Synthetic Coordination and Organometallic Chemistry

Synthetic Coordination and Organometallic Chemistry

edited by

Alexander D. Garnovskii
Rostov State University
Rostov-on-Don, Russia

Boris I. Kharisov
Universidad Autónoma de Nuevo León
Monterrey, Mexico

MARCEL DEKKER, INC.   NEW YORK • BASEL

Library of Congress Cataloging-in-Publication Data
A catalog record for this book is available from the Library of Congress.

ISBN: 0-8247-0880-6

This book is printed on acid-free paper.

Headquarters
Marcel Dekker, Inc.
270 Madison Avenue, New York, NY 10016
tel: 212-696-9000; fax: 212-685-4540

Eastern Hemisphere Distribution
Marcel Dekker AG
Hutgasse 4, Postfach 812, CH-4001 Basel, Switzerland
tel: 41-61-260-6300; fax: 41-61-260-6333

World Wide Web
http://www.dekker.com

The publisher offers discounts on this book when ordered in bulk quantities. For more information, write to Special Sales/Professional Marketing at the headquarters address above.

Current printing (last digit):
10 9 8 7 6 5 4 3 2 1

PRINTED IN THE UNITED STATES OF AMERICA

To Russian chemist Lev Alexandrovich Chugaev (1873–1922)

Preface

Metal complexes* (complex compounds or coordination compounds) are one of the most important groups of chemical compounds and form the basis of coordination chemistry. Although complex compounds date back to the 18th century, coordination chemistry was considered a science only after the formulation of the coordination theory by the Swiss chemist A. Werner at the end of the 19th century [1]. During the 20th century, thousands of metal complexes were obtained, characterized, and widely applied. Much has been written on heir synthesis, structure, and properties [2–13].

Given the huge body of work on this subject, two questions arise: why is it necessary to write a new book dedicated to the synthesis of coordination compounds and, how can it be made different from earlier publications? In *Synthetic Coordination and Organometallic Chemistry* we focus on the role of the ligand in modern synthetic coordination chemistry and to the main synthetic methods (interaction of ligands and metal salts, template and direct synthesis), as well as to the synthesis of complexes with programmed localization of a coordination bond (regioselective synthesis) and structure (polyhedron-programmed synthesis).

Regioselective synthesis is examined mainly on the basis of the conception of competitive coordination and the principle of hard and soft acids and bases [14]. A description of polyhedron-programmed synthesis is given, taking into consideration thin structure compounds to be used as ligands (number, nature, and mutual situation of donor centers; and the presence and character of organic fragments, annealed to a metal-cycle).

The considerable role of synthetic methods—in particular, direct gas-phase [11,15–18] and electrochemical [11,17,19] techniques—in controlled creation of coordination compounds of various types (molecular and η^n-π-complexes, chelates, and homo- and heteronuclear compounds) is discussed. Particular attention is given to the complex compounds having "standard" and "nonstandard" coordination modes of the most widespread (typical) ligands, for example σ(N)- and $\pi(\eta^6$-C)-complexes

* We use the term to distinguish the metal-containing, organic, and H-complexes.

of pyridine or (η^2-O.O) and chelate η^1(O or C) metal-bound β-diketonates [4]. The use of novel methods in synthetic coordination chemistry is discussed, such as microwave [20] and ultrasonic [21,22] activation of the reaction system.

Synthetic methods for obtaining coordination compounds of several classes of ligands are emphasized. Among these ligands, the β-diketones, *o*-hydroxyazomethines and their S,Se,Te-containing analogues, five- and six-member aromatic heterocycles, oximes, phthalocyanines, quinones, and organodithiocarbamates are discussed. Additionally, the peculiarities of complexes of radioactive elements are reviewed, with special attention paid to the corresponding precautions required for the isolation of these compounds.

We are grateful to RSU and the Russian Foundation for Fundamental Research (grants No. 94-03-09731, 96-03-32026, 97-03-33479, 00-15-97320), the UANL (projects PAICyT 1999-2002), and the National System of Researchers of Mexico (SNI) for financial support. We especially thank Professors Juan Manuel Barbarín Castillo and Ubaldo Ortiz Méndez for their help in preparation of this book. It is hoped that the chapters herein will be valuable to professionals in the field and to students.

Alexander D. Garnovskii
Boris I. Kharisov

REFERENCES

1. Werner, A.Z. *Anorg. Chem.* **3**, 267 (1983).
2. *The Chemistry of Coordination Compounds.* (Edit. Beilar, J.). Reinold: New York, 1956.
3. *Modern Chemistry of Coordination Compounds.* (Edit. Lewis, J., Wilkins, R.G.). Interscience Publishers: New York, 1960.
4. *Comprehensive Coordination Chemistry.* (Edit. Wilkinson, G.). Pergamon Press: Oxford, 1987.
5. (a) Cotton, F.A.; Wilkinson, G. *Advanced Inorganic Chemistry.* John Wiley: New York, 1988; (b) Cotton, F.A.; Wilkinson, G.; Murillo, C.A.; Bochmann, M. *Advanced Inorganic Chemistry*, 6th edition. John Wiley: New York, 1999.
6. (a) Davies, J.A.; Hockensmith, C.M.; Kukushkin, V. Yu.; Kukushin, Yu. N. *Synthetic Coordination Chemistry: Principles & Practice.* World Scientific: Singapore, 1996; (b) Kukushkin, Yu. N.. *Chemistry of Coordination Compounds.* High School: Moscow, 1985, 455 pp; (c) Kukushkin, V. Yu.; Kukushkin, Yu. N. *Theory and Practive of the Synthesis of Coordination Compounds.* Nauka: Leningrad, 1990, 260 pp.
7. *Inorganic Reactions & Mechanisms.* Vol. 14 (Formation of Bonds to Transition & Inner-Transition Metals). (Edit. Atwood, J. D.) 1998.
8. Von Zelewsky, A. *Stereochemistry of Coordination Compounds.* John Wiley: New York, 1995.
9. *Synthetic Methods of Organometallic and Inorganic Chemistry.* (Edit. Herrmann, W.A.) Thieme: New York, 1996, 1997.
10. Candlin, J.P.; Tailor, K.A.; Thompson, D.T. *Reactions of Transition Metal Complexes.* Elsevier: Amsterdam, 1968.
11. (a) *Direct Synthesis of Coordination Compounds.* (Edit. Skopenko, V.V.). Ventury: Kiev, 1997, 172 pp; (b) *Direct Synthesis of Coordination and Organometallic Compounds.* (Edit. Garnovskii, A.D.; Kharisov, B.I.) Elsevier Science: Lausanne, 1999, 244 pp.

12. *Coordination Chemistry. A Century of Progress.* (Edit. Kaufman, G.B.). ACS: Washington, D.C., 1994.
13. Skopenko, V.V.; Savranskii, L.I. *Coordination Chemistry.* Libid: Kiev, 1997, 334 pp.
14. Garnovskii, A.D.; Osipov, O.A.; Bulgarevich, S.S. *Russ. Chem. Rev.* **41**(4), 341 (1972).
15. Klabunde, K.J. *Chemistry of Free Atoms and Particles.* Academic Press: New York, 1980.
16. Klabunde, K.J. *Free Atoms, Clusters, and Nanoscale Particles.* Academic Press: New York, 1994.
17. Garnovskii, A.D.; Kharisov, B.I.; Gójon-Zorrilla, G.; Garnovskii, D.A. *Russ. Chem. Rev.* **64**(3), 201 (1995).
18. Kharisov, B.I.; Garnovskii, A.D.; Blanco, L.M.; Garnovskii, D.A.; Garcia-Luna, A. *J Coord. Chem.* **49**, 113 (1999).
19. Garnovskii, A.D.; Blanco, L.M.; Kharisov, B.I.; Garnovskii, D.A.; Burlov, A.S. *J Coord. Chem.* **48**, 219 (1999).
20. Roussy, G.; Pearce, J.A. *Foundations and Industrial Applications of Microwave and Radio Frequency Fields.* John Wiley: New York, 1995.
21. *Active Metals: Preparation, Characterization, Applications.* (Edit. Furstner, A.) VCH:Weinheim, Germany, 1996.
22. (a) Cintas, P. *Activated Metals in Organic Synthesis.* CRC Press: Boca Raton, FL, 1993; (b) Mason, T.J.; Lorimer, J.P. *Sonochemistry: Theory, Applications and Uses of Ultrasound in Chemistry.* John Wiley: New York, 1988, 251 pp; (c) Mason, T.J. *Advances in Sonochemistry.* JAI Press: London, 1990.

Contents

Contributors

Leonor M. Blanco-Jerez, Ph.D. Facultad de Ciencias Químicas Universidad Autónoma de Nuevo León, Monterrey, Mexico

Alexander D. Garnovskii, Dr. Habilitation Institute of Physical and Organic Chemistry, Rostov State University, Rostov-on-Don, Russia

Dmitry A. Garnovskii, Ph.D. Institute of Physical and Organic Chemistry, Rostov State University, Rostov-on-Don, Russia

Boris I. Kharisov, Ph.D. Facultad de Ciencias Químicas, Universidad Autónoma de Nuevo León, Monterrey, Mexico

Miguel A. Méndez-Rojas, Ph.D. Department of Chemistry and Biochemistry, Universidad de Las Américas, Puebla, Mexico

Igor S. Vasilchenko, Ph.D. Institute of Physical and Organic Chemistry, Rostov State University, Rostov-on-Don, Russia

1

Fundamental Concepts of Coordination Chemistry

ALEXANDER D. GARNOVSKII, IGOR S. VASILCHENKO, and DMITRY A. GARNOVSKII

Rostov State University, Rostov-on-Don, Russia

1.1 METAL COMPLEXES, COORDINATION NUMBER, LIGANDS

Metal complexes consist of a central atom (ion) – *metal center* – and the neutral molecules or ions (*ligands*) connected to it [1b–3]. The combination of a complex-former and its ligands produces the *coordination sphere*, which is formed by coordination bonds having a donor–acceptor nature [3–6]. A *coordination bond* is mostly formed as a result of the overlapping of atomic orbitals (*AO*) of ligands, filled with electrons, and vacant *AO* of the central atom. The coordination bonds can be subdivided into two-electron two-center and multicenter bonds. In the first case, the coordination bond is formed by *AO* of the metal and ligand; in the second case, electronic pairs of the bonding molecular orbitals (*MO*) can be delocalized in the area, including various atomic centers [5]. Typical examples of metal complexes with localized coordination bonds are, for example, ammines ***1*** and acetylacetonates ***2***:

$[M(NH_3)_n]X_3$

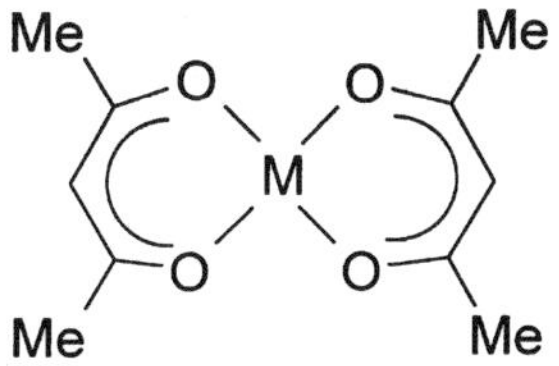

1: M = Co, Cr; X = Cl, NO_3

2: M = Co, Mn, Ni, Cu, Zn

Amongst the complexes with delocalized coordination bonds, the sandwich structures of aromatic (*3*, *4*) [1,2,7,10] and heteroaromatic (*5*, *6*) [11–14] series are widely represented:

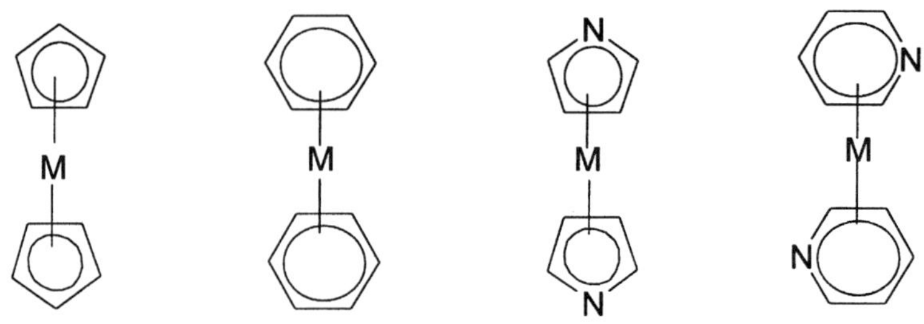

3 *4* *5* *6*

The *coordination number* (*CN*) is the most important characteristic of the central atom, determined by the number of two-electron two-center bonds formed by the *metal center* and ligands. Generally, *CN* coincides with the number of electron-donor centers of the ligands (atoms or π-bonds) with which the central atom is directly coordinated [10,15]. If a ligand has one coordinating fragment (atom, multiple bond), *CN* equals the number of ligand molecules. In the case of the formation of many-center coordination bonds at the expense of delocalized bonds (mostly in five- or six-member aromatic or heteroaromatic ligand systems), *CN* cannot often be determined exactly. As examples of the complexes with σ-bonds and $CN=4$, the acetylacetonates *2* and platinum dichloride *cis*-ammine *7* can be examined. Complex compounds with multielectron multicenter bonds are represented by sandwich structures *3*–*6*. A considerable group of coordination compounds includes complexes where both σ- and π-bonds are simultaneously present, as for instance in compounds *8* and *9*:

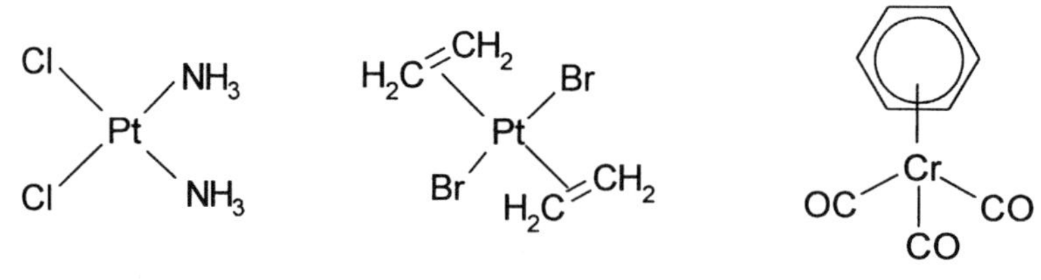

7 *8* *9*

CN is equal to 4 for the complexes *7* and *8*, but for *9* it becomes ambiguous. Thus, in order to calculate *CN* according to the number of donor centers, it could be equal to 4 (*9*) or to 9 (*10*). However, when a benzene formula with localized double bonds is used, *CN* for the same complex becomes equal to 6 (*11*):

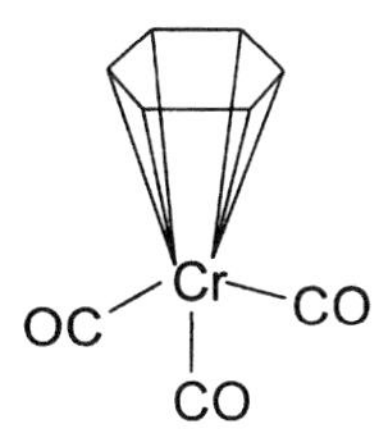

10

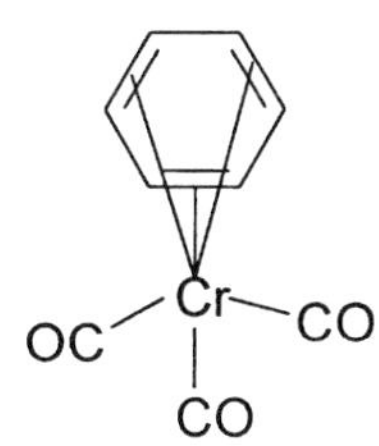

11

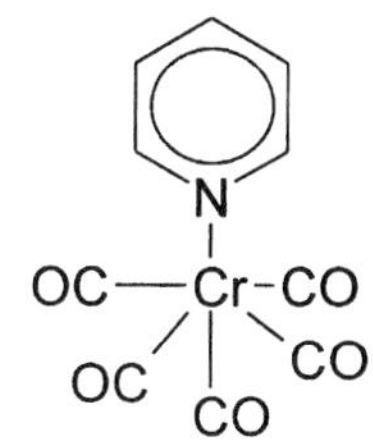

12

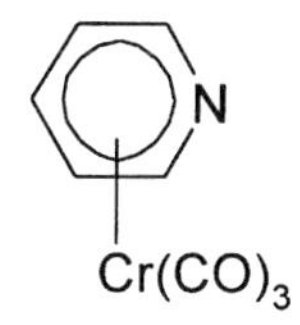

13

CN depends not only on the composition of a coordination compound, but also on the type of σ- and π-bonds present in it. Chromium carbonyl-pyridine complexes are the classic example to illustrate the difficult task of determining unambiguously the coordination numbers. Only σ-bonds are present in $Cr(py)(CO)_5$ and $CN = 6$ (***12***). However, in the σ, π-complex ***13***, as well as in ***9***, the coordination number is quite questionable (compare formulae ***9–11***).

The procedure to determine *CN*, as described above, is correct only for the gaseous phase and for dilute solutions in noncoordinative solvents. In order to determine *CN* for crystalline complexes, it becomes necessary to take into account the possibility of intermolecular coordination arising from the participation of a central atom together with ligand centers coordinately unsaturated [15]. This circumstance leads to the increase of *CN* of the metal center. Thus, intermolecular coordination of type M—O is characteristic for a series of metal acetylacetonates ***2*** [16–18]. So, the formula ML_2 (***2***) does not correspond perfectly to *CN* of the central ion. This magnitude can be changed from 5 to 6 depending on the metal nature. In this respect, it is necessary to note that *CN* can be determined exactly on the basis of x-ray single crystal diffraction data, showing real coordinated donor centers in the crystal of a complex compound. Therefore, for crystal complexes it is expedient to use the concept of "*crystallochemical CN*" which corresponds to the number of atoms, or molecules, present in the crystal cell as next neighbors to the metal center.

Ligands (Latin "*ligo*" – connect) are neutral molecules, ions, and radicals connected with the complex-former [1,2,10,15,19]. The frequently used [20] concept of "*coordinated ligands*" is truly incorrect, since the concept "*ligand*" in itself includes already a coordination. In other words, ligands cannot be uncoordinated. So, it is more accurate to examine the coordination chemistry of inorganic and organic (and sometimes organometallic) compounds used as ligands. For instance, ammonia, carbon monoxide, ethylene, benzene, and cyclopentadiene are chemical compounds themselves (in the *free form*), but not yet ligands. During their connection with a metal, they already participate as ligands, as for instance in the above complexes ***1–13***. The molecules used as ligands in coordination compounds may either conserve their composition unchanged, or change it. The first case is characteristic of the majority of complexes ***1***, ***4***, ***6–13***. However, acetylacetone ***14*** is used to obtain chelates ***2***, cyclopentadiene ***15*** for cyclopentadienyls ***3***, and pyrrole ***16*** to prepare ***5***:

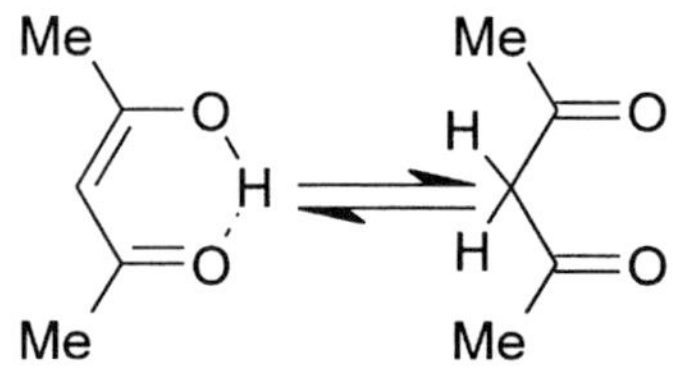
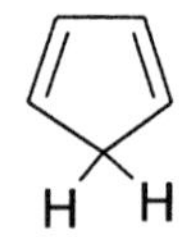
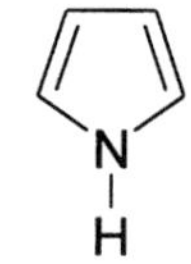

14a *14b* *15* *16*

Deprotonation takes place in all the complexes shown above; the anions participate as ligands and not as neutral initial molecules ***14–16***. Other examples of change of composition and structure, in particular tautomeric forms and conformations of compounds used as ligands, are also known [11,16,21,22]. This principal difference between compounds used as ligands and ligands themselves should always be taken into account. Nevertheless in some cases, due to plain convenience, we have kept unchanged the traditional use of the concept of "*ligands*."

The main feature of ligands is their *dentacity*, and this peculiarity is determined by the number of donor centers in ligand systems, or by the number of chemical bonds between the ligand and the metal complex-former during the formation of a two-center bond. According to this designation, ligands can be divided into *monodentate* (having one donor atom or fragment, for example, π-bond), *bidentate*, *oligodentate*, and *polydentate*. Coordination number is directly related to dentacity. For monodentate ligands, they coincide. In the case of di- and polydentate ligands, *CN* is obtained by multiplying the number of ligands by their dentacity. Thus, for the complexes *7* and ***8***, $CN = 4$ coincides with the number of monodentate ligands (ammonium, chlorine, bromine, ethylene). The same situation is characteristic of compound ***12***, for which *CN* and the number of monodentate ligands are both equal. For the chelate ***2***, two acetylacetonate anions are bidentate ligands. In this respect, the *CN* of complex-formers, given by the product of the number of ligands by their dentacity, equals 4. For the complexes ***3–6*** and ***9–13***, an exact determination of dentacity of ligands, and of *CN* as well, becomes ambiguous.

The η-notation is widely used in modern coordination chemistry to indicate the dentacity of ligand(s) [16,23]. In this notation, upper indexes denote the dentacity of a ligand, so η^1 is a monocoordinated ligand, η^2 dicoordinated, η^3 tricoordinated, etc. These designations are applicable for mono-, di-, tri-, and higher-dentate ligands; they mainly form σ-bonds with the complex-former. Thus, ammonium and all the other monodentate ligands always form η^1-coordinated complexes (for instance, *7*). Bidentate chelating ligand systems form η^2-connected complexes, in the majority of cases [21], as for instance complex ***2***. However, although complexes ***3*** and ***5*** are connected by η^5-mode, and ***4***, ***6***, ***9***, and ***13*** by η^6-mode, the number η is not identical to the dentacity of cyclopentadienyl and pyrrole anions, benzene and pyridine. When several different donor centers are present in the ligands, the donor atoms and dentacity of ligands [24–26] must be indicated. In particular, η^1-N (*17*), η^1-S (***18***), and η^2-N,S (***19***) coordinated ligands are known for the NCS anion:

M-NCS NCS-M M-NCS-M'

17 *18* *19*

Moreover, the indexes k (χ) [16,27] and λ [28] were introduced with the same aim. These indexes are placed before the coordinated atoms of a ligand. The number of coordinated atoms is written over κ [16]: κ^1 means a connection with one donor center, κ^2 with two, κ^3 with three, etc.

If the ligand carries out a bridge function, the letter μ is introduced to specific donor centers of the ligand. Thus, for the bridge thiocyanate ion in ***19***, this is written as η^2(N,S-μ). If the number of metal atoms, connected by the bridge ligand, is more than two, their quantity is designated by a subindex after μ (for example, μ_3-). An upper index is generally used in Russian literature [10]. If a metal complex contains ligands of different types, the bridge ligands are primarily indicated by the name of the complex, followed by the others in alphabetical sequence. Moreover, in order to denote donor centers of di- and polydentate ligands, it was proposed to distinguish the coordinated nucleophilic atoms in the formulae of coordination compounds by means of the indexes æ [27] and Ω [28]. Examples of such a presentation are thoroughly described in a review [29]:

M—NO_2 nitrito-æ*N*
M—O—N=O nitrito-æ*O*
M—NCS thiocyanato-æ*N*
M—SCN thiocyanato-æ*S*

Special attention is also given to the description of names of metal-cyclic structures [28]. If a group of simultaneously coordinated atoms of the chelating ligand is present in the complex, the symbols of all atoms and groups are enumerated, separated by a comma, after the symbol Ω followed by a "-". An upper index number on the right of the symbol of an element indicates its number within the group.

We note that the last described approach is seldom used in coordination chemistry. On the contrary, the η-notation of F.A. Cotton is the most widely applied. The nomenclature of coordination compounds according to IUPAC is fully described in Refs. 3 and 10.

1.2 CLASSIC CONCEPTS OF COORDINATION IN THE SYNTHESIS OF METAL COMPLEXES

1.2.1 Lewis Theory

Considerable progress in the development of theoretical and synthetic coordination and organometallic chemistry was made with the use of electron ideas. Lewis elaborated in 1923 the classic electron theory of acids and bases [30], and used it to explain the coordination ideas of Werner [31] (in Ref. 32, this achievement is ascribed to Sidgwick). A *Lewis acid* (**A**) is a acceptor of the electron pair and a *Lewis base* (**B**) is its donor [33]. In other words, **A** is a species that can form a new covalent bond by accepting a pair of electrons and **B** is a species that can form a new covalent bond by donating a pair of electrons. The fundamental *Lewis acid–base theory* is described by a direct equlibrium [Scheme (1.1)], leading to the formation of the adduct (acid–base complex):

$$A + B \rightleftarrows A{:}B$$

Scheme 1

According to such an approach to the evaluation of the nature of coordination compounds, the coordination (donor–acceptor) bond between the metal center M and each joining group (ligands L) is formed by the electron pair [Scheme (1.2)]:

$$M + :L \rightleftarrows M:L$$

Scheme 2

The processes of exchange (substitution) of acids and bases are similar [Schemes (1.3)–(1.5)]:

$$AB^1 + B^2 \rightleftarrows AB^2 + B^1$$

$$A^1B + A^2 \rightleftarrows A^2B + A^1$$

$$A^1B^1 + A^2B^2 \rightleftarrows A^1B^2 + A^2B^1$$

Schemes 3–5

Qualitative and quantitative aspects of the Lewis theory of acids and bases, and practical applications of Lewis acids, are discussed in a series of monographs [1,4–6,30–46] and reviews [47–49]. The following aspects are taken into account: (a) electronic configuration of acceptors (**A** = M; MX_n are generally metal and boron salts), (b) nature of anions (usually halides), (c) peculiarities of thin structure of donors (**B** are generally the compounds containing N, P, As, Sb; O, S, Se, Te; F, Cl, Br, I atoms): their electronic structure, spatial accessibility, and mutual position of donor centers. Moreover, the nature of X, order of binding of **A** and **B** in formation of adducts of type AB_n, nature of solvents, and evaluation of ΔH or ΔG of the processes (1.1)–(1.5) [31,48] should also be considered.

In relation to the synthetic direction of the present monograph, we would like especially to mention that the theoretical approach of Lewis is the basis of preparative techniques in the synthesis of coordination compounds described later (Chap. 3): interaction of the ligands with metal salts or carbonyls (Sec. 3.3.1) is described by Schemes (1.1) and (1.2), ligand exchange (Sec. 3.3.2.1) by Schemes (1.3) and (1.6), metal exchange (Sec. 3.2.2) by Schemes (1.4) and (1.7):

$$ML^1 + L^2 \rightleftarrows ML^2 + L^1$$

$$M^1L + M^2 \rightleftarrows M^2L + M^1$$

Schemes 6, 7

The reactions (1.2) describe also the "direct" syntheses (especially in the gas phase, Sec. 3.4.1). The equilibrium (1.5) is the basis of synthetic reactions of double exchange (which are not yet well studied because of the small number of reported data); Scheme (1.8):

$$M^1L^1 + M^2L^2 \rightleftarrows M^1L^2 + M^2L^1$$

Scheme 8

It should also be taken into account that a lot of synthetic processes in coordination chemistry are carried out in solvating media. So, the reaction schemes (1.1)–(1.8) should include solvating reagents and consider the possibility of formation of adducts of final complexes with solvent molecules [Scheme (1.9)]:

$$A(=M)(Solv)_m + B(=L)Solv_n \rightleftharpoons AB(=ML)(Solv)_{m+n}$$

Scheme 9

The electron theory of Lewis made a considerable contribution in understanding not only reaction routes, but also reaction mechanisms with participation of Lewis acids and bases [20,31,50]. In particular [31], substitution (exchange) reactions of ligands in octahedral complexes include the acid–base interaction (1.1). Oxidative addition reactions can occur when a complex behaves simultaneously as a Lewis acid and a Lewis base [the metal provides electrons for ligand binding and has vacant coordination sites to accommodate two additional ligands, Scheme (1.10)] [34b]:

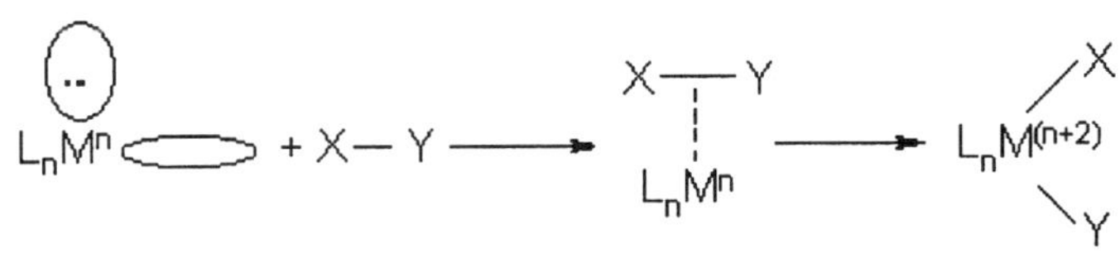

Scheme 10

Interest in the synthesis of Lewis acid and base complexes $MX_n \cdot B_m$ is permanent [51–53] (Secs. 3.1.1.1, 3.1.1.3, 3.2.1, and 3.3.2.3). The advantages of this theory are evident: using the united electronic viewpoint, it helps us to examine a huge series of reactions, in particular the formation and transformations (reactivity) of coordination and organometallic compounds [16,31,50,54].

1.2.2 Principle of Hard and Soft Acids and Bases (HSAB)

The HSAB principle, formulated by Pearson in 1963 [55], has been developed not only by its author [55–73], but also by a series of other researchers, for instance the authors of Refs. 29,31, and 74–90. The generalization of this material is given in some monographs [33,41,91]. The following three statements are the basis of HSAB [55,56].

1. Chemical reactions, in particular complex formation, can be classified as acid–base ones; the resulting products can be examined as complexes of the type Lewis acids and bases.
2. All acids (**A**) and bases (**B**) can be divided into hard (**H**), soft (**S**), and intermediate (**I**).
3. The HSAB principle itself is the following: the acid–base reactions take place in such a way that hard acids (**HA**) prefer to be connected with hard bases (**HB**), meanwhile soft acids (**SA**) react with soft bases (**SB**).

In other words, the HSAB principle emphasizes the preference for hard–hard and soft–soft interactions, and the highest thermodynamic stability of complexes formed as a result of such interactions is achieved. These statements are proved theoretically as well as experimentally. Collective notions of "hardness" and "softness" are related to a series of classification signs summarized in Table 1.

The rows shown below indicate that the hardness of the elements (donor centers of ligands) decreases from left to right:

$$N > P > As > Sb; \qquad O > S > Se > Te; \qquad F > Cl > Br > I$$

Ligands with N, O, F, Cl-donor centers, or compounds containing a combination of these elements, are "hard" bases according to Pearson. On the contrary, all ligand systems, containing elements further to the right of these rows, and their combinations, are "soft" bases. Specific examples of such bases–ligands are represented in Table 2 and in Chap. 2.

The classification of acids was also given by Pearson and is shown in Table 3. In this respect, we emphasize that the "hardness" and "softness" of acids depend considerably on the oxidation number of the metal center, as presented in Table 4.

Hard–soft properties of donors and acceptors were evaluated using the multi-electron perturbation theory [41,92], density functional theory [93,94], and molecular orbital theory [95]. The quantum-chemical basis was assigned to the HSAB principal, in that number in the limits of the ab initio method [96]. The HSAB conception has been widely used to explain various coordination modes in the complexes of di- and polydentate ligands (competitive coordination, Sec. 4.1.1) and in discussions on regioselective synthesis (Sec. 4.1.2). The solvent nature can be an important factor. Taking into account that the proton belongs to hard acids, Table 3, proton-containing solvents frequently interact with the "hard" donor atom of ligand systems and direct the electrophilic attack by the metal towards the more accessible soft nucleophilic centers. So, the most favorable conditions to control the localization mode of a coordination bond with participation of ligands containing "hard" and "soft" donor centers are created when complex-formation reactions are carried out in aprotic nonaqueous solvents.

1.2.3 18-Electron Rule

The 18-electron rule [particular case of the rule of *effective atomic number EAN)*], formulated by Sidgwick [97], defines that, in general, mid-transition metal ions, upon coordination with ligands, aspire to accept a sufficient number of electrons to satisfy the effective atomic number of the closest noble gas, i.e., the complexes formed adopt structures that allow the metal center to attain a closed-shell, 18-valence electron configuration [5,6,97–100]. Classic examples of complexes with 18-electron valence shell are the majority of sandwich complexes of π-arene ligands of the types ***3*** and ***4*** [5–10,16,34] and coordination compounds of heteroarene ligand systems ***5*** and ***6*** [11,14,101–103]. In complexes of the late transition metals, such as Pd(II) and Pt(II), metal centers typically attain 16-valence electron configuration. Ligand substitution reactions in these and similar Ni(II), Au(II), Rh(I), and Ir(I) complexes constitute the most extensively investigated class of ligand substitution reactions [104]. Early transition metals often possess electron-deficient structures, because steric effects limit the coordination number. It has been proposed that reactions of organometallic compounds (the majority of which have 18 electrons in the valence shell of the metal [104]) occur by $16 \leftrightarrow 18$ electron sequences [99,105,106].

Table 1.1 Classification Signs of Hard and Soft Acids and Bases

Acid (acceptor of electrons)			Base (donor of electrons)		
Characteristics	Hard	Soft	Characteristics	Hard	Soft
Polarizability	Low	High	Polarizability	Low	High
Electropositivity	High	Low	Electronegativity	High	Low
Positive charge (oxidation state)	High	Low	Negative charge	High	Low
Size of ions (atoms)	Small	Large	Size of ions (atoms)	Small	Large
Usual types of bond with a base	Ionic, electrostatic	Covalent	Usual types of bond with an acid	Ionic, electrostatic	Covalent
Electrons in exterior orbitals of electron-donor atoms	A small number of such electrons; they are excited with difficulty	Some such electrons; they are excited easily	Vacant orbitals of electron-donor atoms	Practically inaccessible	Accessible

Table 1.2 Pearson's Classification of Bases

Hard bases	Soft bases
H_2O, OH^-, F^-, CH_3COO^-, PO_4^{3-}, SO_4^{2-}, Cl^-, CO_3^{2-}, ClO_4^-, NO_3^-, ROH, RO^-, R_2O, NH_3, RHN_2, N_2H_4	R_2S, RSH, RS^-, I^-, SCN^- (at coordination on the sulfur atom), $S_2O_3^{2-}$, R_3P, R_3As, $(RO)_3P$, CN^-, RNC^-, CO, C_2H_4, C_6H_6, H^-, R^-
Bases occupying an intermediate position	
$C_6H_5NH_2$, C_5H_5N, N_3^-, Br^-, NO_2, SO_3^{2-}, N_2	

Two methods for counting (electron counting formalism) are generally applied: the ionic (charged) and the covalent (neutral) model. In the first method, all of the ligands are removed from the metal and, if necessary, the proper number of electrons are added to each ligand to bring it to a closed valence shell state. In the second method, all of the ligands are removed from the metal, but rather than take them to a closed shell state, we do whatever is necessary to make them neutral. For example, for $CoCp_2$ the sum of electrons is 19 in both cases: ionic model, Co(III) 7 e + $2Cp^-$ 12 e; covalent mode, Co(0) 9 e + $2Cp^{\cdot}$ 10 e [107a]. The methyl group can be represented as an anion CH_3^- (2 e, first model) or neutral radical $CH_3^{\cdot}$ (1 e, second model). The ammonia molecule is a neutral two-electron donor in both cases.

Table 1.3 Pearson's Classification of Acids

Hard bases	Soft bases
H^+, Li^+, Na^+, K^+	Cu^+, Ag^+, Au^+, Tl^+, Hg^+
Be^{2+}, Mg^{2+}, Ca^{2+}, Sr^{2+}, Mn^{2+}	Pd^{2+}, Co^{2+}, Pt^{2+}, Hg^{2+} $MeHg^+$,
Al^{3+}, Sc^{3+}, Ga^{3+}, In^{3+}, La^{3+}	$Co(CN)_5^{2-}$
Nd^{3+}, Ce^{3+}, Gd^{3+}, Lu^{3+}	Pt^{4+}, Te^{4+}
Cr^{3+}, Co^{3+}, Fe^{3+}, As^{3+}, Sn^{3+}	Tl^{3+}, $TlMe_3$, BH_3, $GaMe_3$
Si^{4+}, Sn^{4+}, Ti^{4+}, Zr^{4+}, Th^{4+}, U^{4+}	$GaCl_3$, GaI_3, $InCl_3$
Pb^{4+}, Hf^{4+}, WO^{4+}, Cr^{6+}	R^+, RSe^+, RTe^+
UO_2^{2+}, VO^{2+}, MoO^{3+}	I^+, Br^+, HO^+, RO^+
$BeMe_2$, BF_3, $B(OR)_3$	I_2, Br_2, ICN, etc.
$AlMe_3$, $AlCl_3$, AlH_3	Trinitrobenzene, etc.
RPO_2^+, $ROPO_2^+$	Chloranile, quinones, etc.
RO_2^+, $ROSO_2^+$, SO_3	Tetracyanoethylene
I^{7+}, I^{5+}, Cl^{7+}	O, Cl, Br, I, N, $RO^{\cdot}$, RO_2
RCO^+, CO_2, NC^+	M (metal atoms)
HX (molecules forming hydrogen bonds)	Bulk metals
	CH_2, carbenes
Acids occupying an intermediate position	
Fe^{2+}, Co^{2+}, Ni^{2+}, Cu^{2+}, Zn^{2+}, Pb^{2+}, Sn^{2+}, Sb^{2+}, Bi^{3+}, Rh^{3+}, Ir^{3+}, BMe_3, SO_2, NO^+, Ru^{2+}, Os^{2+}, R_3C^+, Ph^+, GaH_3	

Table 1.4 Dependence of "Hardness" and "Softness" of the Elements on Their Oxidation State[a]

Group						
I	II	VI	VII	VIII		
Cu I[b] (s)[c] II (int)	Zn II (int)	Cr VI–III (h) II (int) 0 (s)	Mn VII–II (h) I, 0 (s)	Fe VI–III (h) II (int) 0 (s)	Co III (h) II (int) 0 (s)	Ni II (int) 0 (s)
Ag II (int) I (s)	Cd II (s)	Mo VI–III (h) 0 (s)	Tc VII–II (h) 0 (s)	Ru VI–III (h) II (int) 0 (s)	Rh III (int) I, 0 (s)	Pd II (s) 0 (s)
Au II (int) I (s)	Hg II (s)	W VI–II (h) O (s)	Re VI–III (h) O (s)	Os VI–III (h) O (s)	Ir III, II (int) O (s)	Pt II (s) O (s)

[a]Group III–V metals are usually hard Pearson acids.
[b]Oxidation state.
[c]s = soft, int = intermediate, h = hard.

Transition metal complexes can be assigned to three classes according to the Δ_0 magnitude: **I** [... 16, 17, 18, 19 ... valence electrons; 18-e rule not obeyed; examples, $Mn(acac)_3$, 16 e; $Co(NH_3)_6^+$, 18 e]; **II** [... 16, 17, 18 valence electrons; 18-e rule not exceeded; examples, $OsCl_6^{2-}$, 16 e; $W(CN)_8^{3-}$, 18 e]; and **III** [18-e rule obeyed; example, $CpMn(CO)_3$] [107a].

This rule has considerable predictive value, in that the composition of many transition metal complexes may be anticipated from combinations of sets of ligands with transition metals of appropriate metal *d*-electron count. Just as organic chemists have their octet rule for organic compounds, so organometallic chemists have the 18-electron rule. And just as the octet rule is often violated, so is the 18-electron rule. Each derives from a simple count of the number of electrons that may be accommodated by the available valence orbitals (one *s* and three *p* for organic chemists; organometallic chemists get five bonus *d*-orbitals in which to place their electons) [107a]. For organometallics of the *f*-elements this procedure is not applicable. Under the actual application of this rule, each ligand is seen as introducing two electrons into the valence shell of a metal. To such kinds of ligands belong most of the above examined simple molecules (for instance, CO, CS, NH_3, H_2O, H_2S) and anions (OH^-, SH^-, Hal^-, NCS^-), as well as PR_3, SeR_2, TeR_2, etc. π-Ligands are seen as donors of electrons; besides, the number of introduced electrons is equated to the number of electrons in the bonding and antibonding π-MO. This number of electrons is evaluated on the energy diagram (see, for example, Refs. 6,91c,107b), according to which the cyclopentadienyl anion introduces five electrons while benzene introduces six.

It can easily be demonstrated that the rule of 18 electrons is effective and valid in stable π-complexes of both aromatic and heteroaromatic compounds (Sec. 2.2.4.1). Examples of the organometallic compounds which obey this rule are

$Ni(CO)_4$, $Mn_2(CO)_{10}$, $V(CO)_4(C_5H_5)$, $FeMe(CO)_2(C_5H_5)$ (all possess 18 e) [107a]. At the same time, it is necessary to take into account that this rule has no minor exceptions as, for example, vanadocene (15 e, M = V, R = H), nickelocene (20 e, M = Ni, R = H) [6,99], $ZrCl_2(C_5H_5)_2$ (16 e), $TaCl_2Me_3$ (10e), WMe_6 (12 e), $Pt(PPh_3)_3$ (16 e), $IrCl(CO)(PPh_3)_2$ (16 e) [107a]. The "apparent exceptions to the 18 electron rule" $W(PhC{\equiv}CPh)_3$, $Zr(BH_4)_4$, Cp_3UR, and UO_2L_6 are examined in Refs. 108 and 109. Square planar compounds are exceptions to the 18-electron rule. The combination of eight metal *d*-electrons and two electrons from each of the four ligands gives a total of 16. These are particularly stable complexes, because with 16 electrons all of the bonding molecular orbitals are occupied, any extra would destabilize the complex [107a]. Recently [110], the 18- and 19-electron complexes $[Fe(NO)(py\text{-}S_4)](BF_4)$ and $[Fe(NO)(py\text{-}S_4)]$, respectively, [py-S_4^{2-} is 2,6-bis(2-mercaptophenylthiomethyl)pyridine(2-)] were reported. Most 17-electron complexes are very liable in comparison to their 18-electron analogues [104]. Other important applications of the 18-electron rule are the possibility of its use for counting paramagnetic complexes (for example, to explain why the 20-electron complex Cp_2Ni is paramagnetic despite having an even number of electrons) and to predict the number of formal metal–metal bonds in polynuclear complexes (up to four metal centers) and polyhedral shapes [104]. A topological electron-counting scheme was also described [111].

Other theoretical considerations for synthetic coordination chemistry (isolobal and isoelectronic analogies, chelate, cis and trans effects, factors affecting the acid/base properties of coordination compounds, bond theories, etc.) are covered in detail in excellent recent monographs [3,34b,104,106] and, for this reason, are not presented in this book.

1.3 TYPES OF COORDINATION COMPOUNDS

1.3.1 General Principles and Classification

There are some definite differences in the various approaches nowadays in use for the classification of coordination compounds [1,2,10,32,112–117]. They are often conflicting, inducing an incorrect view of the state of modern coordination chemistry, which negatively affects its presentation in classrooms (negative effect upon correct evaluation of the state of modern coordination chemistry and its presentation in high school). Practically from the moment of birth of classic coordination chemistry, the nature of ligands was the basis for classifying metal complexes. Werner introduced such concepts as "aqua," "amino," and other types of complexes which had been used [116,118,119] and are still being used [1,10] at present. The classification of complexes on the basis of the type of donor centers of a ligand [32,120] is practically identical to the approach presented above. Ligand character was also taken into account in the classification of three types of Werner complexes [1,2,10]:

1. Acido compounds, containing only deprotonated acid anions in the inner coordination sphere, for example, $K[Ag(CN)_2]$.
2. Molecular complex compounds, containing neutral ligands in the inner coordination sphere, for example, $[Co(NH_3)_6]Cl_3$.
3. Mixed acido–molecular metal complexes, for example, $[Cr(H_2O)_5Cl]Cl_2 \cdot 2H_2O$.

However, these classification approaches are not extensive enough, leaving out many types of complexes now common in modern coordination chemistry. Moreover, they do not allow us to distinguish definite types of complexes, and give very limited ideas about their nature. In fact, the ligands with N-donor center form a very wide range of metal complexes, including in the coordination sphere molecular and atomic nitrogen, ammonia, and amines, as well as heterocyclic N-containing compounds such as pyrrole, azoles, and azines, and even chelate-forming and macrocyclic compounds, such as prophyrins, phthalocyanines, and other ligand systems.

Such a situation was clear, in principle, long ago, and in this respect various classifications of metal complexes were offered according to: coordination number [18,32], oxidation state of the central ion [32], electronic configuration of the metal center [32], nature of the coordination bond [32], character of the electric charge of the coordination sphere (cationic, anionic, and neutral) [116,121,122], number of nuclei (metals) in the complexes [121–123], and dentacity of ligands [121]. However, such a division of complexes comes short of satisfying modern coordination chemistry.

In this respect, a classification was reported whose basis is the nature of ligands together with the characteristic bonding and structural peculiarities of the metal complexes [112,113]. This allowed the identification of four types of coordination compounds: molecular complex compounds (*MCC*); metal-cyclic complex compounds, metal chelates (*MC*); complexes with multicenter coordination bonds (*CMCB*); and di- and polynuclear coordination compounds (*DPCC*). The expediency of such a classification is based on the following considerations.

1. The separation of coordination compounds into four types, as described above: (a) encompasses practically all types of complexes reported in modern coordination chemistry [1,2,10,16,34,124–127]; (b) is clear and informative with respect to a given molecule in a coordination compound as a whole, and does not rely on its individual portions; (c) concentrates its attention on the special features of some types of metal complexes; (d) follows from the modern approach to study the structures of metal complexes [16,34,125,126], together with their systematic nomenclature [122,123].

2. Although the classifications offered earlier have allowed for a "more deep understanding of the nature and variety of coordination compounds" [32] and are still useful at present in some cases, they do not shed vital information on the peculiarities of coordination compounds, which is only found sparingly included, at the most, in one or another of the types used. Thus: (a) a classification based on the coordination number is generally useful for a profound study of stereochemistry, stereodynamics, and isomery of coordination compounds [19,128,129]; (b) a classification based on the oxidation number of the central ion allows for the resolution of problems dealing with the stability of metal complexes [130,131] and with the nature of the coordination bond [1,4–6]; (c) a classification based on the type of donor centers is especially useful in examining the competitive coordination and reactions of ambidentate ligand systems [11,25,26,29,74]; (d) a classification based on the charge helps to characterize the state of some types of complexes, helps to explain the nature of dissociation processes [1,121], and is useful in one of the approaches followed for classifying coordination compounds [123].

1.3.2 Some Types of Metal Complexes

1.3.2.1 Molecular Complex Compounds (*MCC*)

The *MCC* are complex compounds consisting of neutral molecules having a general formula $MX_n \cdot L_pL'_g$ where M is a metal center, X is an anion, L and L′ are ligands, $n > 0$, p and $g = 0, 1, \ldots$). The general formula of *MCC* emphasizes the circumstance that the complexes of this type can be obtained not only by putting together some otherwise separated or independently existing molecules (p and $g = 0$), but also that they can be synthesized by substitution of one ligand by another (different-ligand compounds p, $g \neq 0$).

The general formula for *MCC* does not provide detailed information about the structure of this type of complex. However, after detailed physicochemical studies, it is possible to establish the composition of the coordination sphere in an *MCC*, their stereochemistry, properties, and the peculiarities of the bonds present. In fact, the general formulae $BF_3 \cdot NH_3$, $AlCl_3 \cdot 6H_2O$, and $Fe(CN)_2 \cdot 4KCN$ indicate that these compounds are formed from components capable of independent existence. As a result of their study (electroconductivity and x-ray single crystal diffraction), it was established that: in the first of them the group $[BF_3 \cdot NH_3]$ is netural, in the second the group $[Al(H_2O)_6]^{3+}Cl_3^-$ is cationic, and in the third the group $K_4^+[Fe(CN)_6]^{4-}$ is anionic. A tetrahedral configuration ($CN = 4$) is present in the first complex, while an octahedral configuration ($CN = 6$) is observed in the second and third complexes.

Data available on modern coordination chemistry testifies that *MCC* includes almost all types of classic Werner complexes [1,114,124,132]. This type is broad enough so that the three other groups mentioned above (*MC*, *CMCB*, and *DPCC*) were considered not long ago as "other complex comppounds" with respect to the *MCC* [122] or were even isolated as belonging to "some especial groups" [114]. However, such an approach does not rightfully reflect the state of modern coordination chemistry [16,34,117,124–126,133,134]; this fact was recognized in review articles [112,113] and textbooks [1,2,10].

The peculiarities of classical localized coordination bonds (two-electron and two-center [1,4,5]) are displayed most clearly in *MCC*. Mostly, the elements of the first period of the Periodic Table (C, N, O) participate as electron donors in the formation of such bonds. In complexes of this type, the role of π-dative interactions is significant. These interactions are revealed in coordination compounds of ligands containing the elements of the next periods as donor centers (P, As, Sb; S, Se, Te; Cl, Br, I). We note that the examined complexes are the most successful objects to study the influence of ligand and metal nature on the character of the coordination bond, since, in this case, the factors which could distort this influence (chelate, macrocyclic, and other effects [117,135]) are absent.

An examination of several groups of ligands capable of forming *MCC*, as well as other types of complexes, has been extensively documented [16, vol. 2] and reviewed [112,113], and will be discussed in the next chapter.

1.3.2.2 Metal Chelates (MC)

A general feature of *MC* is the presence of metallocycles in their molecules, whose main role is to produce stabilization effects (chelate, macrocyclic, and other effects) on the complexes [117,135–139]. An outstanding role in the study of this type of coordination compound belongs to the Russian scientist L.A. Chugaev (to whom the present book is dedicated), who formulated the cycles rule which serves as a basis for a better understanding of the chelate effect [1,117].

The type of metal complexes now discussed specifically involve compounds with a definite position of donor centers in relation to the metal center; for instance ***20*** and ***21***, or the conformational nonrigid di- and polydentate ligands illustrated by ***22*** [112,113,117]:

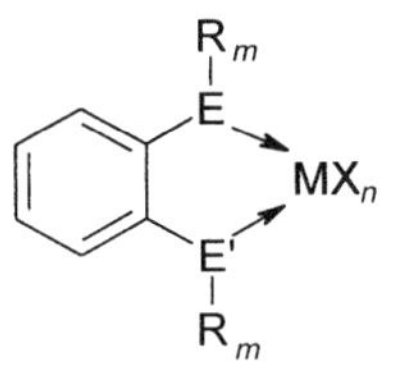

20: E,E' = N, As, P, O, S, Se; m = 1, 2; R - H, Alk, Ar

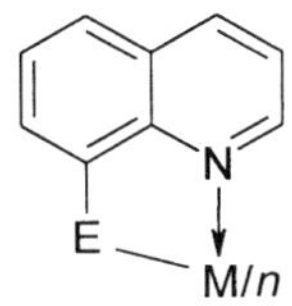

21: E = NR, O, S, Se; R = Ar, Ts

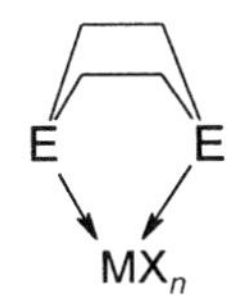

22: E = NR, PR, O, S, Se

Amongst chelates, the molecular metal chelates ***20*** and ***22***, the *inner-complex* ***21***, and the *macrocyclic* and *macromolecular* compounds are worth distinguishing. The concept of "*inner-complex compounds*" (*ICC*), adapted from older publications mentioned elsewhere [117], is used to emphasize their synthesis starting from ligand cycle-forming systems together with inner-molecular H-bond, for instance acetylacetone ***14***. However, this concept is mainly used in German and Russian literature. In general, only the term "*chelates*" is mentioned as including *ICC*.

Chelates and other effects serving as stabilizing molecules for complex compounds have been discussed repeatedly in scientific [116,117] and educational literature [1,2,10] and, in this respect, do not require additional examination.

Chelating ligands include a very wide range of compounds. To this category belong the β-diketones ***14*** [1,2,10,16,117] *o*-hydroxyazomethines ***23*** [16,125,140,141], porphyrins ***24*** [113,142] (see details in Sec. 2.2.4.2), and macrocyclic systems, as for instance crown ethers ***25*** [16,113] (described in Sec. 2.2.4.4), which are the focus of a great many researchers. Other examples on the basics of chelating ligands and their complexes are presented in Chap. 2.

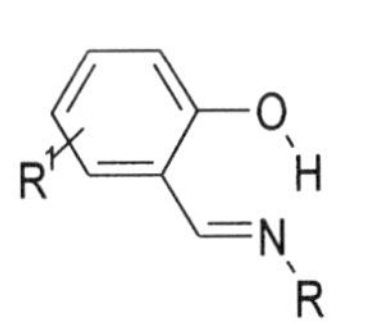

23

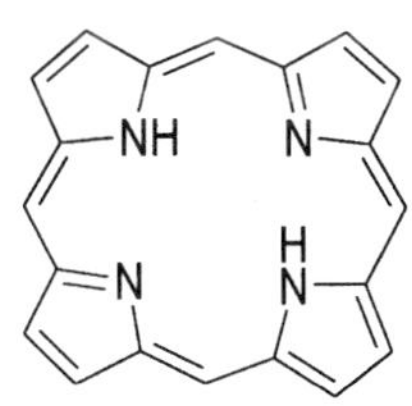

24

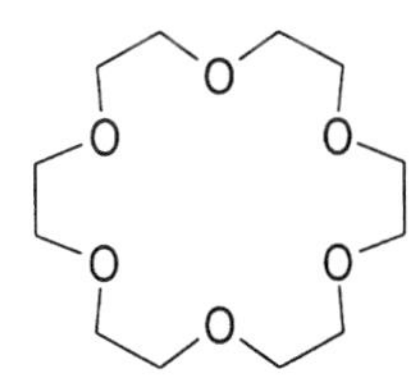

25

1.3.2.3 Complexes with Multicenter Coordination Bonds (*MCCC*)

MCCC are mainly formed from organic and inorganic compounds having delocalized bonds [1,2,5,7–10,112,113]. They are widely represented by η^5-(***3***, ***5***, ***26***) and η^6-(***4***, ***6***, ***9***, ***10***, ***13***) π-complexes and inorganic structures, for example ***27*** [112]:

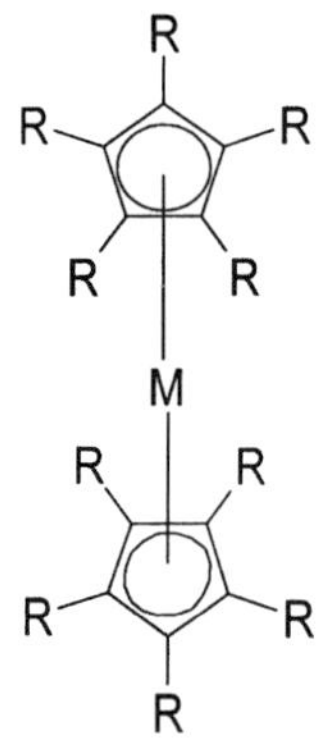

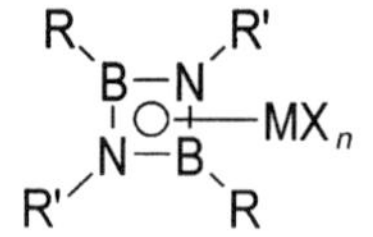

26
R = H,Alk,Ar;
M = Fe,Co,Ni etc.

27

Multidecker sandwiches and coordination compounds, formed by unstable organic particles such as radicals, biradicals, carbenes, and carbynes [7,112,113] (see Sec. 2.2.4.1), also belong to *MCCC*.

1.3.2.4 Di- and Polymetallic Coordination Compounds (*DPCC*)

DPCC may be subdivided into two groups: the first group includes compounds with direct metal–metal bonds, while the second one consists of complexes in which the atoms of metal centers are divided by dredging ligand fragments. *DPCC* with the metal–metal bond are formed by bidentate ligands having ***28*** as general formula, with the orbital parallel, or almost parallel, to their nucleophilic centers [113,143]; amongst them, for instance, are found methylenediamine and the monoanions of hydroxy- and aminoheterocycles [112,113]. Ligands having two or more chelating groups, for instance ***29***, are widely used to obtain *DPCC* [112,113]:

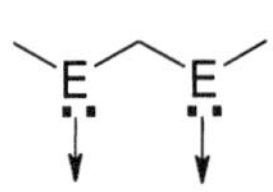

28: E = N, P, As,
O, S, Se

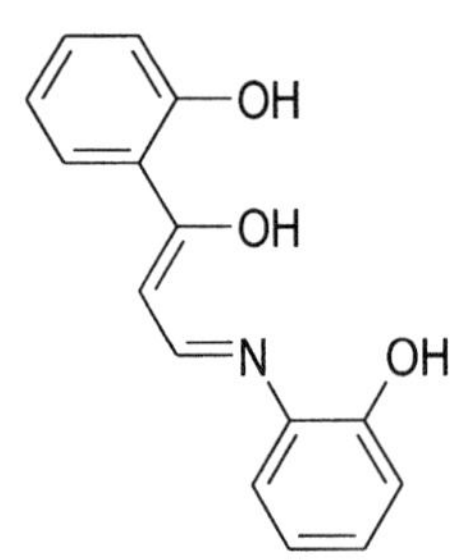

29

Clusters [144–158] are di- and polymetallic homo- and heteronuclear compounds containing a skeleton formed from metal atoms kept together through direct “metal–metal” bonds, or analogous atoms separated by ligand bridges. Clusters are formed by atoms of Groups **I** (H), **IV** (C, Si), **V** (N, P, As, Sb), **VI** (O, S, Se, Te), and **VII** (Cl, Br, I), homo- and heteronuclear molecules, methylene, and many other organic fragments.

DPCC are obtained on the basis of metal-containing ligands, which can be classical Werner complexes (Scheme 1.11), salicylaldiminates [140], and other inner-complex compounds, as well as numerous organometallic compounds [112,113,159]:

$$M(NCX)_m + M^1A_n \rightarrow [M(NCX)_m \rightarrow M^1A_n]$$

$$X = S, Se; A = Hal, NO_3; m,n = 2, 3$$

Scheme 11

Metal-containing polymers, formed on the basis of polymer organic systems and metal-containing monomers [113,133,134], are important complexes among *DPCC*.

Thus, it is possible to observe distinctive and specific characteristics for the types of metal complexes examined.

1. For *MCC* – their formation is via reunion (substitution) of individual neutral components (ligands and mostly metal salts), and the presence of two-center two-electron coordination bonds.
2. For *MC* – the formation of metal-cycles is provoked by enhanced stabilization effects in the complex molecule.
3. For *MCCC* – the presence of multielectron multicenter coordination bonds.
4. For *DPCC* – di- and polynuclear structures.

Noting the paramount importance and the role played by ligands in the synthesis of complexes, we have dedicated the next chapter to their study.

REFERENCES

1. Kukushkin, Yu.N. *Chemistry of Coordination Compounds*. High School: Moscow, 1985, 455 pp.
2. Skopenko, V.V.; Savranskii, L.I. *Coordination Chemistry*. Libid: Kiev, 1997, 334 pp.
3. Rodgers, G.E. *Introduction to Coordination, Solid State, and Descriptive Inorganic Chemistry*. McGraw-Hill: New York, Madrid, Mexico, 1995.
4. (a) Butler, I.S.; Harrod, J.F. *Inorganic Chemistry: Principles and Applications*. Benjamin/Cummings: Redwood City, 1989, 785 pp.; (b) Manku, G.S. *Theoretical Principles of Inorganic Chemistry*. McGraw-Hill: San Francisco, 1988.
5. Douglas, B.E.; McDaniel, D.H.; Alexander, J.J. *Concepts and Models of Inorganic Chemistry*, 2nd Edition. John Wiley: New York, 1992.
6. Minkin, V.I.; Simkin, B.Ya.; Minjaev, R.M. *Theory of Molecular Structure*. Fenix: Rostov-on-Don, 1997, 558 pp.
7. *Comprehensive Oragnometallic Chemistry*. (Edit. Wilkinson, G.; Stone, F.G.A.; Abel, E.W.). Pergamon Press: Oxford, New York, London, 1982, 1st Edition, vols 1–9; 1995, 2nd Edition, vols. 1–13.
8. Lukehart, C. *Fundamental Transition Metal Organometallic Chemistry*. Brooks Cole: Monterey, 1985.
9. Yanamoto, A. *Organotransition Metal Chemistry*. John Wiley: New York, 1986.

10. Kostromina, N.A.; Kumok, V.N.; Skorik, N.A. *Chemistry of Coordination Compounds*. High School: Moscow, 1990, 432 pp.
11. Garnovskii, A.D.; Garnovskii, D.A.; Vasilchenko, I.S.; Burlov, A.S.; Sadimenko, A.P.; Sadekov, I.D. *Russ. Chem. Rev.* **66**(5), 389 (1997).
12. Sadimenko, A.P.; Garnovski, A.D.; Sheinker, V.N.; Osipov, O.A. *Khim. Geterotsykl. Soed.* 1299 (1283).
13. *Comprehensive Heterocyclic Chemistry*. (Edit. Katriitzky, A.R.; Rees, C.W.; Skrives, E.F.V.). Pergamon Press: Oxford, 1984, 1st Edition, vols. 2–6; 1996, 2nd Edition, vols. 2–9.
14. Garnovskii, A.D.; Sadimenko, A.P. *Adv. Heterocycl. Chem.* **72**, 1 (1999).
15. *Chemical Encyclopedia*. Soviet Encyclopedia: Moscow, 1990, vol. 2, 671 pp.
16. *Comprehensive Coordination Chemistry*. (Edit. Wilkinson, G.). Pergamon Press: Oxford, 1987, vols. 1–7.
17. Mehrotra, R.C.; Rohra, R.; Gauer, D.P. *Metal β-Diketonates and Allied Derivatives*. Academic Press: New York, 1978.
18. Shkolnikova, L.M.; Porai-Koshitz, M.A. *Results of Science & Technology. Crystallochemistry*. VINITI: Moscow, 1982, vol. 16, p. 117.
19. Von Zelewsky, A. *Stereochemistry of Coordination Compounds*. John Wiley: Chichester, New York, 1995.
20. (a) Hipp, C.J.; Busch, D.H. In: *Coordination Chemistry*. (Edit. Martell, A.E.). American Chemical Society: Washington, 1978, vol. 2; (b) Black, D. St. In: *Comprensive Coordination Chemistry*. Pergamon Press: Oxford, vol. 1, p. 415; (c) *Reactions of Coordinated Ligands*. (Edit. Braterman, P.S.). Harcover: New York, London, 1986, vol. 1; Plenum Press: New York, 1989, vol. 2.
21. Garnovskii, A.D.; Garnovskii, D.A.; Burlov, A.S.; Vasilchenko, I.S. *Mendeleev Chem. J.* **40**(5/6), Part I, 19 (1996).
22. Minkin, V.I.; Garnovskii, A.D.; Elguero, J.; Katritzky, A.R.; Denisko, O.V. *Adv. Heterocycl. Chem.* **76**, 157 (2000).
23. Cotton, F.A. *J. Am. Chem. Soc.* **90**, 6230 (1968).
24. *The Chemistry of Pseudohalides* (Edit. Golub, A.M.; Keler, H.; Skopenko, V.V.). Elsevier Science: Lausanne, Amsterdam, New York, Tokyo, 1986.
25. Kawagudi, S. *Variety of Coordination Modes of Ligands in Metal Complexes. Inorganic Chemistry Concepts*. Springer-Verlag Chemie: Berlin, 1988, vol. 11.
26. Burmeiser, L. *Coord. Chem. Rev.* **105**, 77 (1990).
27. Sloan, T.E.; Bush, D.H. *Inorg. Chem.* **17**, 2043 (1978).
28. Leich, G.J. *Inorg. Chem.* **17**, 2047 (1978).
29. Garnovskii, A.D. *Koord. Khim.* **6**(12), 1779 (1980).
30. (a) Lewis, G.N. *Valence and Structure of Atoms and Molecules*. Chem. Catalogue Co., Inc.: New York, 1923; (b) Lewis, G.N. *J. Franklin Inst.* **226**, 293 (1938).
31. Basolo, F.; Pearson, R. *Mechanism of Inorganic Reactions. A Study of Metal Complexes in Solution*, 2nd Edition. John Wiley: New York, London, Sydney, 1967.
32. Dai, M.C.; Selbin, J. *Theoretical Inorganic Chemistry*. Reinhold: New York, Amsterdam, London, 1975.
33. (a) Gutmann, V. *Coordination Chemistry in Non-Aqueous Solutions*. Springer-Verlag: Wien, New York, 1968; (b) Gutmann, V. *The Donor–Acceptor Approach in Molecular Interaction*. Plenum Press: New York, 1978.
34. (a) Cotton, F.A.; Wilkinson, G. *Advanced Inorganic Chemistry*. John Wiley: New York, 1988; (b) Cotton, F.A.; Wilkinson, G.; Murillo, C.A.; Bochmann, M. *Advanced Inorganic Chemistry*, 6th Edition. John Wiley: New York, 1999.
35. Luder, W.F.; Zuffanti, S. *The Electronic Theory of Acid and Bases*. Dover Publications: New York, 1946; 2nd Edition, 1961.
36. Bell, R.P. *Acid and Bases*. Methuen: London, 1952; 2nd Edition, 1969.

37. Brigleb, G. *Molecularverbindungen und Koordination Verbindungen in Eizeldarstellung Elektronen-Donator-Acceptor Komplexe*. Springer-Verlag: Berlin, 1961.
38. Mulliken, R.S.; Pearson, W.B. *Molecular Complexes*. Wiley Interscience: New York, 1969.
39. Miskidjyan, S.P.; Garnovskii, A.D. *Introduction to the Modern Theory of Acids and Bases*. Vysha Shkola: Kiev, 1979.
40. (a) Jensen, W.B. *Lewis Acid–Bases Concept*, 2nd Edition. Wiley Interscience: New York, 1980; (b) *Molecular Interactions* (Edit. Ratajczak, H.; Orvill-Thomas, W.J.). John Wiley: Chichester, New York, Brisbane, Toronto, 1981, vol. 2.
41. Garnovskii, A.D.; Sadimenko, A.P.; Osipov, O.A.; Tsintsadze, G.V. *Hard–Soft Interactions in Coordination Chemistry*. Rostov University Press: Rostov-on-Don, 1986.
42. *Selectivities in Lewis Acid Promoted Reactions*. (Edit. Schinzer, D.). Kluwer Academic: Dordrecht, 1989, vol. 289.
43. Santelli, M.; Pons, J.M. *Lewis Acids and Selectivity in Organic Synthesis*. CRC Press: Boca Raton, FL, 1995.
44. *Lewis Acid Reagents: A Practical Approach*. (Edit. Yamamoto, H.). Oxford University Press: Oxford, 1999.
45. Thompson, A.J. In: *Metal Sites in Proteins and Models: Phosphates, Lewis Acids, and Vanadium*. (Edit. Hill, H.A.; Thomson, A.J.; Sadler, P.J.). Springer-Verlag: New York, 1999.
46. Yamamoto, H. *Lewis Acids in Organic Synthesis*. John Wiley: New York, 2000.
47. Guryanova, E.N. *Usp. Khim.* **37**, 1981 (1968).
48. Satchell, R.P.; Satchell, R.S. *Chem. Rev.* **69**, 251 (1969); *Quart. Rev.* **25**, 171 (1971); *Usp. Khim.* ("*Russ. Chem. Rev.*") **42**, 1009 (1973).
49. Jensen, W.B. *Chem. Rev.* **78**, 1 (1978).
50. *Inorganic Reactions & Mechanisms. Vol. 14* (*Formation of Bonds to Transition & Inner-Transition Metals*). (Edit. Atwood, J.D.). 1998.
51. Jolley, J.; Cross, W.I.; Pritchard, R.G.; McAuliffe, C.A.; Nolan, K.B. *Inorg. Chim. Acta* **315**, 36 (2001).
52. Manson, J.L.; Schlueter, J.A.; Geiser, U.; Stone, M.B.; Reich, D.H. *Polyhedron* **20**, 1423 (2001).
53. Sokol, V.I.; Davidov, V.V.; Elekhina, N.A.; Shkylaev, Yu.V.; Sergienko, V.S. *Zhurn. Neorg. Khim.* **46**, 77 (2001).
54. Candlin, J.P.; Tailor, K.A.; Thompson, D.T. *Reactions of Transition Metal Complexes*. Elsevier: Amsterdam, London, New York, 1968.
55. Pearson, R.G. *J. Am. Chem. Soc.* **85**, 3533 (1963).
56. Pearson, R.G. Science **151**, 172 (1966).
57. Pearson, R.G. *Chem. Br.* **3**, 103 (1967).
58. Pearson, R.G. *J. Chem. Educ.* **45**, 581 (1968).
59. Pearson, R.G.; Songstadt, J. *J. Org. Chem.* **32**, 2899 (1967); *Usp. Khim.* **38**, 1223 (1969).
60. Pearson, R.G. *J. Chem. Educ.* **64**, 561 (1987).
61. Pearson, R.G. *Inorg. Chem.* **27**, 734 (1988).
62. Pearson, R.G. *J. Am. Chem. Soc.* **110**, 2092 (1988).
63. Pearson, R.G. *J. Org. Chem.* **54**(6), 1423 (1989).
64. Pearson, R.G. *Coord. Chem. Rev.* **100**, 403 (1990).
65. Pearson, R.G. *Chem. Br.* **27**(5), 444 (1991).
66. Pearson, R.G. *Inorg. Chim. Acta* **198/200**, 781 (1992).
67. Pearson, R.G.; Palke, W.E. *J. Phys. Chem.* **96**(8), 3283 (1992).
68. Pearson, R.G. *J. Molec. Struct.* **255**, 261 (1992).
69. Pearson, R.G. *Struct. Bond.* **80**, 1 (1993).
70. Pearson, R.G. *Chemtracts: Inorg. Chem.* **3**(6), 317 (1991).
71. Pearson, R.G. *Acc. Chem. Res.* **26**(5), 250 (1993).

72. Pearson, R.G. *Inorg. Chim. Acta* **240**(1/2), 93 (1995).
73. (a) Pearson, R.G. *Educ Quím.* **8**(4), 208 (1997); (b) Pearson, R.G. *Chemical Hardness*. Weinheim: Wiley-VCH. 1997. 197 pp.
74. Garnovskii, A.D.; Osipov, O.A.; Bulgarevich, S.B. *Russ. Chem. Rev.* **41**(4), 341 (1972).
75. Werner, H. *Chem. Uns. Zeit.* **1**, 135 (1967).
76. Seyden-Penne, *J. Bull. Soc. Chim. Fr.* 3871 (1968).
77. Ho, T.L. *Chem. Rev.* **75**, 1 (1975).
78. Davies, J.A.; Hartley, F.R. *Chem. Rev.* **81**, 79 (1981).
79. Chermette, H.; Lissilour, R. *Act. Chim.* **4**, 59 (1985).
80. Ermakov, A.A.; Kliment'eva, G.A. *Russ. J. Inorg. Chem.* **36**(8), 1116 (1991).
81. Semenov, S.A. *Russ. J. Inorg. Chem.* **36**(3), 455 (1991).
82. Datta, D. *Inorg. Chem.* **31**, 2797 (1992).
83. Gazquez, J.L.; Mendez, F. *J. Phys. Chem.* **98**(17), 4591 (1994).
84. Gazquez, J.L.; Mendez, F. *J. Am. Chem. Soc.* **116**(20), 9298 (1993).
85. Arland, S. *Coord. Chem. Rev.* **154**, 13 (1996).
86. Hancock, R.D.; Martell, A.E. *J. Chem. Educ.* **73**, 654 (1996).
87. Gazquez, J.L. *J. Phys. Chem. A* **101**(26), 4657 (1997).
88. Roy, R.K.; de Proft, F.; Greerlings, P. *J. Phys. Chem. A* **102**(35), 7035 (1998).
89. Damoun, S.; Van de Woude, G.; Geerlings, P. *J. Phys. Chem. A* **103**(39), 7861 (1999).
90. (a) Gerrard, L.A.; Wood, P.T. *Chem. Commun.* **21**, 2107 (2000); (b) Petrukhin, O.M. *Koord. Khim.*, **28**, 722 (2002).
91. (a) *Hard and Soft Acids and Bases*. (Edit. Pearson, R.G.). Hutchinson and Ross: Stroundsburg, 1973; (b) Ho, T.L. *Hard and Soft Acids and Bases Principle in Organic Chemistry*. Academic Press: New York, 1987; (c) Huheey, J.E. *Inorganic Chemistry. Principles of Structure and Reactivity*, 3rd Edition. Harper & Row: New York, Cambridge, Philadelphia, San Francisco, 1983; (d) Beck, M.; Nagypal, I. *Chemistry of Complexes Equilibria*. Akademia Kiado: Budapest, 1989.
92. Klopman, G. In: *Chemical Reactivity and Reaction Paths*. John Wiley: New York, London, Sydney, Toronto, 1974.
93. (a) Parr, R.G.; Pearson, R.G. *J. Am. Chem. Soc.* **105**, 7512 (1983); (b) Pearson, R.G. *J. Am. Chem. Soc.* **110**, 2092 (1988).
94. Chattaraj, P.K.; Lee, H.; Parr, R.G. *J. Am. Chem. Soc.* **113**, 1854 (1991).
95. Li, Y.; Evans, J.N.S. *J. Am. Chem. Soc.* **117**, 7756 (1995).
96. (a) Pal, S.; Vaval, N.; Roy, R. *J. Phys. Chem.* **97**, 4404 (1993); (b) Chattaraj, P.K.; Von Schleyer, P.R. *J. Am. Chem. Soc.* **116**, 1067 (1994).
97. Sidgwick, N.V. *The Electronic Theory of Valence*. Clarendon Press: Oxford, 1927.
98. Green, M.H.L. *The Transition Elements. Organometallic Compounds*. Methuen: London, 1968, vol. 2.
99. Tolman, C.A. *Chem. Soc. Rev.* **1**, 337 (1972).
100. Astruc, D. *Chem. Rev.* **88**(7), 1189 (1988).
101. (a) *Direct Synthesis of Coordination Compounds*. (Edit. Skopenko, V.V.). Ventury: Kiev, 1997, 172 pp.; (b) *Direct Synthesis of Coordination and Organometallic Compounds*. (Edit. Garnovskii, A.D.; Kharisov, B.I.). Elsevier Science: Lausanne, Amsterdam, London, New York, 1999, 244 pp.
102. (a) Sadimenko, A.P.; Garnovskii, A.D.; Retta, N. *Coord. Chem. Rev.* **126**, 237 (1993); (b) Sadimenko, A.P. *Adv. Heterocycl. Chem.* **78**, 1 (2000).
103. Sadimenko, A.P. *Adv. Heterocycl. Chem.* **80**, 158 (2001).
104. Collman, J.P.; Hegedus, L.S.; Norton, J.R.; Finke, R.G. *Principles & Applications of Organotransition Metal Chemistry*. University Science Books, 1987.
105. *Mechanisms of Inorganic and Organometallic Reactions*. (Edit. Twigg, M.V.). Plenum Press: New York, 1983–1985, vols. 1–3.
106. Atwood, J.D. *Inorganic & Organometallic Reaction Mechanisms*, 2nd Edition. VCH: Weinheim, 1997.

107. (a) Koutsantonis, G. *Organometallic Compounds of the Transition Elements*. Internet site http://www.chem.uwa.edu.au/enrolled_students/Organometallic/contents.html; (b) Orchin, M.; Jaffe, H.H. *The Importance of Antibonding Orbitals*. Houghton Mifflin: Boston, 1967.
108. Laine, R.M.; Moriarty, R.E.; Bau, R. *J. Am. Chem. Soc.* **94**, 1402 (1972).
109. Chu, S.-Y.; Hoffmann, RF. *J. Phys. Chem.* **86**, 1289 (1982).
110. Sellman, D.; Blum, N.; Heinemann, F.W.; Hess, B.A. *Chem. Eur. J.* **7**(9), 1874 (2001).
111. (a) Teo, B.K. *Inorg. Chem.* **23**, 1251 (1983); (b) Teo, B.K.; Longoni, G.; Chung, F.R.K. *Inorg. Chem.* **23**, 1257 (1984).
112. Garnovskii, A.D. *Izv. Vuzov. Khim. i Khim. Tekhn.* **30**(10), 1 (1987).
113. Garnovskii, A.D. *Koord. Khim.* **14**(5), 579 (1988).
114. Grinberg, A.A. *Introduction in Chemistry of Complex Compounds*. Chemistry: Moscow, Leningrad, 1971, 631 pp.
115. Cotton, F.A.; Wilkinson, G. *Bases of Inorganic Chemistry*. Mir: Moscow, 1979, p. 413.
116. (a) Akhmetov, N. *Inorganic Chemistry*. Mir: Moscow, 1973 (in English); (b) Shoeib, T.; Gorelsky, S.I.; Lerer, A.B.P.; Siu, K.S.W.; Hopkinson, A.C. *Inorg. Chim. Acta*, **315**, 236 (2001).
117. *Ross. Khim. Zhurn.* ("*Russ. J. Chem.*") **40**(4/5), 1–189 (1996); *Mendeleev Chem. J.* **40**(4/5), Part I, II (1996) issue dedicated to metal chelates.
118. Werner, A. Z. *Anorg. Chem.* **3**, 267 (1893).
119. Skopenko, V.V.; Grigorieva, V.V. *Coordination Chemistry*. High School: Kiev, 1984, p. 13.
120. *The Chemistry of Coordination Compounds*. (Edit. Beilar, J.). Reinold: New York, 1956.
121. Skorik, N.A.; Kumok, V.N. *Chemistry of Coordination Compounds*. High School: Moscow, 1975, p. 61.
122. Kan, R.; Dermer, O. *Introduction to Chemical Nomenclature*. Chemistry: Moscow, 1983, 224 pp.
123. Kachanova, N.N.; Reshakova, A.A.; Sukhova, A.G. *Koord. Khim.* **12**(7), 867 (1986).
124. *Coordination Chemistry. A Century of Progress*. (Edit. Kaufman, G.B.). ACS: Washington, 1994.
125. *Koord. Khim.* ("*Russ. J. Coord. Chem.*") **19**(5), 331–408 (1993) (issue dedicated to 100 years of Werner's coordination theory).
126. *Chem. Rev.* **93**, 847–1280 (1993) (issue dedicated to 100 years of Werner's coordination theory).
127. *Handbuch der Preparative Anorganische Chemie*. (Herausgegeben von Brauer, G.) Ferdinant Enke Verlag: Stuttgart, 981, Bd. 3.
128. Kepert, D.L. *Inorganic Stereochemistry*. Springer-Verlag: Berlin, Heidelberg, New York, 1982.
129. Ebsworth, E.A.V.; Rankin, D.W.H.; Cradock, S. *Structural Methods in Inorganic Chemistry*, 2nd Edition. Blackwell: Oxford, 1991.
130. Smith, R.M.; Martell, A.E. *Critical Stability Constants*. Plenum Press: New York, 1974–1982, vols. 1–5.
131. Yatsimirsky, K.B.; Kriss, E.E.; Gvyazdovskaya, V.L. *Stability Constants of Metal Complexes with Bioligands*. Nauk. Dumka: Kiev, 1979, 255 pp.
132. Werner, A. *New Views in Inorganic Chemistry*. ONTI: Leningrad, 1936, 506 pp. (translated to Russian).
133. Pomogailo, A.D.; Savostianov, V.S. *Metal-Containing Monomers and Polymers*. Chemistry: Moscow, 1988, 384 pp.
134. Pomogailo, A.D.; Ufyand, I.E. *Macromolecular Chelates*. Chemistry: Moscow, 1991, 303 pp.
135. Martell, A.E.; Calvin, M. *Chemistry of the Metal Chelate Compounds*. Prentice-Hall: Englewood Cliffs, NJ, 1952.

136. Yatsimirsky, K.B.; Kolchinskii, A.G.; Pavlishuk, V.V.; Talanova, G.G. *Synthesis of Macrocryclic Compounds*. Nauk. Dumka: Kiev, 1987, 279 pp.
137. Dashevskii, V.G.; Baranov, A.P.; Kabachnik, M.I. *Usp. Khim.* **52**, 268 (1983).
138. Martell, A.E.; Hancock, R.D. *The Chelate, Macrocyclic and Cryptate Effects*. In: Ref. 4, vol. 1, p 240.
139. Yatsimirsky, K.B. *Mendeleev Chem. J.* **40**(4/5), 1 (1996).
140. Garnovskii, A.D.; Nivorozhkin, A.L.; Minkin, V.I. *Coord. Chem. Rev.* **126**, 1 (1993).
141. Garnovskii, A.D. *Russ. J. Coord. Chem.* **19**, 368 (1993).
142. Gerbeleu, N.V.; Arion, V.B.; Burgess, J. *Template Synthesis of Macrocyclic Compounds*. John Wiley: Weinheim, New York, Chichester, 1999, 565 pp.
143. Fallis, I.A. *Ann. Rep. Progr. Chem. A* **94**, 351 (1998).
144. *Transition Metal Clusters*. (Edit. Johnson, B.F.G.). John Wiley: Chichester, 1980, 681 pp.
145. Gubin, S.P. *Chemistry of Clusters*. Nauka: Moscow, 1987, 263 pp.
146. Dange, J.G. In: Ref. 4, vol. 1, p. 137.
147. *The Chemistry of Metal Cluster Complexes*. (Edit. Shriver, D.F.; Kaesz, H.D.; Adams, D.A.). Verlag Chemie: Weinheim, 1990, 437 pp.
148. Ma, L.; Williams, G.K.; Shapley, J.R. *Coord. Chem. Rev.* **128**(1/2), 261 (1993).
149. Thimmappa, B.H.S. *Coord. Chem. Rev.* **143**, 1 (1995).
150. Grimes, R.N. *Coord. Chem. Rev.* **143**, 71 (1995).
151. Pignolet, L.H.; Aubart, M.A.; Craighead, K.L.; Gould, R.A.T.; Krogstad, D.A.; Wiley, J.S. *Coord. Chem. Rev.* **143**, 219 (1995).
152. Chen, L.; Poë, A.J. *Coord. Chem. Rev.* **143**, 265 (1995).
153. McClunchey, M.J.; Girard, L.; Ruffolo, R. *Coord. Chem. Rev.* **143**, 331 (1995).
154. Lentz, D. *Coord. Chem. Rev.* **143**, 383 (1995).
155. Comstock, M.C.; Shapley, J.R. *Coord. Chem. Rev.* **143**, 501 (1995).
156. Wadepohl, H.; Gebert, S. *Coord. Chem. Rev.* **143**, 535 (1995).
157. Teo, B.K.; Zhang, H. *Coord. Chem. Rev.* **143**, 611 (1995).
158. Halet, J.-F. *Coord. Chem. Rev.* **143**, 637 (1995).
159. Garnovskii, A.D.; Sadimenko, A.P.; Uraev, A.I.; Vasilchenko, I.S.; Garnovskii, D.A. *Russ. J. Coord. Chem.* **26**, 311 (2000) [*Koord. Khim.* **26**, 334 (2000)].

2

Ligands of Modern Coordination Chemistry

ALEXANDER D. GARNOVSKII, IGOR S. VASILCHENKO, and DMITRY A. GARNOVSKII

Rostov State University, Rostov-on-Don, Russia

2.1 GENERAL CONCEPTS

Ligands, as the main part of metal complexes, are the object of a great deal of attention in coordination and organometallic chemistry [1–5]. They are examined in Ref. 1 (vol. 2) according to the nature of donor centers and types of ligand systems, where these donor fragments (atoms or groups of atoms) are included. Additionally to the mentioned aspects, the central place in our further discussion belongs to the problem of various coordination modes of di-, tri-, and polydentate ligands.

The list of ligands covered by Ref. 1 (vol. 2) starts with the exotic mercury ligand (Dean, p. 1). Silicon and a series of metals (Ge, Ti, Pb) (Harrison, p. 15) also reveal ligand properties. Hydrogen and a variety of hydride anion complexes (Crabtree, p. 689), as well as the complexes formed by anions with a carbon-donor center (cyanides, fulminates, etc.) (Sharpe, p. 25) are briefly discussed.

A large number of nitrogen-containing ligands are well described and reviewed, like those containing ammonium and a variety of amines (House, p. 23), also the group involving the triazene and azabutadiene compounds (Vrieze and Van Kotten, p. 189), a number of nitrogen heterocycles (Reedijk, p. 98), polyazamacrocycles (Curtis, p. 899), porphyrins and related tetrapyrrol ligands (Mashiko and Dolphin, p. 813), and the polypyrazolylborates and their related gallium derivatives (Shaver, p. 245).

Oxygen-containing ligands are represented by a variety of bridged atoms, or molecules, or even ions as, for instance, atomic oxygen, the hydroxyl anion, or water

(Burgess, p. 228), also molecular oxygen, and the super- and peroxides (Chill and Tew, p. 315), as well as the alkyl- and aryloxides (Malcolm and Rothwell, p. 335), a diversity of oxyanions (Hathaway, p. 413), the group of β-diketonates (Siedle, p. 365), many carboxylates (Oldham, p. 435), the hydroxyacids (Pedrosa, p. 461), and some complexones (Angeregg, p. 777).

Sulfur participates as a donor center in sulfide complexes (Müller and Diemann, p. 515), dithiolenes (Müller-Westerhoff and Vance, p. 595), and other S-containing ligands (thiosulfates, thiourea, mono-, and dithioketones) (Livingstone, p. 633). Some other complex compounds with Se and Te as donor centers have been described (Berry, p. 661).

An important place among the whole range of ligands is occupied by the di- and polydentate donors with different nucleophilic centers: amongst them can be listed the amide- and nitroderivatives (Chisholm and Rothwell, p. 161), oximes, guanidines, and related compounds (Mehrotra, p. 269), Schiff bases, (Calligaris and Randaccio, p. 715), aminoacids and proteins (Lauril, p. 739), sulfoxides, amides, amino-oxides, and similar ligand systems (Goggin, p. 487).

Missed in older treatises [1], the di- and multicenter donors – both aliphatic and aromatic compounds – are discussed as components of organometallic compounds in an excellent later edition [2]. Ideas about ligands, developed by Kukushkin [3], Kostromina et al. [4], and Skopenko and Savranskii [5], are also of considerable interest. Thus, most attention is devoted to the relatively simple ligands [3]; water molecules and OH^-, amines, N_2H_4, NH_2OH, pyridine, organic nitriles, amine N-oxide, phosphines, arsines, stilbines, trialkylphosphites, phosphinoxides, H_2S and its deprotonated forms, thioethers, mercapthanes, sulfoxides, CO, RNC, alkenes and alkynes (mostly C_2H_4, C_2H_2, and their derivatives), anions: halides, cyanide, thiocyanate, nitrite, nitrate, carbonate, carboxyl, sulfate, and perchlorate.

Despite the variety of these ligand systems, they were divided into only two groups. Ligands in the first group characteristically include the electron-donor atoms with the highest electronegativity (σ-donors), the presence of pairs of electrons and the lack of vacant orbitals becoming typical for all the members of this group, which explains their capability to be donors of electronic density. The donor–acceptor bond formed by these ligands has a considerable ionic component. Examples of these ligands are the following molecules and ions: NH_3, H_2O, ROH, F^-, and Cl^-.

The main feature of the second group of ligands is their low electronegativity and the presence of vacant orbitals, together with a pair of electrons. Such ligand systems are σ-donors and π-acceptors. The coordination bond in coordination compounds has mostly a covalent character. The more representative ligands of this type are CO and NO, and the cyanide (nitrile) and isocyanide (isonitrile) ions. Those ligands capable of forming coordination bonds at the expense of the united π-electronic system belong to a special group. They include unsaturated aliphatic and aromatic hydrocarbons, for instance alkenes, benzene, and its homologues.

In addition to the ligands above, considerable attention is given to more complex ligand systems [4,5]: aromatic and heteroaromatic compounds (heteroarenes) (i.e., five- or six-member cyclic structures with delocalized π-bonds in the ring containing, besides carbon atoms, either N, P, As, O, S, Se, or Te compounds [6–8]), various chelate-forming compounds, such as macrocyclic crown-ethers, cryptands, porphyrins, and phthalocyanines.

In this work we have paid attention to the most frequently reported ligand systems from within the enormous number of ligands used in modern coordination chemistry, as reported in the many pieces of literature dedicated to this topic. In the material that follows we have separated inorganic and organic compounds. In general, the data summarized in monographs [1,9] and our previous reviews, [10–17a] forms the basis for the presentation of this work. It must be said that not only neutral molecules are examined throughout, but also anions. Considerable attention is paid to the reviewed literature of the last decade. Due to the synthetic direction of this book, classic theories on the structure of coordination compounds [17b] are not examined here.

2.2 MAIN INORGANIC AND ORGANIC LIGANDS

2.2.1 Homoelement Ligand Systems

2.2.1.1 Hydrogen

Hydride complexes are present in a broad spectrum of literature, summarized in monographs [18–21], reviews [22–37], and a large series of original articles, for example, Ref. 38–44. The simplest two-electron σ-donor ligand is the hydride H^- anion [3,11,18–27]. It contains a pair of electrons, forming the two-center coordination σ-bond, and can participate both as a terminal and as a bridge ligand.

A considerable number of monohydride mononuclear complexes are known. In these compounds the hydride anion plays the role of a terminal, as for instance in $[HCo(CO)_4]$, $[HFe(CO)_2Cp]$, $[HCo(CN)_3]^{3-}$, $[HIrCl(CO)(PPh_3)_2]$, and more [18a,20,23–27,39]. The H^- anion may also carry out a bridge function, forming clusters with one {as in $[HNb_6I_{11}]$, $[HRu_6(CO)_{18}]^-$, $HOs_{10}C(CO)_{24}]^-$}, two {as in $[H_2Os_3(CO)_4]$, $[H_2Os(CO)_{11}(RC{\equiv}CH)]$}, three {as in $[H_3Os(CO)_{11}(RC{\equiv}CH)]$}, and even four {as in $[H_4Re_4(CO)_{12}]$, $[H_4R_4(CO)_{12}]$} hydride bridges [18a,20,23–27,30]. The bond formed, of the kind M—H—M, is described in MO terms as a three-center two-electron bond [24,29].

In hydride bridge complexes, three-center two-electron MO can be formed from a 1s-AO of a hydrogen anion and the appropriate *d*-orbitals of the metal atom. Those orbitals depend on the particular symmetry and can yield arrangements in the plane $M{\diagup}^{H}{\diagdown}M$ [29]. However, such a description is only qualitative and has to be taken as being oversimplified [29].

A bridge hydride ligand forms "*open*" and "*closed*" three-center two-electron bonds depending on the rate force of the metal–metal interaction [29]. The open bridge bond is shown, for instance, in the complex cation $[(PtEt_3)_2HPt(\mu\text{-}H)PtPh(PEt_3)_2]^+$ presenting a low metal–metal interaction [29], while the closed bridge bond is formed in the complex $[(\mu\text{-}H)W_2(CO)_9NO]$, where the WHW angle is 125.5° and the metal atoms are closer one to the other [29]. Thus, molecules of "closed" hydride complexes are more bent. Bimetallic hydride complexes have been extensively reviewed [24].

The main methods for the synthesis of hydride complexes are sufficiently well discussed, their procedure features are: interaction of complex compounds with hydrogen, oxidative addition of hydrogen-containing molecules, formation of hydride complexes resulting from cleavage of the C—H bond, and protonation of the central atom of certain complexes [18a,28]. Various physical and chemical methods, including x-ray single crystal diffraction [20,23–30], have been applied for the study of the properties and structures of hydride complexes. Additionally, the data on acidity of the common transition metal hydrides (pK_a values in various solvents) are available [21b]. Thus, $HCo(CO)_4$ is a much weaker acid than perchloric, a somewhat weaker one than HBr and H_2SO_4, and approximately equal in acid strength to HCl and HNO_3. Electron-donating substituents decrease acidity.

Complexes in which molecular hydrogen participates as a ligand are also known [5,28,29,34,36,37]. The bonding in them can be seen in Fig. 2.1.

The number of such complexes stable at ambient temperature, discovered before 1992, totaled some 170 compounds [34]. The following complexes belonging to this group have been obtained: $[M(CO)_n(H_2)]$, $[CpM(CO_2)_2(H_2)]$, $[M(CO)_3(PR_2)_3(H_2)]$, $[MH(H_2)(PR_3)_4]^+$, $[M(NH_3)_n(H_2)]^{2+}$, etc., where M = Cr, Mo, W; Ru, Os, Ir; Co, Re, Fe; $n = 4$–5 [28]. One of the first compounds of molecular hydrogen, $[W(\eta^2\text{-}H_2)(CO)_3\{P(i\text{-}Pr)_3\}_2]$, described in Ref. 38, as well as a series of complexes containing cations $[(RCp)_2TaH_2CO]^+$ (R = H, *t*-Bu, $SiMe_3$) [41] and $[(Me_3SiCp)_2NbH_2L]^+$ (L = PMe_2Ph) [42], should be noted.

The main synthetic methods used in the preparation of the coordination compounds mentioned above are: reactions of coordinatively unsaturated complexes with hydrogen, protonation of hydride complexes, and formation of complexes containing molecular hydrogen by a synthesis reaction under reducing conditions.

Coordination compounds of H_2 [28] and those which contain H and H_2 ligands simultaneously [28,44a,c] have been studied by x-ray single crystal diffraction, ^1H NMR (by solid state NMR), and vibration spectroscopy. Before 1993 only six complexes had been studied by the first method. The distances H–H and M–H vary depending on the nature of the complex compounds of hydrogen within the limits 0.75–1.17 Å and 1.50–1.99 Å, respectively [28]. These data are as follows: $[W(CO)_3(P\text{-}i\text{-}Pr_3)_2(H_2)]$, 0.75(16) Å and 1.95(23) Å; $[FeH(H_2)(Cy_2PCH_2CH_2PCy_2)_2]BF_4$ (where Cy is cyclohexyl), 0.87(3) Å and 1.67(2) Å; $[Fe(\eta\text{-}C_5H_3)(CHMeNMe_2)(P\text{-}i;Pr_2)(\eta^5\text{-}C_5H_5)(H_2)Ru(\mu\text{-}H)(\mu\text{-}Cl)_2RuH(PPh_3)_2]$, 0.80(6) Å and 1.50(4) Å; $[Re_2Cl(H_2)(PMePh_2)_4]$, 1.17(13) Å and 1,49(9) Å; $[Re(H_2)H_4(PPh(CH_2CH_2PCy_2)_2]SbF_6$, 1.08(5) Å and 1.98(9) Å; $[RuH(H_2)(I)(PCy_3)_2]$, 1.07(7) Å and [1.605(5) Å and 1.59 (4) Å]. Similar bond lengths were obtained from neutron diffraction [28].

The magnitudes given above testify to an increase in the H–H distance upon coordination, a circumstance already noted [5], since for the H_2 molecule this length is ~ 0.736(10) Å.

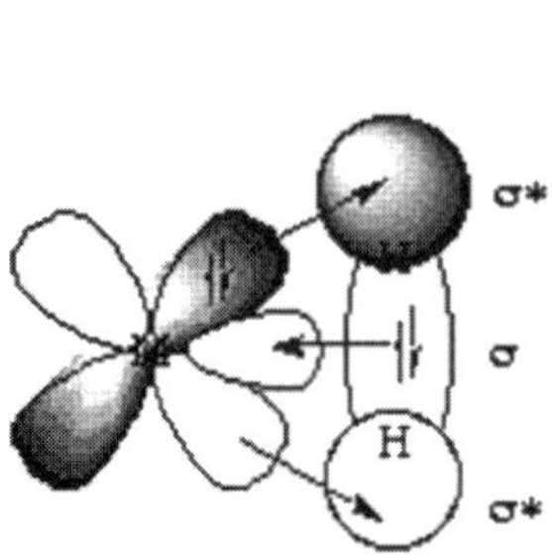

Figure 2.1 Bonding in complexes of molecular hydrogen.

Finally, polyhyride complexes have been fully reviewed [35]. The quantum-mechanical exchange coupling in dihydrogen and polyhydrogen complexes is carried out.

2.2.1.2 Carbon

Carbon participates as a ligand both in the atomic [11,25,26,45] and molecular [11,46] state. Atomic carbon forms part of complex clusters [26,27,45]. Examples of such compounds are the neutral $[Fe_4C(CO)_{13}]$, $[Fe_5C(CO)_{15}]$, $[Ru_5C(CO)_{15}]$, $[Os_5C(CO)_{16}]$, $[Ru_6C(CO)_{17}]$ and the anionic $[Fe_4C(CO)_{12}]^-$, $[Fe_6C(CO)_{16}]^{2-}$, $[Rh_6C(CO)_{15}]^{2-}$, $[CoC(CO)_{18}]^{2-}$ [26]. Structures of these and similar clusters are discussed by several authors [25,26].

It has been shown that the structural unity of isostructural complexes with a general formula $[M_5C(CO)_{15}]$, where M = Fe, Ru, Os, is a square pyramid with each metal atom occupying a vertex. Every metal atom is connected by three terminal carbonyl groups. Hexanuclear clusters form mostly two different types of structures: octahedral and trigonal-prismatic. Thus, a group Ru_6C in the complex compound $[Ru_6C(CO)_{17}]$ is an octahedron, containing the carbon atom in the center. An analogous skeleton structure is typical of the iron cluster $[Fe_6C(CO)_{16}]^{2-}$.

On the contrary, a trigonal-prismatic polyhedron arrangement is presented by the cluster $[Rh_6C(CO)_{15}]^{2-}$. The skeleton structure of the cluster $[Co_8C(CO)_{18}]^{2-}$ is a tetragonal antiprism. The structures of more complex carbide clusters are also discussed, for instance $[Re_8C(CO)_{24}]^{2-}$, $[Os_{10}C(CO)_{24}]^{2-}$, and $[Os_{11}C(CO)_{27}]^{2-}$ [25].

Graphite forms a kind of layer compound [46] of general formula $(M^0)_pC_n$ and $(MX_m)_pC_n$, where M^0 is a zero-valent transition metal, and MX_m are halides of metals belonging to practically all groups of the Periodic Table. Carbon atoms form nets with six-angled cells in the graphite structure, like in benzene. Each carbon atom in such a system has a free π-electron in a *p*-orbital, perpendicular to the plane of the molecule. Thus, the carbon net represents a polydentate aromatic π-ligand system and the metal atoms, arranged in the graphite matrix, can in this way form π-complexes with graphite, or even with graphite nets as shown in Ref. 46.

The synthesis of metal–graphite compounds is carried out by the reduction of layer graphite compounds and metal chlorides by using different reducing agents such as, for instance, solutions of aromatic anion radicals in THF, metallic sodium in liquid ammonia, complex boron and aluminum hydrides, and vapor or liquid potassium. As a result, the complex compounds containing one (Sn, Co, Mn, Cr, Mo, W, Pd) and two (Fe–Mo, Fe–W, or Mo–W) metals were synthesized.

The complexes described above have been studied by various physicochemical methods, including x-ray diffraction, γ-resonance, Röntgen-fluorescent spectroscopy, and EXAFS [46].

2.2.1.3 Elements of Group VA (Nitrogen, Phosphorus, and Arsenic)

Nitrogen

Atomic nitrogen forms complex cluster systems [11,26,27], whilst molecular nitrogen forms a series of adducts with transition metal salts [13,47,48]. The fragment N is a part of the following nitrido cluster anions: $[Fe_4N(CO)_{12}]^-$, $[Os_4N(CO)_{12}]$, $[Ru_5N(CO)_{14}]^-$, $[Co_6N(CO)_{15}]^-$, $[Ru_6N(CO)_{16}]^-$, $[PtRh_{10}N(CO_{21}]^-$, etc. [26]. Heteroligand nitride clusters are also studied, for instance $[HRu_4(N)(CO)_{12}]$,

$[HFe_5(N)(CO)_{14}]$, $[Ir_3(N)(SO_4)_6(H_2O)_3]$, $[Re(N)Cl_2(PPh_3)_2]$, $[Ru_2(N)Cl_8(H_2O)_2]^{3-}$, and $[Mo_3(N)(O)(CO)_4Cp_3]$, whose structures have been discussed in a review [26].

Dinitrogen N_2 takes part as a mono- and bidentate ligand and is bound with metals via three structurally-proven types ***30–32*** (E = N) [13,49–56]:

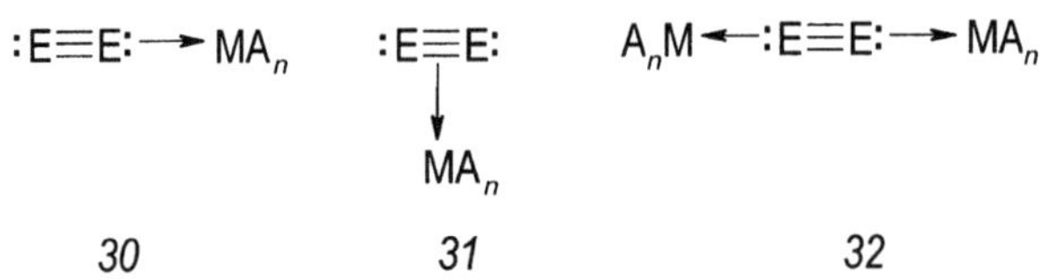

σ-Bonds are formed in the complexes ***30*** with the terminal coordination and ***32*** with a bridge connection of the nitrogen molecule, meanwhile a donor–acceptor interaction of π-type exists in compound ***31***. Additionally to the structures above, the nitrogen molecule can form complexes where N_2 behaves as a π,π-bridge ***33*** [13] or a tetradentate ligand ***34*** [54]:

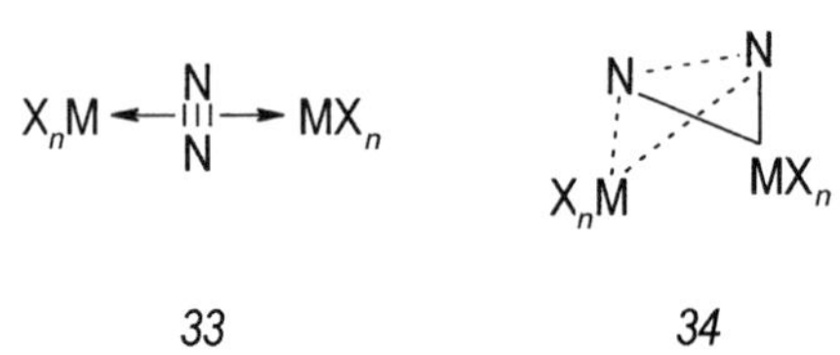

It has been shown that the bonding isomers* of nitrogen complexes of the types ***30*** and ***31*** are in the following equilibrium [13]:

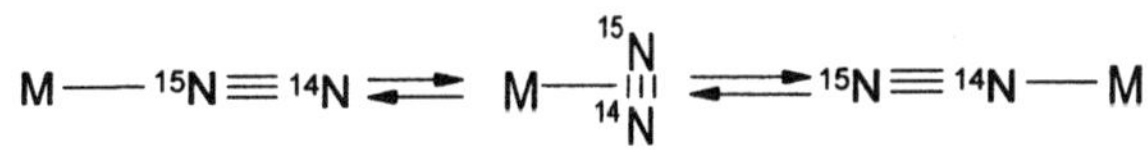

Such an equilibrium was registered in particular for the complex compound $[Re(\eta^5\text{-}(CMe_5)Re(CO)_2N_2]$ by ^{15}N NMR [51]. Other examples of complex compounds with different coordination mode of nitrogen molecules are described particularly in reviews [54–56].

Phosphorus

In a feature that distinguishes this element from nitrogen, phosphorus forms some allotropic modifications, hence the formation of complexes with different schemes for coordination of ligands becomes characteristic. Atomic phosphorus, as well as other elements discussed above, is part of the clusters as in $[Co_6P(CO)_{16}]^-$ [26]. The diphosphorus molecule P_2 participates both as terminal and bridge ligands [1,57–60], forming complexes ***30–32*** (E = P). An example of such compounds is the dimeric complex ***35*** [57], where P_2 has a bridge function:

---

*Bond isomery describes the isomers with different coordination mode of ligands on different donor centers [3,4].

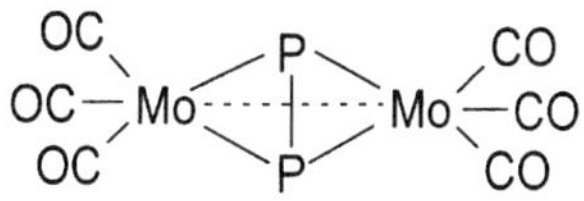
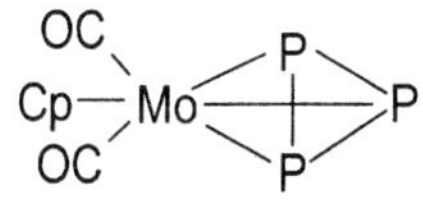
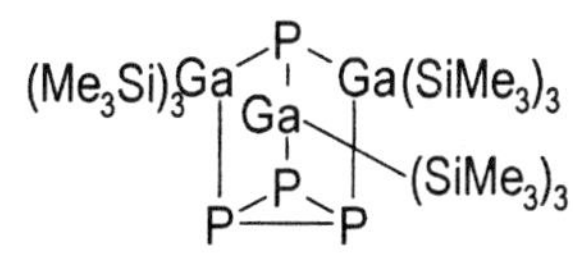

35 *36* *37*

Metal complexes, obtained on the basis of the triphosphorus molecule P_3, are represented by compounds with the η^3-coordinated triangle ligand fragment ***36*** [57]. An illustration of such a ligand fragment is found in the trinuclear gallium cluster ***37*** obtained from the P_4 molecule [61]. During the complex-formation process this molecule eliminates atomic phosphorus which, as well as in the case of other phosphorus clusters [26], is capable of connecting some metal atoms. The tetrahedral P_4 molecule can also act as a monodentate η^1-ligand, forming complexes of the type ***38*** [62]:

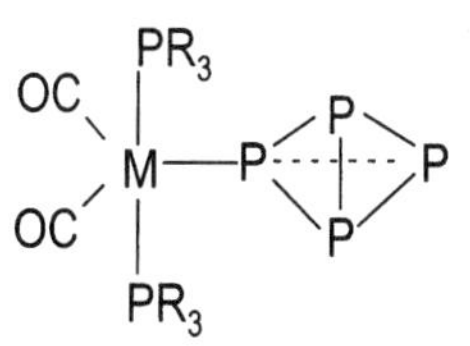
M = Mo, W; R = C_6H_{13}-cyclo

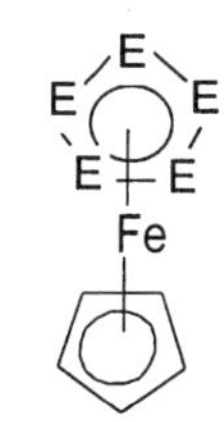
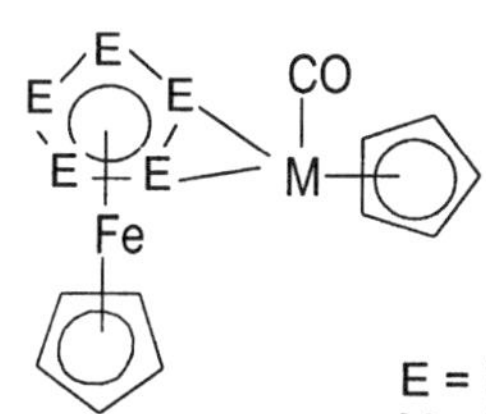

E = P, As
M = Ti, Rh

38 *39* *40*

In the case of pentameric phosphorus (P_5 molecule), complex compounds of the types ***39*** [63] and ***40*** [64–66] are formed. Multicenter $\eta^5(\pi)$-coordination is present in the sandwich ***39***, meanwhile, the unique μ-$\eta^5(\pi)$:$\eta^2(\sigma)$-connection of the ligand exists in ***40***.

The P_6 molecule (hexaphosphabenzene) can participate not only as an $\eta^6(\pi)$-coordinated ligand [67–71] for example in the three-decker complex ***41*** [71], but also as an η^3,$\eta^3(\sigma,\sigma)$-bonding ligand system ***42*** [67]:

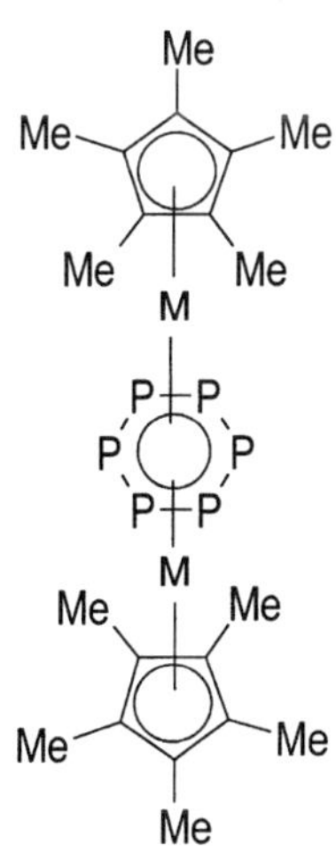
M = V, W

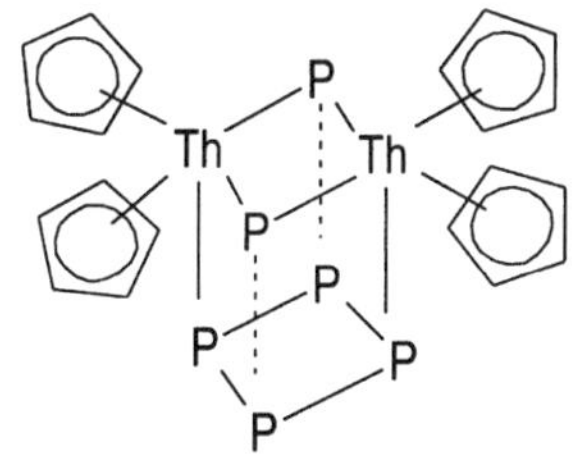

41 *42*

Arsenic

Elemental As forms part of the clusters, as for example in $[Co_3As(CO)_9]$ and $[Fe_3(As)_2(CO)_9]$, whose structures can be represented by a tetrahedron ***43*** and a trigonal bipyramid ***44***, respectively [26]:

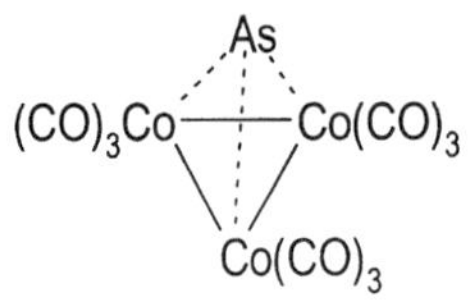

43

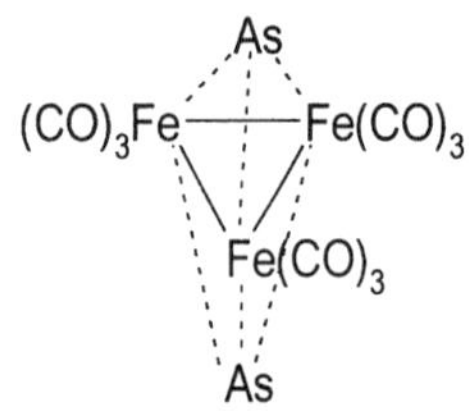

44

In case of the molecule As_2, three types of structures are generally observed, for which the σ-coordination (***45***, ***46***) or π-coordination (***47***) bonds are formed [11,72]:

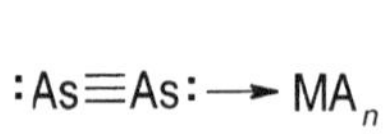

45

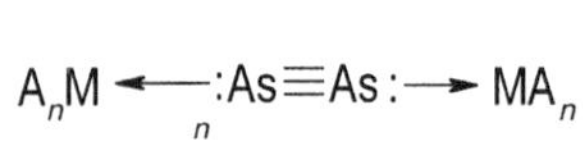

46

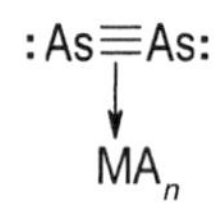

47

The bridge coordination of As_2, of the type ***46*** [72], is especially widespread. The molecules As_3, As_5, and As_6 can participate as ligands too [72]. This is a situation in which the coordination presented by the complexes of Sb and Bi, working as ligands, should be analogous, although there is little reported data available [1,26,73].

2.2.1.4 Elements of Group VIA (Oxygen, Sulfur, Selenium, and Tellurium)

Oxygen

Atomic oxygen participates mainly as a μ_2- or μ_3-bridge ligand [26]. The structures ***48*** $[W_3O_4F_9] \cdot 3H_2O$ and ***49*** $[Os_6O(CO)_{19}]$ [26] are illustrations of these kinds of ligands:

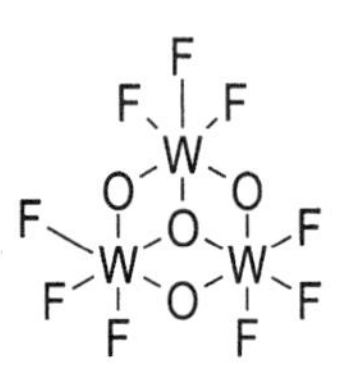

48

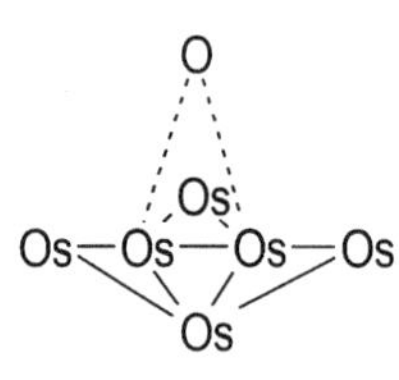

49

For molecular oxygen three modes of coordination with metals are the most widely found: terminal ***50***, chelating ***51***, and bridge ***52*** (E = O) [11,13,74–76]. The possibility of other types of dimeric structures is not excluded [74]:

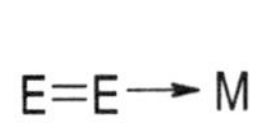

50

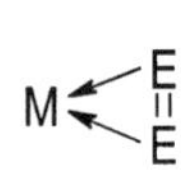

51

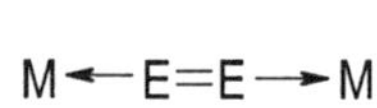

52

A calculation of the complex with oxo bridges by the ab initio method was carried out in a recent work [77].

Sulfur

Numerous clusters with atomic sulfur have been reported [26,78–81]. Amongst them the following compounds are described: $[Rh_{10}S(CO)_{22}]^{3-}$, $[Os_6(S)_2(CO)_{17}]$, $[Rh_{17}(S)_2(CO)_{32}]^{3-}$, $[Mo_3(S)_4Cp_3]$, etc. Carbonyl iron clusters, in which sulfur has a bridge function and is connected mostly with three metal atoms, are well described [81]. The following clusters, structurally characterized, belong to this type of compound: $[Fe_3(\mu_3\text{-}S)\{\mu_3\text{-}CS(CH_2)S\}(CO)_9]$, $[Fe_3(\mu_3\text{-}S)(CO)_9]^{2-}$, $[Fe_3(\mu_3\text{-}S)_2(CO)_9]$, $[Fe_2(\mu_3\text{-}S)Co(CO)_8(NO)]$.

Molecular diatomic sulfur forms part of complex compounds for which the localization modes of a coordination bond are a characteristic feature ***50–52*** (E = S) [11,13,71,78,81], and in fact the same as for those formed by oxygen.

Selenium and Tellurium

The situation close to that described above for sulfur is characteristic for the chalcogenide elements. Atomic selenium [1,26,82–86] and tellurium [86] are part of di-, tri-, and polynuclear clusters, while diatomic molecules of these elements, when acting as ligands, have mostly a bridge function ***52***. Examples of di- and trinuclear monochalcogenide compounds are the complexes ***53*** and ***54*** [82–85]:

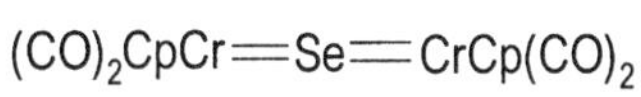

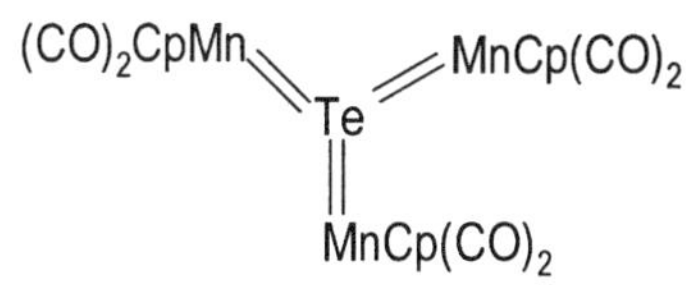

53 54

Bis-chalcogenide coordination compounds have a general formula ***55*** (E = Se, Te; M = Zr, Hf; Cp = C_5H_5, C_5Me_5) [82–85]:

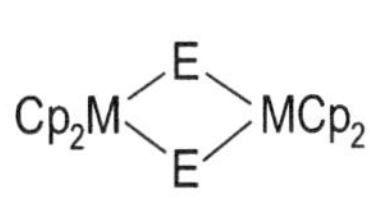

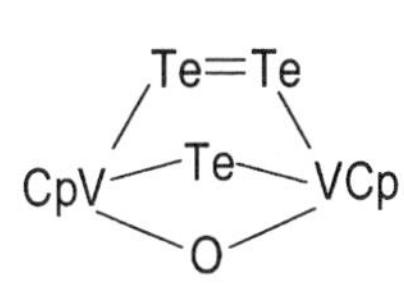

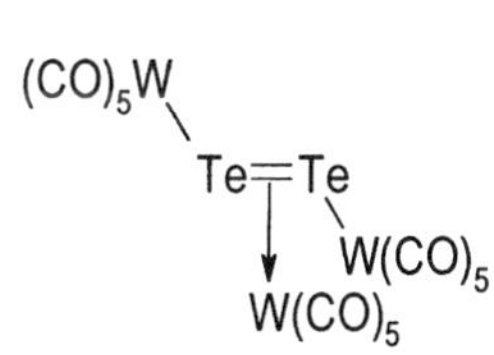

55 56 57

Complexes ***56*** [87] and ***57*** [82–85] are of considerable interest. During the formation of the first compound, both atomic and molecular tellurium participate simultaneously; in the second one, Te_2 carries out the functions of σ-bridge and π-donor ligand [82–85]. The cases are known when two diselenide bridges (for example, ***58***) and pentaselenide η^1-terminal (***59***) ligation occurs [82–85]:

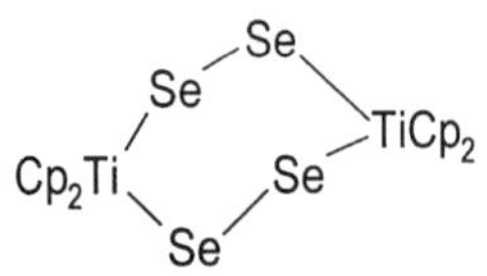

58

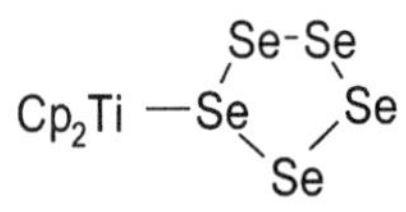

59

The tellurium-containing derivatives (for instance, ***60***) have been obtained and characterized [82–85]. Important exceptions are complexes ***61***, where the Zr$=$E terminal bond (E$=$S, Se, Te) is present as proved by x-ray single crystal diffraction studies [88]:

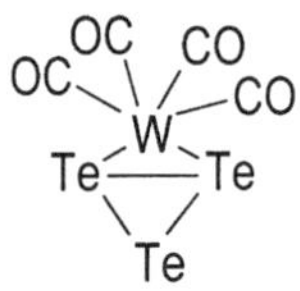

60

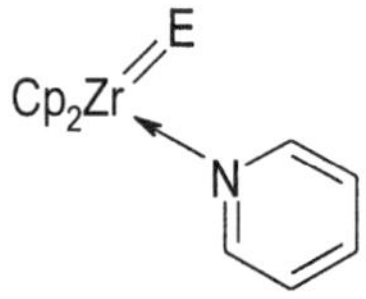

61

Other coordination compounds containing ligand atoms with the chalcogenes are described in several reviews [82–85], one of them [82] reporting data about the transition metal complexes formed mostly by heteroatomic ligands with elements of the main subgroups of both Groups V and VI of the Periodic Table: P, As, Sb (E) and S, Se, Te (E$'$), respectively of the type $E_mE'_n$. It has been observed that the numbers of E and E$'$ atoms (m, n) vary from 1 to 4; the complex compounds thus formed have mono-, di-, and tetranuclear structures, whose ligands have either a terminal, chelating, or bridge function.

2.2.1.5 Elements of Group VIIA (Halogens)

The halide ligands, and particularly those of chloride, are the most propagated ones in modern coordination chemistry. Their complex compounds are represented in classic [89–91] and modern editions [1, vol. 2; 3–5,32,73,92]. Coordination compounds of practically all metals and halogens have been synthesized and characterized. Presentation of the enormous amount of material available for the complexes of this type is beyond the limits of the present book.

The halides participate as terminal and bridge ligands and can be included in inner and external spheres of this kind of complex. A few examples of compounds with such terminal ligands are the boron ammines $[BX_3 \cdot NH_3]$, $[BX_2(NH_3)_2]X$, $BX(NH_3)_3]X_2$, and $[B(NH_3)_4]X_3$ (X$=$F, Cl, Br, I). Halides form part of many mixed-ligand carbonyl complexes, for example, $Fe(CO)_4I_2$ and polyhalide anions with various elements [4]. In this respect, the structure of the anion $[Re_2Cl_8]^{2-}$ provides important information: the Re—Re bond is present and the chlorine atoms have a bridge function [4].

Halide bridges are characteristic of many complexes of the type $M_mHal_nL_p$ [3] and diverse clusters [26]. Not only dimeric $M_2Hal_nL_p$ structures, but also polymeric ones exist in the complexes of this type of bridge. Thus, a "column" configuration with bromine bridges, together with an alternating square-plane and octahedral fragments, are present in the complex molecule $[Pt^{II}(NH_3)_2Br_2][Pt^{IV}(NH_3)_2Br_4]$ [3].

Halide ions can be connected terminally in clusters {for example, $[Os_5C(CO)_{15}I]^-$}, they can also carry out a bridge function {$[ReCl_3)Cl_6]^{3-}$, $[Ru_4(CO)_{13}Cl]^-$} [26], and even be connected to more than two metal atoms (as in $[Pt_4Cl_4Me_{12}]$, where the chlorine atom is connected to two and three Pt fragments [3]). Quantum-chemical calculations (ab initio method) of the halide bridge structures have been carried out extensively [77].

2.2.2 Heteroelement Ligands

2.2.2.1 Dinuclear Molecules

The heteroelement diatomic molecules CO [3–5,11,13,48,93–98], CS [11,13,99,100], and NO [11,13,101–104] are widely found as ligands in modern coordination chemistry. Formation of complexes having the structures ***62–65*** has been proved for carbon mono-oxide [3–5,13]:

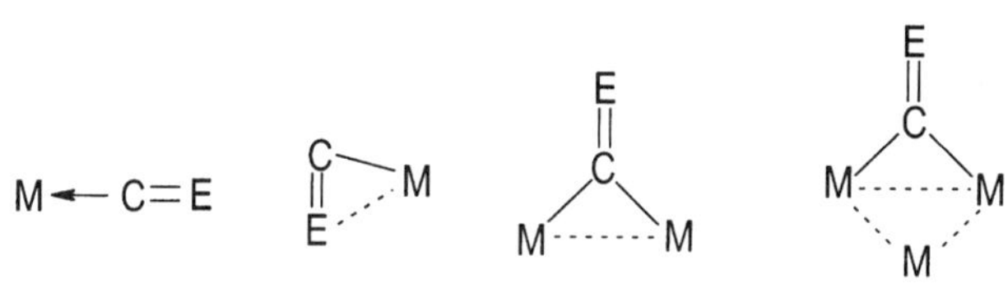

62 *63* *64* *65*

The schematic formula given by ***62*** corresponds to the most frequent mode of localization of the coordination bond (E = O). This mode is present in the majority of metal carbonyls $M(CO)_n$ (details in Sec. 2.2.5.1) and their derivatives $M(CO)_{n-m}L_m$. The properties and structures of the compounds mentioned above have been discussed in detail [3,4,21b,98], and do not require special examination. The comparatively rare type of structure given by ***63*** (E = O) was discovered in the dinuclear complex of *bis*(diphenylphosphine)methane with $Mn_2(CO)_5$. It may be possible in this compound to see the existence of structures with monodentate coordination ***62*** [3], as well as connections between the metal, bonded to carbon atoms, and the oxygen atoms of any of the one to five carbonyl groups in a simultaneous participation ***63*** [4].

Although the coordination modes of the NO group in metal complexes have not been examined in detail [103a], it is clear from numerous structural formulae that they may be described either by (a) terminal η^1-linear ***66a***, (b) η^1-bent ***66b*** (MNO angles are in the range 120–140°, (c) η^1-isonitrosyl ***66c*** and η^2-π-nitrosyl (cyclic side-on) ***66d*** linkage isomers (for monometallic systems), or (d) bridge structures (for polymetallic systems):

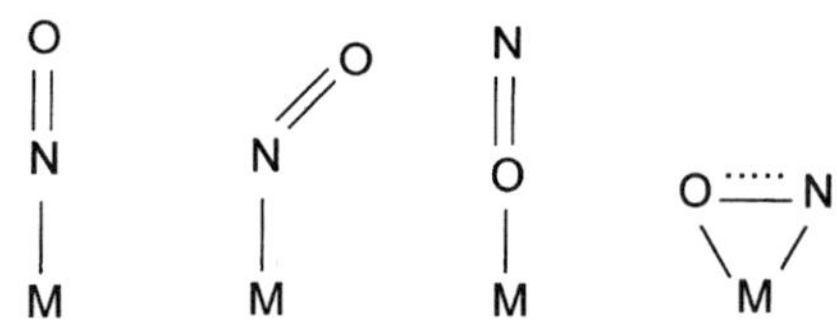

a *b* *c* *d*

66 (monometallic systems)
a - linear nitrosyl, *b* - bent nitrosyl, *c* - linear isonitrosyl,
d - side-on (cyclic)

An example of type ***66b*** is the mononuclear nitrosyl (or it can be a nitrosyl-carbonyl) compound ***67a***. Examples of the bridge structures are binuclear compounds ***67b*** [103a] and the complex $[Pd_2(NO)(CH_3COO)_2]_2 \cdot CH_2Cl_2$ with Pd–NO coordination [103b]:

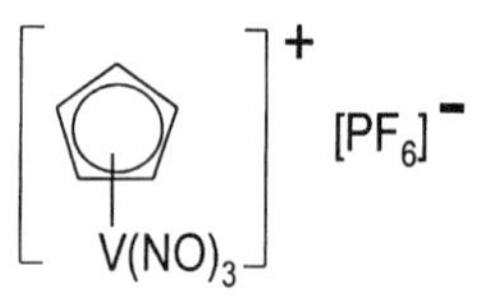

67a

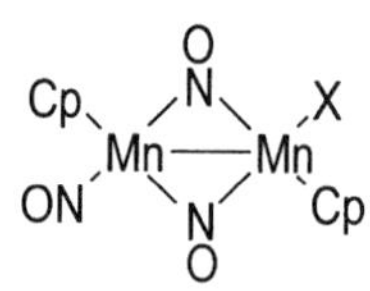

67b: X = Cl, Br, I, NO_2

Homoleptic metal nitrosyl complexes are rare; examples are $Cr(NO)_4$ or $Co(NO)_3$ [21b]. In general, NO ligands are more resistant to ligand substitution than CO groups. Coordinated nitrosyl ligand in complexes can be oxidized, forming nitrite complexes such as, for example, $(Ph_4P)_2[Pd_2(NO)_2Cl_4]$ [103c]. Among the recent investigations on NO complexes, synthesis of hydride and alkyl compounds containing the Cp*Os(NO) fragment and crystal structure of $[Cp^*Os(NO)]_2$ [103d], iron complexes $[Fe(NO)(pyS_4)](PF_6)$, $[Fe(NO)(pyS_4)]$, $[Fe(NO)(pyS_4)](BF_4)$ [pyS_4^{2-} = 2,6-*bis*(2-mercaptophenylthiomethyl)pyridine, which is a pentadentate ligand] [103e], nickel complex CpNi(NO) [103f], and palladium complexes $[Pd_4(O_2CMe)_6(\mu_2\text{-}NO)_2]$ and $[Pd_6(O_2CMe)_8(\mu_2\text{-}NO)_2]$ are described [see also Scheme (3.116), compound ***702***, in Sec. 3.2.1]. Interaction of nitric oxide and organic nitroso compounds with metalloporphyrins and heme [(por)M + NO $\leftrightarrow$ (por)M(NO)] should be especially mentioned [103h]. According to Refs. 103h, i, the above-mentioned linear isonitrosyl and side-on forms can exist as metastable states in model heme–NO compounds, and they are generated by low-temperature photolysis. The best predictor of metal–NO bond geometry in these and other NO–metal complexes is the *Enemark–Feltham* notation [103h, k, l], in which a metal–NO complex is described as an $[M(NO)_x]^n$ system (x is the number of NO ligands and n is the total number of electrons in the metal *d*- and NO π^*-orbitals).

The nitrosonium ion, NO^+, can insert into metal–carbon bonds in organometallic compounds. The synthetic chemistry of nitrosyl complexes and nitrosonium salts is covered in Ref. 92. All complexes of sulfur-containing analogues of nitrosyl ligands have linear M—NS bonds, such as, for instance, in the complex $Cr(CO)_2(NS)$. However, the existence of a bent form is not excluded [21b]. An interesting example of an unusual trinuclear *bis*(thiazyl)rhenium complex $[\{Re(CO)_5\}_3(C_2H_2N_5S_4)][AsF_6]_2$ is reported [103m], which, as well as *trans*-

$[Re(CO)_4Cl(NS)][AsF_6] \cdot CH_3CN$, is produced from *trans*-$[Re(CO)_4(CH_3CN)(NS)][AsF_6]_2$ by its reaction with cesium halides.

2.2.2.2 Trinuclear Molecules

The most widely known and abundant heteroelement triatomic ligands are those of carbon dioxide (CO_2) [11,13,104–109], carbonyl sulfide (COS) [105], carbon disulfide (CS_2) [11,13,105,110,111], and sulfur dioxide (SO_2) [11,13,112].

Five types of fragment structures, typically ***68–72*** (X = Y = O) [104,105], are discussed for carbon dioxide complexes:

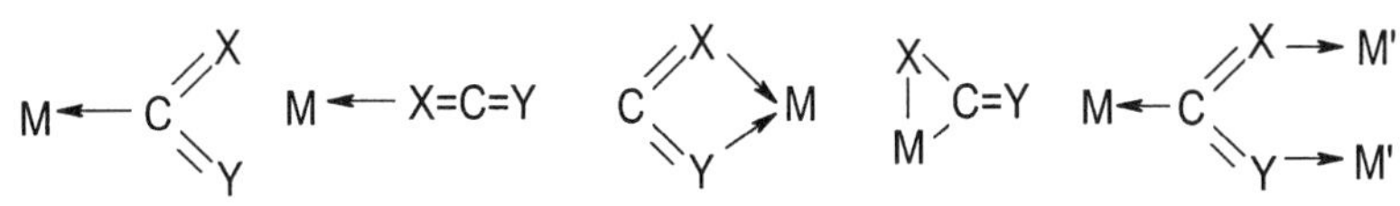

68 *69* *70* *71* *72*

Fragment structure ***68*** is found in the complex compound $[Cp_2Ti(CO_2)PMe_3]$ [105], while structure ***69*** exists in $[Cu(CO)_2(PH_3)_2]$, similarly ***70*** can be seen in both $[(CO)_2Fe(CO_2)(PH_3)_2]$ and $[(PH_3)_2Fe(CO_2)]$, structure ***71*** can be found in complexes with the general formula $M—CO_2M'L_n$ {for example, $[Cp(CO)_2M(CO_2)M'ClCp]$, where M = Fe, Ru and M′ = Ti, Zr; also known, structure ***72*** is present in $[(PEt_3)_2PhPt(CO_2)PtPh(PEt_3)_2]$. X-ray data shows that fragment ***69*** is present in the complex compound (n-Pr-salen)$_2$Co(CO_2)], where salen stands for salicylalethylenediamine. The CO_2 molecule is coordinated through the oxygen atom in the complex $[Re(Me_2AsCH_2CH_2AsMe_2)(\eta^1\text{-}CO_2)Cl]$. The structure *72* [X = Y = O, M = Re(CO)$_5$, M′ = Re(CO)$_4$] is present in the cluster $[Re_3(CO)_{13}(CO_2)]$ [104,105].

Different coordination modes, within those described above, may take place simultaneously in a given complex. Thus, a CO_2 molecule may be connected to iridium to produce a coordination of the kind ***68***, while another CO_2 molecule forms a structure such as that given by ***69*** in the complex compound $[Ir(PMe)_3(CO_2)_2Cl]$ [104]. Structure ***70***, where the corresponding CO_2 coordination is carried out through C and O atoms [104,105], is the most stable amongst the structures shown above. The situation with respect to the coordination of COS and CS_2 ligands is close to that described above. For instance, the structure ***70*** (X = S, Y = O) is observed in the complex compounds $[M(PR_3)_2(CO)_2(COS)]$ (M = Pd, Ir; R = Ph, p-MeC_6H_4) [105]; for the $[Pd(PR_3)_2(COS)_2]$ complexes, a molecule of carbonyl-sulfide is connected by means of a terminal mode ***69*** (through a sulfur atom), while the second one shows the chelating mode ***70*** [112]. In the binuclear complexes $[Mn_2(CO)_9(CS_2)PPh_3]$ and $[Mn_2(CO)_4(CS_2)_2P(Cy)_3LL']$ (Cy is cyclohexyl; L, L′ = CO, $Ph_2PCH_2PPh_2$), carbon disulfide participates as a monodentate S-donor ***69*** (X = Y = S) and bidentate S,S-donor ***70*** (X = Y = S) ligand [111].

Sulfur dioxide forms complexes, and five different modes for metal connection by this ligand have been documented; they are given by *73–77*:

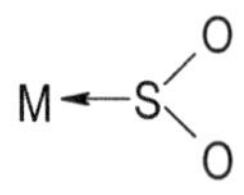

73

η^1-planar

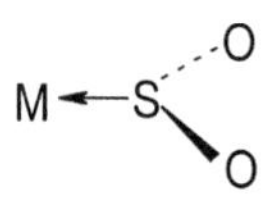

74

η^1-pyramidal

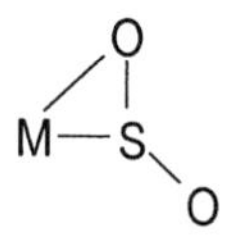

75

η^2-pyramidal

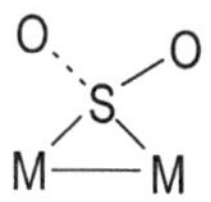

76

η^2-S-bridge

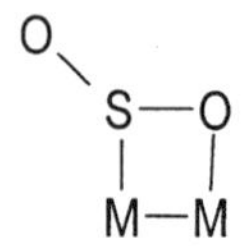

77

η^2-O,S-bridge

An example of a complex compound containing a monodentate S-coordination is ***78*** [13]. The SO_2-ligand is coordinated according to the formula *75* [13], in compounds such as those shown in ***79***:

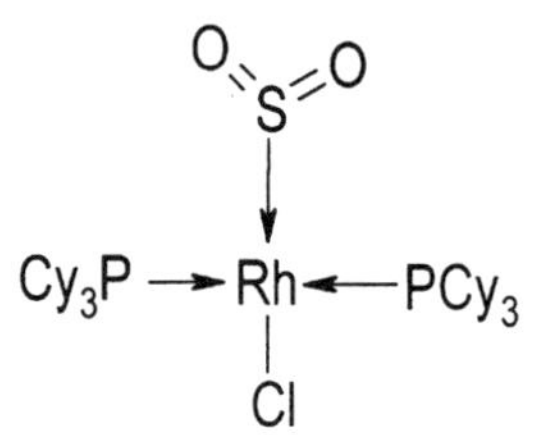

78

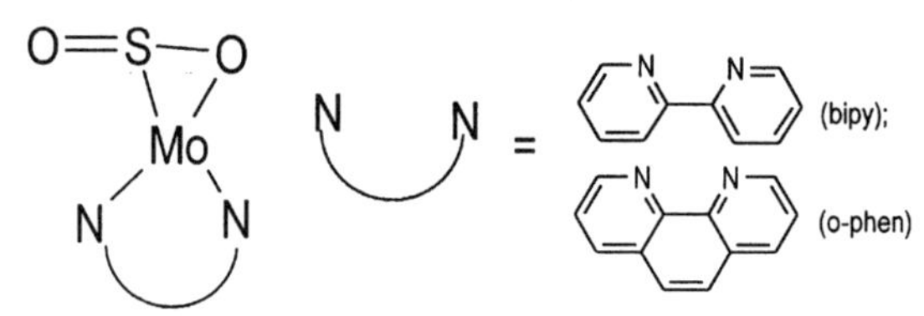

79

It is worth noting that preparative techniques for the synthesis of coordination compounds, formed by ligands such as N_2, O_2, H^-, Hal^-, CN^-, CO, NO, COS, and CS_2, are treated in an excellent handbook [113].

2.2.3 Inorganic Anions

2.2.3.1 Hydroxo- and Hydrosulfide Anions

In addition to the complexes of monoatomic anions described above, those compounds formed by the hydroxyl- [1, vol. 2; 3; 73, vol. 6] and hydrosulfide [1, vol. 2] anionic ligands are the most studied. Two kinds of connections with metals (R = H), either terminal ***80*** or bridge ***81–83***, are a main feature:

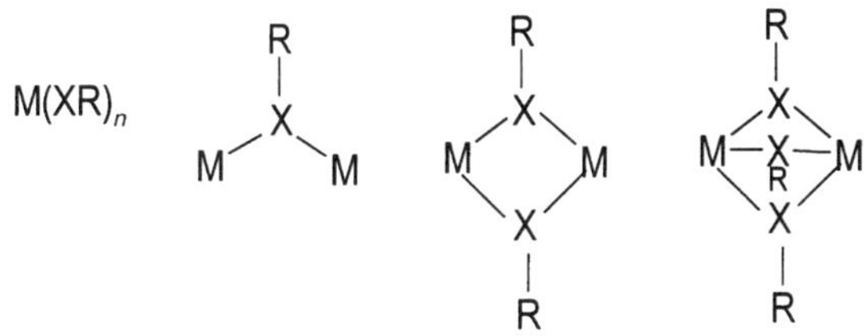

80 *81* *82* *83*

The terminal coordination of the OH^- anion is observed in a series of monomeric metal-hydroxides, for instance in the complexes $\{[(NH_3)_2Pt(OH)L]^+(NO_3)^-\}$, where L = Het; $[(PPh_3)_2MOHL]$, with L = $-HC{=}CCl_2$, $-C(Cl){=}CCl_2$; M = Pt, Pd, and $[CpRu(PMe_3)_2OH]$, where $Cp = \eta^5(\pi)$-C_5H_5 [114a]. Mono- (***81***), di- (***82***), and tri- (***83***) bridged connection (X = O) [3] is observed in many binuclear complexes where the OH^- anion participates. The hydroxo bridge, type ***81***, is observed in the binuclear complex $[(NH_3)_5Cr(OH)Cr(NH_3)_5]Cl_5$ [3] and in the tetranuclear cluster $[Me_{12}Au_4(OH)_4]$. A double hydroxo bridge is characteristic for the cation $[Co_4(OH)_4(NH_3)_{12}]^{6+}$; triple OH binding has been established by x-ray diffraction in the cation $[Co_2(OH)_3(NH_3)_6]^{3+}$. A quantum-chemical evaluation of the stability shown by OH bridges, in particular for $[Cr_2(\mu\text{-}OH)_2(NH_3)_8]^{4-}$ with a double hydroxy fragment, has been given recently [77]. The complexes with plane Cu $(\mu\text{-}OH)_2Cu_2$ binuclear system are synthesized and studied in detail [114b,c].

The same coordination modes as those just seen for OH^- are also characteristic for the SH^- anion ***81–83*** (X = S) [3,115]. In particular, the terminal metal-binding is observed in compounds such as $[HPt(SH)(PPh_3)_2]$ and $[CpNi(SH)(PBu_3)]$ [3]. According to x-ray diffraction data, the same kind of coordination has been determined in compound $[RPt(SH)(Cy_2PCH_2CH_2PCy_2)]$, where R = Ph and Cy = cyclohexyl [115].

Complexes of sulfide anions are widely found; their coordination modes, adapted from Ref. [1], are as follows:

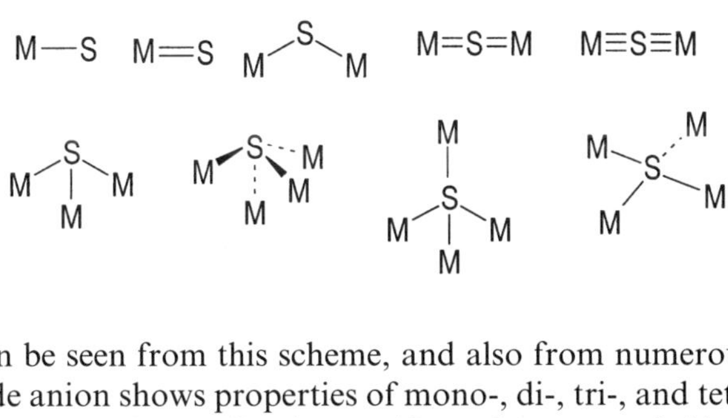

It can be seen from this scheme, and also from numerous reported data [1], that the sulfide anion shows properties of mono-, di-, tri-, and tetradentate ligands. In the last three types of coordination, sulfur takes out a bridge function, as observed for example in complexes $[Pt_2(\mu_2\text{-}S)(PPh_3)_4]$, $[Fe_2(\mu_2\text{-}S)_2(NO)_4]$, $[Au_3(\mu_3\text{-}S)(PPh_3)_3]^-$, $[Fe_4(\mu_3\text{-}S)_3(NO)_7]^-$, etc. [1]. The mixed coordination (both terminal and bridge) of thioanion [1, vol. 2] exists in compounds containing the $[W_3S_9]^{2-}$ and $[W_4S_{12}]^{2-}$ anions.

2.2.3.2 Cyanide Anion

Development of the four structural types shown as ***84–87*** [3,5,112,116–121] is a main feature of CN^-, as has been proved by various physical–chemical methods, including x-ray diffraction:

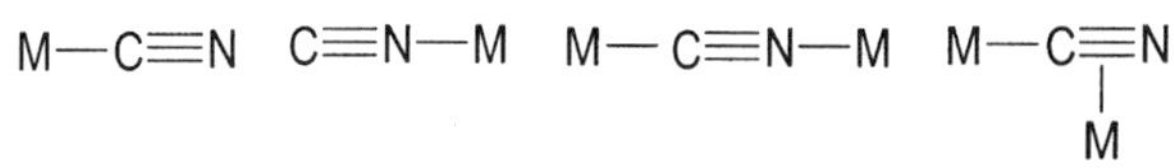

84 *85* *86* *87*

σ,σ-bridge σ,π-bridge

The C—M-coordination ***84*** takes place in the complex anions $[Ni(CN)_2(CO)_2]^{2-}$, $[Ni(CN)(CO)_3]^-$, and $[Mn(CN)(CO)_3NO]^-$ [118], and in a series of polynuclear structures, as for example $[(CO)_4(CN)MnSnPh_2Mn(CN)(CO)_4]$. The CN groups present in structure ***85*** are connected through an N atom in the complex $[Co(NC)_2(trien)ClO_4]$, where trien stands for triethylenetetramine [118].

Bonding isomery with cyanide (M—CN) and isocyanide (M—NC) complexes (see details on M—CN complexes in Sec. 2.2.4.1) taking part [117,118,122] is present in complexes of types ***84*** and ***85***. σ,σ-Cyano bridge structures of type ***86*** are well established for the complex species $[(H_3N)_5Co—N{\equiv}C—Co(CN)_5]$, $[(H_2O)_5Cr—N{\equiv}C—Co(CN)_5]$, $[(R_2AuCNAuR_2)_2]$, $[Ru_6(\mu\text{-}CN)_2(CO)_2]^{2-}$, $[Cr_2(\mu\text{-}CN)(CO)_{10}]^-$, etc. [117,118]. It is highly significant that the two isomers $[(H_3N)_5Co—C{\equiv}N—Co(CN)_5]$ and $[(H_2O)_5Cr—C{\equiv}N—Co(CN)_5]$ are clearly identified [118]. Among heteronuclear complexes, the binuclear compound ***88*** with the moiety Fe—C≡N—Cu is interesting as a biomimetic model of cytochrome C-oxidaze [121]:

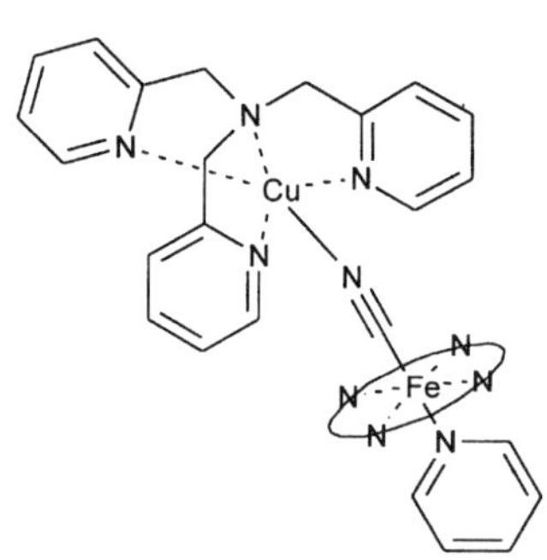

88

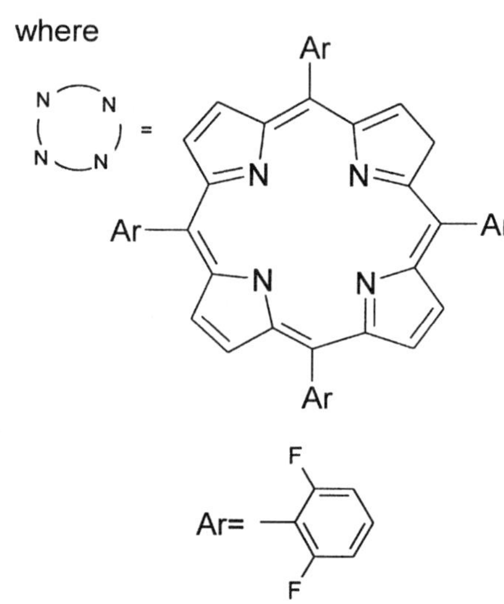

Some complexes with a σ,π-bridge-like connection on C≡N *87* are known [118]. To this category belongs the complex anion $[Cp(CO)_2Mo(\mu\text{-}\sigma,\pi\text{-}CN)Mo(CO)_2Cp]^-$ and cation $[(CO)Rh(\mu\text{-}\sigma,\pi\text{-}CN)(\mu\text{-}CO)Rh(CO)]^+$.

A bridge binding occurs when either a carbon or a nitrogen atom takes part in a coordination with two metal atoms, as shown in ***89a*** or ***89b***, respectively [118]:

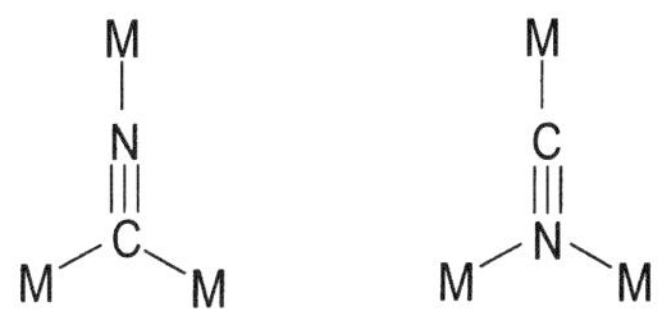

89a *89b*

Fragments of the type ***89a*** (M = Cu) are contained in the polymeric complex $[Cu(CN)NH_3]_n$ and in the compound $[Cu_5(CN)_6(DMF)_4]$. The structure ***89b*** is characteristic for heteronuclear complexes represented collectively by the formula $[Ru_3NCMnX]$, where X = H, NCMn, and $AuPPh_3$ [118].

2.2.3.3 Nitrite Anion

The nitrite anion takes part in numerous complexes for which the bonding isomery is assumed, leading to the eight types of isomers ***90–97*** [3,13,112,116,117,123–126]:

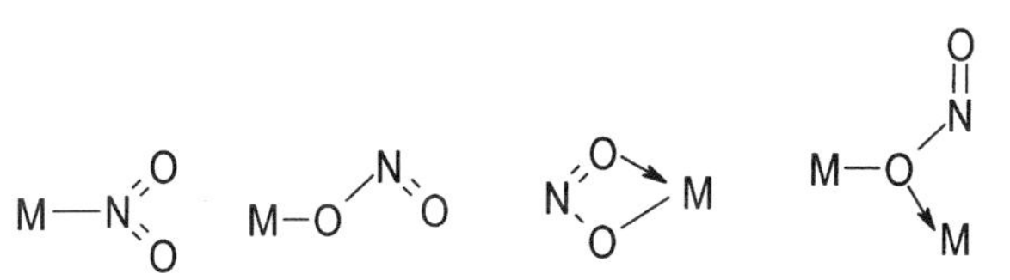

90 *91* *92* *93*

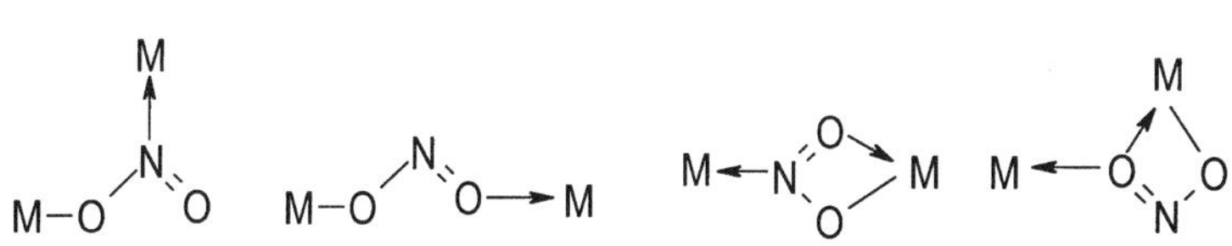

94 *95* *96* *97*

Spatial isomers are also found in several nitrite-anion complex compounds [117]: in an angle, as in ***98***, and trans-planar ***99***:

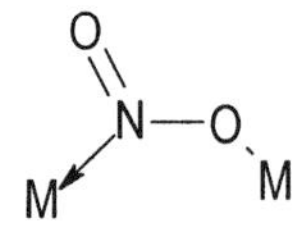

98

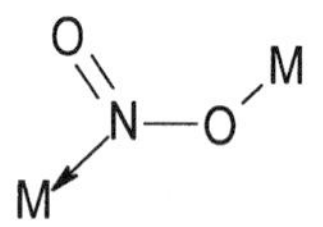

99

Numerous examples of nitrite complexes have been described in Ref. 117.

2.2.3.4 Sulfite and Sulfate Anions

As a ligand, the sulfite anion SO_3^{2-} participates either as S- (***100***) [1, vol. 2; 112,116,117,125], or as O- (***101***) [112,116,117], also as O,O-chelating (***102***) [112,116,117], or even in the form of an O,S-bridge (***103*** [117,125,126]:

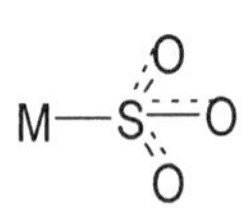

100

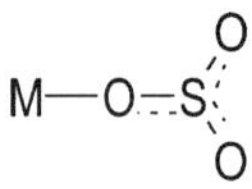

101

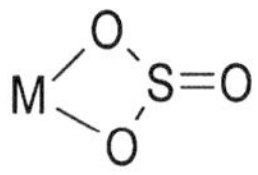

102

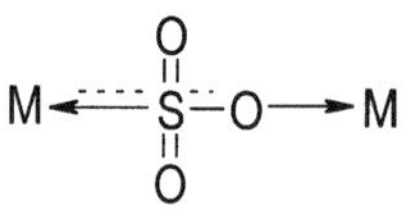

103

There are many cases in which various coordination modes of sulfite anion exist simultaneously in a complex compound. Thus, according to NMR data, the three types of metal binding ***100***, ***101***, and ***103*** are present in the complex anion $[Co^{2+}Co_2^{3+}(SO_3)_6(NH_3)_6]^{4-}$ [125,127].

In the case of the sulfate anion SO_4^{2-}, complexes are described in which the ligand performs a terminal (***104***) or chelating (***105***) function in mononuclear compounds, but the same ligand behaves as a bridge ligand system (***106***) in di- and polynuclear structures [(3,13,128–131]:

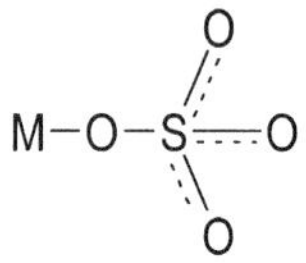

104

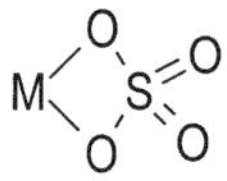

105

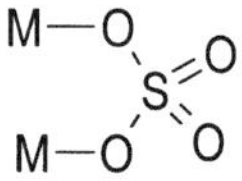

106

Sulfate anions behave simultaneously as terminal ***104*** and bridge ***106*** ligands in complex $K_8[Os_2(\mu\text{-}O)_2(\mu\text{-}SO_4)(SO_4)_4]\cdot 4H_2O$ [129]. The last coordination type (***106***) was also established via x-ray diffraction in the complex compound $H_3K_9[Os_5(\mu_3\text{-}O)_2(\mu\text{-}O)_2(\mu\text{-}SO_4)_8(SO_4)]\cdot 6H_2O$ [130]. Recently [131], complexes $[UO_2SO_4L_2\cdot H_2O]$ and $[UO_2SO_4L'_3]$ were synthesized, where L = N,N-diethylcarbamide and L′ = N,N-dimethylcarbamide; these are the first examples of mononuclear uranyl complexes with bidentate-cyclic sulfate anions. The structures of those compounds mentioned above, as well as several other coordination compounds containing the SO_4^{2-} ligand [13], were all proved through x-ray diffraction data.

2.2.3.5 Pseudohalide Anions

For the complexes of pseudohalide ligands NCX^-, where X = O, S, Se, the 10 most studied coordination modes ***107–116*** are discussed [3,5,13,48,116,117,122,132–137]:

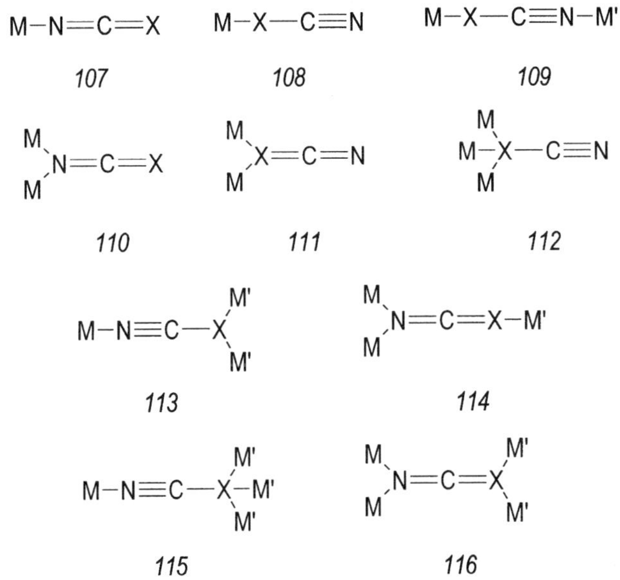

X = O, S, Se

Amongst the fragments above, the structures for the thiocyanate complexes (X = S) given by ***107–109*** are the most abundant, and there is an important amount of crystallographic data dedicated to them [117,122,134–137]. Fragment ***107*** (X = S) is the most frequently found amongst the complexes belonging to this group, and it exists in practically all complexes of hard (Ga [138], In [139], Sn [140,141]) and intermediate (Co [142,143], Mn [144,145], Fe [146–148], Ni [149,150], Cu [151–155]) Pearson acids* [156,157]. There are considerably fewer complexes in which the structure with M—S—C≡N coordination (***108***) is present, and this has been taken as a main feature for the coordination compounds of "soft" metals like mercury or the platinum group elements [13,48,112,116,117,122,136,137,158–161].

Bond isomery [described by Scheme (2.1)] is frequently observed in complexes with the structures ***107*** and ***108*** [48,117,162–164]:

$$\text{M-NCS} \rightleftharpoons \text{M-SCN} \tag{2.1}$$

Moreover, a change of coordination from ***107*** to ***108*** is characteristic for these complexes, in which the main role is given to the nature of other ligands present in the molecule, together with the conditions of synthesis and the solvents used [160,165–167]. In this respect, it is highly indicative of the exchange of configuration

*The classification of metals, their salts, and organometallic compounds, based on the *Principle of Hard–Soft Acids and Bases* (HSAB) of Pearson [112,116,156–158], is discussed in detail in Sec. 1.2.2.

of the NCS group (***117–119***), provoked by the change of nature of the diphosphine ligands of the type $Ph_2P(CH_2)_nPPh_2$ ($n = 1, 2, 3$) [49,167,168]:

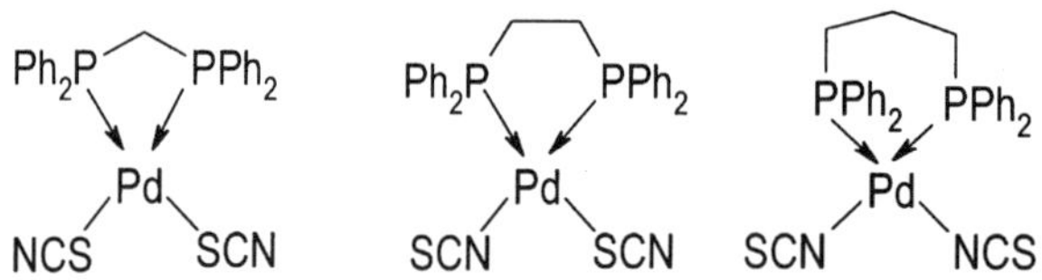

117 118 119

The most important place amongst the thiocyanate complexes belongs to the group of compounds containing the bridge bond M—NCS—M′ (***109***, X=S) [13,112,117,122,135–137]. From amongst recent investigations are worth noting those devoted to the study of thiocyanate complexes of bismuth [169], antimony [170], molybdenum [171], nickel [172,173], copper [174,175], and ruthenium [176], as well as their heterometallic compounds [112,177]. Thus, according to x-ray diffraction data of the complex $[La(NCS)(HMPA)_5(\mu\text{-}NCS)HgCl(SCN)_2]$, where HMPA is hexamethylphosphatriamide, the LaNCSHg binding of the type ***109***, LaNCS binding of the type ***107***, and HgSCN binding of the type ***108*** are all simultaneously present [177].

Other coordination modes in pseudohalide complexes are comparatively rare. Amongst them, we note the structure of complex $[AgLSCN]_n \cdot 0.25L$ (L = bipy), in which two silver ions are simultaneously bound with N atoms from NCS groups [166]. The pseudohalide complexes with simultaneous different kinds of coordination of the NCS group are also rare. In particular, the complex compound $[CuL(HL)_2][Cu(L)(SCN)(\mu\text{-}NCS)]$, where LH is 2-dimethylaminoethanol, contains within its coordination sphere both kinds; S-terminal (***108***) and -bridge (***109***) thiocyanate groups [178].

Other examples of the NCS group coordination in different forms within the same complex are described in a number of monographs [112,122,132,133] and reviews [117,135]. Data on the other modes of localization of the coordination bond is also found in the same sources. In this respect, the structure of $[Hg_2(SCN)_6Co] \cdot C_6H_6$ [135], established by x-ray diffraction studies, is interesting due to the presence of fragment ***113*** (M = Co, M′ = Hg).

Complexes with other pseudohalide groups, cyanato- NCO^- and selenocyanate $NCSe^-$, are also described. For the first one, the M—NCO binding (similar to ***107***) and M—NCO—M (similar to ***109***) are generally typical [117,122]. Side by side with that described above, the fulminate CON^- and thiofulminate CNS^- groups can be examined as pseudohalide ligands. The C—M bonds are generally contained in their complexes [117,122].

In connection with the problems already discussed, concerning the determination of the kind of binding present in a complex compound, we note that the results of recent quantum-chemical calculations of pseudohalide ions NCX^- via ab initio methods testify to the possibility of a chelating kind of binding of these ligands [179]. However, the experimental data which would confirm this conclusion is still absent.

In addition to the information given above, there is reported data on various coordination modes of a series of other inorganic anions. Amongst them are a wide number of anions, such as: NO_3^- [3,4,180,181], CO_3^{2-} [3], $S_2O_3^{2-}$ [117,125,126,182], and ClO_4^- [3], together with some exotic ones such as $[C(CN)_3]^-$ [122], NSO_3^- [117,180], SON^- [117], NSO_2^- [180], S_2CO^{2-} [117], $OSNH^-$ [117], NS_3^- [180], and NX^{2-} (X = Cl, Br, I, Sr, SeR) [183].

2.2.4 Organic Ligand Systems with Monoatomic Electron-Donor Center

2.2.4.1 Ligands with Carbon-Donor Centers

There is no reported data on the ligand systems containing a carbon-donor center in one of the main sources precisely dedicated to ligands [1] (vol. 2 of a seven-volume edition), although some textbooks on coordination chemistry [3–5] contain this material. This situation could be explained as due, mainly, to an artificial separation of metal-containing compounds into two groups: coordination and organometallic compounds. But that is precisely why all organometallic compounds showing σ-and π-localized two-center and π-delocalized M—C bonds are considered to belong to the group of organometallics, as discussed elsewhere [2,21,184].

According to the classical organometallic textbook [21b] (named "*The Bible of Organometallic Chemistry*"), common organometallic ligands with carbon-donor centers are the following: CH_3 (or other alkyl); μ-CH_3 (or other bridging alkyl), example of bonding types M—CH_3—M; η^1-aryl, alkenyl (vinyl), alkynyl (terminal and bound through only one carbon), M—Ph, M—C(R)=CR_2, M—C≡C—R; η^2-alkenyl (vinyl) (terminal and bound through two carbons); μ-η^2-alkenyl (bridging and bound to one metal through both carbons); η^3-allyls (bound through three carbons); η^4-cyclobutadiene (4 e, not aromatic) or acyclic diene (bound through four carbons); η^5-Cp or other cyclic or acyclic dienyl (bound through five carbons); η^6-arene (bound through six carbons); η^7-cycloheptatrienyl [7 e; 8 e as an anion; 6 e as a cation (tropylium cation $C_7H_7^+$ is aromatic, bound through seven atoms]; cyclo-octatetraene (COT, 8 e; not aromatic; forms η^2, η^4, η^6 or $2\eta^4$ complexes; its anion $C_8H_8^{2-}$ is aromatic, 10 e); η^1-acyl (bound through carbon only), M—C(O)—R; η^2-acyl (bound through carbon and oxygen); η^2-ketone (bound through carbon and oxygen); η^2-alkene (bound through both carbons); η^3-cyclopropenyl (3 e); η^2-alkyne (bound through both carbons); μ-η^2-alkyne (bridging and bound to both metals through two carbons); CYR (terminal carbene, where Y is a substituent capable of π-interaction with the carbene carbon), M=CCl_2, M=CPh_2, M=C(OMe)R; CR_2 (terminal alkylidene or carbene, where no substituent is capable of π-interaction with the carbene carbon), M=C(H)—R; μ-CYR or μ-CR_2 (bridging alkylidene or carbene), CR_2M_2; CR, CX (terminal alkylidine or carbyne), M≡CCH_3, M≡CNR_2; μ-CR (bridging alkylidine), CM_3R; CO (terminal or bridging carbonyl, μ-η^2-CO; CNR (isocyanide, isonitrile), etc. The most important π-donor and some selected σ-donor carbon centers are discussed in this section; the synthetic chemistry of π-donors is presented in Sec. 3.1.1.3.

Alkenes (unsaturated hydrocarbons with double bonds) are the oldest and most studied carbon ligands, with interest in their study dating from around 1827* [3,185–189]. An important aspect is the structure of π-alkene and similar π-alkyne complexes. Their structural data is summarized in reviews [186–189] and presented in

*In 1827 the synthesis of the first olephinic complex $K[Pt(C_2H_4)Cl_3]\cdot H_2O$ was reported by the Danish chemist Zeise [3,185].

a multivolume edition [2]. The bonding in alkene complexes is described by the *Dewar–Chatt–Duncanson* model, which provides us with a bonding picture not unlike that seen in carbonyl or phosphine complexes. A σ-type donation from the C=C π-orbital with concomitant π-backbonding into an empty π^* orbital on the ethyelene presents us with a synergistic bonding situation: the greater the sigma donation to the metal, the greater the π-backbonding (Fig. 2.2) [184e]:

According to x-ray diffraction data, for the majority of olephinic complexes the C=C bond of ethylene is disposed perpendicular to the plane of the metal center [184e,190]. In particular, complexes $K[Pt(C_2H_4)Cl_3]\cdot H_2O$, $K[Pt(C_2H_4)Br_3]\cdot H_2O$, and *cis*-$[(C_2H_4)Pt(NH_3)Br_2]$ have such a kind of structure; the closest coordination sphere of these compounds is represented by formulae ***120–123***. The greater the electron density backdonated into the π^*-orbital on the alkene, the greater the reduction in the C=C bond order. An alternative way of stating this would be to say that the hybridization of the alkene carbon changes from sp^2 to sp^3 as backdonation increases. Either formalism describes two limiting structures: a planar olefin adduct and a metallocyclopropane. X-ray crystallographic studies confirm that as the C—C bond length increases, the CH_2 plane is distorted from the ideal planar geometry of an alkene [184e].

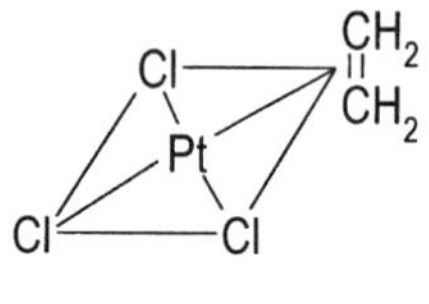

120

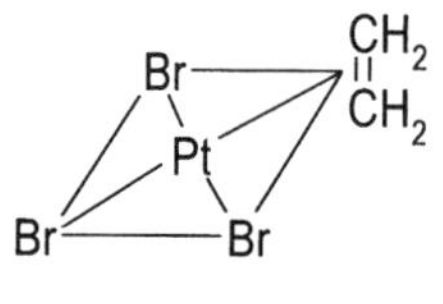

121

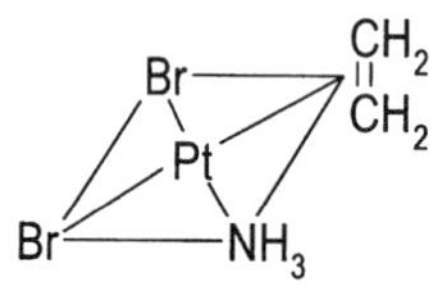

122

An analogous arrangement to the olephinic ligands is also characteristic of dimeric complexes like that shown by ***123*** ([185]:

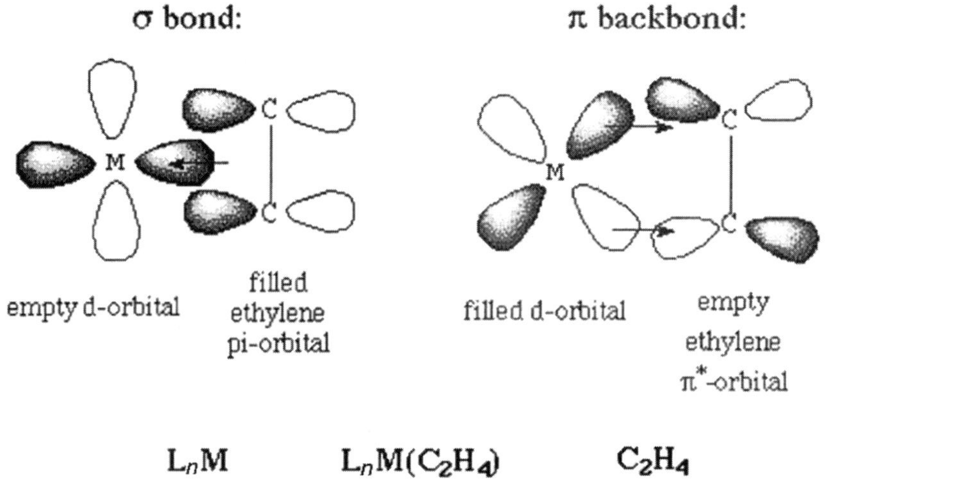

Figure 2.2 Dewar–Chatt–Duncanson model for bonding in alkene complexes. (From Ref. 184e. Reproduced with permission.)

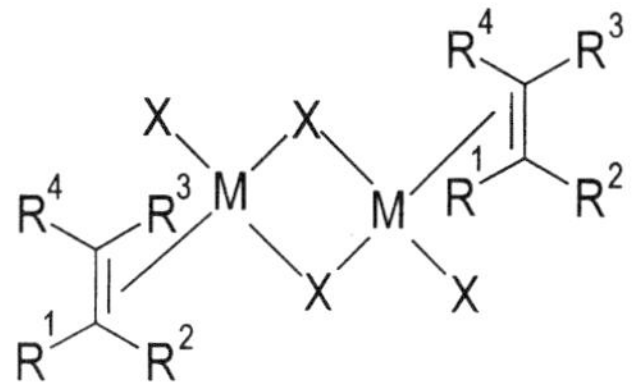

123: R^1,R^2,R^3,R^4 = H, Alk, Ar;
X = Cl, Br, I; M = Pd, Pt

The structures of π-ethyne complexes are more diverse compared to similar olephinic structures. Important modes of ligation of alkynes to single metal atoms and clusters of two to four metal atoms are presented in Ref. 184f. Some authors divide these complexes into three types [190], where:

1. The alkyne is coordinated with one metal atom (nonbridge coordination).
2. The alkyne is coordinated with two metal atoms (bridge coordination).
3. The alkyne is coordinated with three and even more metal atoms.

Nonbridge coordination is confirmed by x-ray diffraction studies, for example in the complex compounds $[(PPh_3)_2PtPhC{\equiv}CPh]$, $[(tert\text{-}C_4H_9C{\equiv}CC_4H_9\text{-}tert)(p\text{-}CH_3C_6H_4NH_2)PtCl_2]$ [190], $Cp_2Ti(\eta^2\text{-}PhC{\equiv}CPh)PMe_3$ [184f], and $Cp_2Mo(PhC{\equiv}CPh)$ [21b]. Diphenylacetylene and related alkynes take part as a bridge function in the complexes $Cp_2Ru_2(CO)_3(\mu\text{-}R_2C_2)$ [21b], $[Co_2(CO)_6(PhC{\equiv}CPh)]$, and $[Ni_2Cp_2PhC{\equiv}CPh)]$ [190]. The same ligand, $PhC{\equiv}CPh$, in $[Fe_3(CO)_9(PhC{\equiv}CPh)]$ is bound with three iron atoms [190]. In a hexanuclear complex $[Pt_2Ag_4\{C{\equiv}C(t\text{-}Bu)\}_8]$, the alkynyl groups act as $\mu\text{-}\eta^1{:}\eta^2$ (σ-Pt, η^2-Ag) ligands [191a,b]. The ethyne ligand takes part simultaneously with four cobalt atoms in the cluster $[Co_4(CO)_{10}(EtC{\equiv}CEt)]$ [190]. Useful information on polynuclear alkyne and acetylide complexes is given in Refs. 191c–i. In particular, the bridge CO and C_2R_2 groups are found in $[(CO)_8C_4(\mu\text{-}C_2R_2)]$ ($R{=}CH_2OH$, CH_2SCMe_3) [191h]. Also, α-bonded complexes with the terminal acetylide anion $L_nM{-}C{\equiv}C{-}R'$ are described [21b], which can undergo either alkylation or protonation with R^+ or H^+ at the β-carbon, forming very stable vinylidene complexes $L_nM{-}C^+{=}C{=}CR(R')$ or $L_nM{-}C^+{=}C{=}CR(H)$. Vinylidene complexes may also be obtained by rearrangement of π-bonded terminal acetylenes.

The bonding of an alkyne to a transition metal atom is similar to that of an alkene, however, an alkyne can act either as a two-electron or a four-electron donor. Alkynes have two sets of mutually orthogonal π-bonds, which can be bound with a metal in a σ-type fashion ***A*** (including a π-backdonation ***B***) and in a π-type fashion ***C*** using an orthogonal metal *d*-orbital (including a small contribution of δ-type backdonation ***D*** (Fig. 2.3) [184e]:

It is possible to depict an $RC{\equiv}CR$ metal binding by several resonance structures including the metallocyclopropane form, M← ||| form, and almost noncontributing form with two M=C bonds. It is noted that alkynes are usually nonlinear when coordinated [184e]. An interesting consequence of the distorted coordination geometry is the fact that small cyclic acetylenes, which are unstable as free molecules, are stabilized by coordination to a metal, for example $CpTa(Me)_2$(benzene) or (cycloheptyne)$Pt(PPh_3)_2$ [21b].

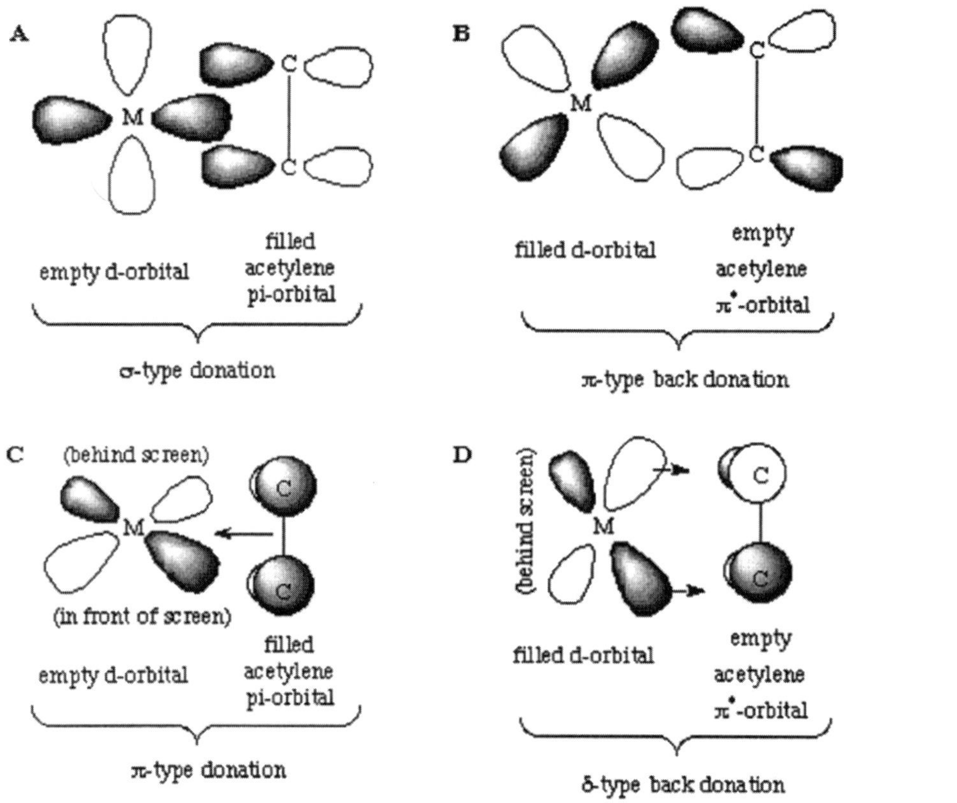

Figure 2.3 Bonding in alkyne complexes. (From Ref. 184e. Reproduced with permission.)

The main conclusions derived from x-ray diffraction studies of alkene and alkyne complexes are the following. The coordinated double bond of type π, which is a two-electron donor, can be expediently examined as one occupied by one coordination place. The coordinated triple bond C≡C can also be a donor of a pair of electrons, occupying one coordination place (nonbridge RC≡CR ligand binding), or a donor of four electrons, occupying two coordination places (bridge coordination type in binuclear complexes). That said, it becomes clear that alkenes are monodentate ligands and alkynes are mono- and bidentate ligands. Moreover, alkynes can carry out the role of ligands of higher dentacity, being found with three and even more metal atoms [189]. Another difference between alkenes and alkynes is that alkynes tend to be more electropositive and therefore tend to bind more tightly to a transition metal than alkenes, frequently substituting them in complexes [184e].

One of the most important carbon π-donors are the *cyclopentadienes* and their heteroanalogues, for instance ***15*** and ***16***. In general, cyclopentadiene itself forms three general types of mononuclear Cp transition metal complexes: Cp_2M ***3*** (symmetric molecules with mutually parallel Cp rings; examples M = Fe, Cr, Ni), Cp_2ML_n ***124*** [*bent metallocenes*, L = H, R, CO, etc. (n = 1–3)], and *half-sandwich* compounds $CpML_n$ ***125*** (n = 1–4) [21b]. In the "bent" sandwich complexes

Cp_2MX, Cp_2MY_2, and $Cp_2M(Y)(X)(Y)$, the two Cp rings are not parallel: the angle between the normals to each ring is less than 180°. Examples of typical bent sandwich complexes are Cp_2FeH^+ (Fe^{IV}, d^4), $Cp_2Mo(CO)$ (Mo^{II}, d^4), Cp_2MoH_2 (Mo^{IV}, d^2), Cp_2TaH_3 (Ta^V, d^0), $Cp_2Zr(Cl)H$ (Zr^{IV}, d^0), etc. The half-sandwich compounds (monocyclopentadienyl complexes) have a "*piano stool*" structure with Cp at the seat and one to four ligands serving as the "legs": $CpCo(CO)_2$, $CpMn(CO)_3$, CpCu(CO), CpNi(NO), $[CpFe(CO)_2]_2$ [184f], etc. There are several additional groups of Cp complexes (as well as their heterocyclic analogues): *tris*- and *tetrakis*-cyclopentadienyl compounds, common for early d^0 transition elements [for example, $Ti(\eta^5\text{-}C_5H_5)_2(\eta^1\text{-}C_5H_2)_2$], Ln and An; tri-, tetra-, and penta-decker sandwiches; "mixed-sandwich" Cp derivatives having general formula CpCr(R) (R is a cyclic aromatic ligand with three to seven carbon atoms) [21b]. Cp* ligand (C_5Me_5) is sterically more demanding than Cp, allowing the isolation of Cp* complexes for which the Cp analogues are unknown or kinetically unstable. In decamethylferrocene, Cp^*_2Fe, the methyl groups are actually tilted above the C_5 plane by 3.4° due to steric interactions. The methyl groups on Cp* are electron donors, so this results in more electron density at the metal than for the analogous Cp complex. This can be observed spectroscopically by looking at carbonyl stretching frequencies of Cp/Cp* carbonyl complexes. The increased donation of the Cp* ligand results in greater π-backbonding, which is apparent from a red shift of approximately 50 cm^{-1} in the IR spectrum. Likewise, electrochemical measurements indicate that Cp* complexes are more easily oxidized than their Cp analogues by approximately 0.50 V [184e].

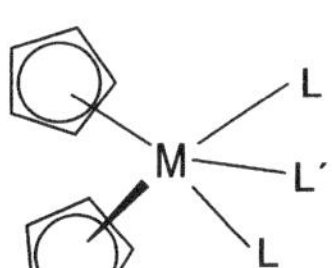

124

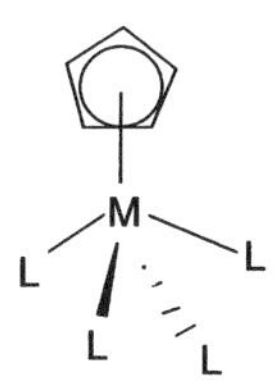

125

On the basis of Cp and its heteroanalogues, a wide series of metallocene structures of types ***3***, ***5***, ***126***, ***127*** [2–5,7,8,13,132,133,192–205] have been obtained. These complexes are typical for Fe, Co, Ni, Mn, V, Cr, and Ru, and are the "standard examples" of complex compounds with multicenter delocalized coordination bonds formed as a result of combined grouping orbitals of the cyclopentadienyl rings with *d*-metal orbitals.

E = NR, PR, AsR, SbR, O, S, Se, Te; R = H, Alk, Ar

126

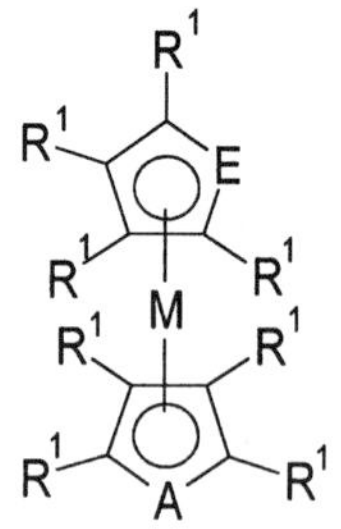

A = CR, E; E = N, P, S; R^1 = H, Alk, Hal

127

The orbital interactions, which determine the stability of sandwich complexes of the *f*-elements (lanthanides), for example, ***128*** [203,206,207a], and three-decker structures, such as ***129***, have an analogous character [2]. The majority of sandwich complexes are diamagnetic. However, paramagnetic compounds of types $[(\eta^5\text{-}C_5Me_5)_2M]^+[TCNE]^-$ (TCNE = tetracyanoethylene) [207b,c] and $[(\text{IsodiCp})_2M]^+[PF_6]^-$ (IsodiCp = isodicyclopentadienyl) [207d] are also reported.

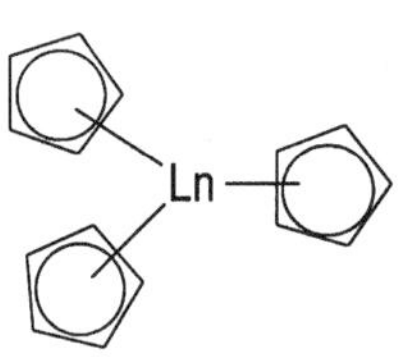

128

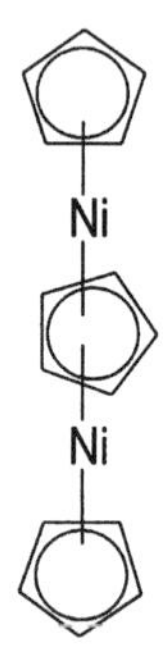

129

η^6-*Arene* (***130***) and *hetarene* (***131***) complexes are formed by π-bonds [2,4,5,7,8]. Chromium, molybdenum, and tungsten compounds of this kind have been studied in great detail [208–210].

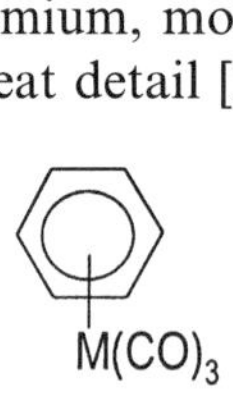

130

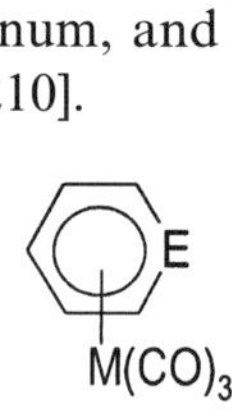

131

E = N, P, As

The following complexes of six-member *hetarenes*, $\eta^6(\pi)$-coordinated with metal, are well studied: those of pyridine ***6*** [211] and its derivatives ***132*** [211–217], pyrimidine ***133*** [8], phosphabenzene (***134***, E = P [2,218,219]), and arsabenzene (***134***, E = As [220]):

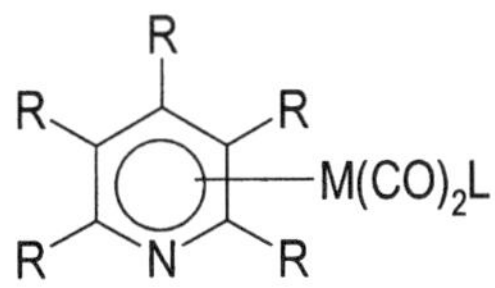

L = CO, PR'$_3$

M = Cr, Mo, W

132

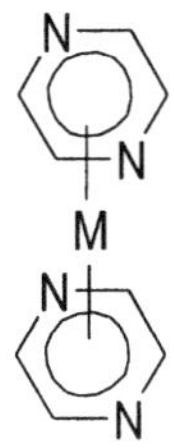

133

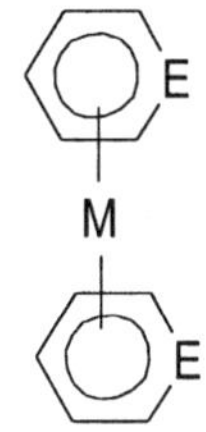

E = P, M = Cr;
E = As, M = Ti, V, Cr

134

Stability of the examined complexes is usually explained by the 18-*electron rule* (details in Sec. 1.2.3) [3,4,21b,184h,221–223]. It has been observed that a large number of coordination compounds are formed by metal-centered radicals containing from seven to 21 electrons [223]. Also important is that many 19-electron compounds are among them, such as, for instance, $[FeCp(\eta^6\text{-arene})]^+(PF_6)^-$, $[CoCp(C_5H_4BR)]$, and $[Co(C_5H_4BR)_2]$, in which the fragments ***135–137*** are contained in due order [223]:

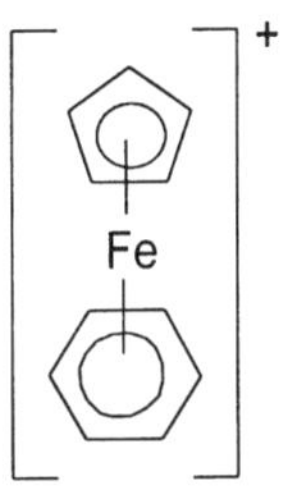

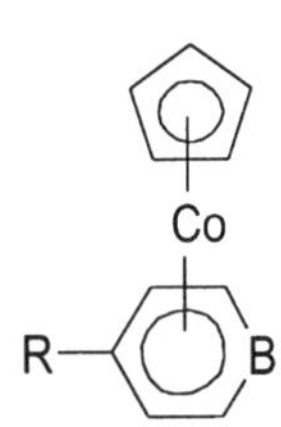

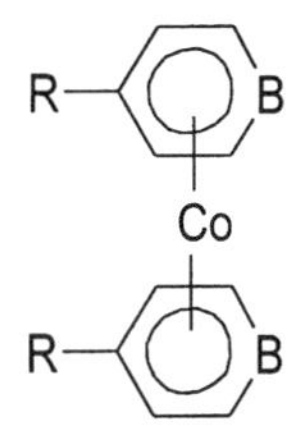

R = Me, Ph

135 *136* *137*

Numerous complexes with π-delocalized coordination bonds are formed from pentadienyl, benzyl, and organic biradicals [224a]. Thus, a considerable number of complexes have been synthesized where a transition metal is bound to a chain formed by five unsaturated carbon atoms, united by the name "*pentadienyl*" ***138***:

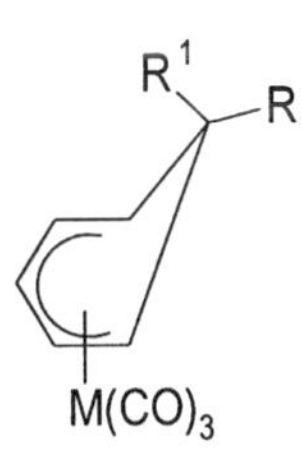

M = Mn, Re

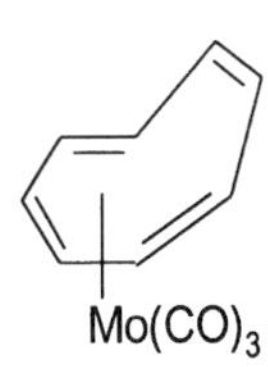

138 *139* *140*

Among these compounds, most importance is placed on complexes where the fragment ***138*** forms a part of the coordinated *ene* ligand, as for example in ***139*** and ***140*** [224a,b].

Although the arene group usually acts as a six-electron donor (η^6-binding), four- and two-electron forms (η^4 and η^2; a π-allyl bond type with the participation of one, or even two, double bonds in the benzene ring) are also known [21b,224a]. η^4-Arene complexes are structurally very interesting because the ring planarity and carbon–carbon bond lengths are strongly distorted by complexation of only four arene carbon atoms. It is as if the metal, requiring only four donor electrons from the arene to acquire 18 valence electrons, has coordinated with the diene portion of a hypothetical "*Kekule* structure." Examples of such complexes are (η^5-Cp)Rh[η^4-$C_6(CF_3)_6$] and (η^6-C_6Me_6)Ru(η^4-C_6Me_6). Also, few η^2-arene complexes exist, for example $(Et_3P)_2Pt[\eta^2$-$C_6(CF_3)_6]$ having four different lengths of the C—C bond in the aromatic ring from 1.35° Å to 1.51° Å (the last one corresponds to the coordinated C—C bond) [21b]. Other examples of such compounds are ***141*** and ***142***, whose structures have been demonstrated by x-ray diffraction [224a]:

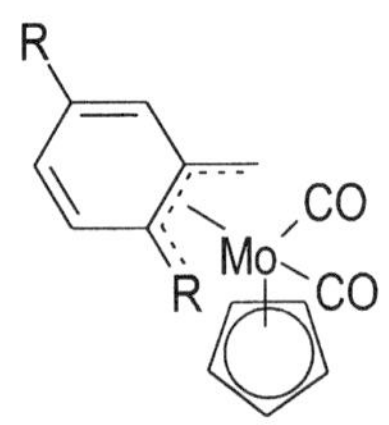

141: R = *i*-Pr

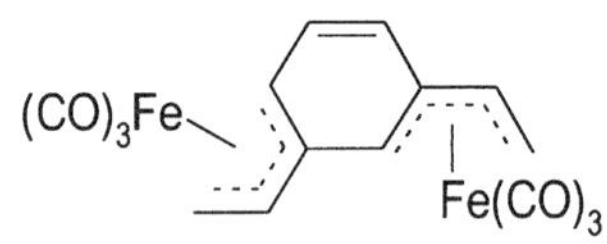

142

Complexes of trimethylenemethane ***143*** and tetramethylene ethane ***144*** biradicals [224a] are typical examples of biradical coordination compounds:

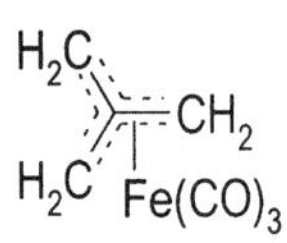

143

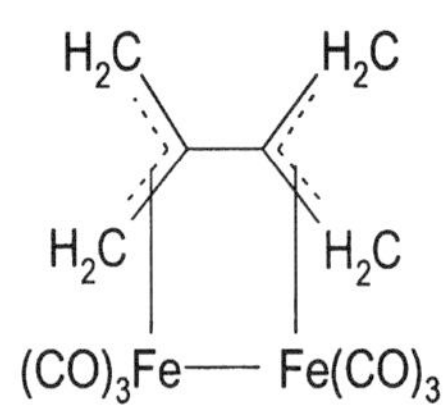

144

Side by side with the examined π-complexes, complex compounds of carbon-containing ligands are represented by structures where M ←: C bonds of σ-type of different multiplicity are present (Fig. 2.4). σ-Organometallic compounds with a single metal–carbon σ-bond are discussed elsewhere [2,21,184] and, for this reason, their detailed examination is not needed. Their classification according to Cotton [184f] includes the following main classes of transition metal alkyls and aryls: (a) mononuclear compounds {such as WMe_6, Cp_2ZrMe_2 [184f], *trans*-$VMe_2(dmpe)_2$ [224c], or $[(C_5Me_5)Os(dmpm)(CH_3)H]^+$ [224d]}, (b) binuclear compounds with M—M bonds ($Li_2Re_2Me_8$), (c) bi- or polynuclear compounds with alkyl bridges, (d) alkylate anions ($[CpLuMe_3]^{2-}$), (e) metallacycles (details in Secs. 2.2.5.1 and 4.1.3, (f) acetylides [$CpCr(CO)_3C{\equiv}CPh$] (see also above in this section), (g) fluoroalkyls and aryls (complexes with CF_3 or C_6F_5 species), and (h) acyls [$(CO)_5MnC(O)Me$] [184f]. Another classification [184h] for the most important organic ligands containing at least two carbon atoms, which coordinate to the transition metal through σ-bonds, is as follows:

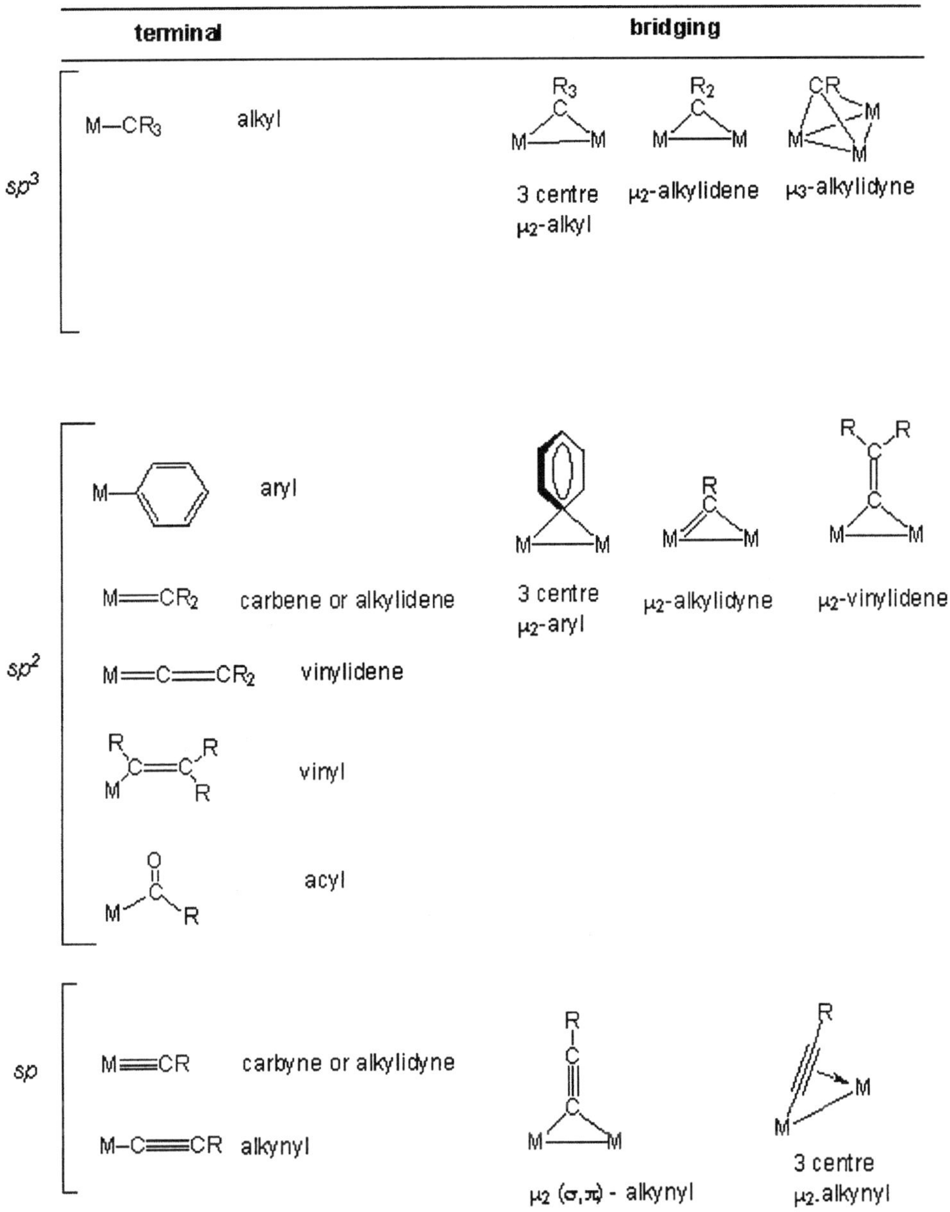

Figure 2.4 Classification of σ-organometallic compounds. (From Ref. 184k. Reproduced with permission.)

Among these compounds, transition metal alkyls are the most widespread. For simple metal alkyls, the M—R bond distance is typically 190 to 220 pm. This is approximately the sum of the covalent radii of carbon and metal, $r_C = 77$ pm and $r_M \sim 120$ pm [184h]. Simple alkyls are simple σ-donors (Fig. 2.5).

Also, alkyls can bridge two metal centers, for example like methyl groups in $(CH_3)_2Al(\mu\text{-}CH_3)_2Al(CH_3)_2$ [184e]. Coordinately saturated complexes are more

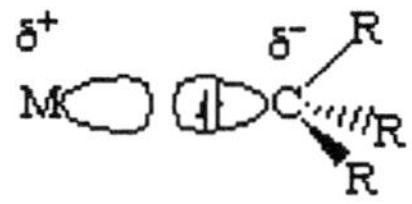

[184h]

Figure 2.5 Bonding in simple alkyl complexes.

stable [for example the stability of $TiMe_4$(bipy) is higher than $TiMe_4$; $MnEt(CO)_5$ is more stable than $MnEt_2$]; thermal stability increases with lack of β-hydrogens [WMe_6, $Cr(CH_2SiMe_3)_4$] [184f,h]. Depending on the capability of the ligand to form multiple bonds, the pure σ-bond can be supplemented by various degrees of π-interaction, metal–carbon multiple bonds [184h].

The synthetic chemistry of complexes of carbon ligands bound prinicipally by σ-bonds is covered in Ref. 21. In particular, transition metal complexes with alkyl ligands are commonly prepared by the following routes: (a) from metal halides ($ZrCl_4$, $TiCl_4$, $NbCl_5$)+organo Li, Mg, Al reagents ($PhCH_2MgCl$, Al_2Me_6, $ZnMe_2$) (final products with M—C bonds); (b) by hydrometallation (alkene insertion into a metal hydride bond) reactions {metal hydride [$(Et_3P)_2(Cl)PtH$]+alekene} [final products $(Et_3P)_2(Cl)Pt(R)$]; (c) by metallate alkylation {carbonylate anion in $[Mn(CO)_5]^-Na^+$ and alkyl halide (MeI), final product $Mn(CO)_4Me$}; (d) by metallate acylation {reaction between carbonylate anion in $[CpFe(CO)_2]^-Na^+$ and alkyl halide (MeI) with intermediate formation of acyl compound $CpFe(CO)_2(COCH_3)$, final product $CpFe(CO)_2Me$}; (e) by oxidative addition {16-e complex $(Ph_3P)_2Pt^0(C_2H_4)$+alkyl halide (CH_3I), intermediate 18-e complex $(Ph_3P)_2Pt^{II}(CH_3)(I)(C_2H_4)$, final product $(CH_3)(I)Pt^{II}(PPh_3)_2$} [21b,184h].

Carbene complexes of transition metals [2,21,225–236] are typical representatives of compounds with a double metal–carbon bond. They are seen as derivatives of a two-covalent carbon in their singlet state [226,232,236]. As a rule, the carbene ligand is an effective σ-donor and a comparatively weak π-acceptor. Formation of a σ-bond M—C takes place via transference of a nonbonding electronic pair with a nucleophilic σ-orbital of the carbenic carbon to the metal atom. Simultaneously, it is also possible to form a π-bond as a result of the interaction of symmetrically appropriate metallic *d*-AO with a vacant electrophilic *p*-orbital of the carbene [236,237]. This situation is a key factor that determines the polarization of most of the carbene complexes according to type ***145*** (Fig. 2.6).

The first carbene complexes ***146*** were prepared by the Nobel laureate E.O. Fisher [238]. At present, several hundred* of this kind of compound have been prepared with practically all transition metals [8,239–244].

Side by side with Fisher carbene complexes ***145–147***, complex compounds of methylene ***148*** (M = Ta, R = H, X = Me, n = 2), diphenylcarbene ***148*** (M = W, R = Ph, X = CO, n = 5) [237], Schrock carbenes such as $Cl_2H(PR_3)_3Ta^V(CHCMe_3)$, and many others, generalized in Ref. 184f, are also described. Schrock carbenes are typically found on high oxidation state metal complexes (early to mid transition metals). These compounds are usually electron-deficient or contain strong π-donating ligands. Contrast this to Fisher carbenes, which are typically low valence, low oxidation state complexes containing strong π-acid (acceptor) ligands

*Since 1979, some 400 stable carbene complexes are known [237].

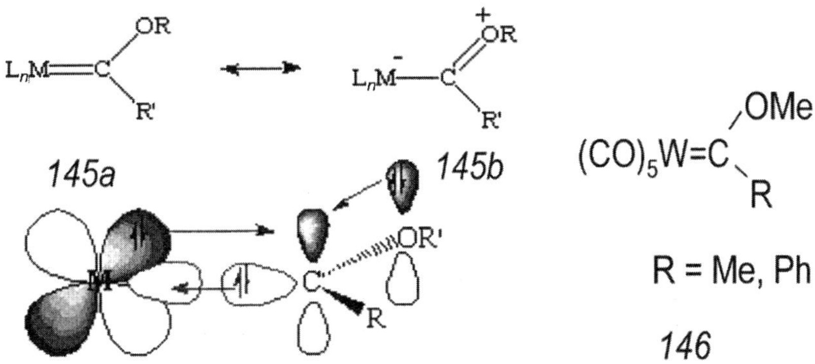

Figure 2.6 Bonding in carbene complexes. (***145*** From Ref. 184e. Reproduced with permission.)

such as CO [184e]. At present, IUPAC recommends calling Fischer and Schrock carbenes *alkylidene complexes*, restricting the term "*carbene*" to a free CR_2 species [184f].

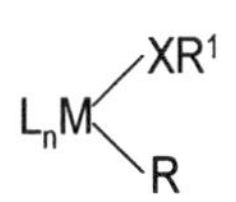
147

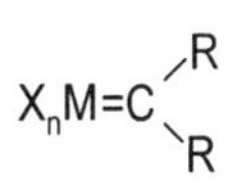
148

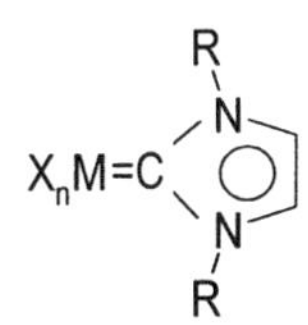
149

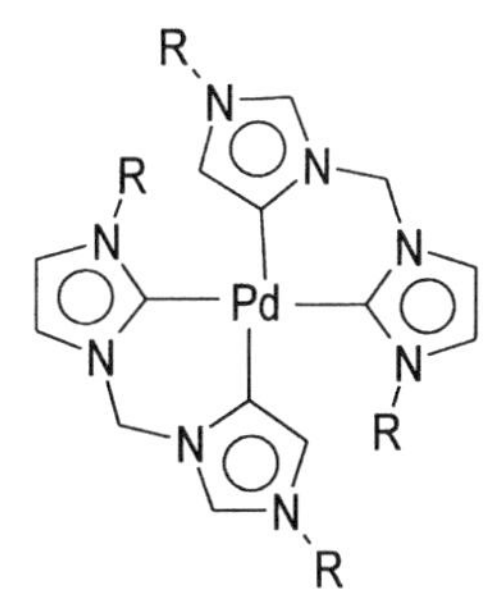
150

Numerous complexes have been obtained from the basis of carbenes of the heterocyclic series; amongst them, the carbenazoles have been studied in detail and are well characterized [8,225,226,239–245]. For instance, complexes ***149*** and ***150*** [8] belong to the group of carbenazoles. A theoretical study (by the quantum-chemical method) of N-heterocyclic carbenes, silylenes, and germanylenes is under way [245]. The carbon atom in the studied complex compounds has a hybridization close to sp^2; according to x-ray diffraction data, the grouping MCXR(R) is planar.

Alkylidenes can be made in many different ways. One of the most common is an α-abstraction reaction. Other methods have also been used, such as reduction of alkyl halide complexes and transfer of an alkylidene from a phosphorane or another metal complex. Reaction of a Wittig reagent ($R_3P{=}CR_2'$) with a metal oxo complex is not a common reaction; apparently the metal–oxo bond strength is too favorable [184e]. In general, synthetic techniques for the preparation and reactivity of carbene complexes have been discussed in numerous reviews [184e,225,226,230,232,234,236] and original articles (for instance, Ref. 246), and no additional examination is required in this work. The structures of carbene complexes do not give rise to doubts, since in many cases they have been proven by x-ray diffraction.

The *isonitrile* (*isocyanide*) complexes can be examined formally as first obtained carbene complexes [236]. This is a class of unsaturated CNR ligands which are isoelectronic to CO. As far back as almost 90 years ago, L.A. Chugaev prepared and characterized $[Pt(CH_3NC)_4]Cl_2$, $[Pt(C_4H_9NC)_2]Cl_2$, $[Pt(CH_3NC)_4]$ $[OC_6H_2(NO_2)_3]_2$, and $[Pt(N_2H_4)(CH_3NC)_8]Cl_2 \cdot 8H_2O$ [247].

Isonitrile coordination chemistry was developed intensively 40 years ago when a general technique for their synthesis by dehydration of N-substituted formamides was proposed [248]. The first monograph on these ligands appeared some years later [249]. In general, isonitrile complexes are prepared by ligand substitution reactions from metal carbonyls and other complexes in the presence of heterogeneous catalysts ($CoCl_2$, Pt) [184f,250].

X-ray diffraction data of isonitrile complexes showed that, additionally to known η^1-terminal ("head on") and μ_2,η^1-bridge binding, μ_2,η^2- and μ_3,η^2-four-electron bridge structures can be formed in transition metal complexes similarly to CO ligands [251]. The terminal isocyanide ligands are bound with transition metals in their complexes by the electronic pair of the carbon atom (η^1-type), which can be represented by two mesomeric structures $^{(-)}M{-}C{\equiv}N^{(+)}{-}R \leftrightarrow M{=}C{=}N{:}{-}R$ [225,253]. In polynuclear complexes, the CNR ligands form bridges [184f], connecting two or three metal centers: through the terminal carbon atom (μ_2,η^1-binding) [254,255], with simultaneous participation of both carbon and nitrogen atoms (μ_2,η^2-binding) [254,255], or fairly rare μ_3,η^2-bridge-type coordination in clusters [256,257]. Coordinated isonitriles are subject to reactions of ligand substitution, insertion, coupling, dealkylation, electrophilic, and nucleophilic attack [184f]. Additional information on isocyanide and related complexes is given in Ref. 258.

Carbyne complexes, for example ***151*** and ***152***, belong to another type of complexes with multiple σ-carbon–metal bond [73,259a–c]:

$[X(CO)_4M{\equiv}CR]$ $\quad$ $[(CO)_5Cr{\equiv}CNR_2]^+[BF_4]^-$

151 $\quad$ *152*

M = Cr, Mo, W;

R = Alk, Ar, NAlk; X = Hal

The main synthetic technique for obtaining carbyne complexes is through the reaction of carbene compounds with metal halides, preferably with those of Group IIIA [259c] as, for example, through reaction (2.2):

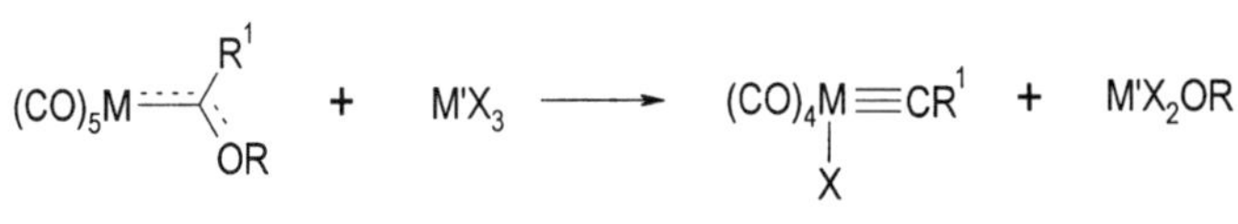

(2.2)

M = Cr, Mo, W; M' = B, Al, Ga; R = Alk, Ar, NR^2; R^1 = Alk; X = Cl, Br, I

The carbene–carbyne rearrangement has been discovered, and it is accompanied by the following transformation equation (2.3) [259c]:

$$\text{carbene} \underset{HCl}{\overset{MX_3}{\rightleftharpoons}} \text{carbyne} \qquad (2.3)$$

Available data on carbene heteroanalogues is summarized in a number of reviews [259d–f].

2.2.4.2 Ligands with Donor Nitrogen Atoms

Organic compounds (and inorganic, for instance NH_3) with nitrogen as donor atom are the most widespread and studied ligands. They are represented by: aliphatic, aromatic, and heterocyclic amines; azoles and azines; bidentate chelating compounds; polyazamacrocycles; polypyrazolylborates and cyclic tetrapyrrol systems (porphyrin, phthalocyanine, etc.). The coordination compounds obtained on this basis occupy a leading position in many classic publications [1,7,89,91,184f], textbooks [3–5,90], monographs [260–268], and reviews [134,269–278]. The increased interest in complexes of this type is explained, first of all, by the fact that the MN_n coordination bonds present in this kind of compound form the portion with the most important biological role [1, vol. 2;266,267,279–284].

Ammonia and Amine Complexes

A series of fundamental monographs and textbooks [1,89–91,285a–d] is devoted to summarizing the enormous amount of material and data available, encompassing the synthesis, properties, and structure of the complexes having formula H_nE, where $n=3$ (E = N, P, As, Sb) and $n=2$ (E = O, S, Se, Te) (see below, Secs. 2.2.4.4–2.2.4.6). The metal complexes generally formed have a general formula $[(H_nE)_mMA_p]$, where M stands for a metal (although B, C, Si, As, Sb, and Te may also participate) of practically all groups, while A could mean any anion from a wide group, but especially one of the halides. In this respect, in our opinion, the x-ray diffraction data on inorganic ammine complex compounds* (E = N) have a good deal of interest [1, vol. 2]. They testify that the bond length M $\leftarrow$: NH_3 varies in the range 1.9–2.6 Å and that it depends on the composition and structure of the many examined complex compounds studied, containing any from a wide number of various metals (Ag, Mg, Cd, V, Co, Fe, Ni, Pd, Pt, Ru, Cu, Cd, Zn).

Among the numerous approaches taken for classifying ammonia organic derivatives – amine ligand systems – the most attractive one is that based on their separation according to the dependence on the number of NH_2 groups [285e]. The mono-, di-, tri-, tetra-, penta-, and hexa-amine ligands have been isolated. The dentacity of these ligands is usually determined by the number of amine groups and their relative position.

Monoamine compounds with the general formula ***153*** take part as monodentate σ-donor ligands and form complexes of type ***154***:

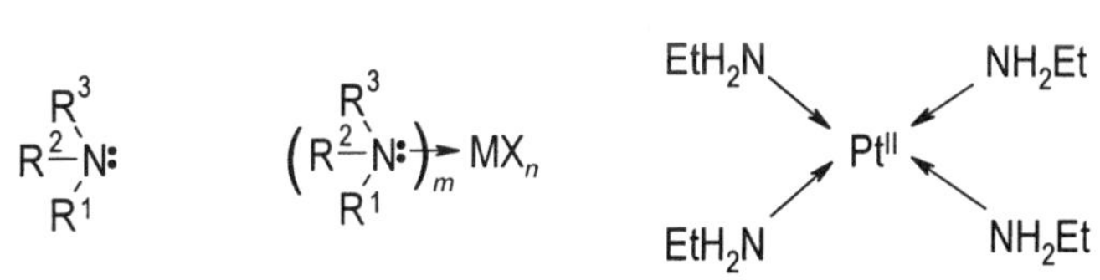

153 *154* *155*

R^1, R^2, R^3 = H, Alk, Ar, AlkAr

---

*In the period between 1977 and 1981, a total of 54 structures of ammine complexes were solved. More recent data can be obtained from the Cambridge bank of x-ray diffraction data.

Complex $[Pt(NH_2Et)_4][PtCl_4]$ is an example of such coordination compounds, in this the cationic fragment has a structure ***155*** [286], according to x-ray diffraction data.

Diamine ligands act generally as bidentate ligands forming chelating or bridge structures, given respectively by ***156*** and ***157***:

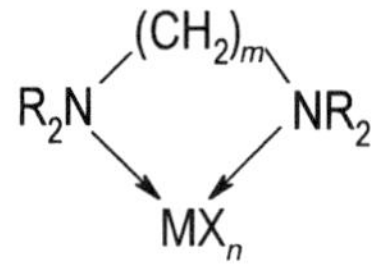

156

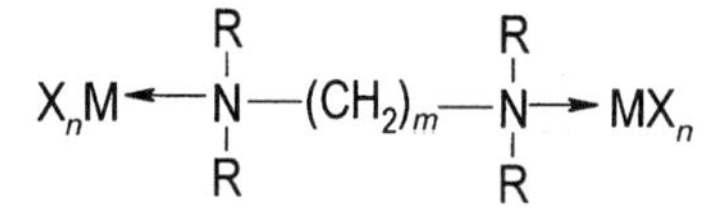

157

The bridge structures are more preferable at $m = 1$; at $m > 1$, and especially when $m = 2$ and 3, the chelating structures are generally formed in a feature known either as the rule of cycles of Chugaev [287], or the chelate effect [288–295]. The most propagated from these ligands is ethylenediamine [296], capable of forming chelating ***158***, ***159*** and chelating-bridge ***160*** coordination compounds, as for example:

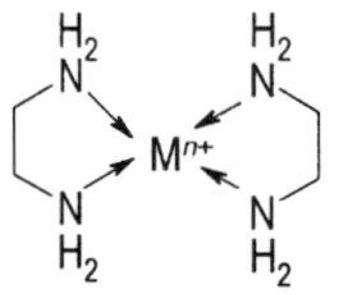

158

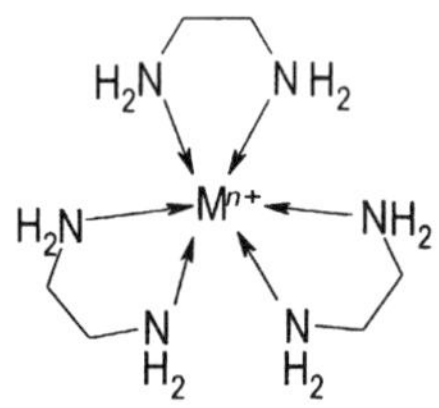

159

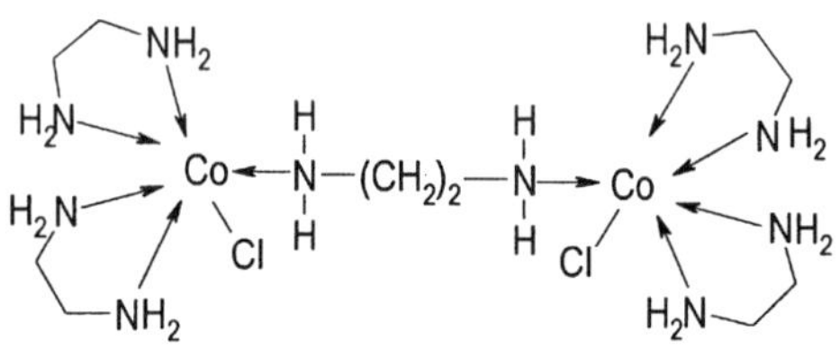

160

In the case of aromatic diamines, the character of metal binding is determined by a mutual position of NH_2 fragments on the benzene nuclei: *O*-phenylenediamine carries out a chelating function (***161***), while the meta- and para-isomers of this ligand rather carry out a bridge function (***162*** and ***163***, respectively):

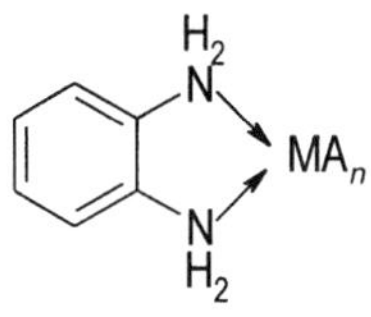

161

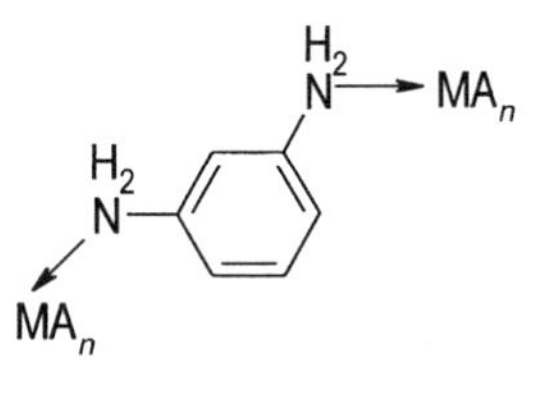

162

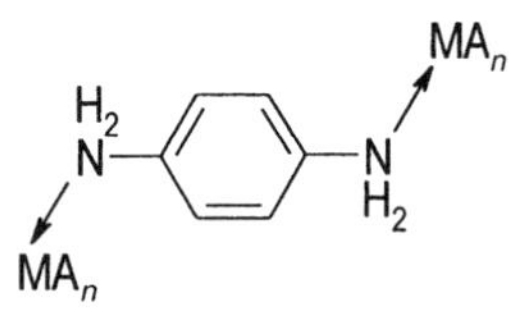

163

o-Phenylenediamine and its N-substituted derivatives take on a chelating function not only in mononuclear, but also in cluster-like structures. An example of such clusters is given by the nickel complex compound $\{[o\text{-}C_6H_4(NH_2)(NHPh)]Ni_4(\mu_3\text{-}OH)(MeCN)_2 \cdot (OOCCMe_3)_2(\mu\text{-}OOCCMe_3)_4\}$ [297]. *o*-Phenylenediamine also participates as a chelate-bridge ligand, as for example structure ***164*** [277]. 1,8-Naphthylenediamine behaves analogously [277]:

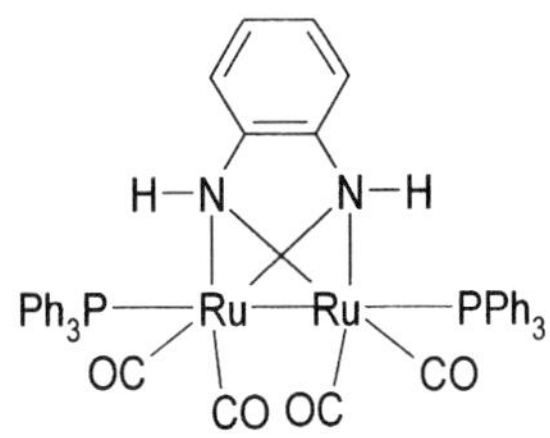

164

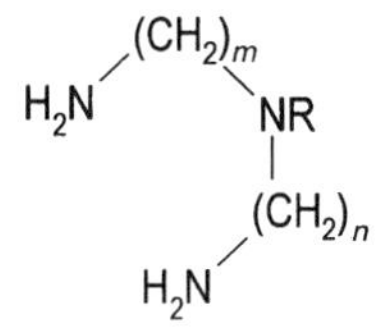

165

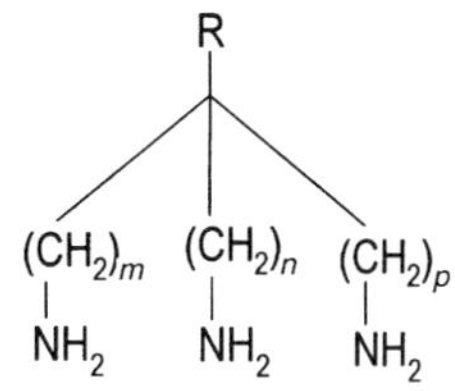

m,n, p = 2-4

166

Most tridentate triamine ligands ***165*** and ***166*** command a high level of interest [285e,286,293–295,297,298]. In general, octahedral complexes have been obtained in which the existence of the three isomeric compounds ***167*—*169*** has been established by x-ray diffraction [285e]:

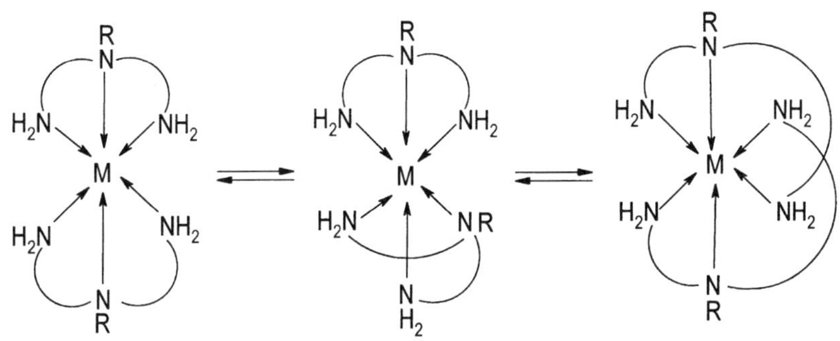

167 (sym-fac) *168 (asym-fac)* *169 (mer)*

For the systems ***165*** ($m = n = 2$), the equilibrium between the isomers has been observed in solution. The ratio between the three species depends on the nature of R [285e].

Tetra-amine ligands are represented by a series of compounds of the types ***170*–*173*** [261,285e,298]:

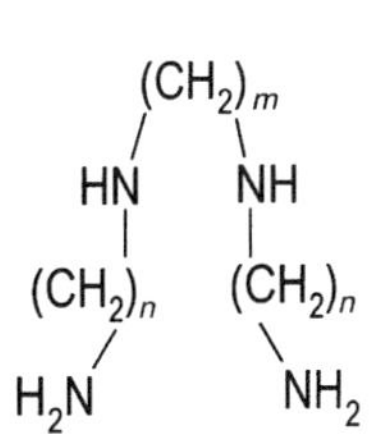

m = 2–4; *n, p* = 2, 3

170

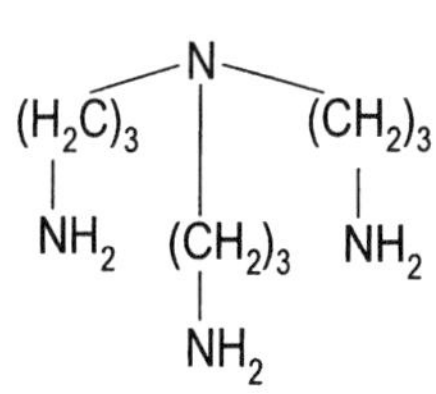

171

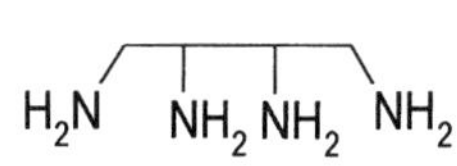

172

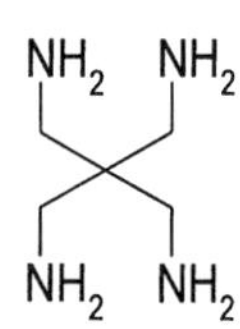

173

The majority of penta-amine ligands studied are the compounds presented as ***174–177*** [285e]:

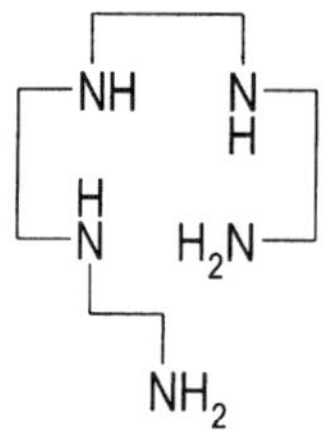

174

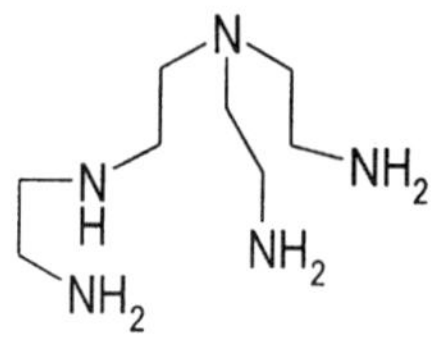

175

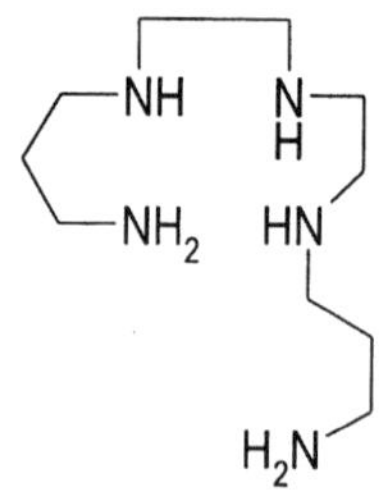

176

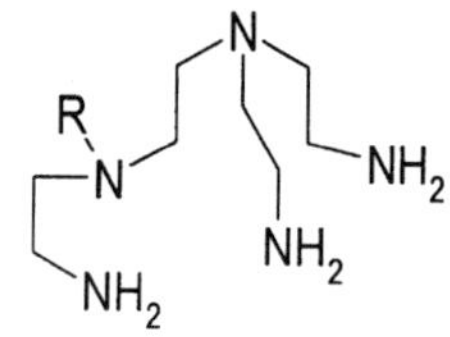

177

The most typical hexa-amine ligands are also compounds, ***178–180*** [285e]:

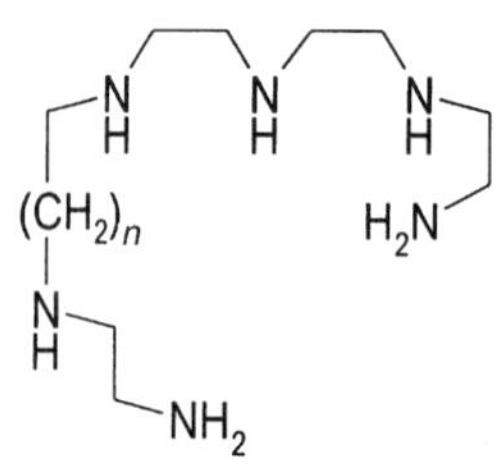

$n = 2, 3$

178

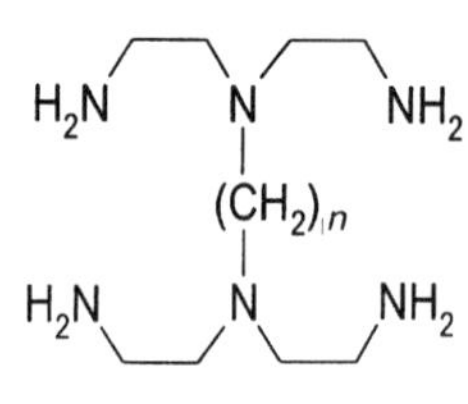

$n = 2–4$

179

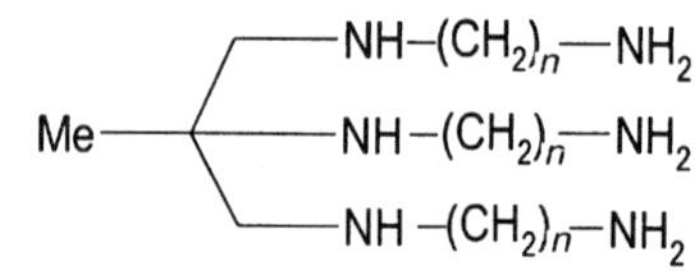

$n = 2, 3$

180

The topology of the complexes of these ligands is discussed extensively in a review [285e] and in a monograph [263]. In addition to the ligands described above, the aniline derivative may participate as tri-, tetra-, and penta-amine ligands, as for example ***181*** and ***182*** [285e]:

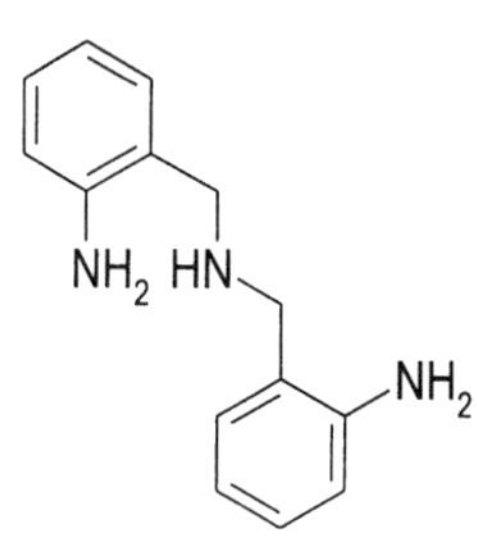

181

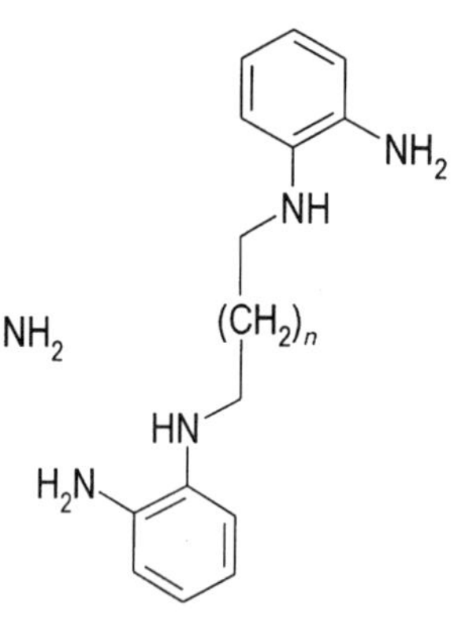

182

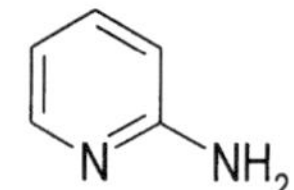

183

Complexes of these ligands have been described [299,300]. The NH_2 group does not possess donor properties in the heterocyclic analogues of aniline – amino- derivatives of nitrogen-containing heterocycles, as for instance in structure ***183*** [301]. Ethyleneimine ligands also take part as donors [302].

Nitrogen-Containing Heterocycles

This group is represented by some types of ligands, amongst which ***184*** (pyrrole), ***185*** (azines), and ***186*** (azoles) are the most widespread [7,8]:

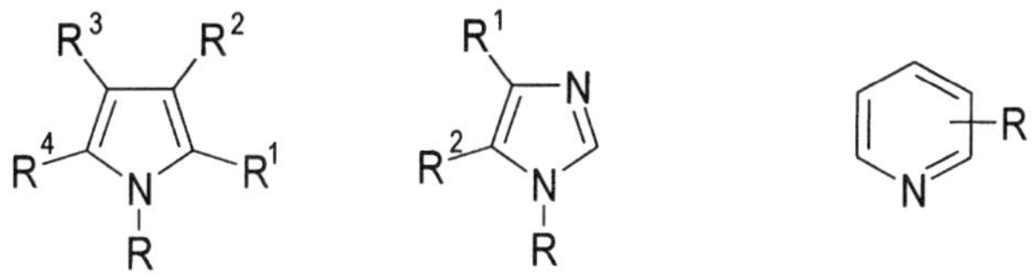

184 *185* *186*

The nitrogen atom contained in a neutral pyrrole molecule (***184***, R = Alk, Ar) is practically deprived of its donor properties [8,302]. In this respect, for nonsubstituted pyrrole (***184***, R = H), the exchange reactions of the proton of the NH group with a metal (***187***) are typical, as well as its participation in the complex formation as an anion $\eta^5(\pi)$-ligand system (see above, Sec. 2.2.4.1):

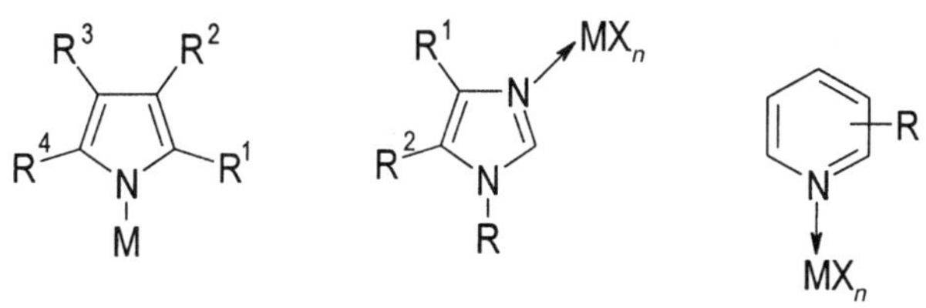

187 *188* *189*

On the contrary, N atoms of azoles 188 and azines 189 have a free pair of electrons (not included in the heteroaromatic π-system), which provides for the existence of a typical coordination bond (two-electron two-center) and thus for the formation of complexes [1, vol. 2;8,303,304].

Coordination compounds of imidazole (***188***: R, R^1, R^2 = H, Alk, Ar) and its benzoderivatives (***188***: R = H, Alk, Ar; R^1, R^2 = C_4H_4-cyclo), as other azoles, were the subject of numerous investigations, whose results were summarized in a series of reviews: pyrazole ***190*** [305–308], imidazole ***188*** [306,309], triazole ***191*** [310], tetrazole ***192*** [311], and isoxazole ***193*** [312,313]:

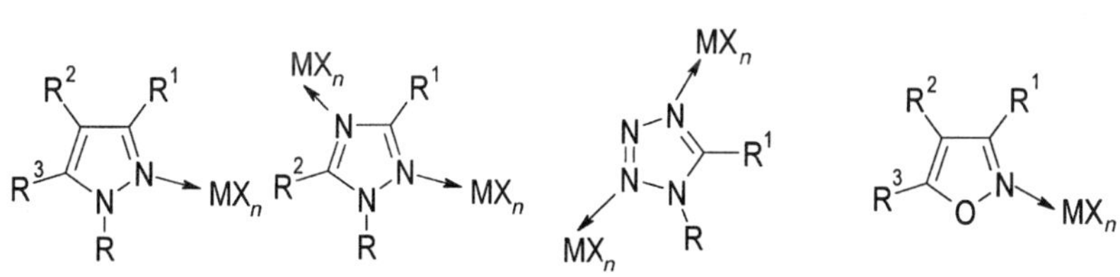

190 *191* *192* *193*

Pyridine ligands ***186*** are monodentate, forming complexes of the kind presented by ***189***. Pyridine is coordinated in the same way in copper acetate ***194*** [314] and iron rhodanide complexes [315], adducts of cobalt complex with *bis*(salicylidene)ethylenediamine ***195*** [316] and nickel chelate, formed by tridentate N,S-donor azomethine ligand ***196*** [317]:

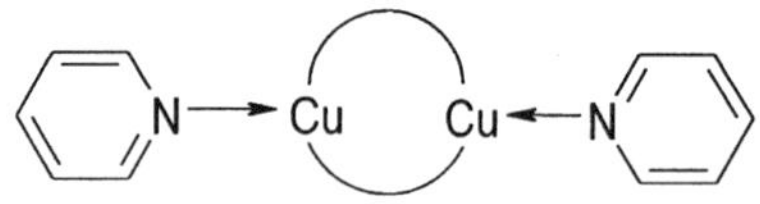

194

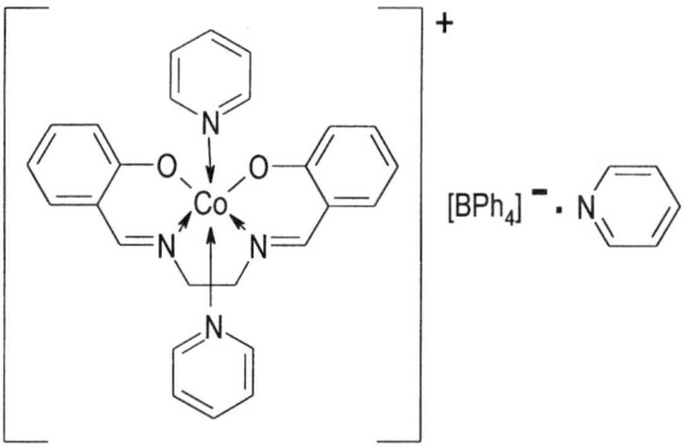

195

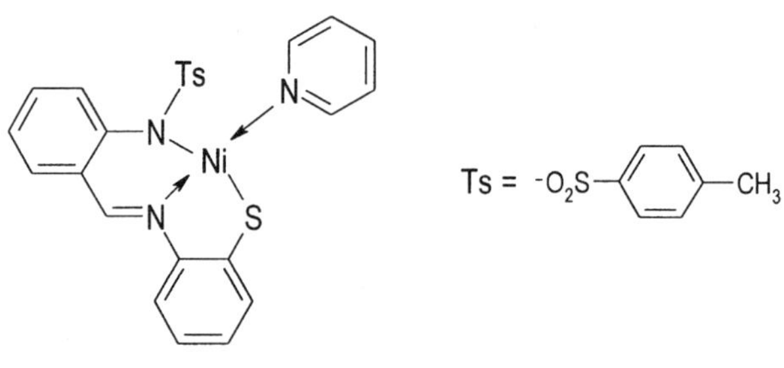

196

Pyridine is mono-N-coordinated in all its adducts with BH_2CN [318], also when combined with the ruthenium–cyclopentadienyl fragment [319], and with a number of chelates which simulate biologically important compounds (vitamin B_{12} [320–323] and corrol iron complex [324]). Complex compounds containing some coordinated pyridine molecules are also of interest [325–328]. Amongst them, for instance, the following complexes have been synthesized: complexes of dirhodium $[Rh_2(MeCOO)_3Py_4](CF_3SO_3)$ and $[Rh_2(MeCOO)_2Py_6](CF_3SO_3)_2$ [325], dicobalt $[Co_2(\mu\text{-}H_2O)(\mu\text{-}Me_3CCOO)(Me_3CCOO)_2Py_4]$, $[Co_2(\mu\text{-}Me_3CCOO)_4Py_4]$ [326], and dinickel $[Ni_2(\mu\text{-}H_2O)(Me_3CCOO)_2(\mu\text{-}Me_3CCOO)Py_4]$, as well as the analogous bipyridine compound $[Ni_2(bipy)_2(\mu\text{-}H_2O)(\mu\text{-}Me_3CCOO)(Me_3CCOO)_2]$ [327].

These dicobalt and dinickel complexes are of interest, since there is an unusual bridge binding of one water molecule at the expense of the two electronic pairs in the oxygen atom. Moreover, a simultaneous terminal and bridge coordination of the carboxyl groups is characteristic in this kind of compound.

For isomeric diazine ligands pyrimidine (1,3-diazine), pyrazine (1,4-diazine), and pyridazine (1,2-diazine), formation of bridge structures is a feature [303,304]. In particular, cobalt [329] and copper [330] pyrazine complexes ***197*** have such structures:

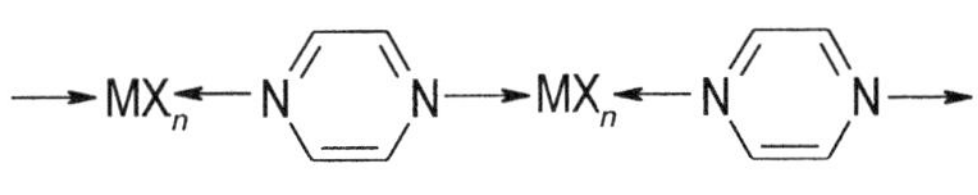

197

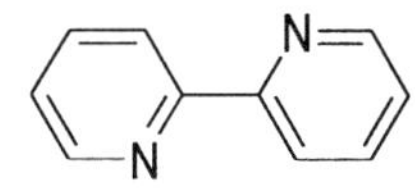

198

Bis-azine systems, for example ***198***, carry out in general a bidentate ligand function and, depending on the position of the nitrogen atoms with respect to one another, may even form chelate ***199*** or bridge ***200–203*** structures [1, vol.2]:

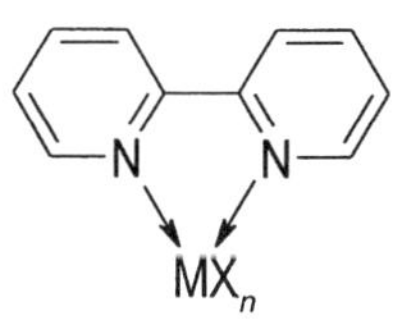

199

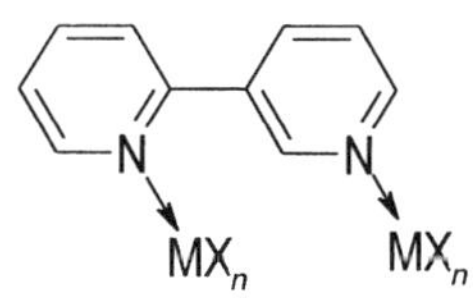

200

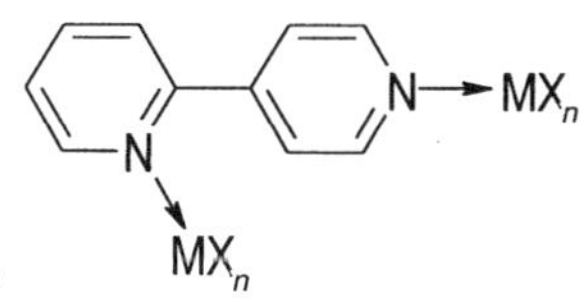

201

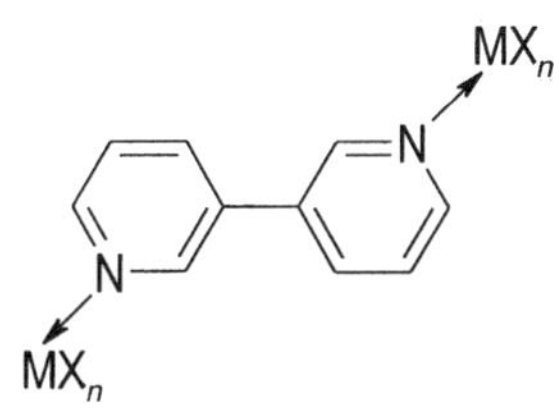

202

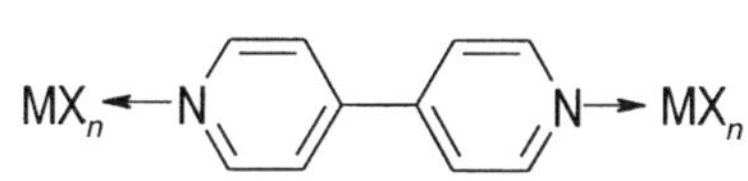

203

Azinodiazines ***204***, *bis*-diazines ***205*** and ***206***, and phenantrolines ***207–209*** [1, vol.2] belong to a similar type of ligands:

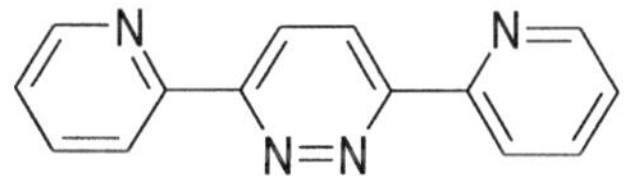

204

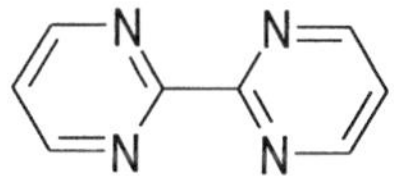

205

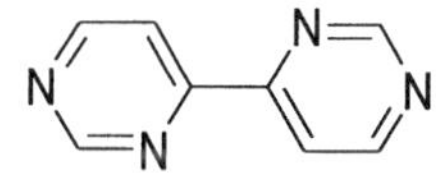

206

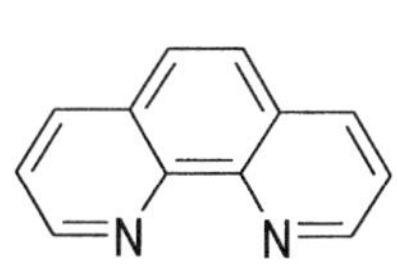

207

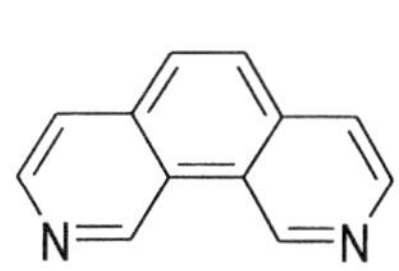

208

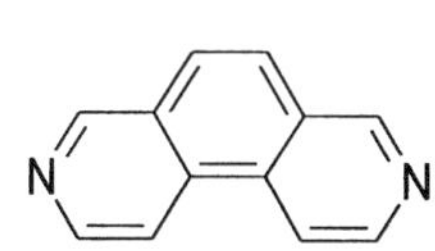

209

α,α′-Bipyridine ***198*** (bipy) and *o*-phenantroline ***207*** (phen) are the most studied among the *bis*-azine ligands. On the basis of these ligands, apart from the complexes of types ***199*** and ***210***, numerous adducts with metal chelates ***211*** have also been obtained [331–333]:

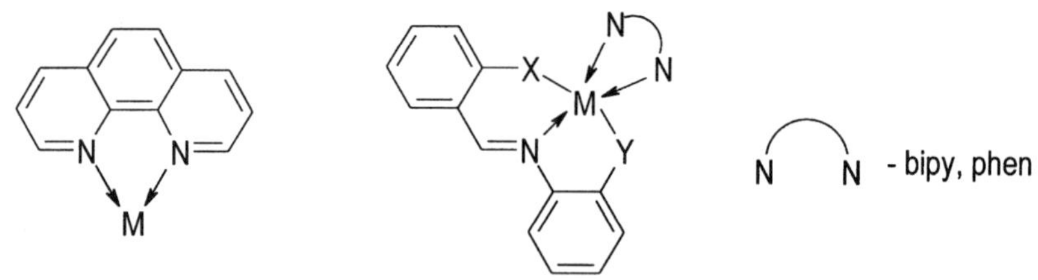

210 *211*

Examples of several other complexes of the type discussed have been prepared on the basis of ligands ***203*** [334] and ***206*** [335], as well as of other di- and oligoazine ligand systems [336].

The most important place among the N-containing ligands is occupied by the amino derivatives of nitrogen heterocycles [8,13,301,337]. On their basis, complexes have been obtained in which the metal center is mostly bound with the endocyclic N-atom of pyridine type ***212a***. At the same time, the coordination of metal with the exocyclic nitrogen atom of the NH_2 group was observed, for example in ***212b*** [44a], or the existence of an equilibrium between the bond isomers ***213a***$\leftrightarrow$***213b*** ("*redox-dependent linkage isomerism*") [337e]:

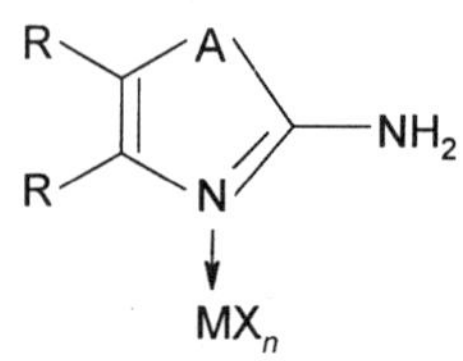

212a

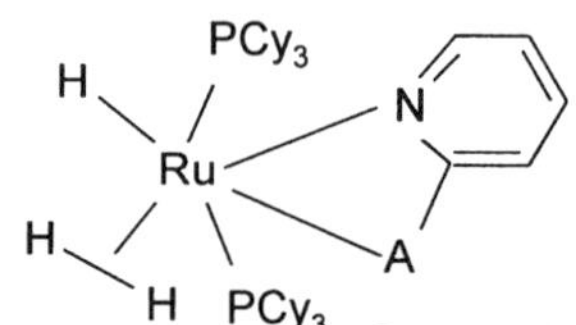

Cy = cyclohexyl
A = NH, O

212b

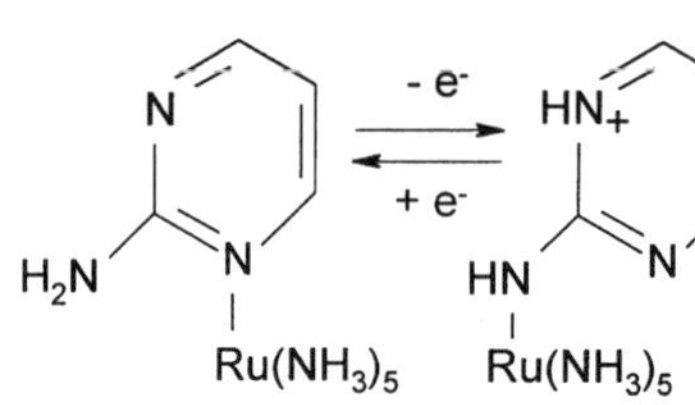

213a *213b*

Chelating Nitrogen-Containing Ligands Forming Inner-Complex Compounds

Ligands with N-donor centers, arranged in spatially convenient positions for metal-cycle formation and containing acidic hydrogen atoms, form inner-complex compounds (ICC). They are represented through various structures in which the coordination unit MN_4 is present in most cases [260,262,298,337–342]. The following ICC containing metal-cycles with a different number of atoms have been reported: four-member cycles {complexes of amidines ***214*** [343–350] and triazenes ***215*** [1, vol. 2;340,341,351]}; five-member cycles {dioxymates ***216*** [1, vol. 2;352–354]}, chelates of arylenediamines ***161*** and ***164*** [298,355], or those formed from aromatic azomethines ***217*** [13,269,270,356–360] and their heteroanalogues ***218*** [32,269,361–369], and also those synthesized from troponimines ***219*** [269,355,370–372]. The chelates of the following ligands contain six-member metallocycles: ICC of azomethines of *o*-tosylaminobenzaldehyde ***220*** [269,340,341,373,374], their cyclic aromatic ***221*** [375–383] and pyrazol ***222*** analogues [384], β-aminovinylimines ***223*** [339,342,385–399], *o*-aminoazocompounds ***224*** [270,341,400–404], and hydrazoneimines ***225*** [270,340,341,405–409] and ***226*** [410–416]:

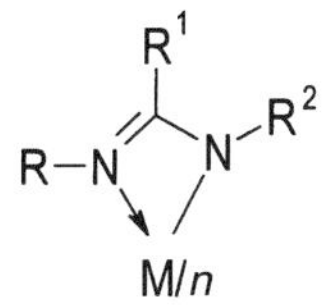

214: R,R^1,R^2 = Alk, Ar

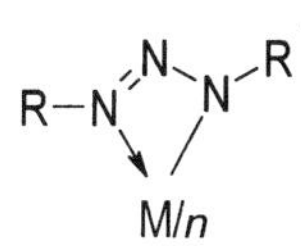

215: R,R^1 = Alk, Ar, Het

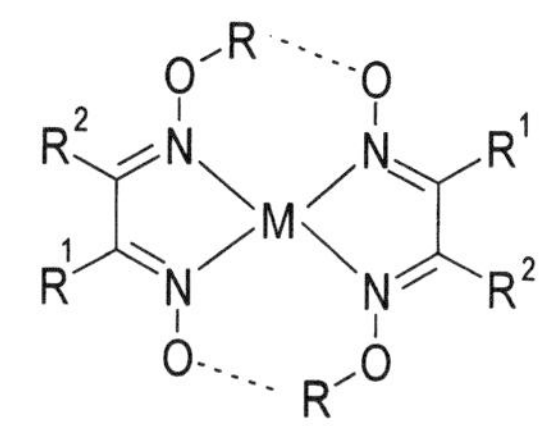

216: R = H, Alk; R^1,R^2 = Alk, Ar; R^1,R^2 = cyclo-$(CH_2)_n$

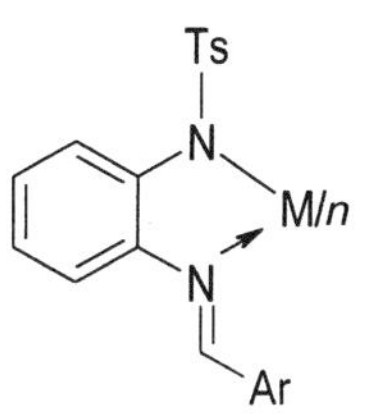

217

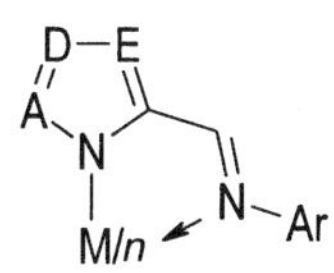

218: A,D,E = N, CR

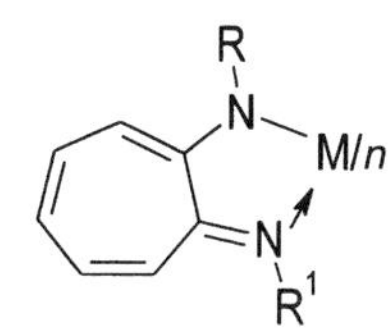

219: R,R^1 = H, Alk, Ar

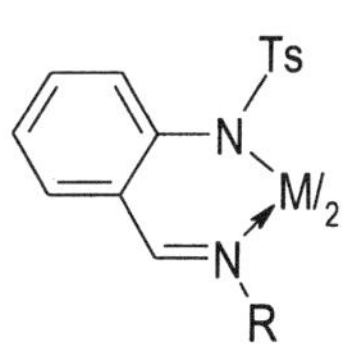

220: R = Alk, Ar

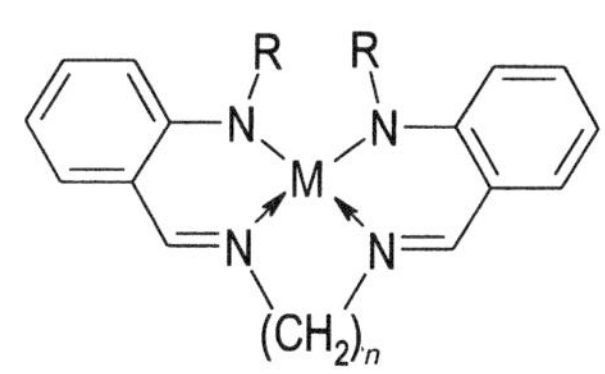

221: R = H, Alk

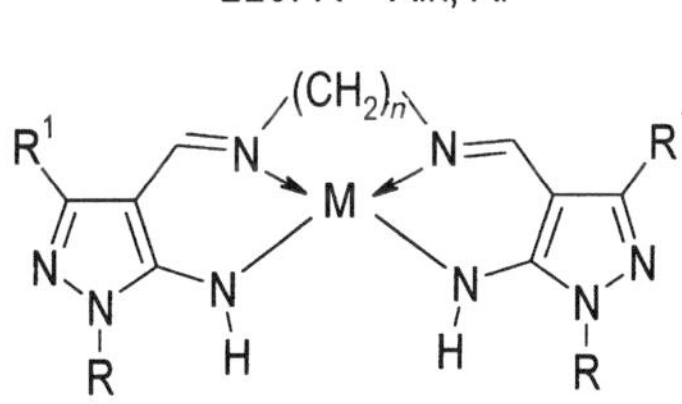

222

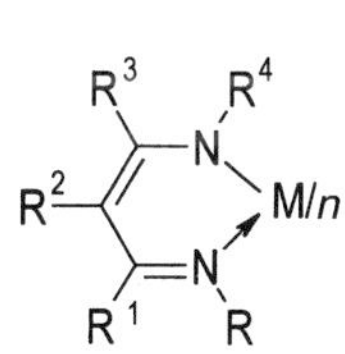

223

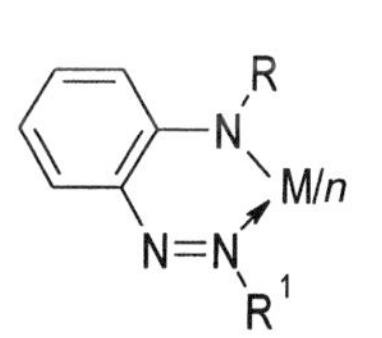

224
R = H, Ts; R^1 = Ar, Het

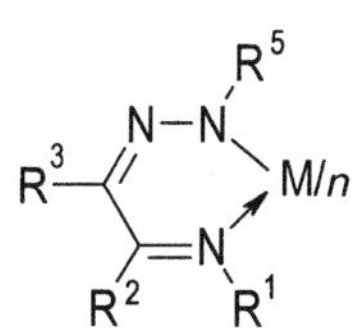

225

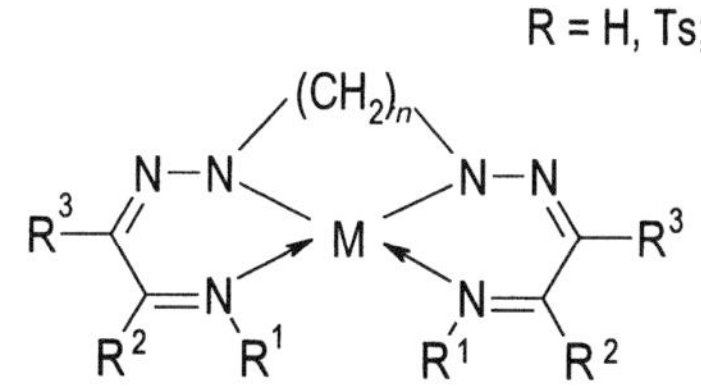

226

Formazanes, whose general structure is given by *227*, form chelates with metal-cycles of different sizes depending on the nature of substituents R^1, R^3, and R^5 [340,342,417,418]:

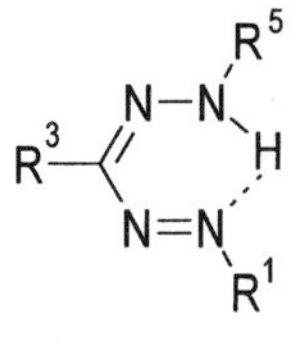

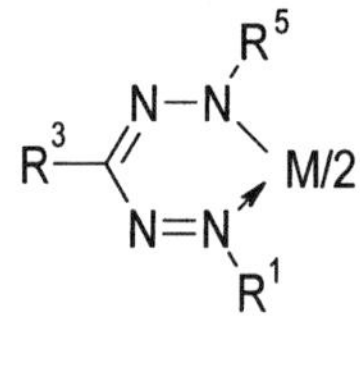

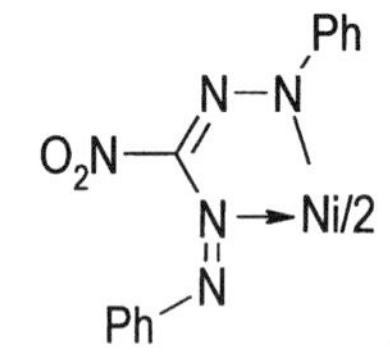

227 *228* *229*

In general, six-member metal-cycles are characteristic of the complexes of aromatic formazanes ***228*** [417–427]. However, a five-member structure ***229*** with the coordination unit MN_4 is formed if $R^3 = NO_2$ [428]. The same structure is observed in the hetarylformazanes ***230*** [340,342,418,420,429–435] and ***231*** [341]:

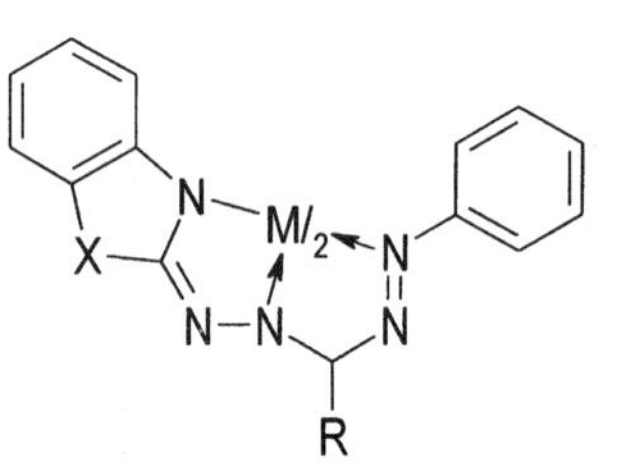

230: X = O, S; M = Ni, Co, Zn, Cd, Hg, Fe

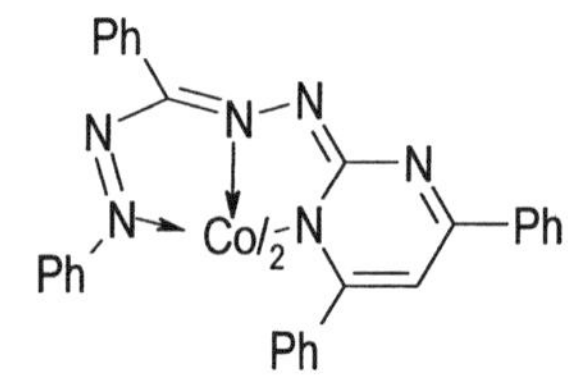

231

Complex ***231*** [341] is an example of a five-member structure. The analysis of structures ***220***, ***223***, ***224***, and ***228*** testifies that, as the number of N atoms increases in a six-member metal-cycle, a transition of configuration from tetrahedral to plane-square takes place [270,340,342]. Another important generalization drawn from the material related to nitrogen-containing ligands is their capacity to form metal-cycles of different sizes, especially typical for formazane ***228–230*** [270,341,342,417,418] and azocompounds ***224*** [340,343,403] complexes. It is worth noting that, on the basis of the ligand systems described above, not only chelating but also bridge structures can be obtained [436].

Ligands of polypyrazolyl-borates (scorpionates) ***232*** and ***233*** (E = B) [437-442] are the focus of great interest in modern coordination chemistry:

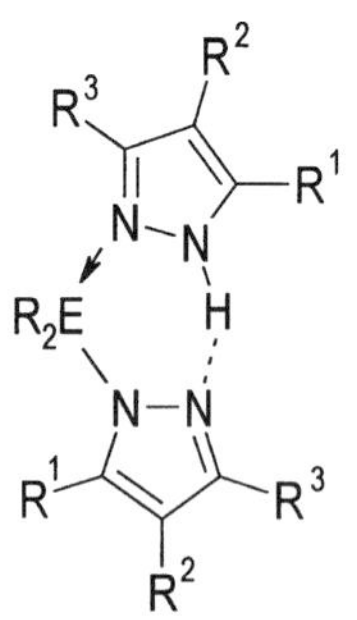

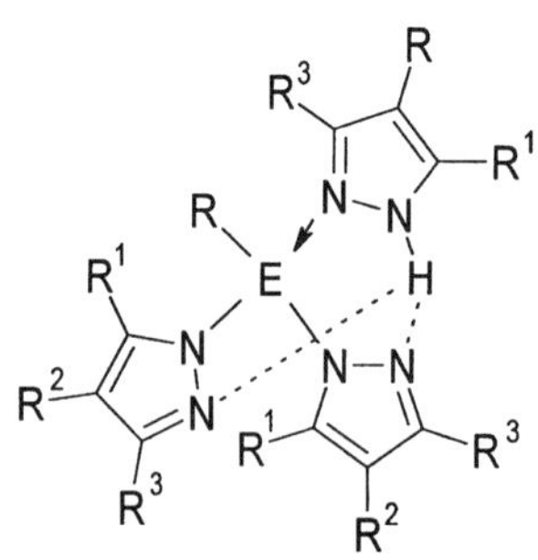

R, R^1, R^2, R^3 = H, Alk, Ar, ArAlk; E=B, Al, Ga

232 *233*

A recent review [440], published 100 years after Werner's coordination theory was first presented, was dedicated to the complex compounds obtained on the basis of polypyrazolyl-borates or scorpionates. Complexes of this kind, an example of which is presented by structure ***234***, have been obtained with practically all *p,d,f*-elements and they can be examined as N-containing analogoues (MN_4 coordination unit) of β-diketonates *2* (MO_4 coordination unit):

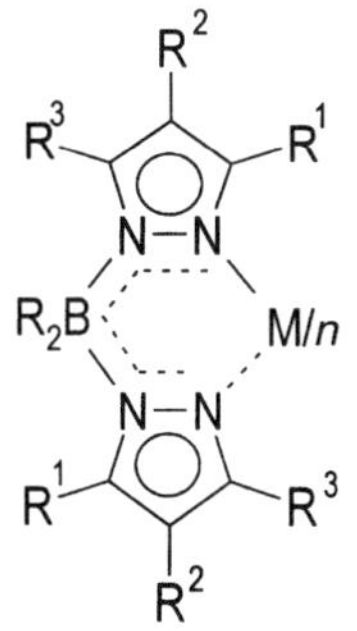

234

The main basic difference between these two wide classes of compounds is described as follows: β-diketonate molecules are associated in the condensed phase (crystalline) and, in a number of cases, also in solution, while the pyrazolyl-borates keep themselves as monomers. In addition to boron, ligands ***232*** and ***233*** may contain aluminum (E = Al) and gallium (E = Ga) [443–445].

Among the coordination compounds obtained on the basis of polypyrazolyl-borates, it is worth emphasizing the copper chelates ***235*** which are still the only biomimetic model of "blue" copper proteins, reproducing all their physical (UV- and EPR-spectral) properties [441,446–448]. Compound ***236*** [449] is also an example of complexes of this kind of system:

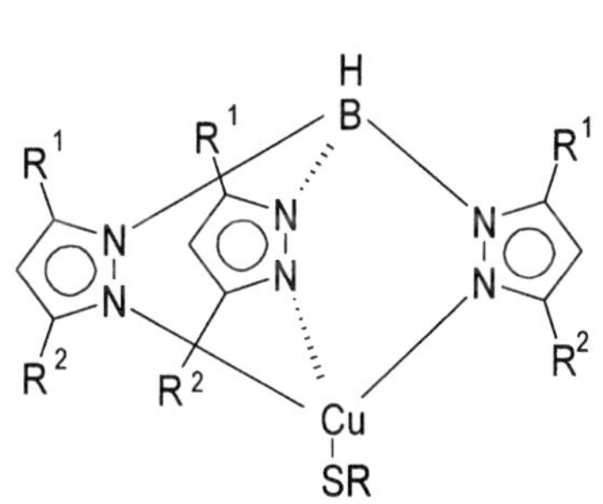

R = Alk, Ar; R^1, R^2 = Alk

235

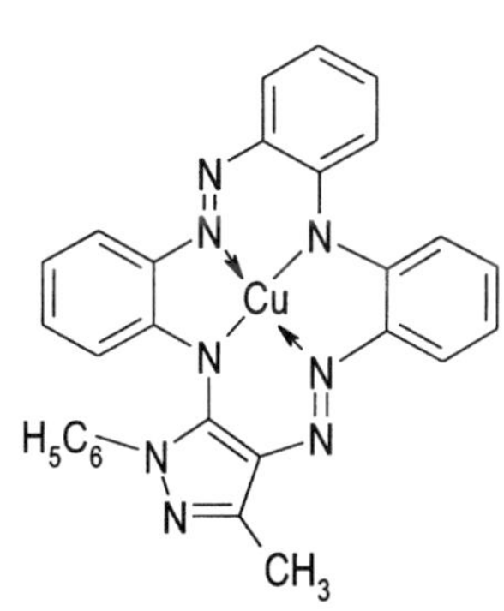

236

Azomacrocyclic ligands, forming MN_n coordination units, occupy a central place amongst N-donors. These ligands and their complexes are discussed in practically all corresponding textbooks and manuals [3–5,91], in numerous monographs [1,7,260–262,266–268,279–281,450–455], and in a number of reviews [271–277,456–467].

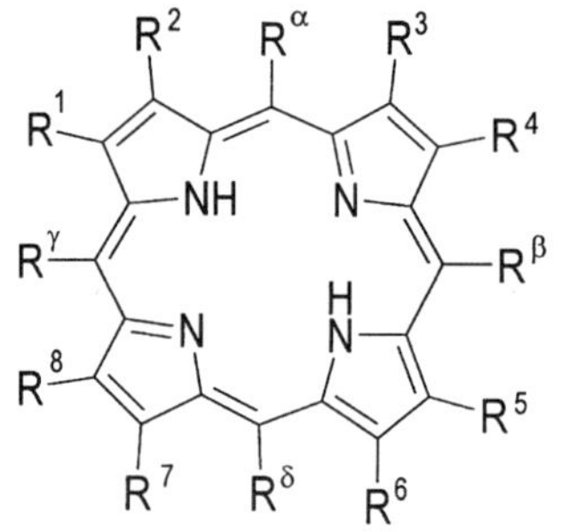

237

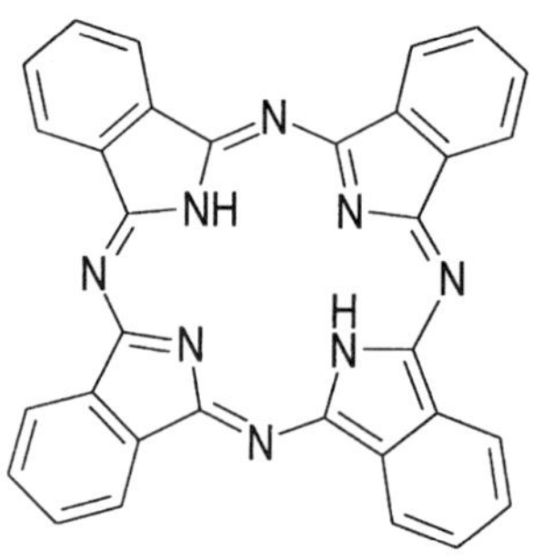

238

The cyclic tetrapyrrol systems belong to this group of compounds, and are considered to be the most important amongst them. On the basis of cyclic tetrapyrrol, metal chelates have been obtained with the coordination unit MN_4 (porphyrins ***237***, phthalocyanines ***238***, and corroles). The transition metal complexes of azamacrocycles are of interest because of the possibility of forming either metal-reduced species or metal-stabilized ligand radical species, in which the electron resides mainly on the macrocyclic ligand [468]. These compounds are also very important from the biological point of view [266,267,279–284,454–466,469]. After taking into account a detailed examination of the chemistry of coordination of azamacrocyclic ligands, together with the enormous amount of data reported in relation to them, it was decided not to discuss this group of complexes in more depth in the present monograph (except phthalocyanines), as also justified by its limited volume.

In addition to the complexes listed above, the following chelates belonging to the class containing an MN_4 unit have been obtained on the basis of: 1,4-diaza-1,3-butadiene [1, vol.2;470], tetra-aza-1,3-butadiene [1, vol.2], penta-azadienes [471], azomethines of 2,5-diformylpyrrole [472,473], cyclic hydrazones and thiosemicarbazones [262], dibenzotetra-aza[14]annulene [474], hetarylaminoazocompounds [8,404,475], azoazine radicals [476], Schiff bases of heterocyclic series [132,133,477], and azoleazines [132,133,478,479].

In conclusion, we note that a broad description of complex compounds of various N-containing ligands, amongst others, has been presented before [480], whose aim was the examination of molecular mechanisms of simulation of coordination compounds.

2.2.4.3 Ligands with Donor Phosphorus Atom

Phosphorus-containing organic compounds are classic subjects in modern coordination chemistry [1]. Phosphines are easy to synthesize and are excellent ligands for transition metals. Although the basicity of phosphorus in the oxidation state III in these ligands is lower than that of N(III), the donor activity of P-donors is sufficiently high [1,480–494]. The nature of the P—M bond and energy characteristics of complex formation with the participation of P(III)-containing ligands are examined in detail in reviews [487] and classic inorganic and organometallic textbooks [21b,184f]. Special attention is paid to the formation of dative bonds (see below), in a manner drastically different to that shown by N-donor compounds, which becomes an important feature for the transition metal complexes of P-donor ligands. At the same time, as follows from Ref. 489a, phosphine complexes contain, besides the transition metals, a variety of hydrides, halides, and organic derivatives of the elements of Groups IIIA (Al, Ga, In), IVA (Ge, Sn), VA (As, Sb, Bi), VIA (Te), and VIIA (I). Transition metal complexes containing trans-spanning diphosphine ligands are described in Ref. 489b.

In a difference of nitrogen, the chemistry of phosphorus is characterized by the possession of weak *pπ-pπ* bonds that lead to the existence of, for instance, $P(OR)_3$ [but not $N(OR)_3$]. Due to the presence of vacant *d*-orbitals of fairly low energy in phosphorus (and As and Sb) atoms (which are absent in the nitrogen atom having no function other than simple donation), these can accept electrons from the filled *d*-orbitals of the metal ion. So, additionally to a σ-bond (as in amine complexes), the phosphines can form an additional bond of the π-type (backdonation). This factor is especially important for the stability of transition metal–P-ligand complexes, where a *dπ–dπ* overlap contributes considerably to the bonding [19,184f]. π-Backbonding is significant among complexes of PF_3 and, to a lesser degree, complexes of phosphites [21b]. The latter, $P(OR)_3$ or $(RO)_2POP(OR)_2$, are often similar to phosphines, but the phosphites tend to be more basic and less sterically hindered [184f]. The empirical ordering of the π-accepting or σ-donating capabilities of phosphines is as follows: $PMe_3 < PPh_3 < P(OMe)_3 < P(OPh)_3 < P(NR_2)_3 < PCl_3 < CO \approx PF_3$ (greater π-acidity →; ← greater σ-donation) [184e]. Another peculiarity of phosphorus coordination chemistry in comparison with nitrogen is the existence of higher coordination numbers 5 and 6, for instance in compounds PPh_5, $P(OR)_5$, or $[P(OR)_6]^-$ [184f] (Fig. 2.7).

The useful classic idea of *cone angles* (θ) as a measure of a phosphine complex's steric bulk [21b,184f,g] is used for a rough definition of the steric demand of monophosphine ligands in their substitution reactions. The rates of phosphine ligand (L) dissociation from NiL_4 or *cis*-$Mo(CO)_4L_2$ correlate very nicely with their θ magnitudes, which suggests a dominant role for steric effects in these processes. Thus, for the dissociation of $Mo(CO)_4L_2$ at 70°C in C_2Cl_4, the change of cone angle from 122° to 141° leads to a considerable increase in the rate of dissociation from 10^{-6} to 1.6×10^{-4}: PMe_2Ph ($\theta = 122°$; rate $< 1.0 \times 10^{-6}\,s^{-1}$), $P(OPh)_3$ (128; $< 10^{-5}$), $PMePh_2$ (136; 1.3×10^{-5}), P(O-*o*-tolyl)$_3$ (141; 1.6×10^{-4}), PPh_3 (145; 3.2×10^{-3}), $PPhCy_2$ (162; 6.4×10^{-2}) [184g]. A useful generalization, as for the σ-donor and π-acceptor properties of phosphines, is that as the average electronegativity of the three groups attached to P increases, the ligand becomes a poorer σ-donor and a better π-acceptor [184f]. Indeed, it is very difficult to separate steric and electronic factors in

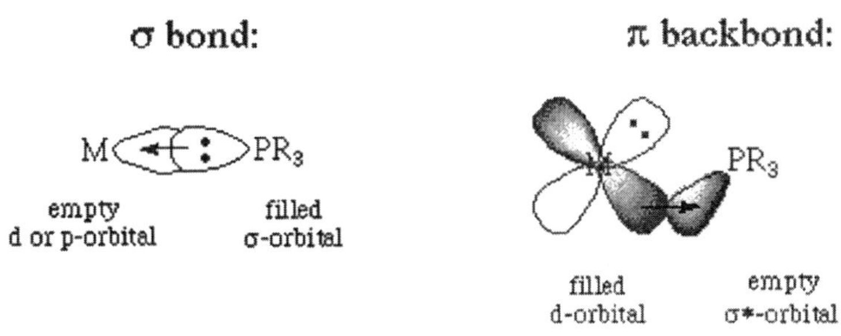

Figure 2.7 Bonding in phosphine complexes. (From Ref. 184e. Reproduced with permission.)

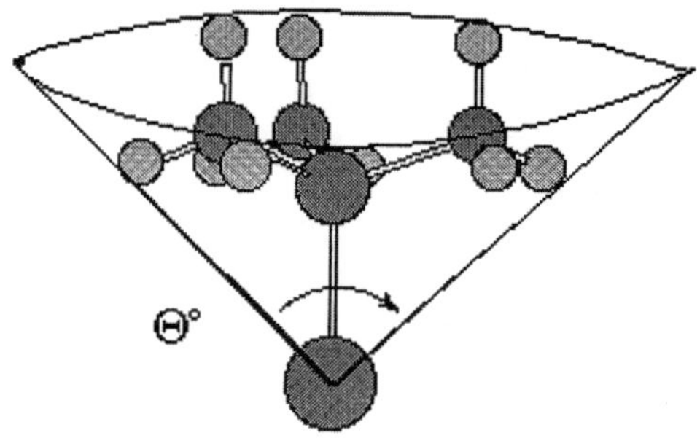

Figure 2.8 Cone angle. (From Ref. 184e. Reproduced with permission.)

dissociation of P-ligands, since, for example, phosphites of the same size as phosphines dissociate more slowly [184g] (Fig. 2.8).

The main phosphorus ligands are shown in Fig. 2.9. Additionally to the homoleptic complexes of phosphine itself (PH_3), its complexes with additional coligands are described, for example the quadruply-bonded *cis*-$Mo_2Cl_4(PH_3)_4$ [489c,d], dicyanoacetylene complex $Pt(PH_3)_2(C_4N_2)$ [489e], and the 14-electron complex $(M(X)(PH_3)_2$ (M = Rh, Ir; X = CH_3, H, Cl) [489f]. The substituted phosphines containing one phosphorus atom form mononuclear complexes of the type $[(R_3P)_m \rightarrow MA_n]$, and binuclear bridge structures as shown by ***239*** [3,481]:

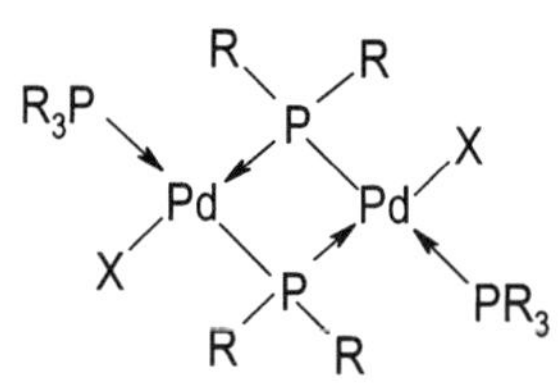

239

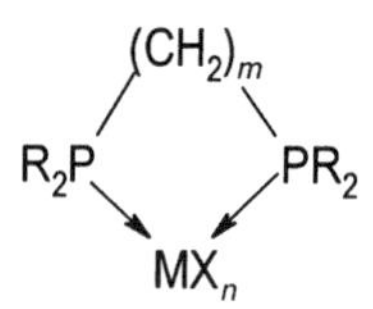

240

Diphosphines are mostly bidentate chelating ligands, forming metal-cyclic structures of type ***240*** [480,481,485,486]. However, in the complexes of *p*-metals, the discussed ligands behave also as a bidentate bridge ***241*** [489a]:

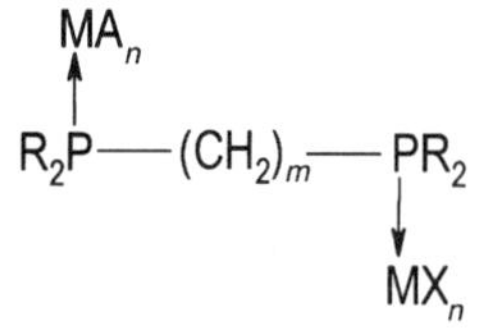

241

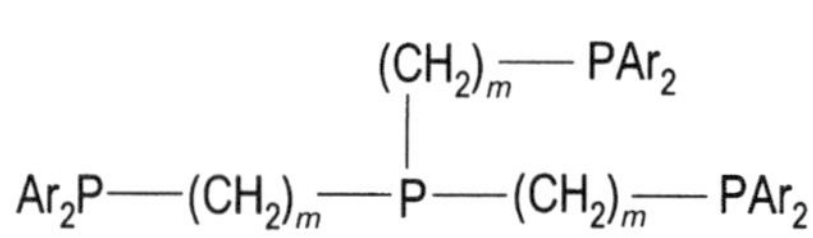

242

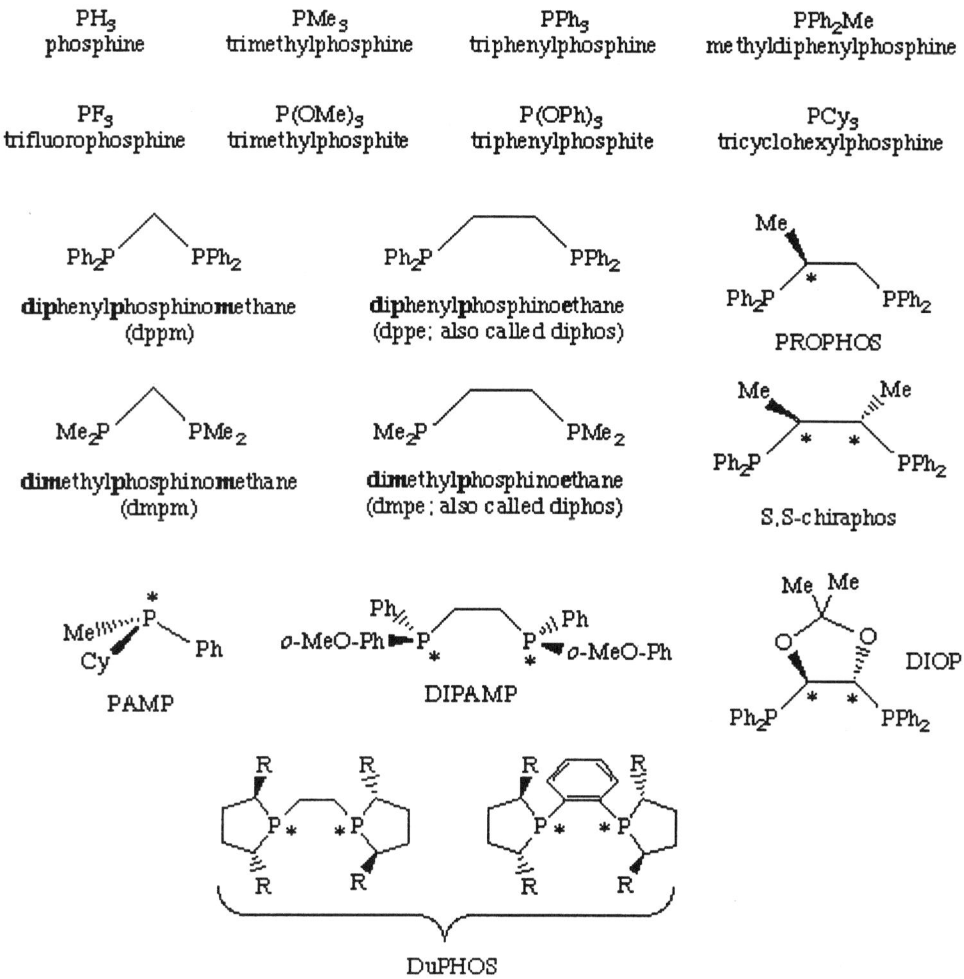

Figure 2.9 Main phosphorous ligands. (From Ref. 184e. Reproduced with permission.)

A number of coordination compounds are known in which this ligand is a bidentate bridge, particularly so for the most simple ligand system of the type discussed, as shown by ***241*** ($m = 1$) [494].

Complexes of tripodal hexa-arylphosphine ligands ***242*** [484], phospharyl-alkynes (R—C≡P) [490], and chiral derivatives of phosphoric acid [495], in all of which the P—M bonds are present, are widely known.

The possibility of using phosphorus-containing heterocycles as ligands [496–499] represents an area of research of great interest. On the basis of these kinds of ligands, complexes of three-member saturated, for example ***243*** [497], and unsaturated ***244*** [498] heterocycles, as well as four-member heterocycles ***245*** [499] have been synthesized:

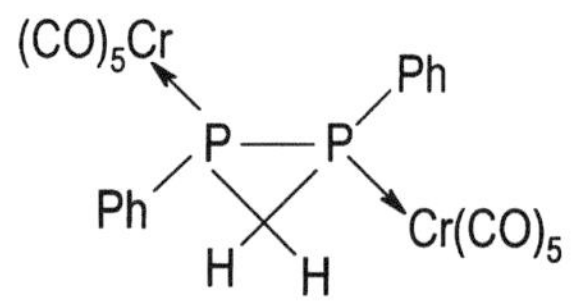 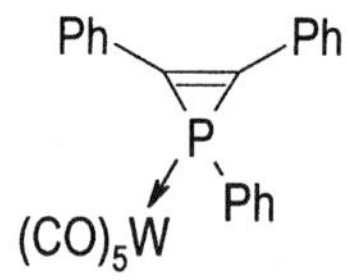 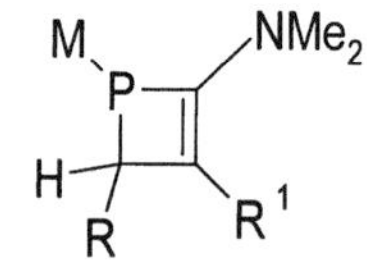

243 *244* *245*: M = Fe(Ru)Cp$(CO)_2$;
R = H, CN, CO_2Me;
R^1 = CN, CO_2Me

Among phosphorus-containing heterocyclic ligands, much attention has been paid to phospholes ***246*** [8,13,204,210,496,500,501] and phosphabenzene ***247*** [2,8,13, 496–499,502]:

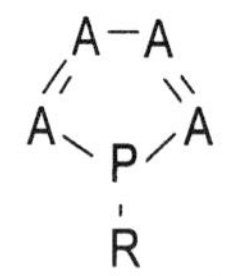 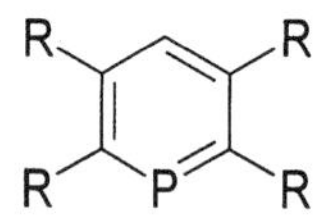

A = CR, P
R = Alk, Ar

246 *247*

These ligand systems, in addition to $\eta^n(\pi)$-complexes ($n = 5$ and 6, respectively), also form coordination compounds by the σ-P—M bond, as shown by ***248*** and ***249*** [496]:

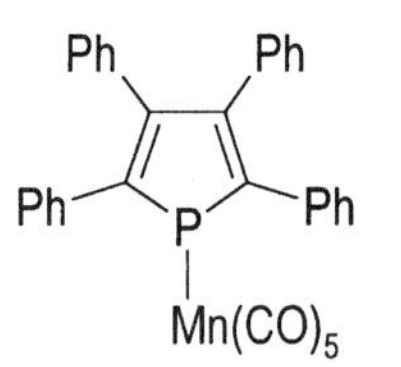 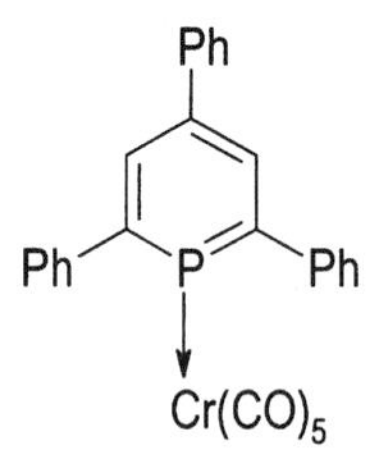

248 *249*

Complexes ***250*** [503] and ***251*** [499] have been prepared, and it is known that the σ- and π-binding exist simultaneously.

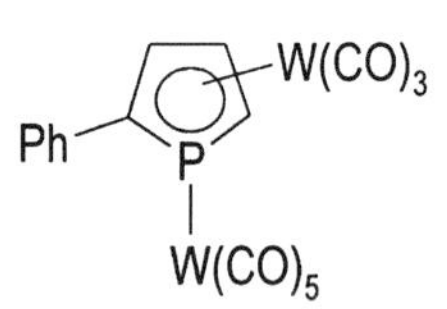 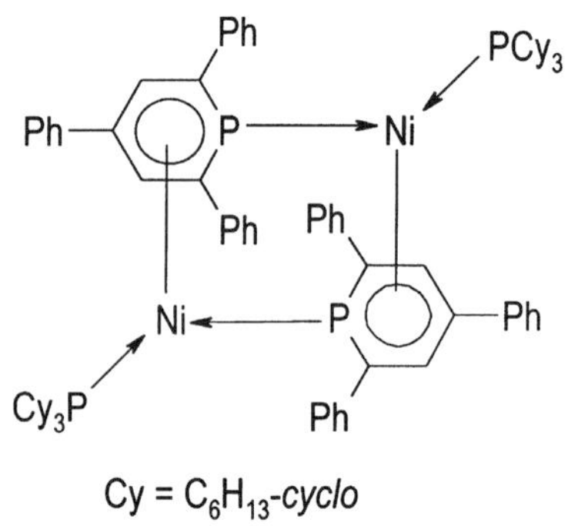

250

Cy = C_6H_{13}-*cyclo*

251

A similar type of binding is also characteristic for the coordination compounds of phospholes containing a few phosphorus atoms in the heterocycle [204,210,504–513]. These compounds form σ-P—M (*252–254*) and σ-P—M, π-Het—M (*255–258*) complexes, as shown below:

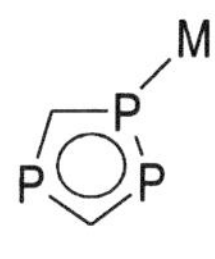

252

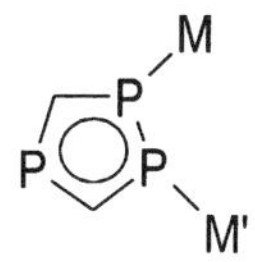

253

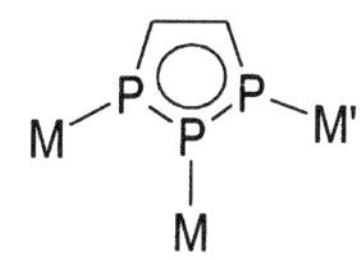

254

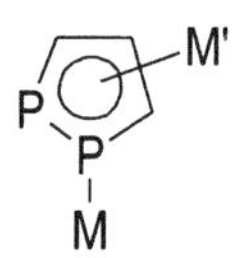

255

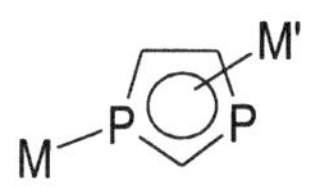

256

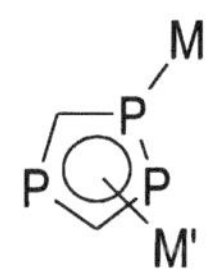

257

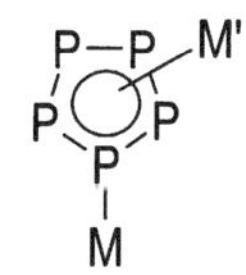

258

π-Coordinated phosphorus analogs of *bis*-imidazole behaves as a σ,σ,P,P-chelating ligand and forms a three-nuclear metal-cyclic structure ***259*** [514]. π(η^6)-Coordinated phosphabenzene [2,515–517] is also a σ-P-donor, forming binuclear complexes of the type shown by ***260*** [518]:

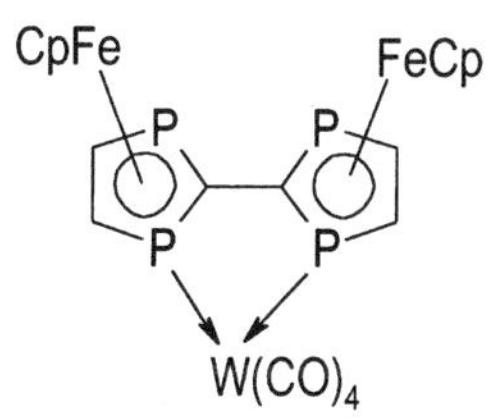

259

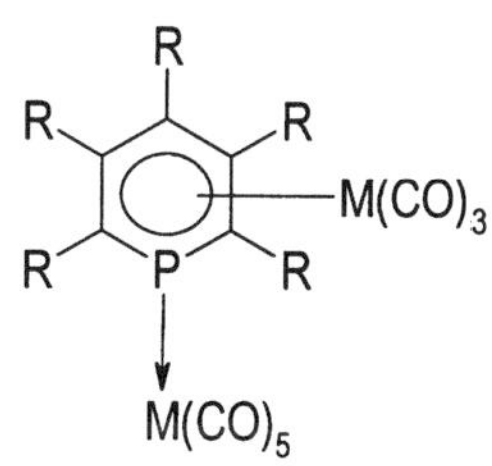

260

Coordination compounds of phosphabenzene ***261*** and α,α-diphosphabenzolyl (the phosphorus analogue of α,α-bipyridine) ***262*** have been synthesized, but they are insufficiently studied as yet [519]:

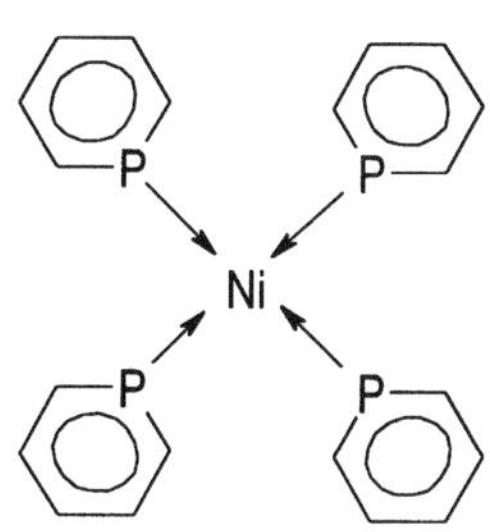

261

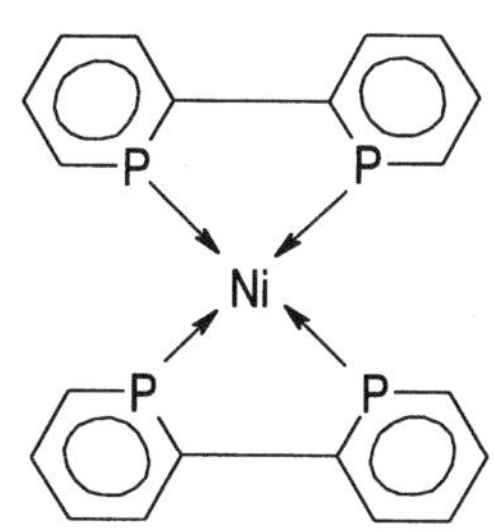

262

On the basis of phosphorus-containing compounds are comparatively rare complexes of cationic ligands have been obtained, some of their representatives are complexes of *bis*-phosphino-*i*-phosphineindolide, for example ***263–265*** [520]:

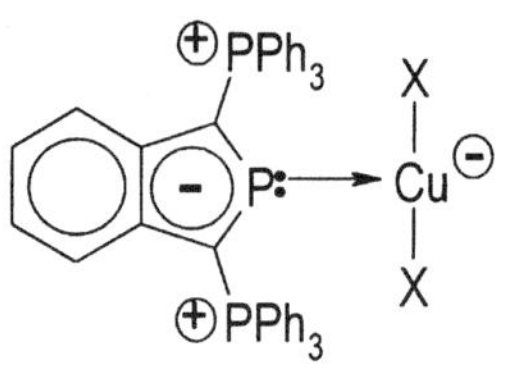

263: X = Cl, Br, I

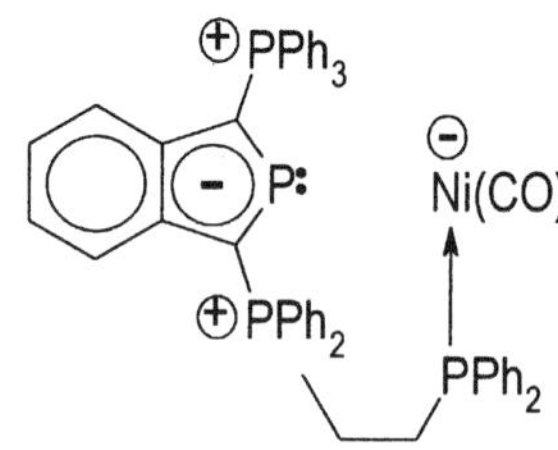

264

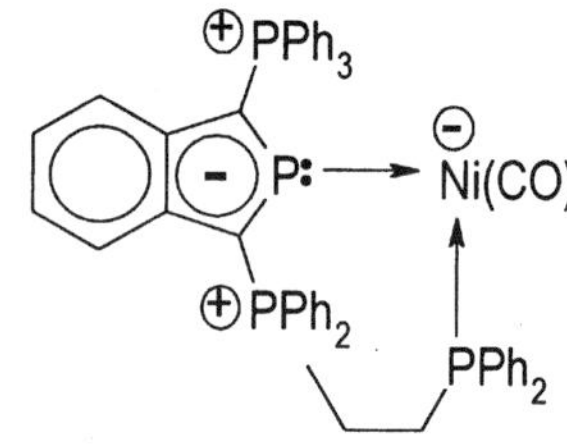

265

Detailed information about As-, Sb-, and Bi-containing compounds used as ligands is found in a number of references [1, vol.2;73, vol.3;63,521].

2.2.4.4 Ligands with Oxygen as Donor Atom

Ligands containing only oxygen atoms as donors include a large group of compounds, the most important of which are alkyl- and aryloxides, organic acids, β-diketones, crown-ethers, and podands. The most classic O-containing ligand, water, is discussed elsewhere [1,184f,285b] (the OH^- group as a ligand is presented in Sec. 2.2.3.1). Alkyl- and aryloxides form, similarly to the hydroxy anions ***80–83*** (X = O, R′ = Alk, Ar), coordination compounds with terminal and bridge ligands [1, vol.2;114,297,522–531]. A terminal M—OR bond is formed in complexes of type ***266***, for instance ***267*** [114]. Cobalt, nickel, and ruthenium [114] participate as central ion in complexes ***266***, in addition to platinum, palladium, and iridium:

$L^1L^2L^3M(OR)$

266

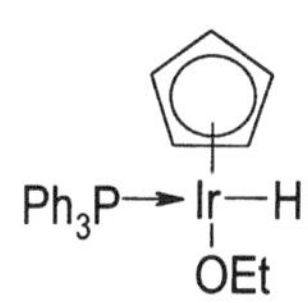

267

The OR groups carry out a bridge function in binculear alkoxide complexes, for example ***268–270*** [522,523]:

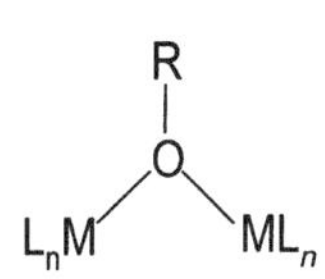

268

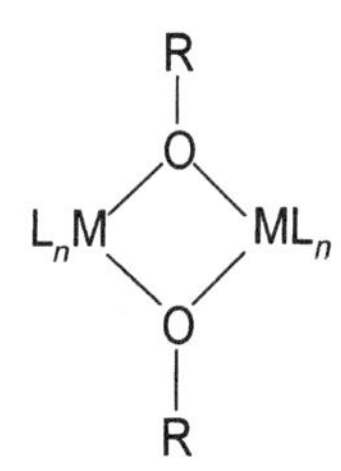

269

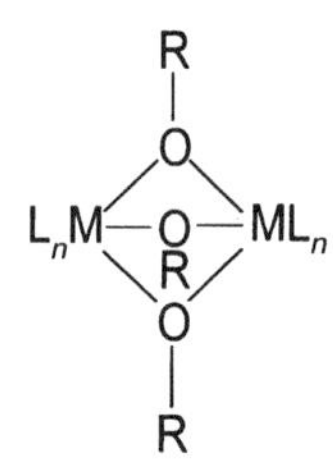

270

The di-OR-bridge structure ***269*** is the most common. It is typical, for instance, for copper complexes (M = Cu) with methoxy (R = Me) and ethoxy (R = Et) bridge fragments and nitrogen-containing ligands (L) [528,529]. The mixed coordination modes are highly typical for the discussed ligands: terminal and bridge, respectively, shown by ***271*** and ***272*** [527, 531]:

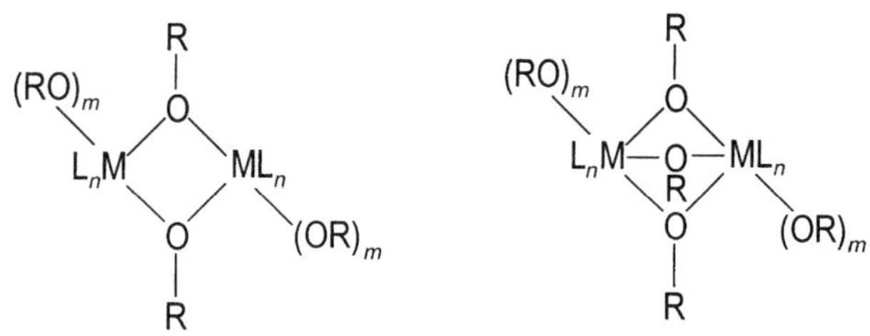

271 *272*

Metal binding of the type shown by ***271*** is formed in the complex compound $[Zr_2(O\text{-}i\text{-}Pr)_4(\mu\text{-}O\text{-}i\text{-}Pr)_2L_2]$, where LH is 2,2,6,6-tetramethylheptane-3,5-dione [531]. Coordination modes of the types ***266***, ***268***, and ***272*** exist simultaneously, according to PMR data, in the complexes $M_3O(OR)_{10}$, where M = W, Th; R = *i*-Pr, *t*-Bu [527]. According to x-ray diffraction data, such a coordination is characteristic for trinuclear clusters with the general formula $[M_3(\mu_3\text{-}O)(\mu_3\text{-}OR)(\mu\text{-}OR)_3(OR)_6]$, where M = Mo, W, U ([531] and references cited therein).

Coordination modes ***273–280*** [297,532–536] exist in complexes of monocarbonic acids:

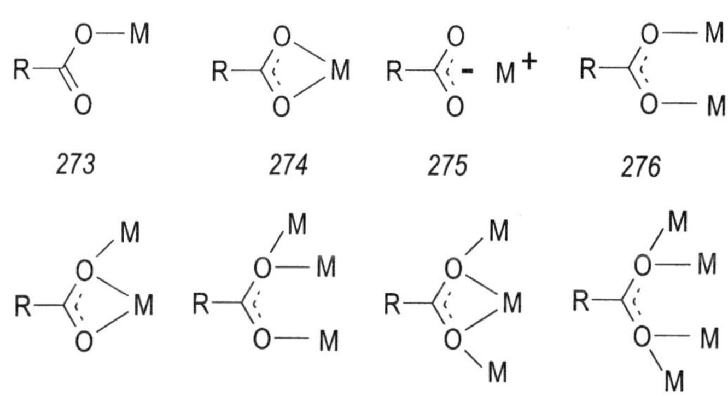

273 *274* *275* *276* *277* *278* *279* *280*

The detailed crystallochemical examination, including systematic stereochemistry of simple monocarboxylates, and the analysis of the factors determining the structural function of a monocarboxylate ligand, have been presented [532].

The widely known types ***273*** and ***274*** were discovered through x-ray diffraction studies, for example in the complexes of tridentate *o*-oxyazomethine ***281*** [536] and in their tetradentate analogues ***282*** [535]:

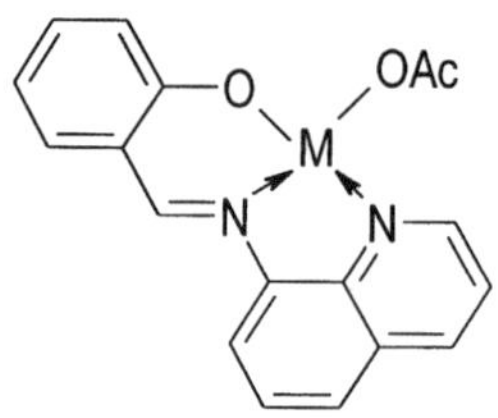

281

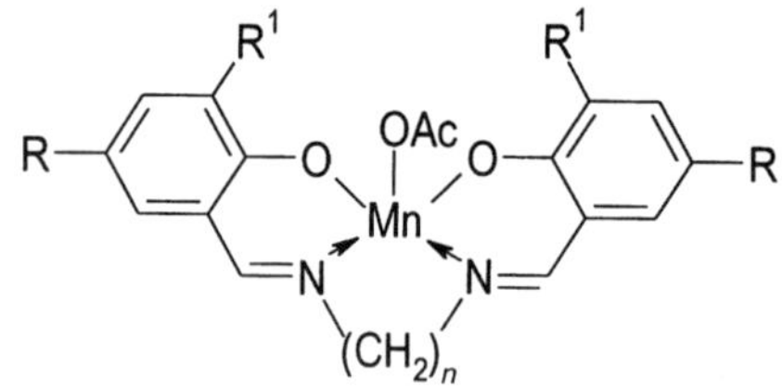

282: $n = 2, 3$; R = H, NO_2; R^1 = H, OMe, Br

Another widespread type of metal binding of carboxylate ligands is that in which the formation of dimers occurs having the structure of a "chinese lantern" ***283*** [4, p.132;537]:

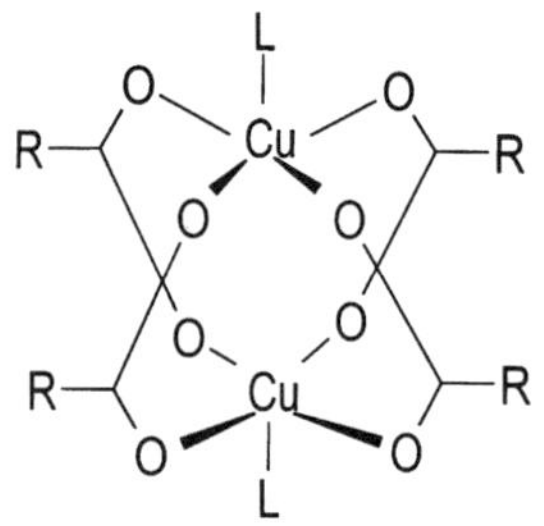

283

L = NR_3, H_2O, azoles, azines

Structures of this type are especially abundant among copper carboxylate complexes ([537] and references cited therein).

β-Diketones are the ones with the most widely repeated type of ligand system and they form mostly ICC of type **2** [1, vol.2;12–14,48,132,133,168,356,538–545].

Additionally to chelating O,O-coordination, the existence is believed of some other structures containing the O—M bond **284–287** [13,14,540]. Thus, according to x-ray diffraction data [540], the monodentate O-coordination ***285*** is present in complexes $Mn(HAA)Br_2$ and $Re(HBA)Cl(CO)_3$, where HAA is acetylacetone and HBA is benzoylacetone. The bidentate-coordinated diketone form of acetylacetone ***286*** ($R = R^2 = Me$, $R^1 = H$) is observed in chelates $[Ni(HAA)_2Br_2]$ and $[Ni(HAA)(H_2O)_2](ClO_4)_2$ [540]:

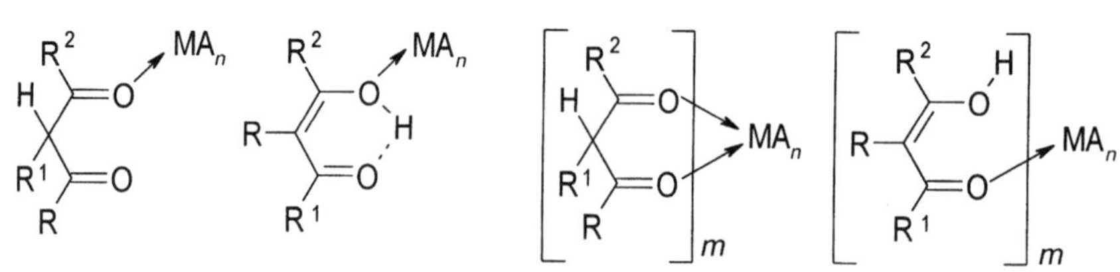

284 285 286 287

In classic β-diketonates **2**, oxygen atoms have free electronic pairs and, in this way, they may provide one additional O-coordination. Such a situation is observed in a series of barium complexes whose fragments are represented by formulae ***288–291*** [13]:

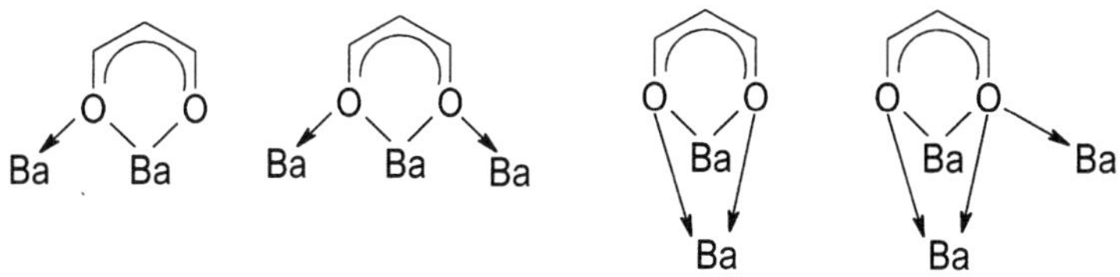

288 *289* *290* *291*

In complex ***288*** the β-diketonate fragment behaves as a tridentate ligand, while in ***289*** and ***290*** it behaves as a tetradentate ligand, changing in ***291*** to a pentadentate ligand system [13,546–550].

β-Carbonylphosphoryl compounds, forming ICC of type ***292*** [551,552], are O,O-donor chelating ligands:

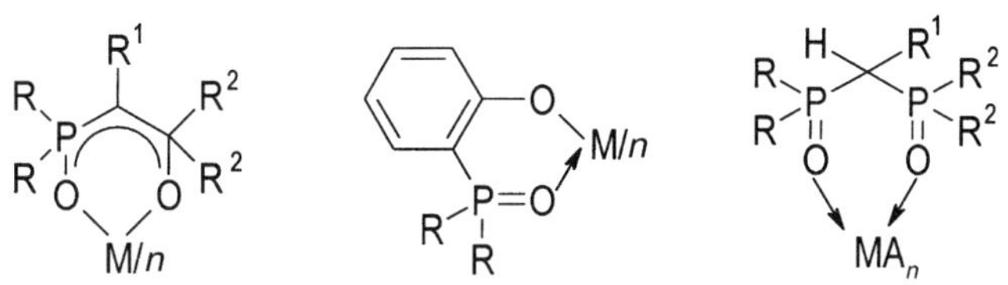

292 *293* *294*

The same ligand types include *o*-phosphoryl phenols and di[alkyl(aryl)diphosphonyl]methanes, for which the complexes of types ***293*** and ***294*** [552] are characteristic. In this respect, it is worth noting that, on the basis of alkyl(aryl)diphosphonylmethanes, molecular chelates ***294*** [552] are formed instead of ICC.

Some oxygen-containing heteroelement compounds, in which other elements are not coordinatively active, could be attributed to O-ligands. This group includes in particular nitrogen, phosphorus, arsenic, and antimony organo-oxides [1,3,11,112]. The most studies amongst them are phosphine oxides, on whose basis complexes, with structure ***295***, of practically all metals have been obtained [1, vol.2;3]:

$$R_3P = O \rightarrow MA_n$$
$$\textit{295}: R = \text{Alk, Ar}$$

The assumption of urea and its derivatives, ***296*** (X = O), as O-donors is quite authorized. Although the possibility of not only O- ***297***, but also N-coordination is assumed, the latter has not been strictly proved yet [13,112]:

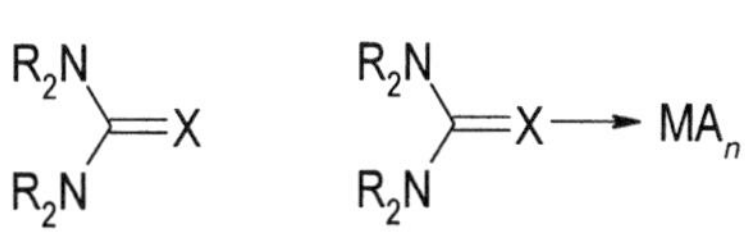

296 *297*

X = O, S, Se, Te

Complexes of O-containing heterocycles are represented by a wide number of coordination compounds of tetrahydrofurane (THF), to yield a structure of type ***298*** [1,2,4,8,73,92,553]:

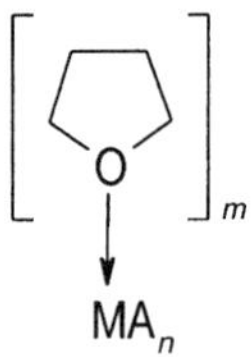

298

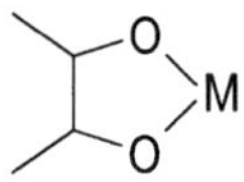

299

A special interest for this type of complex is related to the possibility of their use {for example, [THF · $M(CO)_5$]} for obtaining organometallic derivatives of aromatic and heteroaromatic ligands [2,8], and for the stabilization of unexpected cyclopentadienyl complexes of *p*-metals, for example [$Cp_2M(THF)_2$], where M = Mg, Ca, and [$Cp_2Ca(\mu\text{-}CH_3)_2AlCH_3$] · THF [553]. The oxygen atom in the heteroaromatic system (furane) is practically deprived of its donor properties [7]. However, some lanthanide σ-O-complexes of furane, included in macrocyclic O-containing ligand systems, are known [8,13].

Carbohydrates are also oxygen-containing ligands [1, vol.2;554–559]. It is supposed [557] that the diol function, similar to ***299***, participates as a structural unit in their complexes. It is present both in mono- and dinuclear chelating complexes [557]. Among these coordination compounds, the natural carbohydrate complexes [558] have been isolated in relation to their biological and medical importance (for simulation of enzymes).

Crown-ethers [1,4,5,11,560–572] occupy a special place among oxygen-containing ligands. Examples of these are ***25***, ***300***, and ***301***:

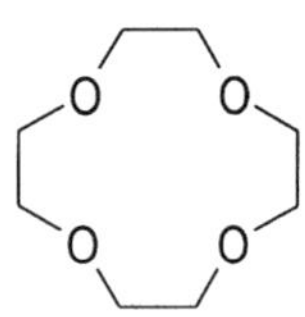

300

12-crown-4

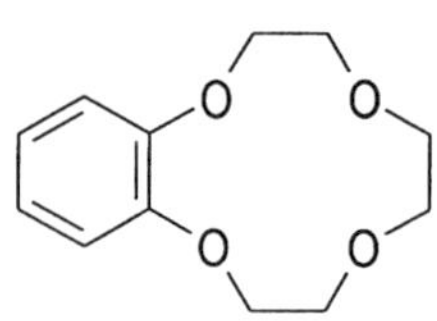

301

benzo-12-crown-4

Other examples of various crown-ethers can be found in textbooks [4,5], monographs [563–568,571], and reviews [569,570,572]. Different classifications are presented and fully examined [564,568], as well as their structures (for example, Refs. 568 and 573), thermodynamics of complex formation [574], and properties and applications [568,571].

The principle of geometric correspondence, formulated by Nobel laureate Pedersen [561], is the basis for explaining the crown-ether complexes formation. Their selectivity depends on the correspondence of the size and form of the ion complex-former to the size of cavity of the macrocycle. However, it was established later that this principle does not have an absolute dependability [571]. Such factors as type of anion, complex composition, and the nature of solvent used during synthesis also have considerable influence [571].

The forces of ion-dipole interactions are mostly the basis for complex formation in these systems. This fact probably helps explain the possibility of complex formation of crown-ethers with cations whose size is bigger than the ether's cavity. In this case, the sandwich structures are formed, or a considerable displacement (going out) of metal cations out of the plane of the oxygen donor atoms [571]. Opened O-containing crown-ether podands [575–578], for instance, ***302*** and ***303***, belong to this structural type:

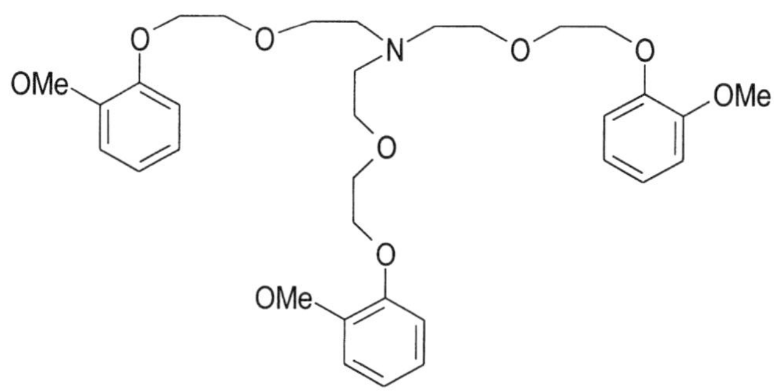

302

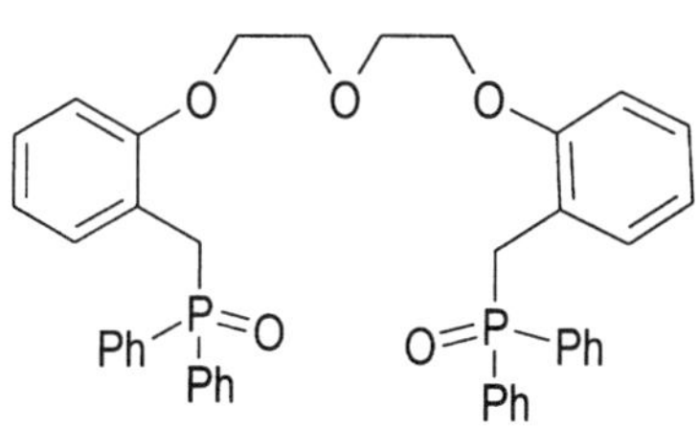

303

The main significance of the examined polycyclic ligands is in the creation of the basis of coordination chemistry for the *s*-elements of Groups IA and IIA of the Periodic Table. Not long ago the above complexes of these elements were only represented by a few single examples; at present, hundreds of them have been described. Another important aspect to be mentioned is the use of crown-ether complexes as biomimetic models of ionophores [579–581].

2.2.4.5 Ligands with Sulfur as Donor Atom

This type of ligand system is represented, in general, by H_2S, sulfides, and disulfides, thioethers, dithiolenes, dithiocarbamates, and many other compounds related to them. The H_2S complexes are not numerous [1,3,582a]. Among them, the following compounds have been described: $AlCl_3 \cdot H_2S$, $TiCl_4 \cdot nH_2S$ ($n=1$, 2), $[Mn(CO)_4(PPh_3)H_2S] \cdot BF_4$ [1, vol.2], $[(CO)_5W(H_2S)]$, $[Ru(NH_3)_5(H_2S)](BF_4)$ and $[RuCl_2(P{-}N)Par_3 \cdot H_2S]$, where P—N is *o*-diphenylphosphino-N,N-dimethylaniline [582a]. The structure presented by ***304*** [582a] is attributed to the last complex:

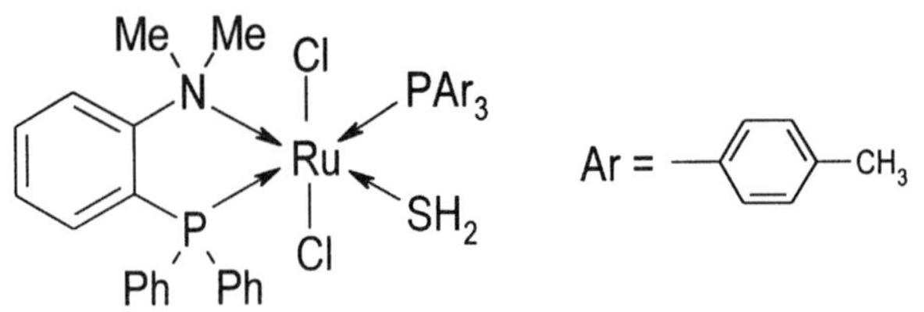

304

Organothiols and organodithiols are deprotonated in reactions of complex formation and participate as anionic ligand systems. Among the complexes containing thiolate SR^- anions, the mononuclear complexes are particularly emphasized [1, chap.16.1], in which the examined ligand behaves as a monodentate [582b]. Such complexes, for instance, are those containing the $[M(SH)_4]^{2-}$ anion (M = Mn, Fe, Co, Ni, Zn, Cd) [582b–584]. Thiolate ligands present in cluster anions, for example in $[Fe_4S_4(SR)_4]^{2-}$, perform the same function [1, vol.2].

Di- and polynuclear complexes with the μ_2-SR ligand are very common. The following conformers [1, vol.2] are four samples of these bridge fragments:

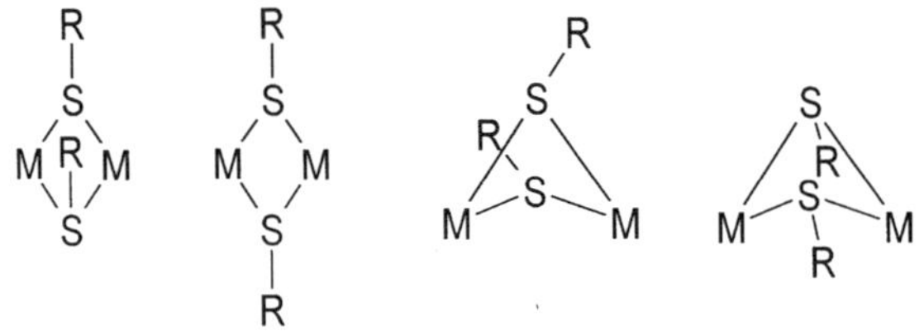

syn *anti* *syn-exo* *syn-endo*

Compound ***305***, $Ni_2(SCH_2Ph)_2L_2$, where $L = S_2CSCH_2Ph$, is an example of a complex containing a di-bridge SR fragment [1, vol.2]. Binuclear molybdenum thiolate-bridge complexes have recently been reviewed [585]. In a series of cases, complex compounds are formed where both the terminal and bridge binding of the sulfur atom of the SR ligand by a metal take place simultaneously. Such a coordination mode is observed, for instance, in the complex anion $[Fe_2(SR)_6]^{2-}$ ***306*** [582b,584]:

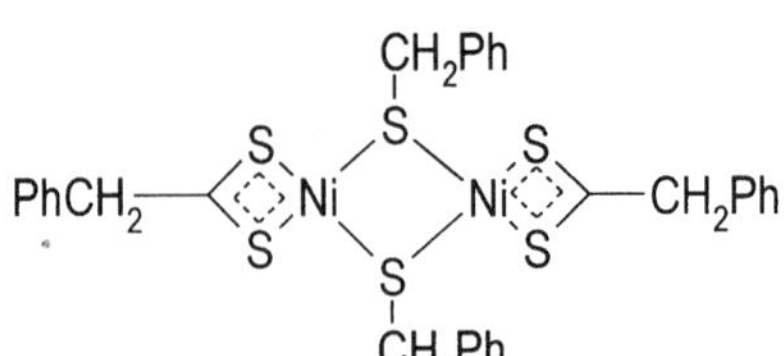

305

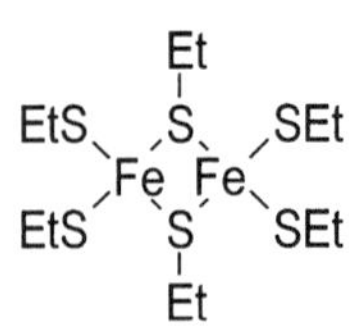

306

Many other complexes containing alkyl and aryl [585] thiolate fragments are also described [132,133,586–588]. It must be noted that not only monothiol compounds [1, vol.2] serve as precursors of this kind of complex, but also organodichalcogenides, in particular disulfides, in which the S—S bond is easily fractured under conditions of electrosynthesis [reaction (2.4)] [132,133]:

$$n/2(\text{R-X-X-R}) \xrightarrow[\text{e/s}]{\text{M}^{\text{o}}} \text{M(XR)}_n \qquad (2.4)$$

Among possible structural types ***307–309***, dithiolates are mostly represented by the structures ***308*** [1, vol.2;132,133,585,587]:

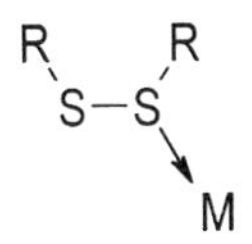

307

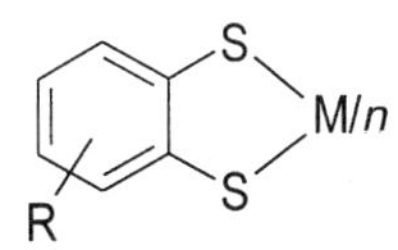

308

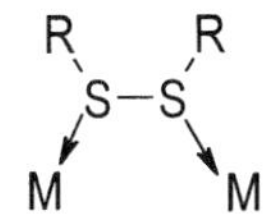

309

Thioethers take part, in general, as terminal (***310***, ***311***) and bridge (***312***) [1, vol.2;589]:

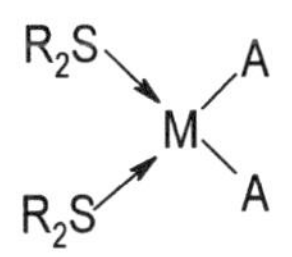

310

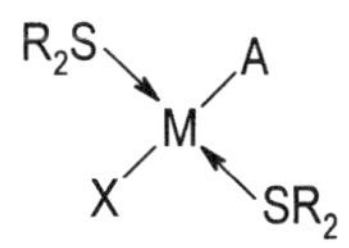

311

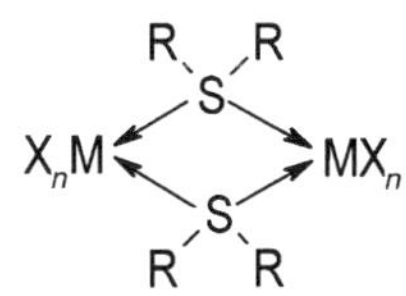

312

Ligands with two or three sulfur atoms also form metal chelates ***313*** and ***314*** [1, vol.2], in addition to the terminal and bridge functions discussed:

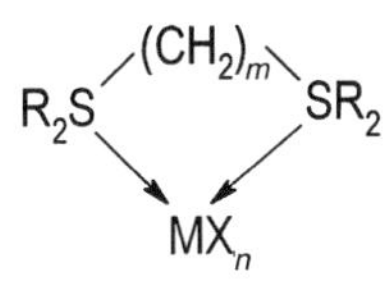

313

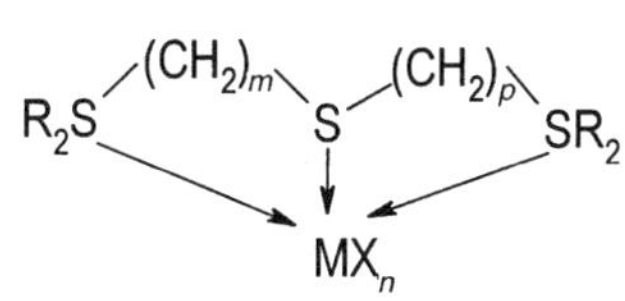

314

Cyclic sulfur-containing ethers are mainly represented by thio-crowns–polythioethers [11,569,590,591]. Typical representatives of this group of compounds are ***315*** (a sulfur analogous to cyclame), ***316*** (a sulfur analogous to 18-crown-6), and ***37*** (an interesting variety of metacyclofane) [568]:

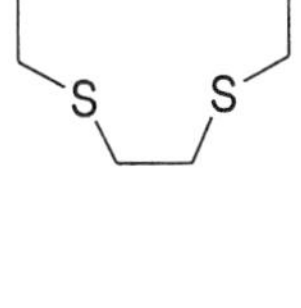

315

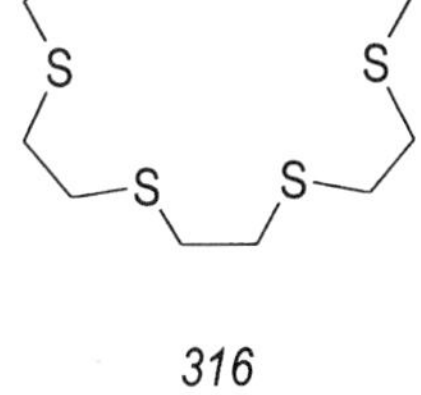

316

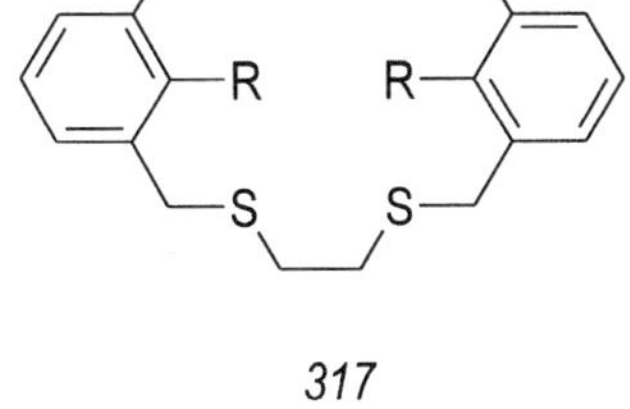

317

We note that, in a marked difference with respect to the oxo-crowns ***300*** and ***301***, sulfur-containing crown-ethers form stable complexes with transition and, especially, "soft" metals [590,591], as discussed later.

Dithiolene complexes are represented by planar monomeric ***318*** and dimeric structures, where the dimerization is carried out at the expense of the formation of either M—M (***319***) or M—S (***320***) intermolecular bonds [1, vol.2, p.515]:

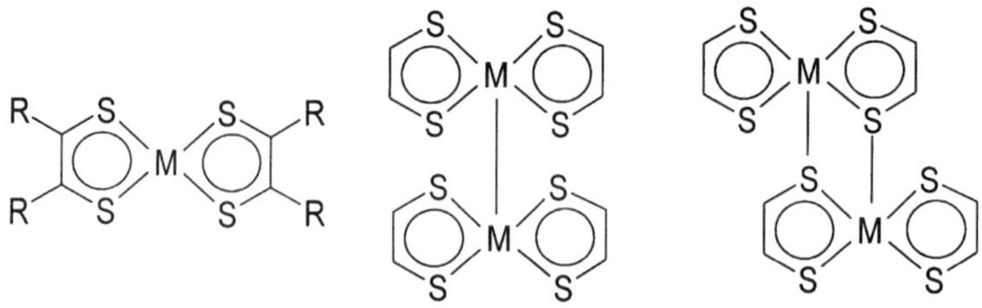

318a 319 320

An example of a complex of type ***318*** is a nickel chelate, while structure ***319*** represents an ICC of platinum and palladium. These complexes are of renewed interest due to their controversial data concerning metal oxidation number (0, +2, or +4), in particular for Ni, Pt, and Pd, which depends on the resonance structures of the ligand systems shown as ***318b*** and ***318c***:

318b 318c

This situation has been widely discussed [592a–c] and has not been finally resolved [592d]. At the same time, there is experimental data (reactions with halogens) in favor of the existence of dithiol structures of type ***318b*** containing M—S bonds (M = Ni, Pd, Pt) with metal oxidation number +4 [592e,f].

Trithiolene complexes are also known, for instance those of chromium, molybdenum, and tungsten [592f], having an octahedral geometry. Structures of the complexes above, as well as of a series of other thiolene (thiol) compounds, have been studied using various physical–chemical techniques (x-ray diffraction, NMR, EPR, and UV spectroscopy). Considerable attention has been devoted to their reducing and photochemical properties [1, vol.2].

The following compounds, each with a general formula included, are widely used as ligands: dithiocarbamates ($R = NR^1_2$; R^1 = Alk, Ar, Het) and ksantogenates ($R = OR^1$; R^1 = Alk, Ar) ***321***, dithiophosphinates ***322*** (E = P; R, R^1 = Alk, Ar), and dithiophosphates ***322*** (E = P; $R = OR^2$; R^1, R^2 = Alk, Ar) [1, vol.2]. The related structure ***322***, representing either As-, Sb-, or Bi-containing ligands (E = As, Sb, Bi), was reviewed recently [593].

321 322

Ligand systems ***321*** and ***322*** form coordination compounds of types ***323*** and ***324***, respectively:

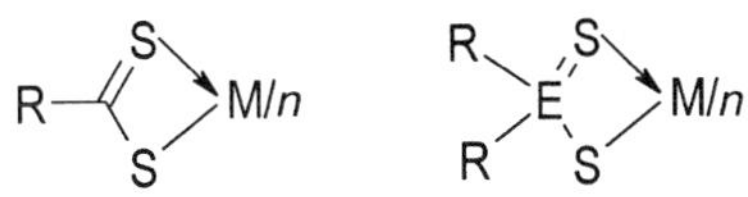

323 324

The greatest research effort is devoted to dithiocarbamate complexes *323* ($R = NR'_2$). Amongst them, the ICC having monomeric, dimeric, and tetrameric structures have been isolated and fully characterized [1, vol.2;132,133,588,594–610]. The groups CS_2^-, as well as PS_2^-, carry a bidentate-cyclic and/or bidentate-bridge role. The special interest placed on this group of compounds is due to the possibility of obtaining, on their basis, a new generation of improved thermochromic coatings [588] and photochromic materials [605]. The ksantogenate complexes *323* ($R = OR^1$; $R^1 = $ Alk, Ar, Het) [1, vol.2;588,611–615] are also the focus of permanent research.

The following structural types are characteristic for the group of dithiocarbamate complexes: *325* and *326* for mononuclear, *327* for binculear [1, vol.2]:

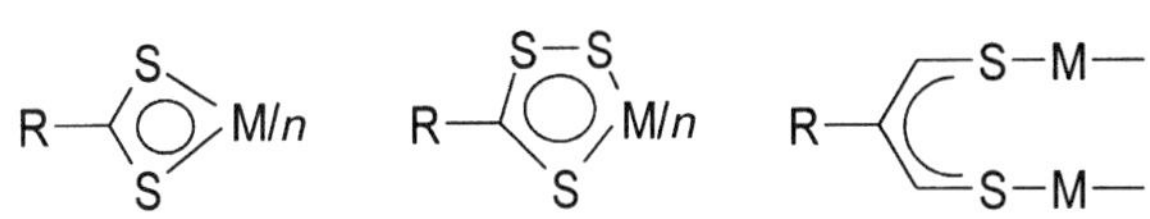

325 326 327

It is known [1,616] that sulfur-containing ligands could help to stabilize the unusual oxidation states of metals. Thus, the direct interaction of metallic copper (see Sec. 3.4.3 on "direct" synthetic methods) with nonaqueous solutions of tetramethylthiuram disulfide [TMTD, $Me_2N{-}C(S){-}S{-}S{-}C(S){-}NMe_2$], in the presence of CI_4 and bipy, has been studied in detail [617]. The synthesized compounds have the general formula $[Cu_nI_m(Me_2NCS_2)_l]A$, where $n = 1$–3, $m = 0, 1, 2, 4$, $l = 1$–3, 5, $A = H_2O$, Cu_2S, and CHI_3. They contain copper in the forms of Cu(I), Cu(II), or rarely Cu(III), according to magnetic data. It should be emphasized, however, that the diamagnetic properties of this kind of complex compound are not sufficient to propose the +3 oxidation state for copper, since the existence of Cu(III) paramagnetic compounds is well known [618]. The Cu^{3+} ion, as also happens with Ni^{2+}, has a d^8-configuration in its electronic shell.

It has been shown that the process of oxidation of copper in these systems takes place through successive stages. Thus, there is an opinion [617] arguing that a one-electron oxidation of Cu^0 by a two-component system [for instance, formamide + bipy or DMF(DMSO)+CI_4], or a one-electron oxidation of Cu(I) to Cu(II) by the thiuram (or its dication), or even one-electron oxidation from Cu(II) to Cu(III), for the system DMSO+CI_4 + TMTD, all take place with the participation of CI_4 or I_2 + $Me_2S \cdot I_2$ as it is formed in the system.

The electrochemical dissolution of copper and other metals (see Sec. 3.4.2 on direct electrosynthesis), in solutions of tetramethylthiuram disulfide and its analogues in nonaqueous solutions of acetone and acetonitrile, was first carried out by Tuck and coworkers [619]. Complexes with a general formula given by $M(R_2NCS_2)_n$ are formed with good yields under these conditions. The proposed mechanism of successive complex formation for this kind of system includes first the electrochemical step followed by the reaction of oxidative addition [reactions (2.5)] [619]:

$$\text{Cathode: } (R_2NCS_2)_2 + 2e^- \rightarrow 2(R_2NCS_2)^-$$
$$\text{Anode: } (R_2NCS_2)^- + Cu \rightarrow Cu^I(R_2NCS_2) + e^-$$
$$\text{then } Cu(R_2NCS_2) + 0.5(R_2NCS_2)_2 \rightarrow Cu^{II}(R_2NCS_2)_2 \qquad (2.5)$$

An additional study on the same system has been reported, including a comparison of direct electrochemical and conventional chemical dissolution of metallic copper in TMTD solutions in various solvents under conditions of simultaneous ultrasonic treatment of the reaction system [133,620]. It has been shown that the system "TMTD–copper–solvent" could serve as a perfect model to study the influence of simultaneous application of ultrasonic treatment (see Sec. 3.5) on the syntheses of complexes of the transition metals in different nonaqueous solutions, by using and testing several techniques [620]. Several other studies on the interaction of copper and iron species with thiruam sulfides have also been reviewed [621].

Among other complexes of S,S-chelating ligands, we note β–dithioketonates ***328***, β-mercaptothioetherates ***329***, and imidodithiophosphinates ***330***, forming ICC with six-member metal-cycles. These coordination compounds are examined in publications summarized in a monograph edition [1, vol.2].

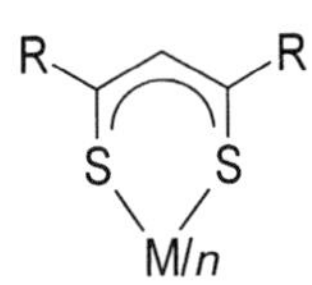

328

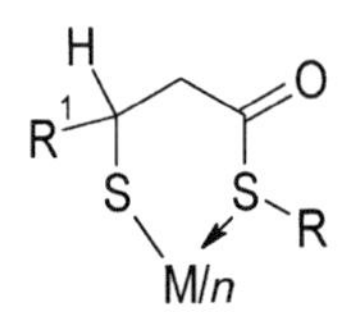

329

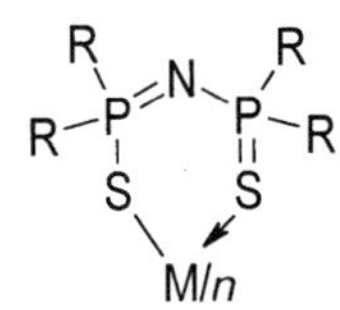

330

Thiophene complexes, containing σ(S)-coordination bond, are represented by a high number of structures [8,13,205b,622,623]. Compound ***321*** [8] is an example of these complexes. η^1(σ-S-Coordination is also characteristic for benzothiophene complexes, for instance ***322*** [8,13,624]:

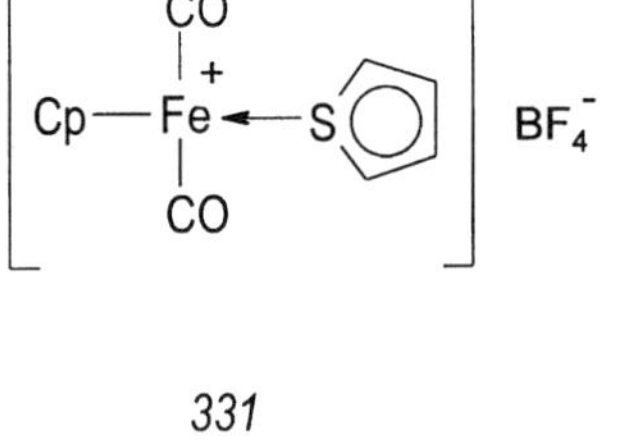

331

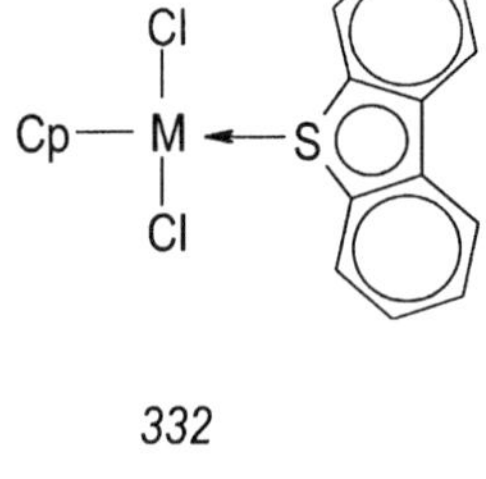

332

Much of the work reported by the Chinese research groups [625–631] is dedicated to x-ray diffraction studies of the transition metal complexes (Ni, Cr) having coordination unit $Ni^{II}N_2S_4$ or $Cr^{III}S_6$, respectively. Other complexes having coordination units N_2S_2, N_2S_4, and N_4S_4 are widely reported [1, vol.2;632] and described in Sec. 2.2.5.4.

2.2.4.6 Ligands with Selenium as Donor Atom

Ligands with electron-donor selenium atoms, similar to S-containing ligand systems, include H_2Se, the selenolate anion (RSe^-), diorganoselenides (R_2S), diorganodiselenides (R—Se—Se—R), diselenolenes (RSe_2^-), diorganoselenocarbamates [$SeC(O)NR_2^-$], and other compounds [1, vol.2;633–636a]. Interaction of salts and metal complexes with H_2Se leads to the formation of coordination compounds containing the fragment HSe, for instance ***333*** and ***334***:

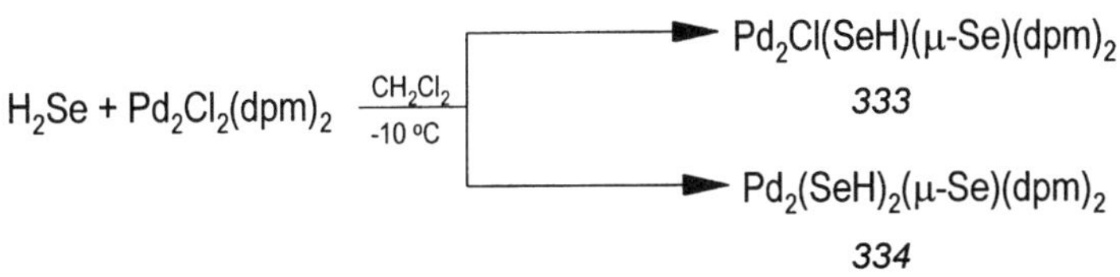

(2.6)

Formation of complex compounds ***333*** and ***334*** is regulated by the reagents' molar ratio: complexes ***333*** and ***334*** are formed for ratios 1:1 and 2:1 (H_2Se:Pd), respectively [636b].

In principle, the metal complexes of organic selenium ligands do not have a considerable difference from their sulfur analogues described in Sec. 2.2.4.5. Thus, the organoselenate anion participates either as a terminal (M—SeR) or bridge (M[μ-SeR]$_2$M) ligand. There is an ample summary of such compounds reported elsewhere [634, table 26, p. 741].

Diorganoselenides are monodentate ligands capable of forming numerous complexes of general formula $MA_n \cdot mR_2Se$ [589,635]. Their complexes of formula [$MA_2(SeR_2)_2$] are the most commonly described, as for example those having M = Pt, Pd; A = Br, I; R = Me, Et, *i*-Am [634].

It seems that the inclusion of diorganodiselenides [634] in the category of monodentate ligands is doubtful. In reality a series of complexes is known where the examined compounds behave as monodentate ligands, for instance ***335*** and ***336*** [16], but at the same time the dimeric structures have been described, for example ***337***, using for an explanation a bidentately-coordinated R_2Se_2 molecule [16,634a,b]:

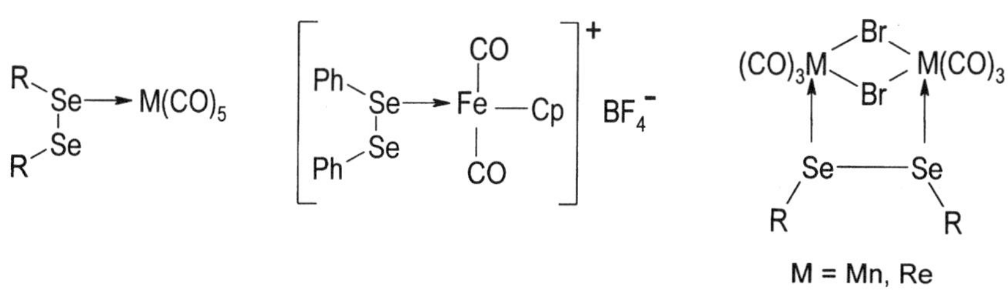

335 *336* *337*

Amongst the diselene complexes, the following structures are presented as an illustration of this kind of compound: ***338*** [637], ***339*** [638–640], and ***340–342*** [641]:

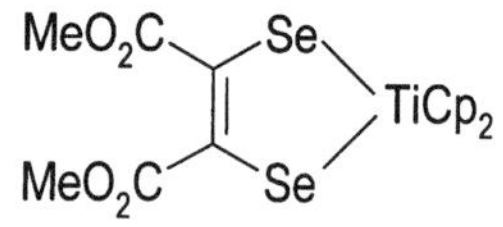

338

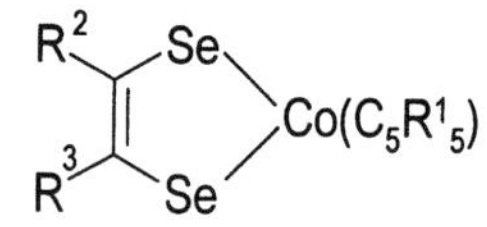

339: R^1 = H, Me; R^2,R^3 = H, Bu, Ph

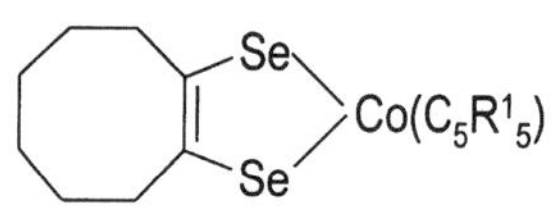

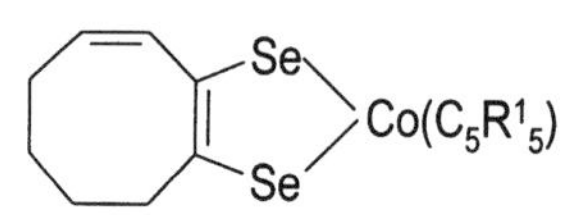

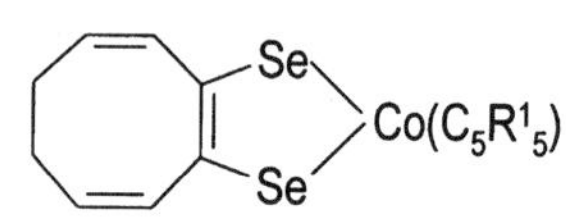

R^1 = H, Me

340 *341* *342*

Optimal synthetic methods for obtaining these complexes have been developed including, for example, the interaction of acetylene derivatives, atomic selenium, and cyclopentadienyl complexes [638,640]. In some derivatives of 1,2,3-selenodiazole, structure ***343*** can be transformed to ***339***, as shown by reaction (2.7) [641]:

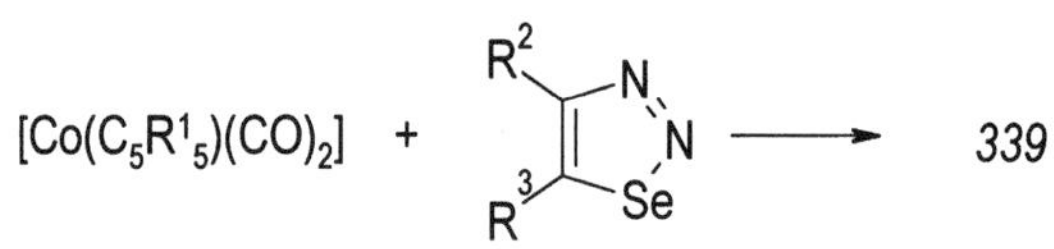

(2.7)

343

Some representatives of chelates containing an Se,Se-ligand environment are known, amongst them the dialkylselenocarbamates ***344*** [588,634,642] and thc β-disclcno-ketonates ***345*** [634]:

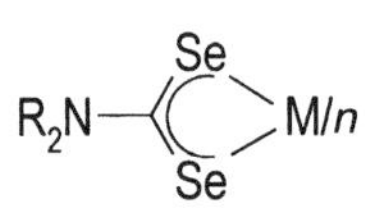

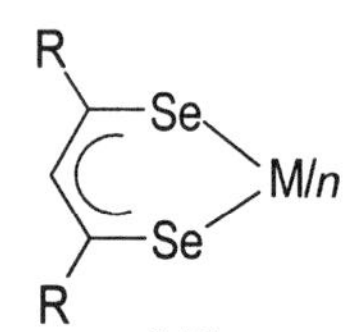

344 *345*

M = Zn, Cd

Other chelating ligands, leading to the formation of complexes with four-, five-, and up to six-member metal-cycles of the type MSe_4, have been fully described [634].

Coordination compounds of selenium-containing heterocycles with an Se—M bond are represented by the complexes of 1,4-diselenane ***346–348*** [634] and selenophene ***349*** [8,643]. Among these complexes, while the Se—M coordination ***350*** was proposed [634] for the complexes on the basis of 1,2,3-benzoselenodiazole, only N—M binding ***351*** was found in these compounds [13,644]:

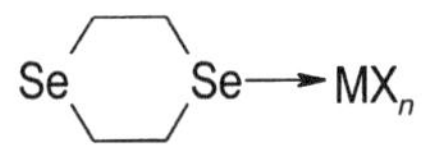

346

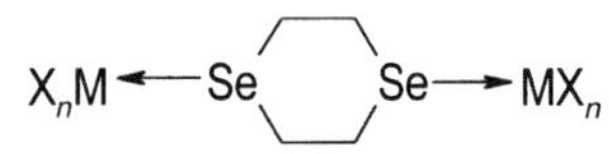

347

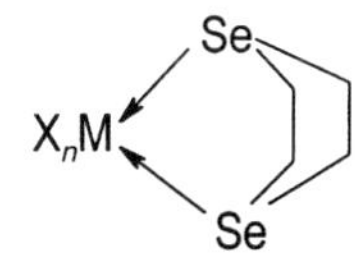

348

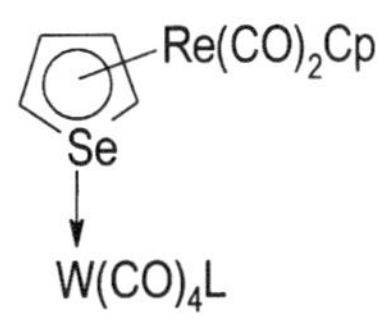

349

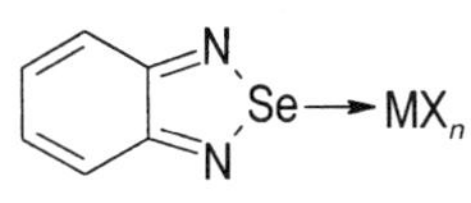

350

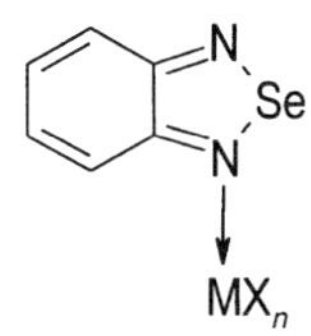

351

On the contrary, selenourea ***296*** (X = Se) is a ligand with selenium as electron-donor center and forms complexes of type ***297*** (X = Se) [112,634].

2.2.4.7 Ligands with Tellurium as Donor Atom

As a result of much research made mostly during the last 20 years, a high number of complexes with tellurium as electron-donor atom have been described [1, vol.2;8,13,16,633–636a,645–649]. Organotellurolate anions carry out, most commonly, a dibridge function, forming *sin*- or *anti*-isomers which are described, for instance, by metal–carbonyl complexes of the types ***352*** and ***353***, respectively [650–652]:

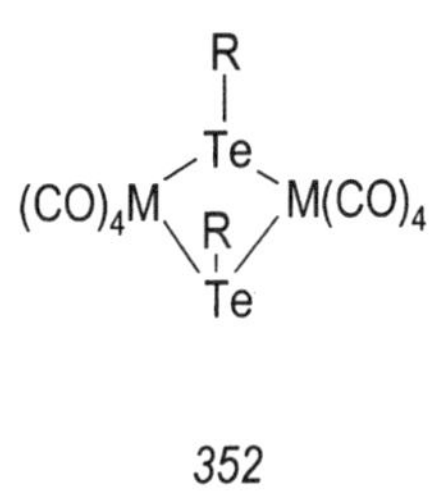

352

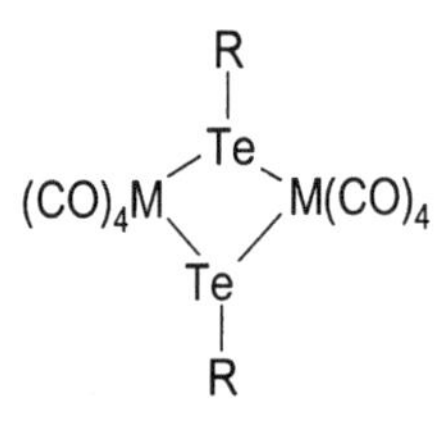

353

One of the most interesting examples is given by a binuclear manganese complex of formula $[Mn_2(CO)_6(TePh)_3]$, where the TePh group participates as a tribridge fragment [653]. At the same time, some complexes with the terminal TeR ligand are known, for instance *cis*-$(PPh_3)_2Ir[Te(2,4,6\text{-}t Bu_3C_6H_2)](CO)$ [654], as well as some others containing mixed-coordinated tellurolate anions ***354*** [1, vol.2]:

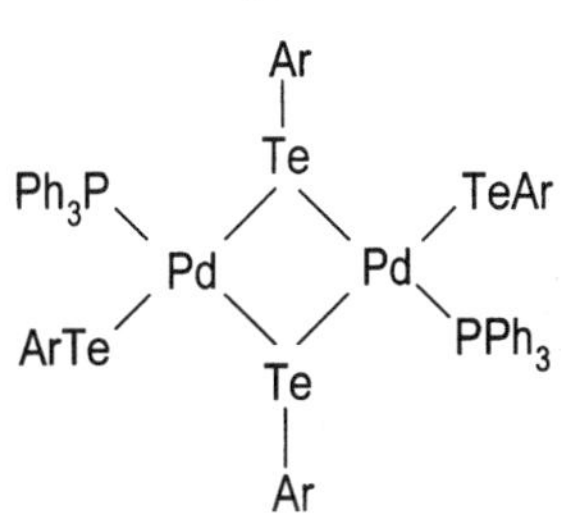

354

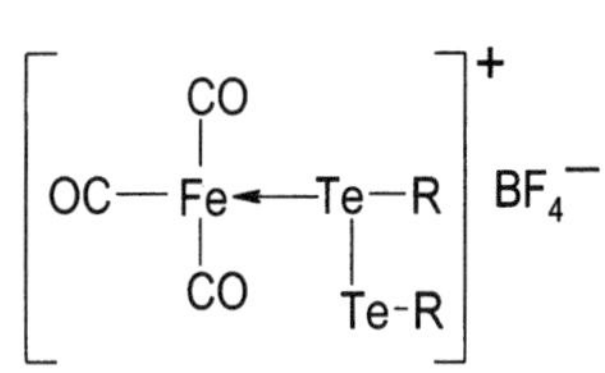

355

Diorganotellurides form complex compounds of the type $(TeR_2)_m \cdot MA_n$, of which compound $[TeR_2 \cdot M(CO)_5]$ (M = Cr; R = $C_6H_4CH_2NMe_2$-*o*) [633–635] is a clear example.

The interaction between diorganoditellurides, metal salts, and carbonyls leads not only to complex compounds of the types ***352–354*** (fracture of Te—Te bonds), but also to adducts of R_2Te_2 [16,646,655]. The examined ligands in these complexes can form either a monodentate ***355*** [16] or a bidentate bridge function (for example, ***356*** [646]). Recently, compound ***357***, having a chelating $ArTe_2Me_2$ ligand, was described [656]:

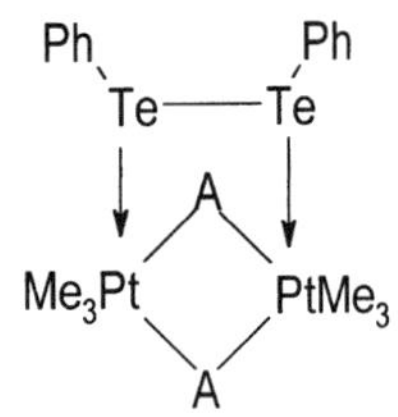

356: A=TePh

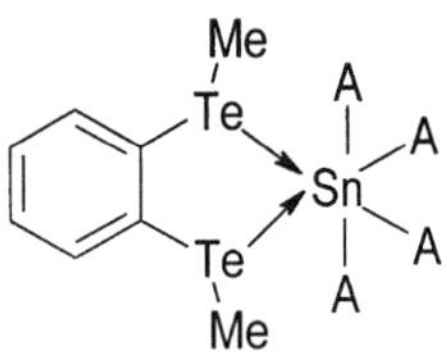

357: A = Cl, Br

The following compounds of tellurium–carbonyl derivatives have been described: metal–carbonyl complexes of tellurobenzaldehyde ***358*** (R = H) and diphenyltelluroketone ***358*** (R = Ph) [16,657,658], tellurourea ***300*** [E = Te; $MA_n = M(CO)_5$; M = Cr, Mo, W, $MnBr(CO)_4$] [16,633,634], and phosphatelluroketone ***359*** [16,659]:

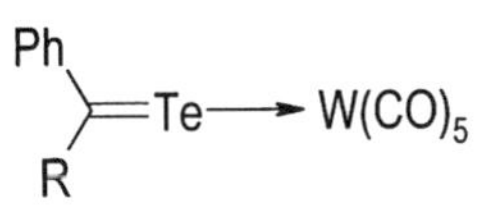

358

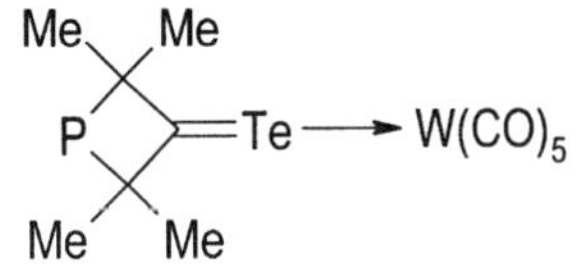

359

The following compounds, obtained on the basis of tellurium-containing saturated ***360*** [16,660,661], unsaturated ***361*** [16,662,663], and aromatic ***362*** [1, vol.2;8,13,16,633,634] heterocycles, are widely present in specialized literature:

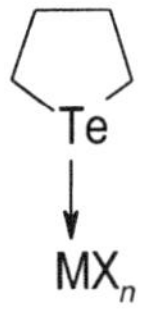

360

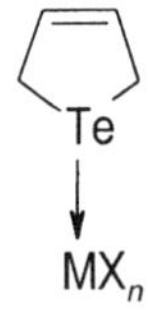

361

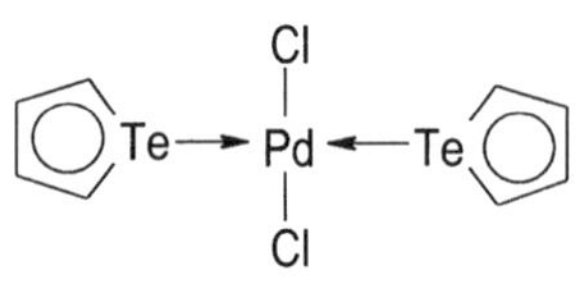

362

Other complexes of heterocyclic ligands with Te-donor center are fully described in reviews [8,13,16,633,634].

2.2.5 Organic Ligands with Different Donor Atoms

2.2.5.1 Carbon-Heteroatomic (N, P, O, S, Te) Ligands

Organic nitriles* (RCN) are the simplest C,N-containing ligands; they form three types of complex compounds represented by ***363–365*** [1, vol.2;3,664-667]:

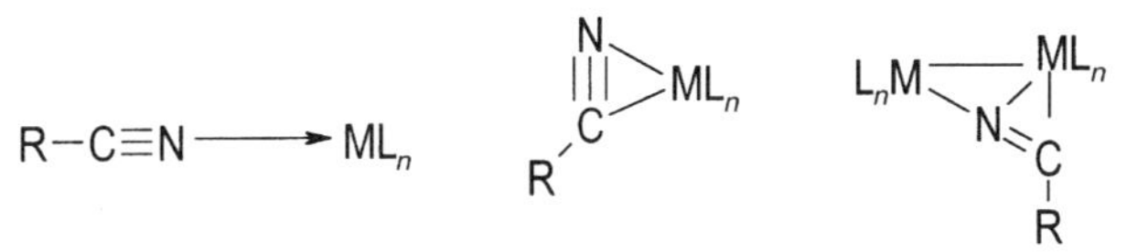

363 *364* *365*

η^1-Coordination (***363***) is the most widespread among the structures shown above when the usual donor–acceptor bond is formed with the participation of the N atom of monodentate organonitrile ligands. The nitrile group of these ligands behaves either as chelate ***364*** or bridge ***365***. The structure ***364*** was found, for instance, in the series of complexes represented by the formula $[M(\eta^5\text{-Cp})_2(\eta^2\text{-NCR})]$, where R = Me, M = Mo [665]. The bridge-bonding ***365*** is characteristic for dimeric and oligomeric cluster-like structures, such as that of the complex compound of formula $[Mn_2(\mu\text{-}\eta^1,\eta^2\text{-NCMe})(CO)_6(Ph_2PCH_2PPh_2)]$ [668].

Among other C,N-donor ligand systems, the diazoalkane complexes [669,670] are widely found. The existence of three types of structures is possible due mainly to their mononuclear compounds, where either the bindings M—C (***366***), M—N (terminal, ***367***) or N—M—N (chelate, ***368***) are present [669,670]:

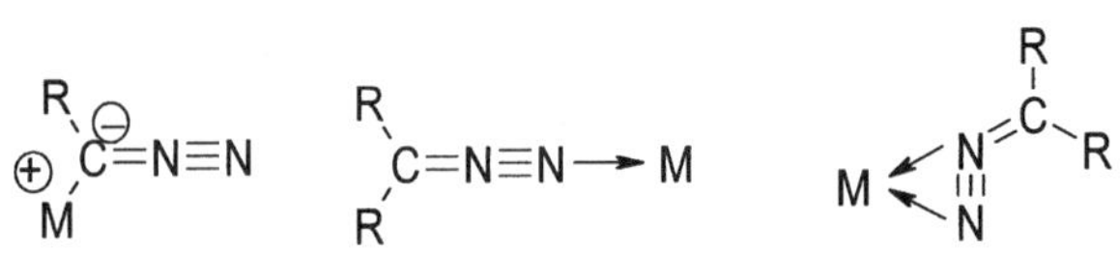

366 *367* *368*

The most frequently reported compounds amongst the types described above are those with η^1-N-coordination (***367***) and η^2-N,N-coordination (***368***) [669,670]. It is possible to attribute as belonging to the first type of complexes $[W\{N_2C(SiMe_3)_2\}(PPh_3)(CO)_4]$ and $[IrCl(N_2C_5Cl_4)(PPh_3)_2]$ which were characterized by x-ray diffraction [669,670]. Chelating N,N-binding is also confirmed by x-ray diffraction in the complex compounds $[CpMo(CO)_2(N_2CPh_2)]$ and $[\{2\text{-}(OMe)C_6H_4O\}_2V(N_2CHSiMe_3)]_2$ [669,670]. Diazoalkyl complexes containing an M—C bond are very common among "soft" metals ("*Hard and Soft Acids and Bases*" principle – see Sec. 1.2.2) (Pd, Rh, Os), as well as for nickel [669,670]. They are represented, for example, by the following complexes: $[PBu_3)_2Pd\{C(N_2)COOEt\}_2]$, $[PEt_3)_3RhC(N_2)SiMe_3]$, and $[Cl(PMe_3)_2NiC(N_2)SiMe_3]$ [669,670]. The possibility of formation of chelate structures with η^2-C—N-coordination is not excluded [669,670]. A wide variety of coordination modes in the binding of diazoalkanes

*Discussion of isonitrile complexes is covered in Sec. 2.2.4.1.

are characteristic for di- and polynuclear complexes [669,670]. The diazopentadiene complexes have also been described [671].

A wide number of possibilities for using heteroatomic nitrogen-containing organic molecules as electron-donor carbon center have been opened with the advent of the cyclometallation reactions, for example (2.8) [8,13,672–689]:

369 *370* (2.8)

E = N, P, O, S

The main stereochemical requirement for ligands of type ***369*** is the arrangement of donor centers in postions which are convenient for the formation of metallacyclic structures ***370*** [689]. Behaving as ligands of the type discussed, most of the C,N-donors take part to form complexes of amines ***371*** [679,689], azomethines ***372*** [13,679,686,690], azocompounds ***373*** [13,679,689,691], aldazines ***374*** [13,676,686,692], oximes ***375***, and their derivatives ***376*** [689]:

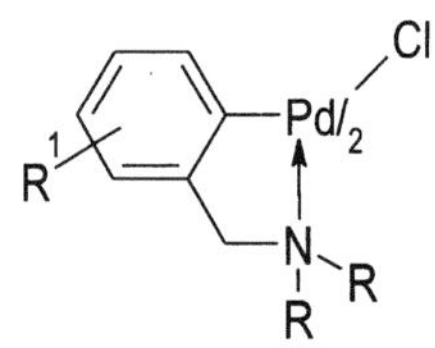

371: R^1 = H, Me; R = Me, Et

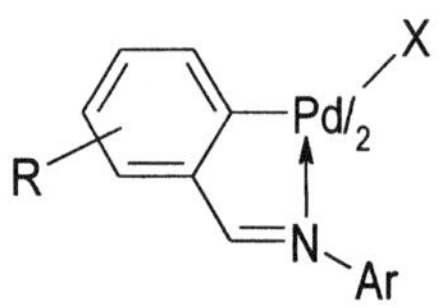

372: X = OAc, Cl, Br, CNS; Ar = $C_6H_4R^1$; R,R^1 = H, Alk, OAlk, Hal, CN, NO_2

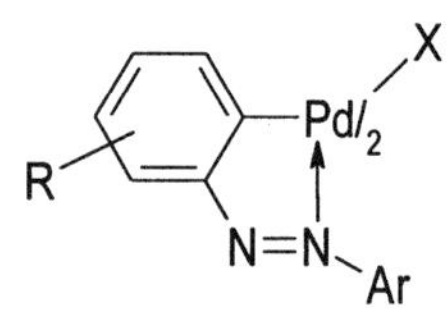

373: X = Hal; Ar = $C_6H_4R^1$; R,R^1 = Alk, OAlk, OCOAlk

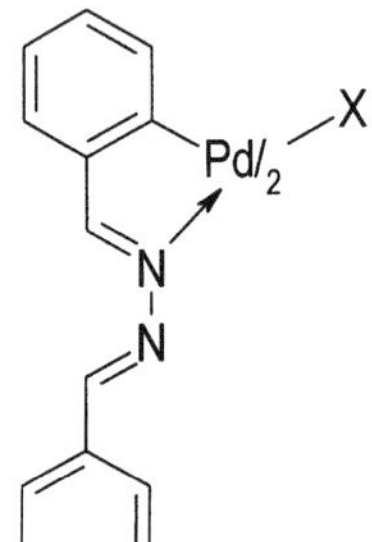

374

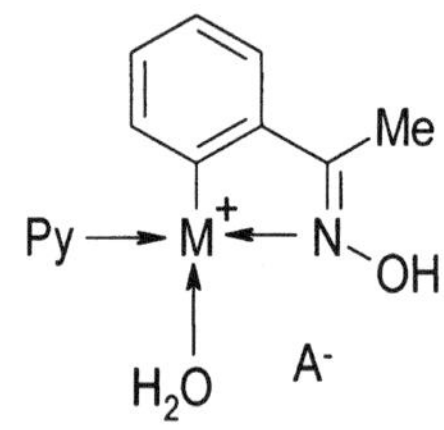

375: A = Cl, Br, SCN

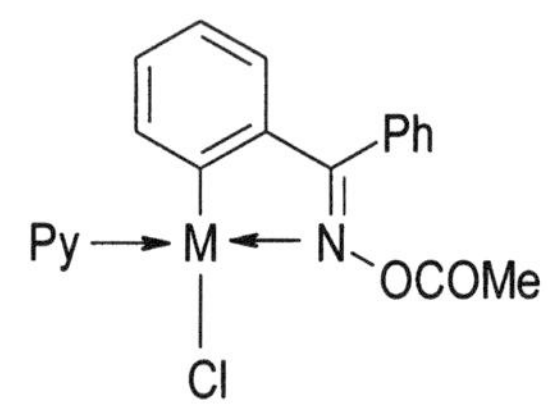

376: M = Pt, Pd

Amongst the last group of complexes, compound ***377*** represents a special interest because its structure testifies that the carbon atom of the cyclopentadienyl fragment of ferrocene may participate in the coordination with the metal (Pt) [689]:

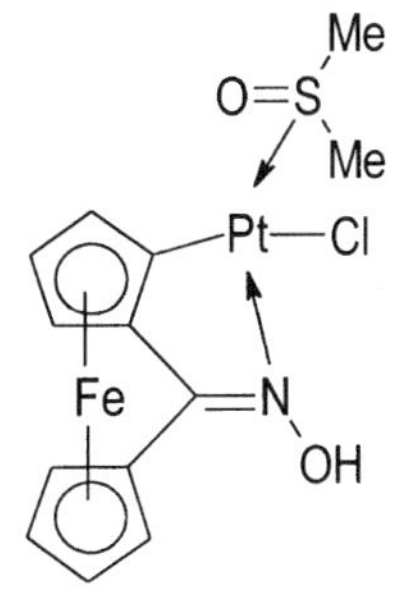

377

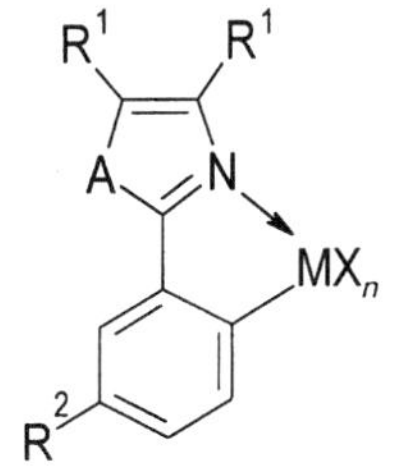

378: A = NR, O, S;
R^1 = H, Alk, Ar, 3,4-benzo;
R^2 = H, Het

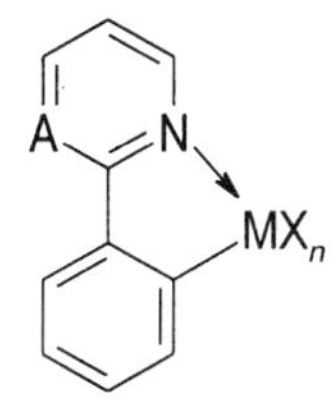

379: A = CH, N

The phenyl derivatives of azoles and azines are widely used as C,N-ligand systems; on their basis the following complexes were obtained: ***378*** (A = NR [13,693,694]; S [679]) and ***379*** (A = CH [695,696]; N [697]). The examples of C,N-coordination of alkyltrimethylsilyl derivatives of pyridine and quinoline have been reviewed [8].

There are considerably fewer complexes containing ligands of the type ***369*** (E = P, O, S). Thus, coordination compounds ***380*** [688], ***381*** [679], and ***382*** [679] have been described:

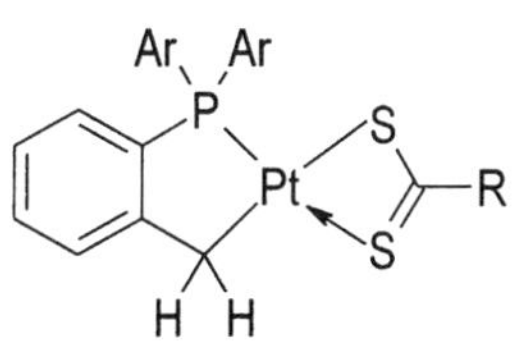

380: Ar = C_6H_4-Me-*o*;
R = NMe_2, OEt

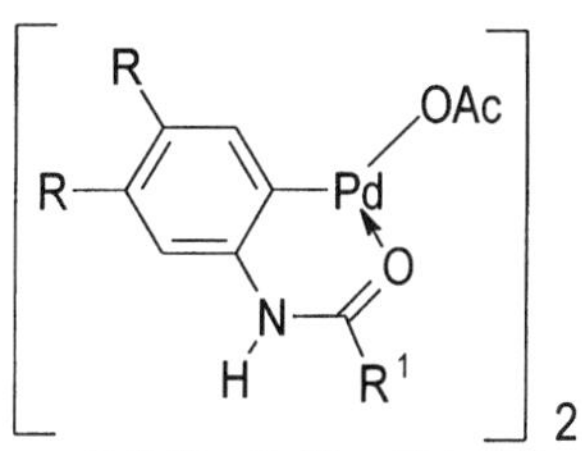

381: R = H, Me, OMe, Cl, COMe, COOEt;
R^1 = Me, Et, *t*-Bu

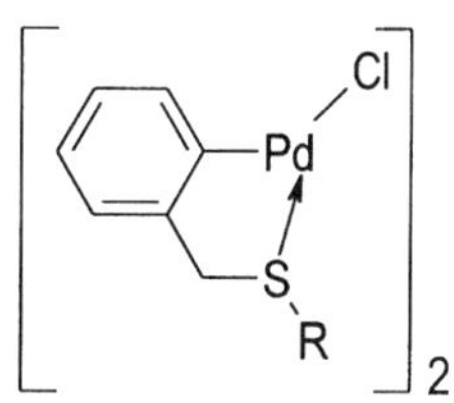

382: R = *i*-Bu, *t*-Bu

A wide range of cyclometallation reactions have been attempted [685]. Depending on the nature of the metal, these reactions were separated into three categories: nucleophilic (Co, Fe, Ru, Rh, Pd, Os, Ir, Pt), electrophilic (Ti, Zr, Hf), and multicentered (Th) cyclometallation [685].

C,P-Coordination is also observed in a series of bi- and trinuclear diphenyldiphosphine complexes containing platinum and palladium [698a]. Some representatives of complexes with C,Te-coordinated ligands have also been described [634].

Carbonyl complexes (Sec. 2.2.2.1) are well presented in all classic monographs and textbooks on organometallic and coordination chemistry (for example, Refs. 21b and 285d) and, so, their detailed coverage is not required in the present book. The bonding in M—CO complexes can be seen in Fig. 2.10 [184h].

They may be mononuclear [$V(CO)_6$, $Fe(CO)_5$] or polynuclear [$Mn_2(CO)_{10}$, $Rh_6(CO)_{16}$, $Os_3(CO)_{12}$] (in particular, mixed-metal $[Ru_2Ir(CO)_9(RC{\equiv}CR')]^-$ and $[HRu_2IR(CO)_9(RC{\equiv}CR')]$ [698b]), anionic {$[Ir(CO)_4]^-$, $[Fe(CO)_4]^{2-}$}, or cationic {$[Re(CO)_6]^+$}, with [$Fe_3(CO)_{12}$] or without [$Ru_3(CO)_{12}$] CO-bridging groups. CO typically bonds in an "end-on" fashion through carbon ($M{=}C{=}O \leftrightarrow {}^-M{-}C{\equiv}$

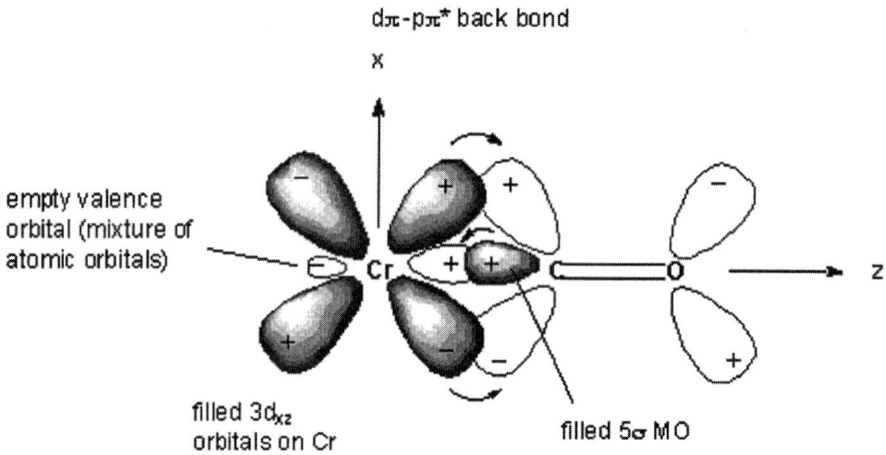

Figure 2.10 Bonding in chromium carbonyl complexes.

O^+; M—C—O angles near 180°). Bridging carbonyls [μ_2- (the most common type; angles between 77 and 90°) or μ_3-] often undergo exchange with terminal carbonyls. A variant of a bridging carbonyl is the "semibridging" carbonyl in which the M—CO—M bond is asymmetric rather than symmetric. In this structural type, the C=O bond is not perpendicular to the M—M bond. Other following coordination modes are also possible: (a) metal atom M^1 forms the bond M^1=C; metal atom M^2 (the same element) is connected with the double bond of the C=O group, i.e., a terminal CO on M^1 is bound to M^2 as an olefin (a σ-donor and a π-acceptor; such complexes are called "four-electron-donor" carbonyls or π-CO); (b) the "end-on", or isocarbonyl, structure M^1=C=O→ M^2 is formed where a terminal carbonyl on metal atom M^1 serves, through oxygen, as a Lewis base toward M^2. Additionally to ***63***, some less common bonding modes exist, such as $\eta^2(\mu_3\text{-C},\mu_2\text{-O})$-CO in niobium cluster $[(CO)_2CpNb][\eta^2(\mu_3\text{-C},\mu_2\text{-O})\text{-CO}]$ [184e]. For a better description of structural and spectroscopic properties of polynuclear carbonyl complexes, the classic concepts of "*carbonyl scrambling*" and "*cluster rotation within CO shells*" are used [184f].

The analogues of metal carbonyls with S, Se, and Te atoms (thiocarbonyl, selenocarbonyl, and tellurocarbonyl complexes) are also known. Because of the instability of CE compounds (E = S, Se, Te), their complexes are prepared indirectly, modifying a coordinated CE precursor (for example, CS_2 or Cl_2CS) [21b]. A general route is the use of dichlorocarbene complexes, for treatment of $Os(PPh_3)_2Cl_2(CO)(CCl_2)$ with HE^-, giving $Os(PPh_3)_2Cl_2(CO)(CE)$. For thiocarbonyl complexes, the ligand may be coordinated in three structural forms: linear terminal M—CS, carbon bridging M[μ-C(S)]M, and "end-to-end" bridging M—C=S—M [21b].

2.2.5.2 Ligands Containing N, P (N, As) as Donor Atoms

Nitrogen- and phosphorus-containing ligands are mainly represented by amino- (***383*** [13,112]) and imino- (***384*** [13,517,699]) phosphines, cyanophosphine derivatives, for example ***385*** and ***386*** [112], and various nitrogen-containing N,P-heterocycles, for example ***387*** and ***388*** [511,517]. Interaction of ***383*** with Lewis acids leads mostly to P-coordinated structures, for example ***389*** [112]:

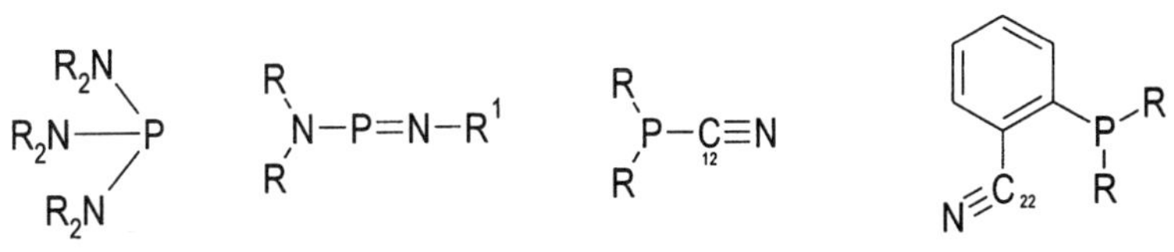

383 384 385 386

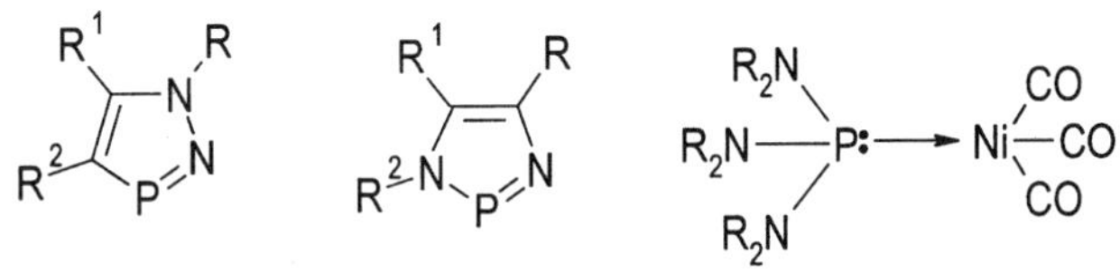

387 388 389

R,R^1,R^2 = H, Alk, Ar

For the iminophosphine complexes, eight coordination modes ***390–397*** have been proposed [13,699]:

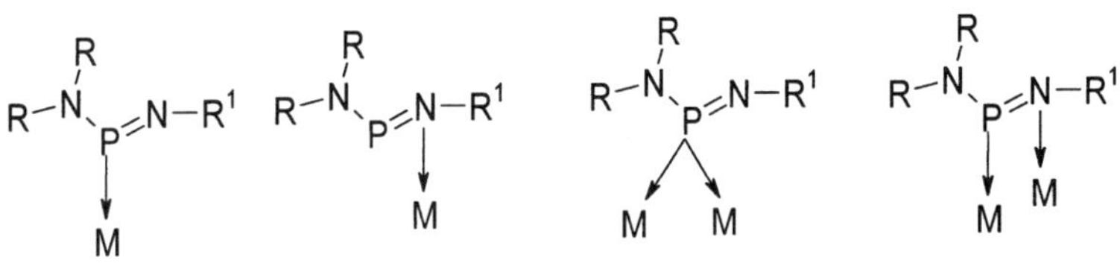

390 391 392 393

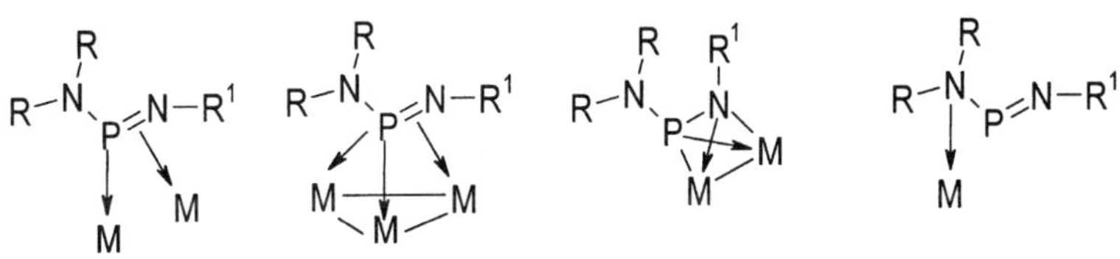

394 395 396 397

Direct x-ray diffraction provides proofs for the formation of structures of type ***390*** which have been obtained for a series of coordination compounds, for instance ***398*** [517]:

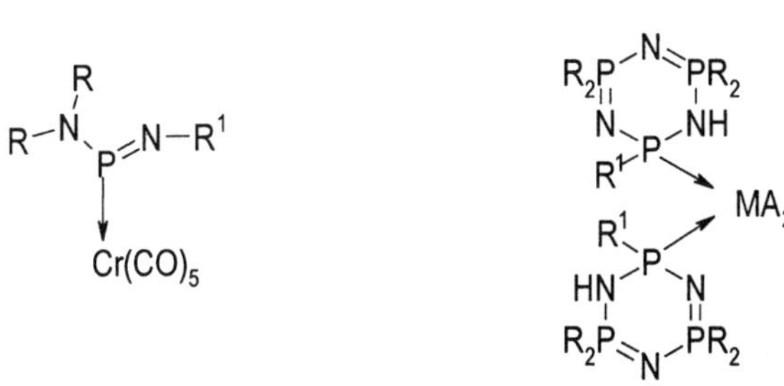

398: R,R^1 = Alk, SiMe$_3$ 399

Compounds with an endocyclic N=P bond (alkyl-, arylphosphazenes) behave similarly, forming complexes of structure ***399*** with a P-coordinated ligand [112]. Information on other phosphazene complexes can be found elsewhere [700].

For the coordination compounds formed from ligands containing the C≡N group (**385** and ***386***), the IR-spectral test is used in order to determine the site where the coordination bond is located. Due to the fact that there are no changes in the position of the absorbance band corresponding to the CN group in the complexes of these ligands, for instance structure ***400***, the conclusion about the localization of the coordination bond on the P atom is correctly made [112]. A similar situation is characteristic for complexes of N,P-ligands containing two phosphorus atoms ***401*** [701]:

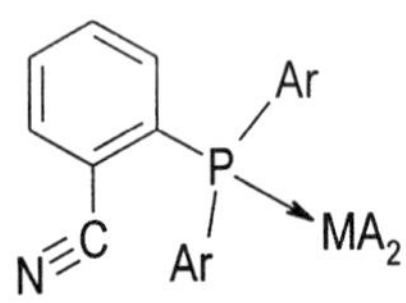

400: M = Pd, Pt; A = Cl, Br

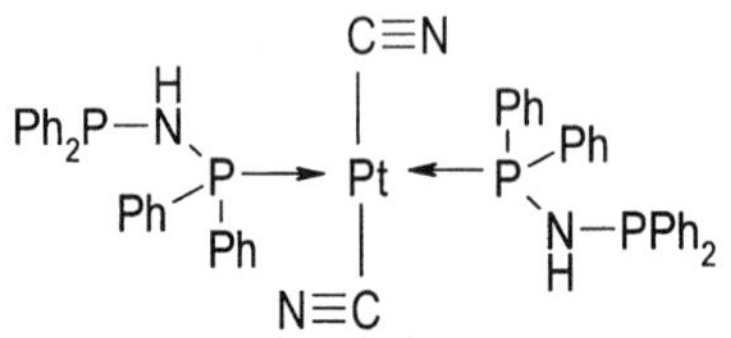

401

Complex compounds for which N,P-containing ligand systems take on the role of chelating ligands are well known. In particular, amongst this kind of complexes are those compounds with donor centers arranged in spatially convenient positions for the formation of metal-cycles, for example ***402*** [112] which belongs to them. According to x-ray diffraction data, a chelate structure is also formed in complex ***403*** [702]:

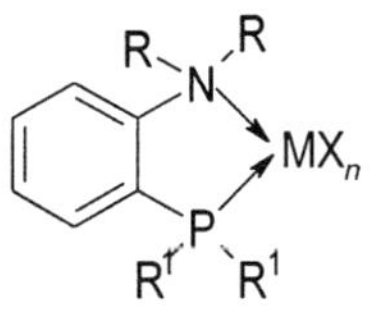

402

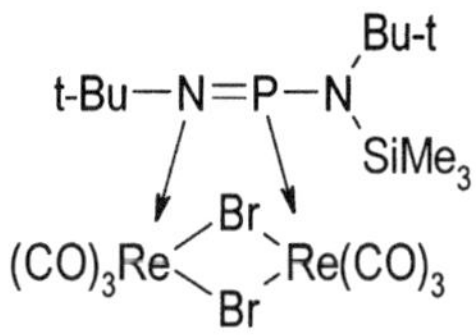

403

Other types of N,P-donor ligand systems (chelate included) were recently reviewed [13,703,704]. In this respect, it is worth noting that the examined ligand systems with N,P,O-donor centers participate in coordination with metals as N,P-bidentate ligands [704–706].

On the basis of N,P-containing heterocycles, the complexes of metal carbonyles have been obtained where mostly a P–metal binding is formed [517]. Amongst these coordination compounds is noted complex ***404***, whose structure was proved through x-ray diffraction studies [517]:

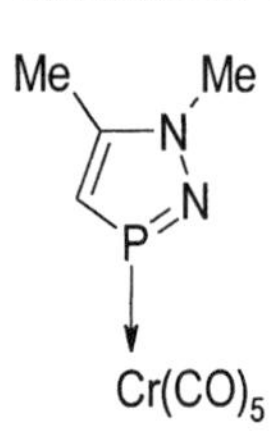

404

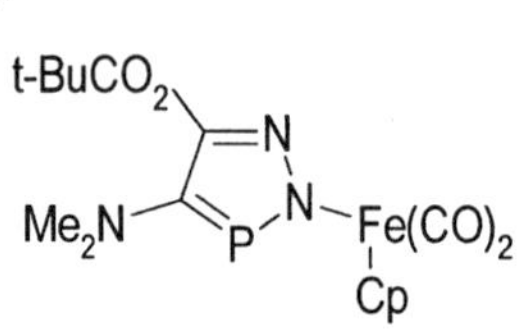

405

Coordination compounds of the type N,P-heterocycles, where the N—M bonds are present, for instance structure ***405***, are well known [707]. The nitrogen–phosphorus analogue of α,α′-bipyridine can be correctly taken as belonging to this type of ligand: chelate complexes, for example ***406***, were obtained on its basis [517]:

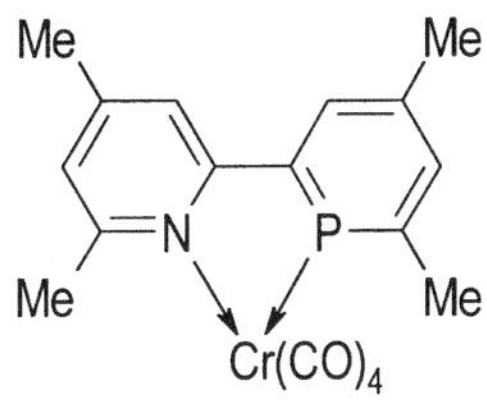

406

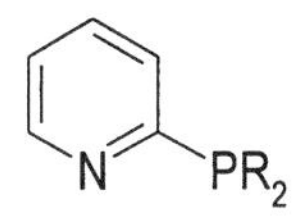

407

Pyridylphosphine compounds ***407*** [708–710] are bidentate N,P-donor ligands. Many complexes containing this ligand system have been obtained by using metal carbonyls as precursors, as well as gold(I) and silver(I) salts. It has been established that, according to the HSAB principle, soft metal carbonyls are coordinated only to a soft phosphorus atom [708,709]. At the same time, the N—M bond can also be formed in binuclear complexes obtained on the basis of coordinated ligands of structure ***407*** [512]:

The complexes of diphosphinamides are widely found, for example ***408*** [703,710,711]:

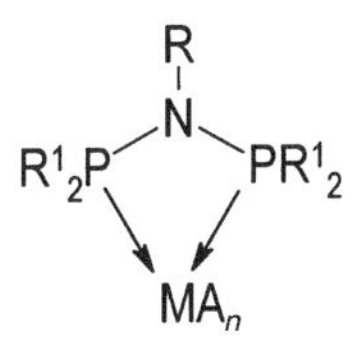

408

409

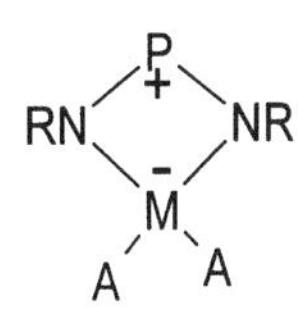

410

Complex compounds obtained on the basis of phosphoranimine ***409*** [712a] and diaminophosphine ***410*** [520] ligands are of considerable interest. Complex compounds of the types ***408*** and ***410*** are represented in general by chelates, and ***409*** by cluster structures. N,P-Bidentate ligands and their rhodium and palladium complexes are described in a recent review [712b].

2.2.5.3 Ligands Containing N and O as Donor Atoms

The range of N,O-containing ligand systems, and their coordination compounds, is too broad to be discussed in the limited possibilities provided by a chapter of this monograph. They include [1]: aminoacids, amides, and hydrazides of carbonic acids, *o*-hydroxyazomethines and their azo analogues, β-aminovinylketones, complexones, 8-hydroxyquinoline and other hydroxyderivatives of heterocycles, peptides, and other fragments of bio-organic ligands. Only the main problems concerning the coordination chemistry of N,O-donors will be discussed, paying particular attention to the not yet sufficiently described ligands belonging to this type, and to the newest data related to the coordination chemistry of the N,O-donors.

The simplest neutral N,O-donor ligand systems are the alkyl(aryl)isocyanates RNCO. As reported [112], little attention is dedicated to these ligands, despite their ambidentate character and their possibility to form complexes $[(RNCO)_n \cdot MA_n]$. The coordination bond in these complexes can be localized in C-, N-, or O-donor centers. This tendency is also observed in a series of RNCO complexes reviewed [713].

Another important aspect of the coordination chemistry of organopseudohalides is that the reactions of cycloaddition and oligo(poly)merization may take place in conditions of complex formation (under interaction with Lewis acids) [112,713].

In this respect, transformation (2.9) is a matter of great interest since the resulting complex compound ***411*** possesses a rare N,N-coordination of substituted urea formed there [13,713]:

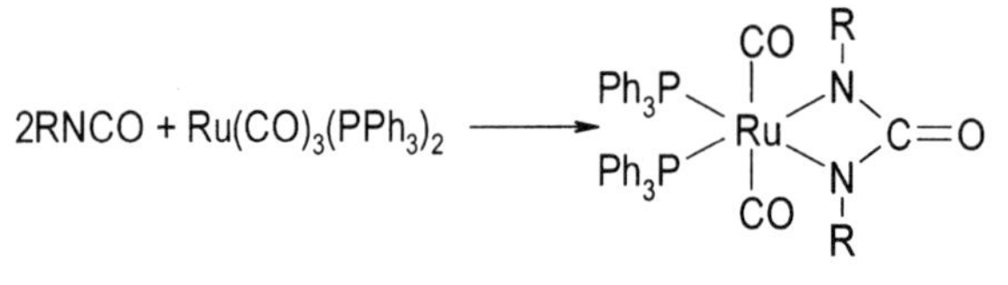

(2.9)

411: R = Alk, Ar

Among other reactions of RNCO complex formation, we note those transformations capable of provoking a modification to the initial ligand systems. Thus, as a result of reaction (2.10), the complex of the benzoxazolone ligand system ***412*** is formed [713] as described in the following:

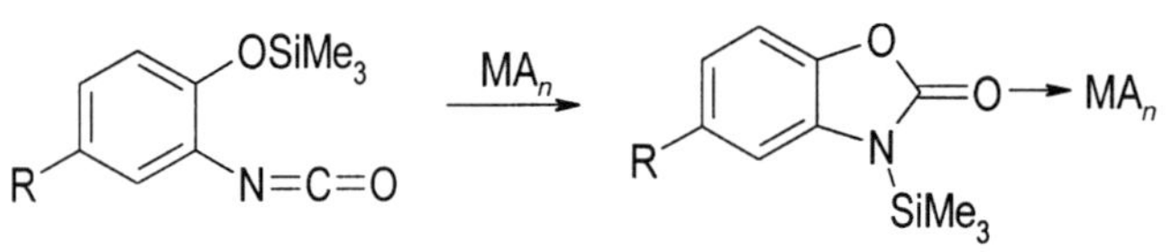

(2.10)

412

R = H, Cl; M = Sb, Sn, Ti; n = 4, 5

Interaction of isocyanates with metal carbonyls [for instance reaction (2.11)] leads to decarbonylation of the ligand and thus, to the formation of a cluster structure ***413*** [713]:

$$PhNCO + Ru_3(CO)_{12} \longrightarrow \underset{413}{Ru_3(CO)_{10}NPh} + 3CO \qquad (2.11)$$

Formation of a metal-cyclic structure such as ***414*** is typical for aminoacids [1,714–722], which is explained by chelate effect. At the same time, complexes have been reported in which aminoacids behave as monodentate ligands with localization of the coordination bond either in N- (***415***) or O- (***416***) donor centres [714]:

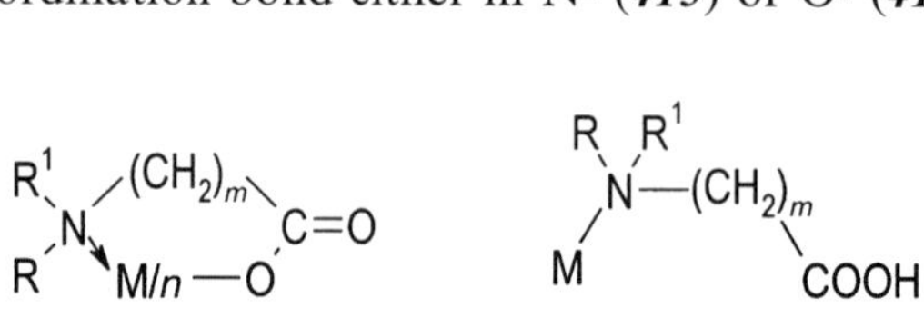

414: $m = 1 \ldots p$; $n = 2$-4; R,R¹ = H, Alk, Ar *415* *416*

We note that, with the lone participation of a carboxyl group in the coordination, the coordination modes examined earlier for the complexes of carbonic acids (Sec. 2.2.5.4, formulae ***276–283***) are possible. Additionally, a mixed binding amongst ligands is observed in many complexes of aminoacids, for example chelates ***277***, ***414***, and O,O^1-bridge ***279*** [714].

Amides of carbonic acids behave, in general, as monodentate O-donor ligands, forming complexes of the type ***417*** [112,722–724]. At the same time, these ligands can participate in the imidoalcohol form, producing chelate structures ***418*** [112]:

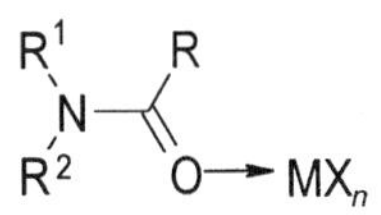

417

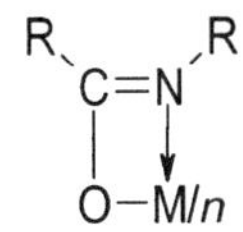

418

A slightly different situation is observed in coordination compounds formed by hydrazides of carbonic acids [112]. In this case, together with complexes ***419*** and ***420*** which are closely related to ***417*** and ***418***, the structure ***421*** is formed, which for the examined compounds has been found to behave as a monodentate N-donor ligand [112]:

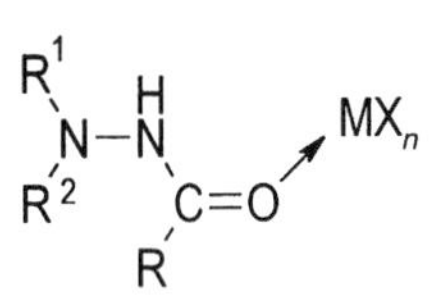

419

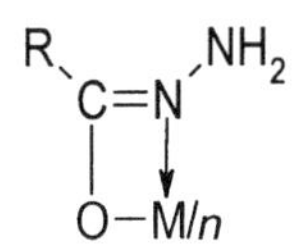

420

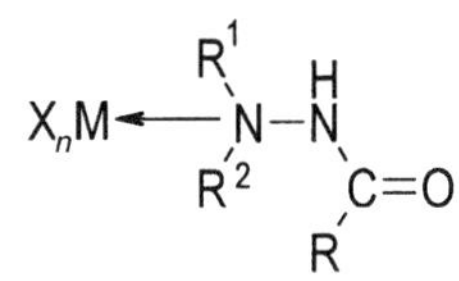

421

o-Hydroxyazomethines ***23*** (R = H, Alk, Ar, Het) are some of the most widespread ligand systems. This could be confirmed not only by the enormous number of publications, but also by their wide generalizations [1,8,12–15,269,270,725–732]. Mostly metal chelates of the type ***422*** have been obtained on the basis of the examined ligands. However, the possibility of the existence of molecular complexes (type ***423***) of *o*-hydroxyazomethines was proved by x-ray diffraction [13,269,270,726f]:

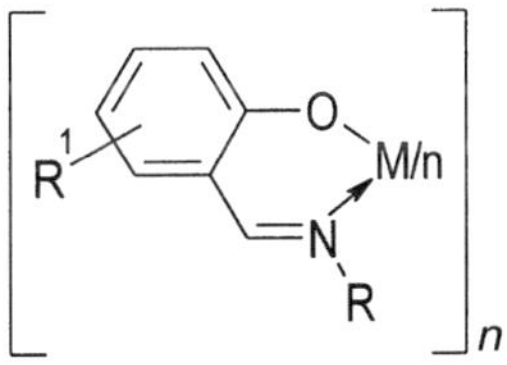

422

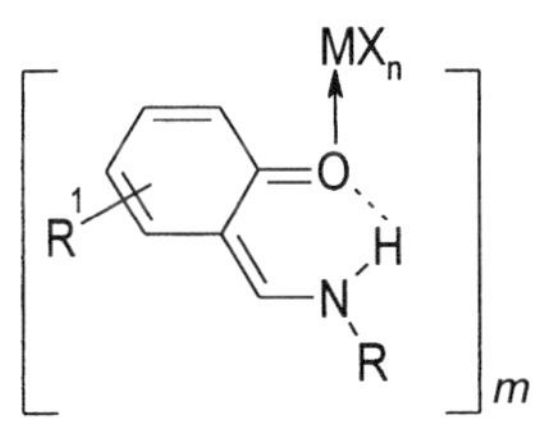

423

Chelates with coordination unit MN_2O_2 are formed on the basis of *o*-hydroxyderivatives of aromatic oxymes (***422***: R = OH) [1, vol.2;733–737], hydrazones (***422***: $R = NR^2_2$, R^2 = H, Alk, Ar) [738–742], semi-, thiosemi-, and selenosemicarbazones (***422***: $R = NHCENH_2$; E = O, S, Se) [260,262,743–749].

We note that the interaction between thio(seleno)semicarbazones of salicylaldehyde and metal salts leads not only to chelates of type ***422*** ($R = NHCENH_2$; E = S, Se), but also to tricyclic structures ***424*** [260,262]. Acylhydrazones of salicylaldehyde

23 ($R=NHCOR^2$, $R^2=$Alk, Ar, Het), additionally to ICC of the type ***422***, may also form chelates with a retained anion and binuclear structures which will be discussed below [741].

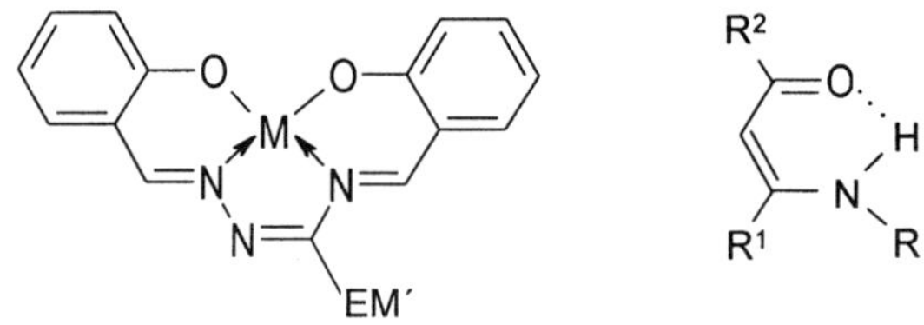

424: E = S, Se
M′= H, NH_4, Na, K

425

β-Aminovinylketones ***425*** occupy an important place among ligands with N,O-donor centers [269,726a,b,f]. On their basis, as well as in the case of *o*-hydroxyazomethines ***23***, not only the chelates with coordination unit MN_2O_2 of the type ***422*** were obtained, but also molecular complex compounds with M—O coordination of the type ***423*** [13,14,269,726c–f].

A broad group of ligands with N,O-donor centers includes *o*-hydroxyarylazo compounds ***426*** [270,401,750,751]:

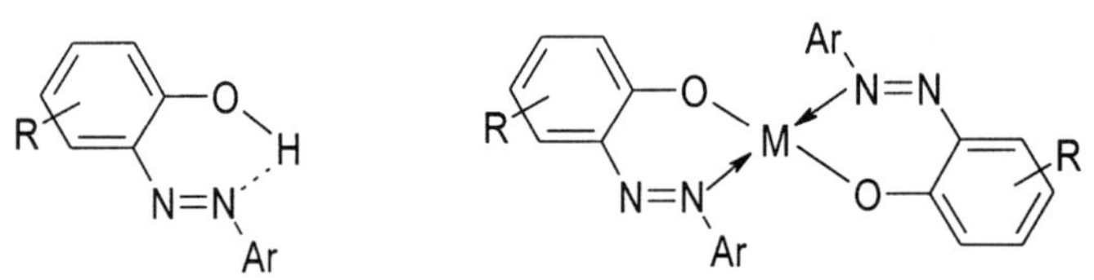

426: Ar = $C_6H_nR^1_{5-n}$;
R, R^1 = H, Alk, Ar

427: M = Ni, Cu, Pt

It is generally accepted that the ICC of nickel, copper, and ruthenium, on the basis of ligands of type ***426***, have trans-planar configuration ***427***, while a tetrahedral polyhedron is characteristic for cobalt and zinc chelates [270,401,751–754]. It was also accepted that a cis-planar configuration for ICC of *o*-hydroxyazo compounds and chelates of *o*-hydroxyazomethines in ***422*** (R = Alk, Ar, Het) is not likely or, at best, only scarcely possible. However, a cis-planar complex ***428*** has recently been prepared and characterized [755]:

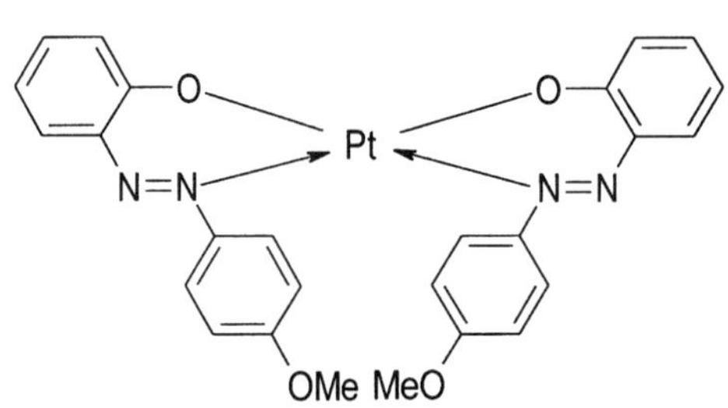

428

Complexones containing amine and carboxyl fragments [756–767] occupy an important place amongst ligand systems with N,O-donor centers. These ligands, in particular, are represented by three types of compounds: monoaminocarbonic ***429***, diaminoalkylcarbonic and diaminotetracarbonic ***430***, and triaminoalkylcarbonic ***431*** acids [761]:

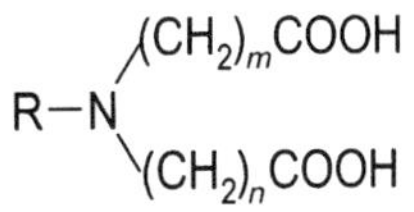

429

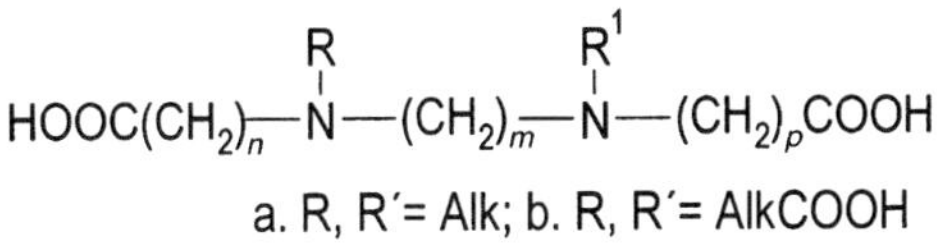

430

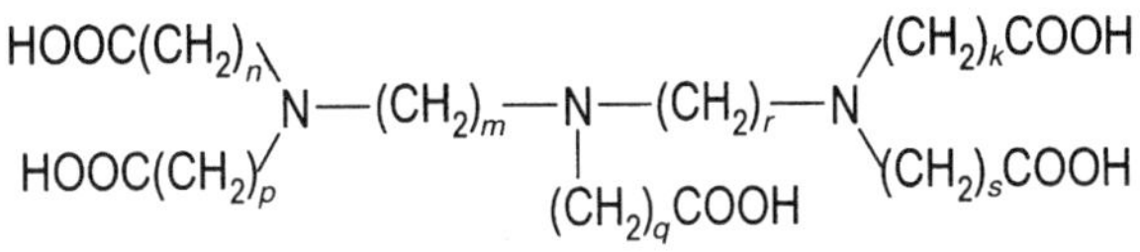

431

The main structural peculiarities of the examined complexones are as follows [756,757,761]. First, both in solution and in crystalline state, they are similar to other aminoacids in which they have a betaine structure, i.e., contain the R_3NH^+ quaternized nitrogen atom and $(CH_2)_mCOO^-$ (deprotonated carboxyl group) fragments. Here in all the complexonates, where a minimum of two carboxyl groups are present per nitrogen atom, one of these groups is deprotonated. The second peculiarity of these complex compounds is the presence of hydrogen bonds, specifically bifurcate ones, under realization of which the NH group proton takes part in the simultaneous binding of two oxygen atoms belonging to carboxyl groups. In the spatial relation, the aminocarboxylate complexones represent a structural template useful for the preparation of metal–chelate cycles, formed through substitution of hydrogen atoms, attached either to imine and/or carboxyl groups, for a metal [756,757,761].

The N,N-, N,O-, and O,O-metal-cyclic structures can be found in these chelates. The first structures are characteristic for di- and triamine ligands (***430a***, ***431***), the second and third ones are formed in the complexes of all three types of ligand systems described above (***429–431***). In principle, the same situation as mentioned before (Sec. 2.2.4.2), for di- and triamine chelates, is observed for N,N-coordination. The O,O-coordination is mostly present in structures formed only by carboxyl groups (compare with the data of Sec. 2.2.4.4). Here the most propagated structures in coordination compounds are ***277***, ***279***, and ***280***, for example, in complexonates of the triamine series [764]. In the case of N,O-metal-binding, the most propagated structures are those with five (***432***) and six-member (***433***) metal-cycles [756–760,762–767]:

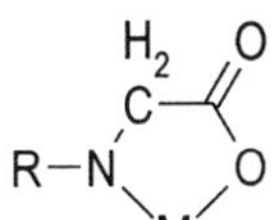

432

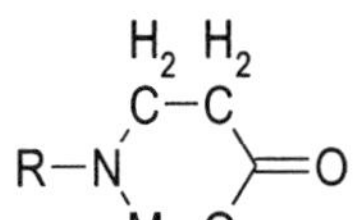

433

The feature of polydentacity shown by the examined ligands allows us to create various structural types in the complexes formed on their basis. Thus, monoaminocarboxyl ligands take part as tridentate (iminoacetates and their analogues) or tetradentate (nitriltriacetates) ligands and may form mononuclear octahedral chelates with a composition ML_2, or metal polymers $[ML]_n$ [756,757]. The existence of octahedral structures (coordination number $CN = 6$) is typical for diamine complexonates, although a considerable number of structures are described where CN varies from 7 to 10 [756,767,766].

Hexacoordinated structures prevail in the case of complexonates of the triamine series [764]. However, a great variety of CN of metals is also observed. Moreover, many coordination modes are observed for the complexones of the triamine series, penta-, hexa-, hepta-, and octa-types with chelate and chelate-bridge function of mono-, di-, tri-, and tetra-deprotonated ligands. As a result, a number of metal-cycles are formed. Thus, maximum dentacity of diethylenetriaminoacetic acid ***431*** ($m = n = p = q = 1$), with respect to a metal atom, equals 8 (3N + 5O), leading to seven metal-cycles (two ethylenediamine and five glycine) being formed [764].

The crystallochemical analysis of complexonates has been carried out and the crystallostructural role of various ligands, on the formation of different coordination modes by metal atoms, is extensively reviewed [761–763]. It is established that complexonate ions of the diamine type are capable of forming 14 different types of coordination [761]. Depending on the coordination mode present, one polydentate diaminocarboxylate ligand has the capability to bind from one to four metal atoms, producing from two to 10 coordination bonds.

Specific complexonate structures are discussed in numerous publications and are well generalized in a number of reviews [756–768]. Taking into account data published elsewhere [764], it is possible to consider that at present more than 30 complexones and 200 complexonates of metals, from practically all the groups of the Periodic Table, have been structurally characterized. The majority of structures have been obtained for complexones of the monoamine series (***429***) and complexonates of various diamine ligands (***430***).

There is a renewed interest in these ligands which can be confirmed in the reviews dedicated to complexonates of the triamine series [764]. Much attention is also paid to complexonates of some metals, in particular to those of ruthenium, as reviewed recently [765].

Among heterocyclic N,O-donors, three groups of compounds, used as ligands, are noted:

1. Crown-ethers with different numbers of N- and O-donor centers and cryptandes [1,450–452,565,568,571,572,769–772];
2. Hydroxyazole and azine compounds [7,8,13,773–779];
3. Nitroxyl radicals and their derivatives [780–790].

N_mO_n-Containing crowns, for example ***434*** [568], differ from macrocyclic polyethers (***303***, ***304***) in their capacity to form complexes not only with *s*- and *p*-metals, but with *d*-metals as well:

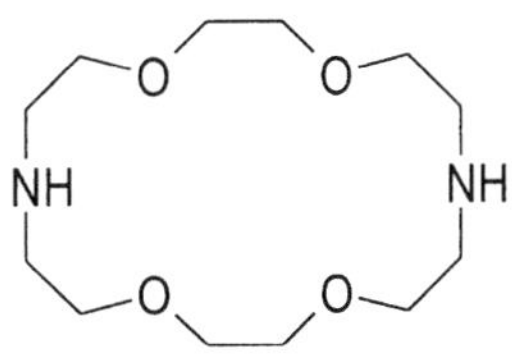

434

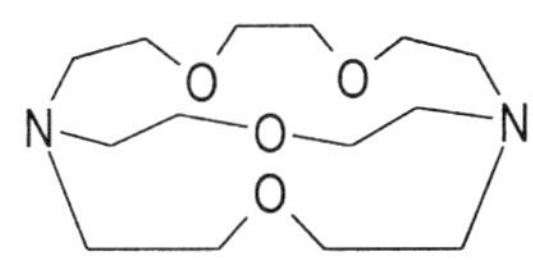

435

Bicyclic structures of "cell" type ***435***, named by *J.M. Lehn* as "*kryptand*,"* form extremely stable complexes, known as kryptates, with various metal ions, hiding them in an inner cavity. Such compounds have an important role in modern supramolecular chemistry [268,581].

Hydroxyazoles mostly form chelate structures, for instance ***436*** [13,773]:

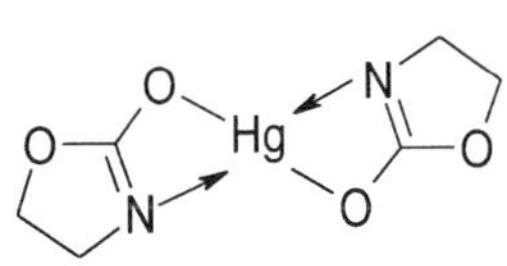

436

437a *437b*

For hydroxyazine coordination compounds, for instance ***437***, different structural types like ***438–444*** are characteristic; they are formed on the basis of the two tautomeric forms illustrated as ***437a*** and ***437b*** [13,774]:

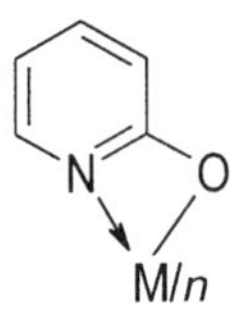

438

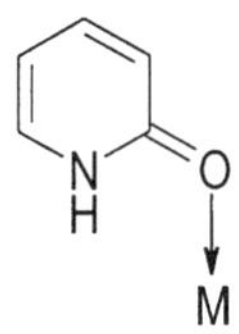

439

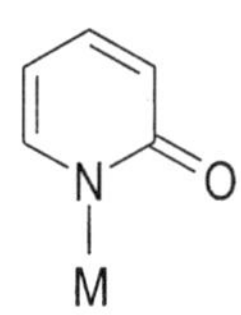

440

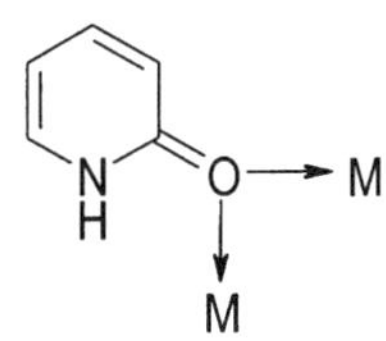

441

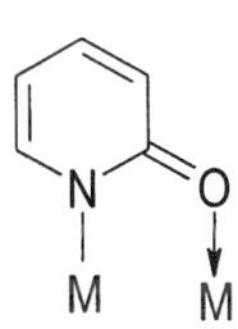

442

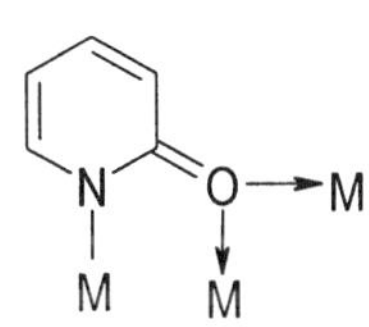

443

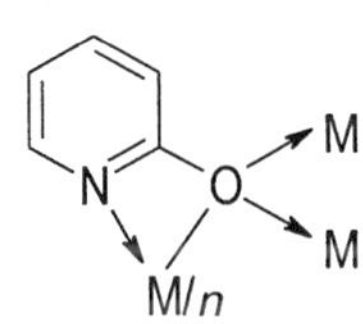

444

Although the chelating-like metal-binding ***438*** has been found to be the most propagated, convincing proofs (in particular, x-ray diffraction data) of structures of the types ***439–444*** have been obtained [8,13].

Amongst other N,O-donor ligand systems are worth noting 8-oxyquinoline, 2-hydroxyphenylbenzazoles, and nitroxyl radicals. The first of them, together with its substituted derivatives, forms the well-known ICC ***45*** [1,8,13,775–778]. The ICC of type ***446*** are characteristic for all 2-hydroxyphenylbenzazole ligands [8,779]:

---

*Hence follows their name – Greek, *kryptos* (concealed, secret).

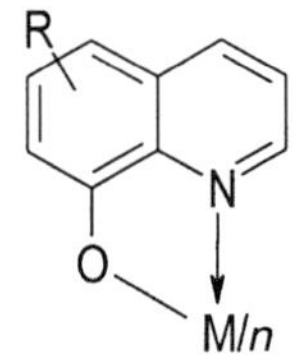

445: $n = 2, 3$; $R = H, Alk, Hal, NO_2$

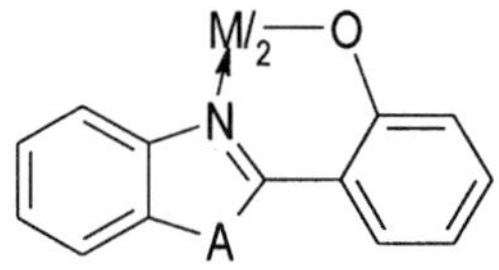

446: A = NR, O, S

Stable nitroxyl radicals of imidazoline [780], which contain chelating groups, form a series of ICC including the coordination units N_2O_2 ***447*** (R = Alk, CF_3, Ph [781–783] or COOEt [784,785], and also $CONH_2$ [786,787]), ***448*** [788], and ***449*** [789]:

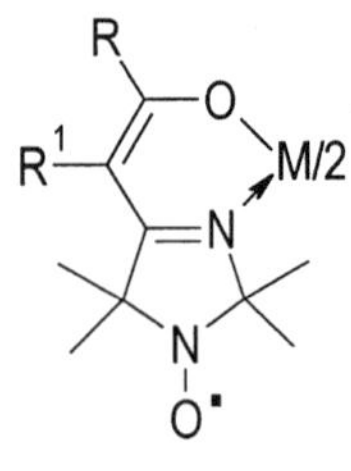

447: R^1 = H, Me

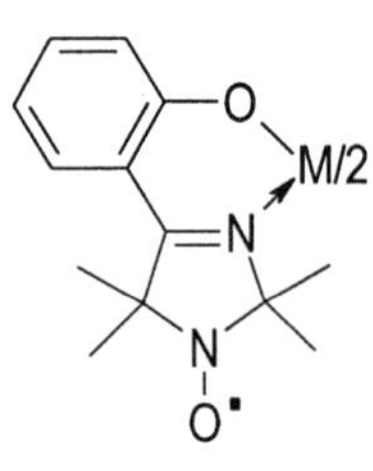

448

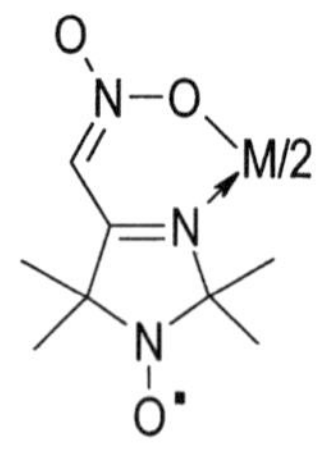

449

It is worth noting that the coordination unit MNO is present in layer polymers containing fragments of deprotonated nitrylnitroxyl radicals, which have an imidazole or benzimidazole substituent in the position 2, as shown by structure ***450*** [790]:

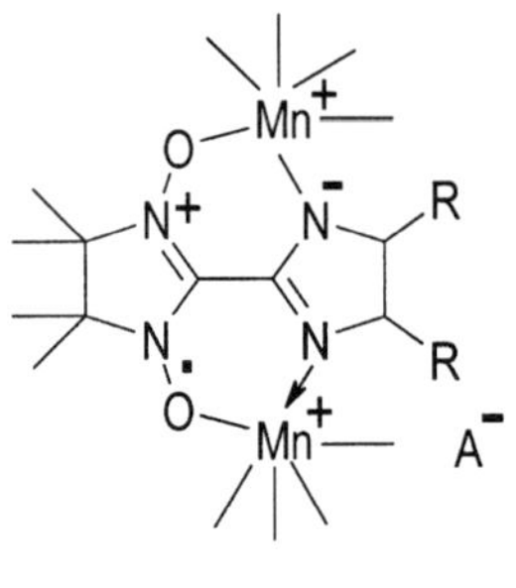

450: R = H, C_4H_4-cyclo; A = ClO_4, BF_4

These complexes, as well as their solvates contained in an alcohol, yield complex compounds of type ***447***; this group of compounds plays an important role amongst molecular ferromagnetics [790].

2.2.5.4 Ligands Containing N and S as Donor Atoms

The simplest organic ligands of this type are mercaptoamines (aminothiols) ***451*** (R = H), their alkyl (aryl) derivatives ***451*** (R = Alk, Ar), and their aromatic analogues ***452*** [1, vol.2]:

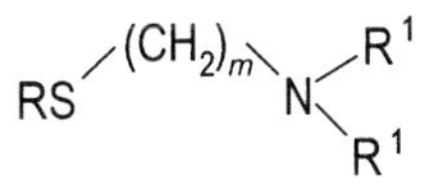

451

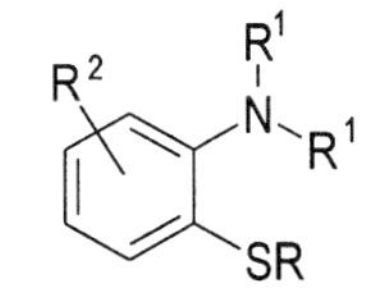

452

R, R^1, R^2 = H, Alk, Ar

Numerous ICC (R = H) ***453***, ***454*** and chelates (R = Alk, Ar) ***455***, ***456*** have been obtained on the basis of these ligands:

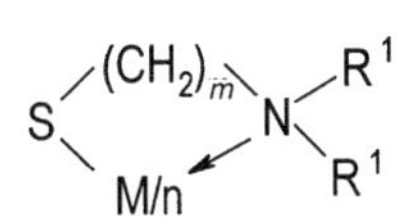

453

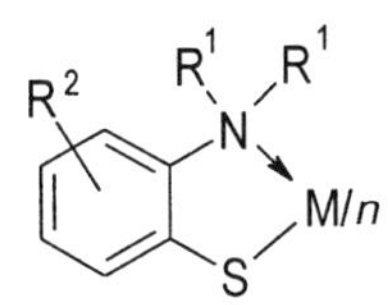

454

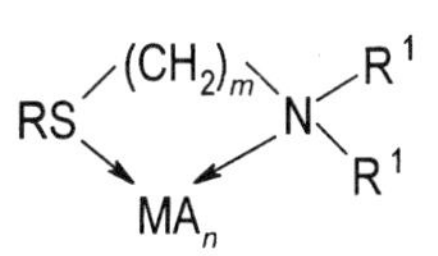

455

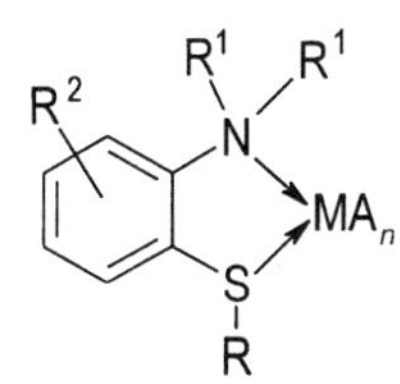

456

The structures of the complexes mentioned above have been determined in a series of cases through x-ray diffraction studies. Thus, nickel ICC of type ***453*** are examples of such complexes (M = Ni, $m = n = 2$, R$'$ = H [791,792], R$'$ = Me [793]). A structural characteristic feature of these chelates is the presence of five-member metal-cycles. Metallocycles of the same size are found in anionic complexes of the types ***457*** (M = K, R = H [794], R = Me [795]; M = Tl, R = Me [796]), ***458*** [797], and ***459*** (M = NEt_4, R = $COCF_3$ [798]; M = $AsPh_4$, R = H [799]):

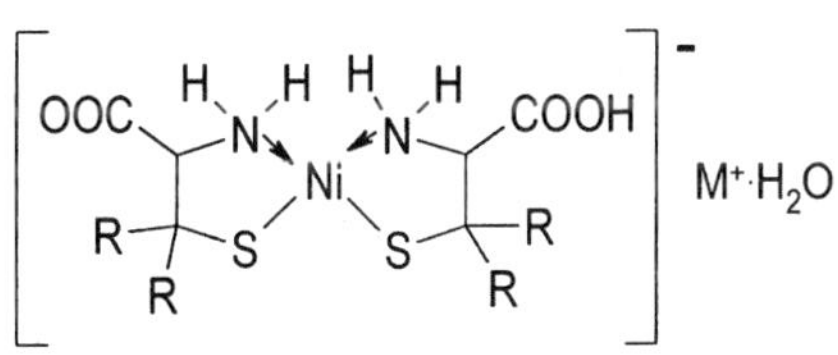

457

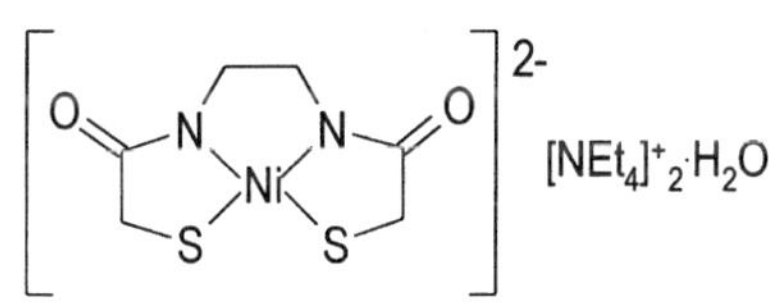

458

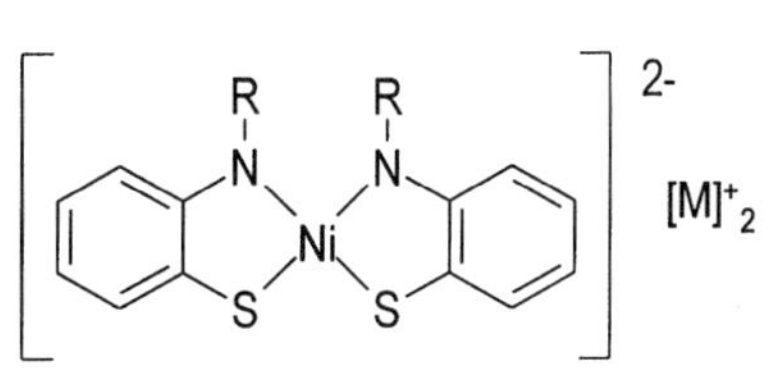

459

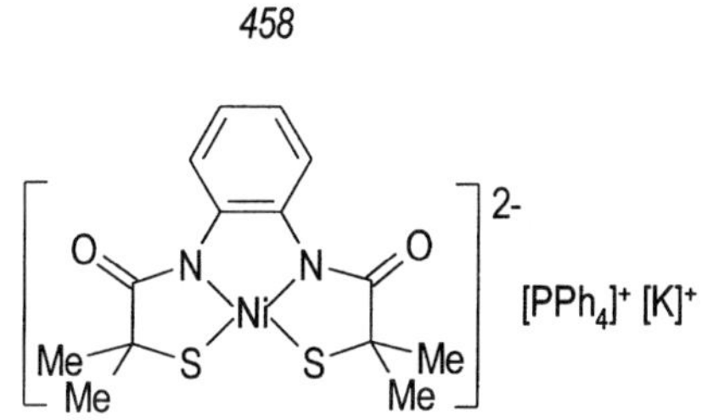

460

The anionic complex ***460***, containing two different cations [800], can also be assumed to belong to the same series.

Data on the structure of mononuclear ICC with saturated six-member metal-cycles of the type MN_2S_2 is as yet unavailable, although such a ligand environment is found in bi- [801] and trinuclear [802] ICC. Another situation is characteristic for the ICC obtained on the basis of ligands which allow the formation of unsaturated metal-cycles [803–822]. Amongst them, many more ligand systems leading to six-member chelates are known than those producing five-member ones. In this respect must be specially emphasized the azomethinic [269,270,448,803,804] and azo [13,270,751,803] compounds.

Precursors of azomethinic complexes with five-member metal-cycles, of the type MN_2S_2, are the benzothiazoline compounds represented by structure ***461*** [337a,805,823,824]; they can easily be transformed into chelates of type ***462*** [806–812] under proper conditions for complex formation (2.12) [9,269,804,823,824]:

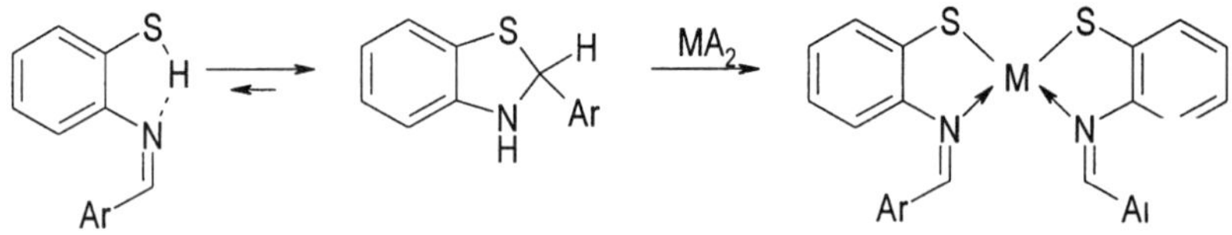

(2.12)

461a *461b* *462*: M = Ni, Pd

The coordination compound ***463*** [813] can also be attributed as belonging to this series of complexes:

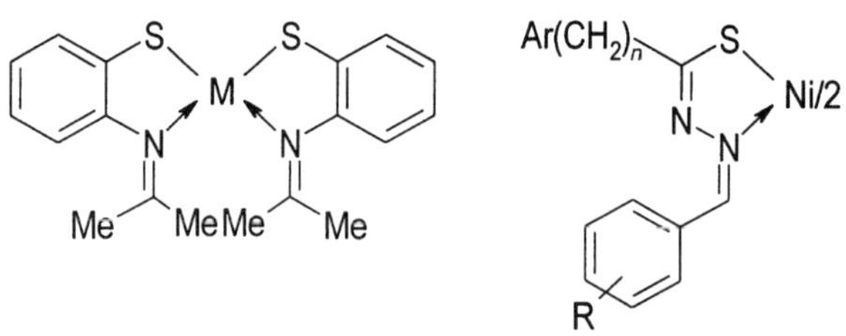

463 *464*: R = H, OMe; *n* = 0, 1

Among other nickel chelates with unsaturated five-member coordination units, we point out the complexes of thioarylhydrazones ***464*** [814–816] and dithiocarbazones ***465*** [817–820] of benzaldehyde and its aryl-substituted derivatives:

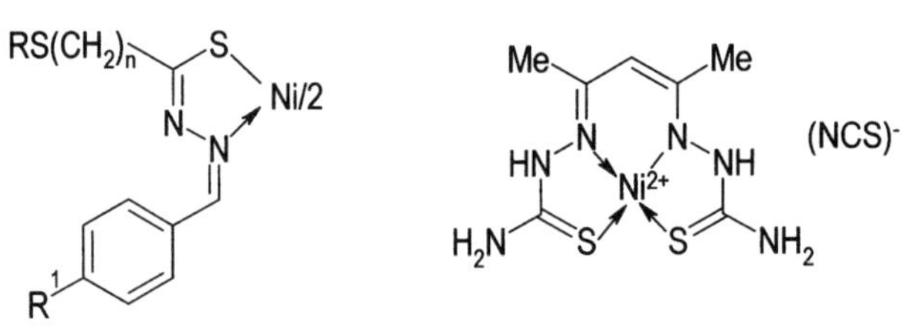

465: R = Me, Ph; R^1 = OMe, NO_2, NMe_2; n = 0, 1 *466*

The complex obtained on the basis of thiocarbazone of 2-furylaldehyde has the structure containing a five-member metallocycle [821]. Two five-member and one six-member metal-cycles are formed in the nickel complex shown by ***466*** [822].

Complex compounds of azomethines containing a six-member N,S-ligand environment are represented by numerous chelates, whose synthesis, physical, and chemical properties, and molecular structures, are reported in a number of reviews [269,270,448,804,823]. Compounds of the types ***467*** and ***468*** are the most propagated amongst this kind of complex:

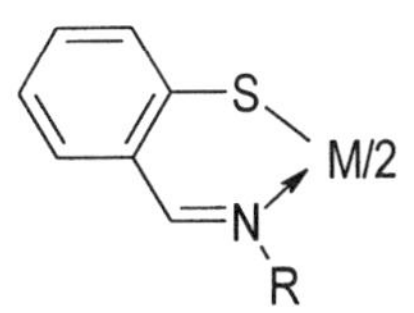

467

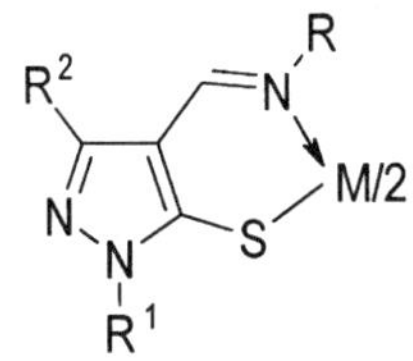

468

Schiff bases of 2-mercaptobenzaldehyde, which can be used as ligands to obtain ICC of type ***467***, are unstable for this purpose. So, the stable compounds ***469*** are mostly used as ligands for the syntheses of chelates ***467***; an alternative way for their synthesis is carried out through the method of template synthesis starting from thiosalicylialdehydates ***470*** (see Sec. 3.3 on template syntheses):

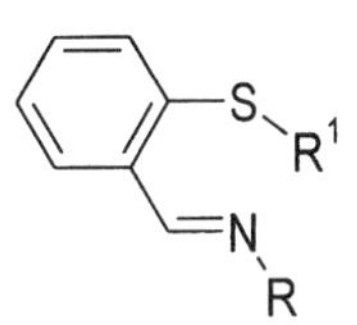

469: R = CN, Alk, Ar;
R^1 = Na, Alk

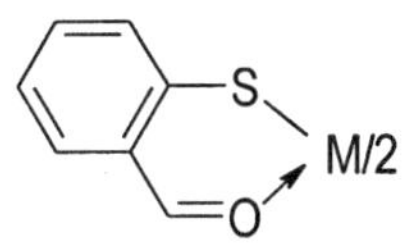

470

As a difference of thioaniles with respect to aromatic aldehydes, analogous hetaryl derivatives are very stable, for example ***471*** (X = S [8,824–826]), and so they are widely used to obtain complexes of type ***468*** (2.13) [8,269,270,804]:

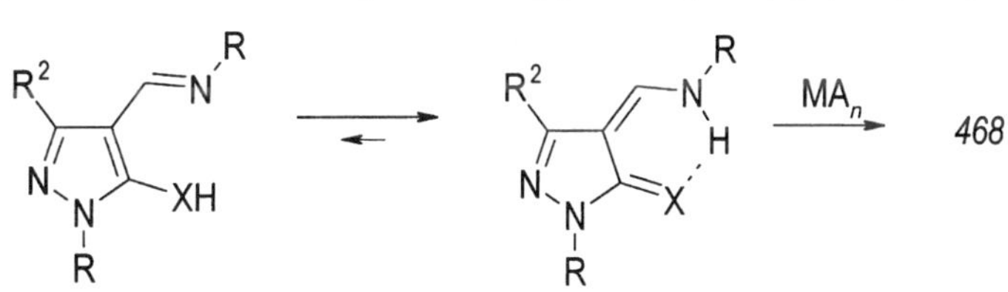

471a *471b* (2.13)

Amongst the chelates shown by ***468***, almost exclusively the nickel(II) complexes were reported [269]. After that early publication some additional reports have appeared which contain structural data of complexes represented by ***472*** ($R = R^1 = H$, $Z = (CH_2)_2$ [827]; $R = R^1 = Me$, $Z = (CH_2)_2$ [828]; R = Me, $R^1 = CHO$, $Z = (CH_2)_3$ [829]) and ***473*** (R = Me, Ph, $n = 3$ [830]; R = Me, $n = 4$ [831]). It has to be mentioned that the x-ray diffraction studies of the complexes belonging to type ***473*** ($n = 2$, 4) were carried out since the early reports [269,832,833].

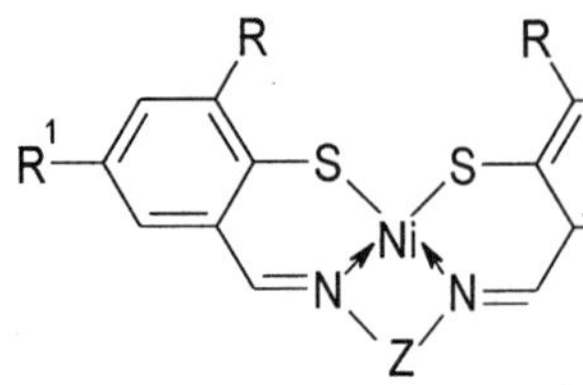

472

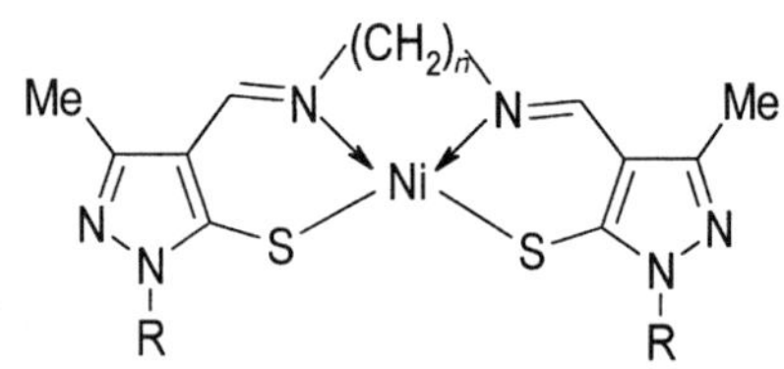

473

Additionally to nickel ICC, chelates of the types ***467***, ***468***, ***472***, and ***473***, containing either cobalt, iron, copper, palladium, zinc, cadmium, or mercury, have also been obtained [269,270,804]. Special interest amongst them is provoked by copper ICC containing an azomethinic N,S-ligand environment. These compounds are one of the several biomimetic models of nonporphyrine metal proteins [448,804,834]. The majority of these ICC have trimetallocyclic structures of the types shown by structures ***474*** [269,448] and ***475*** [269,448,835]:

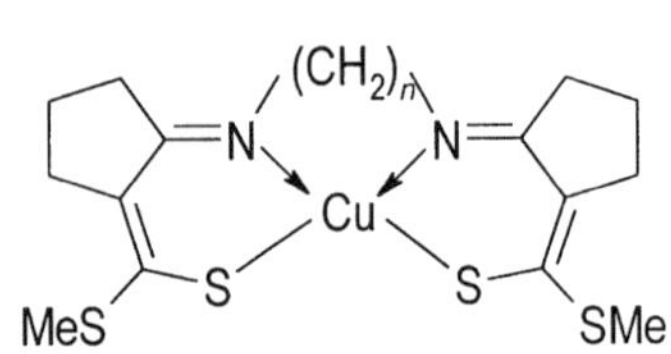

474: $n = 2–4$

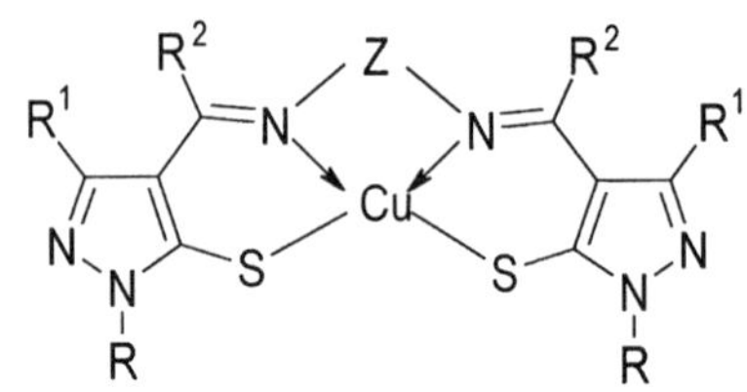

475: R, R^1 = Alk, Ar; R^2 = H, Me, Ph;
Z = $(CH_2)_n$, Ar; $n = 2, 3$

Here the very similar bicyclic chelates of type ***476*** [269,836–838] are poorly presented:

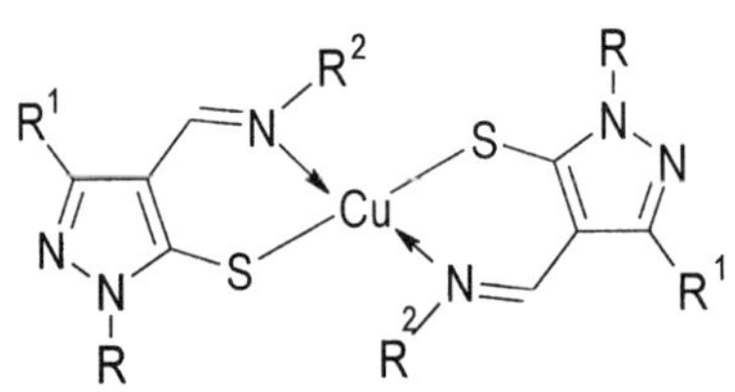

476: R, R^1 = Alk, Ar; R^2 = Alk, Ar, 2-Py, 8-Quin

Iron complexes with an N_2S_2-ligand environment are well exemplified by ICC of the types ***467*** (M = Fe, R = Ar) [269], ***477***, and ***478*** (X = S) [804, 839–841]:

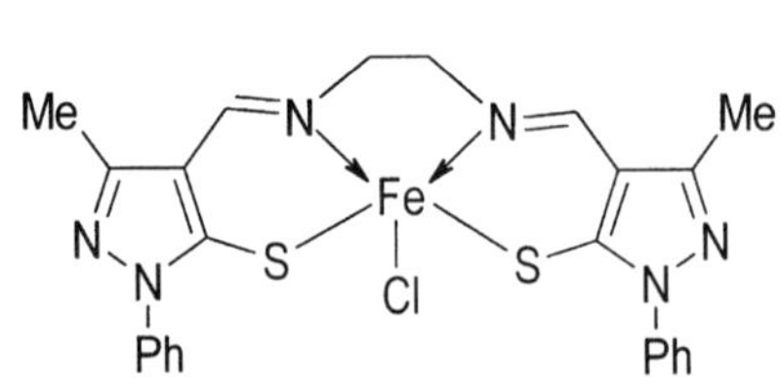

477

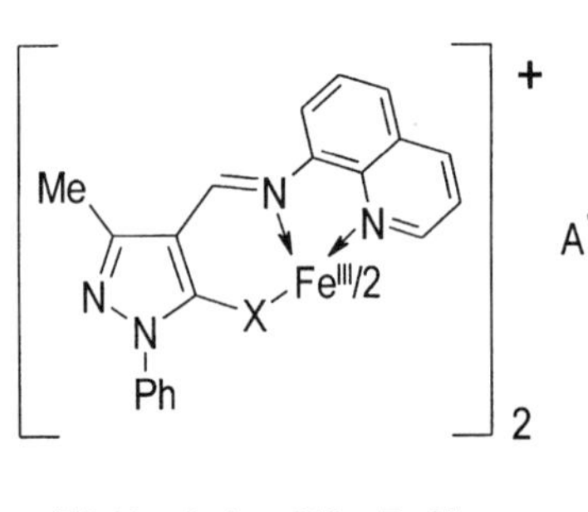

478: X = S, A = ClO_4, $FeCl_4$

Other complexes containing the coordination units FeN_2S_2 [842] and FeN_2S_4 [843–845] are also known. These compounds have due importance in coordination chemistry since they provide a possibility to study the changes of iron spin states in accessible temperature intervals [845]. Moreover, they are permanently used as models of active centers in some iron-containing nonporphyrine metalloenzymes, in particular, nitrylhydrataze [804,846–850].

Complexes belonging to type ***468*** (M = Be, Zn, Cd, Hg, Pb) are very important subjects of study in the field of stereochemical nonrigid tetrahedral structures [851,852]. Using the method of spin labels (a method in dynamic NMR), it is possible to determine the kinetic parameters, separately as intra- or intermolecular processes, for the stereoisomerization reactions in solution.

In a similar way to that discussed above, the six-member coordination unit MN_2S_2 is found in chelates of β-aminovinylthiones shown by ***479*** (R = Bu, R^1 = H, R^2 = Ph [853]; R = R^1 = R^2 = Ph [854]) and by ***480*** [855]:

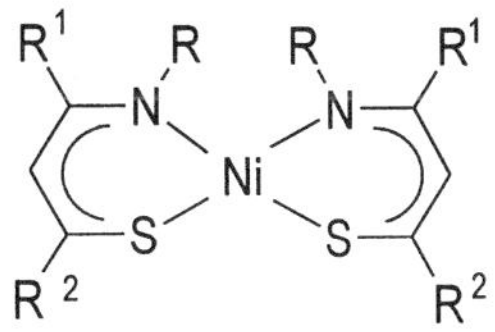

479

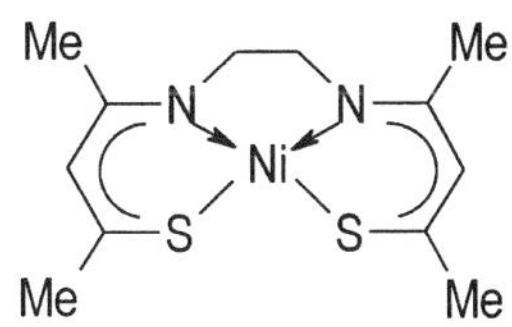

480

Such a coordination mode is also characteristic for the complex compound *bis*[N-(diethyl)aminothiocarbonyl-N′-phenyl-benzamidinato]nickel(II) ***481*** [856]:

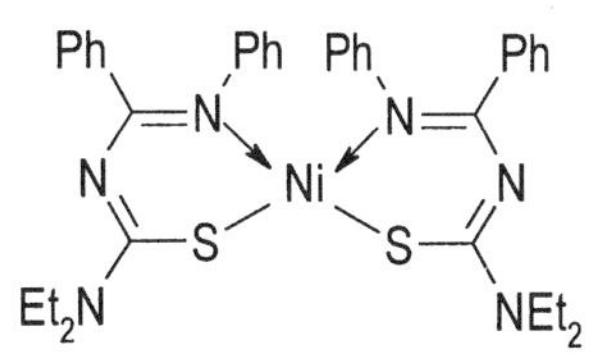

481

In a marked difference with respect to complexes of *o*-hydroxyazo compounds containing a six-member cycle of the type shown by ***427***, the existence of five- [751,857–862] or five- and six-member [751,863] cycles, shown by structures ***482*** and ***483***, respectively, is characteristic for coordination compounds obtained from analogous 2-mercapto derivatives:

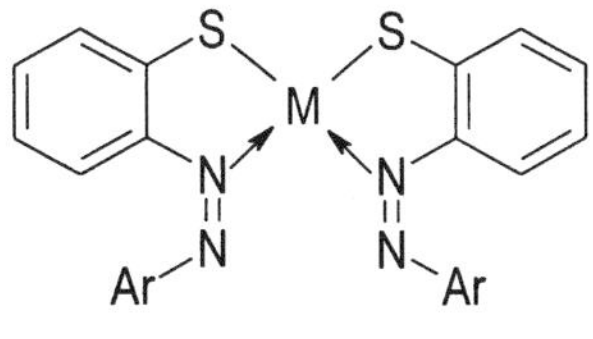

482: M = Ni, Pd, Zn*

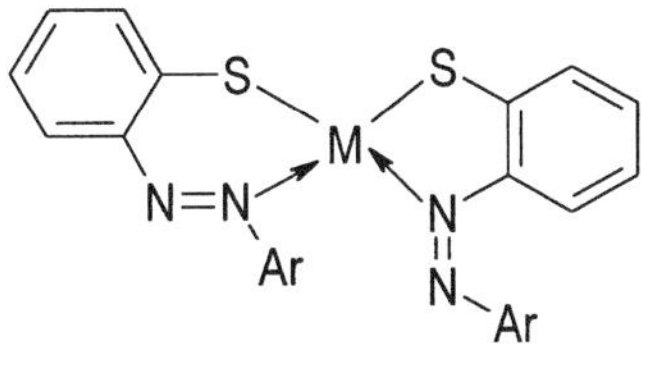

483: M = Pd

*This complex has a tetrahedral structure in its coordination unit.

Complex compounds with an N_2S_2-ligand environment, whose structures are indicated in the usual manner, are obtained on the basis of 2-mercapto derivatives of azoles ***484*** and azines ***485*** [1, vol.2;8,13,864–866] and from 8-mercaptoquinoline ***486*** [1, vol.2;13,777,867–872]:

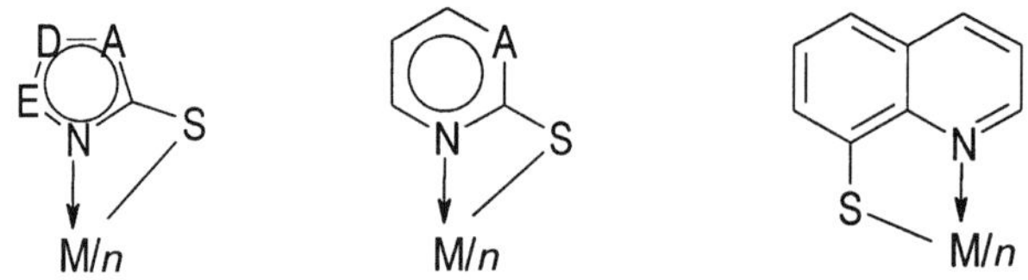

484 485 486

The last one of the series of ligands mentioned above was used for synthesizing compounds of type ***486*** ($M = ER_n$), where E is an element belonging to any of Groups I, III, IV, or V of the Periodic Table [873–877].

A ligand environment, close to ***484–486*** with respect to the nature of donor centers, is found in the complexes of *o*-mercaptophenyl derivatives of oxazoline ***487*** [878]:

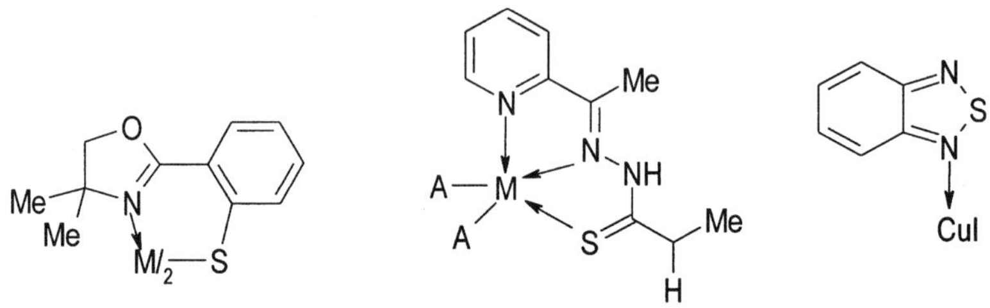

487: M = Zn, Cd, Hg *488*: M = Cd, Hg; A = Cl, Br *489*

Amongst the N,S-donors much attention is being paid to the thiosemicarbazones [260,262,743–746] due to their use as models of active centers in metal proteins, ahead of the derivatives of heterocyclic aldehydes which come in the following order of importance: thiophene [879], izatine [880], and, especially, azines [745,746,881]. Amongst azine complexes are worth noting the coordination compounds of 2-acetylpyridine-4-methylthiosemicarbazone, for instance ***488***, studied by various physical and chemical methods (x-ray diffraction, heteronuclear NMR ^{13}C, ^{195}Pt, ^{199}Hg) [881].

Amongst the N,S-azole ligands [8,13], benzo-2,1,3-thiadiazol deserves a special mention. The controversial question about the possibility of one particular kind of coordination within the complexes formed from benzo-2,1,3-thiadiazol, one in which only one atom of the heterocycle, either N or S, was able to participate [8,13], was solved by x-ray structure studies in favor of the N-coordination ***489*** [882]. However, the examples of the kinds of coordination presented by the sulfur atom in the thiophene cycle (Sec. 2.2.4.5) do not exclude the possibility of M—S binding, particularly when using "softer" (i.e., Pt, Pd) Pearson acids.

2.2.5.5 Ligands Containing N and Se as Donors

The data on complexes of N,Se-donor ligand systems are practically absent in the most extensive issue on coordination chemistry [1]. That can be explained by the difficulties of synthesizing these compounds. Selenourea and its N,N[1]-derivatives can be formally attributed to such ligands; most published work has been devoted to a number of metal complexes belonging to these systems [1, vol.2;112,634]. However, the coordination bond present in known complexes of this ligand is localized only in the selenium atom (***490***):

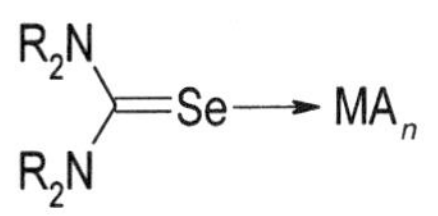

490

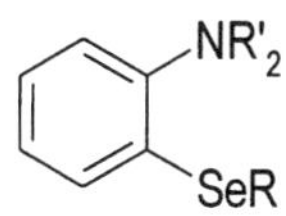

491

The situation with complexes of *o*-selenide ligands ***491***, even though such compounds are supposedly well known, is still unclear [636]. The structural details of coordination compounds of phenoselenoazine ***492*** have not yet been studied in detail [636]:

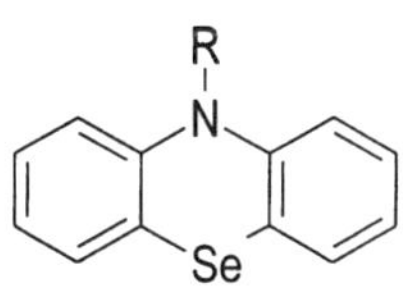

492

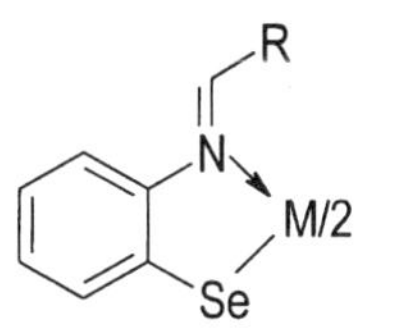

493

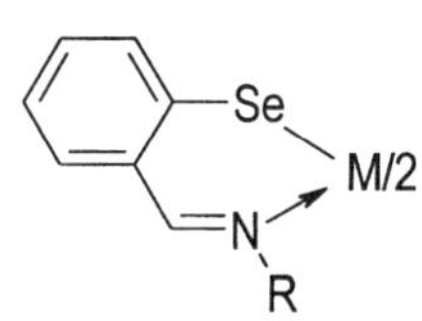

494

Selenium-containing aromatic azomethines are ligands in whose chelates five- (***493***) [269,883] and six-member (***494***) [269,884] metal-cycles are formed. The latter also exist in complexes of β-aminovinylselenoketones ***495*** [885]:

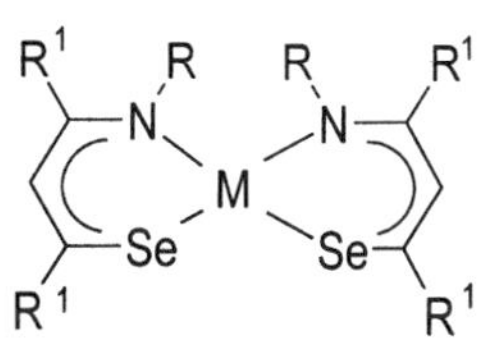

495

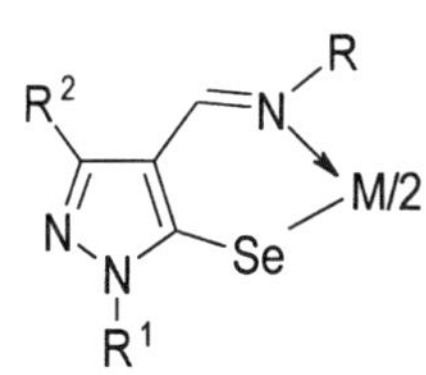

496

Many examples of ICC of the examined type are known amongst complexes of ligands of heterocyclic series [269]. Thus, nickel [886] and copper [887] complexes ***496*** of aminomethylene derivatives of 5-selenopyrazolone ***471*** (X = Se) have been obtained and structurally characterized. A series of coordination compounds has been synthesized on the basis of 2-selenohydrido derivatives of azines ***497*** [8,13,888–894]:

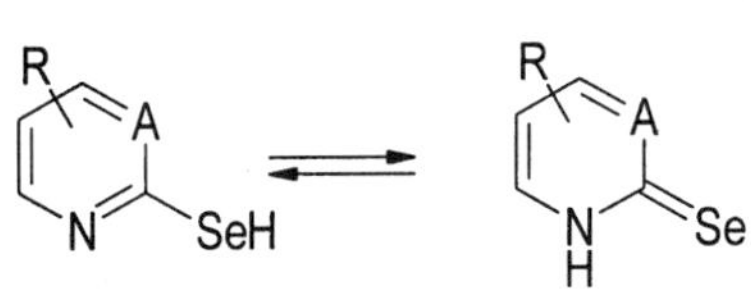

497a *497b*

A = CR[1], N

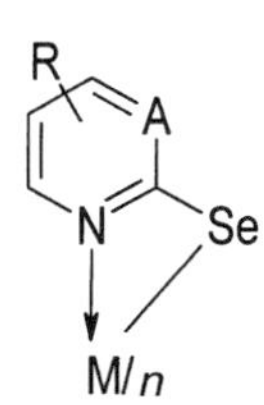

498

Since ligands of type ***497*** are unstable, a plausible way to obtain ICC ***498*** is most frequently by using diselenide derivatives of nitrogen-containing heterocycles, for instance ***499*** [894]. Complexes of type ***498***, either of Li [890], In [891], Sn [890,892], Cu [891], Cd, or Hg have been described [888]. Their structures were verified by x-ray diffraction studies and it was established that, besides the mononuclear complexes (M = Li, n = 1; M = Sn, Cd, Hg, n = 2; M = In, n = 3), the polynuclear ones of formula M_mL_m (M = Cu, m = 4 [891]) are also formed.

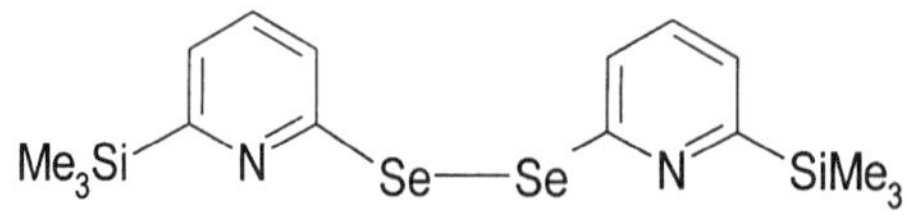

499

Complexes of 8-selenoquinoline ***500*** are well known [895]. We note that the 8-selenocyanate derivative ***501*** is used to obtain complexes of structure ***500***, in a similar way to the application of 8-thiocyanatoquinoline ***502*** for the synthesis of a group of corresponding N,S-complexes ***486***. Recently [895b], the chelates ***500*** (M = Sb, n = 3) were obtained from 8,8-diquinolinediselenide:

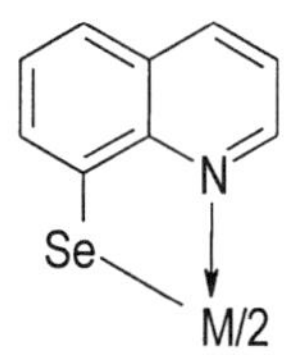

500

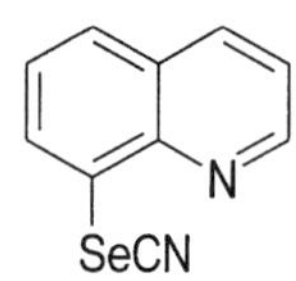

501

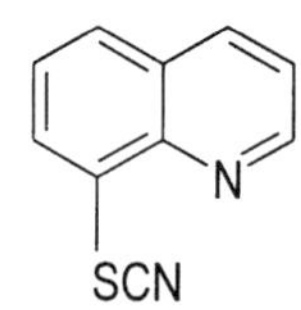

502

The complexes of N,Se-containing azoles ***503*** [13,896] and ***504*** [8,13] have been described. Although the possibility exists for the coordination to be formed through either N or Se, as has been discussed for these ligands, only the first coordination mode has been proved:

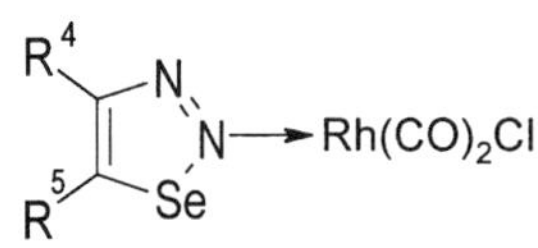

503

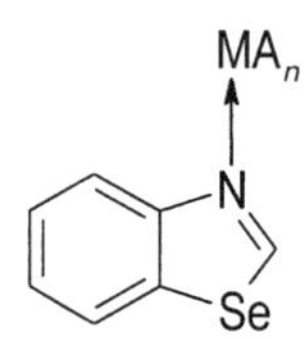

504

The complexes of selenocysteamine of type ***505*** [897], selenosemicarbazide ***506*** [898], selenosemicarbazones ***507*** [899], benzoselenazoles [900], and 2-(2-pyridine)benzoselenophene ***508*** [901] have been comprehensively examined in a monograph [1, vol.2]:

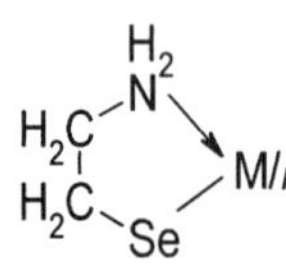

505

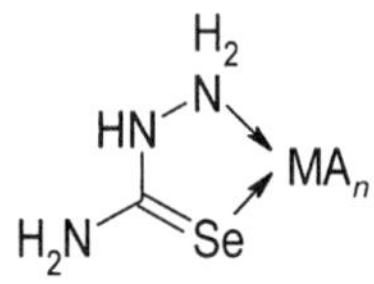

506

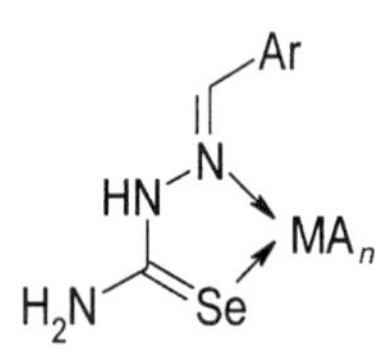

507

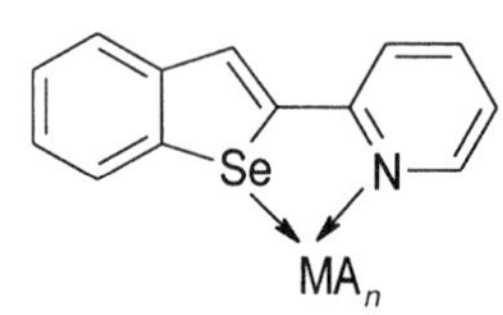

508

In the discussion about the structure of selenosalicylideneiminates ***496*** [884], it was emphasized that the corresponding nickel complex is cis-planar, while the related cobalt and zinc compounds have a tetrahedral structure. Kinetics of enantioisomery in chelates ***496*** (M = Zn, Cd, Pb; R = *i*-Pr) have been studied in detail [884b].

2.2.5.6 Ligands Containing N and Te and Donor Atoms

N,Te-Donor ligands are widely known at present [13,16,902–919], although not long ago [1, vol.2;633,634a] only a few representatives of this group of ligands were known. The most common ligand of this type is tellurourea, whose complexes, showing a structure of type ***509*** with various metals, have been extensively reviewed [16,112,633,634a]:

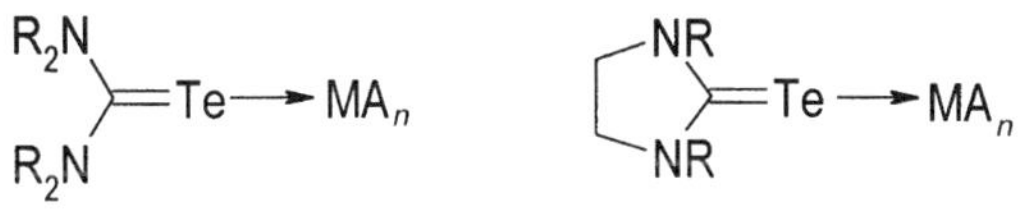

509 *510*

The discussed ligand behaves as a monodentate Te-donor. The cyclic analogue of thiourea reveals similar properties, forming complexes of carbene type ***510*** [633,634a]. The structure of one such complex ***510*** [R = Et, $MA_n = Cr(CO)_5$] has been proved through x-ray diffraction studies [902].

Amongst amine derivatives containing tellurium atoms, where Te can perform as the donor atom, there are a number of compounds with complex-forming characteristics. The following structures exemplify several complex compounds synthesized in this manner: ***511*** [903,904], ***512*** [13,633,634a,903,905,906], ***513*** [636a,907], and ***514*** [903,908]:

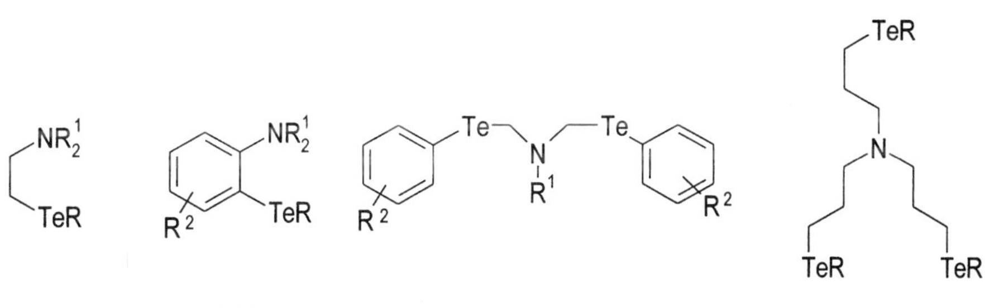

511 *512* *513* *514*

The group of ligands discussed above are mostly of the monodentate Te-donor type, a feature which has an explanation in the use of "soft" Pearson acids (see Sec. 1.1.2) as the sources of metallic ions. Thus, structure ***515*** of the corresponding mercury complex, prepared on the basis of a ligand system ***512*** [905], was established by x-ray diffraction:

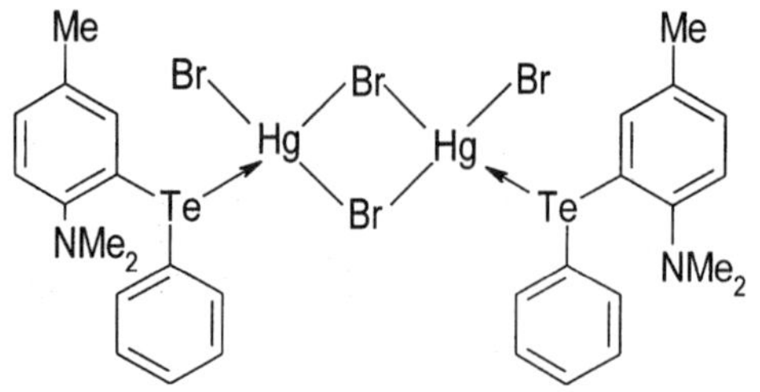

515

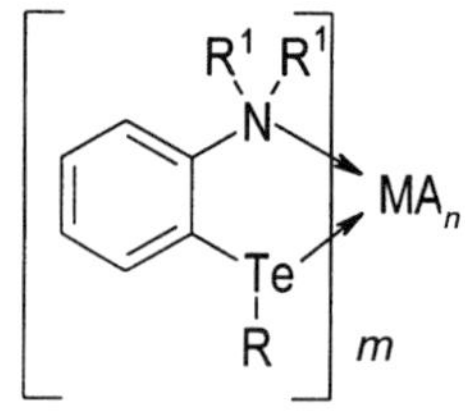

516

At the same time, it is widely accepted [906] that chelate complexes ***516*** can also be obtained from ligands ***512***. The ICC ***517*** [909] belong to the same type of coordination compounds; they are also similar to the *o*-aminothiophenolates ***454***:

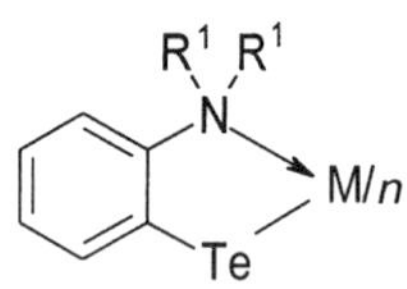

517

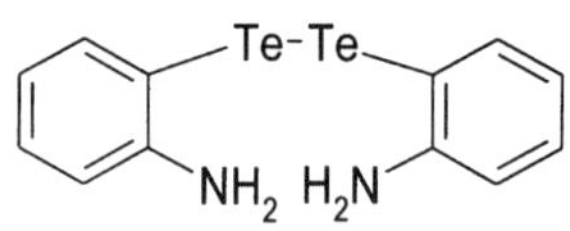

518

Since compounds of structure ***512*** (R = H) are unstable, the ICC ***517*** have been obtained by fracturing the Te—Te bond in diphenylaminoditellurides ***518*** (L) [909]. These compounds ***518*** serve as tetradentate ligands to obtain various complexes, in particular polymeric, with general formula $(MA_nL)_m$ [633,903]. Te-Containing azomethines, capable of forming complexes of types ***519*** [910] and ***520*** [911], could be formally attributed to the potentially N,Te-chelating ligands:

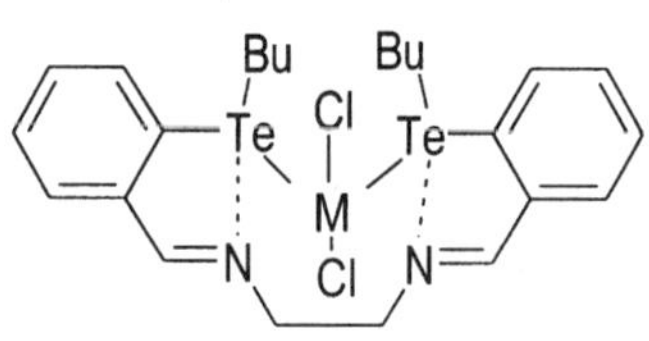

519: M = Pd, Pt

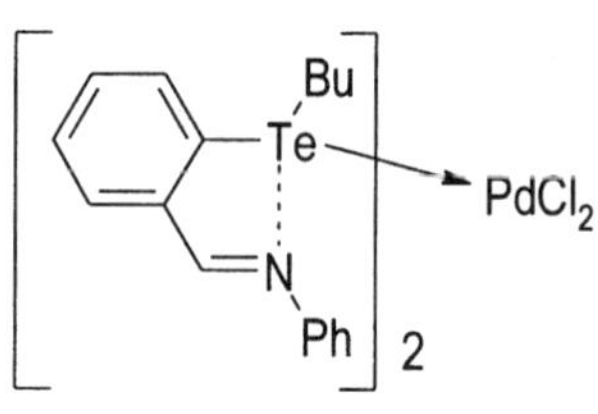

520

According to x-ray diffraction data, the ligands contained in these complexes are coordinated only through tellurium atoms. One of the causes contributing to the decrease in dentacity of the discussed ligands is the formation of an intramolecular Te—N bond [912,913], which decreases the electron-donor activity of the N atom found in the azomethinic bond.

Tellurium-containing nitrogen heterocycles and their derivatives are widely used as ligands. The path for the reaction of coordination of such five-member ligands depends considerably on the nature of both the ligand system itself and the kind of metal carbonyl. Thus, according to x-ray diffraction data and heteronuclear NMR-spectroscopy for the complexes of benzotelluroazole ***521*** with Group VI metal carbonyls, the coordination bond is localized at the Te atom (***522***) [13,16,914,915]:

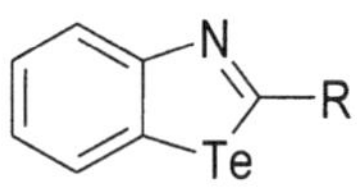
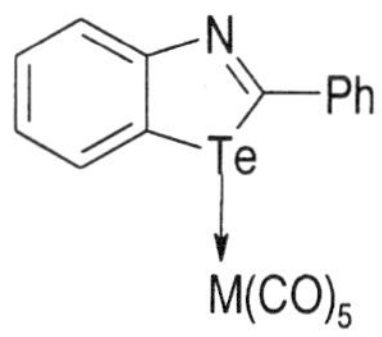
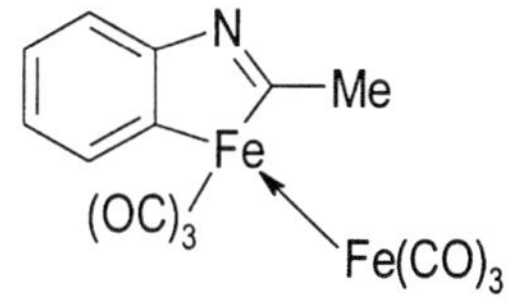

521 *522*: M = Cr, Mo, W *523*

The interaction of that same ligand, type ***521*** (R = Me), this time with $Fe_3(CO)_{12}$, leads to some complexes amongst which a detellurized metal-cyclic compound ***523*** is found [916]. At the same time [reaction (2.14)], benzoisotelluroazole ***524*** forms, together with $Fe_3(CO)_{12}$, the azomethinic compound ***525*** [916]:

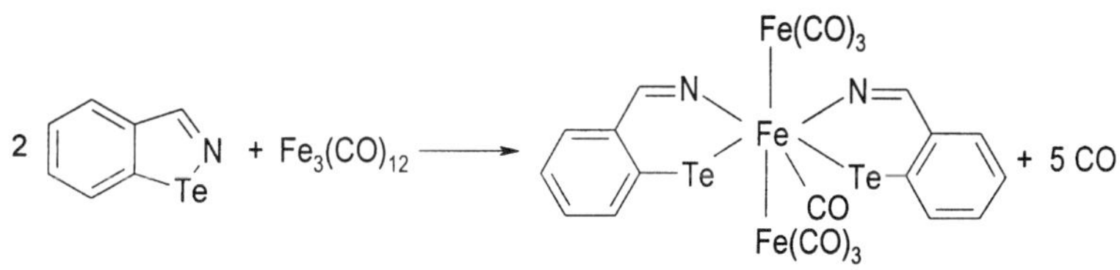

(2.14)

524 *525*

Derivatives of six-member tellurium-containing heterocycles are well represented by the complexes of 10-alkylphenotelluroazines with structure ***526*** [13,16,112,917–919]:

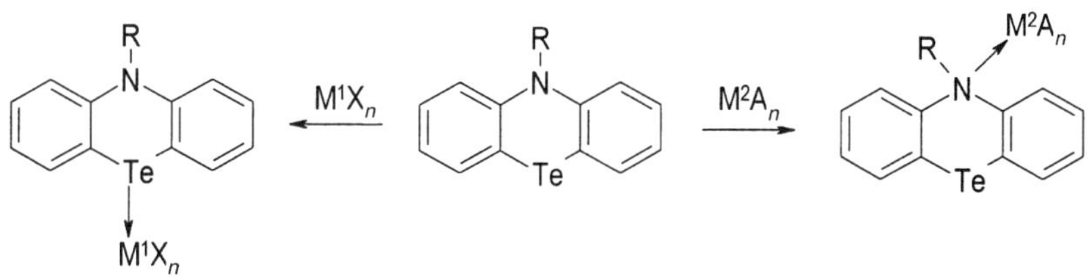

(2.15)

527 *526* *528*

It is accepted that, as a result of reaction (2.15), soft acids {RhCOCl [917], $PdCl_2$, AgX, HgX_2; X = Cl, Br, NO_2, ClO_4 [918,919]} react at the site of the tellurium atom (***527***) while hard acids (TaF_5) are bound to the nitrogen atom (***528***) [919a]. However, due to the absence of any direct proof of the indicated coordination modes for the case of phenotelluroazine ***526***, these conclusions should be considered as preliminary. On the contrary, for the complexes of 2-(2-aryltelluroethyl)pyridine, x-ray diffraction data were obtained which testify that this ligand is a chelating N,Te-donor and forms complexes of type ***529*** [920,921]:

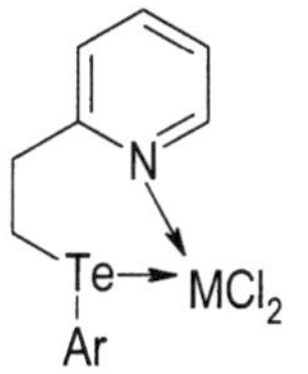
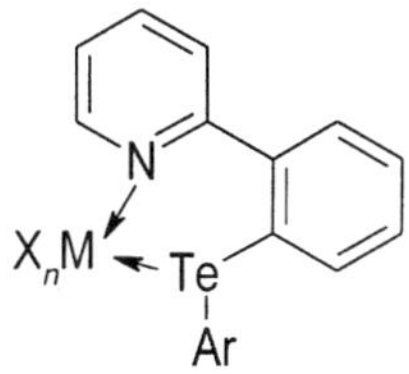
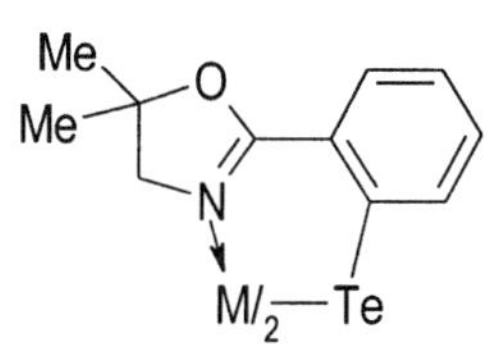

529: M = Pd, Pt; Ar = C_6H_4-OMe-*p* *530* *531*

The examined compound behaves as an N,Te-donor ligand also in analogous cobalt and copper complexes [922]. The N,Te-ligand environment is very likely for complexes ***530*** [923], ***531*** [924], adduct ***532*** [925], and tellurium-containing macrocyclic Schiff bases ***533*** [926]:

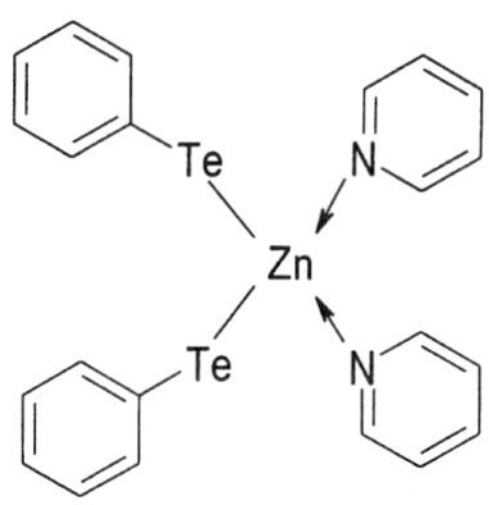

532

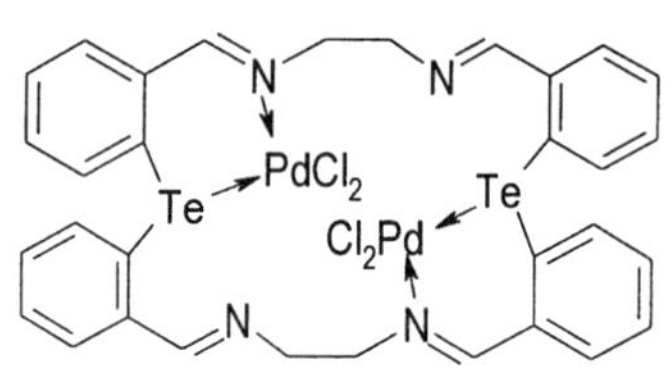

533

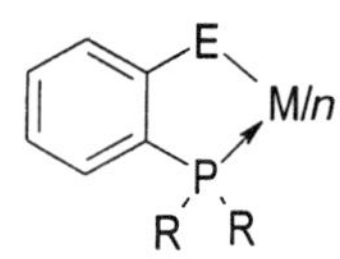

534: E = O, S, Se; R = Alk, Ar

Besides the above ligands containing elements of Groups V and VI, it is worth emphasizing the existence of compounds containing P,O-, P,S-, and P,Se-donor sets, from which metallochelates of type ***534*** were obtained [552].

2.2.5.7 Ligands with Oxygen, Sulfur, Selenium, and Tellurium as Donor Atoms

O,S-Donors are mainly represented by dimethylsulfoxide and its derivatives ***535*** and ***536*** (E = S) [1, vol.2;112,927–933]:

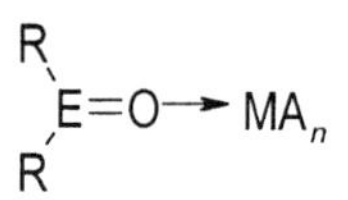

535

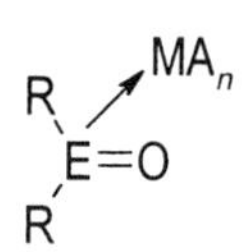

536

On the basis of ***535*** have been synthesized, and studied in detail, coordination compounds with practically all metals [931–933]. Two groups of complex compounds with O-coordinated (***535***) and S-coordinated (***536***) ligands were isolated. A detailed summary of complexes containing sulfoxides in different coordination modes, and the results of their x-ray diffraction studies including the evaluation of structural parameters for M—O and M—S bonds, have been reviewed [933]. It is emphasized [112] that the O-coordination ***535*** is formed with "hard" and intermediate metals, while the S-coordination ***536*** is a characteristic feature for "soft" Pearson acids (see Sec. 1.2.2). For complexes obtained on the basis of organoseleno-oxides ***535*** (E = Se) and organotelluro-oxides ***535*** (E = Te), the hard–hard or soft–soft interactions could also be expected [112]. However, direct proofs of such coordination modes for complexes of type ***534*** (E = Se, Te), described elsewhere [112], are absent in the available literature.

Chelate-forming ligands are widely represented amongst oxygen- and sulfur-containing donors [1, vol.2;112,934]. On their basis the following complexes have been isolated: monothiocarboxylates ***537*** [1, vol.2], thio-oxalates ***538*** [934], and monothio-β-diketonates ***539*** [1, vol.2;935–937]:

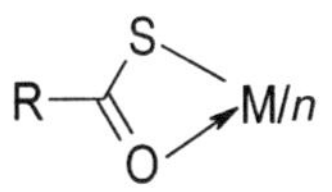

537

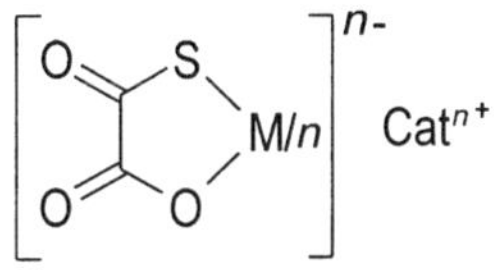

538

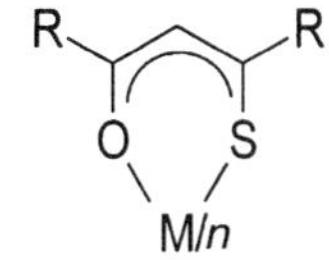

539

N-Benzoylthiourea *540* [935] can also form various chelating ligands:

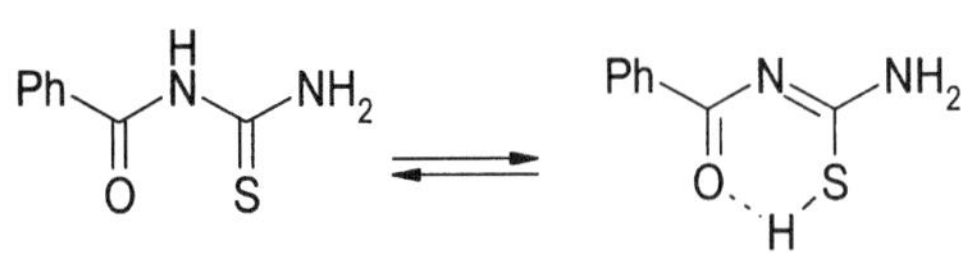

540a 540b

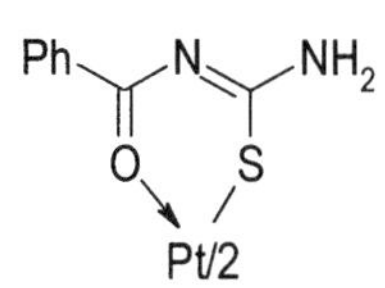

541

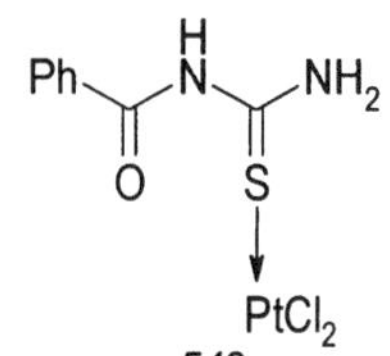

542

However, on the basis of N-benzoylthiourea *540*, not only chelating complexes of type *541* can be synthesized, but also molecular compounds as shown by *542* [938]. In the case of compound *542* (M = Pt) being formed, *540* takes part as a monodentate S-donor ligand. In this respect it has to be noted that other O,S-containing ligands can participate only as S-donors. In particular, dithio-oxalic acid *543* is assumed to belong to this group; complexes of type *544* have been obtained on its basis [1, vol.2]:

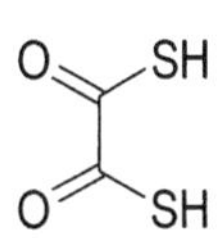

543

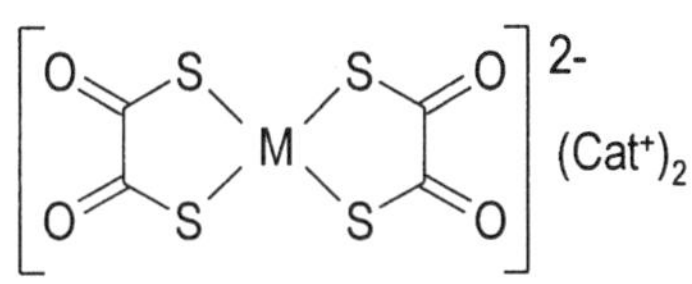

544

Such anionic complexes as *544* are described, in particular, for nickel, palladium, and platinum (M = Ni, Pd, Pt; Cat = K) [1].

As a conclusion, we note that macrocyclic compounds (thiocrown-ethers) are described as belonging to the group of S,O-donors [262,567,939,940]. Also, a number of polyheterodonor ligand systems are known, for example, thiobenzoylhydrazones, containing N,O,S-donor atoms [941].

2.2.5.8 Boron-Containing Ligands

At present, boron-containing compounds and their metal complexes are of considerable interest. Amongst them are found the coordination compounds *545–547* of transition metals containing boryle and borylene ligands [942]:

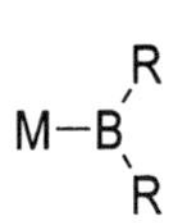

545

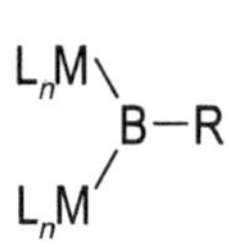

546

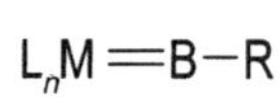

547

Compounds of type *545* are represented, for instance, by the complex compounds of diphenylborylene *548* [943] and phenylene-1,2-dioxyborylene *549* [942,944,945]:

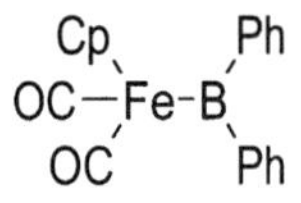

548

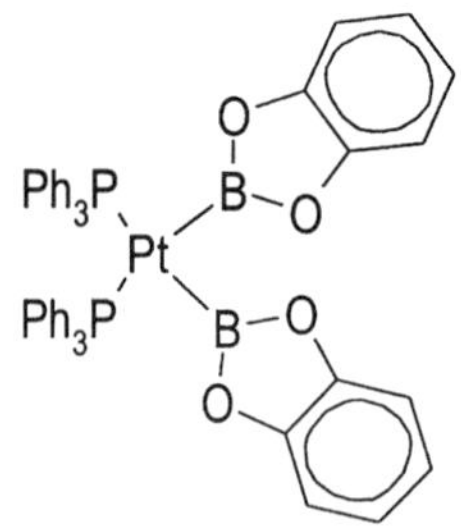

549

550

The compounds, similar to ***549*** [for example, $CpW(CO)_3(BO_2C_6H_4)$], were found to cause C—H activation with the elimination of $R—BO_2C_6H_4$. Coordination compound ***550*** [942,946] is close in its nature to compound ***545***. Binuclear complexes of types ***546***, for instance ***551***, were characterized by x-ray diffraction [942,947]:

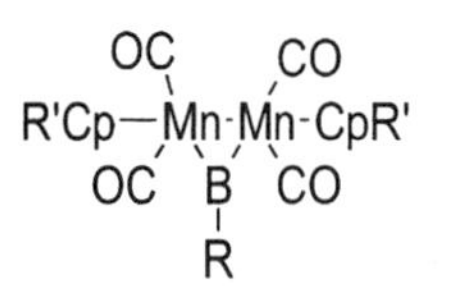

551: R = t-Bu, NMe_2;
R' = H, Me

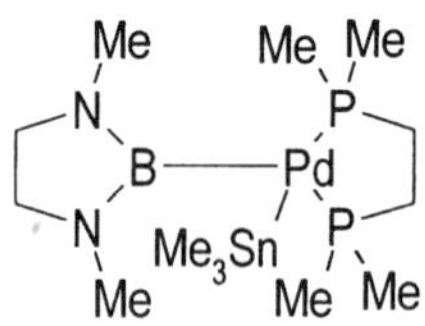

552

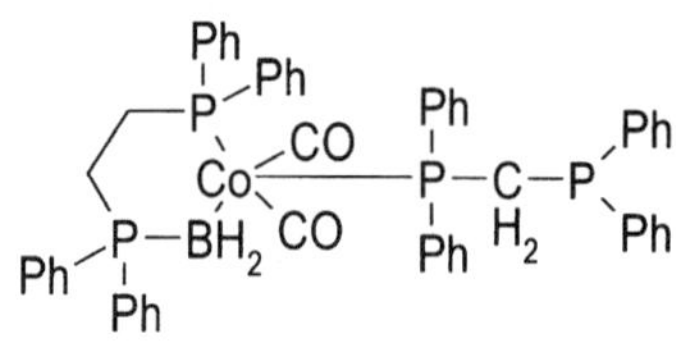

553

Data on coordination compounds of type ***547*** are absent in the literature. The boron atom is connected to the metal in complexes of B,N- (***552*** [942,948]) and B,P-containing (***553*** [942,949–952]) ligands.

The heterocyclic compounds occupy an important place amongst boron-containing ligands [8]. It is well known that borazines, having a six-member ring, form $\eta^6(\pi)$-complexes, for example ***554*** [953]. It was recently shown that σ(B)-metal-coordinated compounds of type ***555*** may also be obtained on the basis of this ligand system [954a]:

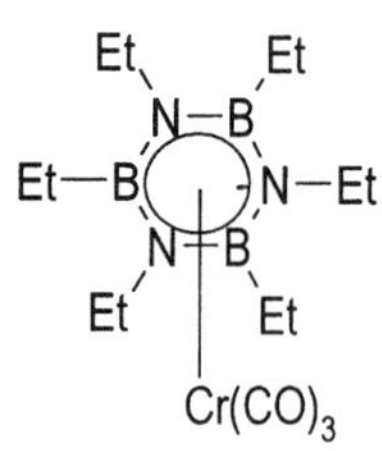

554

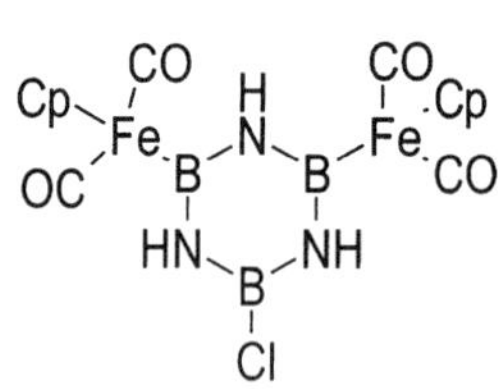

555

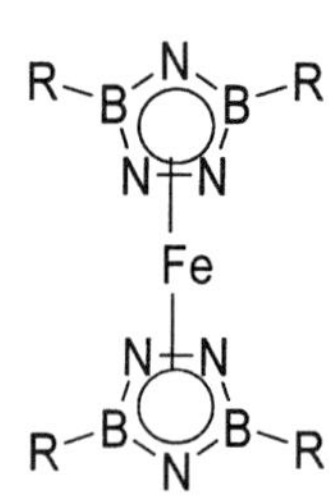

556

On the basis of similar five-member cycles were obtained the $\eta^5(\pi)$-complex compounds of type ***556*** [7, vol.3;8]. In this respect, we note that the $\eta^n(\pi)$-metal-binding shown by ***557–560*** is formed in complexes of five- and six-member boron-containing carbocyclic ligands ($n=5$ and 6, respectively) [8,942,954b]. A variety of similar structures are presented in Ref. 184f.

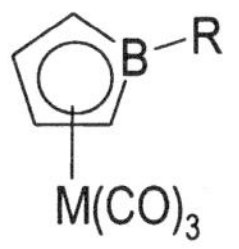
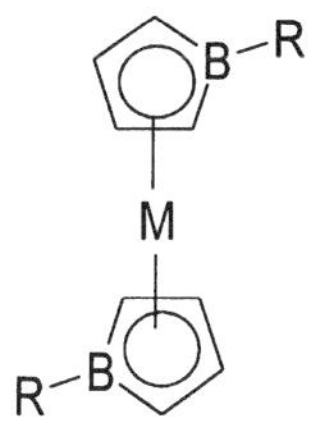
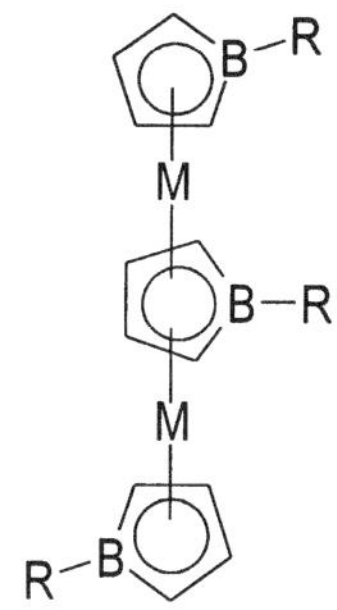
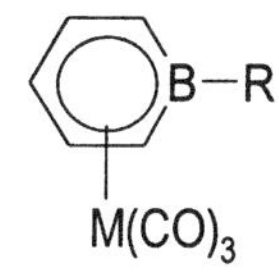

557 *558* *559* *560*

The closopolyhedral hydroborate anions $B_nH_n^{2-}$ [955] ($n=6$ [956], 9 [957], 10 [957–963], 11 [957], 12 [955,964,965]) are ligands of undoubted interest. It is shown, in the data reported above, that this anion can be in the inner or outer sphere of the complex molecule formed. The metal–borane complex ***561*** is an example of the first type of complexes mentioned; it was obtained through the interaction of $M_2B_{10}H_{10}$ (M = K, Ag) with $[Pt(PPh_3)_2I_2]$ in acetonitrile [959]:

$[Pt(PPh_3)_2B_{10}H_{10}]$ $[Pb(bipy)B_{10}H_{10}]$

561 *562*

The closodecaborate anion is part of the inner sphere in complexes **562** too [960]. The examined anion acts as a counterion in complex compounds, containing the substituents ($R=SMe_2$ [961], CH_2CN [962]) in the boron–hydride fragment. The complexes of dodecahydroclosododecaborate anion have similar structures [964,965], for example **563**:

$\{[Pb(bipy)_2]^{2+}(B_{12}H_{12})^{2-}\}_2$ $[(\eta^5\text{-}C_5H_5Ni)B_9H_9]^-[NMe_4]^+$

563 *564*

Probably Pb(II) complexes formed with either the closoanions $B_{10}H_9SMe_2^-$ and $B_{12}H_{11}NEt_3^-$ [960] or the arahno anion $6\text{-}B_{10}H_9CN^-$ [965], both containing amine ligands (bipyridine or dipyridylamine) coordinated with a metal, have a similar structure.

It has been believed that, amongst all the closo anions of type $B_{12}H_{12}^-$, the anion $B_{12}H_{12}^-$ possesses the least capacity to enter the inner sphere of the coordination [964]. This has been assumed from the data reported in a monograph [955].

In addition to that said above, complexes with closopolyhedral hydroborate metal-containing fragments are known, for example ***564***, obtained through the interaction of tetramethylammonium salt of nido-B_9H_{12} anion and nickelocene [957]. An up-to-date covering of carborane and metallacarborane structures and properties is given in Ref. 184f.

2.2.5.9 Other Ligand Systems

The ligand systems mentioned above contain, in general, one or two different donor centers. At the same time, many ligands are known containing three and more donor atoms. N,P,O-Donors of the type ***565*** ([966] and references cited therein) can be attributed to them:

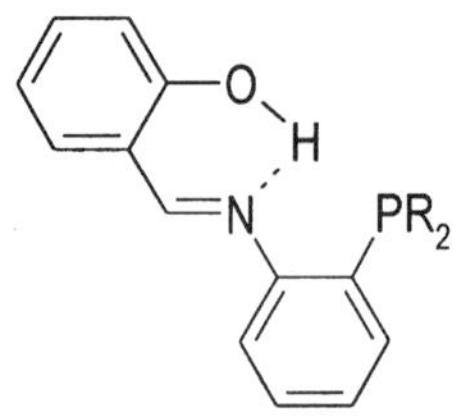

565

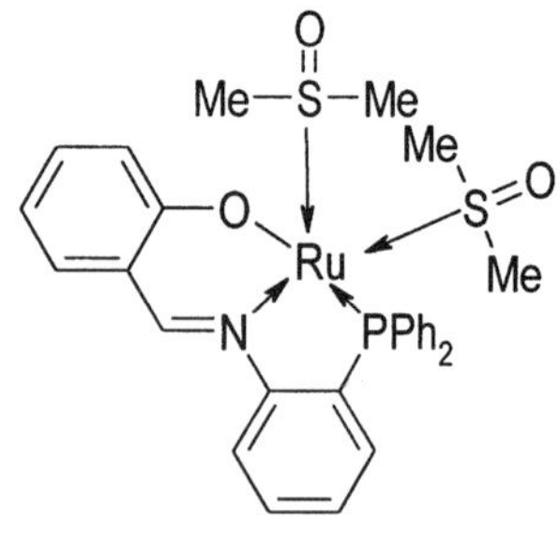

566

On the basis of ***565*** (R = Ph), chelate ***566*** was synthesized and structurally characterized. This chelate contains two metal-cycles and was formed with the participation of three (N, P, O) donor atoms [966].

Ligands having P,N,S and P,N,Se acylic and cyclic combinations can be found in a considerable number of their coordination compounds [967]. The phosphorus–sulfide di-imides, forming complexes with bridge ***567*** or chelate ***568*** structures, belong to the first type of ligands [967]:

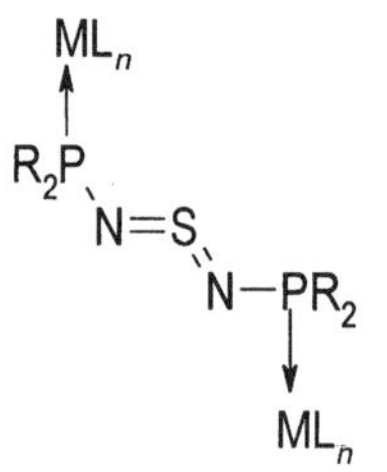

567

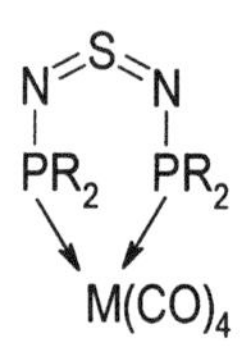

568

Mono- and binuclear structures ***569–571*** can be obtained on the basis of cyclic P,N,S(Se)-donors [967]:

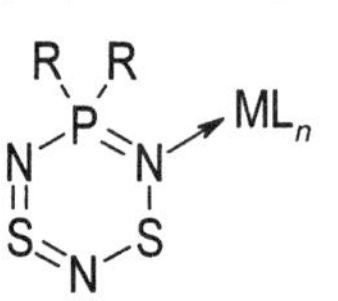

569

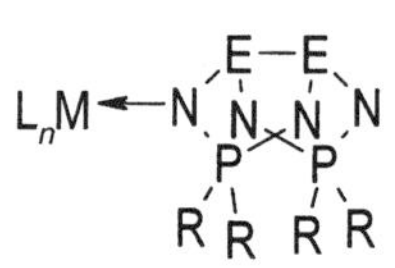

570

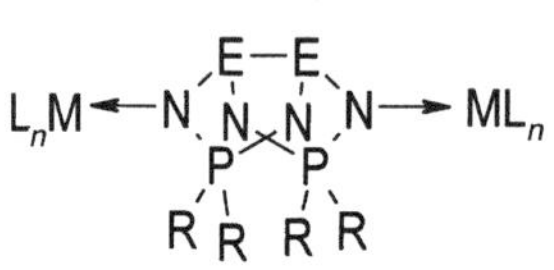

571

E = S, Se; M = Pd, Pt

Such specific ligand systems as catenanes [968], calixarenes [969], fullerenes [970a,b], and cyclodextrines [971], and functionalized dendrimers [972a–d] are of great interest. Thus, on the basis of fullerenes C_{60} and C_{80} were obtained metal–cyclopentadienyl and metal–carbonylcyclopentadienyl compounds with η^1- and η^5-coordination bonds [970b].

To conclude this chapter, we note that the material above gives notions, in general, on the ligands which are the most widely used in modern coordination chemistry. At the same time, it is necessary to take into account the enormous number of known chemical substances (at present, $\sim 10^7$ according to *Chemical Abstracts Service* data), the majority of which possess ligand properties [263].

REFERENCES

1. *Comprehensive Coordination Chemistry*. (Edit. Wilkinson, G.). Pergamon Press: Oxford, 1987, vols. 1–7.
2. *Comprehensive Organometallic Chemistry*. (Edit. Wilkinson, G.; Stone, F.G.A.; Abel, E.W.). Pergamon Press: Oxford, New York, London, 1st Edition, 2nd Edition, 1995, vols. 1–13.
3. Kukushkin, Yu.N. *Chemistry of Coordination Compounds*. High School: Moscow, 1985, 455 pp.
4. Kostromina, N.A.; Kumok, V.N.; Skorik, N.A. *Chemistry of Coordination Compounds*. High School: Moscow, 1990, 432 pp.
5. Skopenko, V.V.; Savranskii, L.I. *Coordination Chemistry*. Libid: Kiev, 1997, 334 pp.
6. Sadimenko, A.P.; Garnovskii, A.D.; Sheinker, V.N.; Osipov, O.A. *Khim. Geterotsykl Soed.* 1299 (1983).
7. *Comprehensive Heterocyclic Chemistry*. (Edit. Katrilitzky, A.R.; Rees, C.W.; Skrives, E.F.V.). Pergamon Press: Oxford, 1st Edition, 1984, vols. 2–6; 2nd Edition, 1996, vols. 2–9.
8. Garnovskii, A.D.; Sadimenko, A.P. *Adv. Heterocycl. Chem.* **72**, 1 (1998).
9. Sadimenko, A.P. *Heterocycl. Chem.* **80**, 158 (2001).
10. Garnovskii, A.D. *Izv. Vuzov. Khim. i. Khim. Tekhn.* **30**(10), 5643 (1987).
11. Garnovskii, A.D. *Koord. Khim.* **14**(5), 579 (1988).
12. Garnovskii, A.D.; Kharisov, B.I.; Gójon-Zorrilla, G.; Garnovskii, D.A. *Russ. Chem. Rev.* **64**(3), 201 (1995).
13. Garnovskii, A.D.; Garnovskii, D.A.; Vasilchenko, I.S.; Burlov, A.S.; Sadimenko, A.P.; Sadekov, I.D. *Russ. Chem. Rev.* **66**(5), 389 (1997).
14. Garnovskii, A.D. *Russ. J. Inorg. Chem.* **43**(9), 1377 (1998).
15. Garnovskii, A.D.; Kharisov, B.I.; Blanco, L.M.; Garnovskii, D.A.; Burlov, A.S.; Vasilchenko, I.S.; Bondarenko, G.I. *J. Coord. Chem.* **46**, 365 (1999).
16. Sadekov, I.D.; Uraev, A.I.; Garnovskii, A.D. *Russ. Chem. Rev.* **68**, 415 (1999).
17. (a) Garnovskii, A.D.; Blanco, L.M.; Kharisov, B.I.; Garnovskii, D.A.; Burlov, A.S. *J. Coord. Chem.* **48**, 219 (1999); (b) Figgis, B.N.; Hitchman, M.A. *Ligand Field Theory and its Applications*. John Wiley: New York, 1999, 376 pp.
18. (a) *Transition Metal Hydrides*. (Edit. Muetterties, E.L.). Marcel Dekker: New York, 1971; (b) *Transition Metal Hydrides*. Advances in Chemistry Series No. 167. (Edit. Bau, R.). American Chemical Society: Washington, 1978.
19. Kukushkin, V.Yu., Kukushkin, Yu.N. *Theory and Practice of the Synthesis of Coordination Compounds*. Nauka: Leningrad, 1990, 260 pp.
20. Dedien, A. *Transition Metal Hydrides*. Verlag Chemie: New York, 1992.
21. (a) Crabtree, R.H. *The Organometallic Chemistry of Transition Metals*. John Wiley: New York, 1994; 3rd Edition, 2001, 556 pp.; (b) Collman, J.P.; Hegedus, L.S.; Norton, J.N.; Finke, R.G. *Principles and Applications of Organotransition Metal Chemistry*. University Science Books, 1987.
22. Bau, R.; Teller, P.G.; Kirtley, S.W.; Koetzle, T.F. *Acc. Chem. Res.* **12**(5), 176 (1979).
23. Venanzi, L.M. *Coord. Chem. Rev.* **43**, 251 (1982).
24. Soloveichik, G.L.; Bulichev, B.M. *Usp. Khim.* **52**, 72 (1983).
25. McParllin, M. *Polyhedron* **3** , 1279 (1984).
26. Nichols, J.N. *Polyhedron* **3**, 1307 (1084).
27. Pearson, R.G. *Chem. Rev.* **85**, 41 (1985).

28. Hlatzky, G.C.; Crabtree, R.H.; Sabo-Etienne, C. *Coord. Chem. Rev.* **65**, 1 (1985).
29. Kukushkin, Yu.N.; Danilova, L.I.; Panina, N.S. *Koord. Khim.* **12**, 1443 (1986).
30. Deutsch, P.P.; Eisenberg, R. *Chem. Rev.* **88**, 1147 (1988).
31. Crabtree, R.H.; Hamilton, D.G. *Adv. Oranomet. Chem.* **28**, 299 (1988).
32. Morris, J.H.; Jessop, P.G. *Coord. Chem. Rev.* **121**, 155 (1992).
33. Crabtree, R.H. *Angew. Chem., Int. Ed. Engl.* **32**(6), 789 (1993).
34. Heinekey, D.M.; Oldham, W.J. *Chem. Rev.* **93** (1993).
35. Lin, Z.; Hall, M.B. *Coord. Chem. Rev.* **135/136**, 845 (1994).
36. Sabo-Etienne, S.; Chaudrecht, B. *Chem. Rev.* **98**, 2077 (1998).
37. Sabo-Etienne, S.; Chaudrecht, B. *Coord. Chem. Rev.* **178–180**, 381 (1998).
38. Kubas, G.J.; Ryan, R.R.; Swanson, B.I.; Vergamini, P.J.; Wasserman, H.J. *J. Am. Chem. Soc.* **106**, 451 (1984).
39. Crabtree, R.H.; Lavin, M.J. *J. Chem. Soc., Chem. Commun.* 794 (1985).
40. Lebouef, J.F.; Lavastre, O.; Leblanc, J.C.; Moise, C.J. *J. Organomet. Chem.* **418**(3), 359 (1991).
41. Sabo-Etienne, S.; Chaudret, B.; Abou El Makarim, H.; Barthelat, J.C.; Doudey, J.P.; Moise, C.; Leblanc, J.C. *J. Am. Chem. Soc.* **116**(20), 9335 (1994).
42. Jalon, F.A.; Otero, A.; Manzano, B.; Vallasenor, E.; Chaudret, B. *J. Am. Chem. Soc.* **117**(40), 10123 (1995).
43. Manzano, B.; Jalon, F.; Matthes, J.; Sabo-Etienne, S.; Chaudret, B.; Ulrich, S.; Limbach, H.H. *J. Chem. Soc., Dalton Trans.* 3135 (1997).
44. (a) Guari, Y.; Ayllon, J.A.; Sabo-Etienne, S.; Chaudret, B.; Hessen, B. *Inorg. Chem.* **37**(4), 640 (1998); (b) Wang, X.F.; Andrews, L. *J. Am. Chem. Soc.* **123**(51), 12899 (2001); (c) Aullon, J.A.; Gervaux, C.; Sabo-Etienne, S.; Chaudrecht, B. *Organometallics* **16**, 2000 (1997).
45. Batsanov, A.S.; Struchkov, Yu.T. *Koord. Khim.* **10**, 603 (1984).
46. Novikov, Yu.N.; Volpin, M.E. *Zhurn. Vses. Khim Obsh. im Mendeleeva* **32**, 69 (1987).
47. (a) Chatt, J.; Dilworth, J.R., Richards, R.L. *Chem. Rev.* **78**, 589 (1978); (b) *New Trends in the Chemistry of Nitrogen Fixation.* (Edit. Chatt, J.; da Camara Pina, L.M.; Richards, R.L.). Academic Press: London, 1980.
48. Yamamoto, A. *Organotransition Metal Chemistry.* John Wiley: New York, 1986.
49. Kawagudi, S. *Variety of Coordination Modes of Ligands in Metal Complexes. Inorganic Chemistry Concepts.* Springer-Verlag Chemie: Berlin, 1988, vol. 11.
50. Pelican, P.; Boca, R. *Coord. Chem. Rev.* **55**, 55 (1984).
51. *Advances in Nitrogen Fixation.* (Edit. Veeger, C.; Newton, W.E.). Nijhoff: Boston, 1991.
52. Shilov, A.E. *New J. Chem.* **16**(1/2), 211 (1992).
53. Hidai, M.; Mizobi, V. *Chem. Rev.* **95**(4), 1115 (1995).
54. Bazhenova, T.A.; Shilov, A.E. *Ross. Khim. Zhurn.* **39**, 50 (1995).
55. Bazhenova, T.A.; Shilov, A.E. *Coord. Chem. Rev.* **145**, 69 (1995).
56. Hidai, M. *Coord. Chem. Rev.* **185/186**, 99 (1999).
57. Vidal, M.; Walker, W.E.; Pruett, R.L.; Schoening, R.C. *Inorg. Chem.* **18**, 129 (1979).
58. Di Vaira, M. *Angew. Chem.* **94**, 338 (1982).
59. Di Vaira, M.; Stopioni, P.; Peruzzini, M. *Polyhedron* **6**, 351 (1987).
60. Scherer, O.J.; Ehses, M.; Wolmerhäuser, G. *Angew. Chem.* **110**, 530 (1998); *Angew Chem., Int. Ed. Engl.* **37**(4), 507 (1998).
61. Uhl, W.; Benter, M. *J. Chem. Soc., Chem. Commun.* 771 (1999).
62. Gröer, T.; Baum, G.; Scheer, M. *Organometallics* **17**(26), 5916 (1998).
63. Scherer, O.J.; Bruck, J.; Wolmerhäuser, G. *Angew. Chem., Int. Ed. Engl.* **26**, 59 (1987).
64. Scherer, O.J.; Bruck, T.; Wolmerhäuser, G. *Chem. Ber.* **121**, 935 (1988).
65. Scherer, O.J.; Blath, C.; Wolmerhäuser, G. *J. Organomet. Chem.* **387**, C21 (1990).
66. Benvenutti, M.H.A.; Hitchcock, P.B.; Nixon, J.F. *J. Chem. Soc., Chem. Commun.* 441 (1996).

67. Scherer, O.J.; Schwalb, J.; Swarowsky, H.; Wolmerhäuser, G.; Kaim, W.; Gross, R. *Chem. Ber.* **121**, 443 (1988).
68. Scherer, O.J.; Werner, B.; Wolmerhäuser, G.; Heckman, G. *Angew. Chem.* **103**, 562; *Int. Ed. Engl.* **30**, 553 (1991).
69. Scherer, O.J.; Wondung, G.; Wolmerhäuser, G. *Angew. Chem.* **101**, 1395; *Int. Ed. Engl.* **28**, 1355 (1989).
70. Scherer, O.J.; Sitzmann, H.; Wolmerhäuser, G. *Angew. Chem.* **97**, 358; *Int. Ed. Engl.* **24**, 351 (1985).
71. Scherer, O.J.; Schwalb, J.; Swarowsky, H.; Wolmershäuser, G.; Kaim, W.; Gross, R. *Chem. Ber.* **121**, 443 (1988).
72. Scherer, O.J. *Angew. Chem.* **102**, 1137; *Int. Ed. Engl.* **29**, 1104 (1990).
73. *Synthetic Methods of Organometallic and Inorganic Chemistry*. (Edit. Herrmann, W.A.). Thieme: Stuttgart, New York, 1996, vols. 1–3, 6; 1997, vols. 4,5,7,8.
74. Savitskii, A.V.; Nelyubin, V.I. *Usp. Khim.* **44**, 214 (1975).
75. Niederhoffer, E.C.; Thimmons, J.H.; Martell, A.E. *Chem. Rev.* **84**, 137 (1984).
76. Nakamoto, K. *Coord. Chem. Rev.* **100**, 363 (1990).
77. Castell, O.; Cabalol, R. *Inorg. Chem.* **38**, 668 (1999).
78. Kubas, G.J.; Wasserman, H.J.; Ryan, R.R. *Organometallics* **4**, 419 (1985).
79. Pasinskii, A.A.; Eremenko, I.L. *Zhurn. Vses. Khim. Obsh. im Mendeleeva* **32**(1), 88 (1987).
80. Gubin, S.P. *Koord. Khim.* **20**, 403 (1994).
81. Katser, S.B.; Porai-Koshits, M.A. *Koord. Khim.* **21**, 243 (1995).
82. Ansari, M.A.; Ibers, J.A. *Coord. Chem. Rev.* **100**, 223 (1990).
83. Ansari, M.A.; McCannachie, J.M.; Ibers, J.A. *Acc. Chem. Res.* **26**(11), 574 (1993).
84. Roof, L.C.; Kolis, J.W. *Chem. Rev.* ***93***, 1037 (1993).
85. Watcher, J. *Angew Chem.* **110**, 783 (1998).
86. Kawachi, A.; Tanaka, V.; Tamao, K. *Chem. Lett.* **1**, 21 (1999).
87. Herberhold, M.; Schrepfermann, M. *J. Organomet. Chem.* **419**, 85 (1991).
88. Howard, W.A.; Parkin, G. *J. Am. Chem. Soc.* **116**(2), 606 (1994).
89. Werner, A. *New Views in Inorganic Chemistry*. ONTI: Leningrad, 1936, 506 pp.
90. Grinberg, A.A. *Introduction in Chemistry of Complex Compounds*. Chemistry: Moscow, Leningrad, 1971, 631 pp.
91. *The Chemistry of Coordination Compounds*. (Edit. Beilar, J.), Reinold: New York, 1956.
92. Davies, J.A.; Hockensmith, C.M.; Kukushkin, V.Yu.; Kukushkin, Yu.N. *Synthetic Coordination Chemistry: Principles and Practice*. World Scientific: Singapore, London, 1992, 452 pp.
93. Chivers, T. *J. Chem. Soc., Dalton Trans.* No. 7, 1185 (1996).
94. Sirkin, V.G. *Metal Carbonyls*. Moscow: Science, 1983.
95. Baird, M.C. *Chem. Rev.* **88**(7), 1217 (1988).
96. Gutler, A.R.; Hanna, P.K.; Vites, J.C. *Chem. Rev.* **88**, 1363 (1988).
97. Stephan, D.W. *Coord. Chem. Rev.* **95**(1), 41 (1989).
98. Aubke, F.; Wang, C. *Coord. Chem. Rev.* **137**, 483 (1994).
99. Butler, J.S. *Pure & Appl. Chem.* **60**, 1241 (1988).
100. (a) Petz, W.; Rehder, D. *Organometallics* **9**, 856 (1990); (b) Rickard, C.E.; Roper, W.R.; Woodgate, S.D.; Wright, L.J. *J. Organomet. Chem.* **623**(1/2), 109 (2001).
101. Feltham, R.; Enemark, J.R. In: *Topic in Inorganic and Organometallic Chemistry*. (Edit. Geoffrey, J.). John Wiley: New York, 1981.
102. Glatfelter, W.L. *Advanced Organometallic Chemistry*. Academic Press: New York, 1985, vol. 24.
103. (a) Richter-Addo, G.B.; Legzdins, P. *Chem. Rev.* **88**(7), 991 (1988); (b) Podberezskaya, N.V.; Balakin, V.V.; Kuznetsova, N.I.; Danilyuk, A.F.; Likholobov, V.A. *Dokl. Akad. Nauk SSSR* **256**, 870 (1981); (c) Bogdanova, L.I.; Pashenko, D.V.; Churakov, A.B.;

Minaeva, N.A.; Stromnova, T.A. *Zhurn. Neorg. Khim.* **46**, 1481 (2001); (d) Brumaghim, J.L.; Priepot, J.G.; Girolami, G.S. *Organometallics* **18**, 2139 (1999); (e) Sellmann, D.; Utz, J.; Heinemann, F.W. *Inorg. Chem.* **38**, 5314 (1999); (f) Boulet, P.; Chermette, H.; Weber, J. *Inorg. Chem.* **40**(27), 7032 (2001); (g) Chiesa, A.; Ugo, R.; Sironi, A.; Yatsimirski, A. *J. Chem. Soc., Chem. Commun.* 350 (1990); (h) Richter-Addo, G.B. *J. Porph. Phthal.* **4**(4), 354 (2000); (i) Fomichev, D.V.; Coppens, P.; Li, T.; Badley, K.A.; Chen, L.; Richter-Addo, G.B. *Chem. Commun.* 2013 (1999); (j) Enemark, J.H.; Feltham, R.D. *Coord. Chem. Rev.* **13**, 339 (1974); (k) Feltham, R.D.; Enemark, J.H. *Top. Stereochem.* **12**, 155 (1981); (l) Westcott, B.L.; Enemark, J.H. In: *Electronic Structure and Spectroscopy*, Vol. II. (Edit. Lever, A.B.P.; Solomon, E.I.). John Wiley: New York, 1999, chap. 7; (m) Ruf. C.; Behrens, U.; Mews, R. *Chem. Commun.* No. 8, 939 (1996).

104. Kolomnikov, I.S.; Lisyk, T.V. *Usp. Khim.* **59**(4), 589 (1990); *Russ. Chem. Rev.* **59**(4), 344 (1990).
105. Pandey, K. K. *Coord. Chem. Rev.* **140**, 37 (1995).
106. Jessop, P.G.; Ikariya, T.; Noyori, R. *Chem. Rev.* **95**, 259 (1995).
107. Leitner, W. *Coord. Chem. Rev.* **153**, 257 (1996).
108. Yin, X.; Moss, J.R. *Coord. Chem. Rev.* **181**, 27 (1999).
109. Walther, D.; Ruben, J.; Rau, S. *Coord. Chem. Rev.* **182**, 67 (1999).
110. Ibers, J.A. *Chem. Soc. Rev.* **11**, 53 (1982).
111. Miguel, D.; Riera, V.; Miguel, J.A. *Organometallics* **10**(6), 1683 (1991).
112. Garnovskii, A.D.; Sadimenko, A.P.; Osipov, O.A.; Tsdintsadze, G.V. *Hard–Soft Interactions in Coordination Chemistry*. Rostov University Press: Rostov-on-Don, 1986, 272 pp.
113. *Hanbuch der Preparative Anorganische Chemie*. (Herausgegeben von Brauer, G.). Ferdinant Enke Verlag: Stuttgart, 1981, Bd. 3.
114. (a) Bryndza, H.E.; Tam, W. *Chem. Rev.* **88**, 1163 (1988); (b) Graford, W.H.; Richardson, H.W.; Wasson, J.R.; Hodgson, D.J.; Hatfield, W.E. *Inorg. Chem.* **15**, 2017 (1976); (c) Ruíz, E.; Alemany, P.; Alvarez, S.; Cano, J. *Inorg. Chem.* **36**, 3683 (1997).
115. Morton, M.S.; Lachicotte, R.J.; Vicic, D.A.; Jones, W.D. *Organometallics* **18**(2), 227 (1999).
116. Garnovski, A.D. *Koord. Khim.* **6**(12), 1779 (1980).
117. Burmeister, J.L. *Coord. Chem. Rev.* **105**, 65 (1990).
118. Fehlhammer, W.P.; Fritz, M. *Chem. Rev.* **93**, 1243 (1993).
119. Zhu, N.; Vahrenrkamp, H. *J. Organomet. Chem.* **472**(1/2), C5 (1994).
120. Bezrukova, A.A.; Khandrakova, V.S.; Petrovskii, P.V.; Rubezhov, A.Z.; Vigodskii, Ya.S.; Butin, K.P.; Magdesieva, G.V.; Piratova, G.A. *Izv. Akad. Nauk SSSR, Ser. Khim.* 968 (1995).
121. Corsi, D.M.; Murthy, N.N.; Young, Y.G.; Karlin, K.D. *Inorg. Chem.* **38**(5), 848 (1999).
122. *The Chemistry of Pseudohalides*. (Edit. Golub, A.M.; Keler, H.; Skopenko, V.V.). Elsevier Science: Lausanne, Amsterdam, New York, Tokyo, 1986.
123. Hitchman, M.A.; Rowbottom, G.L. *Coord. Chem. Rev.* **42**, 55 (1982).
124. Leising, R.A.; Kubow, M.R.; Churchill, M.R.; Buttrey, L.A.; Ziller, J.W.; Takeuchi, K.J. *Inorg. Chem.* **29**, 1306 (1990).
125. Kapandze, Th.Sh.; Tsintsadze, G.V.; Kokunov, Yu.V.; Buslaev, Yu.V. *Polyhedron* **9**(11), 1379 (1990).
126. Das, D.; Laskar, I.R.; Ghosh, A.; Mondal, A.; Okamoti, K.; Chaudehuri, N.R. *J. Chem. Soc., Dalton Trans.* No. 23, 3987 (1998).
127. Gorbunova, Yu.E.; Goryunov, A.B.; Kapanadze, Th.Sh.; Buslaev, Yu.A. *Dokl. Akad. Nauk SSSR, Ser. Khim* **286**, 1128 (1987).

128. Dikareva, L.M.; Zefirov, Yu.A.; Zhilyaev, A.N.; Baranovskii, I.B.; Porai-Koshits, M.A. *Zhurn. Neorg. Khim.* **32**, 118 (1987).
129. Kozmin, P.A.; Larina, T.A.; Surazhskaya, N.D.; Zhilyaev, A.N.; Formina, T.A. *Zhurn. Neorg. Khim.* **35**, 2250 (1990) (p. 1281 of the English version).
130. Zhilyaev, A.N.; Fomina, T.A.; Katser, S.B. *Zhurn. Neorg. Khim.* **43**, 1623 (1998).
131. Mikhailov, Yu.N.; Gorbunova, Yu.E.; Demchenko, E.A.; Serezhkina, L.B.; Serezhkin, V.N. *Zhurn. Neorg. Khim.* **44**, 415 (1999).
132. *Direct Synthesis of Coordination Compounds* . (Edit. Skopenko, V.V.). Ventury: Kiev, 1997, 172 pp.
133. *Direct Synthesis of Coordination and Organometallic Compounds*. (Edit. Garnovskii, A.D.; Kharisov, B.I.). Elsevier Science: Lausanne, Amsterdam, London, New York, 1999, 244 pp.
134. *Chemie der Pseudohalogenide* (Ed. Golub, A.M., Kehler, H., Skopenko, W.W.). VEB Deutscher Verlag der Wissenschaften: Berlin, 1979.
135. Porai-Koshitz, M.A.; Tsintsadze, G.V. In: *Crystallochemistry*. (Series "*Itogi Nauki*"). VINITI: Moscow, 1967, p. 168.
136. Burmeister, J.L. In: *Chemistry and Biochemistry of Thiocyanic Acid and its Derivatives*. (Edit: Newman, A.A.). Academic Press: London, 1975.
137. Kabesova, M.; Boca, R.; Melnik, M.; Valigura, D.; Dunaj-Jurco, M. *Coord. Chem. Rev.* **140**, 115 (1995).
138. Petrosyants, S.P.; Ilyukhin, A.B. *Koord. Khim.* **19**, 839 (1993).
139. Carmel, C.J.; Norman, N.C.; Pember, R.F.; Farrugial, L.J. *Polyhedron* **14**(3), 417 (1995).
140. Charland, I.P.; Khoo, L.E.; Gabe, E.J.; Smith, F.E. *Polyhedron* **8**(15), 1897 (1989).
141. Clarke, B.; Cunningham, D.; Gallsgher, J.F.; Higgins, T.; McArdle, P.; McGinley, J.; Cholcuin, M.N; Sheerin, D. *J. Chem. Soc., Dalton Trans.* No. 17, 2473 (1994).
142. Haanstra, W.Y.; Dricssen, W.L.; Reedij, K.; Frohlich, R.; Krebs, B. *Inorg. Chim. Acta* **185**, 175 (1991).
143. Calabrese, J.C.; Trofimenko, S. *Inorg. Chem.* **31**(23), 4810 (1992).
144. Brooker, S.; McKee, J. *Inorg. Chim. Acta* **173**, 69 (1990).
145. Fabretti, A.C.; Giusti, A.; Sessoli, R. *Inorg. Chim. Acta* **205**(1), 53 (1993).
146. Ozarowski, A.; Shunzhong, Yu.; McGarvey, B.R.; Mislaker, A.; Drake, J.E. *Inorg. Chem.* **30**(16), 3167 (1991).
147. Roux, C.; Zarembowitch, J.; Gallois, B.; Granier, T.; Glaude, T. *Inorg. Chem.* **33**(10), 2273 (1994).
148. Real, J.A.; Munoz, M.C.; Anders, E.; Granier, T.; Gallois, B. *Inorg. Chem.* **33**(16), 3587 (1994).
149. Rakhman, A.F.M.; Salem, G.; Stephens, F.S.; Wild, S.B. *Inorg. Chem.* **29**, 5225 (1990).
150. Pariya, C.; Ghosh, S.; Glosh, A.; Mukherjee, M.; Mukherjee, A.K.; Chaudhuri, N.R. *J. Chem. Soc., Dalton Trans.* No. 3, 337 (1995).
151. Cornelissen, T.P.; De Graaff, R.A.G.; Haasnoot, J.G.; Prins, R.; Reedijk, J.; Beangini-Cingi, M.; Manotti, Langfredi, A.M.; Tiriccio, A. *Polyhedron* **8**, 2313 (1989).
152. Ilyukhin, A.B.; Shkolnikova, L.M.; Poznyak, A.L.; Dyatlova, N.M. *Koord. Khim.* **16**, 811 (1990).
153. McAuliffe, C.A.; Prichard, R.G.; Bermejo, M.R.; García-Vázquez, A.; Macias, A.; Sanmartin, J.; Romero, J.; Sousa, A. *Acta Crystallogr., C* **48**, 1316 (1992).
154. Gliemann, G.; Klement, U.; Stuckl, A.C.; Bolos, Ch.; Manoussanis, G.; Nikolov, G. St. *Inorg. Chim. Acta* **195**(2), 227 (1992).
155. Parker, O.J.; Wilson, B.L.; Breneman, G.L. *Acta Crystallogr., C* **50**, 1681 (1994).
156. Pearson, R.G. *J. Am. Chem. Soc.* **85**, 3533 (1963).
157. Pearson, R.G. *J. Chem. Educ.* **45**, 581 (1968).
158. Garnovskii, A.D.; Osipov, O.A.; Bulgarevich, S.B. *Russ. Chem. Rev.* **41**(4), 341 (1972).

159. Dolenko, G.N.; Polishuk, O.Kh.; Erenin, L.P.; Lukianova, V.A. *Koord. Khim.* **13**, 293 (1987).
160. Raper, E.S.; Clegg, W. *Inorg. Chim. Acta* **180**(2), 239 (1991).
161. Maurya, R.C.; Mishra, D.D.; Pillai, Y. *Synth. React. Inorg. Metal-Org. Chem.* **25**, 139 (1995).
162. Day, M.C.; Selbin, J. *Theoretic Inorganic Chemistry*. Reinhold: New York, Amsterdam, London, 413 pp.
163. Buckingham, D.A.; Clark, C.R.; Liddell, G.F. *Inorg. Chem.* **31**(13), 2090 (1992).
164. Buckingham, D.A. *Coord. Chem. Rev.* **135/136**, 587 (1994).
165. Mashkova, L.; Strubova, K.; Ondreevich, G.; Gazho, Ya. *Koord. Khim.* **13**, 310 (1987).
166. Davidov, V.V.; Sokol, V.I.; Porai-Koshits, M.A.; Zaitsev, B.E.; Bessonov, V.A.; Balebanova, E.V. *Koord. Khim.* **17**, 976 (1991).
167. Garnovskii, A.D.; Garnovskii, D.A.; Vasilhenko, I.S. *Zhurn. Neorg. Khim.* **37**, 1474 (1992).
168. Kawaguchi, S. *Coord. Chem. Rev.* **70**, 51 (1986).
169. Crispini, A.; Errington, R.J.; Fisher, G.A.; Funke, F.J.; Norman, N.C.; Guy Orpen, G.A.; Stratford, S.E.; Struve, O. *J. Chem. Soc., Dalton Trans.* 1327 (1994).
170. Forster, G.E.; Begley, M.J.; Sowerby, D.B. *J. Chem. Soc., Dalton Trans.* 377 (1995).
171. Sarkaz, M. *Transit. Met. Chem.* **17**, 4271 (1992).
172. Monfort, M.; Bastos, C.; Dias, C.; Ribas, J.; Solans, X. *Inorg. Chim. Acta* **218**(1/2), 185 (1994).
173. Monclorf, M.; Ribas, J.; Salans, X. *Inorg. Chem.* **33**(19), 4271 (1994).
174. Koziskowa, Z.; Kozisek, J.; Kabesova, M. *Polyhedron* **9**, 1029 (1990).
175. Vrabel, V.; Kello, E.; Lokaj, J. *Acta Crystallogr. C* **51**, 1286 (1995).
176. Rowley, M.; Steed, J.W.; Tocher, D.A. *Polyhedron* **14**(11), 1415 (1995).
177. Amirkhanov, V.M.; Kapshuk, A.A.; Kramarenko, F.G.; Skopenko, V.V. *Zhurn. Neorg. Khim.* **39**, 1155 (1994).
178. Vasilieva, O.Yu.; Kokozay, V.N.; Skopenko, V.V. *Ukr. Khim. Zhurn.* **10**, 138 (1994).
179. Veszpremm, T.; Pasinszki, T.; Fecher, M. *J. Am. Chem. Soc.* **116**(14), 6303 (1994).
180. Pandey, K.K. *Inorg. Chim. Acta* **182**(2), 163 (1991).
181. Virovets, A.V.; Podberezskaya, N.V.; Lavrenova, L.G.; Bikzhanova, G.A. *Polyhedron* **13**(20), 2929 (1994).
182. Edwards, C.F.; Griffith, W.P.; Williams, D.I. *J. Chem. Soc., Dalton Trans.* 145 (1992).
183. Dehnicke, K.; Strahle, J. *Chem. Rev.* **93**, 981 (1993).
184. (a) Spessard, G.O.; Miessler, G.L. *Organometallic Chemistry*. Prentice-Hall: Englewood Cliffs, NJ, 1996, 561 pp.; (b) Mehrotra, R.C.; Singh, A. *Organometallic Chemistry: A Unified Approach*. John Wiley: New York, 1991, 634 pp. (c) Kegley, S.E.; Pinhas, A.R. *Problems and Solutions in Organometallic Chemistry*. University Science Books, 1986, 322 pp. (d) *Organometallics in Synthesis: A Manual*. (Edit. Schlosser, M.). John Wiley: New York, 1996, 614 pp.; 2nd Edition, 2002; Jordan, R.B. *Reaction Mechanisms of Inorganic and Organometallic Systems*. Oxford University Press: Oxford, 2nd Edition, 1998, 384 pp.; (e) Toreki, R. *The Organometallic HyperTextbook*. Internet site http:www.ilpi.com; (f) Cottom, F.A.; Wilkinson, G.; Murillo, C.A.; Bochmann, M. *Advanced Inorganic Chemistry*. John Wiley: New York, 6th Edition, 1999; (g) Atwood, J.D. *Inorganic and Organometallic Reaction Mechanisms*. VCH: Weinheim, 2nd Edition, 1997, 312 pp.; (h) Koutsantonis, G., *Organometallic Compounds of the Transition Elements*. Internet site http://www.chem.uwa.edu.au/enrolled students/Organometallic/contents.html; (i) Huheey, J.E.; Keiter, E.A.; Keiter, R.L. *Inorganic Chemistry: Principles of Structure and Reactivity*. Benjamin Cummings, 4th Edition, 1997, 964 pp.; (j) *Organometallic Syntheses* (Edit. King, R.B.; Fisch, J.J.). Elsevier Science: Amsterdam, 1988; (k) Elchenbroich, C.; Salzer, A. *Organometallics: A Concise Introduction*. VCH: Weinheim, 2nd Edition, 1992.

185. Herberhold, M. *Metal π-Complexes. Complexes with Mono-olefinic Ligands*. Elsevier Science: Amsterdam, 1972.
186. Albano, V.G.; Natile, G.; Panunzi, A. *Coord. Chem. Rev.* **133**, 67 (1994).
187. Jennings, P.W.; Johnson, L.L. *Chem. Rev.* **98**(8), 2241 (1994).
188. Bruce, M.I. *Chem. Rev.* **98**(8), 2797 (1998).
189. Herndon, J.W. *Coord. Chem. Rev.* **181**, 177 (1999).
190. Gusev, A.I.; Struchkov, Yu.T. *Zhurn. Strukt. Khim.* **59**, 368 (1970).
191. (a) Espinet, P.; Forniés, J.; Martínez, F.; Tomás, M.; Lalinde, E.; Moreno, M.T.; Ruiz, A.J.; Welch, A.J. *J. Chem. Soc., Dalton Trans.* 791 (1990); (b) Ara, I.; Forniés, J.; Gómez, J.; Lalinde, E.; Merino, R.I.; Moreno, M.T. *Inorg. Chem. Commun.* **2**(2), 62 (1999); (c) Went, M.J. *Adv. Organomet. Chem.* **41**, 69 (1997); (d) Rutherford, D.T.; Christie, S.D.R. *Tetrahedron Lett.* **39**, 9805 (1998); (e) Rosenthal, U.; Pulst, S.; Arndt, P.; Ohff, A.; Tillack, A.; Baumann, W.; Kempe, R.; Burlakov, V.V. *Organometallics* **14**, 2961 (1995); (f) Kovács, A.; Frenking, G. *Organometallics* **18**, 887 (1999); (g) Patel, M.M.; Green, J.R. *Chem. Commun.* 509 (1999); (h) Pasynskii, A.A.; Torubaev, Yu.V., Layakhina, A.Yu.; Valiullina, R.I.; Lysenko, K.A. *Zhurn. Neorg. Khim.* **45**, 934 (2000); (i) Kirchbauer, F.G.; Pellny, P.-M.; Sun, H.; Burlakov, V.V.; Arndt, P.; Baumann, W.; Spannenberg, A.; Rosenthal, U. *Organometallics* **20**(25), 5289 (2001).
192. Lukehart, C. *Fundamental Transition Metal Organometallic Chemistry*. Brooks Cole: Monterey, 1985.
193. Rosenblum, M. *Chemistry of Iron Group Metallocenes: Ferrocene, Ruthenocene, Osmocene*. John Wiley: New York, London, 1965.
194. Nesmeyanov, A.N. *Chemistry of Ferrocene*. Nauka: Moscow, 1969.
195. Wilkinson, G. *J. Organomet. Chem.* **100**, 273 (1975).
196. Nesmeyanov, A.N. *Chemistry of σ- and π-Complexes of Iron, Manganese, and Rhenium*. Nauka: Moscow, 1980.
197. Pearson, A.J. *Metallo-Organic Chemistry*. John Wiley: New York, 1985.
198. O'Connor, J.M.; Casey, C.P. *Chem. Rev.* **87**, 307 (1987).
199. Ernst, R.D. *Chem. Rev.* **88**(7), 1255 (1988).
200. Strelets, V.V. *Coord. Chem. Rev.* **114**(1), 1 (1992).
201. Coville, N.J.; Du Plooy, K.E.; Pikl, W. *Coord. Chem. Rev.* **116**, 1 (1992).
202. Koelle, U. *Chem. Rev.* **98**, 1313 (1998).
203. *Metallocenes*. (Edit. Togni, A.; Halterman, R.L.). John Wiley: New York, vol. 1.
204. Mathey, F. *Coord. Chem. Rev.* **137**, 1 (1994).
205 (a) Ashe, A.J.; Al-Ahmad, S. *Adv. Organomet. Chem.* **39**, 325 (1996); (b) Sadimenko, A.P. *Adv. Heterocycl. Chem.* **78**, 1 (2001).
206. Vyazankin, N.S.; Shelokov, P.N.; Kruglaya, O.A. In: *Methods of Elemento-organic Chemistry*. Nauka: Moscow, 1974, p. 905.
207. (a) Schumann, H.; Meese-Marktscheffel, J.A.; Esser, L. *Chem. Rev.* **95**(4), 865 (1995); (b) Kollimar, C.; Courty, M.; Khan, O. *J. Am. Chem. Soc.* **113**, 7994 (1991); (c) Kollimar, C.; Khan, O. *J. Chem. Phys.* **92**, 2988 (1992); (d) Horker, M.A.; Hiller, W.; Kohler, F.H. *Inorg. Chem.* **38**, 2359 (1999).
208. Pannel, K.N.; Kalsotra, B.L.; Parkanyi, C. *J. Heterocycl. Chem.* **15**, 1057 (1978).
209. Bogdanov, G.M.; Bundel, Yu.G. *Khim. Geterosykl. Soed.* 1155 (1983).
210. Sadimenko, A.P.; Garnovskii, A.D.; Retta, N. *Coord. Chem. Rev.* **126**, 237 (1993).
211. Simmons, L.H.; Riley, P.E.; Davies, R.F.; Lagowskii, J.J. *J. Am. Chem. Soc.* **98**, 1044 (1976).
212. Biderman, H.G.; Õfele, K.; Schuhbauerand, N.; Taitelbaum, J. *Angew. Chem., Int. Ed. Engl.* **14**, 639 (1975).
213. Biderman, H.G.; Õfele, K.; Taitelbaum, J. *Z. Naturforsch.* **31b**, 321 (1976).
214. Wucherer, E.J.; Muetterties, E.L. *Organometallics* **6**, 1691 (1987).

215. Eischenbroich, C.; Koch, J.; Kroker, J.; Wunseh, M.; Wassa, W.; Baum, G.; Stork, G. *Chem. Ber.* **121**, 1983 (1988).
216. Uraev, A.I.; Osipov, O.A.; Sadimenko, A.R.; Belskii, V.K.; Magomedova, N.S.; Garnovskii, A.D. *Zhurn. Obsh. Khim.* **57**, 2153 (1987).
217. Uraev, A.I.; Garnovskii, A.D.; Magomedova, N.S.; Belskii, V.K. *Zhurn. Obsh. Khim.* **65**, 1251 (1995).
218. Eischenbroich, C.; Nowotny, M.; Metz, B.; Wassa, W.; Graulich, J.; Biehler, O.K.; Sauer, W. *Angew. Chem., Int. Ed. Engl.* **30**, 547 (1991).
219. Eischenbroich, C.; Bar, F.; Bigler, E.; Manrwald, D.; Novotny, M.; Metz, B. *Organometallics* **12**, 3373 (1993).
220. Eischenbroich, C.; Kroker, J.; Massa, W.; Wünsch, M.M.; Ashe, A.J. *Angew. Chem., Int. Ed. Engl.* **25**, 571 (1986).
221. Green, M.H.L. *The Transition Elements. Organometallic Compounds.* Methuen: London, 1968, vol. 2.
222. Tolman, C.A. *Chem. Soc. Rev.* **1**, 337 (1972).
223. Astruc, D. *Chem. Rev.* **88**(7), 1189 (1988).
224. (a) Kritskaya, I.I. *Usp. Khim.* **41**, 2160 (1972); (b) Winstein, S.; Kaesz, H.D.; Kreiter, C.G.; Friedrich, E.C. *J. Am. Chem. Soc.* **87**(14), 3267 (1965); (c) Morris, R.J.; Wilson, S.R.; Girolami, G.S. *J. Organomet. Chem.* **480**, 1 (1994); (d) Gross, C.L.; Girolami, G.S. *J. Am. Chem. Soc.* **120**, 6605 (1998).
225. Fischer, E.O. *Pure & Appl. Chem.* **30**, 353 (1972).
226. Cardin, D.J.; Cetinkaya, B.; Lappert, M.F. *Chem. Rev.* **72**, 545 (1972).
227. Cotton, F.A.; Lukehart, C.M. *Progr. Inorg. Chem.* **16**, 487 (1972).
228. Cardin, D.J.; Cetinkaya, B.; Doyle, M.J.; Lappert, M.F. *Chem. Soc. Rev.* **2**, 99 (1973).
229. Fischer, E.O. *Angew. Chem.* **86**, 651 (1974).
230. Fischer, E.O. *Adv. Organomet. Chem.* **14**, 1 (1976).
231. Brown, F.J. *Progr. Inorg. Chem.* **27**, 1 (1980).
232. Dötz, K.H.; Fisher, H.; Hofman, P.; Kreissel, F.R.; Schubert, U.; Weis, K. *Transition Metal Carbene Complexes.* Verlag Chemie: Weinheim, 1984.
233. Schubert, U. *Coord. Chem. Rev.* **55**, 261 (1984).
234. Brothers, P.J.; Roper, W.R. *Chem. Rev.* **88**(7), 1293 (1988).
235. (a) *Advances in Metal Carbene Chemistry.* (Edit. Schubert, U.). Kluwer Academic: Dordrecht, 1989; *Advances in Carbene Chemistry* (Edit. Briker, U.H.). Jai Press: Greenwich and Stamford, 1994, vol. 1; 1998, vol. 2; (c) Bourissou, D.; Guerret, O.; Gabbai, F.; Bertrand, G. *Chem. Rev.* **100**, 39 (2000).
236. Nefedov, O.M.; Ioffe, A.I.; Mechnikov, L.G. *Chemistry of Carbenes.* Chemistry: Moscow, 1990, p. 233.
237. Kolesnikov, S.P. *Zhurn. Vses. Khim. Obsh. im Mendeleeva* **24**, 505 (1979).
238. Fischer, E.O.; Massböl, A. *Angew. Chem. Int. Ed. Engl.* **3**, 508 (1964).
239. Fehlhammer, W.P.; Bliss, L.; Sperber, W.; Fuchs, J.Z. *Naturforsch.* **49b**(4), 494 (1994).
240. Herrmann, W.A.; Elison, M.; Fischer, J.; Köcher, C.; Artus, C.R. *Angew. Chem. Int. Ed. Engl.* **34**, 2371 (1995).
241. Kuhn, N.; Kratz, I.; Bläser, D.; Boese, R. *Inorg. Chim. Acta* **238**(1/2), 179 (1995).
242. Herrmann, W.A.; Öfele, K.; Mihalios, D.; Elison, M.; Herdtweck, E.; Priermer, T.; Kiprof, P. *J. Organomet. Chem.* **498**(1), 1 (1995).
243. Kernbach, U.; Ramm, M.; Luger, P.; Fehlhammer, W.P. *Angew, Chem.* **108**, 333 (1996).
244. Francis, M.D.; Hibbs, D.E.; Hursthouse, M.B.; Cameron, J.; Smithies, N.A. *J. Chem. Soc., Dalton Trans.* 3249 (1998).
245. Boehme, C.; Frenking, G. *Organometallics* **17**(26), 5801 (1998).

246. (a) Zhang, S.-W.; Ishii, R.; Takahashi, S. *Organometallics* **16**(1), 20 (1997); (b) Guillaume, V.; Mahias, V.; Lapinte, C. *Organometallics* **19**(7), 1422 (2000); (c) Brumaghim, J.L.; Gross, C.L.; Girolami, G.S. *Chem. Commun.* 953 (1999).
247. Tschugaeff, L.; Teeari, P. *Ber* ("*Berichte der Deutsche Chemische Geselschaft*") **47**, 568 (1914).
248. Ugi, I.; Fetzer, U.; Eholzer, E.; Knupfer, H.; Offermann, K. *Angew. Chem., Int. Ed. Engl.* **4**, 472 (1965).
249. Malatesta, L.; Bonati, F. *Isocyanide Complexes of Metals*. John Wiley: New York, 1969.
250. Pombeiro, A.J.L.; Richards, R.L. *Coord. Chem. Rev.* **104**, 13 (1990).
251. Colton, R.; McCormick, M.J. *Coord. Chem. Rev.* **31**, 1 (1980).
252. Kukushkin, Yu.N. *Koord. Khim.* **11**, 1606 (1985).
253. Crociani, B. In: *Reactions of Coordinated Ligands* (Edit. Braterman, P.S.). Plenum Press: New York, London, 1989, vol. 1, p. 553.
254. Balch, A.L.; Boehm, J.R.; Hope, H.; Olmstead, M.M. *J. Am. Chem. Soc.* **98**, 7431 (1976).
255. Olmstead, M.M.; Hope, H.; Benner, L.S.; Balch, A.L. *J. Am. Chem. Soc.* **99**, 5502 (1977).
256. Rivera, A.V.; Sheldrick, G.M.; Hurthouse, M.B. *Acta Crystallogr., B.* **34**, 1985 (1978).
257. Yamamoto, Y.; Aoki, K.; Yamazaki, H. *Chem. Lett.* 391 (1979).
258. (a) Carvalho, M.F.N.N.; Pombeiro, A.J.L. Schubert, U.; Orama, O.; Pickett, C.J.; Richards, R.L. *J. Chem. Soc., Dalton Trans.* 2079 (1985); (b) Carvalho, M.F.N.N.; Duarte, M.T.; Galvaõ, A.M.; Pombeiro, A.J.L. *J. Organomet. Chem.* **469**, 79 (1994); (c) Rehder, D.; Bottcher, C.; Collazo, C.; Hedelt, R.; Schmidt, H. *J. Organomet. Chem.* **585**, 294 (1999); (d) Tamm, M.; Hahn, F.E. *Coord. Chem. Rev.* **182**, 175 (1999); (e) Cristofani, S.; Leoni, P.; Pasquali, M.; Eisentraeger, F.; Albinati, A. *Organometallics* **19**(22), 4589 (2000); (f) Efendi, L.; Giancarlo, G.; White, A.H. *Inorg. Chim. Acta* **298**(2), 146 (2000); (g) Pombeiro, A.J.L.; Guedes de Silva, M.F.C.; Michelin, R.A. *Coord. Chem. Rev.* 218, 43 (2001); (h) Michelin, R.A.; Pombeiro, A.J.L.; Guedes de Silva, M.F.C. *Coord. Chem. Rev.* **218**, 75 (2001); (i) Valliant, J. F.; Schaffer, P. *J. Inorg. Biochem.* **85**(1), 43 (2001).
259. (a) Fischer, E.O.; Schubert, U. *J. Organomet. Chem.* **100**, 59 (1975); (b) Fischer, H. *J. Organomet. Chem.* **195**, 55 (1980); (c) Kreissl, F.R.; Wolfgruber, M.; Sieber, W.J. *J. Organomet. Chem.* **270**, 4 (1984); (d) Lappert, M.F.; Rowe, R.S. *Coord. Chem. Rev.* **100**, 267 (1990); (e) Grützmacher, H.; Marchand, C.M. *Coord. Chem. Rev.* **163**, 287 (1997); (f) Barrau, J.; Rima, G. *Coord. Chem. Rev.* **178–180**, 593 (1998).
260. Gebreleu, N.V. *Matrix Reactions*. Stiintsa: Kishinev, 1980, 183 pp.
261. Yatsimirski, K.B.; Kilchinskii, A.G.; Pavlishuk, V.V.; Talanova, G.G. *Synthesis of Macrocyclic Compounds*. Nauk. Dumka: Kiev, 1987, 279 pp.
262. Gebreleu, N.V.; Arion, V.B. *Template Synthesis of Macrocyclic Compounds*. Stiintsa: Kishinev, 1990, 373 pp.
263. Von Zelewsky, A. *Stereochemistry of Coordination Compounds*. John Wiley: Chichester, New York, 1996.
264. Garnovskii, A.D.; Vasilchenko, I.S.; Garnovskii, D.A. *Modern Aspects of Synthesis of Metal Complexes. Fundamental Ligands & Methods*. LaPO: Rostov-on-Don, 2000, 354 pp.
265. Ebsworth, E.A.V.; Rankin, D.W.H.; Cradock, S. *Structural Methods in Inorganic Chemistry*. Blackwell: Oxford, 2nd Edition, 1991.
266. Berezin, B.D. *Coordination Compounds of Porphyrins and Phthalocyanines*. Nauka: Moscow, 1978.
267. *The Porphyrins*. (Edit. Dolphin, D.). Academic Press: New York, 1978–1979, vols. 1–7.
268. Lehn, J.-M. *Supramolecular Chemistry*. Verlag Chemistry: Weinheim, 1995.

269. Garnovskii, A.D.; Nivorozhkin, A.L.; Minkin, V.I. *Coord. Chem. Rev.* **126**, 1 (1993).
270. Garnovskii, A.D. *Russ. J. Coord. Chem.* **19**, 368 (1993).
271. Curtis, N.F. *Coord. Chem. Rev.* **3**, 3 (1968).
272. Busch, D. *Usp. Khim.* **38**, 822 (1969).
273. Busch, D. *Pure & Appl. Chem.* **52**, 2477 (1980).
274. Busch, D.H. *Acc. Chem. Res.* **11** (10), 392 (1979).
275. Henrick, K.; Taskes, P.A.; Lindoy, L.F. *Progr. Inorg. Chem.* **33**, 1 (1985).
276. Busch, D.H. *Chem. Rev.* **93**, 847 (1993).
277. Cabeza, Ja.A.; Fernández-Colínas, J.M. *Coord. Chem. Rev.* **126**(1/2), 319 (1993).
278. Sutton, D. *Chem. Rev.* **93**, 995 (1993).
279. *Porphyrins and Metalloporphyrins.* (Edit. Smith, K.M.). Elsevier Science: Amsterdam, 1975.
280. *Vitamin B_{12}* (Edit. Dolpín, D.). Wiley Interscience: New York, 1982.
281. Smith, K.M. In: *Comprehensive Organic Chemistry* (Edit. Barto, D.; Ollis, W.D.). Pergamon Press: Oxford, New York, Toronto, Sydney, Paris, Frankfurt, 1979.
282. Berezin, B.D.; Enikolonov, N.S. *Metalloporphyrins.* Nauka: Moscow, 1988.
283. *Biomimetics.* (Edit. Starikaja, A.; Aksay, I.A.). AIP Press: New York, 1995.
284. *Biocoordination Chemistry.* (Edit. Fenton, D.F.). Paperback Publishers: Oxford, 1996.
285. (a) Mellor, J.W. *A Comprehensive Treatise on Inorganic and Theoretical Chemistry.* Longmans: London, 1931; (b) Cottom, F.A.; Wilkinson, G. *Advanced Inorganic Chemistry.* John Wiley: New York, 1988; (c) Eischenbroich, C.; Salzer, A. *Organometallics: A Concise Introduction.* VCH-Wiley: New York, 2nd Edition, 1992, 495 pp.; (d) *The New Chemistry.* (Edit. Hall, N.). Cambridge University Press: Cambridge, 2000; (e) House, D. In: Ref. 1, vol. 2, p. 23.
286. Brown, K.L.; Hall, D. *Acta Crystallogr., B* **32**, 279 (1976).
287. Chugaev, L.A. *Investigations in the Area of Complex Compounds.* Izd. Akad. Nauk SSSR: Moscow, 1954.
288. Yatsimirsky, K.B.; Kriss, E.E.; Gvyazdovskaya, V.L. *Stability Constants of Metal Complexes with Bioligands.* Nauk. Dumka: Kiev, 1979, 255 pp.
289. Martell, A.E.; Calvin, M. *Chemistry of the Metal Chelate Compounds.* Prentice-Hall: Englewood Cliffs, NJ, 1952.
290. Dashevskii, V.G.; Baranov, A.P.; Kabachnik, M.I. *Usp. Khim.* **52**, 268 (1983).
291. Martell, A.E.; Hancock, R.D. *The Chelate, Macrocyclic and Cryptate Effects.* In: Ref. 1, vol. 1, p. 240.
292. Yatsimirsky, K.B. *Mendeleev Chem. J.* **40**(4/5), 7 (1996).
293. Yatsimirsky, K.B. *Theoret. Exp. Chem.* **16**, 34 (1980).
294. Yatsimirsky, K.B. *Zhurn. Neorg. Khim.* **28**, 2995 (1983).
295. Martell, A.E.; Hancock, R.D.; Montekaitis, R.J. *Coord. Chem. Rev.* **133**, 39 (1994).
296. Pomogailo, A.D.; Ufyand, I.E. *Macromolecular Chelates.* Chemistry: Moscow, 1991, 303 pp.
297. Sidorov, A.A.; Diomidov, S.A.; Nefedov, S.E.; Fomina, I.G.; Danilov, P.V.; Novotriev, V.M.; Volkov, O.G.; Ikorskii, V.I.; Eremenko, I.L. *Zhurn. Neorg. Khim.* **44**, 396 (1999).
298. Black, D.St.C.; Harshorn, A.D. *Coord. Chem. Rev.* **9**, 219 (1973).
299. Tasker, P.A.; Fleiser, E.B. *J. Am. Chem. Soc.* **92**, 7072 (1970).
300. Lau, H.; Gutsche, C.D. *J. Am. Chem. Soc.* **100**, 1857 (1978).
301. Garnovskii, D.A.; Garnovskii, A.D.; Sadimenko, A.P.; Sigeikin, S.G. *Russ. J. Coord. Chem.* **20**, 74 (1994).
302. Von Zelewsky, A.; Barbosa, L.; Schläpfer, C.W. *Coord. Chem. Rev.* **123**(1/2), 229 (1993).
303. Giuffrida, G.; Campagna, S. *Coord. Chem. Rev.* **135/136**, 517 (1994).
304. Szczepura, L.F.; Witcham, L.M.; Takeuchi, K.J. *Coord. Chem. Rev.* **174**, 5 (1998).

305. Trofimenko, S. *Chem. Rev.* **72**, 497 (1972).
306. Garnovskii, A.D.; Osipov, O.A.; Kuznetsova, L.I.; Bogdashev, N.N. *Usp. Khim.* **2**, 177 (1973).
307. (a) Sadimenko, A.P.; Garnovskii, A.D.; Retta, N. *Coord. Chem. Rev.* **126**, 237 (1993); (b) Sadimenko, A.P.; Basson, S.S. *Coord. Chem. Rev.* **147**, 247 (1996); (c) La Monica, G.; Ardizzo, G.A. *Progr. Inorg. Chem.* **46**, 151 (1997); (d) Cosgriff, J.A.; Deacon, G.B. *Angew. Chem., Int. Ed. Engl.* **37**, 286 (1998).
308. Elguero, J. In: Ref. 7, 2nd Edition, vol. 3, p. 1.
309. Sundber, R.J.; Martin, R.B. *Chem. Rev.* **74**, 471 (1974).
310. Moore, S.; Robinson, S.D. *Adv. Inorg. Chem.* **32**, 171 (1988).
311. (a) Popov, I.A. *Coord. Chem. Rev.* **4**, 463 (1969); (b) Palkina, K.K.; Kuzmina, N.E.; Lyakhov, A.S.; Gaponik, P.N.; Bogatikov, A.N.; Govorova, A.A.; Ivashkevich, L.S. *Zhurn. Neorg. Khim.* **46**(10), 1649 (2001).
312. Wakefield, B.J.; Wright, D.J. *Adv. Het. Chem.* **25**, 147 (1979).
313. Munsey, M.S.; Natale, N.L. *Coord. Chem. Rev.* **109**(2), 251 (1991).
314. Antsishkina, A.S.; Sadikov, G.G.; Porai-Koshits, M.A.; Kogan, V.A. *Koord. Khim.* **19**, 64 (1993).
315. Roux, C.; Zarembowitsh, J.; Gallois, B.; Garnier, T.; Claude, R. *Inorg. Chem.* **33**(10), 2273 (1994).
316. Zhu, D.-R.; You, X.-Z.; Shi, X.-H.; Fun, H.-K.; Sivakumar, K. *Acta Crystallogr., C* **52**(5), 1146 (1996).
317. Burlov, A.S.; Antsishkina, A.S.; Romero, J.; Garnovskii, D.A.; Garcia-Vázquez, A.; Sousa, A.; Garnovskii, A.D. *Zhurn. Neorg. Khim.* **40**, 1427 (1995).
318. Das, M.K.; Maiti, P.K. *Inorg. Chim Acta* **172**(1), 35 (1990).
319. Fish, R.H.; Fond, R.H.; Tran, A.; Eduardo, B. *Organometallics* **10**(4), 1209 (1991).
320. Randaccio, L.; Geremia, S.; Dreos-Garlatti, R.; Tauzher, G.; Asaro, F. *Inorg. Chim. Acta* **194**(1), 1 (1992).
321. Geremia, S.; Randaccio, L.; Zangrando, E. *Gazz Chim. Ital.* **122**(6), 229 (1992).
322. Toscano, P.J.; Dreos-Carlatti, R.; Randaccio, L.; Tautzer, G.; Antolini, L. *Inorg. Chim. Acta* **216**(1/2), 125 (1994).
323. Toscano, P.J.; Di Mauro, P.; Geremia, S.; Randaccio, L.; Zangrando, E. *Inorg. Chim. Acta* **217**(1/2), 195 (1994).
324. Vogel, E.; Will, S.; Tilling, A.S.; Nudger, N.; Lex, J.; Bill, E.; Trautwein, A.X.; Weighard, K. *Angew Chem.* **106**, 711; *Int. Ed. Engl.* **33**(7), 731 (1994).
325. Dikareva, L.M.; Golubnichaya, M.A.; Baranovskii, I.B. *Zhurn. Neorg. Khim.* **33**, 2068 (1988).
326. Golubnichaya, M.A.; Sidorov, A.A.; Fomina, I.G.; Eremenko, I.L.; Moiseev, I.I. *Zhurn. Neorg. Khim.* **44**, 1479 (1999).
327. Eremenko, I.L.; Sidorov, A.A.; Nefedov, S.E.; Moiseev, I.I. *Izv. RAN, Ser. Khim.* No. 3, 409 (1999).
328. Fochi, G; Strahle, J.; Gingl, F. *Inorg. Chim. Acta* **30**, 4669 (1991).
329. Carreck, I.W.; Goldstein, M.; McPartlin, E.M.; Unsworth, W.D. *J. Chem. Soc., Chem. Commun.* 1634 (1971).
330. Richardson, H.W.; Wasson, J.R.; Hatfield, W.E. *Inorg. Chem.* **16**, 484 (1977).
331. Labisbal, E., Gómez, C.; García-Vázquez, J.A.; Macías, A.; Romero, J.; Sousa, A.; Englet, U.; Fenton, D.E. *Inorg. Chim. Acta* **203**, 67 (1993).
332. Labisbal, E.; Romero, J.; García-Vázquez, J.A.; Gómez, C.; Sousa, A.; Pritchard, R.; McAuliffe, C.A. *Polyhedron* **13**(11), 1735 (1994).
333. Garnovskii, A.D.; Burlov, A.S.; Garnovskii, D.A.; Vasilchenko, I.S.; Antsishkina, A.S.; Sadikov, G.G.; Sousa, A.; García-Vázquez, J.A.; Romero, J.; Durán, M.; Sousa-Pedrares, A.; Gómez, C. *Polyhedron* **18**, 863 (1999).
334. Newkome, G.R.; Kohli, D.K.; Fronczek, F.R. *J. Am. Chem. Soc.* **104**, 994 (1982).

335. Ernst, S.; Kaim, W. *J. Am. Chem. Soc.* **108**, 3578 (1986).
336. (a) Constable, E.C. In: *Advances in Inorganic Chemistry and Radiochemistry*. (Edit. Emeleus, H.J.; Sharpe, A.G.). Academic Press: London, 1986, vol. 30; (b) Bolder, J.; Gourdon, A.; Ishow, E.; Launau, J.P. *Inorg. Chem.* **35**, 2937 (1996); (c) Patterson, B.T.; Keene, F.R. *Inorg. Chem.* **37**, 645 (1998); (d) Albela, B.; Corbella, M.; Ribas, J.; Castro, I.; Sletten, J.; Stroeckli-Evans, H. *Inorg. Chem.* **37**, 788 (1998); (e) Treadway, J.A.; Meyer, T.J. *Inorg. Chem.* **38**, 2267 (1999); (f) Naiz, R.B.; Lee, K.Y.; Murphy, C.J. *Inorg. Chem.* **38**, 2536 (1999); (g) Chiorboli, C.; Bignozzi. C.A.; Scandola, F.; Ishow, E.; Gourdon, A.; Leunau, J.-P. *Inorg. Chem.* **38**, 2402 (1999); (h) Neve, F.; Crispini, A.; Campagna, S.; Serroni, S. *Inorg. Chem.* **38**, 2250 (1999); (i) Witthew, M.A.; Blake, A.J.; Campness, N.R.; Cooke, P.A.; Hubberstey, P.; Li, W.-S.; Schroder, M. *Inorg. Chem.* **38**. 2259 (1999).
337. (a) Kitamura, H.; Ozawa, T.; Jitsukawa, K.; Masuda, H.; Aoyama, Y.; Einaga, H. *Inorg. Chem.* **39**, 3294 (2000); (b) Castillo, O.; Lugue, A.; Julve, M.; Lloret, F.; Roman, P. *Inorg. Chim. Acta* **315**, 9 (2000); (c) Castillo, O.; Lugue, A.; Roman, P.; Lloret, F.; Julve, M. *Inorg. Chem.* **40**, 5526 (2001); (d) Manson, J.L.; Schlueter, J.A.; Geiser, U.; Stone, M.B.; Reich, D.H. *Polyhedron* **20**, 1423 (2001); (e) La Chance-Galang, K.J.; Maldonado, I.; Galagher, M.; Jian, W.; Prock, A.; Chacklos, J.; Calang, R.D.; Clarke, M. *Inorg. Chem.* **40**, 485 (2001).
338. Lindoy, L.F. *Chem. Soc. Rev.* **14**, 421 (1975).
339. Panova, G.V.; Vikulova, N.K.; Potpov, V.M. *Usp. Khim,* **49**, 1234 (1980).
340. Garnovskii, A.D.; Alexeenko, V.A.; Burlov, A.S.; Kurbatov, V.P.; Lipunova, G.N.; Nedzvetskii, V.S.; Sigeikin, G.I. *Izv. Vissh. Ucheb. Zaved. Ser. Khim. i Khim. Tekhn.* **34**(1), 3 (1991).
341. Garnovskii, A.D.; Alexeenko, V.A.; Burlov, A.S.; Nedzvetskii, V.S. *Zhurn. Neorg. Khim.* **36**, 886 (1991).
342. Garnovskii, A.D.; Kurbatov, V.P.; Lipunova, G.N.; Sigeikin, G.I. *Koord. Khim.* **17**, 1011 (1991).
343. Seveik, J.; Cambel, F. In: *The Chemistry of Amidines and Imidates*. (Edit. Pataj, S.). John Wiley: London, 1975.
344. De Rode, H.W.; Vrieze, K.; von Gustort, E.A.K.; Ritter, A. *J. Organomet. Chem.* **135**, 183 (1977).
345. Crove, D.M.; Van Koten, G.; Ubers, H.J.C.; Vrieze, K.; Niemann, L.; Stam, C.H. *J. Chem. Soc., Dalton Trans.* 717 (1986).
346. Piraino, P.; Bruno, G.; Loschiavo, S.; Laschi, F.; Zanello, P. *Inorg. Chem.* **26**, 2205 (1987).
347. Cotton, F.A.; Matusz, M.; Poli, R. *Inorg. Chem.* **26**, 1472 (1987).
348. Cotton, F.; Feng, X.; Matusz, M.; Poli, R. *J. Am. Chem. Soc.* **110**, 7077 (1988).
349. Yan, C.L.; He, L.P.; Korp, J.D.; Bear, J.L. *Inorg. Chem.* **27**(24), 4389 (1988).
350. Barker, J.; Kilner, M. *Coord. Rev.* **133**, 219 (1994).
351. (a) Moore, D.S.; Robinson, S.D. *Adv. Inorg. Chem. Radiochem.* **30**, 1 (1986); (b) Ruiz, L; Lopez, J.; Rodrigues, V.; Peres, J.; de Arellano, M.C.R.; Lopez, B. *J. Chem. Soc., Dalton Tran.* **2683** (2001).
352. Kukushkin, V.Yu.; Tudela, D.; Pombeiro, A.J.L. *Coord. Chem. Rev.* **156**, 333 (1996).
353. Kukushkin, V.Yu.; Pombeiro, A.J.L. *Coord. Chem. Rev.* **181**, 147 (1999).
354. Simonov, Yu.A.; Bourosh, P.N.; Bologa, O.A.; Mazus, M.D.; Gebreleu, N.V. *Zhurn. Strukt. Khim.* **40**(1), 85 (1999).
355. House, D.A. In: Ref. 1, vol. 2, p. 23.
356. Garnovskii, A.D.; Garnovskii, D.A.; Burlov, A.S.; Vasilchenko, I.S. *Mendeleev Chem. J.* **40**(4/5), Part I, 19 (1996).
357. Garnovskii, A.D.; Olekhnovich, L.P.; Alexeenko, V.A.; Lukova, O.A.; Komissarov, V.N. *Zhurn. Obsh. Khim.* **60**, 1920 (1990).

358. Olekhnovich, R. Ya.; Korobov, M.C.; Lyubchenko, S.N.; Sukholenko, E.V.; Riskina, T.A.; Nivorozhkin, L.E.; Olekhnovich, L.P. *Zhurn. Obsh. Khim.* **62**, 901 (1992).
359. Garnovskii, A.D.; Lukova, O.A., Alexeenko, V.A.; Komissarov, V.N.; Litvinov, V.V.; Kharabaev, N.N. *Zhurn. Neorg. Khim.* **37**, 1299 (1992).
360. Olekhnovich, I.P., Lyubchenko, I.N.; Kurbatov, S.V.; Kogan, V.A. *Ross. Khim. Zhurn.* **40**(4/5), 139 (1996); *Mendeleev J. Chem.* **40**(4/5), 217 (1996).
361. Matsumoto, N.; Asakawa, M.; Nogami, H.; Higuchi, M.; Ohyoshi, A. *J. Chem. Soc., Dalton Trans.* 101 (1985).
362. Castro, J.A.; Vilasandes, J.E.; Romero, J.; García-Vázquez, J.A.; Durán, M.; Sousa, A.; Castellano, E.E.; Zuckerman-Schpector, J. *Z. Anorg. Allg. Chem.* **612**, 83 (1992).
363. Castro, J.A.; Romero, J.; García-Vázquez, J.A.; Castineiras, A.; Durán, M.; Sousa, A. *Z. Anorg. Allg. Chem.* **615**, 155 (1992).
364. Castro, J.A.; Romero, J.; García-Vázquez, J.A.; Sousa, A.; Castellano, E.E.; Zuckerman-Schpector, J. *Polyhedron.* **12**(1), 31 (1993).
365. Castro, J.A.; Romero, J.; García-Vázquez, J.A.; Castineiras, A.; Sousa, A. *Z. Anorg. Allg. Chem.* **619**(3), 601 (1993).
366. Castro, J.A.; Romero, J.; García-Vázquez, J.A.; Sousa, A.; Castellano, F.E.; Zuckerman-Schpector, J. *J. Coord. Chem.* **30**(2), 165 (1993).
367. Castro, J.A.; Romero, J.; García-Vázquez, J.A.; Macias, A.; Sousa, A.; Englert, U. *Polyhedron* **12**(11), 1391 (1993).
368. Castro, J.A.; Romero, J.; García-Vázquez, J.A.; Macias, A.; Sousa, A.; Englert, U. *Acta Crystallogr. C.* **50**, 369 (1994).
369. Castro, J.A.; Romero, J.; García-Vázquez, J.A.; Castineiras, A.; Sousa, A. *Trans. Met. Chem.* **19**, 343 (1994).
370. Davies, W.M.; Roberts, M.M.; Zask, A.; Nakanishi, K.; Nozoe, T.; Lippart, S.J. *J. Am. Chem. Soc.* **197**, 3864 (1985).
371. Davies, W.M.; Zask, A.; Nakanishi, K.; Nozoe, T.; Lippart, S.J. *Inorg. Chem.* **23**, 3737 (1985).
372. Zask, A.; Gonnella, N.; Nakanishi, K.; Turner, C.J.; Iwao, S. *Inorg. Chem.* **25**, 3400 (1986).
373. Zaletov, V.G.; Alexeenko, V.A.; Kogan, V.A.; Garnovskii, A.D.; Osipov, O.A. *Zhurn. Neorg. Khim.* **23**, 1852 (1978).
374. Kwiatkowski, M.; Bandoli, G. *J. Chem. Soc., Dalton Trans.* No. 3, 379 (1992).
375. Green, M.; Tasker, P.A. *J. Chem. Soc. A* 2531 (1970).
376. Green, M.; Tasker, P.A. *Inorg. Chim. Acta* **5**, 791 (1971).
377. Karlsson, R.; Engelhardt, E.M.; Green, M. *J. Chem. Soc., Dalton Trans.* 2463 (1972).
378. Losman, D.; Engelgardt, L.M.; Green, M. *Inorg. Nucl. Chem. Lett.* **9**, 791 (1973).
379. Hall, D.; Waters, T.N.; Wright, P.E. *J. Chem. Soc., Dalton Trans.* 1508 (1973).
380. Bailey, N.A.; McKenzie, E.D.; Worthington, J.M. *J. Chem. Soc., Dalton Trans.* 1363 (1974).
381. Browmaker, G.A.; Waters, T.N.; Wright, P.E. *J. Chem. Soc., Dalton Trans.* 867 (1975).
382. McKenzie, E.D.; Worthington, J.M. *Inorg. Chim. Acta* **16**, 9 (1976).
383. Knapp, S.; Keenan, T.P.; Zhang, X.; Fikar, R.; Potenza, J.A.; Schugar, H.J. *J. Am. Chem. Soc.* **112**(9), 3452 (1990).
384. Nivorozhkin, A.L.; Tofflund, H.; Jorgensen, P.L.; Nivorozhkin, A.L. *J. Chem. Soc., Dalton Trans.* 1215 (1996).
385. Hennig, H. *Z. Chem.* **11**, 81 (1971).
386. Serbina, S.V.; Kurbatov, V.P.; Osipov, O.A. *Zhurn. Neorg. Khim.* **16**, 1089 (1971).
387. Kurbatov, V.P.; Serbina, S.V.; Osipov, O.A. *Zhurm. Obsh. Khim.* **43**, 701 (1973).
388. Kurbatov, V.P.; Serbina, S.V.; Osipov, O.A. *Zhurn. Neorg. Khim.* **18**, 2662 (1973).
389. Torgova, T.V.; Osipov, O.A.; Kurbatov, V.P.; Belonogov, N.A. *Zhurn. Obsh. Khim.* **47**, 896 (1977).

390. Torgova, T.G.; Osipov, O.A.; Kurbatov, V.P.; Visotskii, V.D. *Zhurn. Obsh. Khim.* **49**, 439 (1979).
391. Torgova, T.V.; Kurbatov, V.P.; Osipov, O.A. *Zhurn. Neorg. Khim.* **26**, 120 (1981).
392. Kirichenko, A.V.; Kurbatov, V.P.; Osipov, O.A. *Zhurn. Obsh. Khim.* **55**, 688 (1985).
393, Knorr, R.; Polzer, H.; Bischler, E. *J. Am. Chem. Soc.* **97**, 643 (1975).
394. Knorr, R.; Weiss, A.; Polzer, H.; Bischler, E. *J. Am. Chem. Soc.* **99**, 650 (1977).
395. Knorr, R.; Ruf, F. *J. Am. Chem. Soc.* **101**, 5429 (1979).
396. Knorr, R.; Weiss, A. *Chem. Ber.* **114**, 2104 (1981).
397. Knorr, R.; Ruf, R. *Angew. Chem.* **23**, 350 (1984).
398. Sheldrick, W.S.; Knorr, R.; Polzer, H. *Acta Crystallogr., B* **35**, 739 (1979).
399. Dessy, G.; Fares, S. *Cryst. Struct. Comm.* **8**, 101 (1979).
400. Price, R. *J. Chem. Soc. A.* 2048 (1967).
401. Price, R. In: *Chemistry of Synthetic Pigments.* (Edit. Venkataraman, K.). Chemistry: Moscow, 1946 (1974), vol. 3.
402. Alexeenko, V.A.; Zhuchenko, T.A.; Kogan, V.A. *Zhurn. Obsh. Khim.* **42**, 578 (1972).
403. Aldoshin, S.M.; Alexeenko, V.A.; Atovmyan, L.O.; Dyachenko, O.N.; Kogan, V.A.; Kochin, S.G.; Osipov, O.A. *Koord. Khim.* **1**, 1075 (1975).
404. (a) Garnovskii, A.D.; Burlov, A.S.; Antsyshkina, A.S.; Divaeva, L.N. *Zh. Neorg. Khim.* **41**, 90 (1996); (b) Burlov, A.S.; Antsishkina, A.S.; Sadikov, G.G.; Divaeva, L.N.; Garnovskii, A.D.; Sergienko, V.S. *Koord. Khim.* **26**, 691 (2000).
405. Spacu, P.; Banciu, A.C.; Lipadatu, C.J. *Rev. Roum. Chim.* **18**, 1143 (1973).
406. Banciu, A.C.; Lipadatu, C.J. *Z. Phys. Chem. (BDR)* **97**(3/4), 197 (1975).
407. Balakrisnan, K.P.; Kirshnan, V. *Trans. Met. Chem.* **3**, 337 (1978).
408. Balakrisnan, K.P. *J. Chem. Eng. Data* **25**, 187 (1980).
409. Banciu, A.C. *Rev. Roum. Chim.* **33**, 225 (1988).
410. Potapov, V.M.; Panova, G.V.; Turovets, N.M.; Kasumov, R.D.; Vikulova, N.K. *Zhurn. Obsh. Khim.* **53**, 1353 (1983).
411. Fomicheva, E.B.; Belskii, V.G.; Panova, G.V. *Zhurn. Obsh. Khim.* **56**, 607 (1986).
412. Fomicheva, E.B.; Belskii, V.G.; Panova, G.V. *Koord. Khim.* **15**, 209 (1989).
413. Fomicheva, E.B.; Belskii, V.G.; Panova, G.V. *Koord. Khim.* **15**, 403 (1989).
414. Fomicheva, E.B.; Belskii, V.G.; Panova, G.V. *Koord. Khim.* **15**, 409 (1989).
415. Fomicheva, E.B.; Belskii, V.G.; Panova, G.V. *Koord. Khim.* **15**, 524 (1989).
416. Zakharov, A.N.; Panova, G.V.; Zefirov, N.S. *Zhurn. Obsh. Khim.* **65**, 10 (1995).
417. Buzikin, B.I.; Lipunova, G.N.; Sisoeva, L.P.; Rusinova, L.I. *Chemistry of Phormazanes.* Nauka: Moscow, 1992, 375 pp.
418. Lipunova, G.N.; Rusinova, L.I.; Maslakova, T.I. *Ross. Khim. Zhurn.* **40**(4/5), 148 (1996); *Mendeleev Chem. J.* **40**(4/5), Part II, 230 (1996).
419. Dale, H. *J. Chem. Soc. A* 278 (1967).
420. Bednyagina, N.P.; Postovskii, I.Ya.; Garnovskii, A.D.; Osipov, O.A. *Usp. Khim.* **44**, 1052 (1975).
421. Renkema, W.E.; Lute, C.N.; Stam, C.H. *Acta Crystallogr., B* **35**, 75 (1979).
422. Meuldijk, J.; Renkema, W.E.; Herk, A.M.; Stam, C.H. *Acta. Crystallogr., C* **39**, 1536 (1983).
423. Balt, S.; Meuldijk, J.; Renkama, W.E. *Trans. Met. Chem.* **5**, 357 (1980).
424. Siedle, A.R.; Pignolet, L.H. *Inorg. Chem.* **19**, 2052 (1980).
425. Balt, S.; Klok, C.; Meuldijk, J. *Acta Crystallogr., C* **41**, 528 (1985).
426. Batsanov, A.S.; Struchkov, Yu.T.; Sevego, Kh.M., Zaitsev, B.E.; Ivlev, V.I. *Zhurn. Neorg. Khim.* **32**, 93 (1987).
427. Polyakova, I.N.; Starikova, Z.A.; Sulim, O.P.; Lipunova, G.N. *Kristalogr.* **37**, 653 (1992).
428. Kalia, K.S.; Kumar, A.; Single, M. *Indian J. Chem. Soc.* **20A**, 610 (1981).
429. Batsanov, A.S.; Gambashidze, L.M.; Davidova, C.L.; Rapoport, L.M.; Struchkov, Yu.T. *Koord. Khim* **14**, 94 (1988).

430. Krol, I.A.; Agre, V.M.; Shmelev, L.V.; Stopnikova, M.N. *Koord. Khim.* **14**, 259 (1988).
431. Shmelev, L.V.; Stopnikova, M.N.; Ryabokobilko, Yu.S. *Koord. Khim.* **15**, 1027 (1989).
432. Takhirov, T.G.; Alexandrov, Yu.I.; Lipunova, G.N.; Rusinova, L.I.; Klyuev, N.A.; Kozina, O.A.; Dyachenko, O.A.; Atovmyan, L.O. *Zhurn. Obsh. Khim.* **59**, 2362 (1989).
433. Shmelev, L.V.; Lipunova, G.N.; Kessenih, A.V.; Pervova, I.G.; Polyakova, I.N.; Starikova, Z.A.; Lipunov, I.N. *Koord. Khim.* **19**, 215 (1993).
434. Shmelev, L.V.; Pervova, I.G.; Lipunova, G.N.; Kessenih, A.V.; Polyakova, I.N.; Starikova, Z.A.; Lipunov, I.N. *Koord. Khim.* **19**, 904 (1993).
435. Müller, A.; Drown, D.; Lipunova, G.N. *Z. Naturforsch.* **81b**, 136 (1994).
436. Vrieze, K.; Van Koten, G. In: Ref. 1, vol. 2, p. 189.
437. Trofimenko, S. *Acc. Chem. Res.* **4**, 17 (1971); *Usp. Khim.* **41**, 1660 (1972).
438. Trofimenko, S. *Progr. Inorg. Chem.* **34**, 115 (1986).
439. Trofimenko, S. *Top. Curr. Chem.* **131**, 1 (1986).
440. Trofimenko, S. *Chem. Rev.* **93**, 943 (1993).
441. (a) Kitajima, N.; Moro-oka, Y. *Chem. Rev.* **94**, 735 (1994); (b) Kitajima, N.; Tolman, W.R. *Progr. Inorg. Chem.* **43**, 419 (1995).
442. (a) Parkin, G. *Adv. Inorg. Chem.* **42**, 219 (1995); (b) Reger, D.L. *Coord. Chem. Rev.* **147**, 571 (1996); (c) Etienne, M. *Coord. Chem. Rev.* **156**, 201 (1996); (d) Trofimenko, S. *Scorpionates: The Coordination Chemistry of Polypyrasolyl-Borate Ligands.* Imperial College Press: Wilmington, 1999; (e) Sadimenko, A.P. *Adv. Heterocycl. Chem.* **81**, 167 (2001).
443. Shaver, A. In: Ref. 1, vol. 2, p. 254.
444. Eagle, A.A.; Mackay, M.F.; Vuong, C.G. *Inorg. Chem.* **30**, 1425 (1991).
445. Doyle, R.A.; Daniels, L.M.; Angelici, R.J. *J. Am. Chem. Soc.* **111**, 4995 (1989).
446. Kitajima, N.; Fujisawa, K.; Tanaka, M.; Moto-oka, V. *J. Am. Chem. Soc.* **114**(23), 9232 (1997).
447. Qui, D.; Kilpatrick, L.T.; Kitajima, N.; Spiro, T.G. *J. Am. Chem. Soc.* **116**(6), 2585 (1997).
448. Mandal, S.; Das, G.; Singh, R.; Shukla, R.; Bharadwaj, P.K. *Coord. Chem. Rev.* **160**, 191 (1997).
449. Larin, G.M.; Zvereva, G.A. *Koord. Khim.* **22**, 310 (1996).
450. Yatsimirskii, K.B.; Lampeka, Ya.D. *Physico-Chemical Properties of the Complexes with Macrocyclic Ligands.* Naukova Dumka: Kiev, 1985.
451. *Synthesis of Macrocycles.* (Edit. Izaff, R.M.; Christensen, J.J.). John Wiley: New York, 1987.
452. *The Chemistry of Macrocyclic Ligand Complexes.* Cambridge University Press: Cambridge, 1989.
453. *Vitramin B_{12} and B_{12}-Proteins.* (Edit. Kräutler, B.; Golding, B.T.; Arigoni, D.). John Wiley: New York, 1998.
454. Thomas, A.L. *Phtalocyanines. Research & Applications.* CRC Press: Boca Raton, FL, 1990.
455. *Phthalocyanines. Properties & Application.* (Edit. Leznoff, C.C.; Lever, A.B.P.). Verlag Chemie: Weinheim, 1993, 1996.
456. Chembright, P. *Usp. Khim.* **46**, 1207 (1977); *Coord. Chem. Rev.* **6**, 247 (1971).
457. Berezin, B.D.; Koifman, O.I. *Usp. Khim.* **49**, 2389 (1980).
458. Gilard, R.; Kadish, K.M. *Chem. Rev.* **88**, 1121 (1988).
459. (a) Berezin, B.D. *Zhurn. Neorg. Khim.* **37**, 1260 (1992); (b) Berezin, B.D.; Berezin, M.D.; Berezin, D.B. *Ross. Khim. Zhurn.* **41**(3), 105 (1997).
460. Bencini, A.; Bianchi, A.; Paoletti, P. *Coord. Chem. Rev.* **120**, 51 (1992).
461. Alexander, V. *Chem. Rev.* **95**, 273 (1995).
462. Stuzhin, P.A.; Khelevina, O.G. *Coord. Chem. Rev.* **147**, 41 (1996).
463. Kudrevich, S.V.; Van Lier, J.E. *Coord. Chem. Rev.* **156**, 163 (1996).

464. Takemura, H.; Shinmyozu, T.; Inazu, T. *Coord. Chem. Rev.* **156**, 183 (1996).
465. Gould, R.D. *Coord. Chem. Rev.* **156**, 237 (1996).
466. Reiter, W.A.; Gerges, A.; Lee, S.; Deffo, T.; Clifford, T.; Danby, A.; Bowman-James, K. *Coord. Chem. Rev.* **174**, 343 (1998).
467. Elias, H. *Coord. Chem. Rev.* **1987**, 37 (1999).
468. (a) Marquez, V.E.; Anacona, J.R.; Rodríguez-Barbarin, C. *Polyhedron* **20**, 1885 (2001); (b) Marquez, V.E.; Anacona, J.R. *Polyhedron* **16**, 2375 (1997); (c) Marquez, V.E.; Anacona, J.R. *Trans. Met. Chem.* **25**, 188 (2000); (d) Marquez, V.E.; Anacona, J.R. *J. Coord. Chem.* **49**, 281 (2000).
469. Mashinko, T.; Polphin, D. In: Ref. 1, vol. 2, p. 813.
470. Guillamont, D.; Finger, K.; Hackey, M.R.; Daniel, C. *Coord. Chem. Rev.* **171**, 439 (1998).
471. Beck, J.; Strähle, J. *Angew. Chem.* **97**, 419 (1985).
472. Adams, H.; Bailly, N.A.; Fenton, D.E.; Moss, S. *Inorg. Chim. Acta* **83**, L79 (1984).
473. Adams, H.; Bailey, N.A.; Fenton, D.E.; Moss, S.; Rodríguez de Barbarín, C.O.; Jones, G. *J. Chem. Soc, Dalton Trans.* 693 (1986).
474. Gulter, A.R.; Dolphin, D.J. *Can. J. Chem.* **55**, 3062 (1977).
475. Garnovskii, A.D.; Burlov, A.S.; Antsishkina, A.S.; Divaeva, L.N. *Koord. Khim.* **41**, 90 (1996).
476. Doslik, N.; Sixt, T.; Kaim, W. *Angew. Chem.* **110**, 2521 (1998).
477. Addison, A.W.; Stenhouse, J.H. *Inorg. Chem.* **17**, 2161 (1978).
478. Sieler, J.; Hennig, H.; *Z. Anorg. Allg. Chem.* **381**, 219 (1971).
479. Erbe, J.; Beck, W. *Chem. Ber.* **116**, 3867 (1983).
480. Hay, O.P. *Coord. Chem. Rev.* **126**, 177 (1993).
481. *Transition Metal Complexes of Phosphorus, Arsenic, and Antimoni Ligands.* (Edit. McAuliffe, C.A.). McMillan: London, 1973.
482. Chaudret, B.; Delavaux, B.; Roiblanc, R. *Coord. Chem. Rev.* **86**, 1991 (1988).
483. Davies, J.A.; Dutremez, S. *Coord. Chem. Rev.* **114**(1), 61 (1992).
484. Bianchini, C.; Meli, A.; Peruzzini, M.; Vizza, F.; Zanobini, F. *Coord. Chem. Rev.* **120**, 193 (1992).
485. Saxena, A.K.; Hosmane, N.S. *Chem. Rev.* **93**, 1081 (1993).
486. Anderson, G.K. *Adv. Organomet. Chem.* **35**, 1 (1993).
487. (a) Dias, P.B.; Minas de Piedade, M.E.; Martinho, B. *Coord. Chem. Rev.* **135/136**, 737 (1994); (b) Andersen, N.G.; Keay, B. *Chem. Rev.* **101**, 997 (2001).
488. Quin, L.D. *Coord. Chem. Rev.* **137**, 525 (1994).
489. (a) Norman, N.C.; Pickett, N.L. *Coord. Chem. Rev.* **145**, 27 (1995); (b) Bessel, C.A.; Aggarwal, P.; Marschilok, A.C.; Takeuchi, K.J. *Chem. Rev.* **101**, 1031 (2001); (c) Lledos, A.; Jean, Y. *Chem. Phys. Lett.* **287**, 243 (1998); (d) Demachy, I.; Jean, Y.; Lledos, A. *Chem. Phys. Lett.* **303**, 621 (1999); (e) Nakai, H.; Fukada, S.; Nakatsuji, H. *J. Phys. Chem. A. (Molecules, Spectroscopy, Kinetics, Environment & General Theory)* **101**(6), 973 (1997); (f) Su, M.-D.; Chu, S.-Y. *J. Am. Chem. Soc.* **119**(42), 10178 (1997).
490. Nixon, J.F. *Coord. Chem. Rev.* **145**, 201 (1995).
491. Gimeno, M.C.; Laguna, A. *Chem. Rev.* **97**(3), 511 (1997).
492. Le Froch, P.; Mathey, F. *Coord. Chem. Rev.* **178–180**, 771 (1998).
493. Caminade, A.-M.; Laurent, R.; Chaudert, B.; Majoral, J.P. *Coord. Chem. Rev.* **178–180**, 793 (1998).
494. Casas, J.M.; Fornies, J.; Martínez, T.M.; Rueda, F.; Welch, A.J. *Inorg. Chem.* **38**(7), 1529 (1999).
495. Gavrilov, K.N.; Mikhel, I.S. *Koord. Khim.* **25**, 83 (1999).
496. Weber, L. *Coord. Chem. Rev.* **158**, 1 (1997).
497. Borm, J.; Huttner, G.; Orama, O.; Zsolnai, L. *J. Organomet. Chem.* **282**, 53 (1985).
498. Mathey, F. *Chem. Rev.* **90**(6), 997 (1990).

499. Weber, L.; Kaminski, O.; Stammler, H.-G.; Neumann, B.; Romanenko, V.D. *Z. Naturforsch. B* **48**(12), 1784 (1993).
500. Mathey, F. *New J. Chem.* **11**, 585 (1987).
501. (a) Mathey, F. *Chem. Rev.* **88**, 429 (1988); (b) Cloke, F.G.N.; Hitchcock, P.B.; Nixon, J.F.; Vickers, D.V. *J. Organomet. Chem.* **212**, 635 (2001).
502. Scherer, O.J. In: *Multiple Bond and Low Coordination in Phosphorus Chemistry*. (Edit. Regiitz, M.; Scherer, O.J.). Thieme: Stuttgart, 1990.
503. Ginzburg, A.G.; Batsanov, A.S.; Struchkov, Yu.T. *Metallorg. Khim.* **4**, 854 (1991).
504. (a) Mathey, F. *J. Organomet. Chem.* **400**, 149 (1990); (b) Hitchcock, P.B.; Meidine, M.F.; Nixon, J.F.; Sillett, G.J.D. *J. Chem. Soc., Chem. Commun.* 317 (1990).
505. Ring, B.; Scherer, O.J.; Seckmann, G.; Wolmerhäuser, G. *Chem. Ber.* **125**(5), 1011 (1992).
506. Maigrot, N.; Recard, L.; Charrier, C.; Mathey, F. *Angew. Chem., Int. Ed. Engl.* **31**, 1031 (1992).
507. Hitchcock, P.B.; Matos, R.M.; Nixon, J.F. *J. Organomet. Chem.* **462**(1/2), 319 (1993).
508. Weber, L.; Sommer, L.; Stammler, H.G.; Kolle, U. *Chem. Ber.* **128**(7), 665 (1995).
509. Hitchcock, P.B.; Johnson, J.A.; Nixon, J.F. *Organometallics* **14**(9), 4382 (1995).
510. Detzel, M.; Friedrich, G.; Schrerer, O.J.; Wolmershäuser, G. *Angew. Chem., Int. Ed. Engl.* **34**(12), 1321 (1995).
511. Scherer, O.J.; Detzel, M.; Berg, G. *Phosph. Sulf. Silic.* **109/110**, 13 (1996).
512. Garnovskii, A.D.; Sadimenko, A.P.; Uraev, A.I.; Vasilchenko, I.S.; Garnovskii, D.A. *Russ. J. Coord. Chem.* **26**, 301 (2000).
513. Streubeb, H.-M.; Schiemann, U.; Jones, P.G.; Grunenburg, J.; Schiebel, H.-M.; Gugat, D. *Angew. Chem.* **113**, 2531; *Int. Ed. Engl.* **40**, 2471 (2001).
514. Weber, L.; Sommer, O.; Stammler, H.G.; Neumann, B. *Z. Anorg. Allg. Chem.* **622**(3), 543 (1996).
515. Vahrenkamp, H.; Nöth, H. *Chem. Ber.* **106**, 2227 (1973).
516. Ashe III, A.J.; Butler, W.; Colburn, J.C.; Obu-Arabi, S. *J. Organomet. Chem.* **282**, 233 (1985).
517. Chernega, A.N.; Antipin, M. Yu.; Struchkov, Yu.T. *Zhurn. Strukt. Khim.* **29**(2), 122 (1988).
518. Märkl, G. In: *Miltiplete Bond and Low Coordination in Phosphorus Chemistry* (Edit. Regliitz, M.; Scherer, O.J.). Thieme: Stuttgart, 1990.
519. Le Floch, P.; Ricard, L.; Mathey, F.; Jatand, A.; Amatore, C. *Inorg. Chem.* **34**(1), 11 (1995).
520. Gudat, D. *Coord. Chem. Rev.* **163**, 71 (1997).
521. Champness, N.R.; Levason, W. *Coord. Chem. Rev.* **133**, 115 (1994).
522. Bradley, D.C.; Mehrotra, R.C.; Gaur, D.P.; Singh, A. *Metal Alkoxides*. Academic Press: New York, London, 1978.
523. Mehrotra, R.C.; Agarwal, S.K.; Singh, Y.P. *Coord. Chem. Rev.* **68**, 101 (1985).
524. (a) Mehrotra, R.C.; Singh, V.P.; Tripathi, U.M. *Chem. Rev.* **91**(6), 1287 (1991); (b) *Facets of Coordination Chemistry*. (Edit. Agarvala, B.V.; Munshi, K.N.). World Scientific: Singapore, New Jersey, London, 1993, 253 pp.
525. Healy, M.D.; Power, M.B.; Barron, A.R. *Coord. Chem. Rev.* **130**(1/2), 63 (1994).
526. Pisarevskii, A.P.; Martinenko, L.I. *Koord. Khim.* **20**, 324 (1994).
527. Turevskaya, E.P.; Kozlova, N.I.; Turova, N.Ya.; Belokon, A.I.; Berdiev, D.B.; Kessler, V.G.; Grishin, Yu.K. *Russ. Chem. Bull.* **44**, 735 (1995).
528. (a) Yamashita, M.; Ito, H.; Ito, T. *Inorg. Chem.* **2**, 2102 (1983); (b) Hartmann, E.; Strahle, J. *Z. Anorg. Allg. Chem.* **583**, 31 (1990); (c) Bernardinelli, G.; Kubel-Pollak, A.; Ruttimann, S.; Williams, A.F. *Chimia* **46**, 115 (1992); (d) Lee, S.C.; Holm, R.H. *Inorg. Chem.* **32**, 4745 (1993).
529. (a) Van Albada, G.A.; Lakin, M.T.; Veldman, N.; Spek, A.L.; Reedijk, J. *Inorg. Chem.* **34**, 4910 (1995); (b) Van Albada, G.A.; Smeets, W.J.J.; Spek, A.L.; Reediijk, J. *Inorg.*

Chim. Acta **260**, 151 (1997); (c) Komaei, S.A.; Van Albada, G.A.; Mutikainen, P.; Turpeinen, U.; Reedijk, J. *Eur. J. Inorg. Chem.* **37**, 1577 (1998); (d) Komaei, S.A.; Van Albada, G.A.; Haasnoot, J.G.; Kooijman, H.; Spek, A.L.; Reedijk, J. *Inorg. Chim. Acta* **286**, 24 (1999); (e) Komaei, S.A.; Van Albada, G.A.; Mutikainen, P.; Turpeinen, U.; Reedijk, J. *Polyhedron* **18**(7), 1991 (1999).

530. Goodgame, D.J.L.; Hussain, I.; White, A.J.P.; Williams, D.J. *J. Chem. Soc., Dalton Trans.* No. 17, 2899 (1999).
531. Fleeting, K.A., O'Brien, P.; Otway, D.J.; White, A.J.P.; Williams, D.J.; Jones, A.C. *Inorg. Chem.* **38**(7), 1432 (1999).
532. Porai-Koshitz, M.A. *Results of Science & Technology. Crystallochemistry.* VINITI: Moscow, 1981, vol. 15, p. 30.
533. Mehrotra, R.C.; Bohra, R.C. *Metal Carboxylates.* Academic Press: London, 1983.
534. Hulme, C.E.; Watkinson, M.; Haynes, M.; McAuliffe, C.A.; Prichard, R.G.; Jaboon, N.; Beagley, B.; Sousa, A.; Bermejo, M.R.; Fondo, M. *J. Chem. Soc., Dalton Trans.* 1805 (1997).
535. Watkinson, M.; Fondo, M.; Bermejo, M.R.; Sousa, A.; McAuliffe, C.A.; Pritchard, R.G.; Jaiboon, N.; Aurangzev, N.; Naeem, M. *J. Chem. Soc., Dalton Trans.* No. 1, 31 (1999).
536. Vasilchenko, I.S.; Antsyshkina, A.S.; Burlov, A.S.; Uraev, A.I.; Novorozhkin, A.L.; Garnovskii, D.A.; Sergienko, V.S.; Kurbatov, V.P.; Korshunov, O.Yu.; Garnovskii, A.D. *Russ. J. Inorg. Chem.* **44**, 1205 (1999).
537. Garnovskii, D.A.; Antsishkina, A.S.; Sadikov, G.G.; Kurbatov, V.P.; Eliseeva, A.Yu.; Vasilchenko, I.S.; Garnovskii, A.D. *Zhuirn. Strukt. Khim.* **41**, 573 (2000).
538. Mehrotra, R.C.; Rohra, R.; Gaur, D.P. *Metal β-Diketonates and Allied Derivatives.* Academic Press: New York, 1978.
539. *Problems of Chemistry and Applications of Metal β-Diketonates.* (Edit. Spitsin, V.I.). Science: Moscow, 1982.
540. Shkolnikova, L.M.; Porai-Koshitz, M.A. *Results of Science & Technology. Crystallochemistry.* VINITI: Moscow, 1982, vol. 16, p. 117.
541. *Metal β-Diketonates.* (Edit. Martinenko, L.I.). DGU Press: Vladivostok, 1990, vol. 1; 1991, vol. 2.
542. Martinenko, L.I. *Usp. Khim.* **60**, 1969 (1991).
543. Wong, C.Y.; Woollins, J.D. *Coord. Chem. Rev.* **130**(1/2), 175 (1994).
544. Il'in, E.G.; Buslaev, Yu.A. *Mendeleev Chem. J.* **40**(4/5), Part II, 98 (996).
545. Martinenko, L.I.; Kuzmina, N.P.; Grigorieva, A.N. *Mendeleev Chem. J.* **40**(4/5), Part II, 174 (1996).
546. (a) Rees Jr., W.S.; Carris, M.W.; Hesse, W. *Inorg. Chem.* **30**(24), 4479 (1991); (b) Otway, D.; Rees Jr., W.S. *Coord. Chem. Rev.* **210**, 279 (2000).
547. Rossetto, G.; Polo, F.; Benetollo, F.; Porchia, M.; Zanella, P. *Polyhedron* **11**, 979 (1992).
548. Drozdov, A.A.; Troyanov, S.I. *Polyhedron* **12**, 2877 and 2973 (1993).
549. Glez, A.; Drozdov, A.A.; Troyanov, S.I. *Koord. Khim.* **20**, 922 (1994).
550. Martinenko, L.I.; Kuzmina, I.P.; Grigoriev, A.N. *Koord. Khim.* **21**, 515 (1995).
551. Kabachnik, M.I. *Zhurn. Vses. Khim. Obsh. im Mendeleeva* **29**(3), 32 (1984).
552. Popov, L.D.; Shvetz, A.A.; Kogan, V.A. *Koord. Khim.* **15**, 1299 (1989).
553. Hanusa, T.P. *Chem. Rev.* **93**, 1023 (1993).
554. Poonia, N.S.; Bajaj, A.V. *Chem. Rev.* **79**, 389 (1979).
555. Vano, S. *Coord. Chem. Rev.* **92**, 113 (1988).
556. Angyal, J. *Adv. Carbohydr. Chem. Biochem.* **47**, 1 (1989).
557. Alexeev, Yu.E.; Garnovskii, A.D.; Burlov, A.S.; Zhdanov, Yu.A. *Mendeleev Chem. J.* **40**(4/5), Part II, 242 (1996).
558. (a) Alexeev, Yu.E.; Garnovskii, A.D.; Zhdanov, Yu.A. *Russ. Chem. Rev.* **67**(8), 649 (1998; (b) Zhadonov, Yu. A.; Alexeev, Yu. E. *Russ. Chem. Rev.* **71**(11), 1090 (2002);

(c) Alexeev, Yu.E.; Garnovskii, A.D.; Vasilchenko, I.S.; Zhadanov, Yu. A. *Russ. J. Inorg. Chem.* (Suppl. Rev.) **46**, 235 (2001)).
559. Gyurcsik, B.; Nagy, L. *Coord. Chem. Rev.* **203**, 81 (2000).
560. *Chemical Encyclopedia.* Soviet Encyclopedia: Moscow, 1990, vol. 2, 671 pp.
561. Pedersen, K.D.; Frensdorf, H.K. *Usp. Khim.* **42**, 492 (1973); *Angew. Chem., Int. Ed. Engl.* **11**, 16 (1972).
562. *Synthetic Multidentate Macrocyclic Compounds.* (Edit. Izatt, R.M.; Christensen, J.J.). Academic Press: New York, 1978.
563. Datley, N.K. In: *Synthetic Multidentate Macrocyclic Compounds.* (Edit. Izatt, R.M.; Christensen, J.J.). Academic Press: New York, 1978.
564. *Progress in Macrocyclic Chemistry.* (Edit. Izatt, R.M.; Christensen, J.J.). John Wiley: New York, 1979, vol. 1; 1981, vol. 2.
565. *Coordination Chemistry of Macrocyclic Compounds.* (Edit. Melson, G.A.). Plenum Press: New York, 1979.
566. *Host–Guest Complexes Chemistry.* Springer-Verlag Chemie: Berlin, 1981.
567. Gokel, G.W.; Korzeniowski, S.H. *Macrocyclic Polyether Synthesis.* Verlag Chemie: Berlin, 1982.
568. Hiraoka, M. *Crown Compounds. Their Characteristics & Applications.* Kadansha Ltd: Tokyo, Amsterdam, Oxford, New York, 1982.
569. (a) Izatt, R.M.; Brandshaw, J.S.; Nielsen, S.A.; Lamband, J.D.; Christensen, J.J. *Chem. Rev.* **85**, 271 (1985); (b) Isatt, R.M.; Pawlak, K.; Brandshaw, J.S. *Chem. Rev.* **95**, 2529 (1995).
570. Bogatskii, A.V.; Lukianenko, N.G.; Kirichenko, T.I. *Zhurn. Vses. Khim. Obsh. Im. Mendeleeva* **30**, 487 (1985).
571. Tsivadze, A.Yu.; Varnek, A.A.; Khutorskii, V.E. *Coordination Compounds of Metals with Crown Ethers.* Science: Moscow, 1991.
572. (a) Tsivadze, A.Yu. *Mendeleev Chem. J.* **40**(4/5), Part I, 62 (1996); (b) Bren, V.A. *Usp. Khim.* **70**, 1152 (2001).
573. Truter, M.K. *Struct. Bond.* **16**, 113 (1973).
574. Izatt, R.M.; Eatough, D.J.; Christensen, J.J. *Struct. Bond.* **16**, 161 (1973).
575. Fögtle, F.; Weber, E. *Angew. Chem.* **91**, 813 (1979).
576. Fögtle, F. *Zhurn. Vses. Khim. Obsh. im Mendeleeva* **30**, 500 (1985).
577. Markovich, I.S.; Dziomko, V.M. *Zhurn. Vses. Khim. Obsh. im Mendeleeva* **30**, 562 (1985).
578. *Host–Guest Complexes Chemistry Macrocycles.* (Edit. Vögtle, F.; Weber, E.) Springer-Verlag: Berlin, 1985.
579. Ovchinnikov, Yu.A.; Ivanov, V.T.; Shkrob, A.M. *Membrane-Active Complexones.* Science: Moscow, 1974.
580. Raevskii, O.A. *Usp. Khim.* **59**(3), 375 (1990).
581. *Comprehensive Supramolecular Chemistry.* (Edit. Atwood, J.L.; Davies, J.E.D.; MacNicol, D.A.; Fögtle, F.). Pergamon Press: Oxford, 1996.
582. (a) Mufalige, D.; Ma, E.S.; Rettig, S.J.; James, B.R.; Gullen, W.R. *Inorg. Chem.* **36**(24), 5426 (1997); (b) Hagen, K.S.; Holm, R.H. *J. Am. Chem. Soc.* **104**, 5496 (1982).
583. Koch, S.A.; Maclin, L.E.: Millar, M. *J. Am. Chem. Soc.* **105**, 5944 (1983).
584. Hagen, K.S.; Holm, R.H. *Inorg. Chem.* **23**, 418 (1984).
585. Petilon, F.Y.; Schollhammer, P.; Talarmin, J.; Muir, K.W. *Coord. Chem. Rev.* **178–180**, 203 (1998).
586. Shklover, V.E.; Nagapetyan, S.S.; Struchkov, Yu.T. *Usp. Khim.* **59**, 1179 (1990).
587. Carrillo, D. *Coord. Chem. Rev.* **119**, 137 (1992).
588. Larionov, S.V.; Zemskova, S.M. *Mendeleev Chem. J.* **40**(4/5), Part II, 270 (1996).
589. Murray, S.G.; Martley, F.R. *Chem. Rev.* **81**, 365 (1981).
590. Voronkov, M.G.; Knutov, V.I. *Usp, Khim.* **51**, 1484 (1982).

591. Mori, A.; Kubo, T.; Takeshita, H. *Coord. Chem. Rev.* **148**, 71 (1996).
592. (a) McCleverty, J.A. *Progr. Inorg. Chem.* **10**, 49 (1968); (b) Eisenberg, R.E. *Progr. Inorg. Chem.* **12**, 295 (1970); (c) Runs, R.S.; McAuliffe, C.A. *Adv. Inorg. Chem. Radiochem.* **22**, 303 (1979); (d) Zvereva, G.A.; Larin, G.M. *Koord. Khim.* **24**, 761 (1998); (e) Larin, G.M.; Zvereva, G.A.; Shelokov, R.N. *Dokl. An SSSR 237*, 1107 (1977); (f) Larin, G.M.; Zvereva, G.A. *Dokl. AN SSSR* **2709**, 1404 (1984); (g) Bigoli, F.; Cassoux, P.; Deplan, O.P.; Mercuri, M.L.; Pelligheli, M.A.; Pintus, G.; Serpe, A.; Trogu, E.F. *J. Chem. Soc.*, Dalton Trans, 4639 (2000).
593. Chaudhan, H.P.S. *Coord. Chem. Rev.* **173**, 1 (1998).
594. Yaneff, P.V. *Coord. Chem. Rev.* **23**, 183 (1977).
595. Torn, G.D.; Ludvig, R. *The Dithiocarbamates and Related Compounds*. Elsevier Science: Amsterdam, 1981.
596. Werner, H. *Coord. Chem. Rev.* **43**, 165 (1982).
597. Birko, V.M. *Dithiocarbamates*. Science: Moscow, 1984.
598. Geloso, C.; Kumar, R.; López-Grado, J.; Tuck, D.G. *Can. J. Chem.* **65**, 928 (1987).
599. Zemskova, S.M.; Glinskaya, L.A.; Durasov, V.B.; Klevtsova, R.F.; Larionov, S.V. *Zhurn. Strukt. Khim* **34**(5), 157 (1993).
600. Woo, L.K. *Chem. Rev.* **93**, 1125 (1993).
601. Zeng, D.; Hempden-Smith, M.J.; Alam, T.M.; Rheingold, A.L. *Polyhedron* **13**, 2715 (1994).
602. Jeliazkova, B.G.; Doicheva, M.A. *Polyhedron* **15**(8), 1277 (1996).
603. Jeliazkova, B.G.; Sarova, G.C. *Polyhedron* **16**(23), 3967 (1997).
604. Zemskova, S.M.; Glinskaya, L.A.; Klevtsova, R.F.; Larionov, S.V. *Zhurn. Strukt. Khim.* **37**, 1114 (1996).
605. Plyusnin, V.F., Gorivin, V.P.; Larionov, S.V. *Coord. Chem. Rev.* **159**, 121 (1997).
606. Volkov, V.V.; Il'inchik, E.A.; Volkov, O.V.; Voronina, G.S.; Mikyashev, K.G. *Koord. Khim.* **24**, 902 (1998).
607. Zemskova, S.M.; Prashad, G.; Glinskaya, L.A.; Klevtsova, R.F.; Durasov, V.B.; Tkachev, S.V.; Gromilov, S.A.; Larionov, S.V. *Zhurn. Neorg. Khim.* **43**, 1644 (1998).
608. Varand, V.L.; Glinskaya, L.A.; Klevtsova, R.F.; Larionov, S.V. *Zhurn. Strukt. Khim.* **39**, 300 (1998).
609. Klevtsova, R.F.; Glinskaya, L.A.; Zemskova, S.M.; Larionov, S.V. *Zhurn. Strukt. Khim.* **40**(1), 71 and 77 (1999).
610. (a) Kuzmina, N.P.; Ivanov, R.A.; Ilyukhin, A.B.; Paramonov, S.E. *Koord. Khim.* **25**, 635 (1999); (b) Varand, V.L.; Klevtsova, R.F.; Glinskaya, L.A.; Larionov, S.V. *Koord. Khim.* **26**(11), 869 (2000).
611. Gable, R.; Hockins, B.F.; Winter, G. *Inorg. Chim. Acta* **96**, 151 (1985).
612. Abrahams, B.A.; Hockins, B.F.; Winter, G. *Austr. J. Chem.* **43**, 1759 (1990).
613. Leonova, T.G.; Kirichenko, V.N.; Glinskaya, L.A. *Koord. Khim.* **23**, 590 (1997).
614. Glinskaya, L.A.; Leonova, T.G.; Kirichenko, V.N.; Klevtsova, R.F.; Larionov, S.V. *Zhurn. Strukt. Khim.* **38**(1), 142 (1997).
615. (a) Larionov, S.V.; Glinskaya, L.A.; Leonova, T.G.; Klevtsova, R.F. *Koord. Khim.* **24**, 909 (1998); (b) Leonova, T.G.; Glinskaya, L.A.; Klevtsova, R.F.; Batrachenko, N.I.; Larionov, S.V. *Koord. Khim.* **26**(9), 657 (2000).
616. Larin, G.M. *Mendeleev Chem. J.* **40**(4/5), 112 (1996).
617. Shirshova, L.V.; Korableva, L.G.; Astakhova, A.S.; Lavrentiev, I.P.; Ponomarev, V.I. *Koord. Khim.* **16**(3), 348 (1990).
618. Shirshova, L.V.; Ponomarev, V.I.; Korableva, L.G.; Lavrentiev, I.P. *Koord. Khim.* **16**, 1118 (1990).
619. Geloso, C.; Kumar, R.; Lopez-Grado, J.R.; Tuck, D.G. *Can. J. Chem.* **65**, 928 (1987).
620. Kharisov, B.I.; Blanco, L.M.; Salinas, M.V.; Garnovskii, A.D. *J. Coord. Chem.* **47**, 135 (1999).

621. Victoriano, L.I. *Polyhedron* **19**(22/23), 2269 (2000) (special issue of *Polyhedron*, dedicated exclusively to inorganic and coordination chemistry in Latin America).
622. Angelici, R.J. *Coord. Chem. Rev.* **105**, 61 (1990).
623. Sadimenko, A.P.; Garnovskii, A.D. *XXIII International Conference on Coordination Chemistry* (Book of Abstracts). Florence, Italy, 1998, p. 537.
624. Rao, K.M.; Day, C.L.; Jacobson, R.A.; Angelici, R.J. *Inorg. Chem.* **30**(26), 5046 (1991).
625. Song, B.-L.; Xiong, R.-G., You, X.-Z.; Huang, X.-Y. *Acta Crystallogr., C* **51**, 2258 (1995).
626. Huang, X.-Y.; Xiong, R.-G.; You, X.-Z. *Acta Crystallogr., C* **51**, 2261 (1995).
627. Xiong, R.-G.; You, X.-Z.; Huang, X.-Y. *Acta Crystallogr., C* **51**, 2263 and 2484 (1995).
628. Zuo, J.L.; Xiong, R.-G.; You, X.-Z.; Huang, X.-Y. *Acta Crystallogr., C* **52**, 46 (1996).
629. Xiong, R.-G.; Lui, C.-M.; Li, H.-Z.; You, X.-Z.; Huang, X.-Y. *Acta Crystallogr., C* **52**, 519 (1996).
630. Xiong, R.-G.; Zuo, J.-L.; You, X.-Z.; Huang, X.-Y. *Acta Crystallogr. C* **52**, 521 (1996).
631. (a) Guo, Y.-H.; Xue, Y.-Q.; Xiong, R.-G.; Zuo, J.-L.; You, X.-Z.; Huang, X.-Y. *Acta Crystallogr., C* **52**, 523 (1996); (b) Xiong, R.-G.; Zuo, J.-L.; You, X.-Z.; Huang, X.-Y.; *Acta Crystallogr., C* **52**, 1157 (1996).
632. (a) Roesky, H.W. *Adv. Inorg. Chem. Radiochem.* **22**, 239 (1979); (b) Roesky, H.W.; Anhaus, J.; Sheldrick, W.S. *Inorg. Chem.* **23**, 75 (1984); (c) Chivers, T. *Chem. Rev.* **85**, 341 (1985).
633. Gusling, H.J. *Coord. Chem. Rev.* **42**, 133 (1982).
634. (a) Gusling, H.J. In: *The Chemistry of Organic Selenium and Tellurium Compounds.* (Edit. Pataj, S.; Rappoport, Z.). John Wiley: New York, 1986, vol. 1, p. 679; (b) Atwood, J.L.; Bernal, L.; Calderazzo, F.; Canada, L.G.; Poli, R.; Rogers, R.D.; Veranici, C.A.; Vitali, D. *Inorg. Chem.* **22**, 1797 (1983).
635. Arnold, J. In: *Progress in Inorganic Chemistry*. John Wiley: New York, 1995, vol. 43, p. 353.
636. (a) Hope, E.G.; Levason, W. *Coord. Chem. Rev.* **122**(1/2), 109 (1993); (b) Besenyei, G.B.; Parkanyi, L.; Simendi, L.I.; James, B.R. *Inorg. Chem.* **34**, 6118 (1995).
637. Bolinger, C.M.; Rauchfuss, T.B. *Inorg. Chem.* **21**, 3947 (1982).
638. Miller, E.J.; Landon, S.J.; Brill, T.B. *Organometallics* **4**, 533 (1985).
639. Volhardt, K.P.C.; Walborsky, E.C. *Polyhedron* **7**, 1023 (1988).
640. Kajitani, M.; Ochiai, R.; Kohki, K.; Kavajachi, N.; Akijama, T.; Sugimori, A. *Bull. Soc. Chim. Jpn.* **62**(10), 3266 (1989).
641. Morley, C.P.; Vaughan, R.R. *J. Chem. Soc., Dalton Trans.* No. 5, 703 (1993).
642. O'Bryen, P.; Malik, M.A.; Hursthouse, M.B.; Motevalli, M. *Polyhedron* **11**(1), 45 (1992).
643. Choi, M.G.; Angelici, R.J. *J. Am. Chem. Soc.* **112**, 7811 (1993).
644. Kaim, W.; Kazak, V. *Angew. Chem.* **94**, 712 (1982).
645. Kerhner, D.K.; Basolo, F. *Coord. Chem. Rev.* **79**, 279 (1987).
646. Singh, A.K.; Srivastava, V. *J. Coord. Chem.* **27**, 237 (1992).
647. Hey-Hawkins, E. *Chem. Rev.* **94**(6), 1661 (1994).
648. Sadekov, I.D.; Skopenko, V.V.; Burlov, A.S.; Ivanova, E.I.; Garnovskii, A.D. *Zhurn. Obsh. Khim.* **67**, 1124 (1997).
649. Sadekov, I.D.; Maksimenko, A.A.; Nivorozhkin, V.L. *Usp. Khim.* **67**, 219 (1998).
650. Liaw, W.F.; Ou, D.S.; Li, V.S.; Le, Y.P.; Lee, G.H.; Peng, S.M. *Inorg. Chem.* **34**(14), 3747 (1995).
651. Liaw, W.F.; Chiou, S.J.; Lee, W.Z.; Gene-Hsiang, P.; Peng, S.M. *J. Chin. Chem. Soc. (Taipei)* **43**(1), 29 (1996).
652. Pasinskii, A.A.; Gorubaev, Yu.V.; Eremenko, I.L.; Vegini, D.; Yanovskii, A.I.; Struchkov, Yu.T. *Koord. Khim.* **22**, 369 (1996).
653. Nefedov, S.E.; Kolobkov, B.I.; Pasinskii, A.A.; Eremenko, I.L.; Sadekov, I.D.; Yanovskii, A.I.; Struchkov, Yu.T. *Zhurn. Neorg. Khim.* **37**, 335 (1992).

654. Bonasia, Ph.J.; Arnold, J. *J. Organomet. Chem.* **449**(1/2), 147 (1993).
655. Kaur, R.; Singh, H.B.; Butcher, R.J. *Organometallics* **14**, 4755 (1995).
656. Genge, A.R.J.; Levason, W.; Reid, G. *J. Chem. Soc., Dalton Trans.* No. 23, 4549 (1997).
657. Fisher, H.; Zeuner, S. *J. Organomet. Chem.* **252**, C63 (1983).
658. Fisher, H.; Früh, A.; Troll, C. *J. Organomet. Chem.* **415**(2), 211 (1991).
659. Minoura, M.; Kawashima, T.; Okazaki, R. *J. Am. Chem. Soc.* **115**(15), 7019 (1993).
660. Minoura, M.; Kawashima, T.; Tokitoh, N.; Okazaki, R. *J. Chem. Soc., Chem. Commun.* No. 2, 123 (1996).
661. Al-Rubaie, A.Z.; Al-Obaidi, Y.N.; Yosif, L.Z. *Polyhedron* **9**(9), 1141 (1990).
662. Khanna, P.K.; Singh, H.B. *Trans. Met. Chem.* **16**, 311 (1991).
663. (a) Minoura, M.; Kawashima, T.; Tokitoh, N.; Okasaki, R. *J. Chem. Soc., Chem. Commun.* 123 (1996); (b) *Phosph. Sulf. Silic. Relat. Elem.* **136/137**, 549 (1998).
664. Kukushkin, Yu.N. *Koord. Khim.* **7**, 323 (1981).
665. Kukushkin, Yu.N. *Reaction Capacity of Coordination Compounds.* Chemistry: Leningrad, 1987, 288 pp.
666. Sdorhoff, B.N. *Coord. Chem. Rev.* **23**, 1 (1977).
667. Michelin, R.A.; Mozzon, M.; Bertani, R. *Coord. Chem. Rev.* **147**, 299 (1996).
668. Garcia Alonso, F.J.; Garsia Sand, M.; Ricra, V.; Anillo Abril, A. *Organometallics* **11**, 80 (1992).
669. Mizobe, Y.; Ishii, Y.; Hidai, M. *Coord. Chem. Rev.* **139**, 281 (1995).
670. Dartigunave, M.; Meny, J.; Deydier, E.; Dartigunave, V.; Siebald, H. *Coord. Chem. Rev.* **178–180**, 623 (1998).
671. Clegg, W.; Coles, S.G.; Coppe, E.K.; Mair, F.S. *Angew. Chem.* **110**(6), 841 (1998).
672. Dehand, J.; Pfeffer, M. *Coord. Chem. Rev.* **18**, 327 (1976).
673. Bruce, M. *Angew. Chem., Int. Ed. Engl.* **16**, 73 1977).
674. Abbicht, H.P.; Issleib, R. *Z. Chem.* **17**, 1 (1977).
675. Omae, I. *Coord. Chem. Rev.* **51**, 1 (1983); **53**, 261 (1984).
676. Omae, I. *Coord. Chem. Rev.* **28**, 97 (1979); **32**, 235 (1980); **42**, 245 (1982); **83**, 137 (1988).
677. Constable, E.C. *Polyhedron* **3**, 1037 (1984).
678. Rothwell, I.P. *Polyhedron* **4**, 177 (1985).
679. Ryabov, A.D. *Usp. Khim.* **44**, 253 (1985).
680. Omae, I. *Organometallic Intramolecular-Coordination Compounds.* Elsevier Science: Amsterdam, New York, 1986.
681. Newkome, G.R.; Puckett, W.E.; Gupta, V.K.; Kiefer, G.E. *Chem. Rev.* **86**, 451 (1986).
682. Dunina, V.V.; Zalevskaya, O.A.; Potapov, V.M. *Usp. Khim.* **57**, 434 (1988).
683. Pavlovski, V.; Poznyak, A.L. *Z. Chem.* **29**, 6 (1989).
684. Evans, D.W.; Baker, G.R.; Newkome, G.R. *Coord. Chem. Rev.* **93**, 155 (1986).
685. Ryabov, A.D. *Chem. Rev.* **90**(2), 403 (1990).
686. Hadson, S.A.; Maitlis, P.M. *Chem. Rev.* **93**, 861 (1993).
687. Sokolov, V.I. *J. Organomet. Chem.* **500**, 299 (1995).
688. Fornies, J.; Navarro, M.R.; Sicilia, V.; Villarroja, P. *Organometallics* **15**, 1826 (1996).
689. Ryabov, A.D. *Mendeleev Chem. J.* **40**(4/5), 46 (1996).
690. Baena, M.J.; Espinet, P.; Ros, M.B.; Serano, J.L.; Ezcurra, A. *Angew. Chem.* **105**, 1260 (1993).
691. Ryabov, A.D.; Van Eldik, R. *Angew. Chem.* **106**, 798; *Int. Ed. Engl.* **33**(7), 783 (1994).
692. Casas, J.L.; Gayoso, E.; Vila, J.M.; Perera, M.T.; Gayoso, M. *Synth. React. Inorg. Metal-Org. Chem.* **21**(2), 263 (1991).
693. Ruttimann, S.; Piguet, C.; Bernardinelli, G.; Bocquet, B.; Williams, A.F. *J. Am. Chem. Soc.* **114**(11), 4230 (1992).
694. Ruttimann, S.; Bernardinelli, G.; Williams, A.F. *Angew. Chem.* **105**, 432; *Int. Ed. Engl.* **32**(3), 392, (1993).

695. Ryabov, A.D.; Kuzmina, L.G.; Polyakov, V.A.; Kazankov, G.M.; Rjabova, S.E.; Pfefer, M.; Van Eldik, R. *J. Chem. Soc., Dalton Trans.* No. 6, 999 (1995).
696. Gruter, G.J.-M.; Van Klink, G.P.M.; Akkerman, O.S.; Bickelhaupt, F. *Chem. Rev.* **95**(7), 2405 (1995).
697. Krilova, L.F.; Dikanskaya, L.D. *Metalloorg. Khim.* **4**, 752 and 757 (1991).
698. (a) Casas, J.M.; Fornies, J.; Martínez, F.; Rueda, A.J.; Tomas, M.; Welch, A. *J. Inorg. Chem.* **38**(7), 1529 (1999); (b) Ferrand, V.; Suss-Fink, G.; Neels, A.; Stoeckli-Evans, H. *Eur. J. Inorg. Chem.* **5**, 853 (1999).
699. Markovskii, L.N.; Romanenko, V.D.; Ruban, A.V. *Zhurn. Obsh. Khim.* **57**, 1433 (1987).
700. Browning, C.S.; Farrar, D.H. *J. Chem. Soc., Dalton Trans.* No. 4, 521 (1995).
701. Witt, M.; Roesky, H.W. *Chem. Rev.* **94**(5), 1163 (1994).
702. Schrerer, O.J.; Kerth, J.; Auselmann, R.; Sheldrick, S.W. *Angew. Chem., Int. Ed. Engl.* **22**, 984 (1983).
703. Ly, T.Q.; Woolins, J.D. *Coord. Chem. Rev.* **176**, 451 (1998).
704. Gavrilov, K.N.; Michel, I.S. *Koord. Khim.* **24**, 883 (1998).
705. Gavrilov, K.N.; Michel, I.S.; Teleshev, A.T.; Nifantyev, E.E. *J. Organomet. Chem.* **461**, 229 (1993).
706. Gavrilov, K.N.; Michel, I.S.; Polosukhin, A.I.; Rebrov, A.I.; Cherkaev, G.V.; Timofeeva, G.I. *Koord. Khim.* **23**, 925 (1997).
707. Weber, L.; Kaminski, O.; Stammler, H.G.; Neumann, B. *Organometallics* **14**(2), 581 (1995).
708. Zhang, Z.Z.; Cheng, H. *Coord. Chem. Rev.* **147**, 1 (1996).
709. Borners-Price, S.J.; Bewen, R.J.; Galletit, P.; Healy, P.C.; McKeage, M.J. *Coord. Chem. Rev.* **185/186**, 823 (1999).
710. King, R.B. *Acc. Chem. Res.* **13**, 243 (1980).
711. Balakrishna, M.S.; Reddy, V.S.; Krishnamurthy, S.S.; Nixon, J.F.; Burckett, St.; Laurent, J.C.T.R. *Coord. Chem. Rev.* **129**(1/2), 1 (1994).
712. (a) Dehnicke, K.; Krieger, M.; Massa, W. *Coord. Chem. Rev.* **182**, 19 (1999); (b) Gavrilov, K.N.; Polosuchin, A.I. *Russ. Chem. Rev.* **69**, 661 (2000).
713. Garnovskii, A.D. *Koord. Khim.* **13**, 715 (1987).
714. *Inorganic Biochemistry*. (Edit. Einhorn, H.). Elsevier Science: Amsterdam, Oxford, New York, 1973.
715. Williams, D. *Metals of Life*. Mir: Moscow, 1975.
716. Sigel, H.; Martin, R.B. *Chem. Rev.* **82**, 385 (1982).
717. Pettit, L.D.; Bezer, M. *Coord. Chem. Rev.* **61**, 97 (1985).
718. Hartley, F.R. *Coord. Chem. Rev.* **67**, 1 (1985).
719. Corradi, A.B. *Coord. Chem. Rev.* **117**, 45 (1992).
720. Yankovidis, A.; Hadjilianidis, N. *Coord. Chem. Rev.* **135/136**, 17 (1994).
721. Appleton, T.G. *Coord. Chem. Rev.* **166**, 313 (1997).
722. Aldrich-Wright, J.R.; Vega, R.S.; Williams, P.A. *Coord. Chem. Rev.* **166**, 361 (1997).
723. Clement, O.; Rapko, B.M.; Hay, B.P. *Coord. Chem. Rev.* **170**, 203 (1998).
724. Berthet, J.C.; Ephritikhine, M. *Coord. Chem. Rev.* **178–180**, 83 (1998).
725. Holm, R.H.; Ewerett, G.W.; Chakravorty, A. *Progr. Inorg. Chem.* **7**, 83 (1966).
726. (a) Sergienko, V.S.; Garnovskii, A.D.; Abramenko, V.L.; Porai-Koshitz, M.A. *Koord. Khim.* **13**, 1695 (1987); (b) Korshunov, O.Yu.; Uraev, A.I.; Sherbakob, I.N.; Antonova, I.A.; Kurbatov, V.P.; Garnovskii, A.D. *Zhurn. Neorg. Khim.* **45**, 1491 (2000); (c) Jones, D.; Roberts, A.; Cavell, K. *J. Chem. Soc., Dalton Trans.* 255 (1998); (d) Abramenko, V.L.; Sergienko, V.S.; Churakov, A.V. *Koord. Khim.* **26**, 922 (2000); (f) Garnovskii, A.D.; Vasilchenko, I.S. *Usp. Khim.* **71**(11), 1064 (2002).
727. Maslen, H.S.; Waters, T.N. *Coord. Chem. Rev.* **17**, 137 (1975).
728. Daul, C.; Schlapfer, C.M.; Von Zelewsky, A. *Struct. Bond.* **36**, 137 (1979).

729. Yamada, S.; Takeuchi, A. *Coord. Chem. Rev.* **43**, 187 (1982).
730. Calligaris, M.; Randaccio, L. In: Ref. 1, p. 715.
731. Syamal, A.; Mauriya, M.R. *Coord. Chem. Rev.* **95**, 183 (1989).
732. Collinson, S.R.; Fenton, D.E. *Coord. Chem. Rev.* **148**, 19 (1996).
733. Chakravorty, A. *Coord. Chem. Rev.* **13**, 1 (1974).
734. Keeney, M.E.; Osseo-Assere, K.; Woode, K.A. *Coord. Chem. Rev.* **59**, 141 (1984).
735. Singh, R.P.; Singh, N.K. *Asian J. Chem. Rev.* **2**, 62 (1991).
736. Gerasimchuk, N.N.; Bowman-James, K. In: *Encylopedia of Inorganic Chemistry*. John Wiley: Chichester, 1994, vol. 5, p. 2254.
737. Lalia-Kantouri, M.; Hartophilles, M.; Jannakoudakis, P.D.; Voutsas, G.P. *Z. Anorg. Allg. Chem.* **621**, 645 (1995).
738. Parpiev, N.A.; Yusupov, V.G.; Yakimovich, S.I.; Sharapov, Kh.T. *Acylhydrazones and their Complexes with Transition Metals*. PhAN: Tashkent, 1988, 164 pp.
739. Kogan, V.A.; Zelentsov, V.V.; Larin, G.M.; Lukov, V.V. *Complexes of Transition Metals with Hydrazones. Physico-Chemical Properties and Structures*. Science: Moscow, 1990.
740. Lukov, V.V.; Levchenkov, S.I.; Kogan, V.A. *Zhurn. Neorg. Khim.* **42**, 617 (1997).
741. Lukov, V.V.; Kogan, V.A.; Levchenkov, S.I.; Zavada, L.V.; Lisenko, K.A.; Shishkin, O.V. *Zhurn. Neorg. Khim.* **43**, 421 (1998).
742. Koh, L.-L.; Kon, O.L.; Loh, K.W.; Long, V.C.; Rantord, J.D.; Tan, A.L.C.; Tjan, V.V. *J. Inorg. Biochem.* **72**, 155 (1998).
743. Randhey, S.B.; Kauffman, G. *Coord. Chem. Rev.* **63**, 127 (1985).
744. West, D.X.; Randhey, S.B.; Sonawane, B. *Struct. Bond.* **76**, 1 (1991).
745. Liberta, A.E.; West, D.X. *Biometals* **5**, 125 (1992).
746. (a) West, D.X.; Liberta, A.E.; Pandhye, S.B.; Chikate, R.C.; Sonawane, P.G.; Kumbhar, A.S.; Yerande, R.G. *Coord. Chem. Rev.* **123**(1/2), 49 (1993); (b) Casas, J.C.; García-Tasende, M.S.; Sordo, J. *Coord. Chem. Rev.* **209**, 197 (2000).
747. Sergienko, V.S.; Abramenko, V.L.; Ilyukhin, A.B.; Porai-Koshitz, M.A.; Garnovskii, A.D. *Koord. Khim.* **18**, 585 (1992).
748. Jirgi, B.S.; Tembe, G.L.; Sonar, M.H.; Murty, A.S. *Trans. Met. Chem.* **18**, 171 (1993).
749. Dorokhova, N.I.; Burlov, A.S.; Amarskii, E.G.; Etmetchenko, L.N.; Garnovskii, A.D. *Koord. Khim.* **20**, 387 (1994).
750. Kogan, V.A.; Osipov, O.A.; Sherbak, S.N. *Dokl. Akad. Nauk SSSR* **181**, 1416 (1968).
751. Diyachenko, O.A.; Atovmyan, L.O.; Aldoshin, S.M.; Tkachev, V.V. *Zhurn. Strukt. Khim.* **19**, 829 (1978).
752. Mukherjee, A.K.; Mukherjee, D.P.K. *Acta Crystallogr., C* **42**, 793 (1986).
753. Choudhury, S.B.; Ray, D.; Chakravorty, A. *J. Chem. Soc., Dalton Trans.* No. 1, 107 (1992).
754. Sui, K.; Peng, S.M.; Bhattacharya, S. *Polyhedron* **18**(5), 631 (1999).
755. Kogan, V.A.; Kochin, S.G.; Antsyshkina, A.S.; Sadikov, G.G.; Garnovskii, A.D. *Mendeleev Commun.* 82 (1999).
756. Polinova, T.N.; Porai-Koshitz, M.A. *Results of Science & Technology. Crystallochemistry*. VINITI: Moscow, 1984, vol. 18, p. 64.
757. *Zhurn. Vses. Khim. Obsh. im Mendeleeva* **29**, 1–120 (1984) (issue dedicated to complexones).
758. Radanovic, D.J. *Coord. Chem. Rev.* **54**, 159 (1984).
759. Dyatlova, N.M.; Temkina, V.Ya.; Potapov, K.I. *Complexones and Metal Complexonates*. Chemistry: Moscow, 1988.
760. Shkolnikova, L.M.; Porai-Koshitz, M.A. *Usp. Khim.* **59**, 1111 (1990).
761. Porai-Koshitz, M.A.; Serezhkin, V.N. *Zhurn. Neorg. Khim.* **39**, 1109 (1994).
762. Porai-Koshitz, M.A.; Serezhkin, V.N. *Zhurn. Neorg. Khim.* **39**, 1967 (1994).
763. Serezhkin, V.N.; Polinova, T.N.; Porai-Koshitz, M.A. *Koord. Khim.* **21**, 253 (1995).

764. Ilyukhin, A.B.; Sergienko, V.S. *Mendeleev Chem. J.* **40**(4/5), Part II (1996).
765. Chatterjee, D. *Coord. Chem. Rev.* **168**, 273 (1998).
766. Porai-Koshitz, M.A. *Sov. Sci. Rev., Sec. B, Chem.* **10**, 91 (1987).
767. Douglas, B.E.; Radanovic, D.J. *Coord. Chem. Rev.* **128**(1/2), 139 (1993).
768. Mederos, A.; Domínguez, S.; Hernándes-Molina, R.; Sánchez, J.; Brito, F. *Coord. Chem. Rev.* **193–195**, 857 (1999).
769. Zhang, X.X.; Izatt, R.M.; Bradshaw, I.S.; Karkowiak, K.E. *Coord. Chem. Rev.* **174**, 179 (1998).
770. Graf, E.; Hosseini, M.W. *Coord. Chem. Rev.* **178–180**, 1193 (1999).
771. Formica, M.; Fusi, V.; Micheloni, M.; Poutellini, R.; Romani, P. *Coord. Chem. Rev.* **184**, 347 (1999).
772. Lukjanenko, N.G. *Ukr. Khim. Zhurn.* **65**(9), 21 (1999).
773. Callad, L.H.; Goodgame, D.L.M.; Hill, S.P.W.; Williams, D.J. *J. Chem. Soc., Dalton Trans.* No. 7, 1003 (1993).
774. Rawson, J.M.; Winpenny, R.E.P. *Coord. Chem. Rev.* **139**, 313 (1995).
775. Bankovskii, Yu.A.; Belskii, V.K.; Pech, L.Ya.; Ashaks, Ya.B. *Zhurn. Neorg. Khim.* **38**, 1988 (1993).
776. Kuzmina, L.G.; Kukina, G.A.; Yashaks, Ya.B.; Ivanova, N.A.; Bankovskii, Yu.A.; Efimenko, I.A. *Zhurn. Neorg. Khim.* **40**, 1817 (1995).
777. Alexandrov, G.G.; Sergienko, V.S.; Yashaks, Ya.B.; Leeis, A.E.; Pech, L.Ya.; Bankovskii, Yu.A.; Ivanova, N.A.; Efimenko, I.A. *Zhurn. Neorg. Khim.* **42**, 1820 (1997).
778. McKinney, T.M.; Fanwick, P.E.; Walton, R.A. *Inorg. Chem.* **38**(7), 1548 (1999).
779. Burlov, A.S.; Kuznetsova, L.I.; Volbushko, N.V.; Korshunov, O.Yu.; Garnovskii, A.D. *Zurn. Obsh. Khim.* **68**, 496 (1998).
780. Volodarsky, L.B.; Reznikov, V.A.; Ovcharenko, V.I. *Synthetic Stable Nitroxides.* CCR: London, 1994.
781. Ikorskii, V.N.; Ovcharenko, V.I.; Vostrikova, K.E.; Pervukhina, N.V.; Podberezskaya, N.V. *Zhurn. Neorg. Khim.* **37**, 1177 (1992).
782. Ikorskii, V.N.; Romanenko, G.V.; Sigurova, M.K.; Ovcharenko, V.I.; De Pontu, F.L.; Re, P.; Reznikov, V.A.; Podoplelov, A.V.; Sagdeev, R.Z. *Zhurn. Strukt. Khim.* **35**(4), 76 (1994).
783. Ovcharenko, V.I.; Vostrikova, K.V.; Podoplelov, A.V.; Sagdeev, R.Z.; Romanenko, G.V.; Ikorskii, V.N. *Polyhedron* **13**(19), 2781 (1994).
784. Vostrikova, K.E.; Ovcharenko, V.I.; Romanenko, G.V.; Ikorskii, V.N.; Podberezskaya, N.V.; Reznikov, V.A.; Volodarskii, L.B. *Zhurn. Neorg. Khim.* **37**, 1755 (1992).
785. Ovcharenko, V.I.; Vostrikova, K.E.; Podoplelov, A.V.; Romanenko, G.V.; Ikorskii, V.N.; Reznikov, V.A. *Polyhedron* **16**(18), 1279 (1997).
786. Ovcharenko, V.I.; Shvedernikov, Yu.G.; Musin, R.N.; Ikorskii, V.N. *Zhurn. Strukt. Khim.* **40**(1), 36 (1999).
787. Burdikov, A.B.; Ovcharenko, V.I.; Pervukhina, N.V.; Ikorskii, V.N.; Grigoriev, I.A.; Kirilyuk, I.A.; Ovcharenko, I.V.; Musin, R.N. *Zhun. Neorg. Khim.* **44**, 1131 (1999).
788. Ovcharenko, V.I.; Fokin, S.V.; Romanenko, G.V.; Ikorskii, V.N.; Reznikov, V.A.; Podoplelov, A.V. *Zhurn. Strukt. Khim.* **38**, 751 (1997).
789. Ovcharenko, V.I.; Ikorskii, V.N.; Romanenko, G.V.; Reznikov, V.A.; Volodarskii, L.B. *Inorg. Chim. Acta* **187**, 67 (1991).
790. Ovcharenko, V.I.; Sagdeev, R.Z. *Usp. Khim.* **68**, 381 (1999).
791. Suades, J.; Solans, X.; Font-Altaba, M. *Polyhedron* **3**(11), 1227 (1984).
792. Carrillo, D.; Gouzert, P.; Jeannin, Y. *Polyhedron* **8**(23), 2837 (1989).
793. Girling, R.L.; Amma, E.L. *Inorg. Chem.* **6**, 2009 (1967).
794. Baidya, N.; Ndreu, D.; Olmstead, M.M.; Mascharak, P.K. *Inorg. Chem.* **30**(10), 2448 (1991).

795. Baidya, N.; Olmstead, M.M.; Mascharak, P.K. *Inorg. Chem.* **30**(20), 3967 (1991).
796. Müller, A.; Johannes, K.U.; Straube, M.; Krickmeyer, E.; Bogge, N.Z. *Anorg. Allg. Chem.* **619**(6), 1037 (1993).
797. Kruger, H.-J.; Peng, G.; Holm, R.H. *Inorg. Chem.* **30**, 734 (1991).
798. Ueyama, N.; Tuji, T.; Okamura, T.; Nakamura, A. *Acta Crystallogr., C* **54**, 1424 (1998).
799. Liaw, M.-C.; Lee, C.-H.; Peng, S.-M. *Bull. Inst. Chem. Acta Sin.* **40**, 23 (1993).
800. Hanss, J.; Kruger, H.-J. *Angew. Chem., Int. Ed. Engl.* **37**(3), 360 (1998).
801. Porai-Koshitz, M.A.; Gilinskaya, E.A. *Crystallochemistry. Results of Science.* Institute of Academy of Sciences of USSR: Moscow, 1966.
802. Barerra, H.; Suades, J.; Perocand, M.C.; Brianso, J.L. *Polyhedron* **3**, 839 (1984).
803. Kogan, V.A.; Kharabaev, N.N.; Osipov, O.A.; Kochin, S.G. *Zhurn. Strukt. Khim.* **22**(1), 126 (1981).
804. Nivorozhkin, A.L.; Uraev, A.I.; Burlov, A.S.; Garnovskii, A.D. *Mendeleev Chem. J.* **40**(4/5), 255 (1996).
805. Valters, R.E.; Fülöp, F.; Korbonits, D. *Adv. Heterocycl. Chem.* **66**, 1 (1996).
806. Kochin, S.P.; Garnovskii, A.D.; Kogan, V.A.; Sushko, T.G. *Zhurn. Neorg. Khim.* **14**, 1428 (1969).
807. Diyachenko, O.A.; Atovmyan, L.O.; Aldoshin, S.M.; Kogan, V.A.; Kochin, S.G.; Osipov, O.A. *Izv. Akad. Nauk SSSR, Ser. Khim.* 2130 (1975).
808. Kawamoto, T.; Kushi, Y. *Chem. Lett. Chem. Soc. Jpn.* 297 (1992).
809. Kawamoto, T.; Kushi, Y. *Chem. Lett. Chem. Soc. Jpn.* 893 (1992).
810. Kawamoto, T.; Kuma, H.; Kushi, Y. *J. Chem. Soc., Chem. Commun.* 2121 (1996).
811. (a) Kawamoto, T.; Kuma, H.; Kushi, Y. *Bull. Chem. Soc. Jpn.* **70**(7), 1599 (1997); (b) Kawamoto, T.; Isoroki, N.; Kuma, H.; Kushi, Y. *Inorg. Chim. Acta* **265**(1/2), 163 (1997).
812. Ercan, F.; Ulku, D.; Ancin, N.; Ostas, S.G.; Tuzun, M. *Acta Crystallogr., C* **52**(5), 1141 (1996).
813. Müller, A.; Johannes, K.U.; Plass, W.; Bogge, H.; Krahn, E.; Schneider, K.Z. *Anorg. Allg. Chem.* **622**(10), 1765 (1996).
814. Zelenin, K.N.; Khorseeva, L.A.; Alexeev, V.V.; Toshev, M.T.; Dustov, Kh.V. *Zhurn. Obsh. Khim.* **60**, 2348 (1990).
815. Toshev, M.T.; Yusupov, V.G.; Saidov, S.O.; Dustov, Kh.V.; Zelenin, K.N.; Khorseeva, L.A.; Alexeev, V.V.; Parpiev, N.A. *Zhurn. Neorg. Khim.* **37**, 621 (1992).
816. Yusupov, V.G.; Toshev, M.T.; Saidov, S.O.; Dustov, Kh.V.; Zelenin, K.N.; Parpiev, N.A.; Alexandrov, G.G. *Zhurn. Neorg. Khim.* **37**, 1039 (1992).
817. Tian, Yu.-P.; Duan, C.-Y.; Lu, Z.I.; You, X.-Z.; Fun, H.-K.; Sivakumar, K. *Trans. Met. Chem.* **21**(3), 254 (1996).
818. Fun, H.-K.; Sivakumar, K.; Tian, K.-P.; Lu, Z.-L.; You, X.-Z. *Acta Crystallogr., C* **52**, 1143 (1996).
819. Tian, Yu.-P.; Dung, C.-V.; You, X.-Z.; Fun, H.-K. *Acta Crystallogr., C* **52**, 2150 (1996).
820. Bian, Z.-W.; Duan, C.-V.; Mak, T.C.W. *Jiegou Xuanxue (J. Struct. Chem.).* **16**, 307 (1997).
821. Puranik, V.G.; Tavale, S.S.; Row, T.N.G.; Umpathu, P.; Budhar, A.P. *Acta Crystallogr., C* **43**, 2303 (1987).
822. Leovac, V.M.; Cesl'jewic, V.I.; Argay, G.; Kalman, A.; Ribar, B. *J. Coord. Chem.* **34**(4), 357 (1995).
823. Minkin, V.I.; Garnovskii, A.D.; Elquero, J.; Katritzky, A.R.; Denisco, O.V. *The Isomerism of Heterocyclic Compounds.* Academic Press: San Diego, London, Boston, 2000, vol. 76, p. 157.
824. Uraev, A.I.; Nivorozhkin, A.L.; Tylchenko, L.S.; Kurbatov, V.P.; Lysenko, K.A.; Antipin, M. Yu.; Vlasenko, V.G.; Garnovskii, A.D. Int. Conf. "New Approaches in Coord. and Organomet.Chem." Nizhnii Novogorod, Russia, 2002, p. 183.

825. Bren, V.A.; Minkin, V.I. *Izv. Vissh. Ucheb. Zaved. Khim. i Khim. Tekhnol.* **25**, 663 (1982).
826. Litvinov, V.P.; Vaisburg, A.F.; Mortikov, V.Yu. *Chem. Heterocycl. Comp.* **28**(1), 5 (1992).
827. Yaramura, T.; Tadokoro, M.; Tanaka, K.; Kurodo, R. *Bull. Chem. Soc. Jpn.* **66**(7), 1984 (1993).
828. Henderson, R.K.; Bouwman, E.; Reedijk, J.; Powell, A.K. *Acta Crystallogr., C* **52**(11), 2696 (1996).
829. Christenen, A.; Jensen, H.S.; McKee, V.; McKenzie, C.J.; Munch, M. *Inorg. Chem.* **36**(26), 6080 (1997).
830. Adhikhari, B.; Anderson, O.P.; La Cour, A.; Hazell, R.; Olsen, C.E.; Toftlund, H. *J. Chem. Soc., Dalton Trans.* No. 23, 4539 (1997).
831. La Cour, A.; Findeisein, M.; Hansen, K.; Hazell, R.; Hennig, L.; Olsen, C.E.; Pedersen, L. *J. Chem. Soc., Dalton Trans.* 2045 (1997).
832. Nivorozhkin, A.L.; Konstatinovskii, L.E.; Nivorozhkin, L.E.; Minkin, V.I.; Takhirov, T.G.; Diyachenko, O.A.; Tagiev, D.B. *Izv. Akad. Nauk SSSR, Ser. Khim.* 327 (1990).
833. Nivorozhkin, A.L.; Nivorozhkin, L.E.; Minkin, V.I.; Takhirov, T.G.; Diyachenko, O.A. *Polyhedron* **10**, 179 (1991).
834. Toftlund, H.; Becher, J.; Olsen, P.H.; Pedersen, J. *Israel J. Chem.* **25**, 56 (1985).
835. Knoblauch, S.; Benedix, R.; Ecke, M.; Gelbrich, T.; Sieler, J.; Somoza, F.; Hennig, H. *Eur. J. Inorg. Chem.* **8**, 1393 (1999).
836. Mistryukov, A.E.; Vasilchenko, I.S.; Sergienko, V.S.; Nivorozhkin, A.L.; Kochin, S.G.; Porai-Koshitz, M.A.; Nivorozhkin, L.E.; Garnovskii, A.D. *Mendeleev Commun.* 30 (1992).
837. Uraev, A.I.; Nivorozhkin, A.L.; Bondarenko, G.I.; Lisenko, K.A.; Korshunov, O.Yu.; Vlasenko, V.G.; Vasilchenko, I.S.; Shuvaev, A.T.; Kurbatov, V.P.; Antipin, M.Yu.; Garnovskii, A.D. *Dokl. Akad. Nauk.* **367**, 67 (1999).
838. Uraev, A.I.; Nivorozhkin, A.L.; Bondarenko, G.I.; Lysenko, K.A.; Korshunov, O.Yu.; Vlasenko, V.G.; Shuvaev, A.T.; Kurbatov, V.P.; Antipin, M.Yu.; Garnovskii, A.D. *Izv. Akad. Nauk. Serv. Khim.* **11**, 1892 (2000).
839. Uraev, A.I.; Vlasenko, V.G.; Kharisov, B.I.; Blanco, L.M.; Shuvaev, A.T.; Vasilchenko, I.S.; Garnovskii, A.D.; Elizondo, N.V. *Polyhedron* **19**, 2361 (2000).
840. Nivorozhkin, A.L.; Uraev, A.I.; Bondarenko, G.I.; Antsishkina, A.S.; Kurbatov, V.P.; Garnovskii, A.D.; Turte, K.I. *J. Chem. Soc., Chem. Commun.* No. 18, 1711 (1997).
841. Uraev, A.I.; Nivorozhkin, A.L.; Antsishkina, A,S.; Korshunov, O.Yu.; Bondarenko, G.I.; Vasilchenko, I.S.; Kurbatov, V.P.; Garnovskii, A.D. *Dokl. Akad. Nauk.* **356**, 212 (1997).
842. Forde, C.E.; Lough, A.J.; Morris, R.H.; Ramachandran, R. *Inorg. Chem.* **35**(10), 2747 (1996).
843. Fallon, G.; Gatehouse, B.M. *J. Chem. Soc., Dalton Trans.* 1344 (1975).
844. Shoner, S.C.; Bornhardt, D.; Kovacs, J. *Inorg. Chem.* **34**(18), 4517 (1995).
845. Zelentsov, V.V. *Mendeleev Chem. J.* **40**(4/5), Part II, 132 (1996).
846. Sugiura, V.; Kuwachara, J.; Nagasava, T.; Vamada, H. *J. Am. Chem. Soc.* **109**, 5848 (1987).
847. Nelson, M.J.; Jin, H.; Turner, I.M.; Grove, G.; Scarrow, R.C.; Brennan, B.A.; Que, L. *J. Am. Chem. Soc.* **113**(18), 7072 (1991).
848. Kobayashi, M.; Magasawa, T.; Vamada, H. *Trends Biotechnol.* **10**, 402 (1992).
849. Jin, H.; Tunner Jr., I.M.; Nelson, M.J.; Girbiel, R.Y.; Doan, E. *J. Am. Chem. Soc.* **115**(12), 5290 (1993).
850. Yamada, H.; Kobayashi, H. *Biosci. Biotech. Biochem.* **60**, 1391 (1996).
851. Minkin, V.I.; Nivorozhkin, L.E.; Korobov, M.C. *Usp. Khim.* **63**, 303 (1994).
852. Minkin, V.I.; Nivorozhkin, L.E. *Mendeleev Chem. J.* **40**(4/5), Part II, 8 (1996).

853. Gerloch, D.A.; Holm, R.H. *J. Am. Chem. Soc.* **91**, 3457 (1968).
854. Kraudelt, Z.; Schilde, U.; Uhlemann, E.-Z. *Kristallogr.* **213**, 175 (1998).
855. Attansio, D.; Dessy, G.; Fares, V.; Pennesi, G. *Mol. Phys.* **40**, 269 (1980).
856. Sieler, J.; Richter, R.; Seyer, L.; Lingvist, O.; Andersen, L. *Z. Anorg. Allg. Chem.* **515**, 41 (1984).
857. Kochin, S.G.; Antsishkina, A.S.; Sadikov, G.G.; Burlov, A.S.; Kogan, V.A.; Garnovskii, A.D. *Dokl. Akad. Nauk.* **355**, 777 (1997).
858. Aldoshin, S.M.; Alexeenko, V.A.; Atovmyan, L.O. *Koord. Khim.* **1**, 1075 (1975).
859. Dyachenko, L.A.; Atovmyan, L.O.; Aldoshin, S.M. *J. Chem. Soc., Chem. Commun.* 105 (1975).
860. Aldoshin, S.M.; Dyachenko, L.A.; Atovmyan, L.O. *Izv. Akad. Nauk SSSR, Ser. Khim.* 2405 (1976).
861. Dyachenko, L.A.; Aldoshin, S.M.; Atovmyan, L.O. *Koord. Khim.* **3**, 1753 (1977).
862. Kogan, V.A.; Kochin, S.G.; Antsishkina, A.S.; Sadikov, C.G.; Garnovskii, A.D. *Mendeleev Commun.* 239 (1097).
863. Dyachenko, L.A.; Atovmyan, L.O.; Aldoshin, S.M. *Izv. Akad. Nauk. SSSR, Ser. Khim.* 2147 (1976).
864. Rapper, E.S. *Coord. Chem. Rev.* **61**, 115 (1985).
865. Rapper, E.S. *Coord. Chem. Rev.* **129**(1/2), 91 (1994).
866. Rapper, E.S. *Coord. Chem. Rev.* **153**, 199 (Part I) (1996); **165**, 475 (Part II) (1997).
867. Henkel, G.; Krebs, B.; Schmidt, W. *Angew. Chem.* **104**, 1380 (1992); *Int. Ed. Engl.* **31**(10), 1366 (1992).
868. Bankovskii, Yu.A; Berzinya, I.P.; Belskii, V.K.; Yanson, G.E. *Zhurn. Neorg. Khim.* **39**, 608 (1994).
869. Kuzmina, L.G.; Kukina, G.A.; Leeis, Ya.E.; Grap, S.R.; Bankovskii, Yu.A.; Efimenko, I.A. *Zhurn. Neorg. Khim.* **41**, 209 (1996).
870. Kuzmina, L.G.; Kukina, G.A.; Leeis, Ya.E.; Grap, S.R.; Bankovskii, Yu.A.; Efimenko, I.A. *Zhurn. Neorg. Khim.* **41**, 216 (1996).
871. Ashkas, Ya.V.; Bankovskii, Yu.A.; Yanson, G.E. *Zhurn. Neorg. Khim.* **44**, 765 (1999).
872. Mikhailov, O.V. *Koord. Khim.* **25**, 143 (1999).
873. Pent, H.; Praecel, U.; Huber, F. *Acta Crystallogr., C* **42**, 1138 (1986).
874. Furmanova, N.G.; Struchkov, Yu.T.; Rokhlina, E.M.; Kravtsov, D.N. *Zhurn. Strukt. Khim.* **21**, 87 (1990).
875. Kuzmina, L.G.; Dvortsova, N.V.; Burtseva, S.Yu.; Porai-Koshitz, M.A.; Smislova, E.I.; Grandberg, K.I. *Metalloorg. Khim.* **3**, 364 (1990).
876. Kuzmina, L.G. *Zhurn. Neorg. Khim.* **37**, 1773 (2992).
877. Kuzmina, L.G. *Metalloorg. Khim.* **5**, 744 (1992).
878. Mugesh, G.; Singh, H.B.; Butcher, R.J. *Eur. J. Inorg. Chem.* **8**, 1229 (1999).
879. García-Tojal, J.; García-Orad, A.; Serra, J.L.; Pizarro, J.L.; Lezama, L.; Arriortua, M.I.; Rojo, T. *J. Inorg. Biochem.* **75**(1), 45 (1999).
880. Rodríguez-Argüelles, M.C.; Sanzez, A.; Ferrari, M.B.; Fava, G.G.; Pelizzi, C.; Pelosi, G.; Albertini, R.; Lunghi, P.; Pinelli, S. *J. Inorg. Biochem.* **73**(1/2), 7 (1999).
881. Bermeio, E.; Carbollo, R.; Castineiras, A.; Domínguez, R.; Liberta, A.E.; Maichle-Mössmer, C.; Salberg, M.M.; West, D.X. *Eur. J. Chem.* 965 (1999).
882. Antsishkina, A.S.; Tokarskaya, S.A.; Tsveniashvili, V.Sh.; Ostryakova, V.N.; Ukhin, L.Yu.; Porai-Koshitz, M.A.; Garnovskii, A.D. *Koord. Khim.* **15**, 214 (1989).
883. Garnovskii, A.D.; Minkina, L.S.; Sakhashik, L.V.; Kurbatov, N.P.; Osipov, O.A. *Zhurn. Obsh. Khim.* **41**, 1884 (1971).
884. (a) Kurbatov, V.P.; Osipov, O.A.; Minkin, V.I.; Nivorozhkin, L.E.; Minkina, L.S. *Zhurn. Neorg. Khim.* **19**, 2191 (1974); (b) Minkin, V.I.; Nivorozhkin, L.E.; Konstantinovskii, L.E.; Korobov, M.S. *Koord. Khim.* **3**, 174 (1977).

885. Nivorozhkin, L.E.; Korobov, M.S.; Minkin, V.I. *Zhurn. Org. Khim.* **14**, 347 (1978).
886. Takhirov, T.G.; Diyachenko, O.A.; Tagiev, D.V.; Nivorozhkin, A.L.; Nivorozhkin, L.E.; Minkin, V.I. *Koord. Khim.* **17**, 505 (1991).
887. (a) Uraev, A.I.; Nivorozhkin, A.L.; Bondarenko, G.I.; Korshunov, O.Yu.; Vasilchenko, I.S.; Antsyshkina, A.S.; Sadikov, G.G.; Yanovskii, A.I.; Vlasenko, V.G.; Shuvaev, A.T.; Turte, K.I.; Brashoveanu, N.D.; Kurbatov, V.P.; Garnovskii, A.D. *XVI Mendeleev Meeting on General and Applied Chemistry*. Memories, Science: Moscow, 1998, p. 312; (b) Uraev, A.I.; Nivorozhkin, A.L.; Garnovskii, A.D. *Abstr. XX. International Chugaev Conference on Coordination Chemistry*, Rostov-on-Don, Russia, Rostov State University: Rostov-on-Don, 2001, p. 444.
888. Cheng, Y.; Emge, T.J.; Brennan, J.G. *Inorg. Chem.* **33**(7), 3711 (1994).
889. Khashis, D.Y.; Buretea, M.; Emge, T.J.; Brennan, J.G. *J. Chem. Soc., Dalton Trans.* No. 1, 45 (1995).
890. Cheng, Y.; Emge, T.J.; Brenan, J.G. *Inorg. Chem.* **35**, 342 (1996).
891. Cheng, Y.; Emge, T.J.; Brennan, J.G. *Inorg. Chem.* **35**, 7339 (1996).
892. Chopra, N.; Damude, L.C.; Dean, P.A.W.; Vital, J.J. *Can. J. Chem.* **74**(11), 2095 (1996).
893. Romero, J.; Duran, M.; García-Vázquez, J.A.; Castineiras, A.; Sousa, A.; Cristianes, L.; Zubieta, J. *Inorg. Chim. Acta* **255**, 307 (1997).
894. Rodríguez, A.; Romero, J.; García-Vázquez, J.A.; Duran, M.; Sousa-Pedrares, A.; Sousa, A.; Zubieta, J. *Inorg. Chim. Acta.* **284**, 133 (1999).
895. (a) Secido, E.; Fujiwara, I. *Talanta* **19**, 647 (1972); (b) Pech, L.Ya.; Bankovskii, Yu.A.; Belskii, V.K.; Silin, E.Ya.; Ashaks, Ya.V.; Zavodnik, V.E. *Zhurn. Neorg. Khim.* **45**(6), 940 (2000).
896. Iretskii, A.I.; Petrov, M.A.; Kukushkin, Yu.N.; Shakhmuratov, E.B.; Batsanov, A.S. *Metalloorg. Khim.* **4**, 1314 (1991).
897. Segiura, V.; Kikuchi, T.; Tanaka, H. *Adv. Chem. Res.* **191**, 293 (1980).
898. Adejumobi, A.J.; Goddard, D.R. *J. Inorg. Nucl. Chem.* **39**, 910 (1977).
899. Vedyanu, A.N.; Gebreleu, N.V. *Zhurn. Neorg. Khim.* **21**, 3319 (1976).
900. (a) Beck, W.; Weis, J.C.; Wiezorek, J. *J. Organomet. Chem.* **30**, 89 (1971); (b) Weis, J.C.; Beck, W. *J. Organomet. Chem.* **44**, 325 (1972).
901. Bark, L.S.; Brandon, D. *Talanta* **14**, 759 (1967).
902. Lappert, M.F.; Martin, T.R.; McLaughlin, G.M. *J. Chem. Soc., Chem. Commun.* 635 (1980).
903. Singh, A.K.; Srivastava, V. *J. Coord. Chem.* **27**(4), 237 (1993).
904. Jain, V.K.; Chaudhury, S.; Bohra, R. *Polyhedron* **12**(19), 2377 (1993).
905. Khandelwal, B.L.; Singh, A.K.; Srivastava, V.; Poveley, D.C.; Smith, G.W. *Polyhedron* **9**(17), 2041 (1990).
906. Srivastava, V.; Singh, A.K.; Khanderwal, B.L. *Trans. Met. Chem.* **17**, 177 (1992).
907. Singh, A.K.; Srivastava, V.; Khandelwal, B.L. *Polyhedron* **9**(6), 851 (1990).
908. Singh, A.K.; Srivastava, V. *J. Coord. Chem.* **21**(3), 269 (1990).
909. Sadekov, I.D.; Abakarov, G.M.; Kuren, S.G.; Garnovskii, A.D.; Minkin, V.I. *Zhurn. Obsh. Khim.* **56**, 2168 (1986).
910. Al-Salim, N.; McWhinnie, W.R.; Hamor, T.A. *J. Chem. Soc., Chem. Commun.* 453 (1986).
911. Garnovskii, A.D.; Antsishkina, A.S.; Vasilchenko, I.S.; Korshunov, O.Yu.; Sadikov, G.G.; Maximenko, A.A.; Sadekov, I.D. *Koord. Khim.* **41**, 1297 (1996).
912. Sudha, N.; Singh, H.B. *Coord. Chem. Rev.* **135/136**, 469 (1994).
913. McWhinnie, W.R.; Sadekov, I.D.; Minkin, V.I. *Sulfur Rep.* **18**, 295 (1996).
914. Garnovskii, A.D.; Uraev, A.I.; Antsishkina, A.S.; Sadekov, I.D.; Minkin, V.I. *Koord. Khim.* **22**, 512 (1996).

915. Garnovskii, A.D.; Sadekov, I.D.; Antsyshkina, A.S.; Sadikov, G.G.; Borodkin, G.S.; Uraev, A.I.; Maksimenko, A.A.; Vasilchenko, I.S.; Borodkina, I.G.; Sergienko, V.S.; Minkin, V.I. *Koord. Khim.* **25**, 767 (1999).
916. Badyal, K.; McWhinnie, W.R.; Hammor, T.A.; Chen, H. *Organometallics* **16**(14), 3194 (1997).
917. Garnovskii, A.D.; Abakarov, G.M.; Sadekov, I.D.; Minkin, V.I.; Cherkasova, T.G.; Varshavskii, Yu.S. *Koord. Khim.* **10**, 234 (1984).
918. Sadekov, I.D.; Abakarov, G.M.; Panov, V.B.; Ukhin, L.Yu.; Garnovskii, A.D.; Minkin, V.I. *Khim. Geterotsykl. Soed. 757* (1985).
919. (a) Sadekov, I.D.; Minkin, V.I.; Garnovskii, A.D. *Sulfur Rep.* **4**, 63 (1985); (b) Singh, A.K.; Sharma, S. *Coord. Chem. Rev.* **209**, 49 (2000).
920. Khalid, A.; Khandelwal, B.L.; Singh, A.K.; Singh, T.P.; Padmanabhan, B. *J. Coord. Chem.* **31**(1), 19 (1994).
921. Khalid, A.; Singh, A.K. *Polyhedron* **16**(1), 33 (1997).
922. Khandelwal, B.L.; Khalid, A.; Singh, A.K. *J. Coord. Chem.* **36**(3), 207 (1995).
923. Mbogo, S.A.; McWhinnie, W.R.; Lobana, T.S. *Inorg. Chim. Acta* **172**(2), 211 (1990).
924. Mugesh, G.; Singh, H.B.; Butcher, R.J. *J. Organomet. Chem.* **577**(2), 243 (1999).
925. Bochmann, M.; Bwembya, G.C.; Powell, A.K.; Song, X. *Polyhedron* **14**(23/24), 3495 (1995).
926. Menon, S.C.; Singh, H.B.; Patel, R.P.; Kulshreshtha, S.K. *J. Chem. Soc., Dalton Trans.* No. 7, 1203 (1996).
927. Norbury, A.H. *Adv. Inorg. Chem. Radiochem.* **17**, 232 (1975).
928. Kukushkin, Yu.N. *Koord. Khim.* **4**, 1170 (1978).
929. Critchlow, P.B.; Robinson, S.D. *Coord. Chem. Rev.* **25**, 69 (1978).
930. Koppikaz, D.K.; Sivapullaiach, P.V.; Ramakrishnan, A. *Struct. Bond.* **34**, 135 (1978).
931. Kagan, H.B,.; Ronan, B. *Rev. Heteroatom. Chem.*7, 92 (1992).
932. Kukushkin, V.Yu. *Coord. Chem. Rev.* **139**, 375 (1995).
933. Calligaris, M.; Carugo, O. *Coord. Chem. Rev.* **153**, 83 (1996).
934. Dietzch, W.; Strauch, P.; Hoyer, E. *Coord. Chem. Rev.* **121**, 43 (1992).
935. Cox, M.; Darkeu, J. *Coord. Chem. Rev.* **7**, 29 (1971).
936. Livingstone, S.E. *Coord. Chem. Rev.* **7**, 59 (1971).
937. Uhlemann, E. *Z. Chem.* **11**, 401 (1971).
938. Kukushkin, Yu.N.; Sibirskaya, V.V.; Vorobiev-Desyatovskii, N.V.; Hoyer, E.; Beier, L.; Avetikyan, G.V.; Strukov, V.V.; Pogareva, V.G. *Zhurn. Obsh. Khim.* **51**, 167 (1981).
939. Buter, J.; Kellog, R.M. *J. Chem. Soc., Chem. Commun.* 466 (1980).
940. Buter, J.; Kellog, R.M. *J. Organomet. Chem.* **46**, 4481 (1981).
941. Larin, G.M.; Yusupov, V.G.; Karimov, M.M.; Parpiev, N.A. *Koord. Khim.* **21**, 855 (1995).
942. (a) Braunschweig, H. *Angew. Chem.* **110**, 1883 (1998); (b) Braunschweig, H.; Colling, M. *Coord. Chem. Rev.* **223**, 1 (2001).
943. Hartwig, J.F.; Huber, S. *J. Am. Chem. Soc.* **115**, 4908 (1993).
944. Iverson, C.N.; Smith, M.R. *J. Am. Chem. Soc.* **117**(15), 4403 (1995).
945. Lesley, G.; Nguyen, P.; Taylor, N.J.; Marder, T.B.; Scott, A.J.; Clegg, W.; Norman, N.C. *Organometallics* **15**, 5137 (1996).
946. Backer, R.T.; Overall, D.W.; Calabrese, J.C.; Westcott, S.A.; Taylor, N.J.; Williams, I.D.; Marder, T.B. *J. Am. Chem. Soc.* **112**, 9399 (1990).
947. Braunschweig, H.; Ganter, B. *J. Organomet. Chem.* **545/546**, 163 (1997).
948. Onozawa, S.; Hatanaka, Y.; Sakamura, T.; Shimada, S.; Tanaka, M. *Organometallics* **15**(26), 5450 (1996).
949. Elliot, D.J.; Levy, C.J.; Ruddephatt, R.J.; Holah, D.G.; Hughes, A.N.; Magnuson, V.R.; Moser, I.M. *Inorg. Chem.* **29**, 5014 (1990).
950. Linti, G.; Nöth, H.; Bolborn, K.; Paine, R.T. *Angew. Chem., Int. Ed. Engl.* **29**, 682 (1990).

951. Drieß, M.; Frankhauser, P.; Pritzkow, H.; Sielbert, W. *Chem. Ber.* **124**, 1497 (1991).
952. Paine, P.T.; Nöth, H. *Chem. Ber.* **128**, 343 (1995).
953. Huttner, G.; Krig, B. *Angew. Chem., Int. Ed. Engl.* **10**, 499 (1971).
954. (a) Braunschweig, H.; Kollann, C.; Müller, M. *Eur. J. Inorg. Chem.* 291 (1998); (b) Putzer, M.A.; Rogers, J.S.; Bazan, G.C. *J. Am. Chem. Soc.* **121**(35), 8112 (1999).
955. Kuznetsov, N.T.; Solntsev, A.K. *Chemistry of Inorganic Hydrides.* (Edit. Kuznetsov, N.T.). Moscow: Science, 1990.
956. Lagun, V.L.; Katser, S.V.; Orlova, A.M.; Kuznetsov, N.T. *Koord. Khim.* **18**, 365 (1992).
957. Leyden, R.N.; Sullivan, B.P.; Baker, R.T.; Hawthore, M.F. *J. Am. Chem. Soc.* **100**, 3758 (1978).
958. Gill, J.T.; Lippard, S.J. *Inorg. Chem.* **14**, 751 (1975).
959. Gaft, Yu.L.; Ustinyuk, Yu.A.; Borisenko, A.A.; Kuznetsov, N.T. *Zhurn. Neorg. Khim.* **28**, 2234 (1983).
960. Malinina, E.A.; Solntsev, K.A.; Butman, L.A.; Kuznetsov, N.T. *Koord. Khim.* **15**, 1039 (1989).
961. Kuznetsov, N.T.; Solntsev, K.A. *Koord. Khim.***17**, 1157 (1991).
962. Orlova, A.M.; Sivaev, I.B.; Lagun, V.L.; Katser, S.B.; Solntsev, K.A.; Kuznetsov, N.T. *Koord. Khim.* **19**, 116 (1993).
963. Orlova, A.M.; Sivaev, I.B.; Lagun, V.L.; Katser, S.B.; Solntsev, K.A.; Kuznetsov, N.T. *Koord. Khim.* **22**, 119 (1996).
964. Lagun, V.L.; Orlova, A.M.; Katser, S.B.; Solntsev, K.A.; Kuznetsov, N.T. *Koord. Khim.* **20**, 431 (1994).
965. Lagun, V.L.; Solntsev, K.A.; Katser, S.B.; Orlova, A.M.; Kuznetsov, N.T. *Koord. Khim.* **20**, 504 (1994).
966. Bhattacharyya, P.; Loza, M.L.; Parr, J.; Slawin, A.M.Z. *J. Chem. Soc., Dalton Trans.* No. 17, 2917 (1999).
967. Chivers, T.; Hilts, R.W. *Coord. Chem. Rev.* **137**, 201 (1994).
968. Fujita, M.; Ogura, K. *Coord. Chem. Rev.* **148**, 249 (1996).
969. Wieser, C.; Dieleman, C.B.; Matt, D. *Coord. Chem. Rev.* **165**, 93 (1997).
970. (a) Kepert, D.L.; Clare, B.W. *Coord. Chem. Rev.* **155**, 1 (1996); (b) Rubin, V. *Eur. J. Chem.* **3**, 1009 (1997).
971. Rizzarelli, E.; Vecchio, G. *Coord. Chem. Rev.* **188**, 343 (1999).
972. (a) Gudat, D. *Angew. Chem., Int. Ed. Engl.* **36**, 1951 (1997); (b) Caminade, A.M.; Lauren, T.R.; Chaudrecht, B.; Majoral, J.P. *Chem. Rev.* **178–180**, 793 (1998); (c) Majoral, J.P.; Caminade, A.M. *Chem. Rev.* **99**, 845 (1999); (d) Beletskaya, I.P.; Chuchukinm, A.V. *Russ. Chem. Rev.* **69**, 639 (2000).

3

Main Methods of the Synthesis of Coordination Compounds

ALEXANDER D. GARNOVSKII, IGOR S. VASILCHENKO, and DMITRY A. GARNOVSKII

Rostov State University, Rostov-on-Don, Russia

BORIS I. KHARISOV

Universidad Autónoma de Nuevo León, Monterrey, Mexico

Analysis of the data presented in Chap. 2 testifies that at present considerable progress is being achieved (compare with Ref. 1) in the creation and selection of ligands, especially those containing donor atoms of the elements of Periods II (O, N), III (P, S, Cl), IV (As, Se, Br), and V (Sb, Te, I). Undoubtedly, not only the donor atoms above can serve as an active part of molecules participating as ligands, but also any of their quantitative and qualitative combinations in various mutual positions.

Using various synthetic methods elaborated and used in modern coordination chemistry, it is possible to control the obtaining of ligands with widely varied dentacity, that is the number of compounds with mono-, di-, and polychelating fragments forming macro-oligocyclic structures with metal-cycles of different size.

3.1 INTERACTION OF LIGANDS WITH METAL SALTS OR CARBONYLS

This method, also named "*immediate (direct) interaction of ligands and sources of metal center*" [2], is the oldest and most widespread, in most cases the safest, preparative method. It is comparatively accessible with, in general, high yields of final products – metal complexes. Precisely as a result of its use, the majority of Werner coordination compounds with the simplest ligands – water (aqua complexes), ammonia (amino complexes), halides, β-diketones, etc. – have been obtained. At present, the above method allows us to obtain practically all types of complexes (see Sec. 1.2) with all types of ligands (see Chap. 2).

The complex-formation reactions (Sec. 1.2.1) which take place using this method can be described (without taking into consideration solvation processes), in general, by simple schemes (3.1) and (3.2):

$$mL + MX_n \rightarrow L_mMX_n \tag{3.1}$$

$$mL + M(CO)_n \rightarrow L_mM(CO)_{n-m} + mCO \tag{3.2}$$

The first type of reaction is observed with participation of the ligands without hydrogen atoms, capable under usual conditions of being substituted on a metal ion, for example (3.3)–(3.5):

$$2RCN + MX_2 \rightarrow [M(RCN)_2X_2] \tag{3.3}$$

$$2PR_3 + MX_2 \rightarrow [M(PR_3)_2X_2] \tag{3.4}$$

$$2py + MX_2 \rightarrow [M(py)_2X_2] \tag{3.5}$$

The reactions (3.2) take place with a proton-donor, especially chelate-forming ligands, for example (3.6), (3.7):

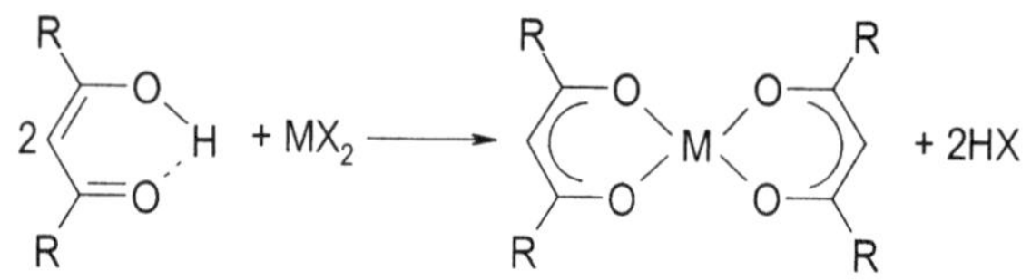

(3.6)

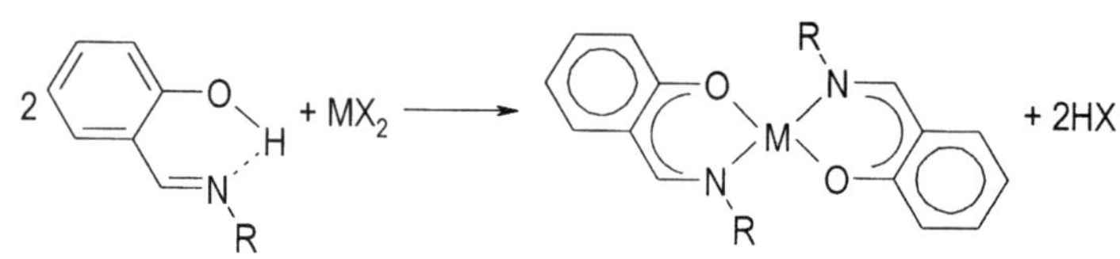

(3.7)

3.1.1 Selection of Ligands

A wide bibliography including a series of fundamental editions, for instance Refs. 1 and 3–6, is dedicated to this theme, covering also the synthesis of ligands and metal complexes. In this respect, the old, but capacious and useful right up to the present time, review article [7] is worthy of special attention. In this publication, geometric aspects of ligand selection, including the influence of the nature of donor atoms, their mutual positions in ligand systems, and the dentacity of the latter, are examined.

At present, the ligand selection is being carried out mainly to create complexes of definite type, control of their stereochemistry and stereodynamics of coordination compounds, their physicochemical and practically useful properties.

3.1.1.1 Molecular Complexes

Among N-donor ligands, ammonia, aliphatic, aromatic, and heterocyclic amines are widely used as ligands at present (see Sec. 2.2.4.2). Amine complexes are obtained by direct (immediate) interaction of ligands and metal salts, for example (3.8) [8]:

$$2NH_3 + Cu(NO_3)_2 \xrightarrow{H_2O} [(NH_3)_2Cu(NO_3)_2] \tag{3.8}$$

The interaction between monodentate aliphatic and aromatic amines ***156*** takes place according to the same scheme, as well as diamine ligands, the spatial positions of the donor N atoms in which exclude the possibility of formation of chelate structures, for instance ***160***, ***165***, ***166***. In this respect, we note the synthesis of liquid amine complexes (diethyl- and triethylamine) with molybdenum dioxodichloride in hexane medium (3.9) [9]:

$$m\,R^1R^1RN + MoO_2Cl_2 \longrightarrow [(R^1R^1RN)_mMoO_2Cl_2] \qquad (3.9)$$

$R = H, R^1 = Et, m = 2; R = R^1 = Et, m = 1$

Many publications are devoted to the synthesis of nitrile complexes, carried out by the immediate (direct) interaction of RCN and MX_n, mostly in the absence of a solvent [10, p. 95]. A series of N-donors, N-containing heterocyclic donors, whose complexes frequently model biologically important objects (Sec. 2.2.4.2), should be mentioned apart. The following compounds belong to this type: azoles ***188***, azines ***189***, and their amino derivatives ***572***. Their interaction with metal salts takes place usually without a solvent with the use of liquid heterocyclic ligands, for example pyridine [10, ch. 4, p. 107; 11], in alcohol or alcohol-aqueous mediums in cases of crystalline ligands (3.10)–(3.12). The specific conditions are presented in the literature, cited in Sec. 2.2.4.2.

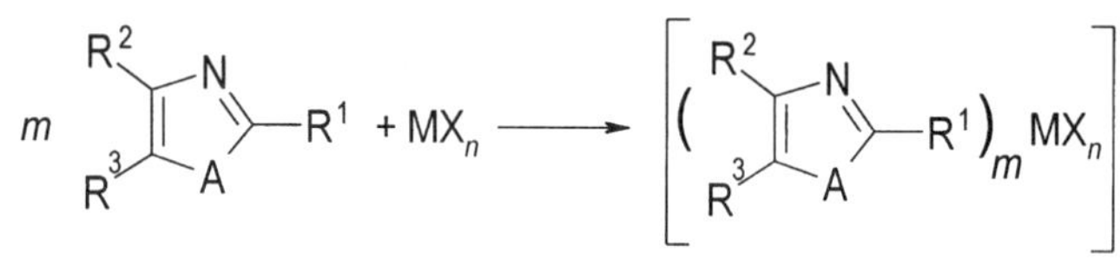

(3.10)

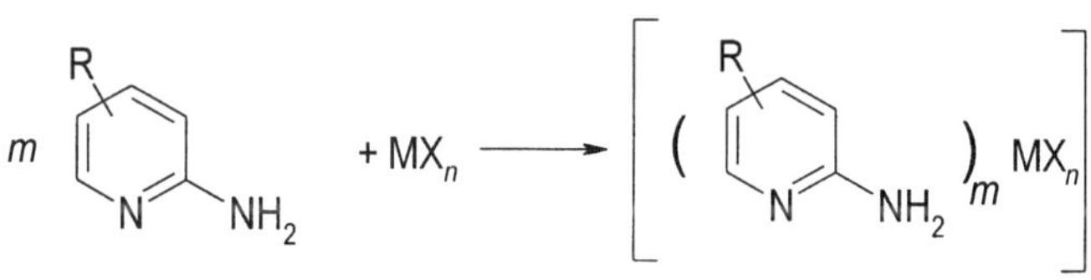

(3.11)

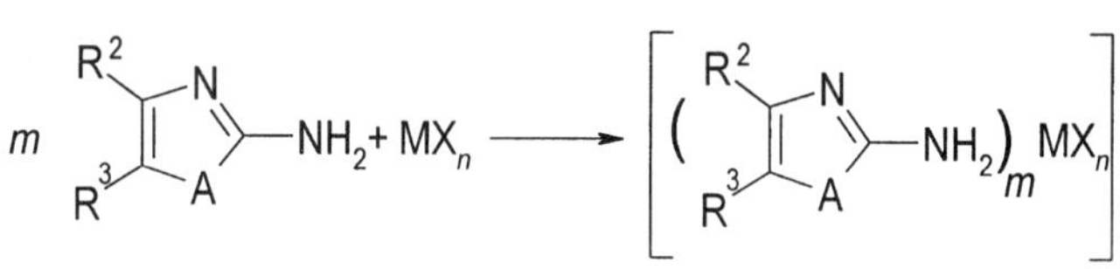

(3.12)

$A = NR^4$, O, S, Se; R, R^1, R^2, R^3 = H, Alk, Ar; X = Hal, NO_3, NCS, ClO_4, BF_4, SO_4; R^5, R^5COO; R^5 = Alk, Ar; $m = 1–6$

In this respect, the syntheses of a wide series of 2-amino-5-nitrothiazole ***572*** complexes with various metal(II) salts are representative. These reactions were carried out in methanol medium and led to adducts of type ***573*** [12] (3.13):

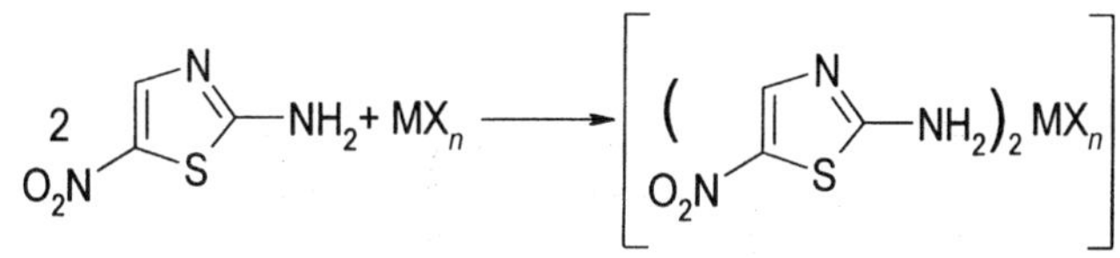

(3.13)

572 573

M = Co, Cu, Cd; X = Cl, NO_3, NCS, SO_4; *n* = 1, 2

The phosphines with one P-donor atom reveal properties close to those of amines and form mononuclear coordination compounds ***574*** or binuclear structures ***239*** (Sec. 2.2.4.3) and ***575*** (3.14):

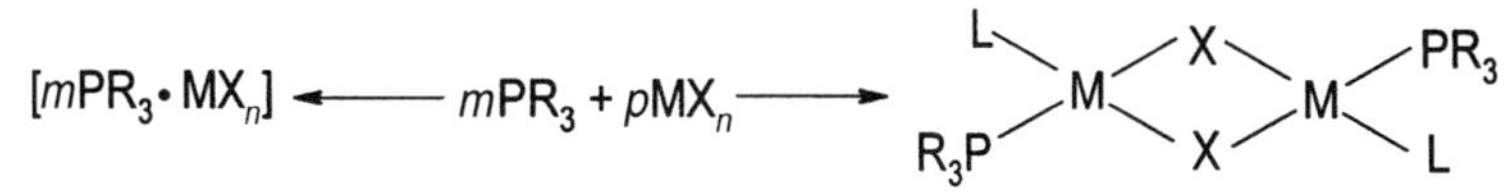

(3.14)

574 M=Pd, Pt 575

The simplest O-containing ligands are represented by water, alcohols, aldehydes, ketones, dioxane, organic acids, and their ethers. Dissolution of metal salts is used to prepare aqua complexes. Dissolving crystallohydrates of metal salts, aqua-containing coordination (in many cases, mixed-ligand) compounds are formed too. The examples of complexes of alcohols, acids, and dioxane are represented in a monograph [10, p. 95]. They could be illustrated by reactions (3.15), (3.16), and (3.17), respectively:

$$6MeOH + Zn(ClO_4)_2 \rightarrow [Zn(MeOH)_6](ClO_4)_2 \quad (3.15)$$

$$6AcOH + ZnX_2 \rightarrow [Zn(AcOH)_6]X_2 \quad (3.16)$$

$$X = ClO_4^-, NO_3^-, BF_4^-$$

$$6C_4H_8O_2 + Zn(ClO_4)_2 \rightarrow [Zn(C_4H_8O_2)_6](ClO_4)_2 \quad (3.17)$$

$$C_4H_8O_2 = \text{dioxane}$$

Among the O-containing heterocyclic ligands, tetrahydrofuran (THF) ***376*** is the subject of permanent interest [13]. As a result of its interaction (3.18) with metal salts, the molecular solvent complexes of the type ***298*** (Sec. 2.2.4.4) and ***577*** are formed, and can be used as precursors to obtain coordination compounds with other ligand systems:

(3.18)

576 577

The adducts ***579*** are obtained by direct (immediate) interaction (3.19) of chalcogen-ethers ***578*** – thioethers (Sect. 2.2.4.5), their selenium (Sec. 2.2.4.6), and tellurium (Sec. 2.2.4.7) analogues with metal salts:

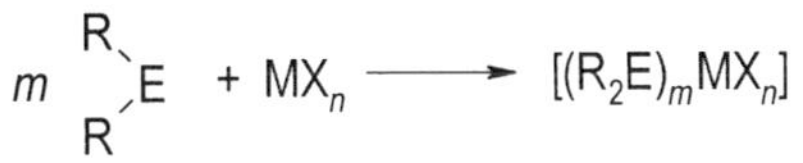

(3.19)

578 579

Interaction of chalcogen-containing heterocycles ***580*** (E = S, Se, Te) with metal salts yields molecular complexes of the type ***581*** (3.20):

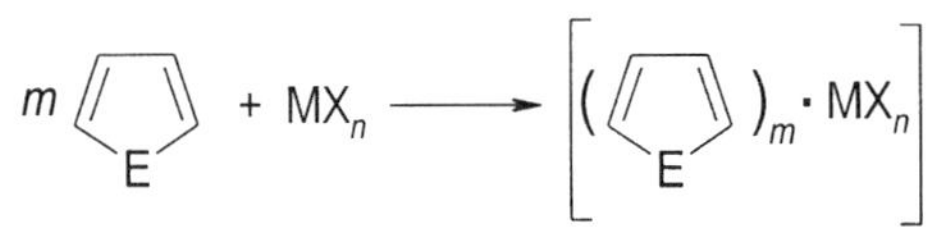

(3.20)

580 581

The readers can be acquainted with specific MX_n-ligand complexes and synthetic conditions of (3.18)–(3.20) in the references, cited in Secs. 2.2.4.4–2.2.4.7.

Among heteroatomic donors (Sec. 2.2.5), used as ligands for obtaining molecular complexes, the organosulfoxides ***582*** are widely used (Sec. 2.2.5.7) [1,10]. They are ambidentate O,S-donor ligands and, in this relation, reacting with hard and intermediate (MX_n) and soft ($M'X_n$) Pearson acids (3.21), they form molecular complexes with O (***583***)- and S (***584***)-coordinated ligands:

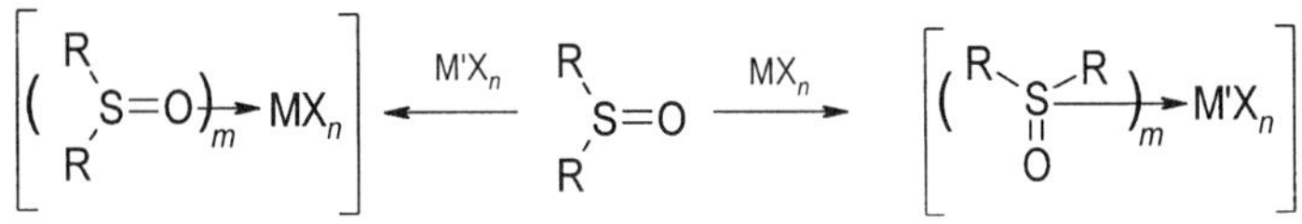

(3.21)

583 582 584

Reference 10 (Chap. 4) contains a large number of dimethylsulfoxide complexes, obtained by the method of direct (immediate) interaction of components. An interesting example of such a synthesis (3.22) is reported in Ref. 14a:

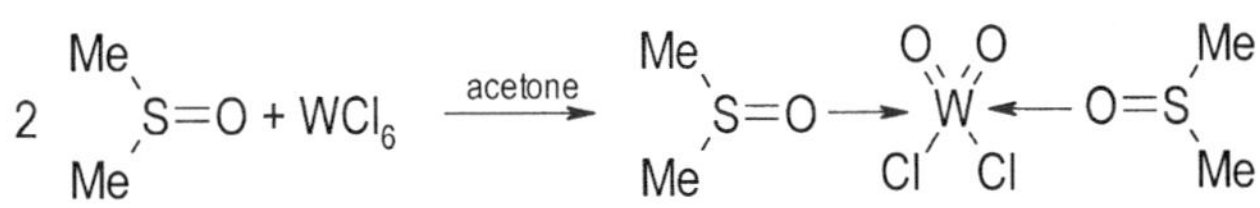

(3.22)

The described syntheses are carried out with the use of not only metal salts, but also metal hydroxides in the presence of acids [for example, (3.23)] [10]:

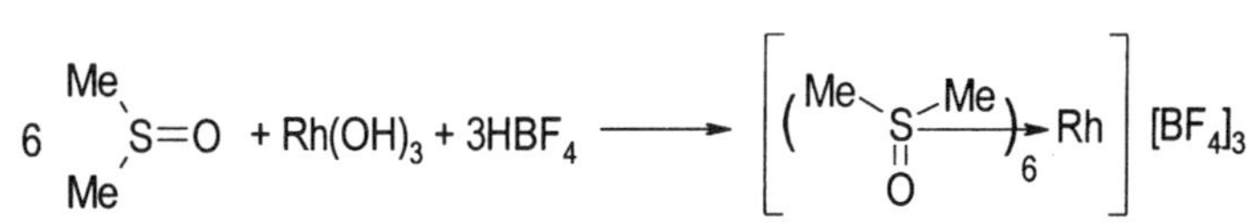

(3.23)

We note that complex salts are used for preparation of DMSO complexes with soft metals [for instance, (3.24)] [10]:

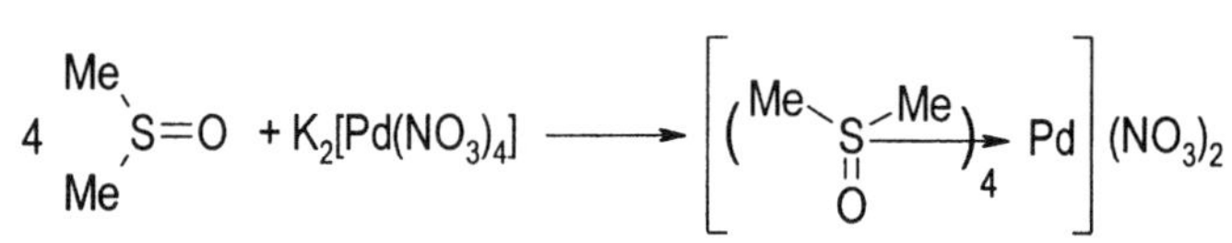

(3.24)

In the majority of cases, the ligand itself is used as a solvent in the synthesis of dimethylsulfoxide complexes.

EXPERIMENTAL PROCEDURES

Example 1

Synthesis of Complexes of Isoxazole (Is) with Titanium(IV) Chloride [14b]

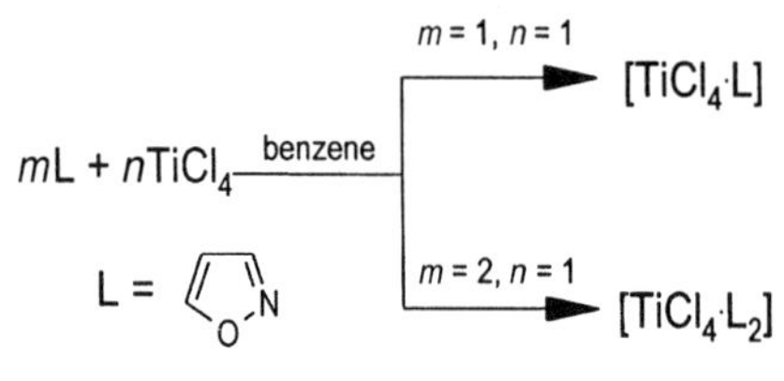

$TiCl_4Is$

A solution of 0.828 g (1.2 mmol) of Is in 15 mL of benzene was added slowly with stirring to a solution of 0.901 g (4.8 mmol) of $TiCl_4$ in 25 mL of the same solvent. After standing for 48 hr at room temperature, the mixture was concentrated and cooled in a refrigerator, and the yellow crystalline solid formed was filtered off, washed with benzene, and dried in vacuo.

$TiCl_4(Is)_2$

A solution of 0.883 g (12.8 mmol) of Is in 15 mL of benzene was added slowly with stirring to a solution of 1.21 g (6.4 mmol) of $TiCl_4$ in 25 mL of the same solvent. The yellow precipitate that formed (exothermically) almost immediately was filtered off, washed with benzene, and dried in vacuo.

Example 2

Synthesis of Complexes of 2-Amino-1,3,4-Thiadiazole (L) [14c,d]

$$ZnCl_2 + 2L \xrightarrow{EtOH} ZnCl_2L_2$$

A hot solution of 1.065 g (30 mmol) $ZnCl_2$ in 25 mL of ethanol was added with stirring to the hot solution of 0.606 g (6 mmol) of L. Mixture was stirred under refluxing for 10 hr, filtered off, and allowed to stand overnight. The formed yellow precipitate was filtered off and recrystallized from ethanol and dried in air. Yield 79%.

$$6L + 4CuClO_4 \xrightarrow{MeOH} [Cu_4L_6](ClO_4)_4 \cdot 2MeOH$$

$Cu(ClO_4)_2 \cdot 6H_2O$ (22.2 mg, 0.06 mmol) was reduced with copper wire under ethylene in methanol (10 mL). After the replacement of ethylene to argon atmosphere, the resultant Cu(I) solution was added to the L (18.2 mg, 0.18 mmol). The colorless solution was sealed in a 7-mm glass tube and allowed to stand at room temperature for a week. Colorless plate crystals were collected. Yield 16 mg (20%).

$$6L + 4AgClO_4 \xrightarrow{MeCOMe} [Ag_4L_6](ClO_4)_4$$

$AgClO_4$ (41.5 mg, 0.20 mmol) and L (20.2 mg, 0.202 mmol) were reacted in acetone (10 mL) under Ar. The colorless solution was sealed in a 7-mm glass tube together with the layered *n*-pentane. After standing at room temperature for a week, colorless brick crystals were obtained. Yield 10 mg (14%).

Example 3

Synthesis of $WO_2Cl_2 \cdot 2DMSO$ (3.22)[14a,e]

0.5 g of WCl_6 was dissolved in 50 mL of acetone and refluxed in order to remove the HCl, after that 1 mL of DMSO was added. The resultant solution was cooled to 0°C. Colorless precipitate was filtered off, washed with acetone, and dried in the dry argon stream. Yield 78%.

3.1.1.2 Metal Chelates

It is known that the general structural characteristic of chelating ligands is a particular position of the donor atoms when the metal-cyclic polyhedra of coordination compounds are formed. The chelating N-donors are widely represented by di-(***582***), tri- (***165***, ***166***,) tetra- (***170–173***), and penta- (***174–177***) amines (Sec. 2.2.4.2). Among diamine ligands, ethylenediamine (en) and its C- and N-substituted derivatives ***585*** are the most used [1]:

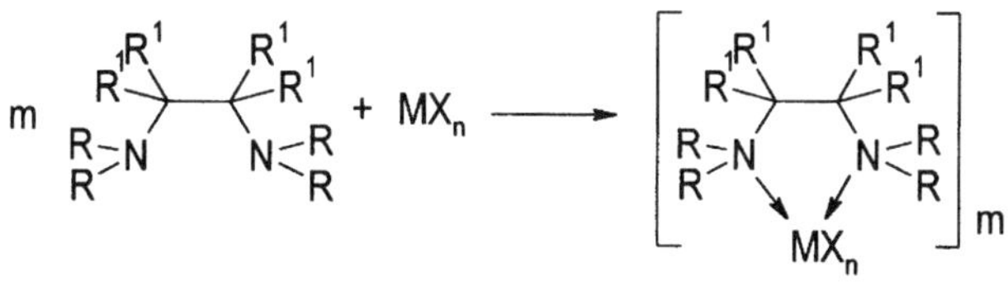

585 586

(3.25)

Their interaction (3.25) with metal salts yields metal chelates ***586***. Thus, the reaction of ***585*** ($R = R^1 = H$) with $Ni(NO_3)_2 \cdot 6H_2O$ and KSeCN in water leads to the formation of *trans*-$[Ni(en)_2(NCSe)_2]$, whose structure was proved by x-ray diffraction [15].

The diamine ligands of the type ***587*** have considerable importance for the formation (3.26) of chelates with metal-cycles of different size ***588***:

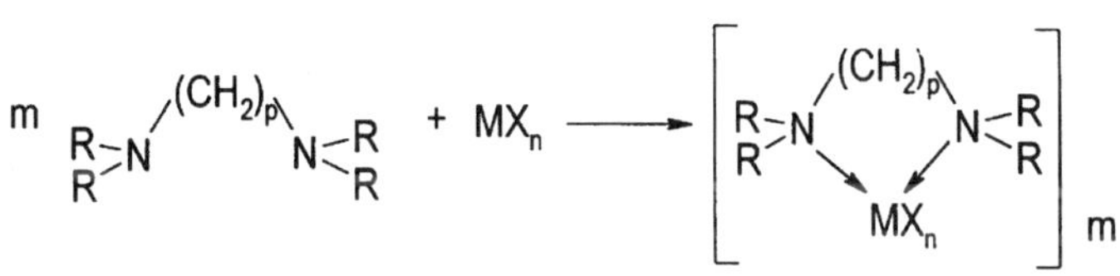

587 588

(3.26)

The complex compounds with five-member ($p = 2$) and six-member ($p = 3$) metal-cycles are the most propagated (widespread) in the series of chelates, obtained on the basis of ***587***.

Metal chelates with different sized metallocycles and coordination units MN_4 have been created (Sec. 2.2.4.2) on the basis of amidines ***214***, triazenes ***215***, dioximes ***216***, azomethines ***217–222***, β-aminovinylimines ***223***, *o*-aminoazocompounds ***224***, hydrazoneimines ***225***, and formazanes ***226***, ***228***, ***229***. In the majority of cases, the syntheses of these complexes have been carried out by interaction of the ligands above and metal salts (mostly metal acetates) in alcohol medium (methanol,

ethanol). The syntheses (3.27), (3.28) [16], (3.29) [17], and (3.30) [18] are representative in this respect:

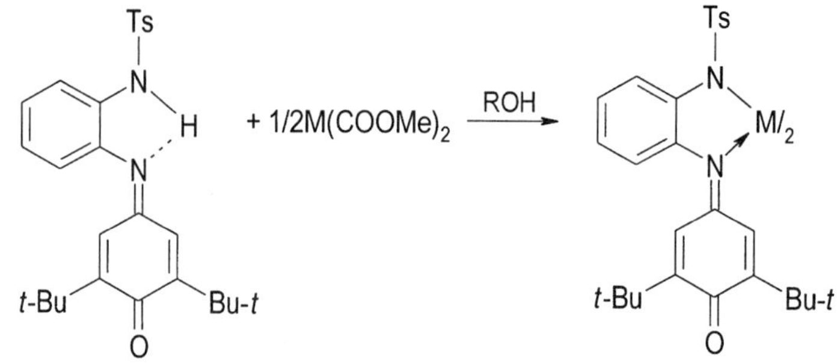

Ts = $SO_2C_6H_4Me$-*p* *589* (3.27)

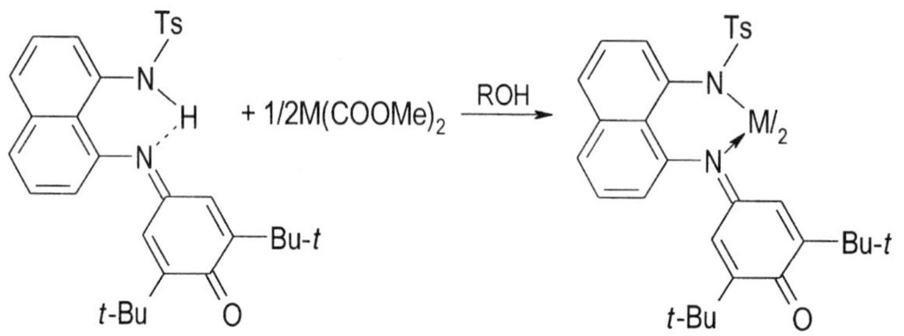

590

M = Co, Ni, Cu(ROH), Pd(MeCOMe) (3.28)

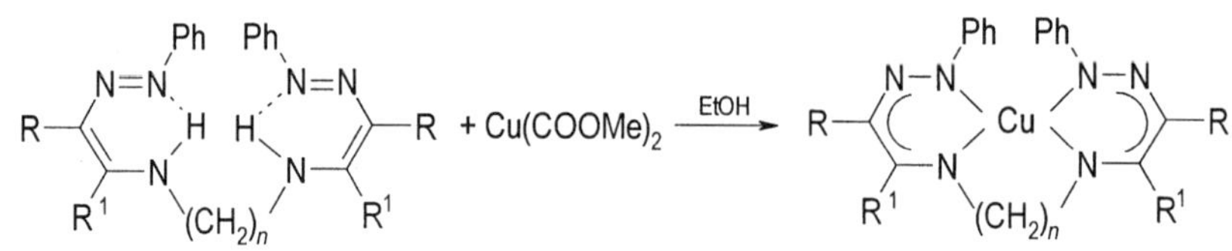

591

(3.29)

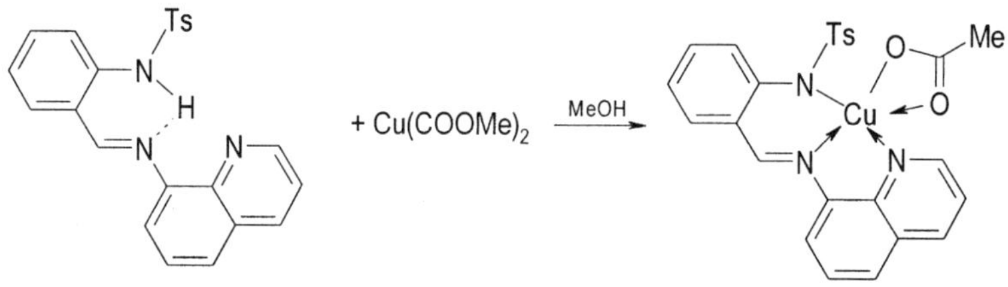

(3.30)

592

There are five-member metal-cycles in the chelates ***589***, six-member metal-cycles in ***590*** and ***591***, and both five- and six-member cycles in ***592***.

As ligands, forming complexes with five- and six-member metal-cycles, the following systems take part: di- (***198***, ***205–207***; see complexes ***210***), tri- (***212***), and tetra- (***213***) azines, *bis*-azoles ***593***, and azinoazoles (***594***, ***595***):

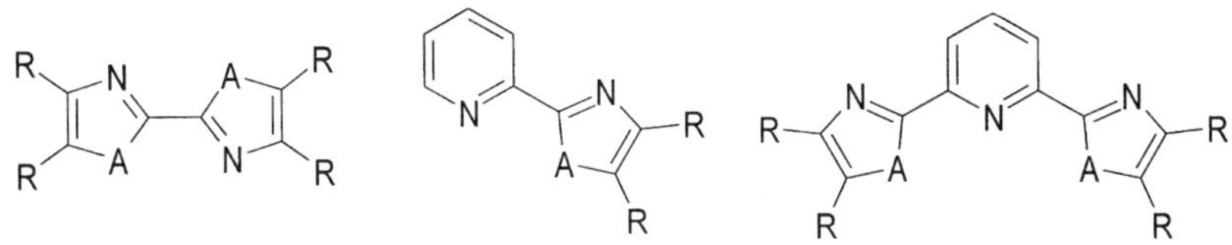

593 *594* *595*

A = NR[1], O, S; R, R[1] = H, Alk, Ar

It is necessary to mention that S-*trans*-conformations (trans position of N atoms of pyridine type ***198***, ***204***, ***593–595***) are characteristic; they change to *cis*-configuration under MX_n coordination [19–21]. Examples of such syntheses [(3.31), (3.32)] are described in Refs. 19 and 20:

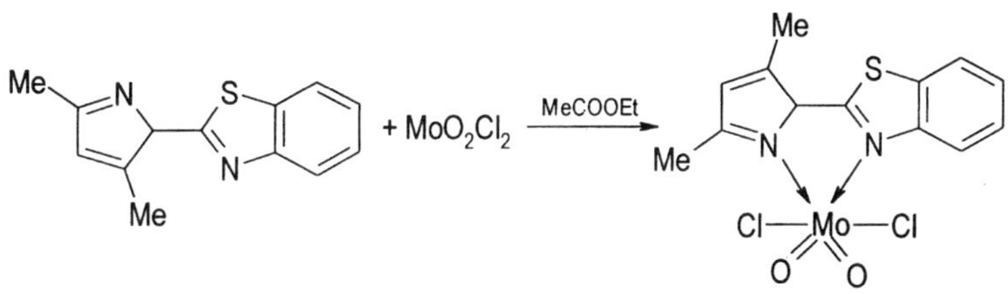

596 *597*

(3.31)

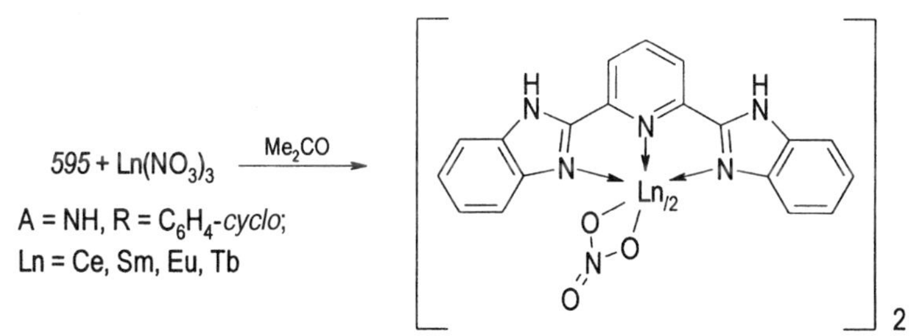

598

(3.32)

Among chelating polycyclic N,N-donors, the porphyrins ***237*** should be especially emphasized due to the important biological role of their metal complexes (Sec. 2.2.4.2). For synthesis of porphyrin complexes, the direct (immediate) interaction of ligands and source of metal ions is used. The example is the interaction of tetraphenylporphyrin ***237*** ($R^{\alpha,\beta,\gamma,\delta} = Ph$; $R^1 \div R^8 = H$) with $AlCl_3$ in high-boiling solvents (pyridine, phenol), leading (3.33) to formation of a macrocyclic chelate ***599***, containing four six-member cycles [22]:

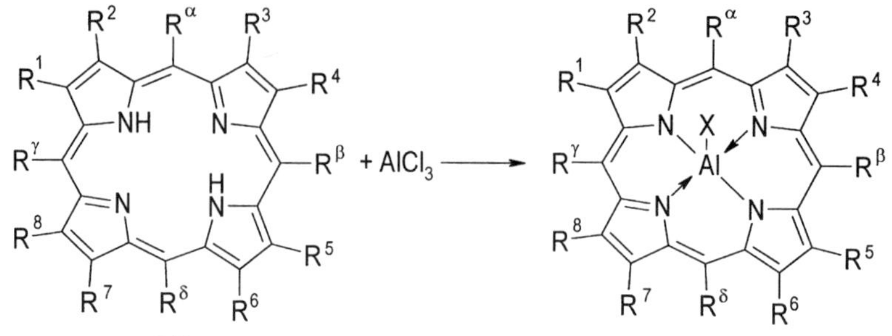

240 599

X = Cl (Py);
X = OH (PhOH)

(3.33)

Much attention is paid to metal chelates with the coordination units MN_n, obtained by direct (immediate) interaction of polypyrazolylborates (***232***, ***233*** (E = B)) with metal salts (Sec. 2.2.4.2).

Among chelating P- and As-donors, the diphosphines and diarsines with a general formula ***600*** are the most propagated. They interact (3.34) with metal salts forming metal-cyclic structures ***601*** (Sec. 2.2.4.3):

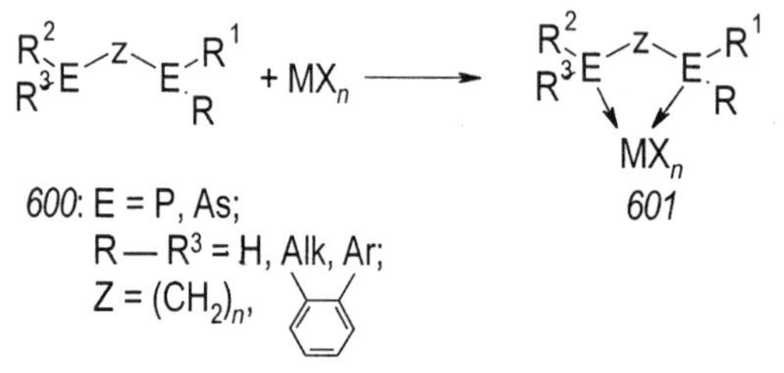

(3.34)

Such syntheses are illustrated by the reaction (3.35) [11]:

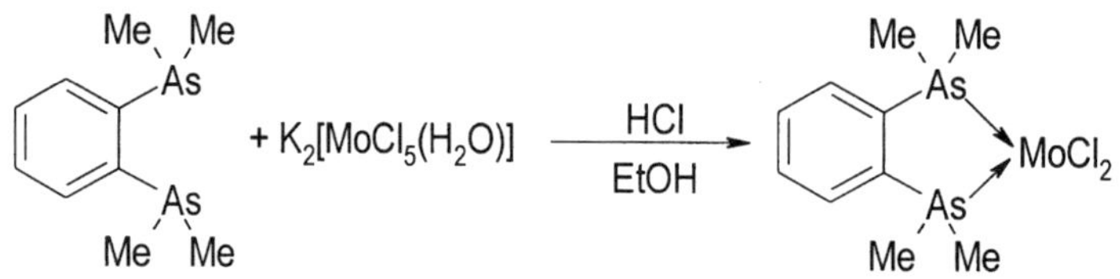

(3.35)

β-Diketones and crown-ethers (Sec. 2.2.4.4) are among the most frequently used O_m-donors. For obtaining β-diketonates by direct (immediate) interaction, the reactions of the examined ligands and metal salts in aqueous or aqua-organic solutions are used (3.36) [23 and references therein]:

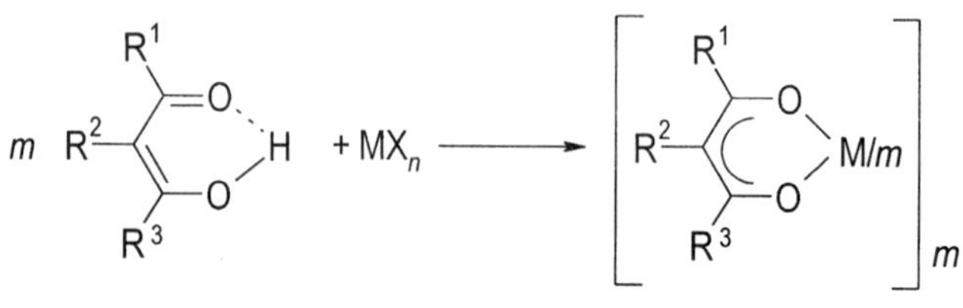

(3.36)

602 603

This transformation, comparatively simple, is complicated by solvolysis processes, leading to the formation of adducts and decomposition (especially hydrolysis) of the final products. Owing to the above, inert nonaqueous solvents are applied. An example of such reactions is the interaction of benzoyltrifluoracetone ***602*** ($R^1 = Ph$, $R^2 = H$, $R^3 = CF_3$) with $WOCl_4$ in benzene, yielding complex ***604*** (3.37) [24]:

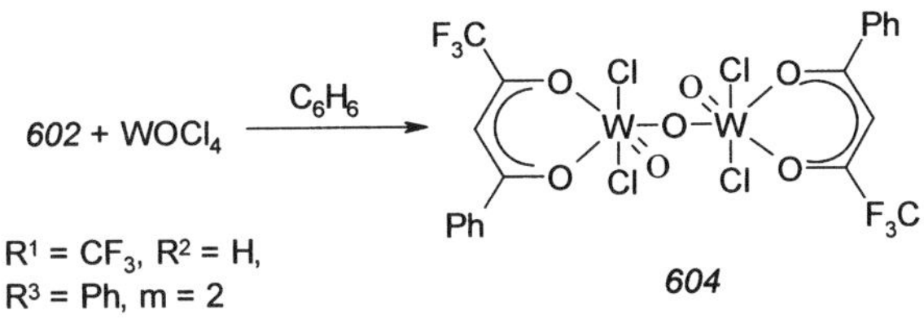

(3.37)

The method of immediate interaction of ligands and metal salts is a general way to obtain O_m-crown-ether complexes, for example (3.38):

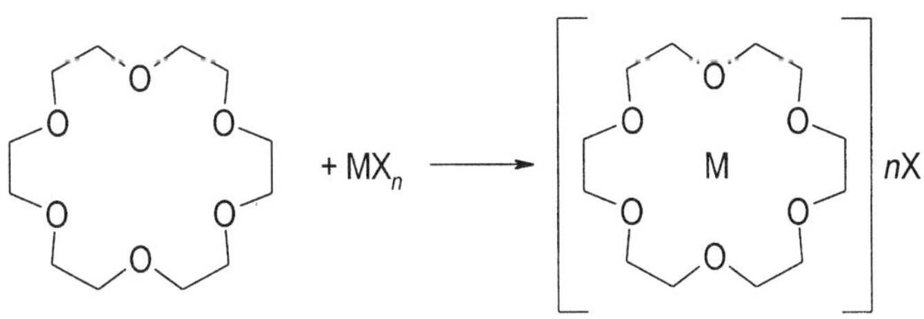

(3.38)

To carry out this transformation, a mixture of appropriate solvents or no solvent is used [25]. The final complexes are isolated by filtration of precipitate from reaction mixture, cooled or partially evaporated under low pressure.

Among crown-ether complexes, the coordination compounds of *s*-metals (for example, Na and K) are the most studied; these compounds are biomimetic models in the processes of metal transportation [26]. The interaction between crown-ethers and Lewis acids, containing metals of Groups III–VI, is well studied [27–40]. It is established by x-ray diffraction that, additionally to inner-cavity complexes, the products of opening and destruction (hydrolysis) of crown-ether rings are formed. Thus, the interaction of 18-crown-6 with $ZrCl_4$ in a mixture of toluene with THF (heating for 24 hr) takes place according to the scheme (3.39) [28]:

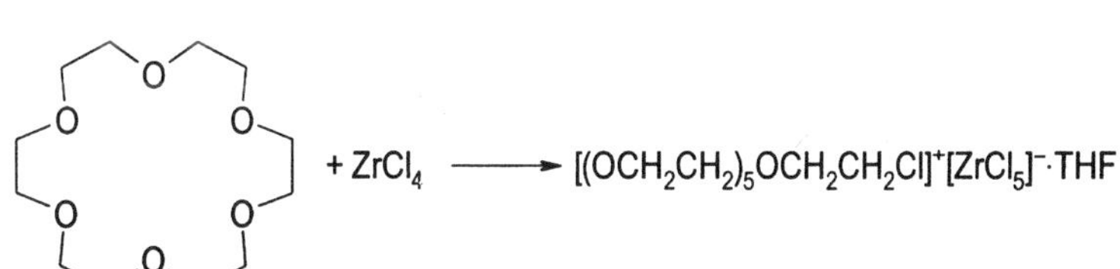

(3.39)

Reaction of 15-crown-5 with $TaCl_5$ in dry acetonitrile yields complexes of various products of decomposition of the crown-ether with the most surprising one $[(18\text{-crown-6})(H_3O)]^+[TaCl_6]^-$. The authors of Ref. 35 believe that formation of this complex most probably is related with destruction of 15-crown-5, growth of the chain of the opened polyether, and its recyclization, leading to a new ligand 18-crown-6. The complex formation of 15-crown-5 with $TaCl_5$ and alkali metal chlorides in acetonitrile results [36] in inner- (Li, Na) and outer-cavity (K, NH_4) compounds $[(15\text{–crown-5})LiMeCN][TaCl_6]$, $([15\text{-crown-5})Na][TaCl_6]$, and [(15-

crown-5)Li(MeCN)$_2$][$Cl_5TaOTaCl_5$]. The last complex was isolated and studied by x-ray diffraction [36]. It is also shown that, during this interaction, part of the molecules of the macrocyclic ligand are destroyed, forming water, which is responsible for hydrolysis processes, in particular, of $TaCl_5$.

References 37–40 are dedicated to the interaction of crown-ethers with uranyl diacetates. It is established [39] that, as a result of the interaction of 18-crown-6 with $UO_2(CH_3COO)_2 \cdot 2H_2O$ in ethanol medium, the following complexes are formed: [(18-crown-6)$_{0.5}$$UO_2(CH_3COO)(OH)(H_2O)$], [(18-crown-6)$UO_2(CH_3COO)(H_2O)_4$], and [(18-crown-6)$UO_2(CH_3COO)(OH)(H_2O)_3$]. The hydrolytic processes were observed in this process too.

The reaction between 15-crown-5 (L) and the same complex-former in toluene is also of undoubted interest [40]. The complex $[L(H_5O_2)(UO_2Cl_3)(H_2O)_2]$ is formed in the first stage; after elimination of HCl, the supramolecular structure $L_{16}[UO_2Cl_2(H_2O)_3]_{16}$ is formed.

Syntheses of chelate complexes of chalcogen-containing ligands are described in Secs. 2.2.4.5 (MS_n: ***308***, ***313***, ***315–317***, ***328–330***), 2.2.4.6 (MSe_n: ***340–342***, ***345***, ***348***) and 2.2.4.7 (MTe_n: ***357***).

Chelates with C,N-ligand environment are obtained mostly as a result of cyclometallation reactions [direct (immediate) interaction of organic compounds with *o*-arranged CH-fragment and N-donor center (Sec. 2.2.5.1)]. As a result of this transformation, described by a general scheme (2.8), the C,N-metal chelates are formed, for example ***371–379***. Metal acetates and halides, as well as metal complex salts $M'_2[MCl_4]$ (M = Pd, Pt; M′ = K, Li), are used as sources of metal centers [41–48]. Alcohols (methanol, ethanol), acetone, dioxane, chloroform, acetic acid and, in some cases, aqua-organic media are used as solvents [41]. The following examples [(3.40), (3.41)] can illustrate this transformation:

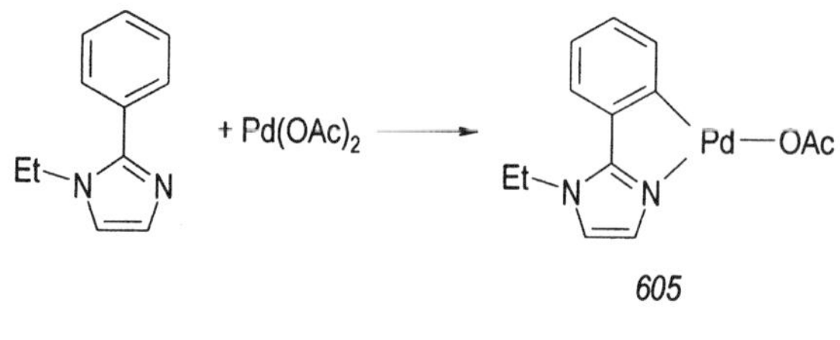

(3.40)

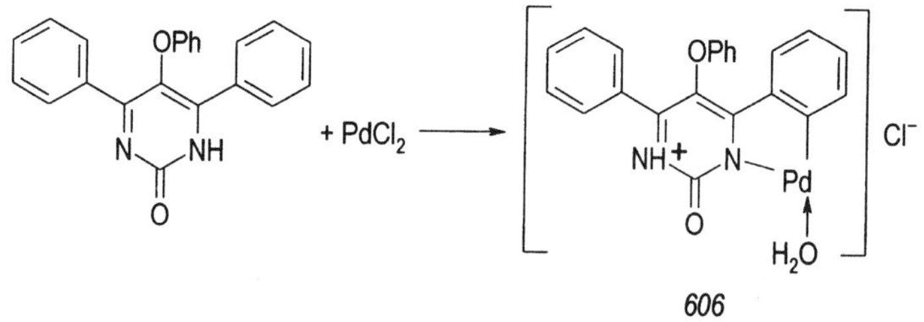

(3.41)

As shown in Sec. 2.2.5.1, the majority of reported data on cyclometallation is related to metals of the platinum group, for example ***606*** [49]. In this respect, the cyclomercuration, described by reactions (3.42) and (3.43) which results in mercury complexes ***607*** and ***608*** [50], represents a definite interest:

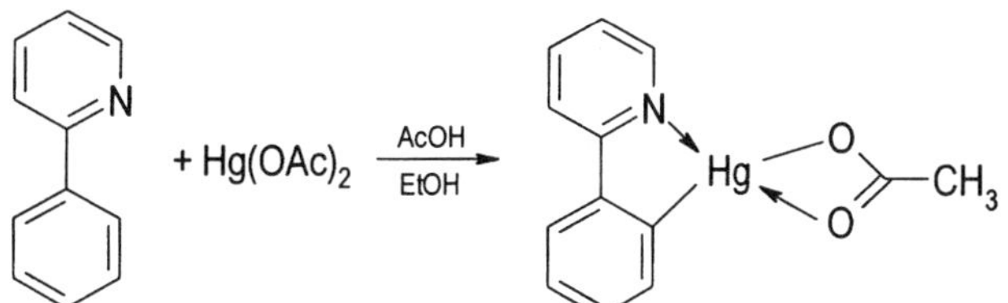

607 (3.42)

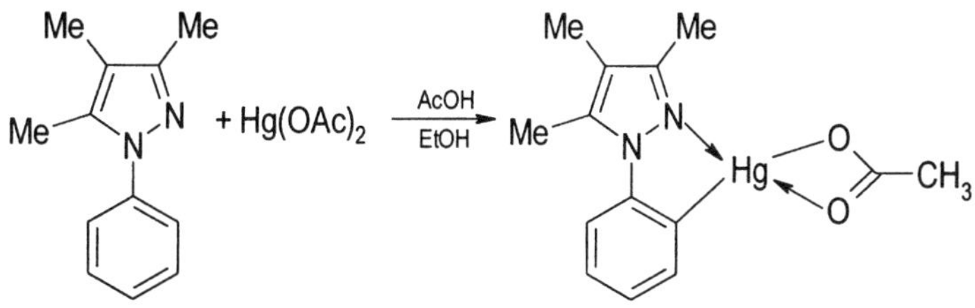

608 (3.43)

There is a reason to consider (Sec. 2.2.4.4) that the acetyl anion carries out a chelate function (***274***) in final products. However, the corresponding x-ray data of these complexes are absent.

An interesting reaction (3.44), which can be examined as C,N-cycloplatination, is reported in a monograph [51, p. 159]:

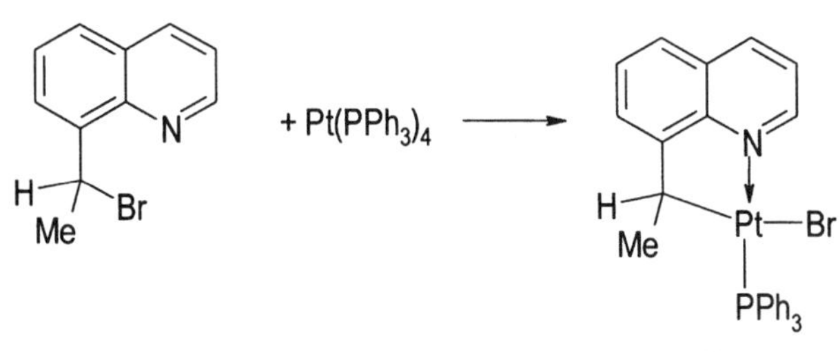

609 (3.44)

An important place among chelating ligands belongs to N,O-donors (Sec. 2.2.5.3) which, similarly to discussed above with N,N-donor ligand systems, allow us to create programmed metal-cycles of different size (Sec. 2.2.5.3: ***414***, ***418***, ***422***, ***424***, ***427***, ***428***, ***432***, ***433***, ***436***, ***438***, ***445***, ***446–449***). To obtain metal chelates with four-member chelate unit, the interaction of, for example, 2-hydroxy derivatives of azines and metal salts is used (3.45) [52]:

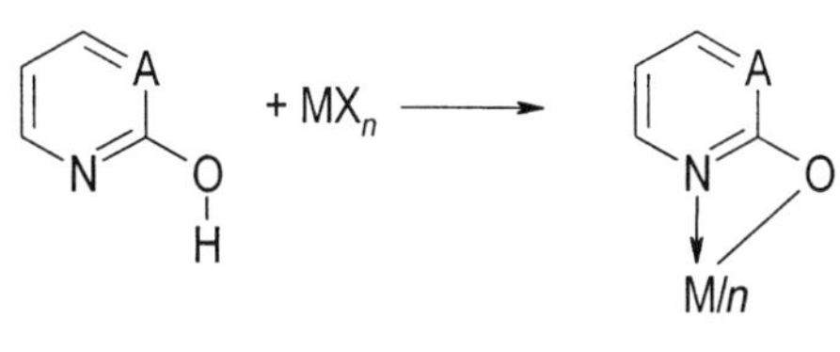

(3.45)

Additionally to 8-hydroxyquinoline, forming complex compounds of the type ***445***, convenient ligands for the synthesis of five-member metal-cycles are arylidene-*o*-aminophenols ***610***. Their interaction (3.46) with metal acetates in methanol results in complexes ***611*** [53–56]:

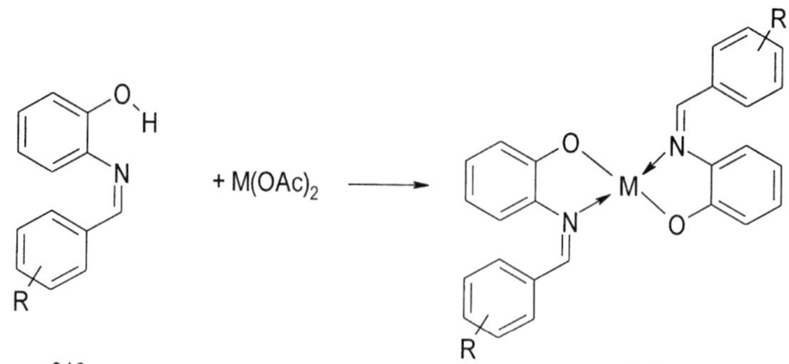

610 611

(3.46)

The majority of complexes with six-member metal-cycles are represented by chelates of the type ***422***, obtained according to the scheme, analogous to (3.46), with participation of salicylidenanilines and their derivatives. Among the latter, we note the tetradentate ligands ***612*** (LH_2), whose interaction (3.47) with manganese acetate leads to chelates MnL(OCOR) ***282*** [57]:

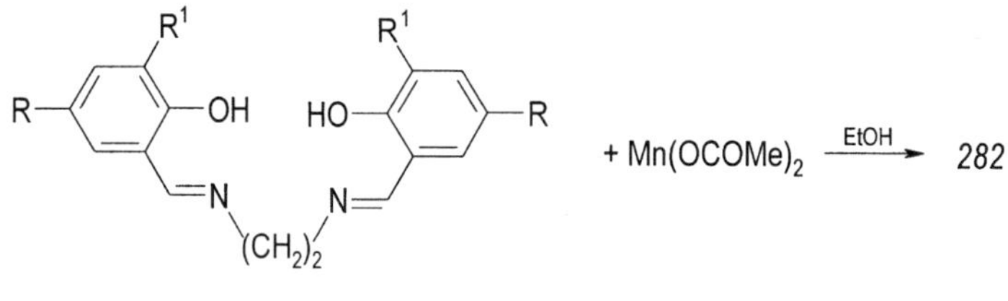

612

(3.47)

We note that, additionally to metal acetates, other metal sources are applied to obtain chelates of the type ***422***. In this respect, the reaction (3.48), carried out in a recent research [58], is representative:

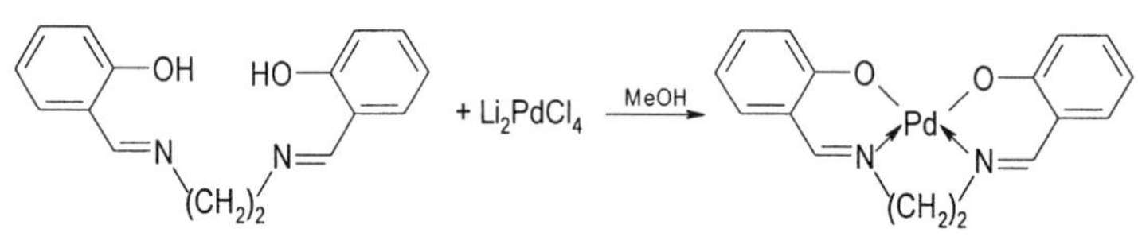

(3.48)

Additionally to the above complexes, six-member metal-cycles are formed in the chelates obtained as a result of interaction of *o*-hydroxyazocompounds, having a general formula ***426***, and 2-hydroxybenzazoles, mostly with metal acetates in methanol medium. These syntheses are described, for example, by the reactions (3.49) [59] and (3.50) [60]:

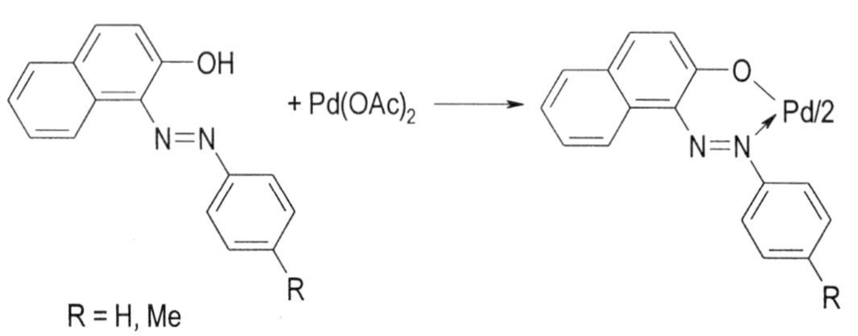

(3.49)

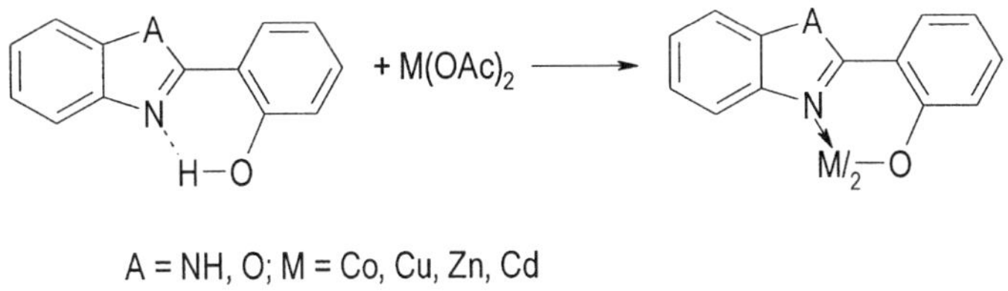

(3.50)

N,E-Containing ligands (E = S, Se, Te), especially those with free EH group, different from similar N,O-donors, are low-stable compounds. This fact complicates their direct interaction with metal salts. Therefore, the discussed method is applied in the case of ligand systems, containing ER fragments ***491***, ***511***, ***513***, and allows us to obtain chelates of type ***455***, ***507***, ***516***, ***519***, ***529***, ***530***, ***531***, ***533***, ***534***. Besides, the method of immediate (direct) interaction is applied under use of nitrogenchalcogen-containing chelating ligands, existing in stable tautomeric forms. These ligands are ***461b*** and ***471b***; on their basis, the coordination compounds ***462*** (2.12), ***463***, ***468***, ***472***, ***473***, ***475–478*** have been synthesized. Their specific synthetic techniques are represented in the literature, cited in Secs. 2.2.5.4, 2.2.5.5, and 2.2.5.6.

At the same time, some case are known when it is possible to carry out complex-formation reactions with N,S-donor chelating systems having a free SH group, for example (3.51) [61], (3.52) [62], (3.53) [63], and (3.54) [64]:

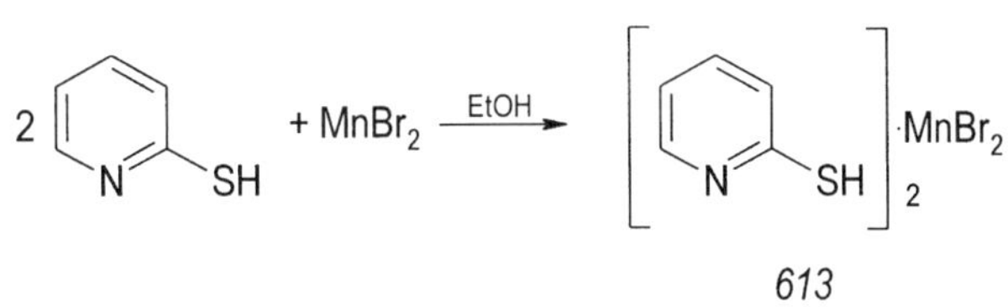

(3.51)

The complex ***613*** is not structurally characterized, so it is not clear whether a metal-cycle exists in this compound. An analogous situation with $Mn(OAc)_2$ leads to a polymer structure with chelate-bridge MnN_2S_2 fragments (x-ray diffraction data [61]).

The compound ***614*** has an interesting alterdentate ligand system (because of a possibility to form chelate cycles on different N atoms), which forms a chelate ***615*** having five-member metal-cycle [62]:

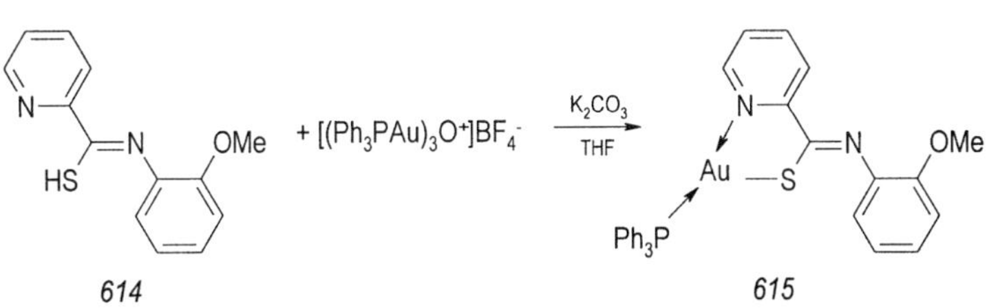

(3.52)

The metal-cycles, analogous by their linkage, exist in the coordination compound ***616***, obtained (3.53) on the basis of thiosemicarbazone of thiophene aldehyde [63]:

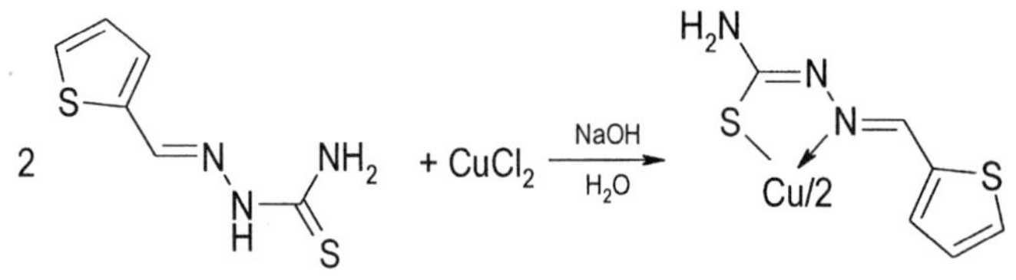

(3.53)

616

We pay special attention to the reaction (2.12). If we take into account the tautomeric equilibrium of thiazoline and mercaptoazomethine forms, it is possible to consider the ligand ***461*** as a compound with free SH group in this case also. Among the syntheses (2.12), we note the recently reported reaction of 2-phenylbenzthiazoline ***617*** with cobalt and nickel acetates (3.54), resulting in structurally characterized chelates of the type ***618*** [64]:

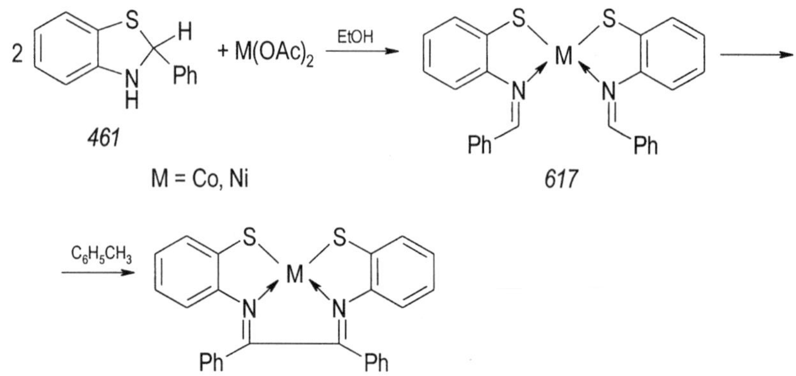

(3.54)

618

It is necessary to mention that the complexes ***617*** (M = Co, Ni, Zn) were synthesized earlier by heating the same reactants in methanol [65]. Temperature conditions play a considerable part in the syntheses above. Thus, at elevated temperature the chelate ***617*** is easily transformed to ***618***.

Another example, confirming the dependence of the type of formed complexes on temperature in the examined synthetic reactions, is the transformation (3.55) [66]:

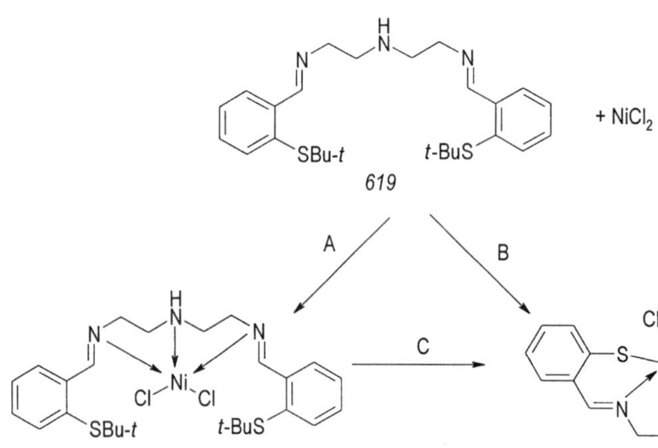

(3.55)

620 621

Route A takes place at 0°C and route B under boiling of the reaction mixture. According to the opinion of the authors of Ref. 66, the synthesis of the complex compound ***621*** takes place (routes A and C) through the intermediate formation of the chelate ***620*** (A), and it is accompanied by isolation of the final product, having one azomethinic group hydrolyzed and reduced. These processes, especially in the case of low-stable hydrolytic ligands of the type ***462*** [53], should be taken into account in the syntheses with use of azomethinic ligands.

The synthetic method of obtaining coordination compounds of the examined type was elaborated, based on the destruction of the disulfide bond by action of metal carbonyls [53]. One example of such a synthesis is the transformation (3.56), leading to hard-accessible and low-stable complex ***622*** [67]:

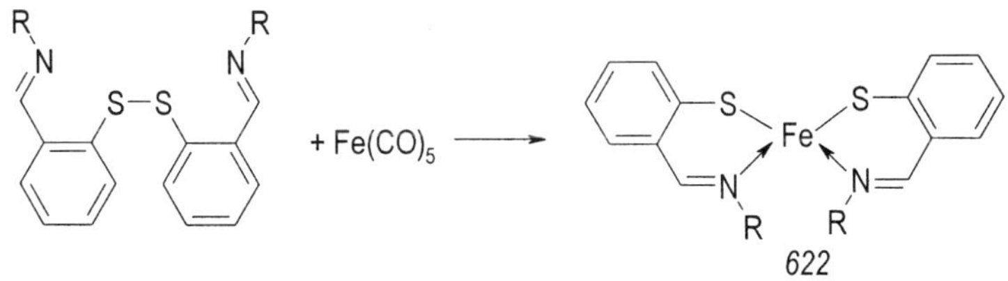

(3.56)

Examples of syntheses of N,Se- (3.57) [68] and N,Te-containing (3.58) [69], (3.59) [70a] chelates are the following reactions:

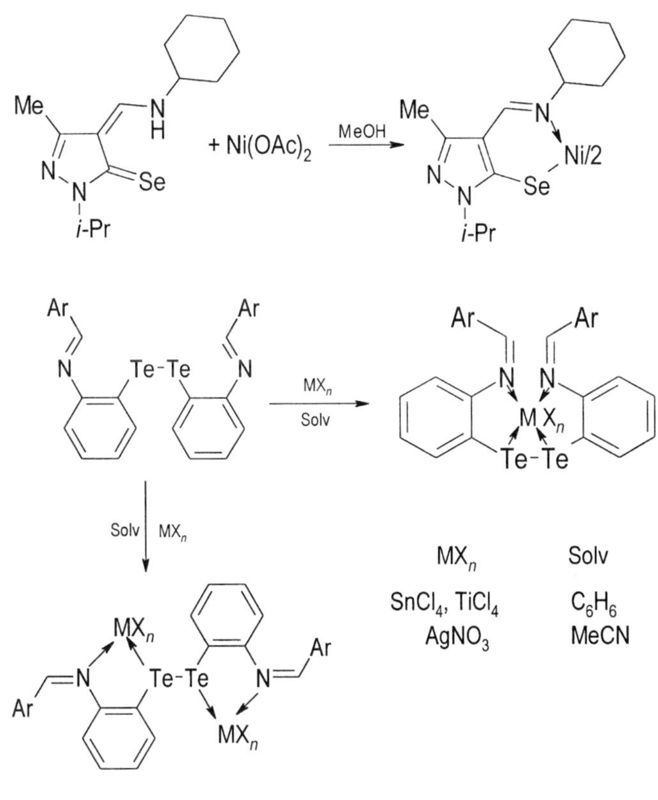

(3.57)

(3.58)

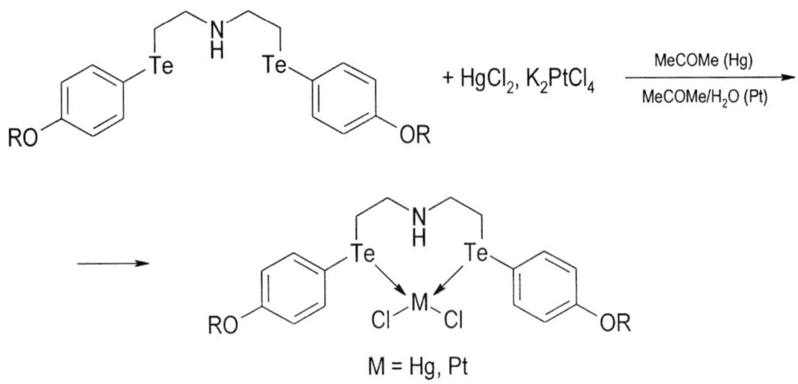

(3.59)

The data presented in this section and in Chap. 2 can help open the way to selecting chelating ligands and synthesizing on their basis chelates with various donor atoms and controlled linkage of metal-cycles.

EXPERIMENTAL PROCEDURES

Example 1

Synthesis of cis- and *trans*-Isomers of Platinum(II) Chelates of *o*-Hydroxyazocompounds [70b]

Synthesis of *bis*(2-hydroxy-4′-methylazobenzeato)platinum(II) was performed by mixing a 2-hydroxy-4′-methylazobenzene (0.42 g, 2 mmol) solution in methanol, a KOH (0.14 g, 2.5 mmol) solution in methanol, and a K_2PtCl_4 (0.42 g, 2 mmol) solution in DMSO. The precipitate (a mixture of *cis*- and *trans*-isomers) was dissolved in the minimally possible volume of toluene, and the hot solution was filtered through a thin bed of silica gel for column chromatography. The adsorbed product was eluted with 50–80 mL of hot toluene. After evaporation and cooling the filtrate and the eluate, crystals of the trans complex of platinum were formed (m.p. 254–255°C), yield 40–45%. Subsequent elution from the sorbent with hot chloroform (80–120 mL) followed by evaporation of the collected fraction to 20 mL and the addition of 5 mL of methanol resulted in the precipitation of crystals of the cis complex [m.p. 243–244°C (decomp.)], yield 4–6%.

Example 2

Synthesis of $Cu(acac)_2$ [71]

$Cu(acac)_2$ was obtained by coupling the water (50 mL) solution of $Cu(MeCOO)_2 \cdot H_2O$ (5 mmol) and water-ethanol (10 mL) solution of Hacac (10 mmol) under stirring. The resulting blue precipitate was filtered, dried, and purified by vacuum sublimation at 180°C. Yield 85%.

Example 3

Synthesis of Cycloaurated Complexes of 1-Ethyl-2-phenylimidazole [72]

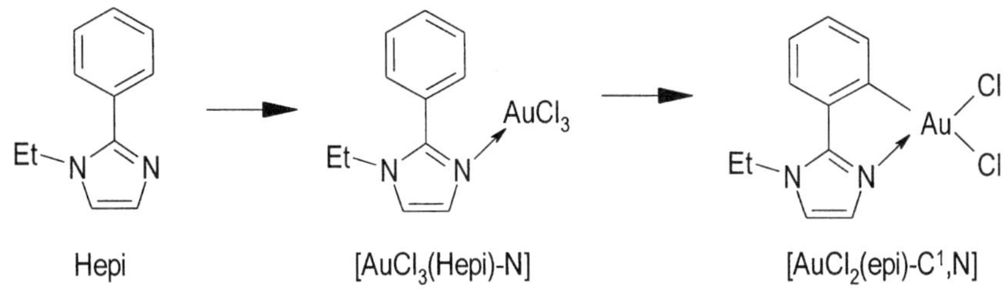

$[AuCl_3(Hepi)]$

An ethanol (5 mL) solution of Hepi (0.522 g, 3.03 mmol) was added to a solution of $HAuCl_4 \cdot 4H_2O$ (0.610 g, 1.48 mmol) in the same solvent (7.5 mL). The resulting mixture became a yellow suspension after stirring at room temperature for 1 hr. After 1 hr the yellow precipitates were collected and washed thoroughly with ethanol to give complex $[AuCl_3(Hepi)]$ (0.6667 g, 95%), m.p. 162°C.

$[AuCl_2(epi\text{-}C^1,N)]$

A dichloromethane (25 mL) solution of $[AuCl_3(Hepi)]$ (0.341 g, 0.716 mmol) was added to a dichloromethane (25 mL) suspension of $AgBF_4$ (0.141 g, 0.724 mmol). The resulting mixture was refluxed for 6 hr and then the volatile materials were removed in vacuo. The residue was extracted with acetonitrile to remove unreactive adduct $[AuCl_3(Hepi)]$. The extract was evaporated to dryness and the residue extracted with dichloromethane. The dichloromethane extract was dried in vacuo and the resulting white residue washed successively with a small amount of acetonitrile and diethyl ether to give complex $[AuCl_2(epi\text{-}C^1,N)]$ (0.063 g, 20%), m.p. 270°C (decomp.).

Example 4

Synthesis of the Chelates *589* and *590*

Chelates ***589*** (3.27) and ***590*** (3.28) [16] were obtained by refluxing the methanol (or ethanol in the case of ***590***) solutions of corresponding ligands and metal acetate in a 2:1 ratio for 15 min. Polycrystalline precipitates were filtered out, washed by corresponding solvent, and dried in vacuo at 150°C. Yields 40–50%.

Example 5

Synthesis of Tetraphenylporphyrinic Complexes of Aluminum(III) (3.33) [22]

Chloride of tetraphenylporphyrinic complex of aluminum(III) ***599*** (X = Cl; R^1–R^8 = H; R^α–R^γ = Ph) was obtained by refluxing the solution of H_2TPP and $AlCl_3$ (1:6) in pyridine (10 mL at 1 g $AlCl_3$) for 6 hr. After cooling 100 mL of water and 30 mL of acetic acid were added to the reaction mixture. The precipitate was filtered, washed with warm water, and dissolved in a minimal amount of $CHCl_3$. This solution was purified by column chromatography (Al_2O_3, $CHCl_3$) twice. The red zone was collected and the solvent was evaporated in vacuo. Yield 61%.

Hydroxide of tetraphenylporphyrinic complex of aluminum(III) ***599*** (X = OH; R^1–R^8 = H; R^α–R^γ = Ph) was obtained by reaction of H_2TPP and $AlCl_3$ (1:6) in boiling phenol (500 K) for 6 hr. The reaction mixture was cooled to room temperature, dissolved in $CHCl_3$, washed more than once with warm water, and poured into a column with Al_2O_3. Chromatographic purification was performed analogously to the previous complex. Yield 81%.

Example 6

Synthesis of Solvate of μ-Oxo-*bis*(benzoyltrifluoroacetonato)ditungsten(VI) Dioxotetrachloride with Benzene $[\{WOCl_2(CF_3C(O)CHC(O\text{-})C_6H_5))\}_2(\mu\text{-}O)]2C_6H_6$ (3.37) [23]

To a dredge of 1.74 g (0.005 mol) of $WOCl_4$ in 20 mL of hot benzene the solution of 1.08 g (0.005 mol) benzoyltrifluoroacetone in 10 mL of the same solvent was added dropwise under stirring. The reaction mixture was refluxed on the water bath for 3 hrs to remove HCl, after that the part of solvent was distilled out, and 10 mL of hexane were added to the mixture cooled to the room temperature. The red-brown crystals were formed after 2–3 hrs. They were filtered, washed with hexane and dried in the stream of dry argon. M.p. 101–103°C.

Example 7

o-Palladation of 4,6-Diphenyl-5-phenoxy-1,2-dihydropyrimidine-2-on (YH_2) ***606*** (3.41) [49]

1 mmol (0.18 g) of $PdCl_2$ and 1 mmol (0.34 g) of YH_2 were added to 10 mL of water and 40 mL of acetone and the resulting mixture was refluxed for 2 hr under stirring. After cooling the yellow precipitate was filtered, washed with water and acetone, and dried at 105°C. Yield 80%.

Example 8

Synthesis of Copper(II) Complexes of N-(quinolyl-8)-[2-N-tozylamino(hydroxy)] Benzilideneimine [18]

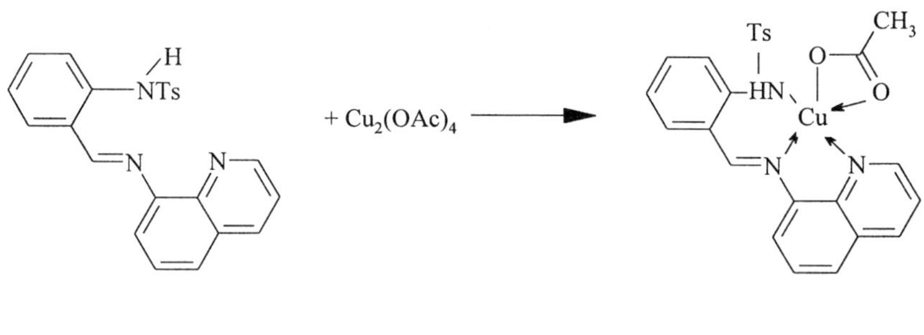

Complexes were synthesized by coupling the corresponding ligands and copper(II) acetate in 1:1 molar ratio in methanol for 5 min. Dark precipitates were filtered, washed with cold methanol, and dried in vacuo at 150°C. Yields about 75%.

Example 9

Synthesis of Metal Chelates of 4-Aminomethylenic Derivatives of 5-Thio(seleno)pyrazoles (2.13), (3.57) [68,73a]

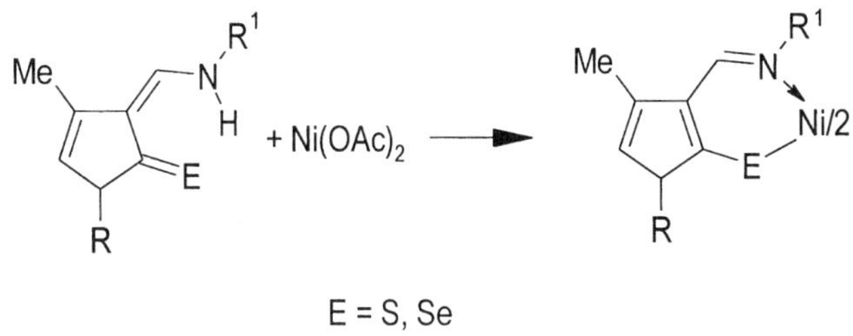

Complexes were prepared by refluxing (5–10 min) equimolecular quantities of nickel acetate with the respective 4-aminomethylene derivatives of 1,3-substituted 5-thio(seleno)-N-pyrazoles in ethanol solution. All the complexes thus obtained were purified by repeated crystallization from benzene.

Example 10

Synthesis of *bis*[*o*-Phenylene-*bis*(dimethylarsin)]molybdenum(II) Dichloride (3.35) [11]

All operations should be performed in the absence of air. *o*-Phenylene-*bis*(dimethylarsin) (DAS) (1.4 g) was dissolved in 50 mL of free-from-air ethanol in a three-funnel flask. A solution of 0.62 g of $K_2[MoCl_5(OH)_2]$ in ~ 5 mL of 6 N HCl was added. After that the flask was vacuumed more than once and filled by the free-from-air nitrogen. After the pressures were levelled the flask was carefully heated at 50°C for 24 hr. The bright-yellow precipitate was filtered, washed with a small amount of diluted HCl, free-from-air ethanol, and dried in vacuum. Yield 0.5 g.

Example 11

Synthesis of [1,2-*bis*-Salicylideneiminopropaneato]palladium(II) (3.48) [58]

A solution of 0.24 mmol of Li_2PdCl_4 in 10 mL of methanol was added to the solution of 0.24 mmol of 1,2-*bis*-salicylideneiminopropane in 20 mL of the same solvent. After heating the yellow precipitate dropped out. The precipitate was filtered, washed with methanol, diethyl ester, and dried in air. Yield 80–85%.

Example 12

Synthesis of Nickel(II) Complex with 2(2′-N-phenylaminonaphthylazo)-1-octylbenzimidazole (HL) [73b]

A solution of $Ni(CH_3COO)_2 \cdot 4H_2O$ (0.248 g, 1 mmol) in 10 mL of CH_3OH was added to a solution of HL (0.95 g, 2 mmol). The mixture was boiled for 1 hr. After cooling, the formed crystals were filtered, recrystallized from CH_3OH, and dried in a vacuum oven at 150°C. Yield 75%.

Example 13

Synthesis of [Ph(pz^{tBu})Bt^{tBu}]Zn(CH$_3$)$_2$ {[Ph(pz^{tBu})Bt^{tBu}}=phenyl(3-*t*-butylpyrazolyl)*bis*{(*t*-butylthio)methyl)borate} [73c]

$Zn(CH_3)_2$ (0.3 mL, 0.6 mmol) was added to [Ph(pz^{tBu})Bt^{tBu}]H (250 mg, 0.6 mmol) in toluene (20 mL). The solution was stirred for 1 hr, and then the solvent was removed under reduced pressure. The resulting white solid was washed with 15 mL of pentane and purified by re-crystallization from CH_2Cl_2/Et_2O. Yield 274 mg (92%).

3.1.1.3 π-Complexes

The method of direct (immediate) interaction of reactants is widely used for obtaining complexes with multicenter coordination bonds (π-complexes). Using it, some methods to prepare *ethylene complexes* have been elaborated. Thus, the interaction of ethylene with chloroplatinatodipotassium in the presence of $SnCl_2 \cdot 2H_2O$ in nitrogen atmosphere (3.60) yields monohydrate of potassium trichloro (η-ethylene)platinate(II) (salt of Zeise) [11]:

$$C_2H_4 + K_2[PtCl_4] \xrightarrow[SnCl_2,\ N_2]{HCl/H_2O} K[PtCl_3(\eta^2\text{-}C_2H_4)]H_2O + KCl \qquad (3.60)$$

Other methods for obtaining complexes of ethylene and other alkenes include ligand substitution reactions, reduction of a higher valent metal in the presence of an alkene, and synthesis from alkyl and related species [reductive elimination, of an allyl or hydride, for example; hydride abstraction from alkyls; protonation of sigma-allyls; from epoxides (indirectly)] [74a].

Such *d*-metals as Cu(I) [but not Cu(II)], form a variety of compounds with ethenes, for example $[Cu(C_2H_4)(H_2O)_2]ClO_4$ (from Cu, Cu^{2+}, and C_2H_4) or $Cu(C_2H_4)(bipy)^+$. It is necessary to mention that, of all the metals involved in biological systems, only copper reacts with ethylene [74b]. Such homoleptic alkene complexes can be useful intermediates for the synthesis of other complexes. The olefin complexes of the *metals in high formal oxidation states* are electron deficient and therefore inert toward electrophilic reagents. By contrast, the olefin complexes of the *metals in low formal oxidation states* are attacked by electrophiles such as protons at the electron-rich metal–carbon σ-bonds [74c].

Ethylene derivatives of organophosphines and organoarsines [75a] are used as σ, π-chelating donors. Thus, 4-pentenediphenylphosphine reacts (3.61) with platinum halides in boiling chloroform, forming complexes of the type ***623***:

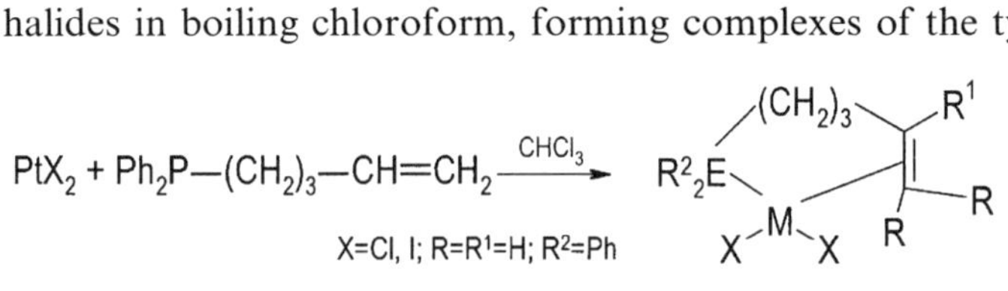

(3.61)

623

E=P, As

The π-attached olefins in the π-complexes containing coordinated hydrogen at the same molecule are very reactive toward other ligands and can be transformed to alkyls by intramolecular insertion of a hydride to these olefins. This is an obligatory step in olefin hydrogenation which is used in catalytic processes. Thus, [Cp*Fe

$(PMe_3)(H)(C_2H_4)]$ in the presence of PMe_3 forms $[Cp^*Fe(PMe_3)_2(Et)]$, where PMe_3 drives the insertion reaction to the alkyl form by filling the vacant coordination site. Other examples are the transformations $[PtL_2H(C_2H_4)]Cl \rightarrow PtL_2Cl(C_2H_5)$, $CpRu(Me_3P)H(C_2H_4)]^+[PF_6]^- \rightarrow [Cp(Me_3P)(C_2H_5)]^+[PF_6]^-$, and $[W(CO)_3(\eta^2\text{-}H_2)(PR_3)_2] \rightarrow [W(CO)_3(H)_2(PR_3)_2]$. Such reactions are typical for η^2-coordinated olefin complexes of Pd(II), Pt(II), Fe(II), etc. [19,74c] and are widely used in many important catalytic processes (olefin hydrogenation, hydroformylation, polymerization, etc.).

The *alkyne* complexes can be obtained, in particular, directly from $RC{\equiv}CR$ and metal carbonyls, for instance $Co_2(CO)_8$, or through ligand substitution reactions, for example from $Mo(CO)(RC_2R')(S_2CNMe_2)_2$ and the same ligand RC_2R' [74c]. The last reaction takes place by a CO dissociation mechanism. The heterobimetallic dihydride $Cp(CO)_2Re(\mu\text{-}H)Pt(H)(PPh_3)_2$ with alkynes produces $(alkyne)Pt(PPh_3)_2$, among other products [75b]. Like olefins, coordinated actylenes are readily attacked by nucleophiles. Intermolecular attack gives trans addition whereas migratory insertions afford cis addition. Thus, stable cationic iron–alkyne complexes $[CpFe(CO)(L)(RC{\equiv}CR)]^+$ undergo reaction with a wide variety of nucleophiles $[Nuc^- = PhS^-, CN^-, {}^-CH(COOEt)_2, L = PPh_3, P(OPh)_3]$, to give stable σ-alkenyl complexes $CpFe(CO)(L){-}C(R){=}C(R){-}Nuc$ resulting from trans-attack of the nucleophile on the coordinated acetylene. Similarly, the reaction of methanol (in the presence of $AgPF_6$) with coordinated distributed acetylenes (generated in situ from $PtClL_2(CH_3)$ and $RC{\equiv}CR$) affords trans σ-bonded vinyl ether complexes $L_2Pt(R)C{=}C(R)OCH_3$ [74c].

Allyl-type transition metal π-complexes can be prepared from metal salts or carbonyls $[NiBr_2, PdCl_2, Ni(CO)_4, HCo(CO)_4]$ with Gringard alkenes, alkenes, or their halides ($CH_2{=}CHCH_2MgBr$, $CH_2{=}CHCH_2Cl$, etc.) [74a,c]. These complexes are represented by a number of examples, especially for noble metals. Thus, in an (η^3-allyl)palladium chloride dimer $(H_2CCHCH_2)Pd(\mu_2\text{-}Cl)_2Pd(H_2CCHCH_2)$, the plane of the allyl group is not perpendicular to the Pd_2Cl_2 plane, but rather is at an angle of 111.5° in it. The $\eta^3 \rightarrow \eta^1$ transformation in allyl complexes, forming σ-allyl group, can take place, thus generating a vacant coordination site on the metal. In the presence of strongly coordinating ligands, such as phosphines or DMSO, the nucleophiles readily attack the η^3-allyl ligand. For instance, the same (η^3-allyl) palladium chloride dimer reacts with ${}^{(-)}CH(COOR)_2$ and PPh_3 yielding $(PPh_3)_2Pd(\eta^2\text{-}CH_2{=}CH{-}CH_2CH(COOR)_2$. In addition to reactions with nucleophiles, (η^3-allyl)palladium chloride complexes are readily oxidized to give conjugated dienes or allylic alcohols, acetates, or chlorides, depending on the oxidant. For example, steroidal η^3-allyl complexes are converted to conjugated enones when oxidized by *m*-chloroperbenzoic acid, CrO_3 in DMF, or oxygen with hν irradiation [74c].

Diene and, especially *cyclodiene* complexes (together with *arene* compounds) are the most explored organometallic ligands. The above-mentioned method of direct (immediate) interaction of reactants is also used to prepare their complexes. As a source of metal center, $PtCl_2$ (3.62) and $K_2[PtCl_4]$ (3.63) are applied in reactions with cyclo-octadiene-1,5 (COD) [11]:

$$C_8H_{12} + PdCl_2 \xrightarrow[EtOH]{HCl/H_2O} PdCl_2(\eta^4\text{-}C_8H_{12}) \tag{3.62}$$

$$C_8H_{12} + K_2[PtCl_4] \xrightarrow[MeCOOH]{H_2O} PtCl_2(\eta^4\text{-}C_8H_{12}) \quad (3.63)$$

In both formed complexes, cyclo-octadiene-1,5 behaves as a chelate-forming π,π-ligand ***624*** [75c, p. 94]. Reduction of ***624*** (M = Pt) (COD)$PtCl_2$ in the presence of COD gives $Pt(COD)_2$, which reacts with ethene to form $Pt(C_2H_4)_4$ and PhC≡CPh to yield $Pt(C_2Ph_2)_2$ [74b]:

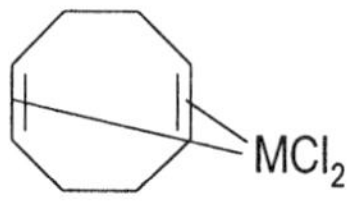

624

The *cyclopentadiene* (Cp) and its analogue C_5Me_5 (Cp*) ligands are undoubtedly at the heart of synthetic organometallic chemistry. There are many methods for preparing their η^5-complexes of the type ***3*** [75d], most of which rely on the acidity of Cp ($pK_a = 9$) and combine salts of its anion with substitution-labile metal halides. A particular high-yield method involves reaction between TlCp and transition metal halides [74c]. Also, the method of exchange of alkali metal in cyclopentadienyl ring to metal center is generally used [11]. However, some cases are known where the method of direct (immediate) interaction of ligands and metal salts in the presence of secondary amines (amine method) is applied for these goals [11]:

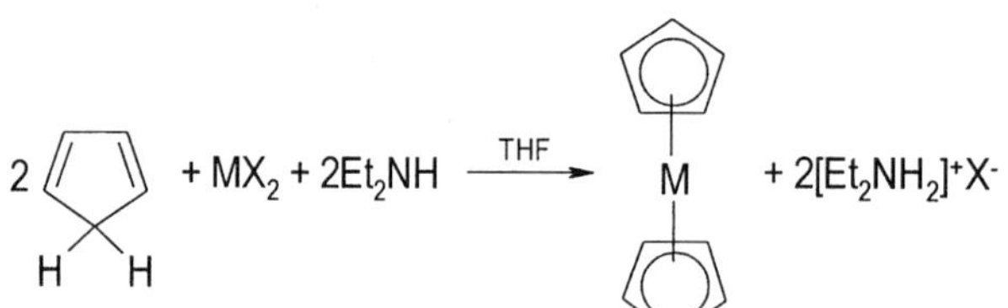

(3.64)

The reaction of the synthesis of ferrocene (M = Fe) from nonaqueous iron (II) chloride is carried out in THF [11]. $FeCp_2$ is lithiated by BuLi, yielding $(\eta^5\text{-}C_5H_4Li)_2Fe$, which by further treatment with CO_2 and H^+ yields functionalized ferrocenes $CpFe(C_5H_4COOH)$ (the same compound can be produced by interaction of Cp_2Fe with CO_2 in the presence of $AlCl_3$) and $Fe(C_5H_4COOH)_2$ [74b,c]. This is a common use of *n*- or *t*-butyllithium in hexane, benzene, or ethers to carry out metal–hydrogen or metal–halogen exchanges [74b]. The auration of the metallated Cp ligands in $[Fe(C_5H_4Li)_2]$ with $[AuCl(PPh_3)]$ yielding $[Fe(C_5H_4AuPPh_3)_2]$ is an example of ion exchange at a site remote from the metal center [10]. Electrosynthesis (Sec. 3.4.2) of $FeCp_2$ by anodic dissolution of iron in TlCp solution was reported. Similarly, MCp_2 (M = Mn, Ni) were prepared by dissolving these elemental metals in THF or py solutions of M′Cp (M′ = Li, Na, K) [10]. Nickelocene (M = Ni) can also be readily obtained from the reaction of Cp^- with Ni(II) salts in dimethyl ether of ethyleneglycol. It is a green paramagnetic 20-electron sandwich complex (two unpaired electrons). It forms mostly 18-electron products, for instance, reacting with (allyl)MgBr gives $CpNi(\eta^3\text{-allyl})$ [74b]. Similarly to nickel salts, $CoCl_2$ reacts with CpH and Et_2NH yielding Cp_2Co (3.64) [74b].

The coordinated ligands in Cp complexes (as well as other π-complexes) are subject to reactions of various types, in particular nucleophilic attack reactions.

Nucleophilic additions to 18-electron cationic transition-metal complexes of unsaturated hydrocarbons are discussed in detail in Ref. 74c. In particular, the rules, applied only for reactions of this type of complex under kinetic control, are offered. The unsaturated hydrocarbon ligands are classified as *even* or *odd*, according to their ligand η, and *closed* (cyclically conjugated ligands) or *open* (acyclic ligands). It is stated that *even* systems are more reactive than *odd* systems and *open* systems are more reactive than *closed* systems. For *even open* polyenes, nucleophilic attack at a terminal carbon atom is always preferred. For *odd open* polyenyls, attack at a terminal carbon occurs only if $[M^+]$ is a strong electron-withdrawing group. These rules, cited in Ref. 74c, are nicely confirmed in numerous reactions of nucleophilic attacks to $\eta^{2,6}$-coordinated complexes, in particular, by reaction of $[Mo(\eta^6\text{-}C_6H_6)(\eta^3\text{-}allyl)(\eta^4\text{-}1,4\text{-}butadiene)]^+$ with CN^-, which is carried out through attack exclusively at the η^4-ligand and exclusively at a terminal position. Another example is the attack of RLi (R = Ph, Me, Et) to the $[CpFe(C_6H_6)]^+$ cation, which takes place on the complexes arene (not the Cp group) and yields $[CpFe(\eta^6\text{-}C_6H_5R)]$ [74c]. Cyclohexadiene complexes are alkylated by a range of carbanions. For instance, the interaction of $PdCl_2$ with C_6H_8 and a nucleophil leads to formation of η^4-coordinated metal complexes [74c].

Oxidative addition reactions of Cp complexes are represented, in particular, by hydrogenolysis by heterolytic dihydrogen activation. Thus, $Cp^*_2U(Cl)(R)$ interacts with H_2 and toluene at $-78°C$, giving $Cp^*_2U(Cl)(H)$ which can eliminate H_2 yielding $[Cp^*_2UCl]_3$. Another example is the hydrogenation of CpZr(X)(R) forming CpZr(X)(H). Addition of RX to the strong reductant $Cp^*_2Zr^{II}L_2$ at room temperature gives $Cp^*_2Zr(X)(R)$ and $Cp^*_2Zr(X)_2$ [74c].

Insertion reactions of CO into a π-complex $[CpFe(PPh_3)(CO)(R)]$ (R = Me, Et) lead to $[CpFe(PPh_3)(CO)\{C(O)R\}]$. Different mechanisms for this process were offered, including alkyl migration and CO insertion steps. Such alkyl–carbonyl migratory insertion reactions can be accelerated using Lewis acids and related inorganic compounds, such as, for example, $NaPF_6$ or amphoteric ligands, such as $R_2PNRAlR_2$. Thus, a nice interaction of $[CpFe(CO)(Me)(PPh_2{-}CH_2{-}R)]$ (where R is an N,O-containing analogue of crown-ethers) with R′Nc, $NaPF_6$ in CH_2Cl_2 yields $[CpFe(RNC)\{C(O)Me\}PPh_2{-}CH_2{-}R'')][PF_6]$, where R″ is R with Na^+ ion coordinated by N and O atoms of the ether. The *electrophilic insertion* reactions of Cp complexes take place with SO_2, $(RSO_2)NSO$, $(RSO_2N)_2S$, S_8, etc. For example, $Cp(CO)_3W{-}CH_2CH_2Ph$ reacts with SO_2 giving an electrophilic insertion product $Cp(CO)_3W{-}\{S(O)_2{-}CH_2CH_2Ph\}$. Similar interaction of $CpFe(CO)_2Me$ with SO_2 yields $CpFe(CO)_2(SO_2Me)$ [74c].

The *rhenium complexes* $(C_5R_5)ReO_3$ (R = Me, Et, etc.) which, as well as $(CH_3)ReO_3$, belong to a very important class of cyclopentadienyl compounds, are intensively studied at present and widely used in various catalytic organic processes. Cp^*ReO_3 ($CpReO_3$ does not exist) is a precursor for many organometallic compounds, such as, for example, Cp^*ReCl_3Et, Cp^*ReOR_2, $Cp^*_2Re(RC{\equiv}CR')$, Cp^*ReCl_4, Cp^*ReMe_4, $Cp^*Re(CO)_3$, etc. [74b]. For well-characterized $Cp^*Re(O)Me_2$ and $Re(O)(I)MeC{\equiv}CMe)_2$, the oxo group is a much less reactive oxygenating agent than such groups in later transition metals. An example of oxygenation of π-complexes is the interaction of Cp^*ReO_3 with PPh_3 and O_2 leading to a μ-oxido dimer $Cp^*Re(O)(\mu_2\text{-}O)_2Re(O)Cp^*$ [74c]. Among other rhenium Cp complexes, the reaction of $[ReH(CO)(NO)(\eta^5\text{-}Cp)]$ with triphenylcarbenium tetrafluoroborate in CH_2Cl_2 at −78°C is reported, which yields $[Re(CO)(Ph_3CH)(NO)(\eta^5\text{-}Cp)][PF_6]$ containing an uncommon η^2-arene ligand $\eta^2C_6H_5{-}CHPh_2$ [75e]. Study of associative substitution mechanisms for rhenium complexes showed that multiple equivalents of the strong donor PMe_3 can cause *reversible* η^5 to η^3 to η^1 slippage and even complete displacement of the Cp group, such as, for example, in the case of the reaction of $[CpRe(NO)(PMe_3)(CH_3)]$ with PMe_3 at different temperatures yielding complexes $[(\eta^1\text{-}Cp)Re(PMe_3)_3(CH_3)(NO)]$, $[(\eta^3\text{-}Cp)Re(PMe_3)_2(CH_3)(NO)]$, $[Re(PMe_3)_4(CH_3)(NO)]^+ (Cp)^-$, and other products [74c].

For *p*-metals, π-coordination is found in Al—Cp compounds, for example Cp_2AlMe with two η^2-bonded Cp rings (in the solid state) and $(\eta^5\text{-}Cp^*)_2Al^+$ cation [74b]. The data on a number of Cp and Cp* complexes of almost all transition metals with many coligands and their reactions have been reported. Among them, the synthesis of di-Gringard reagent $[Ti(CH_2MgBr)_2Cp_2]$ through the reaction of $Cp_2 TiCl_2$ with $CH_2(MgBr)_2$ is described. The titanium-containing Gringard reagent reacts with 0.5 molar equivalents of $SiCl_4$ to produce the compound $Cp_2Ti\{(\mu\text{-}CH_2)_2\}Si\{(\mu\text{-}CH_2)_2\}TiCp_2$. In a similar manner, Cp_2TiCl_2 reacts with one equivalent of $ClMg(CH_2)_3CH{=}CH_2$ yielding $[TiCl\{(CH_2)_3CH{=}CH_2\}Cp_2]$ [10]. In general, Cp_2TiCl_2 is the principal starting material in obtaining other titanium π-complexes, such as $CpTiCl_3$, $Cp_2Ti(\eta^3\text{-allyl})$, $Cp_2Ti(NR_2)_2$, Cp_2TiMe_2, $Cp_2Ti(\mu\text{-}Cl)_2TiCp_2$, etc. [74b]. A similar titanium compound $[TiCl(CH_2Ph)Cp_2]$ reacts with $NaBPh_4$ in CH_3CN generating the solvent complex $[Ti(CH_2Ph)Cp_2(NCCH_3)][BPh_4]$ [75f]. Cp_2Ti reacts with $AlMe_3$ forming $Cp_2Ti(\mu\text{-}Cl)(\mu\text{-}CH_2)AlMe_2$, which acts as a CH_2 transfer agent [74b].

Many Cp complexes, for instance well-known classic π-complex Cp_2ZrCl_2 and $[Cp^*Cr^{III}(CH_2Ph)(THF)_2]^+[BF_4]^-$, can be used in catalytic processes. The first one, Cp_2ZrCl_2, can be transformed to $Cp_2Zr(CO)_2$ by interaction with Mg and CO [74b]. The molybdenum complex $[Mo(CO)_3(CH_2Cl_2)Cp][PF_6]$, containing dichloromethane acting as a monodentate σ-donor ligand, above 15°C dissociates in CH_2Cl_2 to yield a compound $[Mo(CO)_3Cp(FPF_5)]$ containing a coordinated FPF_5^- anion [10]. It should be mentioned that Cp* complexes of such metals as Os may contain multiple metal–metal bonds. Thus, Cp^*OsH_5 reacts with C_6H_6 under hν irradiation giving $Cp^*Os(\mu_2\text{-}H)_4OsCp^*$ with triple bond Os≡Os [74b].

For lanthanide series, a lot of catalytically active hydrido derivatives of the Cp and Cp* lanthanide compounds have been described, for example $Cp_2(THF)Ln(\mu_2\text{-}H)_2Ln(THF)Cp_2$. The most simple Ln complexes are prepared from $LnCl_3$ (Sc, Y, La, Ce, Pr, Nd, Sm, Gd, Dy, Er, Yb) and NaCp giving Cp_3Ln compounds, which react with various ligands L {L = R_3P, RNC, THF, $(EtO)_3PO$} yielding Cp_3LnL. The first unequivocal example of the reaction of CH_4 homogeneously to form an M—C bond was the interaction between Cp_2^*LuR (R = H, CH_3) and CH_4 yielding $Cp_2^*LuCH_3$ [74b].

In *arene* complexes, for example in (η^6-arene)chromium tricarbonyl derivatives, the arene ring C—H bonds have enhanced acidity. These complexes, similarly to Cp complexes (see above), are treated with BuLi, and the resulting lithio (η^6-arene)chromium tricarbonyl complexes react with a variety of electrophiles. For instance, reacting with BuLi, $(C_6H_5Y)Cr(CO)_3$ (Y = H, OMe, F, Cl) forms $(C_6H_5Li)Cr(CO)_3$, and further reaction with CO_2, MeX, PhCHO, MeCOMe, etc. gives *o*-products $(o\text{-}C_6H_4YR)Cr(CO)_3$ [74c].

Arene π-complexes readily take part in ligand substitution reactions. Thus, $[CpV(\mu_2\text{-}C_6H_6)VCp]$ reacts with $C_6H_3(CH_3)_3$ at 110°C with substitution of benzene ligand by $C_6H_3(CH_3)_3$ yielding $[CpV\{\mu_2\text{-}C_6H_3(CH_3)_3\}VCp]$, free benzene, and $CpVC_6H_3(CH_3)_3$. The CO ligands can be substituted by PPh_3 and other ligands, for example $[(\eta^5\text{-indenyl})Rh(CO)(PPh_3)]$ is prepared from $[(\eta^5\text{-indenyl})Rh(CO)_2]$ [74c]. Metal carbonyls can serve as primary materials for obtaining arene complexes, for instance, (toluene)$Mo(CO)_3$ is prepared from $Mo(CO)_6$ and toluene. The classical $[Cr(C_6H_6)_2]^{n+}$ complexes can be prepared by reduction of $CrCl_3$ with Al and $AlCl_3$ in benzene [74b]. A stereospecific reaction of the complex of *o*-methoxyacetophenone $\{o\text{-}C_6H_4(OMe)(C(O)Me)\}Cr(CO)_3$ with Gringard reagents RMgX yields $\{o\text{-}C_6H_4(OMe)(C(R)(OH)(Me)\}Cr(CO)_3$. The steric efficiency of this process depends on the nature of the anion, but specificity is rather high [74c]. Similarly to some Cp complexes, the synthesized arene compounds may contain multiple metal–metal bonds. Thus, the interaction of $(C_6H_5PR_2)Mo(R_2PhP)_3$ and $WCl_4(PPh_3)_2$ gives a compound with a quadruple bond Mo—W $(R_2PhP)_2(Cl)_2MoW(Cl)_2(PPhR_2)_2$ [74b].

To prepare other π-complexes (Sec. 2.2.4.1), the methods of ligand exchange and especially direct gas-phase synthesis are widely used, which will be discussed in Secs. 3.2.1 and 3.4.1 below. For π-complexes of the radioactive elements, see Section 5.3.

EXPERIMENTAL PROCEDURES

Example 1

Synthesis of (Pentamethylcyclopentadienyl)bromo(1,5,-cyclooctadiene)osmium(II), $(C_5Me_5)Os(COD)Br$ [75g]

To a slurry of $(C_5Me_5)_2Os_2Br_4$ (0.62 g, 0.64 mmol) in ethanol (40 mL) was added 1,5-cyclooctadiene (0.82 mL, 6.7 mmol). The solution was refluxed for 90 min, and the solution color changed to a clear orange and an off-white precipitate formed. The solvent was removed under vacuum, the residue was extracted with diethyl ether (4–30 mL), and the extracts were filtered. The filtrates were combined, concentrated to ca. 50 mL, and cooled to −20°C to afford orange crystals. Additional crops of crystals were obtained by further concentrating and cooling the supernatant. Yield 0.48 g (74%).

Example 2

Synthesis of *bis*(Pentamethylcyclopentadienyl)($\mu,\eta^3{:}\eta^3$-2,3-dimethylenebuta-1,4-diyl)tetrachlorodiruthenium, $Cp^*_2Ru_2(\mu,\eta^3{:}\eta^3\text{-}C_6H_8)Cl_4$ [75h]

A Fisher–Potter bottle charged with a cold (−78°C) slurry of $[Cp^*RuCl_2]_2$ (0.57 g, 0.93 mmol) in CH_2Cl_2 (30 mL) was first evacuated and then charged with propadiene (5.0 mmol). The resulting mixture was then warmed to room temperature. After the mixture had been stirred for 3 hr, the dark red solution was filtered, concentrated to ca. 3 mL, and treated with diethyl ether (ca. 2 mL). The resulting mixture was cooled to −20°C to afford red crystals. Yield 0.22 g (34%), m.p. 170°C (decomp.).

Example 3

Synthesis of *bis*(pentamethylcyclopentadienyl)(μ-chloro)(μ-methylene)(ethyldimethylsilyl)diruthenium(III) $Cp^*_2Ru_2(\mu\text{-}CH_2)(SiMe_2Et)(\mu\text{-}Cl)$ [75i]

An orange slurry of $[Cp^*RuCl]_4$ (0.25 g, 0.23 mmol) in diethyl ether (25 mL) at 0°C was treated with $Mg(CH_2SiMe_2Et)_2$ (3.8 mL of a 0.075 M solution in diethyl ether, 0.29 mmol). The solution color turned dark red after ~ 30 sec. The solution was stirred at 0°C for 4.5 hr, the solvent was removed under vacuum, and the residue was extracted with cold pentane (2 × 50 mL) at 0°C. The combined extracts were concentrated to 40 mL, filtered, concentrated further to 15 mL, and cooled to −20°C to afford dark red crystals of the product. Yield 0.14 g (50%), m.p. 164°C.

Example 4

Synthesis of η^3-Allyl(15-COD)palladium(II) Tetrafluoroborate [75i]

The dimer di-μ-dichloro-*bis*(η^3-allyl)dipalladium (1.83 g, 5.0 mmol) and silver tetrafluoroborate (1.95 g, 10.00 mmol) in CH_2Cl_2 (50 mL) were stirred for 15 min. 1,5-COD (2.0 mL) was added and stirring continued for a further 2 min. The mixture was filtered and the residue washed with CH_2Cl_2 (2 × 10 mL). Ether (150 mL) was added to the combined filtrate and washed to give a white or grayish precipitate. It was filtered, washed with ether (3 × 50 mL), and dried in air. The solid was dissolved in CH_2Cl_2 (75 mL), and this solution was repeatedly filtered through a tightly packed plug of cotton until the solution was absolutely clear. Addition of ether (120 mL) precipitated the product, which after filtration, washing with ether, and drying in vacuo was obtained as an off-white solid. Yield 2.63 g (77%).

Example 5

Synthesis of $[CpFe(CO)\{\eta^2\text{-}(PPh_2)_2NH\}]$ [75k]

A solution of $[CpFe(CO)_2I]$ (0.335 g, 1.17 mmol) and dppa $\{NH(PPh_2)_2\}$ (0.45 g, 1.17 mmol) in dry toluene (30 mL) was boiled under reflux for 1.5 hr. The yellow solid formed was filtered off and washed with cold toluene and ether. The complex was dissolved in the minimal amount of CH_2Cl_2, chromatographed by Kieselgel 60, and eluted with acetone. The solution was vacuum-concentrated and the complex precipitated adding diethyl ether. Yield 0.7 g (90%).

Example 6

Synthesis of $[Pt(C_2H_4)(dppf)]$ [dppf = 1, 1′-*bis*(diphenyl-phosphine)ferrocene] [75l]

$[Pt(COD)_2]$ (0.097 g, 0.23 mmol) was added in small portions to 15 mL of petroleum ether saturated with ethylene at 0°C. A solution of dppf (0.33 g, 0.24 mmol) in 15 mL of diethyl ether was then added, and the mixture stirred for 1 hr at 25°C in an Ar atmosphere. The resulting yellow solution was reduced to a small volume, affording a precipitate by bubbling ethylene through the solution at 0°C, which was washed with cold petroleum ether (3 × 5 mL), saturated with ethylene, and dried under reduced pressure for 3 hr. Yield 0.167 g (93%).

Example 7

Synthesis of *meso*-Ethylenebis(4,7-dimethyl-1-indenyl)dimethyl Zirconium [75m]

To a solution of *meso*-ethylenebis(4,7-dimethyl-1-indenyl) $ZrCl_2$ (2.19 g), in 50 mL of THF, 6.5 mL of MeLi 1.6 M solution in diethyl ether was added dropwise at −20°C in 10 min. The solution was stirred for 15 min at −20°C, then allowed to warm slowly to room temperature and stirred for 2 hr. An orange suspension was obtained, dried in vacuo, and extracted with hexane in Soxlet. The extract was concentrated, cooled at −20°C for 12 hr, and then filtrated. Yield 0.74 g (36.5%).

Example 8

Synthesis of $[Pt\{\eta^3\text{-}C(C{\equiv}CPh){=}C(Ph)AuPPh_3\}(dppf)][O_3SCF_3]$ [dppf = 1, 1′-*bis*(diphenyl-phosphine)ferrocene] [751]

A stirred solution of $[Pt(\eta^2\text{-}PhC{\equiv}C{-}C{\equiv}CPh)(dppf)]$ 0.075 g, 0.078 mmol) in THF (5 mL) was treated with $[Au(PPh_3)][O_3SCF_3]$ (0.080 mmol) in THF (16 mL) at 5°C. The resulting orange solution was stirred for 5 min at 5°C, and 10 min at room temperature. The volume was reduced to 5 mL under vacuum, and fast addition of ether (10 mL) afforded yellow precipitate, which was washed with ether (3 × 5 mL) and dried under reduced pressure for 10 hr. Yield 0.106 g (87%).

Example 9

Synthesis of *bis*-Mesitylene Iron(II) Hexafluorophosphate $[\{\eta^6\text{-}C_6H_3(CH_3)_3\}_2Fe][PF_6]_2$ [75n]

Anhydrous $FeCl_3$ (2.5 g) was placed in a "quickfit" round-bottomed flask equipped with a magnetic stirrer bar, and covered with about 30 mL of cyclohexane. The required amount of $AlCl_3$ and Al powder and then a slight excess of mesitylene were added. (It is best to finely powder the $AlCl_3$ in a dry mortar and pestle before addition to the reaction flask.)

The mixture was stirred vigorously whilst heating to reflux in an oil bath for about 2 hr. Then the mixture was cooled to room temperature and the solvent carefully decanted from the solid product. To the remaining solid a slurry of KPF_6 in water was added. The mixture was stirred vigorously, making sure that all the dark-colored material was dislodged from the sides of the reaction flask. Then the mixture was filtered and washed with water and then a little ethanol, and dried at the pump to leave a pale orange/tan-colored solid.

Example 10

Synthesis of Dicarbonyl(η^5-cyclopentadienyl)methyliron $(\eta^5\text{-}C_5H_5)Fe(CO)_2CH_3$ [75n]

A round-bottomed flask (25 mL) containing a magnetic stirrer bar and fitted with a reflux condenser was flushed with nitrogen for 5 min. The condenser was removed and then mercury (5 g) was added. After the beginning of stirring, metallic sodium (320 mg, 13.9 mmol) was added in small portions. The condenser was replaced and flushed with nitrogen for 1 min before sealing the condenser with a screw cap. The amalgam was allowed to cool and solidify. The condenser was removed and dry tetrahydrofuran (2.8 mL) and a solution of cyclopentadienylirondicarbonyl dimer (500 mg, 1.41 mmol) in dry tetrahydrofuran (2.8 mL) were added. The condenser was replaced, flushed with nitrogen for 1 min, and sealed. It was stirred at room temperature overnight under reflux to prevent loss of tetrahydrofuran. When the reaction was complete, idomethane (1.24 mL, 19.9 mmol) was added dropwise via a syringe over a 30-min period. The mixture was stirred in a sand bath for 2–3 hr at 40°C. The solvent was removed by rotary evaporation and the residue flushed with nitrogen. The dry residue was transferred to a sublimation vessel and sublimed at 50°C under vacuum onto a water-cooled probe. Yield 74%.

Example 11

Synthesis of π-Acetylene Tetracobaltododecacarbonyl Clusters – “Butterflies” $[(CO)_8Co_4(\mu\text{-}CO)_2](\mu\text{-}C_2R_2)$ (R = CH_2OH, CH_2SCMe_3) [75o]

$[(CO)_8Co_4(\mu\text{-}CO)_2][\mu\text{-}C_2(CH_2OH)_2]$

A solution of $Co_2(CO)_8$ (0.21 g, 0.6 mmol) and $[Co_2(CO)_6][\mu\text{-}C_2(CH_2OH)_2]$ (0.23 g, 0.6 mmol), obtained according to Ref. 75p, in CH_2Cl_2 (40 mL) was stirred with reflux for 5 hr. Then the reaction mixture was evaporated to dryness; unreacted $Co_2(CO)_8$ was extracted with hexane (45 mL). The solid was dissolved in CH_2Cl_2 (30 mL); the violet solution was evaporated to half of its initial volume. The formed cystals were recrystallized from CH_2Cl_2-hexane mixture. Yield 0.04 g (10%).

$[(CO)_8Co_4(\mu\text{-}CO)_2][\mu\text{-}C_2(CH_2SCMe_3)_2]$

A brown solution of $Co_2(CO)_8$ (0.4 g, 1.2 mmol) in hexane (10 mL) was added to a red solution of $[Co_2(CO)_6][\mu\text{-}C_2(CH_2SCMe_3)_2]$, obtained according to Ref. 75p, in hexane (20 mL) and the mixture was heated at 50°C. The reaction progress was observed chromatographically. The resulting product appeared as a blue spot ($R_f = 0.5$ in the mixture benzene:hexane 1:5), together with the formation of brown $Co_4(CO)_{12}$ as a product of thermal decomposition of $Co_2(CO)_8$. The reaction mixture was eluated on SiO_2 [$Co_4(CO)_{12}$ by hexane; the product by ether]. The ether eluate was evaporated to 10 mL, then hexane (5 mL) was added. Then the mixture was kept at −10°C for 24 hr. The blue crystals were separated and dried in vacuo. Yield 0.09 g (12%).

For other procedures for π-complexes see Secs. 3.2.1, 3.2.2, 3.4.1, and 5.3.

3.1.1.4 Di- and Polymetallic Complexes

Atoms, simplest ions, and molecules; di- and polychelating compounds take part as bridge-type ligands in these complexes. Di- and polynuclear structures of the simplest ligands, in terms of the number of cluster structures, are reported in the literature, cited in Sec. 1.2.2.4.

As described in Sec. 2.2.4.1, the acetylene hydrocarbons frequently have a bridge function [76,77a]. An example of their use is the synthesis of π-complexes of propargyl alcohol with copper(I) chloride (3.65) [77a]:

$$CuCl + HC{\equiv}CCH_2OH \longrightarrow [2CuCl \cdot HC{\equiv}CCH_2OH] \qquad (3.65)$$

According to x-ray diffraction data, the molecule of the prepared complex is stabiized by intermolecular hydrogen bonds of the type $\equiv C{-}H{-}Cl$ and $R{-}OH{-}O(H)R$ ($R = H_2C{-}C{\equiv}CH$), leading to the formation of the tridimensional structure $[Cu_{12}Cl_{12}(HC{\equiv}C{-}CH_2OH)_6]_n$ [77a]. The compounds with μ_2-η^1 bridging acetylides, such as $(PMePh_2)_2Cu(\mu_2\text{-}\eta^1\text{-}C{\equiv}CPh)_2Cu(PMePh_2)_2$, can also be prepared [77b]. Among other binuclear organometallic compounds, the heteronuclear carbonyl complexes with direct metal–metal bonds should be mentioned, such as $Cp_2MeZr{-}RuCp(CO)_2$ {prepared from $Cp_2ZrMeCl$ and $K[Ru(CO)_2Cp]$}, $[(CO)_4Co\text{-}Rh(CO)(PEt_3)_2]$ {obtained from $Na[Co(CO)_4]$ and $[RhCl(CO)(PEt_3)_2]$}, $[WCl_2(CO)_2(\eta^5\text{-}C_5R_5)]_2$ {the compound with a W=W bond, prepared from $[W(\mu\text{-}H)(CO)_2(\eta^5\text{-}C_5R_5)]_2$ and $SOCl_2$ (R = H, Me)}, and $[Me_3Sn{-}Mo(CO)_3Cp]$ {synthesized from Me_3SnCl and $Na[Mo(CO)_3Cp]$} [10]. Heteronuclear compounds can be obtained by the electrochemical dissolution of a sacrificial anode (Sec. 3.4.2) or by cryosynthesis (Sec. 3.4.1). Thus, dissolution of metallic cadmium in solution of Ph_3SnBr and bipy in nonaqueous solvents gives $[CdBr(SnPh_3)(bipy)]$ [10].

A large number of bridge N-donor structures exist (Sec. 2.2.4.2), for example ***157***, ***160***, ***162***, ***163***, ***164***, ***191***, ***197***, and ***200–203***. In this respect, it is necessary to note Ref. 78, where it is shown that amidines form (3.66), additionally to chelate structures of the type ***625***, the bimetallic complexes ***626*** with a bridge function of the ligand:

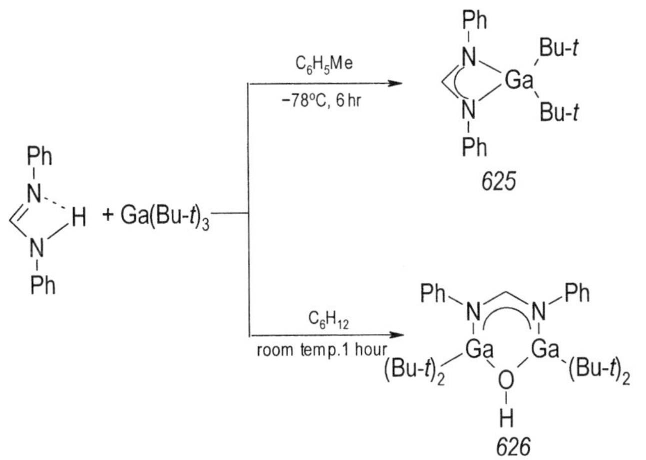

(3.66)

The *bis*(benzimidazole) derivative ***627*** also takes part as a bridge ligand in the complexes ***628*** [20,79] and ***629*** [79], prepared by the method of direct (immediate) interaction (3.67):

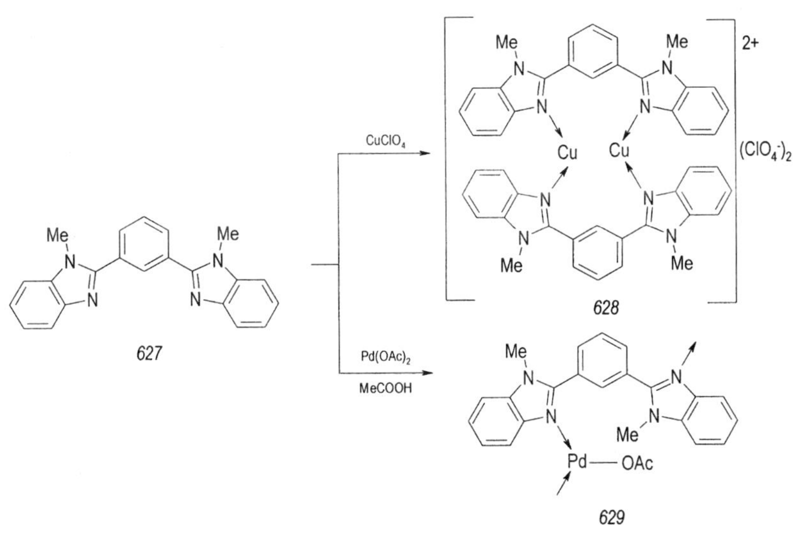

(3.67)

The structures of products are proved by x-ray diffraction. It is established that dimer ***628*** or trimer ***629*** are formed due to the intermolecular M—N bonds.

The ideas on the syntheses of di- and polynuclear complexes with phosphorus bridge by the method of direct (immediate) interaction of ligands and metal source can be received from the literature, summarized in Sec. 2.2.4.3, and related to the coordination compounds ***239***, ***241***, and ***253***. In this respect, we note that chelates can also participate as a metal source, for example in the transformation (3.68) [80]:

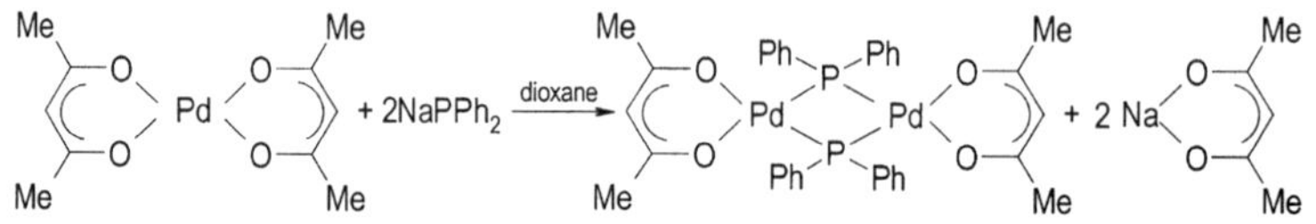

(3.68)

The binuclear complexes with oxygen-containing bridges are widely represented (Sec. 2.2.4.4: ***269–272***). We note that, although they have been obtained by the method of direct (immediate) interaction, oxygen or air, water, alcohols, and crystallohydrates of salts were used as O-donor ligands. Thus, a very unpredicted dimeric O-bridge complex of oxotungsten(VI) β-diketonate ***630*** was obtained as a result of transformation (3.69) [81]:

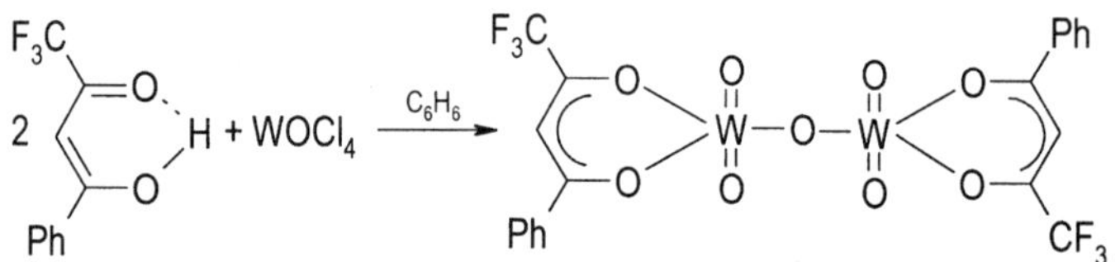

630

(3.69)

The interaction (3.70) between 2-aminopyrimidine with $Cu(ClO_4)_2 \cdot 6H_2O$ results in a binuclear complex ***631***, containing a two hydroxo-bridge fragment [82]:

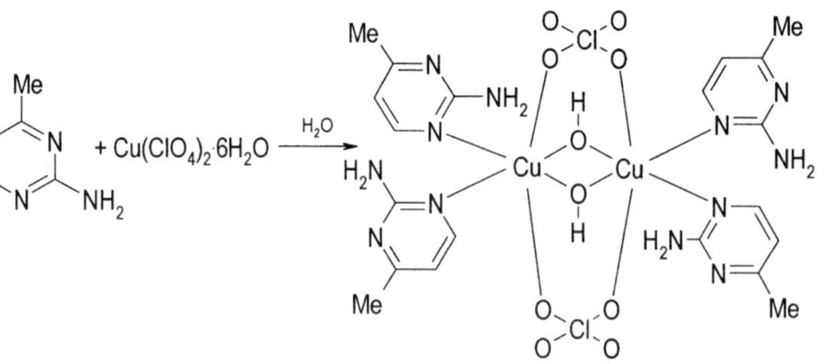

631

(3.70)

Alcoxo-bridges OR (R = Me, Et) are formed by the interaction of 2-amino-4-pycoline ***632*** [83] and 2,4,6-triamino-1,3,5-triazine (melamine) ***633*** [84] with copper(II) nitrate and perchlorate in alcohols:

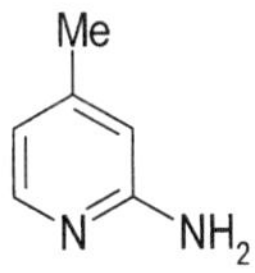

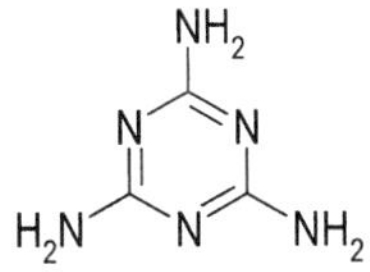

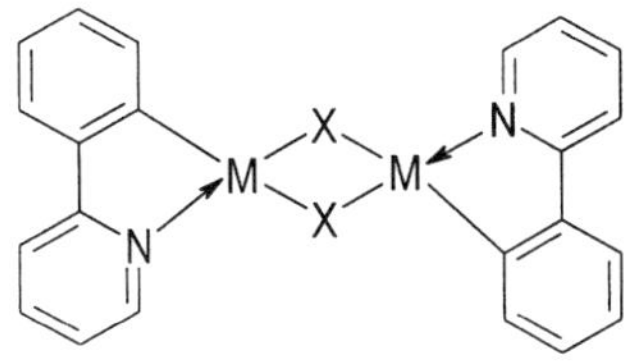

632 *633* *634*: X = OAlk, Hal

Anions of salts frequently participate as bridge ligands in reactions of direct (immediate) interaction of reactants. Dimeric complexes ***634***, described in Sec. 2.2.5.1 and containing acetate, halide, and pseudohalide bridges, are mostly formed in such syntheses. For instance, acetate bridges are formed in cyclometallation reactions, discussed above (Sec. 2.2.5.1), with participation of chelating N,C-donors and $M(OAc)_2$ in acetic acid medium. Here the acetate group participates mostly as a linear bridge.

Another type of acetate bridge ***283*** exists, in general, in binuclear copper complexes, synthesized from various ligands and $Cu(OAc)_2 \cdot H_2O$ in alcohols or aqueous–alcohol mixtures (Sec. 2.2.4.4). Although these reactions with participation of copper(II) acetate formally belong to reactions of direct (immediate) interaction of reactants, $Cu(OAc)_2 \cdot 2H_2O$ itself has the structure of a "*chinese lantern*" ***283*** ($L = H_2O$) and, therefore, the complex formation in this case can be considered as an exchange of ligands (see Sec. 2.2.4.4). The complexes obtained on the basis of caopper acetate, in the majority of cases, possess antiferromagnetic properties [85,86].

It is necessary to mention that the bridge structure, according to x-ray diffraction and magnetochemical data, exists in the anionic complex $K_4\{VO[O_2CCH_2C(O)(CO_2)CH_2CO_2]\} \cdot 6H_2O$; such compounds are rare among carboxylate complexes. This paramagnetic compound was obtained by interaction of citric acid and ammonium metavanadate in the presence of KOH [87].

Complex compounds with halide bridges are prepared by immediate interaction of unsaturated hydrocarbons with metal salts (Sec. 2.2.4.1). Their examples are classic π-complexes of the type ***123*** which are characteristic for d^8-metals [75]. Such complexes are also formed by the method of ligand exchange. The bridges of this type are widespread in products of cyclometallation reactions with the use of metal halides (Sec. 2.2.5.1): ***371–374***, ***381***, ***382*** [41,46,48]. An example of such a synthesis is the reaction of arylhydrazones of 2-oxopropionic aldehyde and benzoylformaldehyde, as well as butadiene-2,3-dione ***635***, with palladium dichloride, leading (3.71) to dipalladium complexes ***636*** [88]:

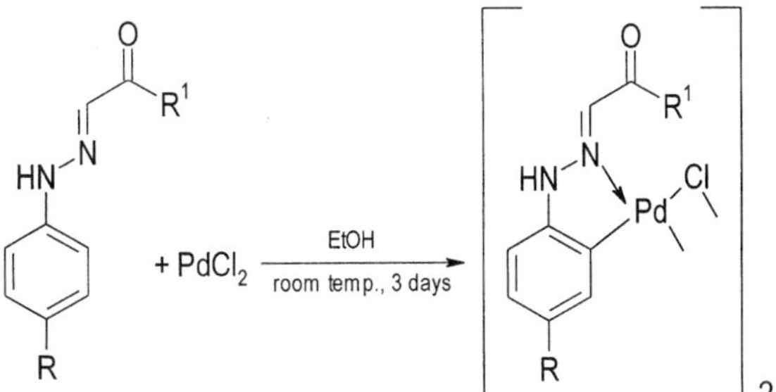

(3.71)

635 *636*

R = H, Br, NO_2; R^1 = Me, Ph

The possibility of formation of dimeric structures should be taken into account in the syntheses of metal–halide complexes of other ligands. In this respect, the transformation (3.72) is interesting and representative, in which not only monomeric mononuclear ***637***, but also dinuclear ***638*** complexes were isolated and their structures determined by x-ray diffraction [89]:

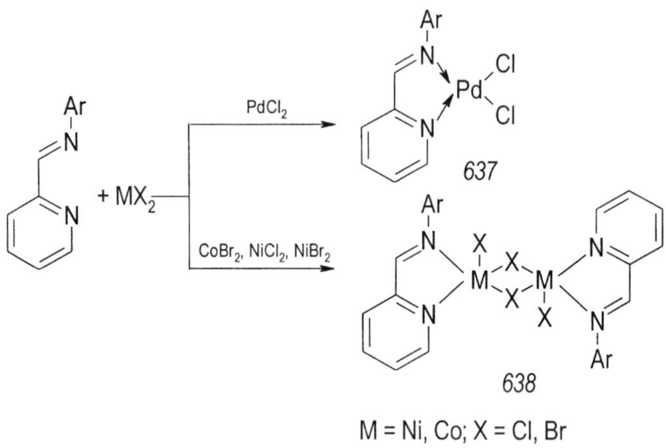

(3.72)

The coordination compounds of bridge chalcogen-containing ligands have been described (Sec. 2.2.4.5: S-donors, formulae ***305***, ***306***, ***309***, and ***312***; Sec. 2.2.4.6: Se-donors, formulae ***337*** and ***347***; Sec. 2.2.4.7: Te-donors, formulae ***352–354***, ***356***). Such complexes have been synthesized, for example, on the basis of disulfides ***639*** (E = S), diselenides ***639*** (E = Se), and ditellurides ***639*** (E = Te), according to the transformation (3.73):

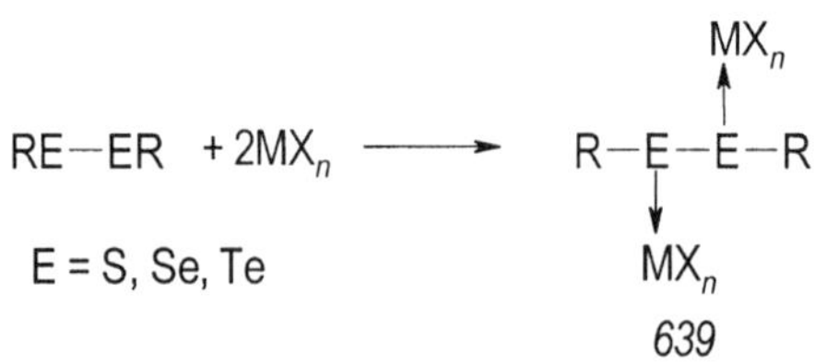

(3.73)

Coordination compounds of the type ***639*** are represented mainly by metal–carbonyl compounds, prepared by the method of ligand exchange, and will be discussed below. We note that some cases are known where some different bridge fragments can exist in complexes obtained by the method of immediate interaction of reagents. Thus, the compound ***631*** has hydroxy- and perclorate bridges; nitrate and alkoxide bridges are formed in the complexes of the above ligands ***632*** and ***633***. We note that, in the case of bridge ClO_4^- and NO_3^- anions, the metals are bound through the oxygen atoms of these bridge fragments [82–84]. The bridge pseudohalide structures are formed as a result of interaction of ligands with $M(XCN)_n$, where X = S, Se (Sec. 2.2.3.5): formulae ***109***, ***113–116*** [90,91].

Dimetalcyclo-forming ligand systems are represented by ligands of tri- and higher dentacity. A permanent interest in these compounds is related mainly with the possibility of creation on their basis of bi- and polynuclear complexes with anomaly magnetic properties [85,86]. Tridentate azomethines of type ***640***, an important group among such ligands, react (3.74) with metal salts yielding dimeric structure ***641*** [92 98]:

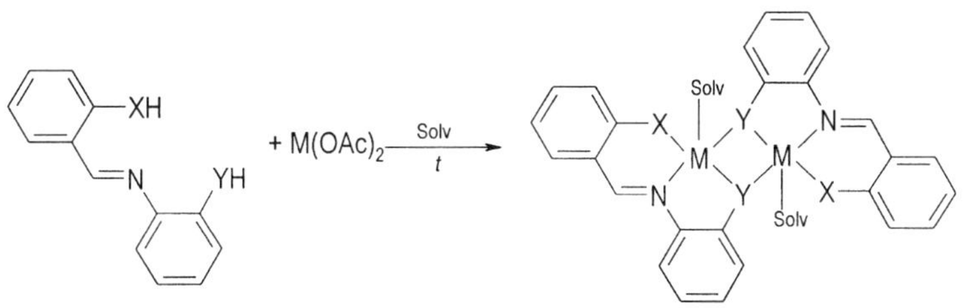

(3.74)

640: X, Y = NTs, O, S

641

The complexes of type ***641*** were obtained according to the same technique by interaction of ligands ***640*** with Co, Ni, Cu, Zn, and Cd acetates in methanol. In case of palladium acetate, acetone was used as solvent [95]. The principal circumstance is the fact that, together with well-known and studied binuclear complexes ***641*** (X = Y = O) [92,93], a series of novel binuclear complexes (X = NTs, Y = O, S) has been obtained and, in some cases, structurally characterized [94,96,98]. They possess antiferromagnetic properties, which are especially characteristic for Cu, Ni, and Co complexes with M_2S_2bridge. Their magnetic moments at 78 K have considerably decreased magnitudes: 1.45 B.M. ($M = Co^{2+}$), 0.45 B.M. ($M = Ni^{2+}$), and 0.41 B.M. ($M = Cu^{2+}$) [95].

To obtain di- and polynuclear structures, other types of azomethinic ligands have been used, for instance, types ***642*** [99] and ***643*** [100]:

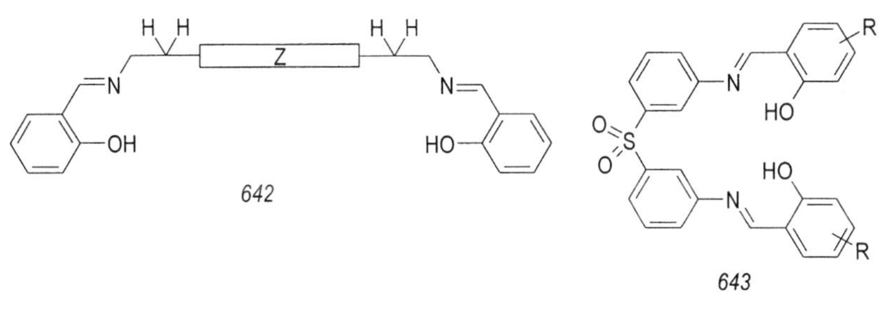

642

643

Dimeric complex was prepared by interaction of ***642*** ($Z = 1,4\text{-}C_6H_4$) with copper acetate in methanol and structurally characterized [99], meanwhile the reaction of ***643*** with the same salt in ethanol yields, according to x-ray diffraction data, tetranuclear supramolecular structures [100].

To synthesize binuclear complexes, the interaction (3.75) of acylhydrazones of salicylaldehyde ***644*** with metal salts is widely used [101–118]:

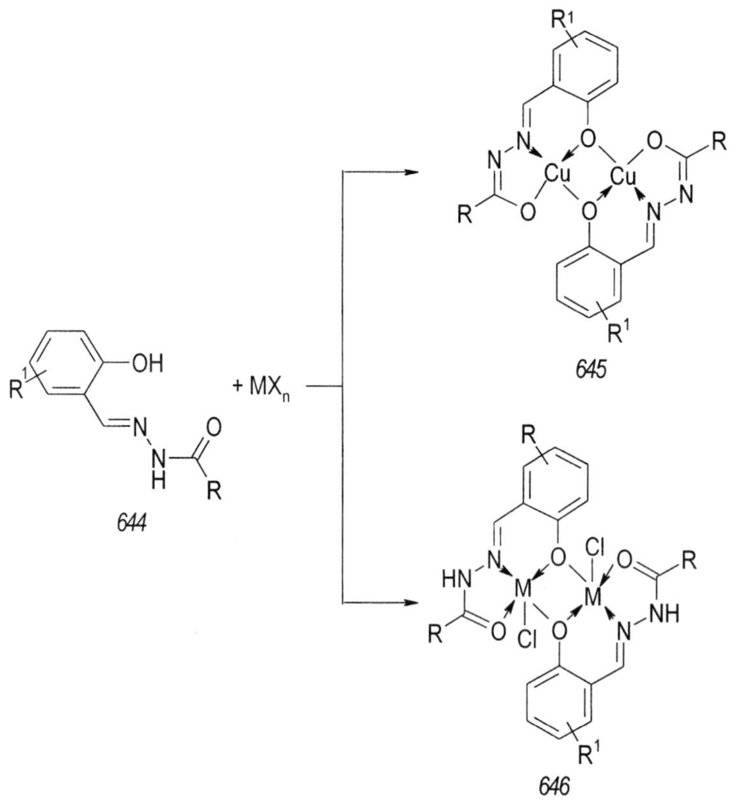

(3.75)

Copper(II) acetate in methanol was used for preparation of complexes ***645*** [107,112,113]. Syntheses of coordination compounds with the retained anion of the salt, for example ***646***, are carried out with metal chlorides [114–116], nitrates [111–113,115], and perchlorates (Cu[103–105,107,110–118], Co[115,118], Mn[115], Cr [115,118], Fe[115], and Ni[115,118]) in alcohols (methanol, ethanol), ethylacetate (EA), dimethylformamide (DMF), as well as in mixtures EtOH—H_2O, EA—EtOH, and DMF—EtOH. One of the isolated complexes, $[Pd_2R_2(\mu\text{-}ClO_4)_2(PPh_3)_2]$ [R = 2-(6-chloropyridyl)], was the first example of a dinuclear complex where two perchlorate ligands bridge the metal centers [10]. Addition of stoichiometric amounts of pyridine to its solution leads to bridge cleavage and the generation of the mononuclear pyridine complexes $[PdR(ClO_4)(py)(PPh_3)]$.

The structures of the examined complexes are proved by x-ray diffraction, for example Refs. 115–117. Magnetochemical properties of these complexes, especially those of copper (mostly antiferromagnetic compounds), are reported in Refs. 104, 111, and 112.

The method of immediate interaction of ligands and metal acetates was also used to obtain binuclear complexes with a bridge M_2O_2 fragment on the basis of benzoyl- (***647***[119]), salicyl- (***648*** [120]), and benzthiazolyl- (***649***[121]) hydrazones of *o*-tosylaminobenzaldehyde:

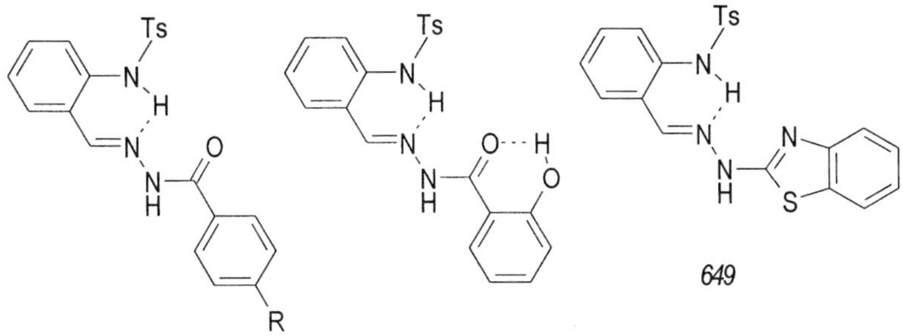

647 *648*

According to the opinion of the authors of Ref. 120, both di-(Zn^{2+}) and trinuclear (Cu^{2+}) chelates can be formed in case of use of ***648***.

Among the ligands used in the method of immediate interaction of reactants for preparation of binuclear complexes with M_2O_2 fragment are *bis*(hydrazones) [106,122,123] and *bis*(alkylisothiosemicarbazones) [124] of 4-substituted 2,6-diformylphenol ***650***. Their interaction (3.76) with metal salts leads to binuclear structures ***651***, containing as the bridge various donor atoms in dependence on the salt anion: the acetates form M_2O_2 (X = OMe) and the chlorides and bromides form MOX (X = Cl, Br) coordination units:

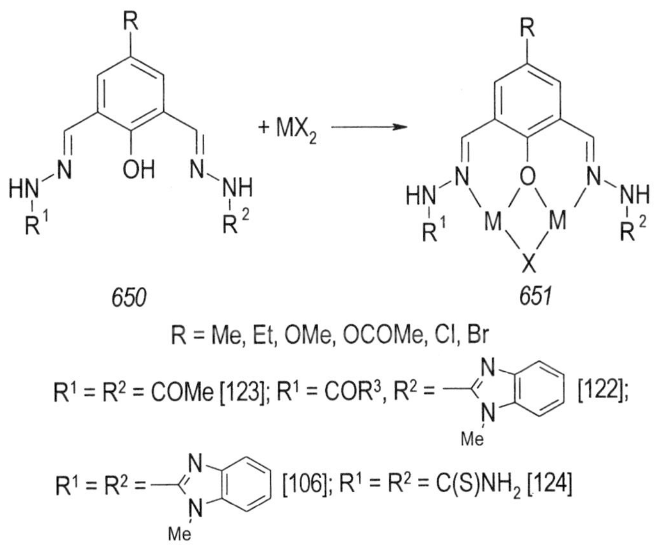

(3.76)

Additionally to ***640*** (Y = S), thioarylhydrazones of salicylaldehydes ***652*** are applied to obtain binuclear complexes with chalcogen-containing bridge by the method above [125]:

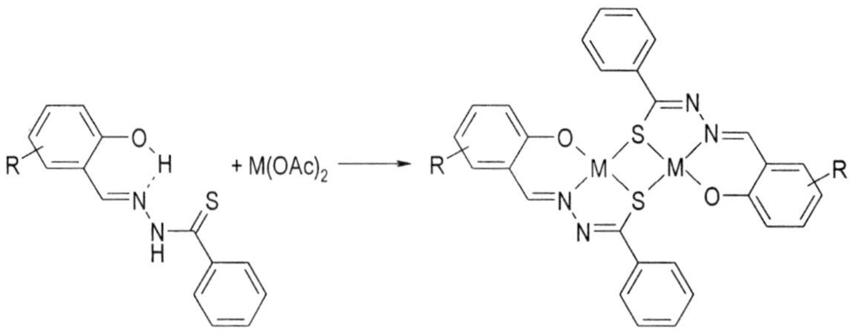

652 653

R = H, NO_2, Br

(3.77)

The interaction (3.77) of these ligands with copper(II) acetate (M = Cu) leads to the formation of a dimer ***653***, possessing unexpected ferromagnetism [126] (compare with Refs. 85 and 86). The explanation of this fact is given in Ref. 127.

Sources of binuclear chelates with M_2X_2 (X = S, Se) fragments are aminochalcogenide compounds, whose chlorhydrates are presented by formulae ***654*** [128]. The syntheses (3.78) with their participation yield complexes ***655*** [128]:

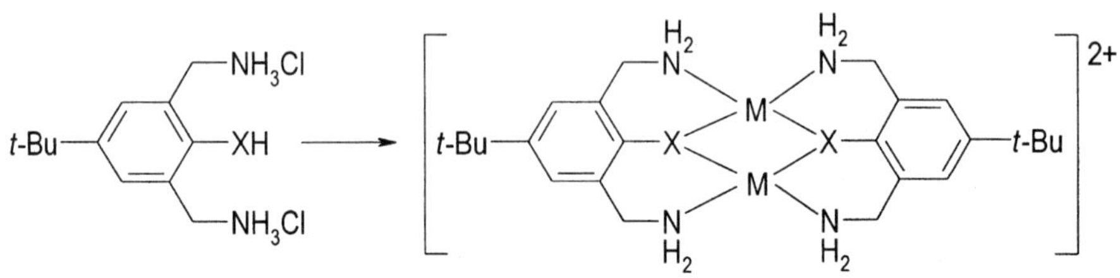

654 655

X = S, Se

(3.78)

Other examples of binuclear structures with oxygen, chalcogen, and mixed-element metal-bridge fragments are reported in Refs. 93, 127, and 129.

One of the most important synthetic routes to produce di- and polynuclear complexes is the immediate interaction of metal-containing ligands and metal salts or carbonyls [130a]. However, in this case, the reactions of coordinated ligands practically take place, so such transformations will be discussed in Sec. 3.3.2.3, dedicated to template synthesis.

EXPERIMENTAL PROCEDURES

Example 1

Synthesis of $[(bipy)(CO)_3Re^I(4,4'\text{-bipy})Ru^{II}(NH_3)_5]^{3+}(PF_6)_3^-$ (bipy = 2,2′-bipyridine; 4,4′-bipy = 4,4′-bipyridine) [130b]

The $[(bipy)(CO)_3Re^I(4,4'\text{-bipy})Ru^{II}(NH_3)_5]^{3+}(PF_6)_3^-$ was prepared as a tetrahydate by stirring $[Re(CO)_3(bipy)(4,4'\text{-bipy})](PF_6)_3^-$ (80 mg, 0.1 mmol) in Me_2CO (10 mL) under Ar and adding, after 1 hr, $[Ru(NH_3)_5(H_2O)](PF_6)_2$ (54 mg, 0.1 mmol), followed by continuous stirring under Ar for 2 hr in the dark. Ether (100 mL) was then added to precipitate the complex. The solid was dissolved in MeOH (2 mL) and sorbed onto a column of Sephadex LH-20 (3 × 10 cm) and eluted with MeOH. The second fraction was collected, rotoevaporated to 5 mL, cooled to room temperature, and precipitated with ether (100 mL). The solid was filtered, washed with ether, and dried in vacuo over P_4O_{10}. Yield 100 mg (78%).

Example 2

Synthesis of $[Cu_2(dppe)_2I_2] \cdot 0.2CH_2Cl_2$ [dppe = 1,2-*bis*(diphenylphosphanyl)ethane] [130c]

A suspension of CuI (0.239 g, 1.26 mmol) in CH_2Cl_2 (2 mL) was treated with a solution of dppe (0.5 g, 1.26 mmol) in CH_2Cl_2 (3 mL). After addition of $CHCl_3$ (5 mL) and stirring for 30 min at room temperature, ether was layered onto the colorless solution, which was kept overnight at −15°C. The white precipitate was filtered off, washed with cold ether, and dried in vacuo. Yield 0.85 g (57%).

Example 3

Synthesis of $[Cp'_2TaSeHW(CO)_5]$ (Cp′ = *t*-BuC_5H_4) [130d]

A mixture of 90 mg (0.18 mmol) of $[Cp'_2TaSeH]$, 0.2 mmol of $[W(CO)_5(THF)]$, and 15 mL of THF was stirred at room temperature for 30 min. After evaporation of the solvent, the residue was chromatographed on SiO_2 (column 10 cm, d 2.5 cm). Eluting with toluene, a red–brown band was obtained, which was recrystallized from toluene/pentane forming dark brown crystals. Yield 100 mg (65%).

Example 4

Synthesis of $Cp_2NbH_2(SnMe_3)$ [130e]

0.502 g (2.22 mmol) of $[CpNbH_3]$ and 0.442 g (2.22 mmol) of Me_3SnCl were mixed in 70 mL of ether, and then 3 mL of NEt_3 was added. The mixture was stirred overnight, then filtered, and the residue was washed with 50 mL of ether. All volatiles were then removed in vacuo from the combined filtrate and washings to leave a light beige powder. Yield 0.730 g (85%).

Example 5

Synthesis of $[Cu_2L(ClO_4)_4]$ {L is a Schiff Base Cryptand Derived from the Direct "2+3" Cyclocondensation of *tris*(2-aminoethyl)amine (tren) and 2,6-*bis*[formylphenoxymethyl]pyridine (bfpp)} [130f]

To prepare the cryptand $L \cdot H_2O$, a solution of tren (4 mmol) in MeOH (20 mL) was added to a warm solution of bfpp (6 mmol) in MeOH (300 mL) and refluxed for 2 hr. The hot solution was filtered and left to stand in a refrigerator overnight. The resulting suspension was filtered to leave a pale yellow solid. Yield 54%.

The prepared cryptand (0.5 mmol) was dissolved in hot ethanol–acetonitrile mixture (1:1, 50 mL) and added to $Cu(ClO_4)_2 \cdot 6H_2O$ (1 mmol) in acetonitrile (20 mL). On mixing, a blue solution was produced from which small blue crystals of $[Cu_2L(ClO_4)_4]$ were isolated after letting the solution stand for 18 hr. Yield 53%.

Example 6

Synthesis of $\{[Pt_2Ag_8\{C{\equiv}C(t\text{-}Bu)\}_8(OClO_3)_2(acetone)_2](O_2ClO_2)_2\}_n$ [130g]

Stirring a solution of $[Pt_2Ag_4\{C{\equiv}C(t\text{-}Bu)\}_8]$ in diethyl ether with $AgClO_4$ (1:4 molar ratio) at room temperature for 5 hr produces an intense yellow solid. Its crystallization from acetone–

hexane affords light yellow crystals of $\{[Pt_2Ag_8\{C{\equiv}C(t\text{-}Bu)\}_8(OClO_3)_2(acetone)_2](O_2ClO_2)_2\}_n$. Yield 68%.

Example 7

Synthesis of $[\{Co_2(CO)_6\}_2MeSiC_2(C_{13}H_8)C_2SiMe_3]$ [130h]

A solution of $[Co_2(CO)_8]$ (0.19 g, 0.56 mmol) and 2,7-*bis*(trimethylsilylethynyl)fluorene (0.10 g, 0.28 mmol) in *n*-hexane (20 mL) was stirred at 0°C and the reaction progress was monitored by IR spectroscopy and spot TLC. After 1 hr the solution turned dark brown and completion of the reaction was revealed from TLC and the disappearance of the bridging carbonyls due to $[Co_2(CO)_8]$. The solvent was then removed in vacuo and the residue chromatographed on silica TLC plates using hexane as eluent. The mayor reddish-band was isolated to afford the product as a dark red microcrystalline solid. Yield 0.13 g (51%).

Example 8

Synthesis of $Rh_2(O_2CCF_3)_4(Ph_2C_2)_2$ [130i]

The compound below was prepared by heating a mixture of unligated dirhodium(II) tetrakis(trifluoroacetate) (0.066 g, 0.10 mmol) and diphenylacetylene (0.036 g, 0.20 mmol) at 65–85°C for 3 days in a sealed evacuated Pyrex ampoule. Dark green block crystals were grown in both "hot" and "cold" zones of the tube. Yield 0.058 g (57.2%).

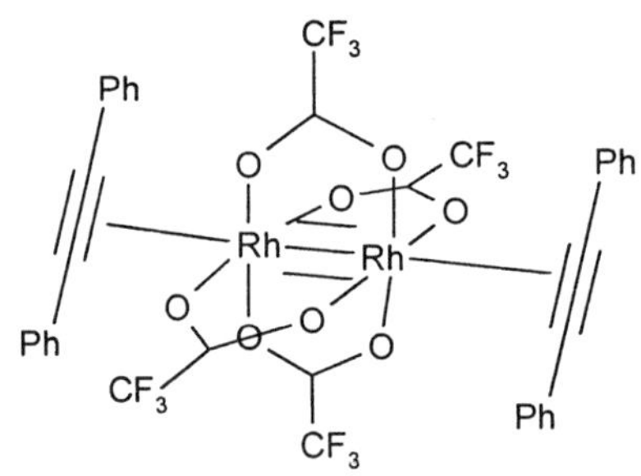

3.1.2 Choice of Metal Source

It is well known that metals are the most important part of coordination compounds. A reflection of this circumstance is the fact that, in the most fundamental of issues, related to coordination chemistry, the material is grouped according to metals [1,3,4,11,131]. In recent years a number of reviews on coordination compounds of Group I–VIII metals have been published. Group I elements – Ag [132,133], Au [134], Cu [135]. Group II element – Be [136]. Group III elements – Al [137–140], Ga [141], In [142,143], Tl [144], Sc [145,146], Y [147]. Group IV elements – Ge [148], Sn [148,149], Hf [150,151], Zr [150,151]. Group V elements – V [152,153], Nb and Ta [154]. Group VI elements – Cr [155–157], Mo [158,159], W [160]. Group VII elements – Mn [161], Tc [162,163], Re [164–166]. Group VIII elements – Fe [167–170], Co [171–173], Ni [174–176], Ru [177], Rh [178–180], Pd [181], Os [177,182,183], Pt [184], lanthanides [185–188], and actinides [185,186,188].

The data above testify that, as well as in previous decades [1], the main attention is paid to coordination chemistry of *d*-elements. At the same time, in the

1990s the interest in *s*-metals, which form complexes mostly with crown-ethers [25,189–198], has decreased. We note that during recent years the most important place in coordination chemistry has belonged to *p*-metal complexes, especially those of aluminum [140] and gallium [141,199,200]. In this respect, the synthesis (3.79) of the complexes of gallium chloride with benzotriazole (HL) is representative [200]. This interaction allowed us to obtain structurally characterized compounds, whose composition depends on the reactants' ratio:

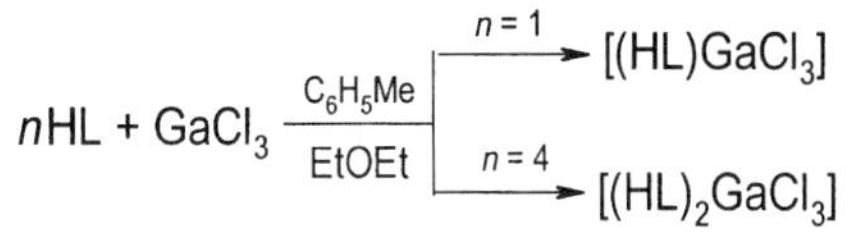

(3.79)

A change of salt anions allows us to carry out controlled syntheses of definite types of coordination compounds. Thus, metal halides are widely used to prepare molecular (Sec. 3.1.1.1) and π-complexes (Sec. 3.1.1.3). Metal acetates are mostly applied in the syntheses of metal chelates and, especially, inner-complex compounds (Sec. 3.1.1.2). Di- and polynuclear structures are formed under use of the anions mentioned above, which, in some cases, determine (see Sec. 3.1.1.4) a type and donor centers of bridge fragments. At the same time, the mentioned approach to choice of salts of metal complex-formers has many exceptions, although it is useful.

The choice of metal halides, especially those of Groups III–V, in the syntheses of molecular complexes is related with their high Lewis acidity and so their capacity to form stable adducts (Sec. 1.2.1). However, this does not mean that molecular complexes cannot be obtained on the basis of weaker Lewis acids, for instance nitrates or acetates of transition metals.

Metal acetates are widely used, because they create an optimal magnitude of pH $\approx$ 5–6 for obtaining ICC. At the same time, in these syntheses, halides (chlorides) or nitrates of transition metals can also be used by adjusting the pH of the reaction medium to weakly acidic by addition of alkaline agents, mostly alcohol or aqueous–alcohol solutions of NaOH or KOH, as well as their alcoholates. In such cases, the danger of formation of hydroxides of metal complex-formers appears; this fact complicates considerably the obtaining of pure final reaction products.

Metal oxides can be used as metal sources. A classic example of such syntheses is the preparation of dioxomolybdenum(VI) *bis*-acetylacetonate in the absence of a solvent (3.80) [11]:

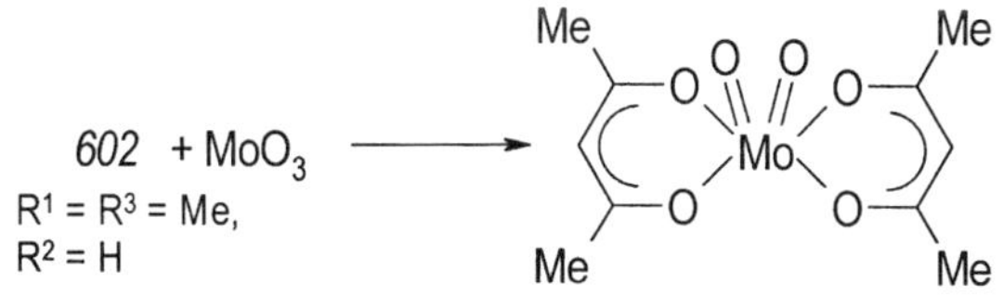

(3.80)

Additionally to the example above, metal oxides are widely applied to prepare complex compounds in so-called "*ammonium synthesis*" [201–204] (Sec. 3.4.3.2). The sense of this method is described by Eq. (3.81):

$$MO + 2NH_4X + 2L \rightarrow ML_2X_2 + NH_3 + 2H_2O \quad (3.81)$$

The oxides of divalent copper, zinc, cadmium, and lead are used most frequently as MO. N-Donor bases (mostly ammonia and pyridine) and O-donor solvents (DMF,

DMSO, CH_3OH) take part as ligands. If, additionally to ammonium salts, other salts are simultaneously used (3.82), the anionic complexes can be formed, for example:

$$MO + 2NH_4SCN + KSCN \rightarrow K[M(SCN)_3(NH_3)] + NH_3 + H_2O \qquad (3.82)$$

Use of metal hydroxides opens some possibilities in the discussed syntheses. Thus, as a result of interaction (3.83) of β-ketoimines ***656*** with divalent nickel, copper, and cobalt hydroxides in acetone solution, the β-ketoiminates ***657*** were obtained [205]:

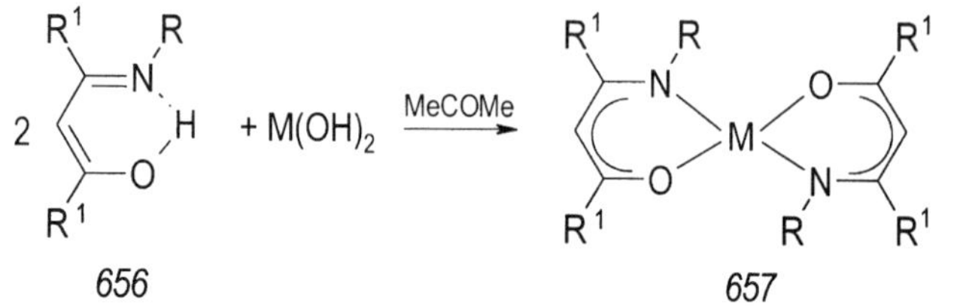

(3.83)

Among other metal sources, the carbonyls should be emphasized [Sec. 3.1.1.2; for example, Scheme (3.56)] and other metal complexes (Sec. 3.2.1), in particular chelates [Schemes (3.112)–(3.116)]. In the majority of cases with $M(CO)_n$ use, an exchange of CO groups to other ligands takes place; such reactions will be discussed in Sec. 3.2.1.

3.1.3 Selection of Solvents

Much attention has been paid to the role of solvents in a series of monographs [3,10,11,206–210] and reviews [211–221]. In general, nonaqueous solvents are applied in modern coordination chemistry. This is explained by the comparatively low solubility of a majority of organic compounds, used as ligands, in water (see Chap. 2).

To synthesize coordination compounds of definite type, the following properties of a solvent should be taken into account, additionally to its capacity to dissolve ligands and metal sources: physical properties (mainly dielectric constant, dipole moment, boiling and melting points) and chemical ones (solvation activity, including acid–base characteristics, specific interaction) [208,222–229].

Selection of solvents for obtaining molecular complexes is carried out taking into account the solubility of metal salts and ligands, which depends on their nature. Synthesis of metal halide adducts and metal organohalides of Group IIIA–VIA elements is generally made in aprotonic solvents with low dielectric characteristics ($\varepsilon < 15$, $\mu < 2D$) and donor numbers ($n < 20$). Among them, the aliphatic and aromatic hydrocarbons, as well as their halide derivatives, are the most common. Thus, benzene has been widely used in the syntheses of complexes of chlorides of Groups III–V metals with amines, azomethines, azoles, and azines (Sec. 2.2.4.2). Dichloromethane was used to prepare adducts of tungsten hexahalides with various ligands (NH_3, NH_2R, NR_3, Py, PR_3) [230].

In a difference from the above, the majority of metal halides of secondary subgroups of Group I, II, and VII and lanthanides are low-soluble in the examined solvents, so synthesis of their complexes is carried out in comparatively high-polar solvents (water, alcohols, and aqueous–alcohol mixtures). To carry out syntheses in water, the corresponding conditions, necessary to obtaining soluble derivatives of organic compounds (for example, halide hydrates of amines or N-containing heterocycles), should be fulfilled. In particular, the complexes of cobalt, nickel, and copper chlorides with 5-aminotetrazole ***658*** were obtained [231].

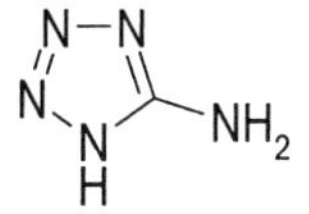

658

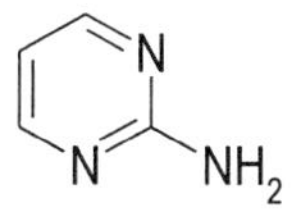

659

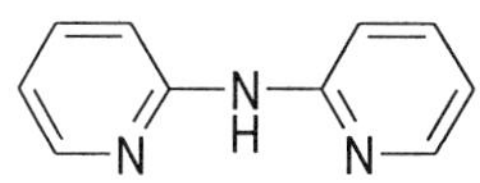

660

The complexes of Cu(I) and Cu(II) bromides with 2-aminopyrimidine (L) ***659*** were synthesized under practically the same conditions [232]. As a result, complexes with different compositions were isolated and structurally characterized: Cu(I) – LCuBr, LCu_2Br_2, and $(LH)CuBr_2$; Cu(II) – L_2CuBr_2, $(LH)CuBr_3$, $(LH)_2CuBr_4$, and $(LH_2)CuBr_4$. It is shown that the Cu—N bond with participation of the endocyclic nitrogen atom of the pyridine cycle exists in their bridge structures (except L_2CuBr_2). The bridges are formed not only at the expense of bromine atoms in Cu(II) complexes, but also both N-atoms of 2-aminopyrimidine.

Complex compounds $LCuCl_2$ of copper dichloride with *bis*(pyridine)amine ***660*** [11] and $LMnCl_2$ of manganese dichloride with bipy [11] were prepared in ethanol. Alcohol solutions are widely used in the syntheses of complexes of lanthanide salts [233]. At the same time, for a series of halides, for example those of Group V and VI metals, the complex-formation reactions are accompanied by alcoholysis processes. Thus, the interaction of niobium pentachloride with azomethinic ligands (LH) in methanol or ethanol leads to halide–alcoxide complexes $NbOCl_3(OR)_2 \cdot LH$ (R = Me, Et) [211,234]. Molybdenum and tungsten halides and oxohalides are also exposed to alcoholysis in the described conditions [211].

Metal nitrates form molecular complexes mostly in aqueous, alcohol, and aqueous–alcohol mediums. A series of such syntheses is described for ***658*** and nitrates of divalent cobalt, nickel, copper, lead, and mercury [231]. The nitrate of *bis*(α,α'-bipyridine)silver $[Ag(bipy)_2]NO_3$ is formed in aqueous–alcohol medium [11].

Metal salts with other cations are synthesized in a similar manner. Among these studies, the obtaining of coordination compounds of copper(II) in methanol with N_2S_2 ligand environment should be mentioned. These complexes are of permanent interest due to the modeling of active centers of "blue" copper proteins on their basis (see Sec. 2.2.5.4) [235–237]. Such complexes were obtained, in particular, by interaction of divalent copper perchlorate and tetrafluoroborate with very exotic ligands ***661*** and ***662*** in methanol [238]:

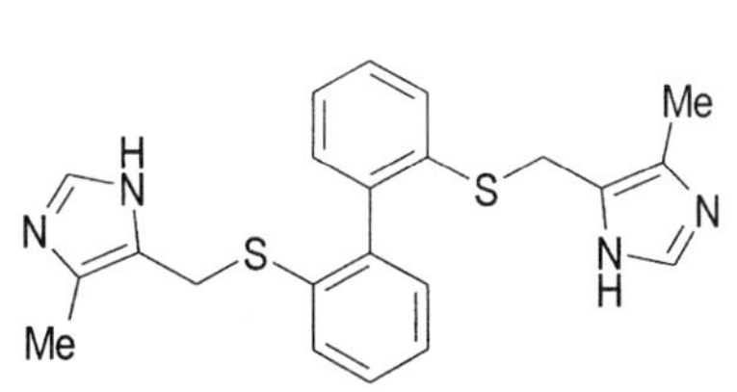

661

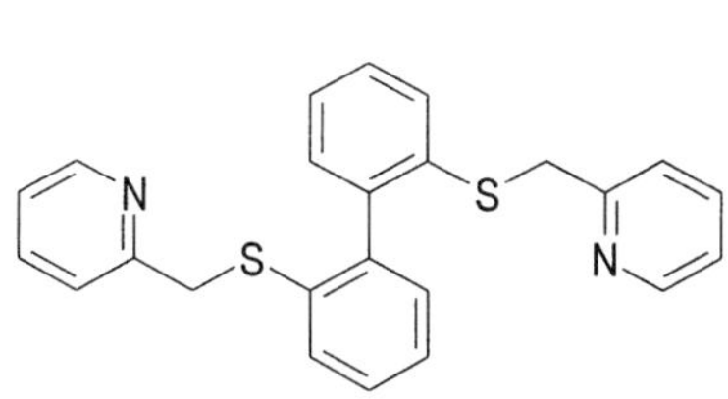

662

Among other O-donor solvents, diethyl ether, THF, and acetone are applied to synthesize molecular complexes [10,211,221]. However, in the majority of cases these reactions take place through the stage of formation of solvent complexes [10,206] and, in this respect, they will be discussed in the next chapter.

Solvents having comparatively high dielectric constant, providing easy dissociation of both EH-containing ligands (E = N, O, S) and metal salts, are widely used to synthesize metal chelates and especially inner-complex compounds. Taking into account the dielectric constant, water would be the most successful solvent. However, the low solubility of the majority of organic chelating compounds, used as ligands, limits an application sphere of aqueous medium in reactions to obtain ICC. As an example of synthesis of metal chelates in water, the interaction (3.84) of 2-(hydroxylamino)propanohydroxamic acid ***663*** with nickel acetate yields the complex ***664*** [239]:

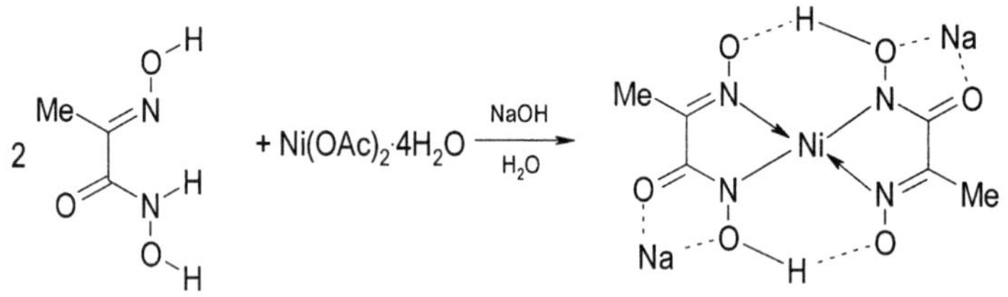

(3.84)

663 *664*

The important synthetic route for obtaining ICC in aqueous medium is a controlled creation of water-soluble chelating ligands [240–242]. In particular [242], mono- (***665*** and ***666***) and di- (***667*** and ***668***) sulfo-substituted azomethinic and azo compounds belong to this type of ligand:

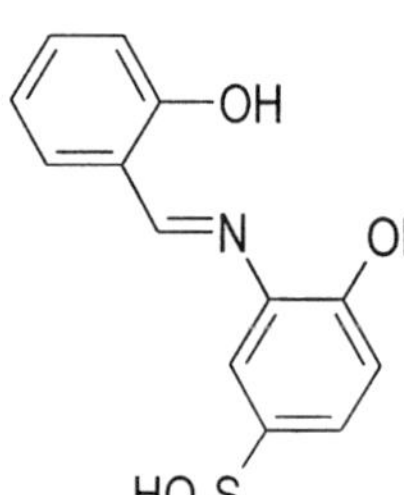

665

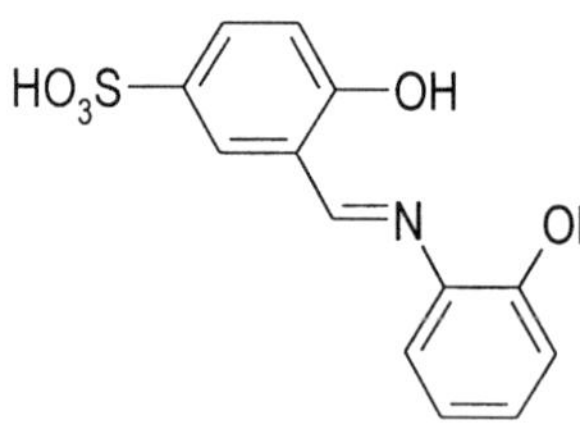

666

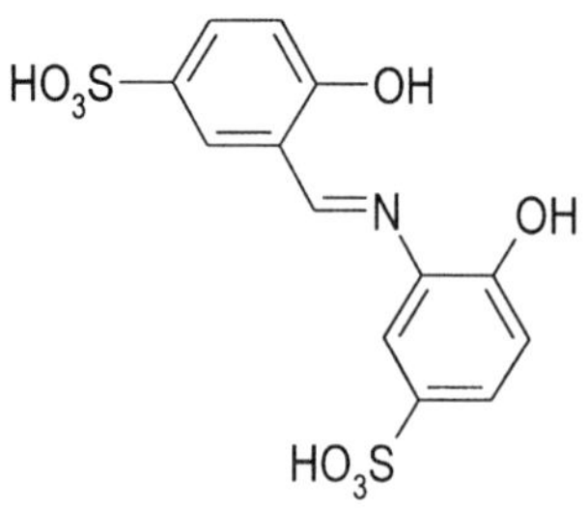

667

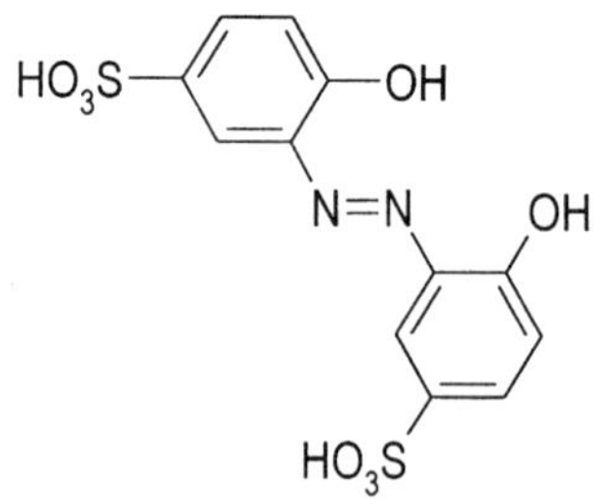

668

Most chelate complexes have been obtained in alcohols, whose dielectric constants are between 10 and 33 [223,224]; methanol is the most useful, among others ($\varepsilon = 32.2$). Additionally to the ICC obtained in CH_3OH and indicated in Sec.

3.1.1.2 and reviews [211,221], we have the nickel(II) complexes with N-mesitylhydrazone of phenylcyanoglyoxal ***669*** (3.85) [243], *o*-hydroxyazomethine of 1-alkyl-2-formylbenzimidazole ***670***, ***671*** (X = O) [244], and its N-tosylamino (X = NTs) and mercapto (X = S) analogues (3.86) [245]:

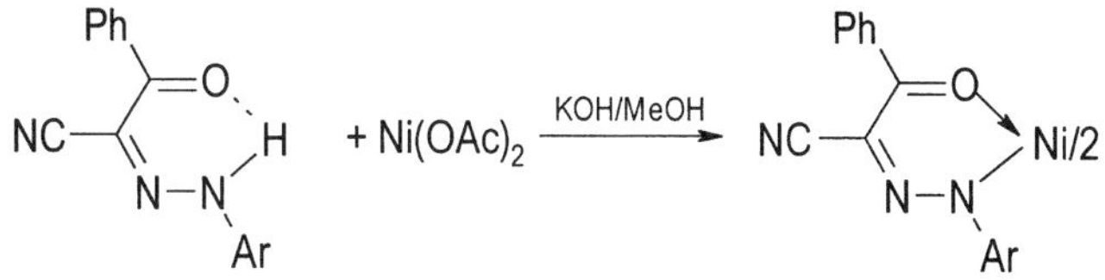

669

(3.85)

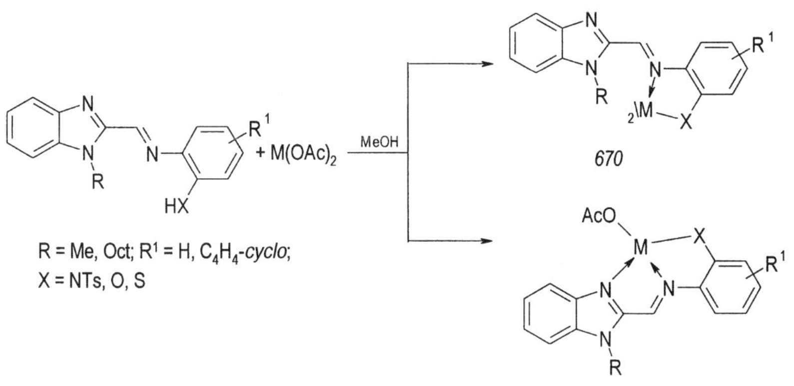

670

671

(3.86)

The complex ***669*** is interesting due to the fact that, according to x-ray diffraction data, the six-member metal-cycle (but not five-member cycle ***672***) exists in its structure [243]:

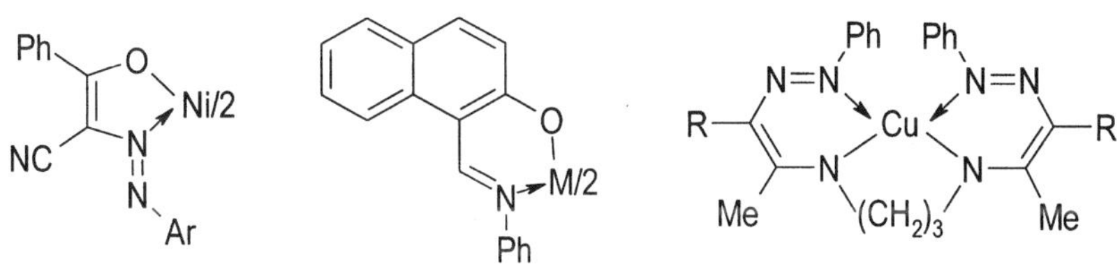

672 673 674: R = H, COMe

The composition of the complexes ***670*** and ***671*** is determined [244,245] by the nature of ligand (LH) and metal: ML_2 (X = NTs, M = Co, Cd, Zn; X = S, M = Co, Cu), MLOAc (X = O, M = Cu, Zn; X = NTs, M = Ni, Cu; X = S, M = Ni, Zn).

Ethanol was used, for example, in preparation of zinc complex of 2-hydroxy-1-(N-phenylaminomethyl) ***673*** [246] and copper complexes of hydrazoneimines of the type ***674*** [247], whose structures were determined by x-ray diffraction. The complexes of salicylidenthiocarbonylhydrazones of the type ***675***–***677*** were synthesized in the same solvent [248]. The complexes ***675*** are formed from metal acetates [Cu(II)], ***676*** from nitrates [Ni(II)], and ***677*** from chlorides [Fe(III)]:

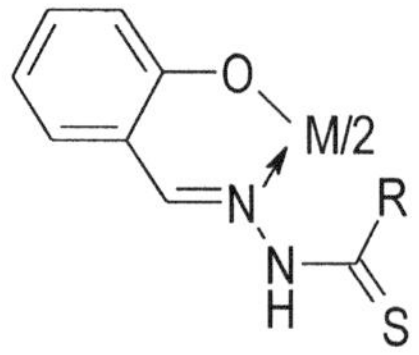

675

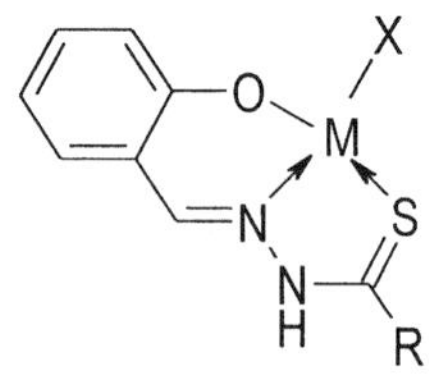

676

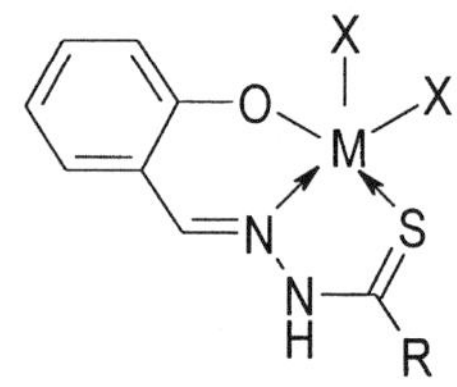

677

Mixtures of the examined solvents are frequently used to obtain ICC. Thus, an aqueous–ethanol mixture was used for the preparation of ***678*** from 5-(2-hydroxyphenyl)-3-phenyl-1,2,4-oxadiazole and copper(II) acetate [249]:

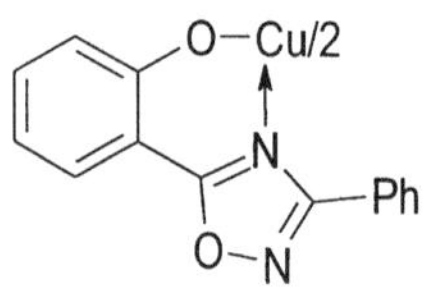

678

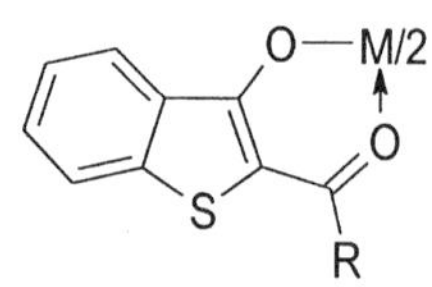

679

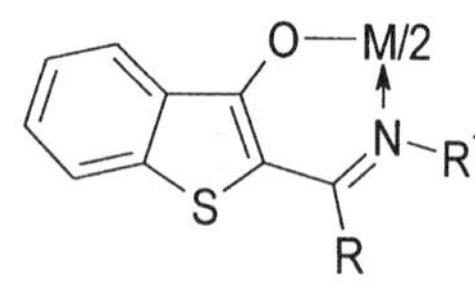

680

A series of complexes of 2-carbonyl derivatives of 3-hydroxybenzo[b]thiophene ***679***, its imines ***680*** ($R' = Ph$) [250] and hydrazones ***680*** ($R' = NR_2R_3$) [251], as well as pyperidine adduct of nickel complex ***681*** [252], have been isolated in methanol–ethanol mixture:

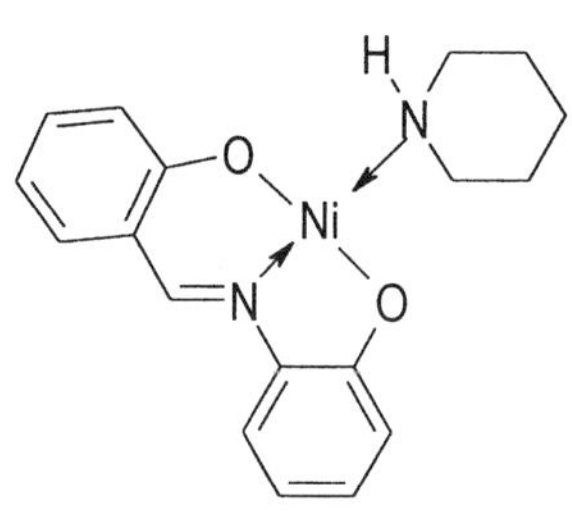

681

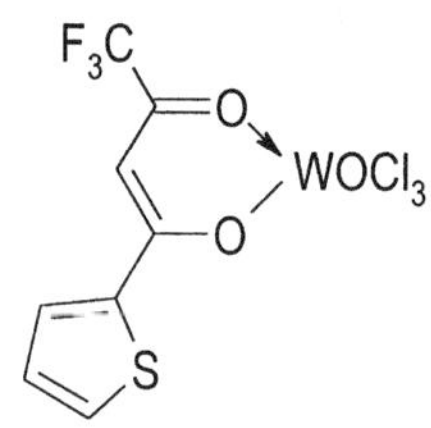

682

In comparison with the material discussed above, the obtaining of β-diketonate ***682*** from $WOCl_4$ and tenoyltrifluoracetone in a mixture of hexane and benzene is highly unexpected [253].

Selection of solvents for isolation of di- and polynuclear complexes (Sec. 3.1.1.4) is practically the same. In this respect, the formation of these compounds from low-dentate ligands in aqueous–alcohol mixtures should be noted. Thus, the tetranuclear complex Ba_4L_8 was obtained in water from bidentate dipivaloylmethane (LH) [254]. A series of polynuclear complexes, containing OH-, S-, and Cl-bridged fragments, was isolated (3.87) in the same solvent (water) starting from formally N,S-bidentate ethylenethiourea ***683*** ($R = H—L^2$) and formally tridentate 2,5-*bis*-N,N′-(alcoxyethylene)thiourea ***683*** ($R = CH_2OR^1$, $R^1 = Et—L^2$; $R = Bu—L^3$) and rhodium trichloride [255]:

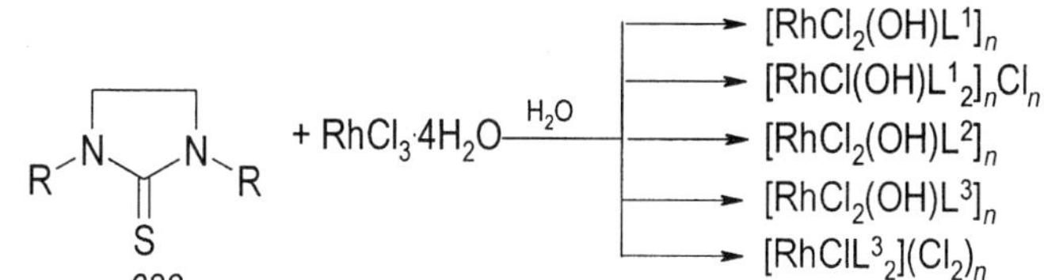

(3.87)

Binuclear structures were also obtained (3.88) from N-2-benzthiazolylhydrazone of salicylaldehyde ***684*** in methanol–dioxane mixture [256a]:

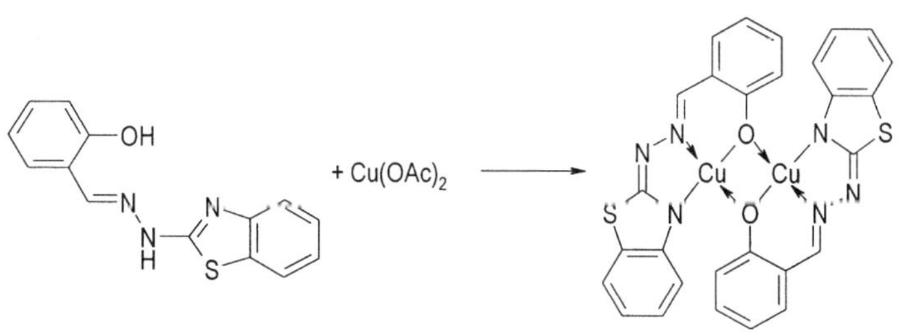

(3.88)

Summarizing the material, we note that the synthesis of di- and polynuclear structures can also be carried out in aprotic solvents with low dielectric constant. The above described reaction (3.72), yielding complexes ***637***, confirms this fact.

EXPERIMENTAL PROCEDURES

Example 1

Synthesis of ***630*** $[\{WOCl_2(CF_3C(O)CHC(O-)C_6H_5)\}_2(\mu\text{-}O)]\cdot 2C_6H_6$ (3.69) [81]

The solution of 1.08 g (5 mmol) benzoyltrifluoroacetone in 10 mL of benzene was added dropwise under stirring to the suspension of 1.74 g (5 mmol) of $WOCl_4$ in 20 mL of hot benzene. The mixture was refluxed for 3 hr to remove HCl, after which part of the solvent was evaporated and 10 mL of hexane was added under stirring. Red–brown crystals were precipitated after 2–3 hr. They were filtered off, washed with hexane, and dried in the dry argon stream.

Example 2

Synthesis of Di-μ-Hydroxo-*bis*(μ-perchlorato-O,O′)-*bis*[*bis*(2-amino-4-methylpyrimidine)dicopper(II) ***631*** (3.70) [82]

2-Amino-4-methylpyrimidine (10 mmol) was dissolved in 25 mL of water. $Cu(ClO_4)_2\cdot 6H_2O$ (10 mmol) was also dissolved in 25 mL of water. The Cu(II) salt solution was then added slowly to the ligand solution, preventing any precipitation, filtered to remove any solids, and after 1 week the blue crystals separated. Yield about 32%.

Example 3

Synthesis of Dinuclear and Tetranuclear Copper(II) Complexes of 2,6-Diformyl-4-methylphenol-di(benzoylhydrazone) [256b]

Ligand (H_3L)

To an ethanolic solution (35 mL) of 2,6-diformyl-4-methylphenol (1 mmol) was added benzoylhydrazide (2 mmol) in ethanol (10 mL). To this solution was added one drop of concentrated HCl and the mixture was heated at ca. 60°C for 1 hr and allowed to stand overnight. The precipitate was filtered off, washed with ethanol, and dried in air. Yield ca. 80%.

Dinuclear Complex [$Cu_2L(OCH_3)$] · DMF

An aqueous solution (10 mL) of copper(II) acetate monohydrate (2.2 mmol) was added to a methanolic solution (40 mL) of H_3L (1 mmol). The solution was allowed to stand for several hours at room temperature. The crude complex precipitated was collected by suction filtration and dried. Yield 90%. The complex was recrystallized from hot methanol–dimethylformamide (1:1) to give dark green crystals of formula [$Cu_2L(OCH_3)$] · DMF. Yield ca. 30%.

Tetranuclear Complex [{$Cu_2L(OH)$ · DMF}$_2$]

The crude complex was recrystallized from a hot dimethylformamide solution to give dark green crystals of [{$Cu_2L(OH)$ · DMF}$_2$]. Yield ca. 30%.

Example 3

Synthesis of *bis*[2-(2-Hydroxyphenyliminomethyl)-1-(4-methylphenylsulfonamido)phenolato]-dinickel(II)-tetramethanol ***641*** (3.74) [97]

[$Ni_2L_2(MeOH)_4$] was obtained by the slow addition of 2-(2-hydroxyphenyliminomethyl)-1-(4-methylphenylsulfonamido)benzene (0.36 g, 1 mmol) in methanol (15 mL) to a solution of nickel acetate (0.248 g, 1 mmol) in the same solvent (10 mL). After refluxing for 1 hr, the brown crystalline precipitate was filtered and washed with methanol.

Example 4

Synthesis of *bis*(α,α′-Bipyridine)silver Nitrate [$Ag(bipy)_2$]NO_3 [11]

$$2\text{Bipy} + \text{AgNO}_3 \rightarrow [\text{Ag(bipy)}_2]\text{NO}_3$$

A hot solution of 16.9 g of $AgNO_3$ in water–ethanol mixture was added to the hot solution of 31.2 g of α,α′-bipyridine in the same solvent. The precipitate was filtered off and recrystallized from hot aqueous enthanol.

Example 5

Synthesis of *trans*-Dinitrodiaqua-*bis*(1-phenyltetrazole)copper(II) [256c]

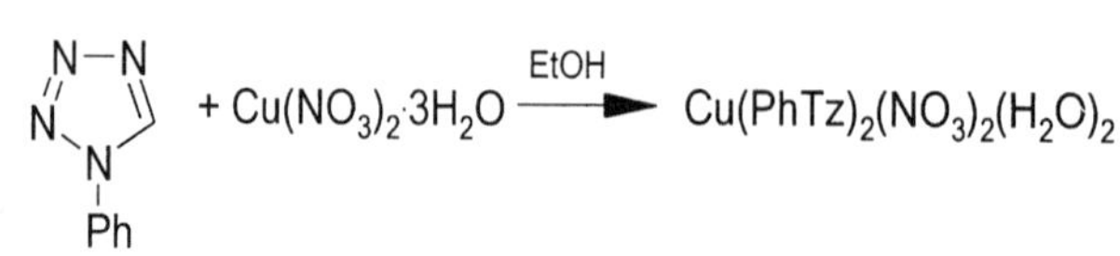

$Cu(NO_3)_2 \cdot 3H_2O$ (0.002 mol, 0.48 g) was dissolved under heating in 7–10 mL of ethanol. The warm solution of PhTz in 5–7 mL of ethanol was added to the solution of $Cu(NO_3)_2$. The surplus of solvent was evaporated in a water bath. A blue precipitate occurred after cooling the solution. The precipitate was filtered off, washed twice with cold ethanol, and dried in vacuo over $Mg(ClO_4)_2$. Yield 90%.

Example 6

Complex of Tenoyltrifluroacetone with Tungsten(VI) Oxochloride [253]

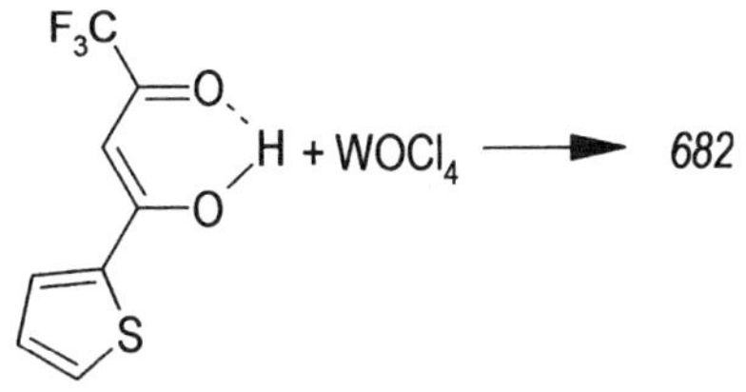

Benzene (2 mL) was added to the suspension of 0.684 g (2 mmol) of $WOCl_4$ in 10 mL of dry hexane. The mixture was heated and the warm (35–40°C) solution of 0.444 g (2 mmol) of tenoylacetone in 10 mL of hexane added dropwise slowly in the boiling mixture. After heating for 0.5 hr the solution was filtered and allowed to stay overnight. Dark needles were filtered off, washed with hexane, and dried in vacuo.

Example 7

Synthesis of $[Cu(C_7H_6N_4)_5(BF_4)]BF_4$ [256d]

A solution of $Cu(BF_4)_2 \cdot 6H_2O$ (1.38 g, 4 mmol) in ethanol:ether (1:2) mixture (10 mL) was added to a solution of 1-phenyltetrazole (3.5 g, 24 mmol) in the same mixture of solvents (15 mL). When the solid began to be formed (5–7 min), ether (13 mL) was added. The sediment was filtered and washed with the ethanol:ether (1:4) mixture. Yield 1.75 g.

3.2 EXCHANGE REACTIONS IN THE SYNTHESIS OF COORDINATION COMPOUNDS

This group of reactions, which is widely used in synthetic coordination chemistry, includes the exchange of metals and anions [10,11,206].

3.2.1 Ligand Exchange

A large number of reported data is dedicated to the exchange (reactions of substitution) of ligands [3,10,11,51,74,206,257–262]. The reaction kinetics and mechanisms are examined in detail there; special attention is paid to the ligand exchange in plane-square and octahedral complexes and the exchange reactions in organometallic compounds are discussed. The factors influencing the direction and velocity of exchange reactions are discussed: nature of metal, ligands, and solvents, role of exiting groups, stereochemical aspects, various effects (trans, cis, chelating, etc.) [51].

Detailed examination of this enormous problem is beyond the limits of this monograph and its synthetic direction. So, we have intended to highlight several aspects of the use of reactions of ligand exchange as a safe preparative method to produce coordination compounds.

The well-known capacity to replace one ligand by another one was used for the preparation of metal complexes as far back as the nineteenth century. Such reactions can be illustrated by two examples [(3.89), (3.90)] [11]:

$$Ni(H_2O)_6]Cl_2 + 6NH_3 \rightarrow [Ni(NH_3)_6]Cl_2 + 6H_2O \qquad (3.89)$$

$$[Co(NH_3)_5Cl]Cl_2 + NaNO_3 \rightarrow [Co(NH_3)_5NO_2]Cl_2 + NaCl \qquad (3.90)$$

The reactions of basic hydrolysis [51,259,261,262], for example (3.91), can also be attributed to the same type of synthetic transformation:

$$Co(NH_3)_5Cl]^{2+} + OH^- \rightarrow [Co(NH_3)_5OH]^{2+} + Cl^- \qquad (3.91)$$

The reactions above take place mostly in aqueous medium and, so, they are limited by using, in general, water-soluble complexes and substituting ligands. The transition to coordination chemistry of nonaqueous solutions [10,206,208–210,212–221] provided the possibility of wide use in syntheses of complexes of various organic ligand systems and exchanging organic compounds by this method. Practically, the majority of reactions, taking place in nonaqueous organic coordinative solvents, represent the processes of ligand exchange.

Among the molecular complexes obtained by this route, the coordination compounds of metal carbonyls and various heterocycles (L) are widely represented [5]. The reactions proceeding in this case can be described by the following transformation (3.92):

$$M(CO)_n + m\text{Solv} \rightarrow M(CO)_{n-m}(\text{Solv})_m + mCO \qquad (3.92a)$$

$$M(CO)_{n-m}(\text{Solv})_m + mL \rightarrow M(CO)_{n-m}L_m + m\text{Solv} \qquad (3.92b)$$

Thus, the transformation (3.93) of hexacarbonyls of Group VI metals with azoles and their benzoanalogues (L) in THF (Solv = THF) results in complexes of the type ***684*** and ***685*** [263,264]:

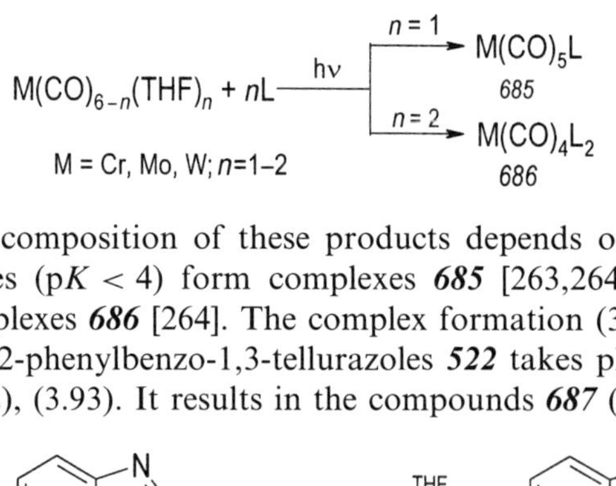

(3.93)

The composition of these products depends on the basicity of ligands: low-basic azoles (pK < 4) form complexes ***685*** [263,264], high-basic azoles (pK > 4) yield complexes ***686*** [264]. The complex formation (3.94) with participation of 2-methyl- and 2-phenylbenzo-1,3-tellurazoles ***522*** takes place according to the reaction types (3.92), (3.93). It results in the compounds ***687*** (Sec. 2.2.5.6) [265]:

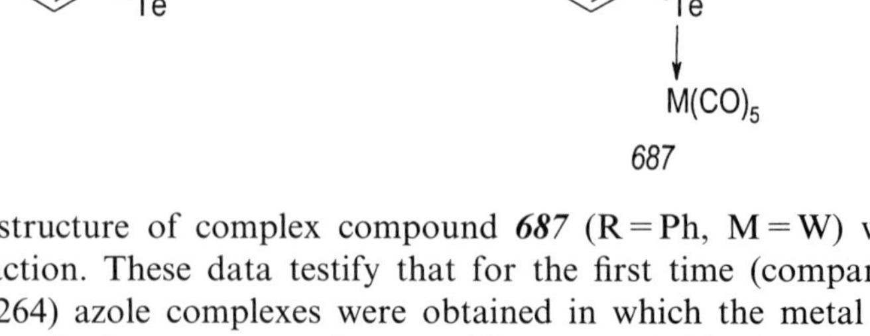

687

(3.94)

The structure of complex compound ***687*** (R = Ph, M = W) was proved by x-ray diffraction. These data testify that for the first time (compare with Refs. 5, 263, and 264) azole complexes were obtained in which the metal coordinates with an endocyclic donor center of nonpyridine type. Evidently, the interaction (3.95) of

2,4,6-triphenylphosphabenzene with tungsten hexacarbonyl takes place according to the same scheme [5]:

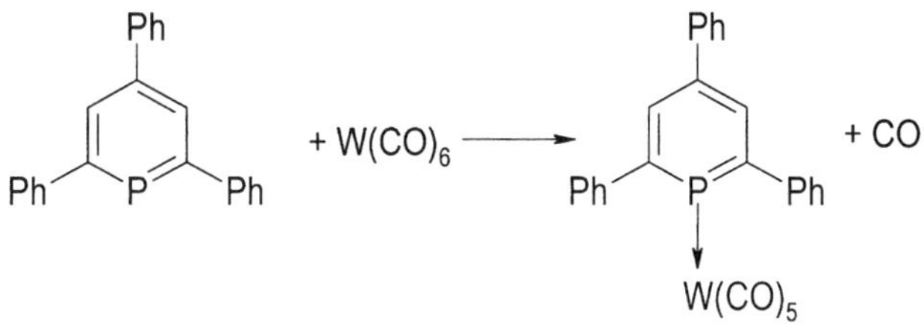

688 (3.95)

The ligand-exchange syntheses of the examined type are applied to obtain complexes of metal carbonyls and carbonyl cyclopentadienyls with tellurium-organic ligands (Sec. 2.2.4.7) [265–268]. Thus, the complexes ***689*** were isolated (3.96) from heated or UV-irradiated THF solutions, containing the adducts $M(CO)_5 \cdot THF$ and diorganotellurides [269,270]:

$$M(CO)_5 \cdot THF + R_2Te \xrightarrow[t,\ h\nu]{THF} \underset{689}{M(CO)_5 \cdot TeR_2} + THF \tag{3.96}$$

Telluroketone reacts (3.97) with $W(CO)_5 \cdot THF$ to form the coordination compound ***690*** [271]:

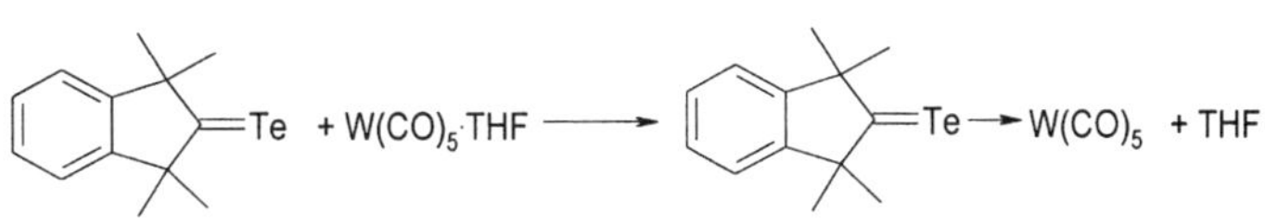

690 (3.97)

The reactions (3.98) of Group VI metal pentacarbonyls with diorganoditellurides also belong to the discussed type of syntheses [268]:

M=Cr, W (3.98)

In this case (compare with Ref. 268), the Te—Te bond remains. This bond is not destroyed also under interaction of diphenylditellurides with cyclopentadienyl–carbonyl complexes of manganese (3.99a) [272] and iron (3.99b) [273]:

$$[CpMn(CO)_3] + Ph_2Te_2 \longrightarrow [CpMn(CO)_2]Te_2Ph_2 \tag{3.99a}$$

$$[CpFe(CO)_2THF]^+(BF)_4^- + RTeTeR \xrightarrow[\sim 20^\circ C]{CH_2Cl_2} [CpFe(CO)_2Te_2R_2]^+(BF)_4^-$$

$$R = Me, Ph \tag{3.99b}$$

The reactions of metal–carbonyl complexes, containing more than one coordinated THF molecule, in a series of cases lead to incomplete exchange of these ligands. In this respect, the following synthesis (3.100) of the radical complex of fluorene ***691*** is representative [274,275]:

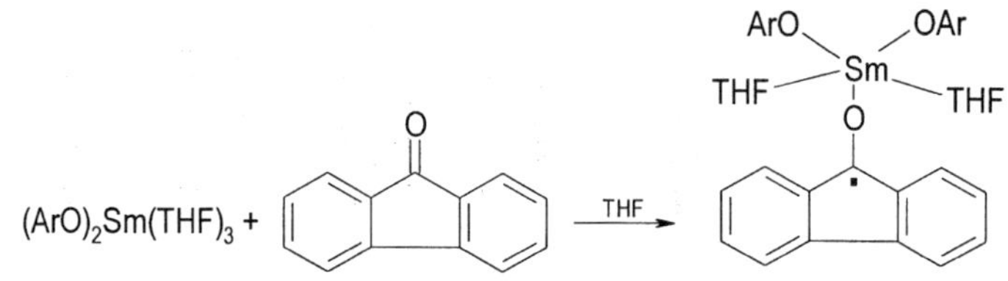

691

(3.100)

Other complexes of metal carbonyls also participate in ligand-exchange processes. In particular, the acetonitrile complexes are applied for these goals (3.101) [265,267]:

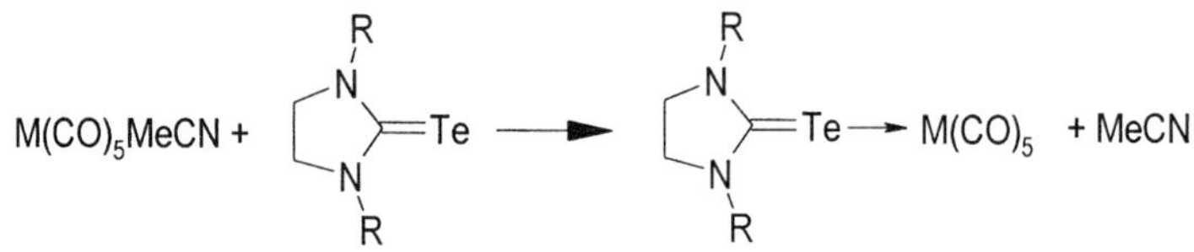

R = Cr, Mo, W; R = Me, Et

(3.101)

Metal etherates were used to obtain complexes by the ligand-exchange method in a series of transformations (3.102):

$$MX_n(ROR)_m + mL \rightarrow MX_nL_m + mROR \quad (3.102)$$

The use of etherate of boron trifluoride is well known for these syntheses [11]. Etherate of gallium trichloride is also applied (3.103) [199]:

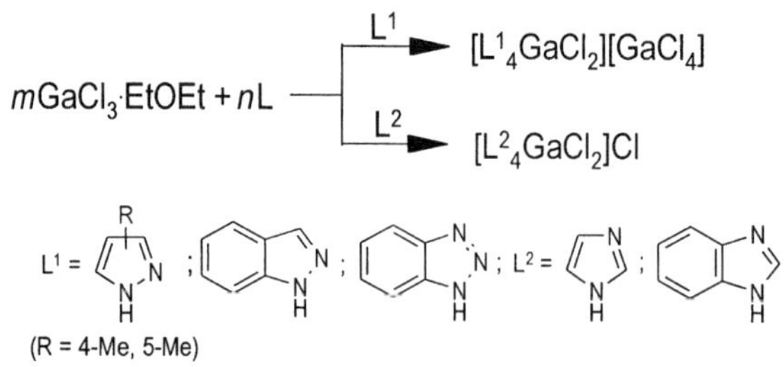

(3.103)

Ligand exchange has found a wide application in the syntheses of metal chelates. Metal-cycles containing C,N-donor atoms are formed as a result of cyclometallation reactions using ligand exchange. For example, the complex $[Pt(DMSO)_2Cl_2]$, which is a universal cycloplatination reactant, takes part in such a process (3.104) [276]:

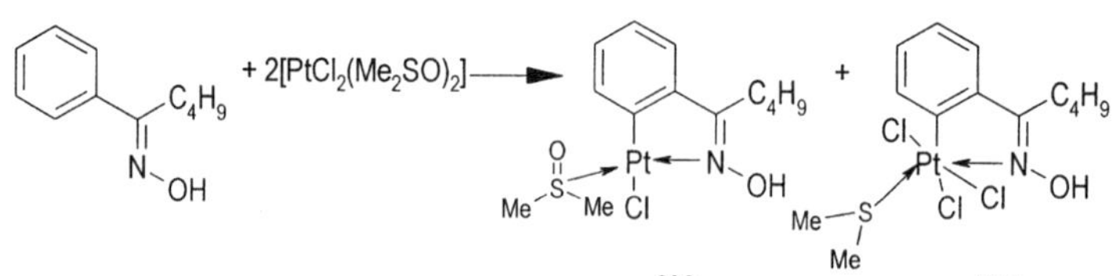

692 693

(3.104)

The complex ***692*** is a "standard" reaction product (Sec. 2.2.5.1); meanwhile, the formation in this reaction of structurally characterized coordination compound ***693*** with deoxygenated DMSO molecule is highly unexpected [277].

The ligand exchange was used to create chelate complexes with P,N- ***694*** (3.105) [278] and As,As-donor atoms (3.106):

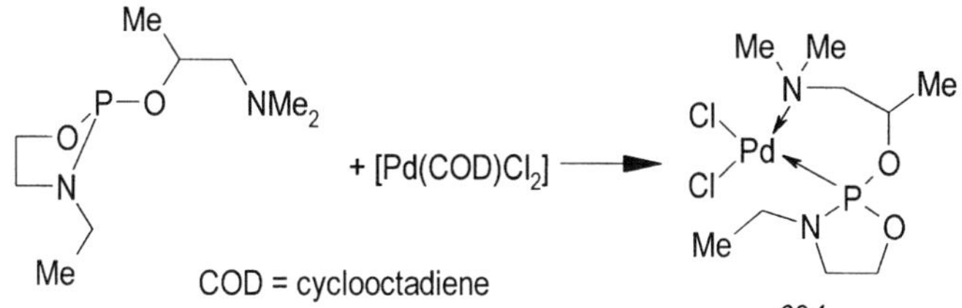

694

(3.105)

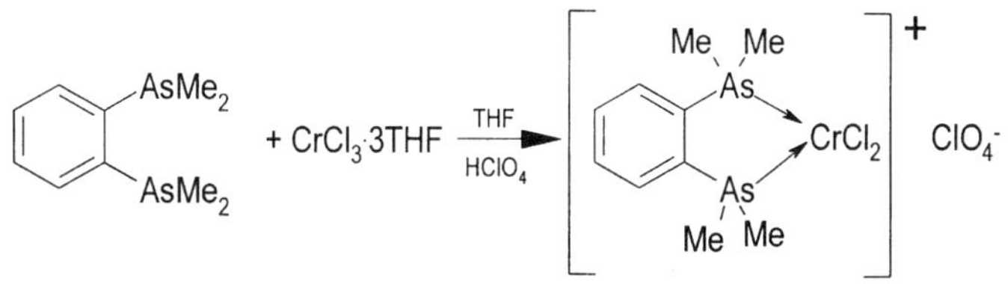

695

(3.106)

The reactions of ligand exchange have frequently been applied to synthesize chelates with the coordination unit containing N,O-donor centers. The complexes of titanium [280], thorium, and uranium [281] tetrachlorides with *bis*(salicylidene)alkylene-amines, obtained in THF solution, are evidently formed through THF complexes of the type MX_n(THF). An example of such syntheses is the transformation (3.107) [280]:

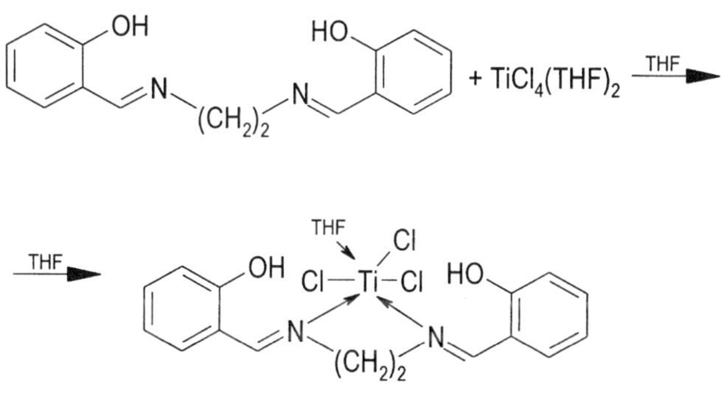

696

(3.107)

Carrying out such syntheses, it is necessary to take into account not only the high donor activity of THF, but also its capacity to reduce some metal ions (3.107). In particular, obtaining the Ti(III) adduct ***696*** from $TiCl_4$ (3.107) with participation of THF testifies to this.

In addition to the above, a series of other molecular complexes are applied as precursors for obtaining chelates on the basis of N,O-donor ligands. Thus, gallium trichloride etherate was used (3.108) to synthesize the ICC of 2-α-hydroxybenzyl-benzimidazole [199]:

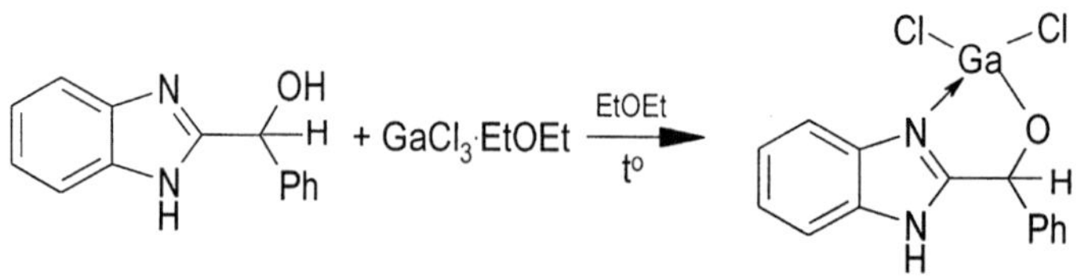

(3.108)

With the objective of obtaining palladium chelates, the ligand-exchange reactions with the participation of nitrile complexes of Pd halides are widely used. Thus, coordination units PdN_4 and PdN_2O_2 are formed as a result of the transformation (3.109) with participation of 2-tosylamino(hydroxy)phenylbenzazoles [282]:

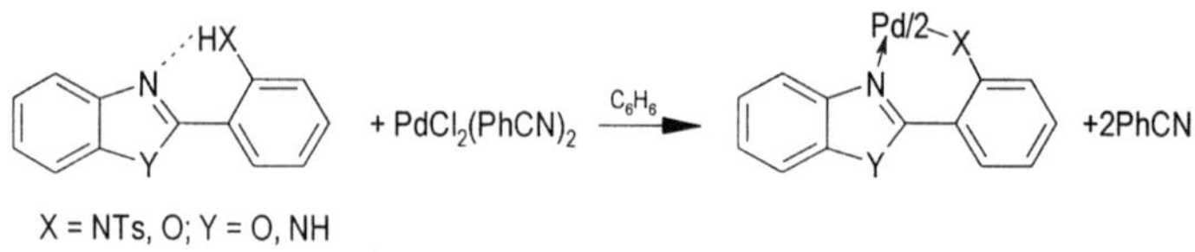

(3.109)

The method of ligand exchange has been used for preparation of metal chelates, containing N,S- ***697*** (3.110) [283] and N,Te-donor atoms ***698*** (3.111) in the coordination unit:

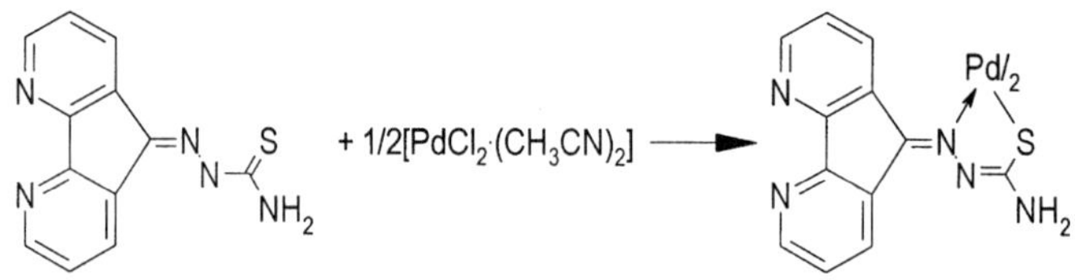

(3.110)

697

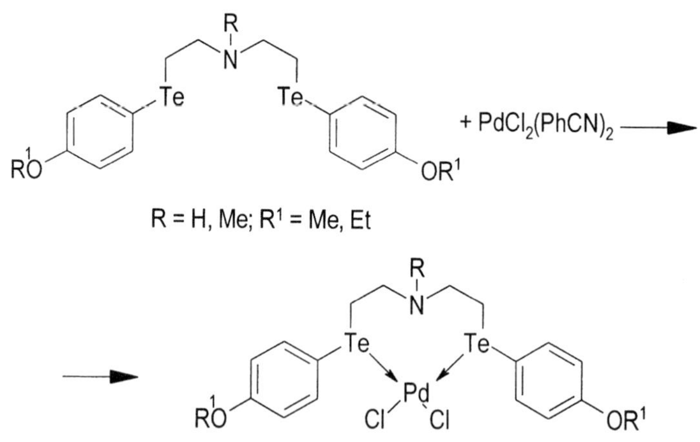

(3.111)

698

The chelate exchange (3.112) is a considerable and important part in the discussed syntheses [2,259] (HCh = chelating agent):

$$MCh_n + nHCh' \rightarrow MCh_n' + nHCh \qquad (3.112)$$

Highly widespread sources of metal complex-former (MCh_n) in this case are metal acetylacetonates. These precursors were used for obtaining ICC with coordination units MN_6 ***699*** (3.113) [285,286]; MN_2O_4 ***700*** (3.114) [287], ***701*** (3.115) [288]; MN_3O_2 ***702*** (3.116) [289]; $MN_2O_2S_2$ ***703*** (3.117) [290]; and MN_3O_2S ***704*** (3.118) [291]:

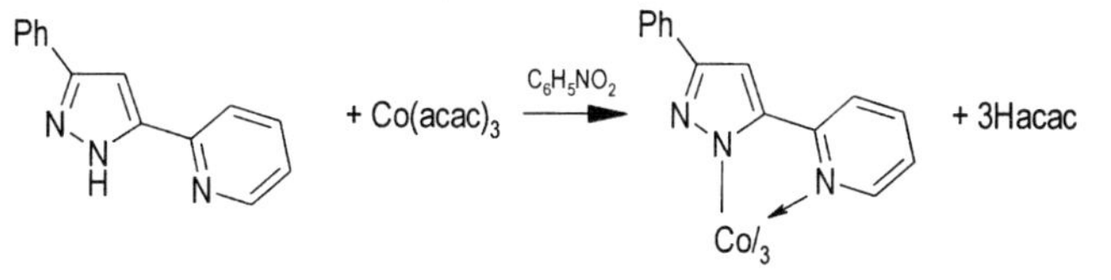

699

(3.113)

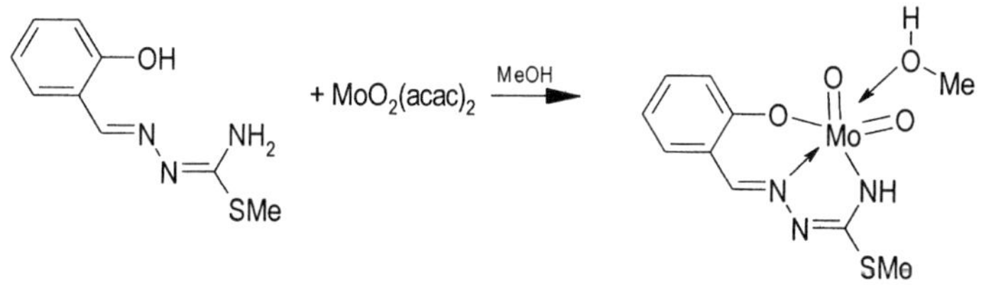

700

(3.114)

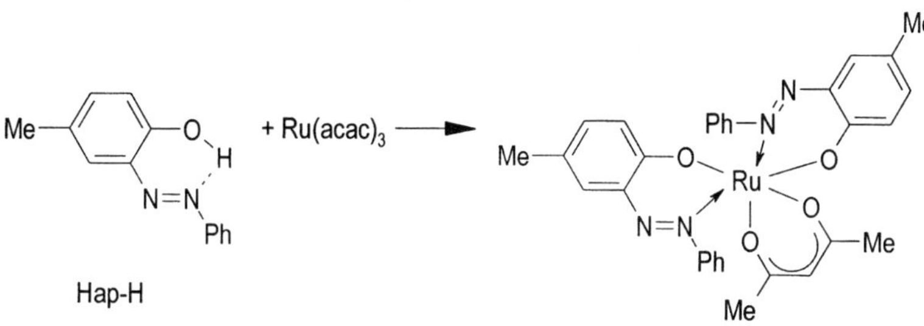

701

(3.115)

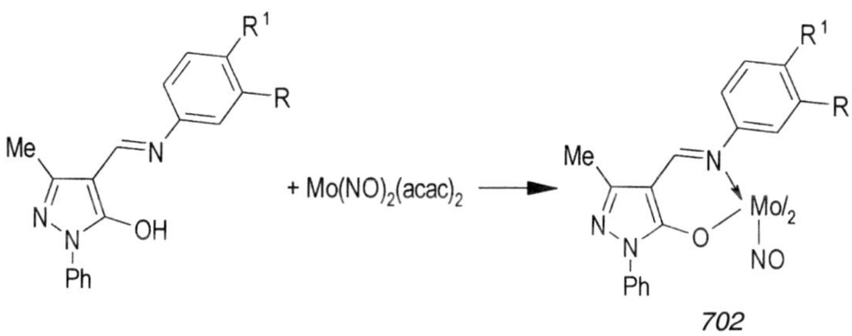

702

(3.116)

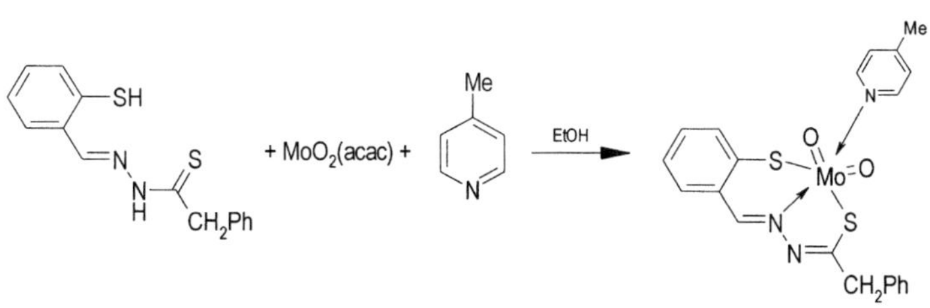

703

(3.117)

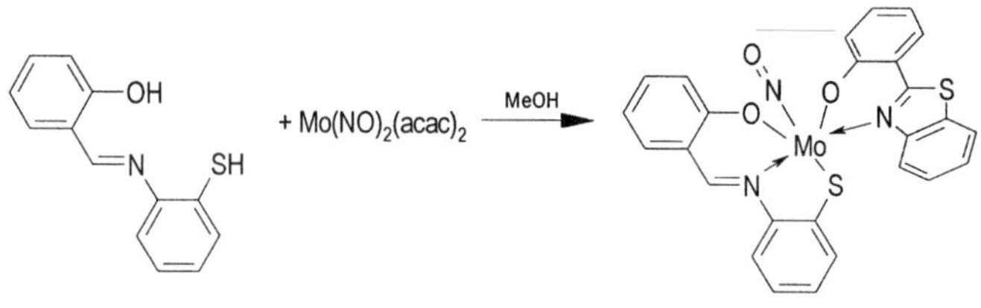

704

(3.118)

The last complex is interesting by the fact that it contains different ligands in the coordination sphere – salicylidenmercaptoanilinate and 2-(hydroxyphenyl)benzthiazole ligands [291].

The ligand-exchange method is the most used method in the synthesis of π-complexes. The examples of such compounds, obtained from metal carbonyls and cyclic unsaturated hydrocarbons, are represented in a classic issue [11]. Among these ligand-exchange syntheses, we mention here the transformations (3.119)–(3.121):

$$M_2(CO)_{10} + C_{10}H_{12} \longrightarrow 2M(\eta^5\text{-}C_5H_5)(CO)_2 + 6CO + H_2 \quad (3.119)$$
$$M = Co, Re$$

$$Ru_3(CO)_{12} + 3C_5H_6 \xrightarrow[3\text{–}4\ hr]{C_6H_{12}} 3Ru(\eta^5\text{-}C_5H_5)(H)(CO)_2 + 6CO \quad (3.120)$$

$$V(CO)_6 + C_7H_8 \longrightarrow V(\eta\text{-}C_7H_7)(CO)_3 + 3CO + 1/2H_2 \quad (3.121)$$
$$C_7H_8 = \text{cycloheptatriene}$$

This synthetic method is applied for obtaining aromatic [11] and heteroaromatic [5,292] π-complexes. Thus, the replacement of carboxyl groups leads to the following η^6-π-Ar–metal carbonyl complexes:

$$Cr(CO)_6 + C_6H_5Cl \xrightarrow[\sim 145°C,\ 20\ hr]{diglym} Cr(CO)_3(\eta^6\text{-}C_6H_5Cl) + 3CO \quad (3.122)$$

$$Mo(CO)_3(MeCN)_3 + C_6H_5Me \longrightarrow Mo(CO)_3(\eta^6\text{-}C_6H_5Me) + 3MeCN \quad (3.123)$$

The complexes of η^5- and η^6-coordinated heterocycles were obtained by the reactions (3.124) and (3.125) [5,293,294]:

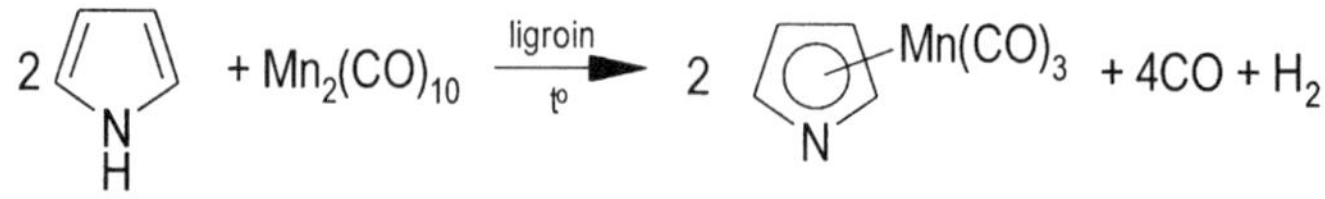

(3.124)

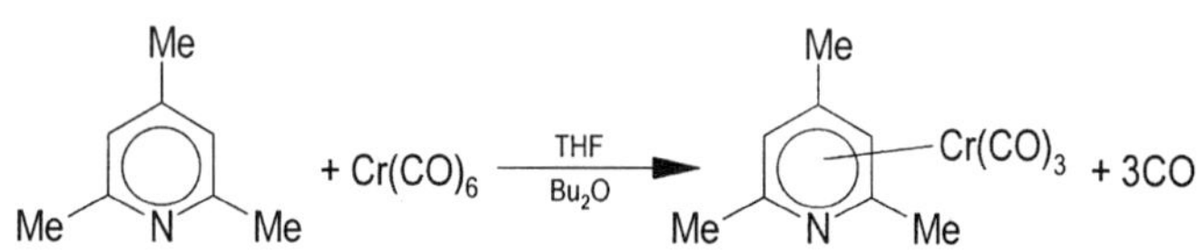

(3.125)

The cases of transformations of σ-metal–carbonyl complexes to π-complexes in the processes of ligand-exchange reactions (3.126), (3.127) are of high interest [5]:

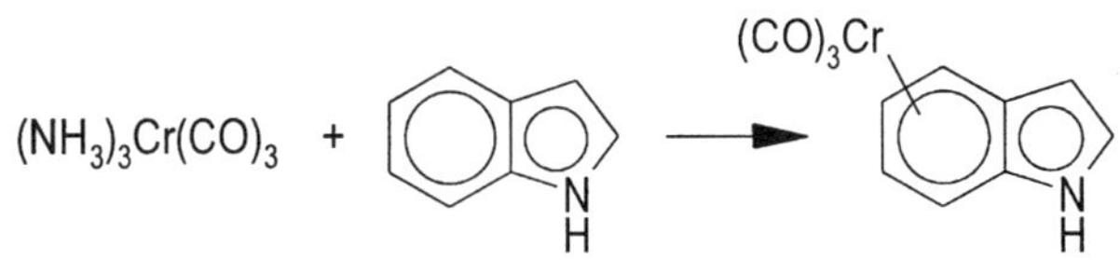

(3.126)

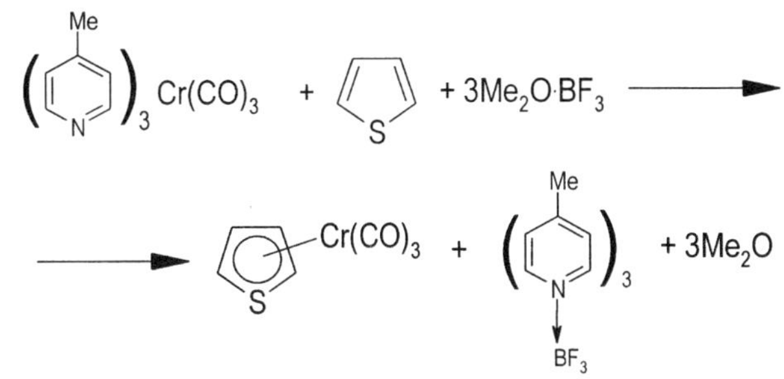

(3.127)

Among molecular complexes recently obtained by the ligand-exchange method we emphasize the coordination compounds of diorganoditellurides. Thus, the interaction (3.128) of molybdenum hexacarbonyl with diphenyltelluride in THF leads to the formation of a binuclear complex ***705*** [272]:

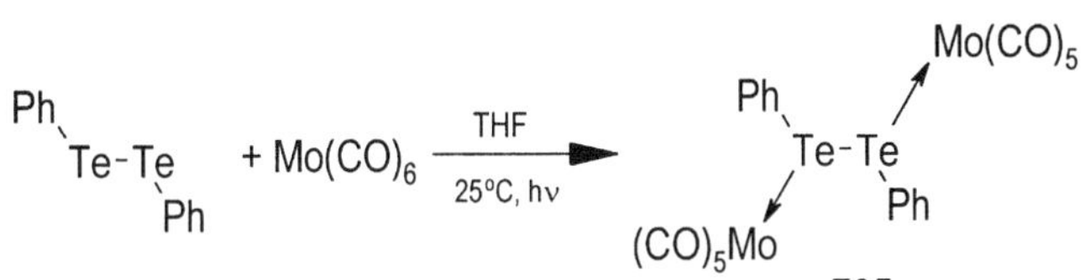

705

(3.128)

The Te—Te bond is conserved in these conditions. However, in more hard conditions (boiling in THF), the destruction of this bond takes place that also leads (3.129) to binuclear structures ***706*** at the expense of formation of two TePh bridges [265–268]:

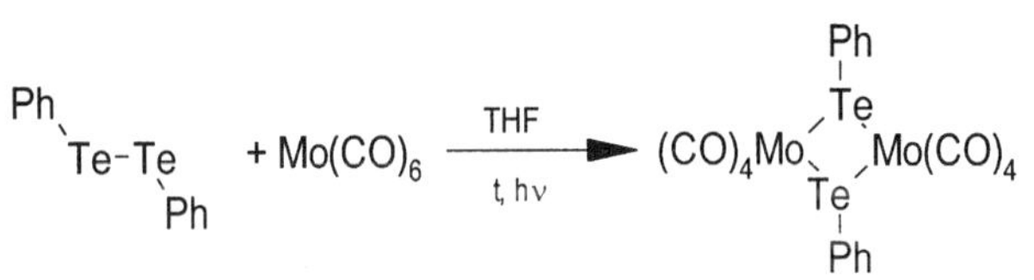

706

(3.129)

The ligand-exchange reactions are the basis of the syntheses of binuclear complexes, formed at the expense of halide bridges. In this respect, the ligand exchange, leading (3.130) to C,N- and C,P-cyclopalladium complexes, is representative [295]:

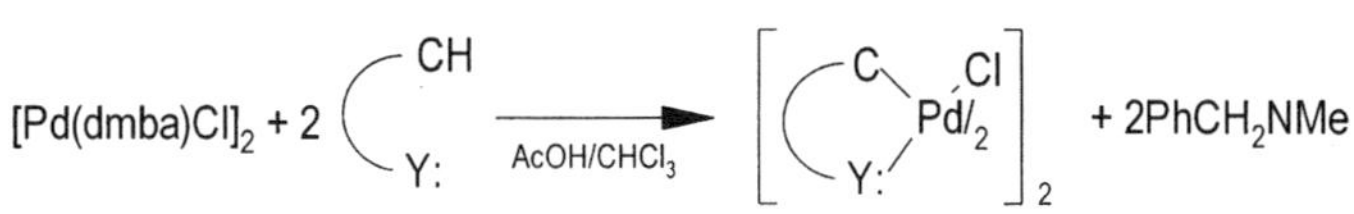

(3.130)

As a result of an exchange reaction (3.131) in acetone or THF in the presence of NEt_3, the binuclear complex ***707*** was obtained. The dimerization of fragments in it is carried out by the O-bridge [296]:

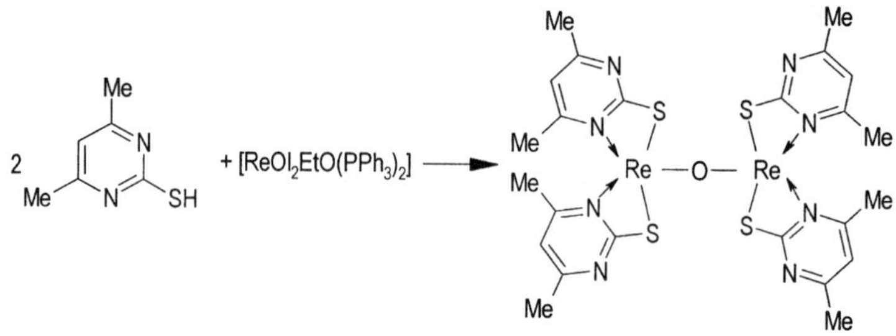

(3.131)

707

The mixed-ligand chelate complex with CPS_2 ligand environment was isolated as a result of the transformation (3.132) [297]:

$$[\mathrm{Pt\{CH_2C_6H_4P(\mathit{o}\text{-tolyl})_2\}(THF)_2]ClO_4} + \mathrm{AgS_2CR} \longrightarrow$$
$$[\mathrm{Pt\{CH_2C_6H_4P(\mathit{o}\text{-tolyl})_2\}(S_2CR)]} + \mathrm{AgClO_4} + 2\mathrm{THF} \qquad (3.132)$$

The ligand-exchange reactions are widely used in the syntheses of binuclear metal–carbonyl, η^5-cyclopentadienyl, and η^6-arene complexes [11, vol. 6]. Examples of these type of transformations are reactions (3.133)–(3.135) [11, vol. 6]:

$$2\mathrm{Fe(CO)_5} + \mathrm{C_{10}H_{12}} \xrightarrow[24\ \mathrm{hr}]{140-160^\circ\mathrm{C}} [\mathrm{Fe}(\eta^5\text{-}\mathrm{C_5H_5})(\mathrm{CO})_2] + 6\mathrm{CO} + \mathrm{H_2} \qquad (3.133)$$
$$\mathrm{C_{10}H_{12}} = \text{dicyclopentadiene}$$

$$2\mathrm{Mo(CO)_6} + \mathrm{C_{10}H_{12}} \xrightarrow{135-145^\circ\mathrm{C}} [\mathrm{Mo}(\eta^5\text{-}\mathrm{C_5H_5})(\mathrm{CO})_3]_2 + 6\mathrm{CO} + \mathrm{H_2} \qquad (3.134)$$

$$2\mathrm{V(CO)_6} + \mathrm{C_6H_3Me_3} \longrightarrow [\mathrm{V(CO)_4}(\eta^6\text{-}\mathrm{C_6H_3Me_3})][\mathrm{V(CO)_6}] \qquad (3.135)$$

The same synthetic method was used in preparation (3.136) of cyclo-octene complexes of trivalent rhodium [11]:

$$2\mathrm{RhCl_3}\cdot 3\mathrm{H_2O} + 4\mathrm{C_8H_{14}} \xrightarrow[\mathrm{N_2},\ 35^\circ\mathrm{C},\ 1-2\ \mathrm{days}]{\mathrm{EtOH/H_2O}} [\mathrm{RhCl}(\eta^2\text{-}\mathrm{C_8H_{14}})_2]_2 \qquad (3.136)$$

The synthetic route (3.137), leading to trinuclear complex ***708*** at the expense of phenyl fragments of 2,4,6-triphenylpyridine, is very interesting [5,298]:

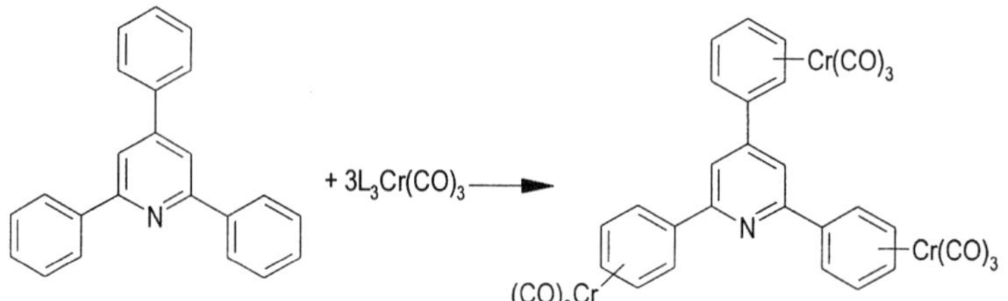

(3.137)

708

EXPERIMENTAL PROCEDURES

Example 1

Synthesis of 2-Phenyl(methyl)benzo-1,3-tellurazolopentacarbonyltungsten(chromium)(VI) ***687*** (3.94) [299,300]

Chromium and tungsten carbonyl complexes with 2-phenyl- (L) and 2-methylbenzo-1,3-tellurazole (L′) were prepared in an argon atmosphere as follows. $Cr(CO)_6$ (0.660 g, 3 mmol) or $W(CO)_6$ (1.05 g, 3 mmol) in 100 mL of dried THF saturated with argon was exposed to a mercury lamp until the evolution of 3 mmol of CO. After irradiation, the corresponding benzotellurazole (3 mmol) was added to the solution, and the mixture was stirred for 1 hr. The solution was filtered off, and the solvent was removed in a rotary evaporator. The solid residue was placed in a sublimation apparatus to remove unreacted metal carbonyl in a high vacuum and then recrystallized twice from hexane.

Example 2

Synthesis of Complexes of Diphenylditelluride with Chromium and Tungsten Pentacarbonyls (3.98) [268]

$(CO)_5W(Ph_2Te_2)$

The solution of 0.18 g (0.5 mmol) of $W(CO)_6$ and 0.2 g (0.5 mmol) of Ph_2Te_2 in 20 mL of THF was irradiated for 1 hr under stirring and water cooling in the stream of argon. Then the mixture was concentrated to a third of the volume under vacuum, and the light petroleum was added. The mixture was concentrated to a half of the volume again and allowed to stay at −18°C for 12 hr. Orange crystals were filtered off, washed with pentane, and dried. Yield 0.14 g (40%).

$(CO)_5Cr(Ph_2Te_2)$

Obtained in a similar way. Yield 60%.

Example 3

Synthesis of *trans-bis*(Diphenyltelluride)-*cis*-dichloro-*cis*-dicarbonylruthenium(II) [301a]

$$Ph_2Te + [RuCl_2(CO)_3] \xrightarrow{CH_2Cl_2} [RuCl_2(CO)_2(TePh_2)_2]$$

Ph_2Te (170 mg, 0.60 mmol) was added to a suspension of 65 mg (0.13 mmol) of $[RuCl_2(CO)_3]_2$ in CH_2Cl_2. The reaction mixture was stirred overnight during which time a yellow solution was formed. The solvent was removed by evaporation, and the orange–yellow precipitate washed with small portions of diethyl ether. The precipitate was dissolved in benzene. Upon slow evaporation of the solvent, orange crystals of $[RuCl_2(CO)_2(TePh_2)_2] \cdot 1/2C_6H_6$ were formed. Yield 51% (110 mg).

Example 4

Synthesis of Dichloro[2,6-*bis*(1-methylethyl)-N-(2-pyridinylmethylene)phenylamine]palladium(II) (3.72) [89]

The ligand (0.76 g, 2.87 mmol) was dissolved in 10 mL of dichloromethane and added to a solution of $[(COD)PdCl_2]$ (0.80 g, 2.80 mmol in 40 mL of CH_2Cl_2). Stirring was continued for 24 hr at ambient temperature at which time precipitation occurred. The precipitated yellow powder was collected by filtration, washed with diethyl ether (2 × 20 mL), and dried in vacuo. Yield 1.07 g (96%).

Example 5

Synthesis of Dioxo[R-salicylidene-S-alkylisothiosemicarbazonato]molybdenum(VI) ***700*** (3.114) [287]

Synthesis of the above complexes was performed by refluxing the solution of molybdenylacetylacetonate and corresponding ligand (1:1) in methanol for 0.5 hr and subsequently cooling the reaction mixture to 0°C. Yield ca. 85–90%.

Example 6

Synthesis of *bis*[2-Phenylazo-4-methylphenolato]rutheniumacetylacetonate ***701*** (3.115) [288]

[Ru(acac)$_3$] (100 mg, 0.25 mmol) and Hap-H ligand [see its formula in (3.115); 320 mg, 1.5 mmol] were taken in 20 ml of ethylbenzoate. The mixture was heated at 160°C with continuous passage of nitrogen to remove the volatile acetylacetone. Heating was continued for 6 hr and the color of the solution gradually became brown. The solvent was then evaporated under reduced pressure. The crude product was dissolved in hexane and subjected to chromatography on a silica gel (60–120 mesh) column. On elution with hexane and 2:1 hexane: benzene mixture as eluent, yellow and light red bands came out and were rejected. With a 1:4 hexane:benzene mixture as the next eluent, a deep brown band came out. This was collected and evaporation of the eluate gave [Ru(acac)(ap-H)$_2$] as a dark brown microcrystalline solid. Yield 65% (102 mg).

Example 7

Synthesis of μ-oxo-*bis*{oxo-*bis*(4,6-Dimethylpyrimidine-2-thiolate)rhenium(V)} ***707*** (3.131) [296]

Method (a)

Solid *trans*-[ReOI$_2$(EtO)(PPh$_3$)$_2$] (0.4 mmol) was added to a solution of Me$_2$pymSH [see its formula in (3.131); 0.8 mmol] containing 2.4 mmol of N(C$_2$H$_5$)$_3$ in 15 cm^3 of methanol (acetone or THF). The mixture was refluxed under stirring for 0.5 hr. On heating, a green–olive microcrystalline solid precipitated immediately. The product was collected, washed with methanol (acetone or THF), and dried in vacuum. Yield ca. 65%.

Method (b)

Trans-[ReO$_2$(py)$_4$]Cl · 2H$_2$O (0.58 mmol) was dissolved in 10 cm^3 of methanol (acetone or THF). After addition of solid Me$_2$pymSH (1.16 mmol), the solution was refluxed upon stirring for 15 min. On heating a green–olive microcrystalline solid separated immediately. The product was filtered, washed with methanol, and dried in vacuo. Yield ca. 66%.

Example 8

Synthesis of {(*i*-Pr)$_2$PC$_2$H$_4$P(*i*-Pr)$_2$}Pd(CO)$_2$ [301b]

A colorless solution of (d^{i}ppe)-Pd(C$_2$H$_4$) (397 mg, 1.00 mmol) in pentane (5 mL) was treated with CO at 20°C. The solution instantaneously turned light red and eventually beige (after

2 min). At −78° colorless crystals separated, from which the mother liquor was removed by cannulation. The solid was washed with cold pentane and dried under vacuum at −30°C (causing a superficial tinge of yellow). Yield 310 mg (73%); d.p. 42°C [R = *t*-Bu, yield 390 mg (81%)].

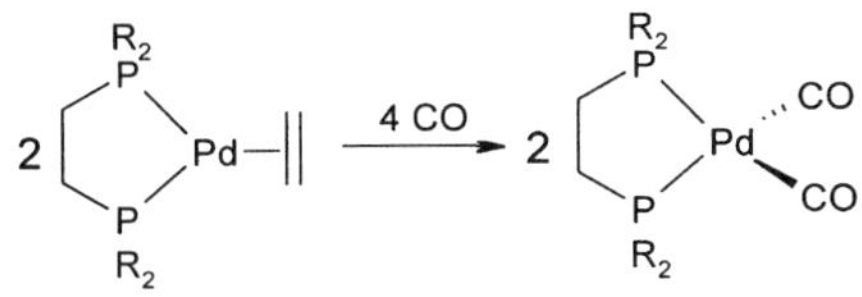

R= *i*-Pr, *t*-Bu

Example 9

Synthesis of Dichloro(9,10-dihydro-9,10-ethenoanthracene)platinum(II) · 0.5(toluene) [301c]

To 0.315 g (0.536 mmol) of $[(C_2H_4)_2PtCl_2]_2$ (Zeise's dimer) dissolved in 25 mL of diethyl ether and 5 mL of toluene was added 0.219 g (1.073 mmol) of 9,10-dihydro-9,10-etheneoanthracene. The solution was stirred vigorously under N_2 for 6 hr. Upon concentration of the resulting solution, a bright orange solid was produced. Filtration yielded 0.46 g of the desired product with 0.5 equiv. of toluene cocrystallized for an 85% yield.

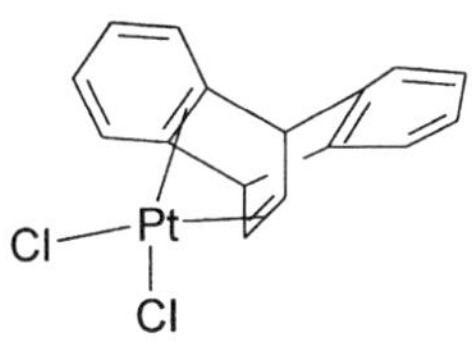

Example 10

Synthesis of $(CH_3)_2Sn(i\text{-}C_3H_7OCS_2)_2$ [301d]

A solution of i-$C_3H_7OCS_2K$ (3.48 g, 0.02 mol) in 40 mL of water was added to a solution of $(CH_3)_2SnCl_2$ (2.20 g, 0.01 mol). The mixture was stirred for 30 min. The formed white solid was filtered, washed with water, and dried in air. The dry solid was dissolved in hexane, then the solution was filtered, and hexane was completely evaporated in air. The crystalline solid was dried to constant weight in vacuo. Yield 2.64 g (62%).

3.2.2 Exchange of Metals

The idea of this method is an exchange of mainly akali metals (M) to transition ones (M′) in complexes of different types (molecular complexes and chelates):

$$M_nL_m + pM' \rightleftharpoons M'_pL_m + nM \qquad (3.138)$$

It was accepted as far back as 40 years ago [2] that the possibilities of metal exchange are not applied for preparative goals. The further development of synthetic coordination chemistry has shown that this route is an important way among other methods to synthesize coordination compounds.

The metal-exchange method is especially widely used in syntheses of chelate complexes starting from the ligands, which are low-stable in free state. Such ligands are, in particular, chalcogen-containing organic compounds. On the basis of their salts with alkali metals, a series of chelate structures with five- (3.139) [302] and six-member (3.140)–(3.142) [303–307] metal-cycles, containing N,X-donor centers (X = S, Se; Secs. 2.2.5.4 and 2.2.5.5), have been isolated:

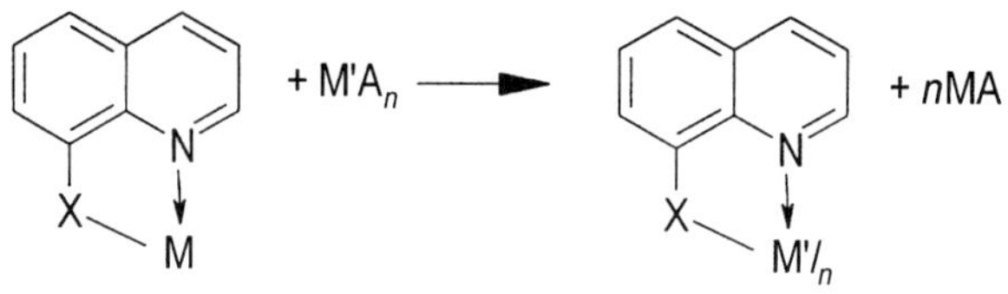

(3.139)

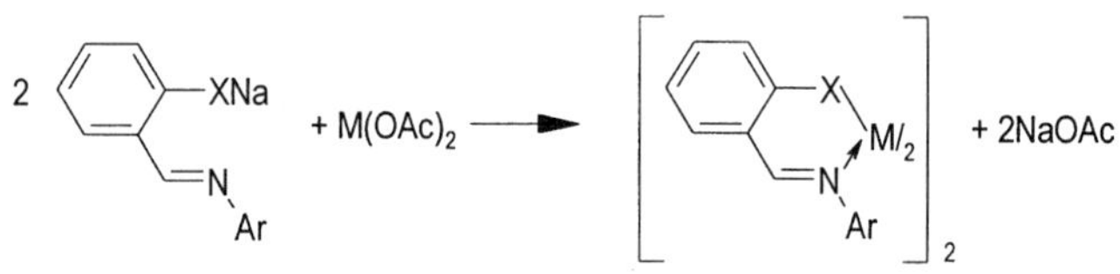

X = S [303], Se [304]

(3.140)

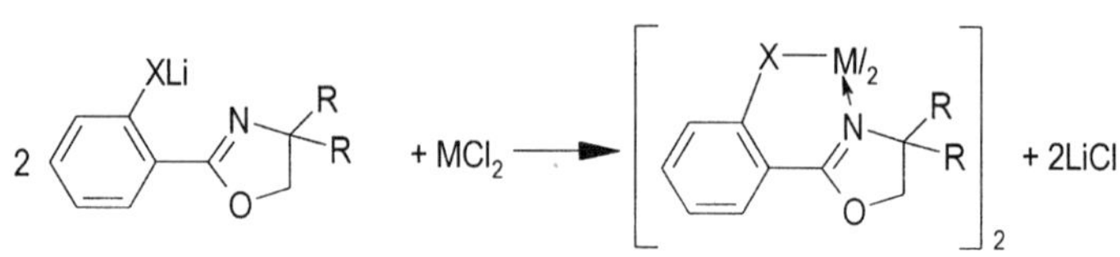

709

X = S [305], Se [306], Te [307];
M = Zn, Cd, Hg

(3.141)

Other metal-exchange syntheses are also known, leading to N,Te-chelate structures, for example (3.142) [308] and (3.143) [309]:

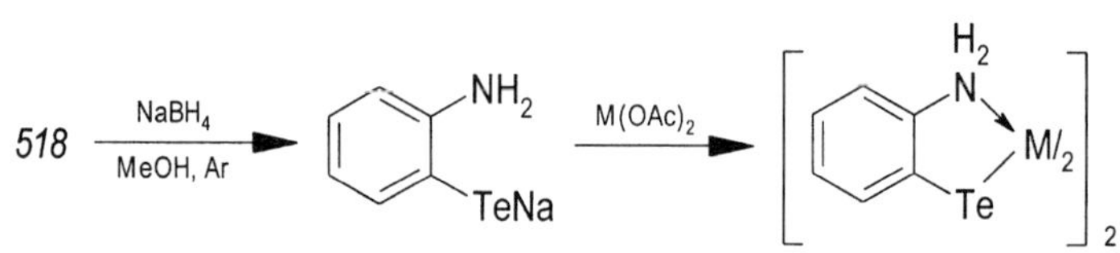

M = Zn, Cd

(3.142)

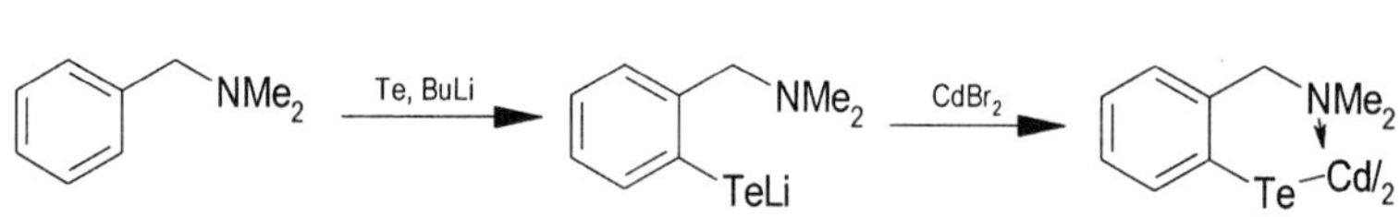

(3.143)

The chelating S,S-donors, applied in the examined syntheses, are widely represented (see Sec. 2.2.4.5) by alkali metal salts of dialkyldithiocarbamates and alkylxantogenates (3.144) [310]:

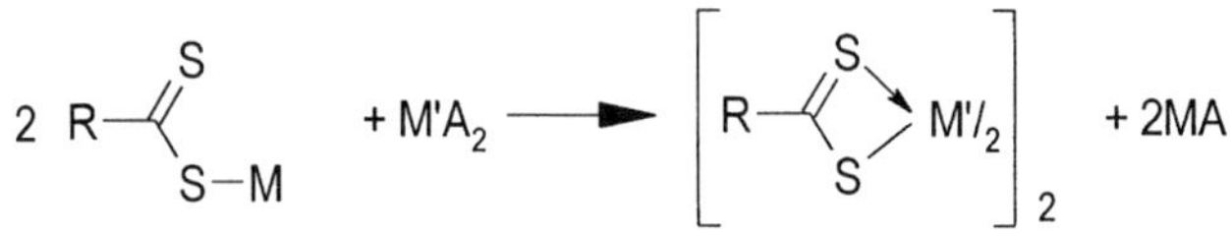

(3.144)

$R = NR^1_2, OR^1; R^1 = Alk$

The method above is used in the synthesis (3.145) of complexes with Te_2P_2-ligand environment [311]:

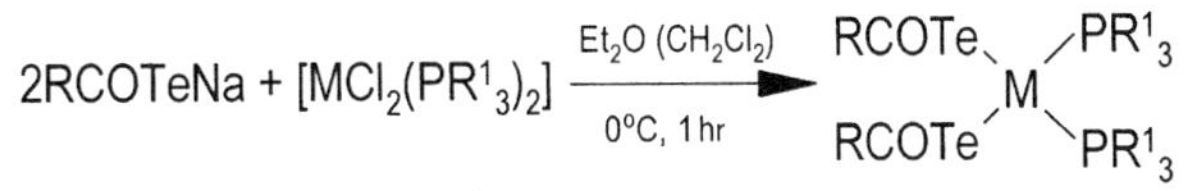

(3.145)

M = Pd, Pt; R = C_6H_4-Me-*p*; R^1 = Et, Ph

The metal-exchange method has found a preparative application for synthesis of ICC starting from ligands containing comparatively weak-acid NH fragments. Such ligand systems are, for example, polypyrazolylborates (see Sec. 2.2.4.2, formulae ***232–235***). Frequently, on the basis of compounds of the types ***232*** and ***234***, their salts of alkali metals are obtained, which are precursors for further preparation of metal chelates ***233*** and ***235***. In this respect, the recently reported [312] metal-exchange synthesis (3.146) of the mixed-ligand complex ***710*** is interesting and representative. This transformation is carried out from potassium salt of *tris*(phenylpyrazolyl)borate, salicylidenimines, and copper acetate in CH_2Cl_2 at room temperature:

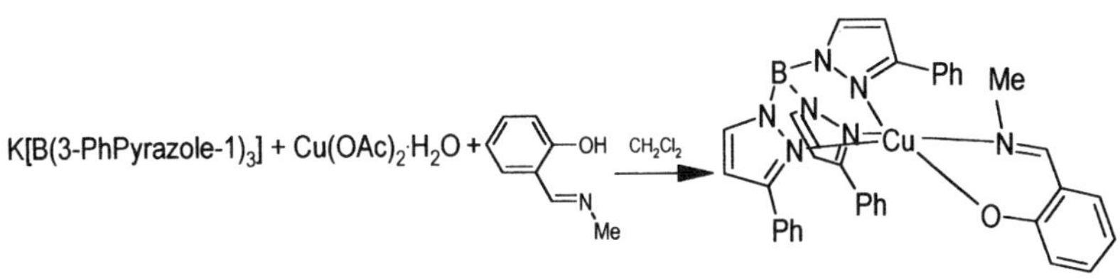

(3.146)

710

The first representatives of $\eta^5(\pi)$-complexes of azoles (compare with Ref. 5) – mixed ligands coordination compounds of the type ***711*** – were obtained via potassium salts of C-substituted pyrazoles (3.147) with high yields ($\sim$ 70%) [313]:

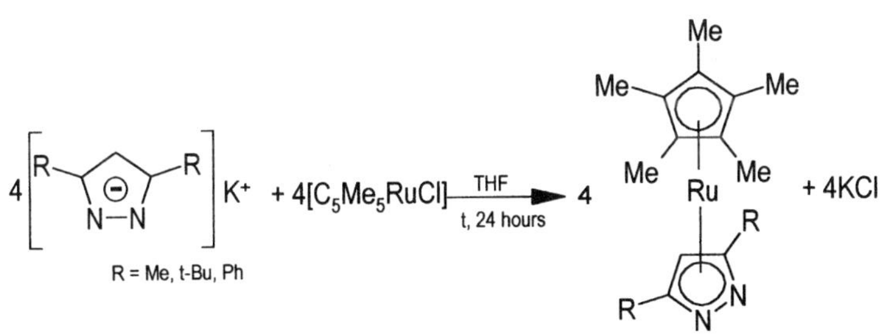

(3.147)

711

The same synthetic approach was used in classic syntheses of $\eta(\pi)$-complexes of various metals [11]. Such syntheses are carried out in two steps: preparation, for example, of cyclopentadienyl derivatives of alkali metals with their further application in the syntheses (3.148), (3.149) of sandwich η^5-complexes of mainly transition metals:

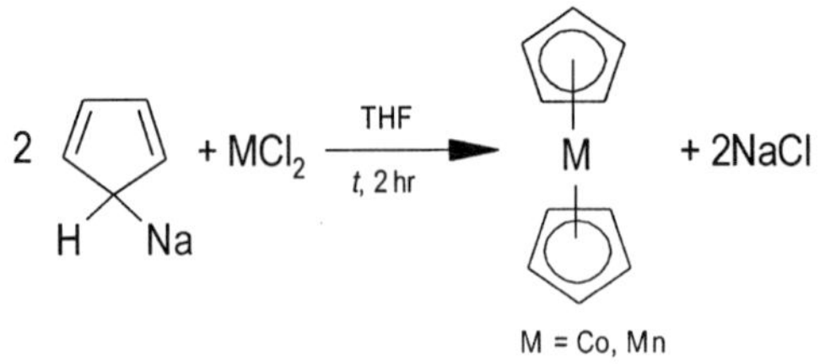

(3.148)

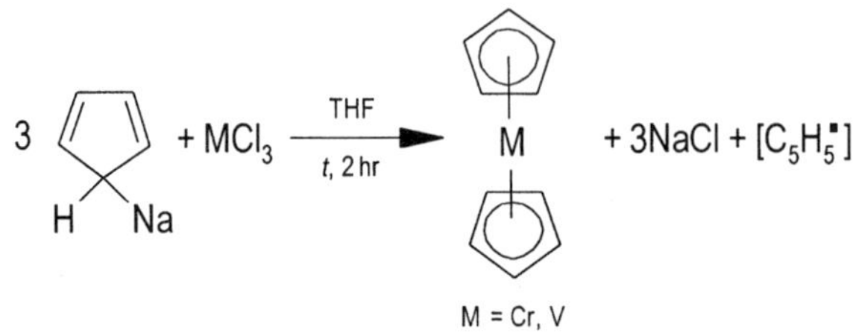

(3.149)

The obtaining of η^1-cyclopentadienyls of nontransition metals is carried out according to a similar scheme, for example (3.50) [11]:

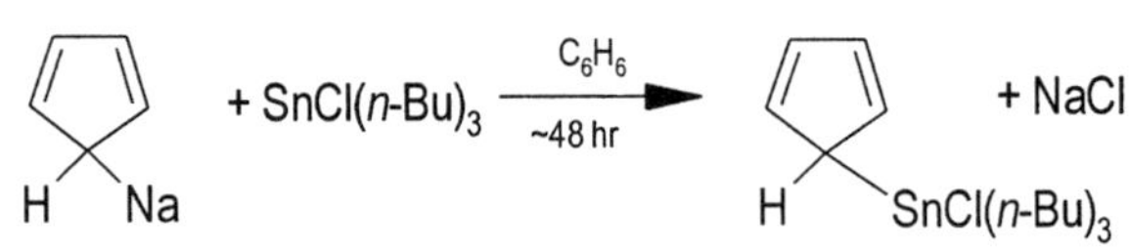

(3.150)

The metal-exchange method has been used practically in all syntheses of complexes of closopolyhedral hydroborate anions (Sec. 2.2.5.8). We note that the reactions of double exchange are also applicable to the synthetic aspect (3.151) [314a]:

$$\mathrm{ML} + \mathrm{M'L'} \rightarrow \mathrm{M'L} + \mathrm{ML'} \qquad (3.151)$$

However, the preparative possibilities of this approach are not yet clear.

To conclude this section, we note that the reactions of anions, which are not part of a coordination sphere, were known long ago [11]. The preparative methods of classic Werner complexes, for instance (3.152) and (3.153), have been developed on their basis:

$$[\mathrm{Co(NH_3)_6}]\mathrm{Cl_3} + 3\mathrm{HNO_3} \longrightarrow [\mathrm{Co(NH_3)_6}](\mathrm{NO_3})_3 + 3\mathrm{HCl} \qquad (3.152)$$

$$[\mathrm{Co(NH_3)_6}]\mathrm{Cl_3} + 3\mathrm{KI} \longrightarrow [\mathrm{Co(NH_3)_6}]\mathrm{I_3} + 3\mathrm{KCl} \qquad (3.153)$$

Evidently, such synthetic transformations should also be characteristic for other cationic complexes, containing complicated organic ligands in a coordination sphere.

EXPERIMENTAL PROCEDURES

Example 1

Synthesis of *tris*(4-Methyl-8-mercaptoquinolinato)iridium(III) of the type ***486*** (3.139) [302b]

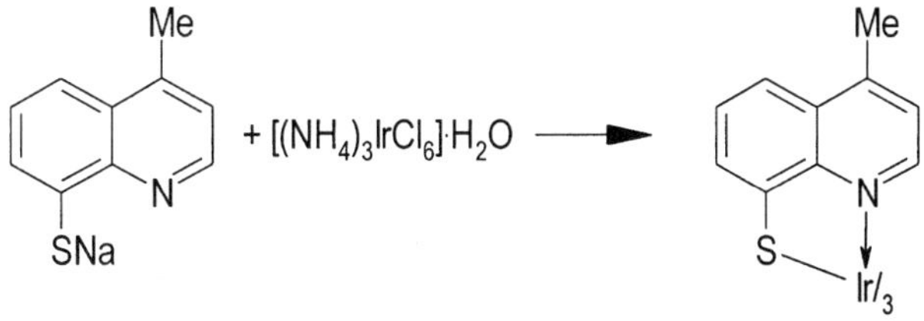

Tris(4-methyl-8-mercaptoquinolinato)iridium(III) was obtained by interaction of a solution of 0.7 g of 4-methylthioxynatosodium in 100 mL of water with a solution of 0.4 g of $(NH_4)_3IrCl_6 \cdot H_2O$ in 20 mL of the same solvent under heating in a water bath for 30 min. The formed precipitate was filtered off, washed with water, ethanol, and dried in air. Monocrystalline samples were obtained by recrystallization from the chloroform–ethanol mixture.

Example 2

Synthesis of [η^5-3,5-Dimethyl(phenyl,*tert*-butyl)pyrazolato] (η^5-pentamethylcyclopentadienyl)ruthenium(II) ***711*** (3.147) [313]

Treatment of $[\{C_5(CH_3)_5\}RuCl]_4$ with 3,5-dimethylpyrazolatopotassium, 3,5-diphenylpyrazolato(tetrahydrofuran)potassium, or 3,5-di-*tert*-butylpyrazolatopotassium in tetrahydrofuran afforded (η^5-3,5-dimethylpyrazolato)(η^5-pentamethylcyclopentadienyl)ruthenium(II) (71%), (η^5-3,5-di-*tert*-butylpyrazolato)(η^5-pentamethylcyclopentadienyl)ruthenium(II) (72%), or (η^5-3,5-diphenylpyrazolato)(η^5-pentamethylcyclopentadienyl)ruthenium(II) (71%), as dark green, pale yellow, or dark brown crystalline solids, respectively.

Example 3

Synthesis of *bis*(Cyclopentadienyl)manganese Mn $(C_5H_5)_2$ (3.148) [11]

$$MnCl_2 + 2NaC_5H_5 \longrightarrow Mn(C_5H_5)_2 + 2NaCl$$

A freshly prepared solution of 88.1 g (1.0 mol) of NaC_5H_5 in 500 mL of THF was introduced into a two-necked flask; then 62.9 g (0.5 mol) of anhydrous $MnCl_2$ was added by portions. After a 12-hr boiling with reflux the solvent was removed in vacuo and the dark brown product was dried in high vacuum to form a powder. The dark red crystal $Mn(C_5H_5)_2$ was obtained by its high-vacuum sublimation at ~ 150°C. Yield 60–70 g (65–72%). The success of the synthesis is determined by absolutely waterless initial reagents and an absence of oxygen in the air.

Example 4

Synthesis of η^1-Cyclopentadienyl(tri-*n*-butyl)tin $Sn(\eta^1\text{-}C_5H_5)(n\text{-}C_4H_9)_3$ (3.150) [11]

$$SnCl(n\text{-}C_4H_9)_3 + NaC_5H_5 \longrightarrow Sn(\eta^1\text{-}C_5H_5)(n\text{-}C_4H_9)_3 + NaCl$$

Powdered NaC_5H_5 (88.1 g, 1.0 mol) was suspended in ~ 1.4 L of absolute benzene, free of thiophene, and treated in four portions with 299.4 g (247.7 mL, 0.92 mol) of $SnCl(n\text{-}C_4H_9)_3$. The formed suspension, which was slightly self-heated during this stage, was stirred for 2 days at room temperature. The unreacted NaC_5H_5 and NaCl were then allowed to precipitate. The obtained yellow solution was filtered through glass filter G3, covered additionally with glass wool. In case of a brown meat-jelly mass washed in the filter, it is recommended not to filter by pressure, otherwise the filter pores are rapidly filled, but to centrifuge the reaction mixture in some portions before filtration. The residue on the filter (or after centrifugation) was multiply washed with benzene (total volume 150 mL). The filtrate (or centrifugate, respectively) was evaporated at 30–50°C in vacuo; the formed yellow oil was distilled in vacuo. $Sn(\eta^1\text{-}C_5H_5)(n\text{-}C_5H_9)_3$ was distilled at 102–104°C (1.5×10^{-2} mm Hg) in the form of a pale yellow oil. Yield 222–248 g (68–76%).

Example 5

Synthesis of $(Cp^{3i})_2Co$ [314b]

A suspension of KCp^{3i} (0.78 g, 3.39 mmol) in 30 mL of THF was cooled to −78°C in a dry ice/acetone bath. Over the course of 10 min, 50 mL of a THF solution of $CoCl_2$ (0.20 g, 1.50 mmol) was added from a dropping funnel; during this time, the reaction mixture turned dark green and then black. The reaction was stirred at −78°C for 1 hr; the dry ice/acetone bath was then removed, and the reaction was stirred for an additional 18 hr, while it warmed to room temperature. The THF was removed under vacuum, and the black residue was extracted with 50 mL of hexanes and filtered through a glass frit to remove a bluish precipitate. Removal of the hexanes under vacuum and fractional sublimation of the remaining black solid (60–110°C, 10^{-6} Torr) gave 0.42 g (63% yield) of $(Cp^{3i})_2Co$ as brown powder (m.p. 136–138°C).

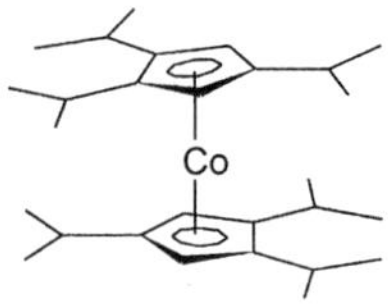

Example 6

Synthesis of $[(C_5Me_5)Yb]_2(C_8H_8)$ [314c]

$YbI_2(THF)_2$ (5.00 g, 8.7 mmol) and KC_5Me_5 (1.53 g, 8.7 mmol) were stirred in 8 mL of THF and formed a cloudy purple suspension. To this mixture, a solution of $K_2C_8H_8$ (0.80 g, 4.4 mmol) in THF was added dropwise, causing the color to change to deep brown. After 4 hr of stirring, the reaction mixture was centrifuged to remove white insoluble material. THF was removed from the red–brown supernatant by rotary evaporation, leaving $[(C_5Me_5)Yb(THF)]_2(C_8H_8)$ as a red–brown powder (3.15 g, 81%). In a tube fitted with a high-vacuum stopcock, the compound (3.15 g, 3.6 mmol) was heated to 30°C at 10^{-7} Torr on a vacuum line. The color changed from a red–brown to bright green within 3 hr. The material was brought into a THF-free glovebox and extracted with 10 mL of toluene. Removal of toluene afforded $[(C_5Me_5)Yb]_2(C_8H_8)$ (2.20 g, 83%) as a bright green powder.

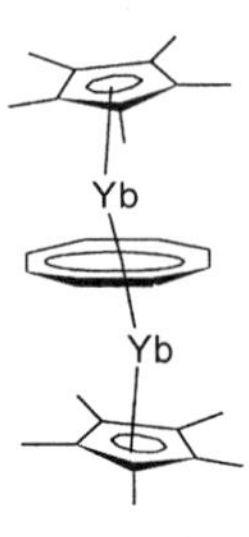

Example 7

Synthesis of $PtX(PR_3)\{CR''(PPh_2{=}NC_6H_4R'_4)_2\text{-}C,N\}$ ($PR_3 = PEt_3$, PMe_2Ph; X = Cl, Br; R′ = CH_3, OCH_3, R″ = H, Me) [314d]

To a stirred solution of freshly prepared $Na[CR''(PPh_2{=}NC_6H_4R'_4)_2]$ (0.45 mmol) in 30 mL of THF was added a solution of $Pt_2X_4(PR_3)$ (0.23 mmol) in 10 mL of THF. After 4 hr the cloudy yellow solution was filtered through a glass filter (G4), and the residue (precipitated NaX) was washed with 10 mL of THF. The combined clear yellow filtrates were evaporated to dryness, leaving an oily residue. Addition of cold pentane (40 mL) resulted in the solidification of the product, which was washed with pentane (2 × 20 mL) and dried in vacuo, yielding a yellow powder (84–95%).

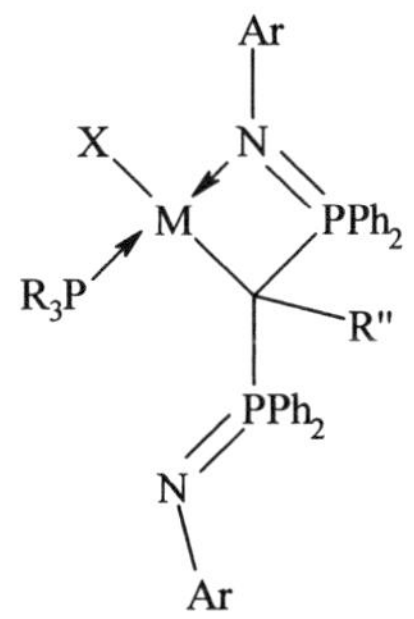

3.3 TEMPLATE SYNTHESIS

Template synthesis has long since [2,314a,315–325] foreseen two aspects: formation of the ligand system on a metal matrix and the reaction of mainly organic compounds, connected with the central atom (ion) (reactions of coordinated ligands).

3.3.1 Syntheses on a Metal Matrix

This preparative method includes reactions between the components of the reaction system, which take place in the presence of a metal salt in appropriate solvents at optimal temperature. It is convenient, since it does not require a preliminary special synthesis of compounds used as ligands. These syntheses are especially appropriate and widely used in the case when the ligand system is hard-accessible, low-stable, or exposed to undesirable transformations.

At the same time, this method has a series of disadvantages. Among them, we note the possibiity of contamination of the final product not only by the excess of one of the reactants [2], but also by complexes of the components of the ligand system. So, to carry out strictly template synthesis experiments, it is necessary to take into account a comparative stability of coordination compounds, obtained on the basis of initial components-precursors and the ligand itself. Not only the thermodynamic characteristics of complex-formation processes should be taken into consideration [326,327], but also the influence of solvolysis processes (especially hydrolysis) and the type of atmosphere (air oxygen).

The synthesis on a metal matrix is mainly applied to prepare chelate complexes. The majority of template synthesis reactions were known long ago, allowing their use for the isolation of hundreds of metal–chelate structures. However, the interest in these syntheses has not disappeared at present. We fix our attention on several examples, illustrating this tendency.

As far back as 1927, as a result of the interaction (3.154) between 1,2-dicyanobenzene and copper(I) cyanide in pyridine at 200°C, the synthesis of the first phthalocyanine ***712*** (R = H) was carried out and studied in detail in 1934 [318,322,328–333]:

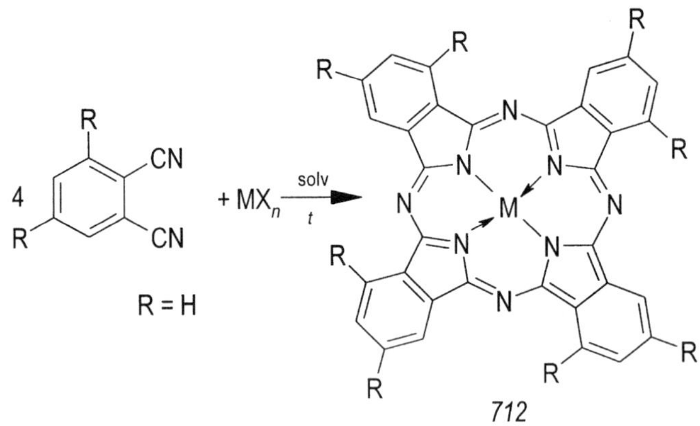

(3.154)

This scheme in principle is also conserved at present [318,319,328,332–334]. Thus, the structural isomers of octa-*tert*-butylphthalocyanines ***712*** (R = *t*-Bu) were obtained in 1997 according to a similar scheme (3.154): $MX_n = Ni(OAc)_2 \cdot 4H_2O$, Solv = $C_5H_{10}OH$, $t = 139°C$] [334]. A more detailed description of phthalocyanine formation from various precursors is presented in Sec. 5.1.

Starting from the mid-1930s, the template syntheses of porphyrines and corrines have been developed [318], which are widely applied at present to isolate metal complexes of these terpyrrol and terpyrrolidine systems. Examples of these reactions are the transformations (3.155):

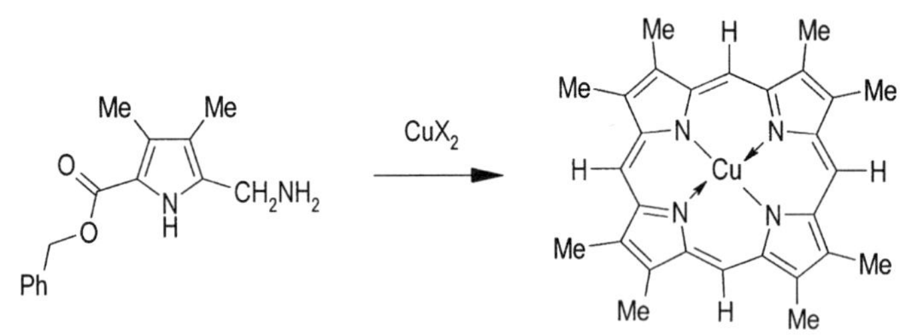

(3.155)

The complexes of octaphenyltetraazaporphyrin were obtained by heating (270–280°C) diphenylmaleinodinitrile and metal chlorides (MCl_n; $n = 3, 4$; M = Ga, In, V, Lu, Sn, Ti, Zr) [335]. We note that the application of template synthesis allowed us to obtain the vitamin B_{12} and a series of other metal–corrinoide systems [318,336].

The template syntheses found application to prepare ICC of Schiff bases and their analogues with coordination units MN_4 (3.156) [337–339]:

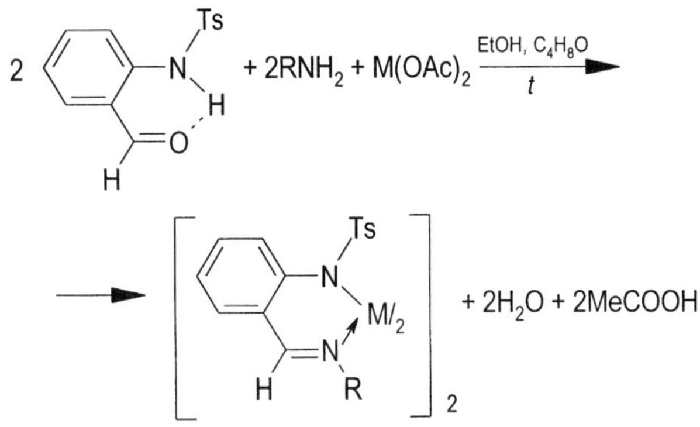

(3.156)

Ts = $SO_2C_6H_4CH_3$-*p*; R = Alk, Ar, $(CH_2)_nOH$; n = 2, 3; M = Co, Ni, Cu, Zn

Such transformations have been carried out in a metal matrix with use of S-alkylthiosemicarbazides, *o*-aminobenzaldehyde ***713*** (3.157) [340], and acetylacetone ***714*** (3.158) [341,342]:

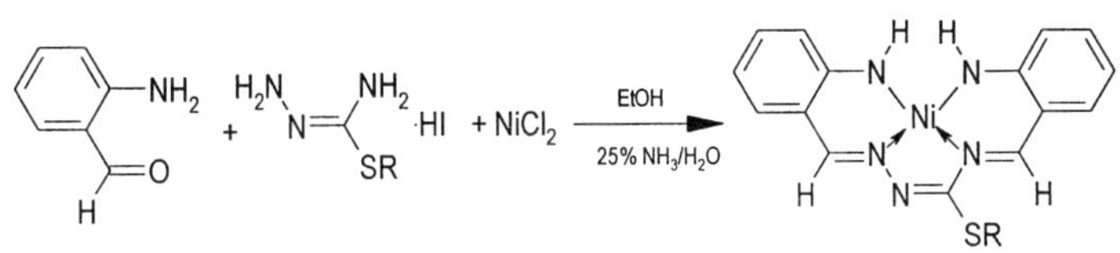

713

(3.157)

Two isomers of the complex ***714***, whose structures were proved by x-ray diffraction, were isolated from the reaction (3.158) mixture.

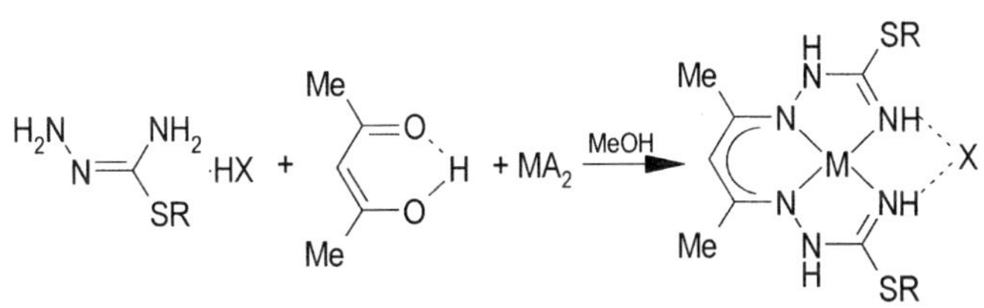

714

(3.158)

M = Ni [341], Co [342]; X = I, NO_3; A = Cl, OAc;
R = Me, Et, Pr, Bu, CH_2Ph

The transformation (3.159), which is carried out in methanol solution of pyperidine and leads to molecular metal chelates ***715***, should be mentioned [343]:

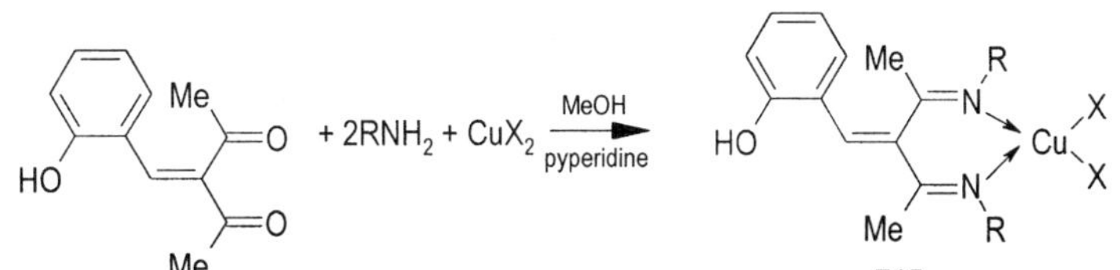

715

(3.159)

R = Me, C_6H_4Cl; X = OAc, Cl, Br

Template syntheses of complexes of macrocyclic ligands with the coordination units, containing N atoms, are examined in detail [316–318,322]. An interesting example of such transformations is the synthesis on the basis of 2,5-diformylpyrrole (3.160), leading to a complex with different di-imine fragments (1,2- and 1,3-diaminopropanes) [318]:

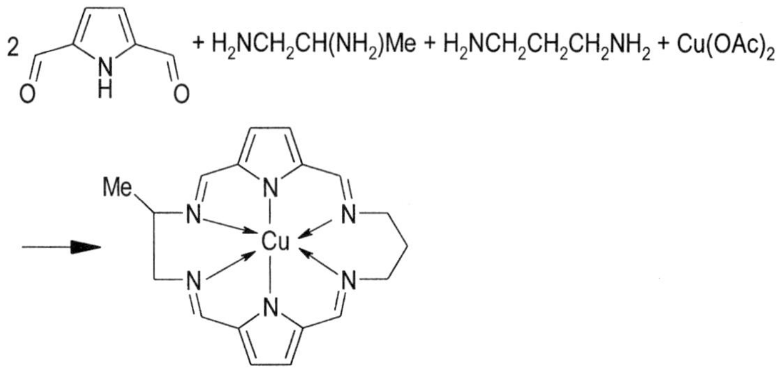

(3.160)

As a result of the examined transformation, in our opinion, the formation of complexes having equal di-imine fragments would be expected. However, the structure of this coordination compound is proved by x-ray diffraction and does not provoke doubts.

Other dicarbonyl components, widely used in the examined syntheses, are 2,6-diformyl(acetyl)pyridines, for instance (3.161) [318]:

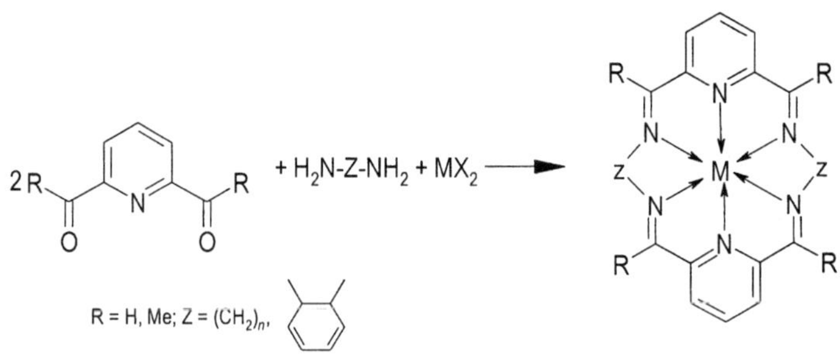

(3.161)

Among chelating O-containing donors, the crown-ethers are the most frequently used in template syntheses [25,189–194,318,344]. Such transformations are described by the general reaction (3.162) [344]:

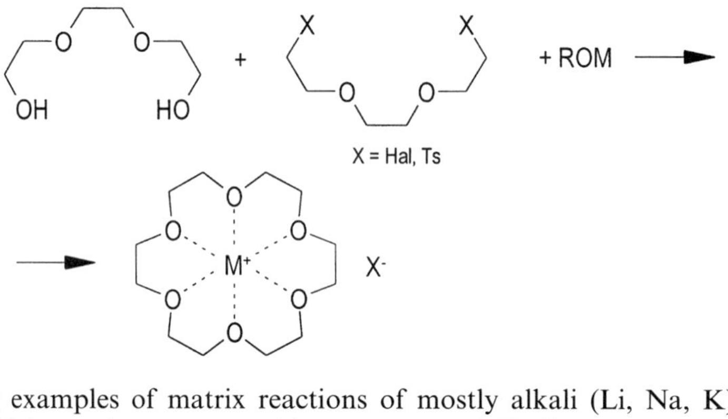

(3.162)

Other examples of matrix reactions of mostly alkali (Li, Na, K) and alkali-earth (Ba, Mg) metals are reported in a monograph [318].

The reaction (3.163) is as an example of such matrix synthesis [345]:

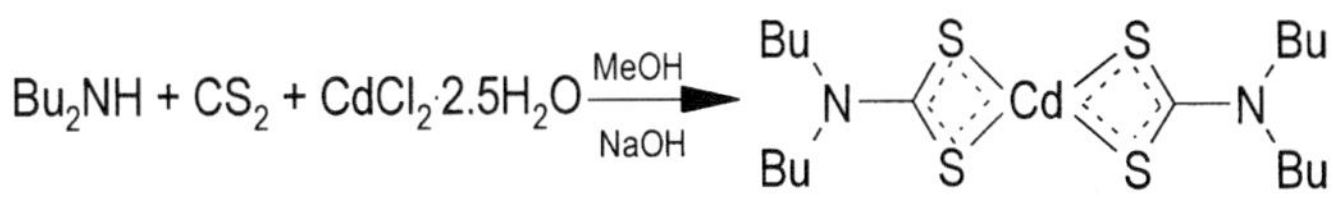

(3.163)

This is a pathway to the S-containing crowns and their complexes [346].

Template syntheses starting from the components of ligand systems and metal salts were widespread in the 1930s to prepare chelates having six-member coordination unit MN_2O_2. In particular, the reaction (3.164) has become classic [2]:

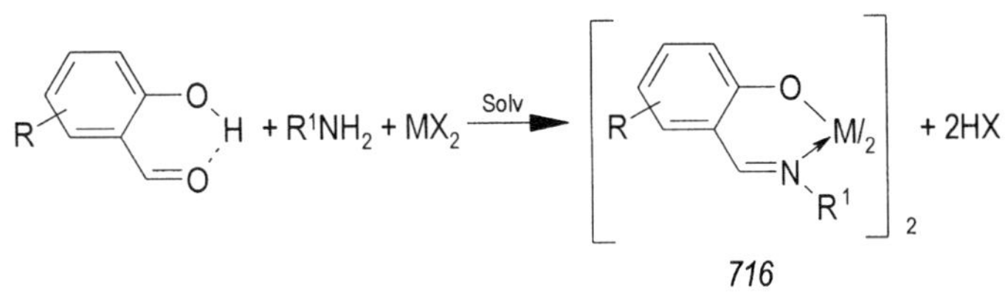

(3.164)

The most favorable conditions for these reactions are created using metal acetates (X = OAc) and methanol. However, some works are known, for example Ref. 2, where copper nitrate is used in the presence of NaOH solution. Alkyls, aryls, or hetaryls are mainly used as R^1. Salicylideniminates ***716*** ($R' = H$, Alk, Ar, Het) of divalent transition metals are formed as a result of the reactions above [53,54].

At the same time, numerous examples of other ligands in similar syntheses have been reported, where oximes (R = OH), hydrazones ($R^1 = NR^2R^3$, R^2, $R^3 = H$, Alk, Ar, Het, COR^4, R^4 = Alk, Ar, Het), semi- and thio(seleno)semicarbazones (R^1 = NHCENR^2, E = O, S, Se; R^2 = H, Alk, Ar) are used. In case of thiosemicarbazide ($R^1 = NHCSNH_2$), not only chelates of the type ***716***, but also metallotricyclic compounds of the type ***717*** are formed (3.165) [316,318,347–350]:

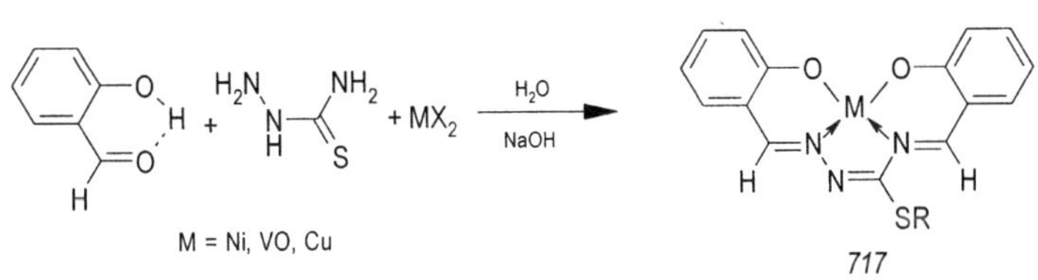

(3.165)

In a difference from the above transformations, attempts to carry out a template synthesis for isolation of azomethinic chelates, having five-member metal-cycle MN_2O_2 were unsuccessful (3.166) [53,351,352]:

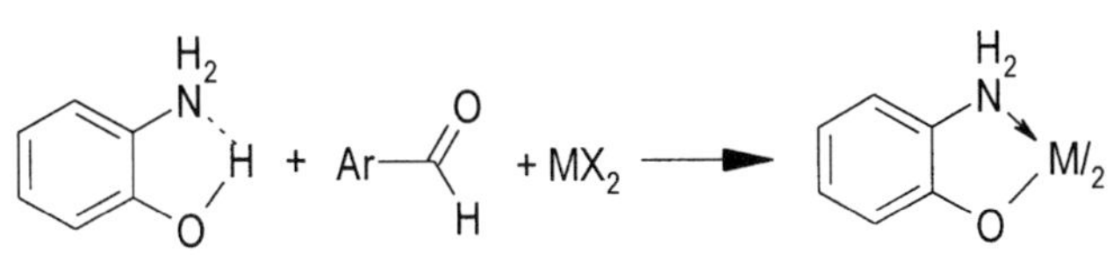

(3.166)

The interesting variant of template synthesis, yielding (3.167) the different-ligand complex ***718***, is examined [314a]:

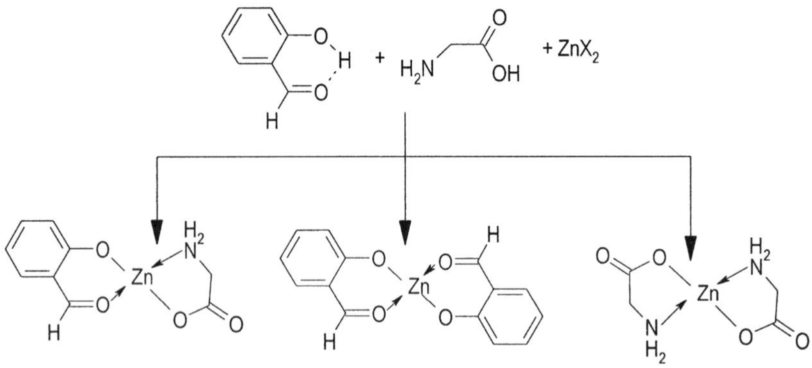

718 719 720

(3.167)

However, the preparative possibilities of this transformation are not yet clear. In our opinion, additionally to the reported product ***718*** [314a], the complexes ***719*** and ***720*** can be formed and so contaminate the different-ligand complex.

The discussed variant of the template synthesis has also found a wide application for obtaining macrocyclic complexes with N,O-donors [315–318]. In this relation, we note the syntheses on a metal matrix on the basis of polyfunctional dicarbonyl compounds, oxadi- and oxapolyamines [318,322]. An example of such syntheses is the reaction (3.168):

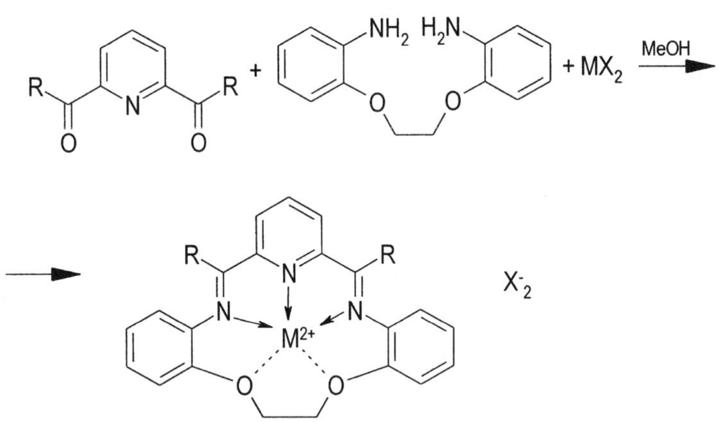

(3.168)

R = H, Me; M = Mn, Zn; X = NO_3, ClO_4

Template synthesis is the most convenient technique for obtaining chelate complexes having a nitrogen–chalcogen ligand environment. In this case, a joint use of the methods of matrix synthesis and metal exchange is especially effective because of the instability of ligands. A combination of these synthetic methods allows us to obtain hard-accessible (3.169) ICC of *o*-mercaptobenzaldehyde ***721*** (X = S) [303] and its selenium analogue ***721*** (X = Se) [304]:

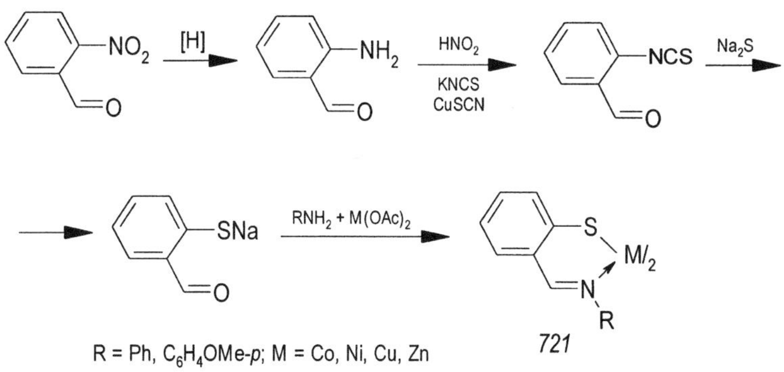

(3.169)

In definite cases, such a template transformation is carried out directly from stable derivatives of mercaptobenzaldehyde, for instance (3.170) [353]:

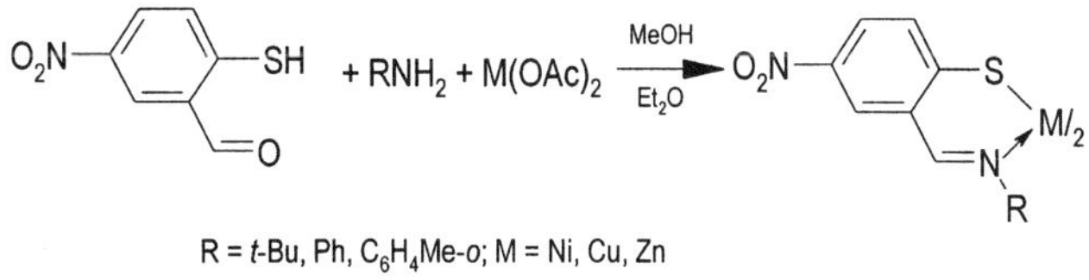

(3.170)

The azomethinic N,S-ligand environment is formed as a result of the template transformation (3.171) [354]:

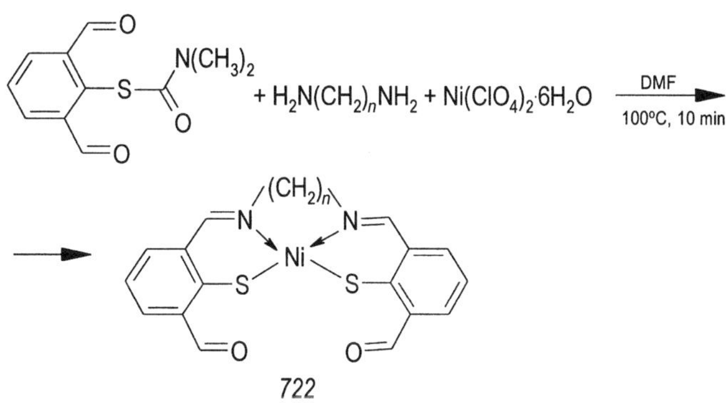

(3.171)

Also, related complexes are available by the following interaction (3.172) [355]:

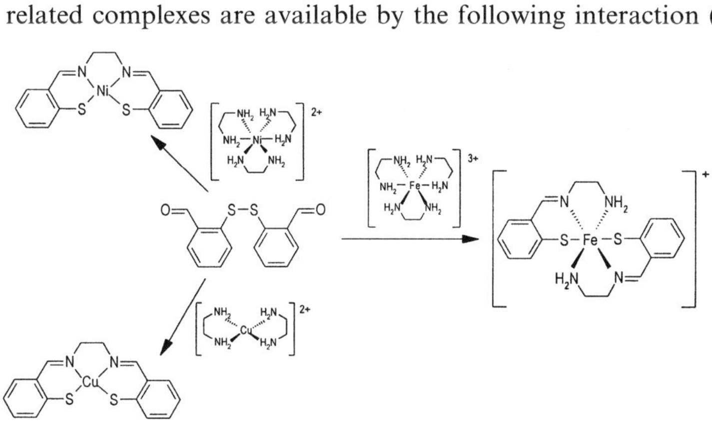

(3.172)

As well as in the above-described template reaction (3.168), the 2,6-diacetylpyridine is widely used for preparation of complexes of macrocyclic ligands having N,S-ligand environment [318]. An example of such syntheses is the reaction (3.173):

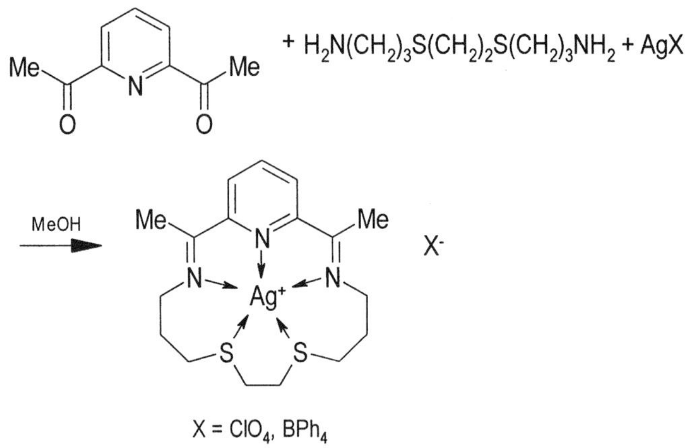

(3.173)

Some examples of other combinations of donor centers in coordination compounds of macrocycles, obtained as a result of template syntheses of the examined type, are also known, for example (3.174) [318,356] and (3.175) [318,357]:

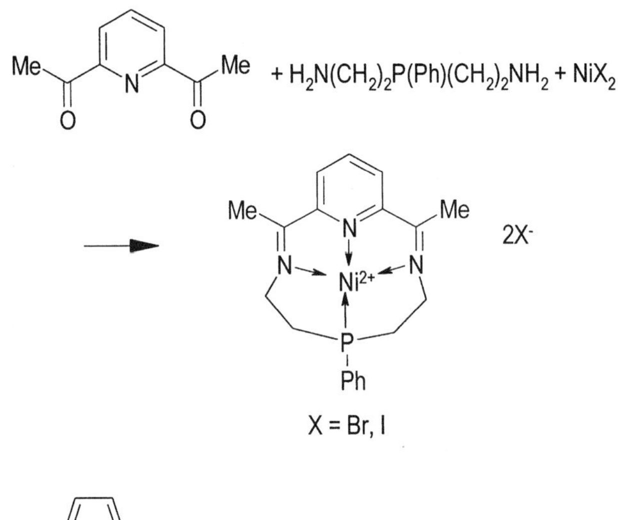

(3.174)

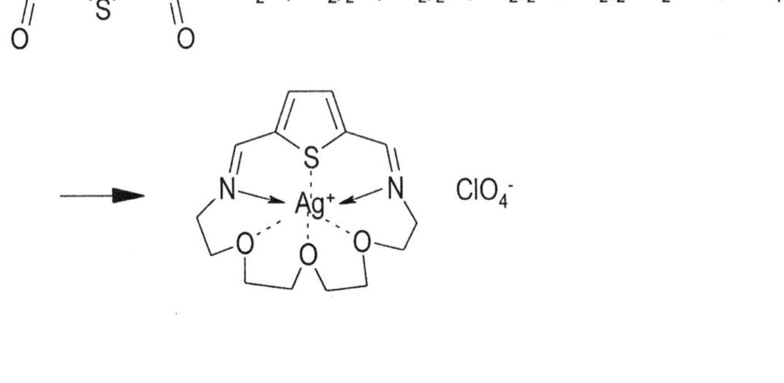

(3.175)

The template syntheses of macrocyclic ligands, containing N,P-, P,P-, S,S-, and O,E-atoms (E = S, Se, Te) [318,358] as donor centers have been reported. The possibility of formation of binuclear structures on a metal matrix using fragments of crown-ethers is shown [318]. However, a detailed examination of these syntheses is beyond the limits of the present monograph.

EXPERIMENTAL PROCEDURES

Example 1

Synthesis of ***715*** (3.159) [343]

Methanolic solutions of salac (2.04 g, 10 mmol, 20 mL) and aniline (*p*-chloraniline, $MeNH_2$), (20 mmol) were mixed. To the resulting solution, 0.5 mL of piperidine was added and the solution mixture was kept stirred for 4 hr. Then a methanolic solution of $CuBr_2$ [$CuCl_2$, $Cu(OAc)_2$] (10 mmol) was added. A dark solid complex, instantly separated, was filtered, washed with methanol, and dried in air. The complexes were recrystallized from acetonitrile/chloroform.

Example 2

Synthesis of Di[isobutyldithiocarbamate]cadmium(II) (3.163) [345]

i-Bu_2NH (17 mL, 0.1 mol) was mixed with 17 mL of MeOH, and the mixture inserted into a flask refrigerated by ice–NaCl mixture. CS_2 (6.0 mL, 0.1 mol) was added to the cooled mixture under vigorous stirring. After that the solution of 4.0 g (0.1 mol) of NaOH in 100 mL of MeOH was dropped under cooling and stirring. The resultant light yellow solution was portionally added to the solution of 11.4 g (0.05 mol) of $CdCl_2 \cdot 2.5H_2O$ in the same solvent under cooling and stirring. White precipitate was filtered off, washed with water, and dried in air. Crude product could be recrystallized from the $CHCl_3$:EtOH (1:1) mixture. Yield 24.4 g (94%), m.p. 172°C.

Example 3

Synthesis of ***721*** (3.169) [303]

o-Rhodanbenzaldehyde (0.02 mol) carefully mixed with 5 g of crystalline $Na_2S \cdot 9H_2O$ was added to 20–25 mL of *n*-butanol, and the mixture was heated in a water bath under stirring for 10 min, after which the precipitate of NaSCN was slowly filtered off through the lay of anhydrous Na_2SO_4. To warm filtrate the 0.02 mol of corresponding amine was added, and the solution was heated for 5 min, and the ethanolic solution of 0.01 mol of metal acetate was added dropwise. Precipitate forming after cooling was filtered, washed with ethanol, and recrystallized from *n*-butanol.

Example 4

Synthesis of [1,2-*bis*(2-Mercapto-3-formyl-1-benzylideneimine)ethanoato]nickel(II) ***722*** (3.171) [354]

1,2-Ethanediamine (0.149 g, 2.48 mmol) and $Ni(ClO_4)_2 \cdot 6H_2O$ (0.3937 g, 1.07 mmol) were dissolved in 10 mL of DMF. 2-(N,N-Dimethylcarbamoylthio)-5-methylisophthaldehyde (0.3641 g, 1.45 mmol) in 5 mL of DMF was added and the mixture was heated to ca. 100°C for 10 min. The yellow–brown Ni(eftp) crystallized on standing at room temperature. Yield 0.2746 g, 85%.

Example 5

Synthesis of Metal Complexes on the Basis of 1,2-Bisthiosalicylideneiminatoethane (tsalen) (3.172) [355]

(a) Ni(tsalen)

The $[Ni(en)_3]Cl_2$ reactant was synthesized by the addition of ethylenediamine in a 3:1 molar ratio to a solution of $NiCl_2 \cdot 6H_2O$ in methanol, resulting in the precipitation of $[Ni(en)_3]Cl_2$ as a light purple solid. $[Ni(en)_3]Cl_2$ (220 mg, 0.930 mmol) was dissolved in 100 mL of methanol in a 250-mL round-bottomed flask equipped with a reflux condenser, and a solution of 1.53 g (5.58 mmol) of 2,2′-dithiodibenzaldehyde (DTDB) in 20 mL of CH_2Cl_2 was added to the refluxing solution. The reaction mixture turned brown in 8–10 min; refluxing was continued for 1 hr until a golden precipitate appeared. This precipitate was collected by filtration, washed with methanol and ether, and recrystallized from CH_2Cl_2 to give 249 mg of Ni(tsalen) as a golden microcrystalline material (0.680 mmol, 75% yield).

(b) Cu(tsalen)

The $[Cu(en)_3]Cl_2$ reactant was synthesized in the same manner as $[Ni(en)_3]Cl_2$, using a 2:1 molar ratio of ethylenediamine and $CuCl_2 \cdot 2H_2O$ and producing a dark blue–purple solid. $[Cu(en)_3]Cl_2$ (520 mg, 2.04 mmol) was dissolved in 100 mL of methanol and a solution of DTDB (1.12 g, 4.09 mmol) in 20 mL of CH_2Cl_2 was added to the refluxing solution. After 1 hr, the reaction mixture was allowed to cool to room temperature, resulting in the deposition of a dark purple precipitate. This precipitate was collected by filtration, washed with methanol and acetonitrile, and recrystallized from CH_2Cl_2 to give Cu(tsalen) (384 mg, 1.06 mmol, 50% yield).

(c) Fe(Haetsaln)$_2$

The $[Fe(en)_3]Cl_3$ reactant was synthesized in the same manner as $[Ni(en)_3]Cl_2$, with a 3:1 molar ratio of ethylenediamine to $FeCl_3 \cdot 6H_2O$ producing a brown solid. $[Fe(en)_3]Cl_3$ (584 mg, 1.71 mmol) was suspended in 100 mL of methanol and a solution of DTDB (2.81 g, 10.3 mmol) in 20 mL of CH_2Cl_2 was added to the refluxing solution. After 30 min, the reaction mixture had turned dark and a dark precipitate had formed on the bottom of the flask. After standing at room temperature, x-ray quality crystals formed from the reaction solution. The bulk was recrystallized from CH_2Cl_2–methanol to give Fe(Haetsaln)$_2$ (461 mg, 1.03 mmol, 60% yield).

3.3.2 Modification of Ligand Systems

This synthetic method includes various chemical reactions of ligands and allows us to obtain new coordination compounds of many types on the basis of metal complexes. It became known long ago [2,314a,319,320] as an integral part of template synthesis and at present is examined as one of its variants [318]. Modification of ligands is the most important part of the problem of the reaction capacity of coordination compounds [130,320,321].

The reactions of the following coordinated organic and inorganic compounds are examined in monographs [321]; unsaturated hydrocarbons, nitrogen, and CO molecules; amines, imines, nitriles, and isonitriles; cyanides and cyanites; nitro- and nitrosyl groups; amides and phosphines; water, hydroxides, and ethers; sulfides, thiol fragments, thioethers, organosulfoxides, thiourea, and thioamides; halides and pseudohalide ligands. The reaction capacity of aromatic [314,315] and heterocyclic [359,360] compounds, connected with metals, is presented separately.

The transformations, related to an increase of cycle number in macrocyclic systems, are important in reactions of coordinated ligands [2,318,319,322]. The majority of these reactions have been used in preparative goals. Their detailed description is beyond the limits of the present monograph because of a high number of reported literature (hundreds of publications). So, we will discuss several examples which have been of interest during many years.

3.3.2.1 Template Synthesis of Chelates of Azomethines and Related Ligands

The following two synthetic techniques are applied for obtaining complexes of aliphatic, aromatic, and heterocyclic imines in the limits of the discussed variant of the template synthesis: interaction of coordinated mono- and diamines with mono- and dicarbonyl derivatives (method A) and reactions of complexes of *o*-hydroxyaldehydes (ketones) with amines and diamines (method B). The simplest transformation of this type is represented by the reaction (3.176) [53,54]:

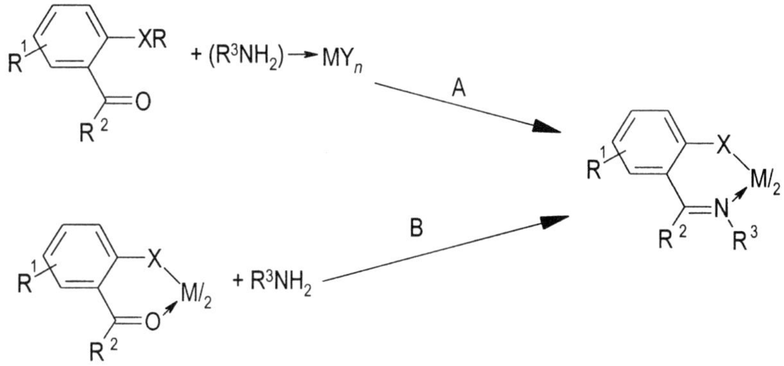

(3.176)

723

X = O, S, Se; R = H, Na, K; R^1 = H, 5-NO_2; R^2 = H, Alk; R^3 = Alk, Ar, Het

An example of the synthesis, taking place according to method A, is the reaction (3.177) [2]:

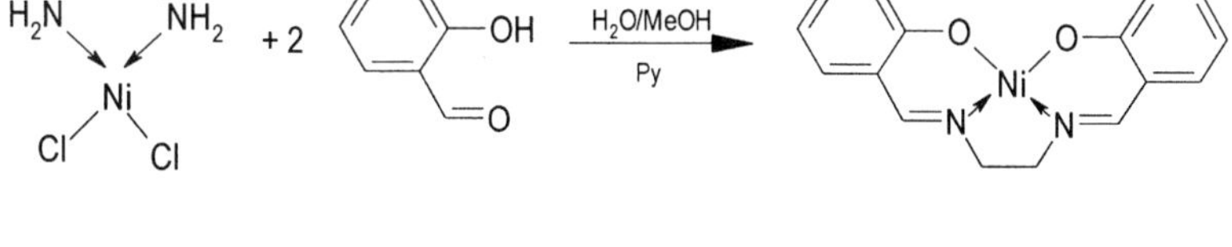

(3.177)

Preparation of chelates according to the method B (aldehydate method) is represented by numerous examples, known as far back as from the middle of the nineteenth century [315] and widely reported in modern coordination chemistry. Among them, the reactions (3.178) (products ***724*** [361], ***725*** [362], and ***726*** [121] and (3.179) (product *727* [363]):

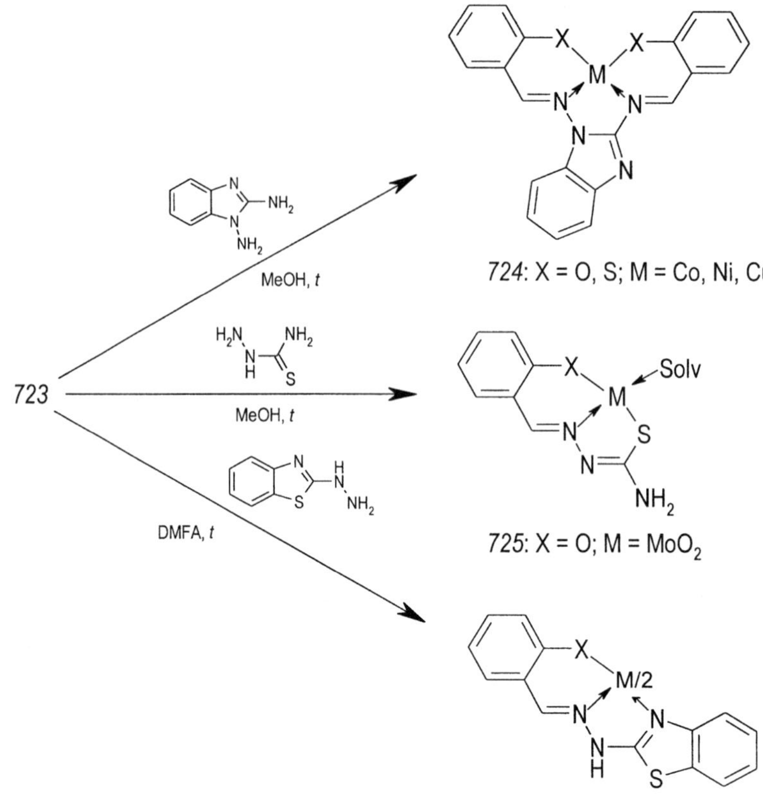

724: X = O, S; M = Co, Ni, Cu, Zn

725: X = O; M = MoO_2

726: X = O; M = Ni

(3.178)

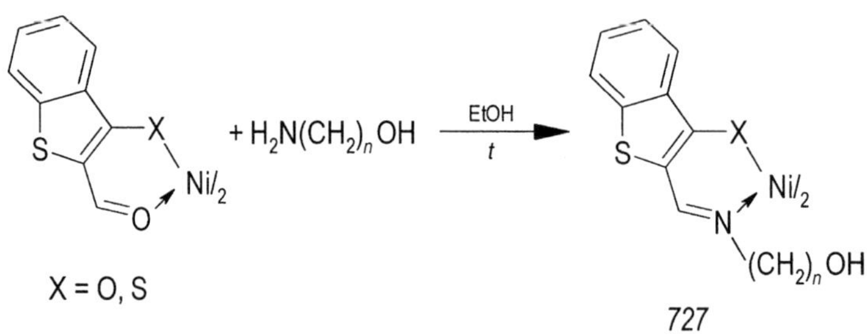

X = O, S

727

(3.179)

Such syntheses are widely used for obtaining complexes of macrocyclic ligands [316–319]. For example, the interaction of metal acetylacetoniminates, containing electron-acceptor substituents in γ-positions of the ligand, with ethylenediamine (3.180) [318,364–366]:

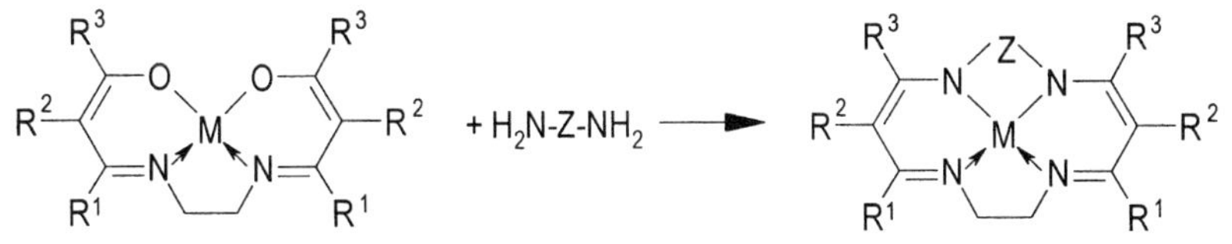

(3.180)

R^1, R^2 = H, Me, Ph; R^2 = COMe, COOEt;
Z = $(CH_2)_2$, *o*-phen; M = Ni, Cu

The aldehydate method is especially useful in the preparation of hard-accessible complexes of azomethinic derivatives of *o*-mercaptobenzaldehyde (3.176) (method B, X = S). Using it, additionally to the transformation (3.179) (product *727*, X = S) [361], the following cycle of preparative reactions (3.181) and (3.182) was carried out, yielding the products ***728*** (R = Ar) [367], ***728*** (R = 2-Py) [368], ***729*** [367,369,370], ***730*** [371], ***731*** [369,372], and ***732*** [369,373]:

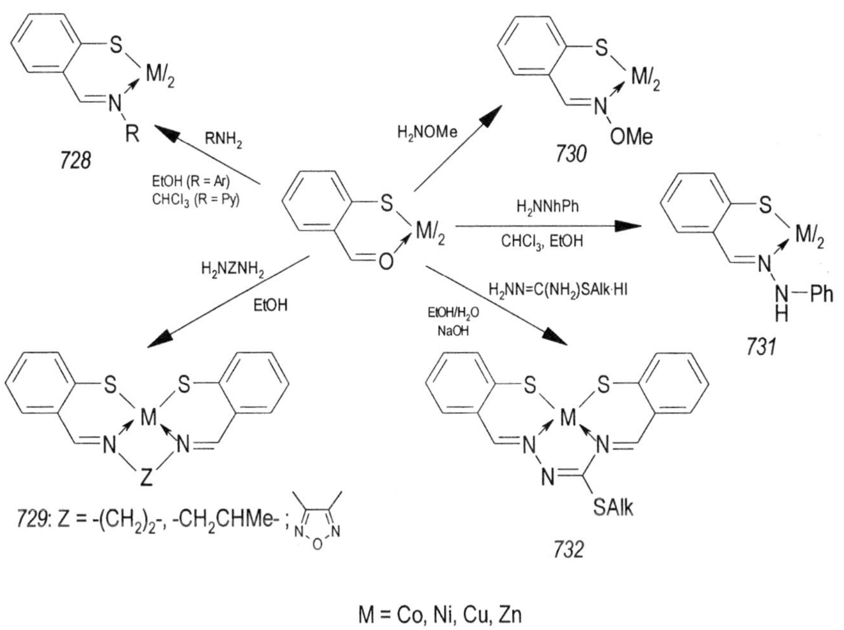

M = Co, Ni, Cu, Zn

(3.181)

Additionally to the above, the nickel thiosalicylaldehydate and its nitroanalogue was used to obtain (3.182) a series of other chelates ***733***, ***734*** (R′ = H) [372], and ***735*** (R′ = NO_2) [374]:

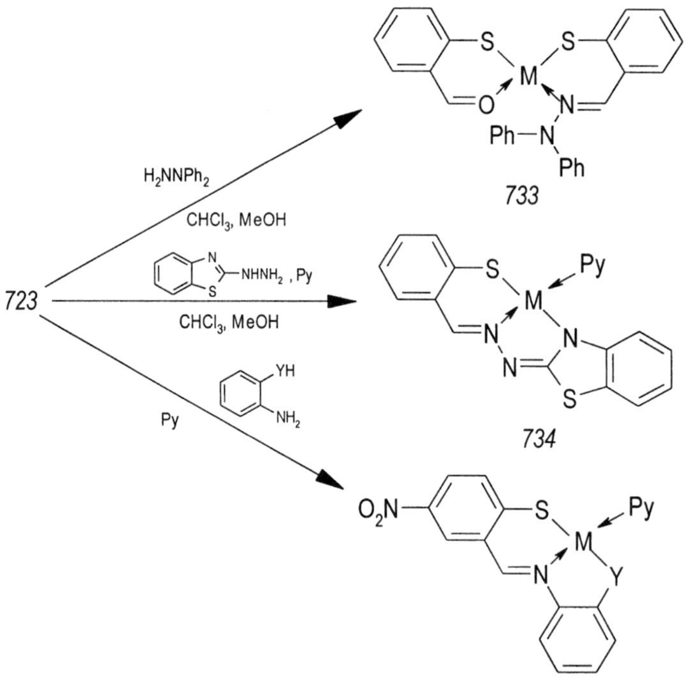

(3.182)

In relation to the synthesis of chelates of chalcogen–nitrogen-containing ligands of the types *724*, *727* (X = S), *729–732*, the possibility of a further modification of their ligand systems is opened by the S-alkylation. Although this transformation was known long ago (3.183) [375], these ligands are permanently applied in template reactions of macrocyclization [318], for instance using *bis*-alkyling agents (3.184) [376] and (3.185) [377]:

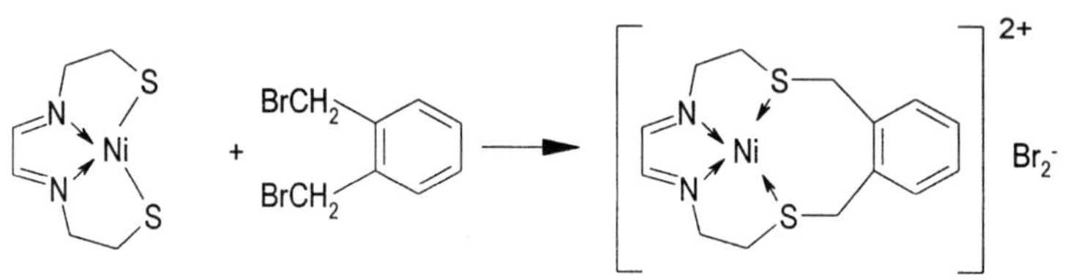

(3.183)

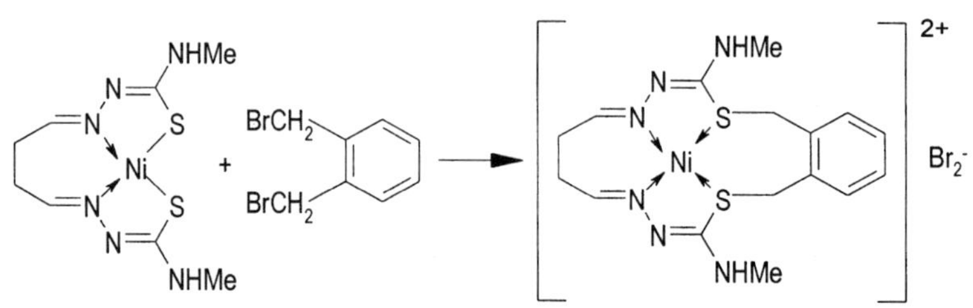

(3.184)

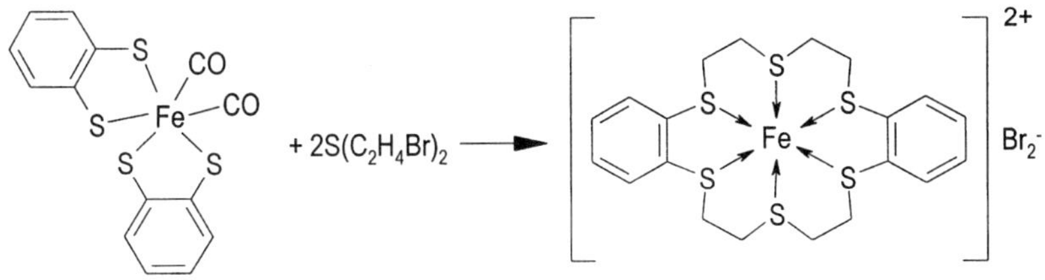

(3.185)

In our opinion, the perspectives of such an alkylation, as well as the possibility of S-alkylation of templately synthesized chelates, discussed in the present section, are highly attractive.

EXPERIMENTAL PROCEDURES

Example 1

Synthesis of *bis*[5-Nitrothiosalicylideneiminato]nickel(II) (3.176) [374]

A solution of 2.1 mmol of corresponding amine in pyridine was added to the solution of 1 mmol of *bis*(5-nitrothiosalicylaldehydato)nickel(II) *723* ($R^1 = 5\text{-}NO_2$, $R^2 = H$, $X = S$) in methanol. The mixture was stirred at room temperature for 1 hr, after which it was heated to boiling. Crystals precipitated after cooling were filtered off, carefully washed with hot methanol, and dried in air. Yield 50–60%.

Example 2

Synthesis of ***724*** (3.178) [361]

To the boiling solution of 1 mmol of metal acetate in 10 mL of methanol consequently were added a solution of 0.244 g (2 mmol) of salicylaldehyde in 5 mL of methanol and a solution of 0.148 g (1 mmol) of 1,2-diaminobenzimidazole in 10 mL of the same solvent. After refluxing for 2 hr the mixture was cooled, and precipitated crystals were filtered off, washed with hot methanol, and dried in air at 120°C. Yield ca. 60–80%.

Example 3

Synthesis of *bis*[2-(N-Hydroxyethyl)carbimin-3-hydroxybenzo[b]thiofenato]nickel(II) Ethanolamine ***727*** (3.179) [363]

The ethanolic solution of excess of ethanolamine (4 mmol) was added to a solution of 1 mmol of nickel chelate of 2-formyl-3-hydroxybenzo[b]thiophene in 15–20 mL of the same solvent. The mixture was refluxed for 2 hr. After that the precipitated complex was filtered off, washed with ethanol, and dried in air. The compound was recrystallized from the chloroform/hexane mixture.

Example 4

Synthesis of Nickelthiosalicylhydrazonates [372]

Bis(thiosalicylidene-N-phenylhydrazonato)nickel(II) ***731*** *(3.181)*

Obtained by coupling equimolar quantities of *bis*(thiosalicylaldehydato)nickel(II) and phenylhydrazine in chloroform on heating. Crude product was recrystallized from acetone. Yield 85%.

(Thiosalicylaldehydato)(thiosalicylidene-N,N-diphenylhydrazonato)nickel(II) ***733*** *(3.182)*

Synthesized analogously with N,N-diphenylhydrazine in $CHCl_3$ or ethanol. Precipitate recrystallized from benzene. Yield 56%.

Adduct [Thiosalicylidene-N-(benzthiazolyl-2)]pyridinatonickel(II) ***734*** *(3.182)*

Obtained by heating the equimolar quantities of *bis*(thiosalicylaldehydato)nickel(II) and 2-benzthiazolylhydrazine in pyridine without additional purification. Yield 45%.

3.3.2.2 Template Reactions of Oximes and Their Metal Complexes

During the last century, oximes have been used as ligands of various coordination compounds [378–388]. In particular, they are used in template syntheses [316,318,328,384,389–391], among which we emphasize the reactions of coordinated oximes (3.186), (3.187) [384,392], (3.188) [393], and (3.189) [394,395]:

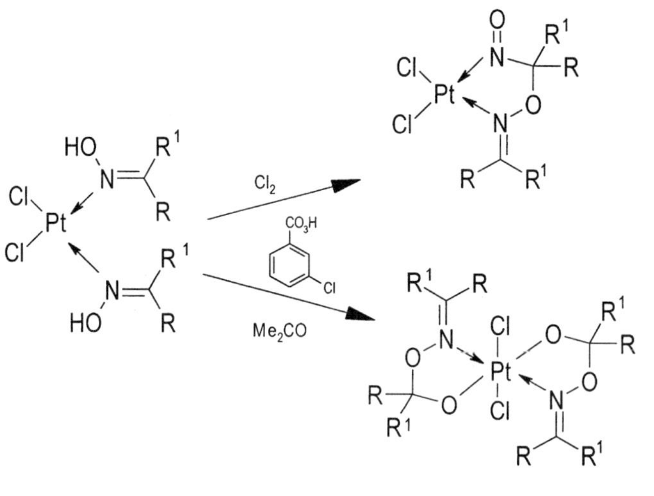

(3.186)

(3.187)

$R = R^1 = Me$; $R,R^1 = (CH_2)_4, (CH_2)_5$

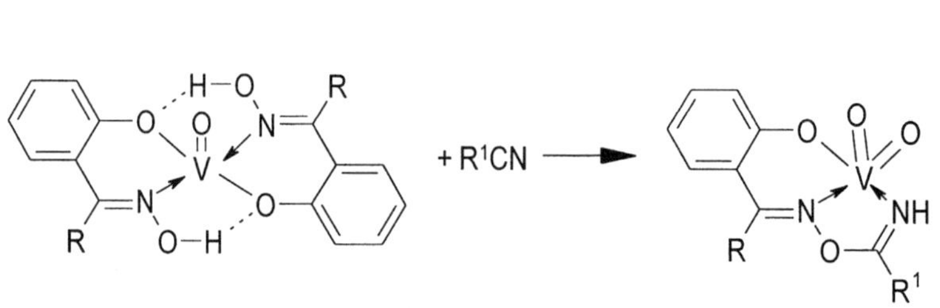

(3.188)

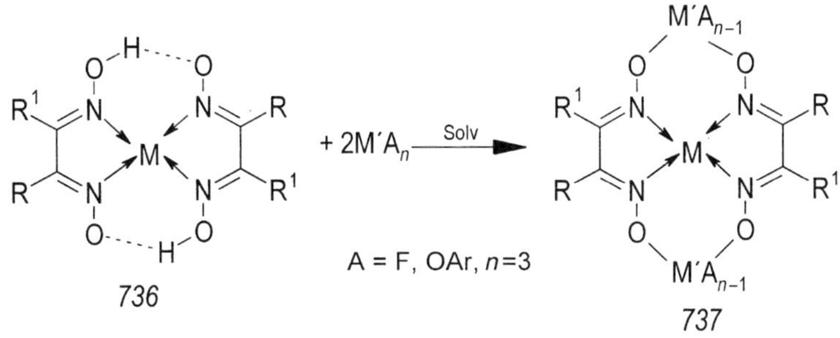

(3.189)

The transformation *736*→*737* (3.189) was discovered in 1962 [394] in the example of interaction of nickel *bis*(dimethylglyoximate) with $BF_3 \cdot OEt_2$ by cooling in CH_2Cl_2 and yielded a new type of compound *737* (R = R′ = Me, M = Ni, M′ = B, A = F, n = 3). The majority of trinuclear compounds have been prepared according to the reaction scheme (3.190) without intermediate isolation of chelates of the type ***736*** [396–412]:

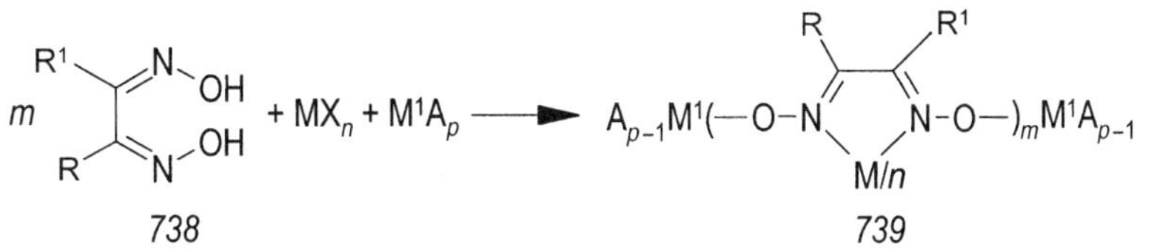

(3.190)

In the transformations above, not only classic dioximes ***738***, but also their analogous alicyclic and aromatic derivatives, as well as oxime-hydrazones with one or two OH groups, were used as oxime component [389,390]. The highest number of reported data was obtained from the example of template syntheses of iron clatrochelates [318,322,389,390]. Thus, the template reaction (3.190) has been widely applied for isolation of iron clatrochelates of type ***739*** on the basis of iron(II) glyoximate with inorganic and organic boron [318,396–403], germanium [404], tin [405–408], and antimony [409] derivatives. Scheme (3.190) reproduces a general result of the reaction: obtaining compounds of the type ***739***. However, judging by the reported techniques, the obtained clatrochelates can also be produced through preliminary formation of the ICC ***736*** [synthesis of type (3.189)] or the products of replacement of hydrogen atoms in ***738*** to MA_{p-1}. The reaction routes do not depend on the order of mixture of reactants, in particular, in the template synthesis of macrobicyclic boron-containing cyclo-octadionedioximates of divalent iron (3.191) [399]:

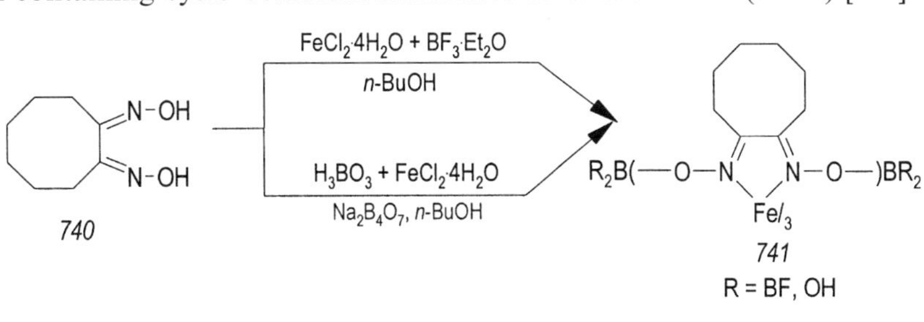

R = BF, OH

(3.191)

Template reactions with participation of mono-oximehydrazone ***742*** as an oxime component take place analogously to (3.190) [410]:

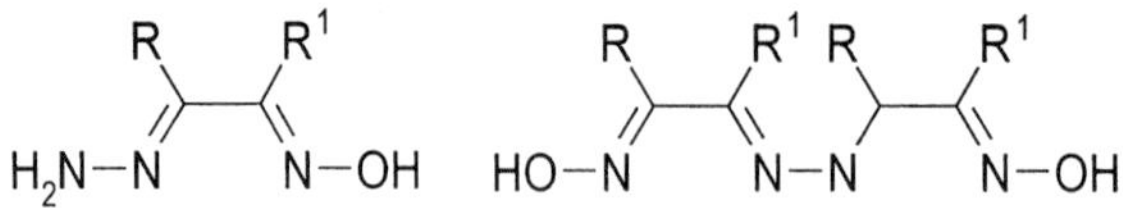

742 743

Thus, diacetylmono-oximehydrazone ***742*** ($R = R^1 = Me$, L^1H) and cyclohexanedione-1,2-mono-oximehydrazone ***742*** ($R\text{-}R^1 = cyclo\text{-}C_4H_8$, L^2H) react with $FeX_2 \cdot mH_2O$ and SnA_4 ($X = A = Cl, Br$; $m = 4, 6$) in *n*-butanol yielding [410] clatrochelates having compositions $[FeL^1{}_3(SnA_3)]$ and $[FeL^2{}_3(SnA_3)]$. The stirring of L_2H, $Fe(BF_4) \cdot 6H_2O$, and $PhB(OH)_2$ for 1 hr yields the complex $[FeL^2{}_3(BPh)(BF_4)]$, which, being treated with triethylorthoformiate in acetonitrile, leads to the coordination compound $[FeL^2{}_3(BPh)(HCOC_2H_5)_3](BF_4)$, whose structure was proved by x-ray diffraction [410].

In a difference from the above, the clatrochelates containing bis-deprotonated fragments ***743*** were obtained by the method (3.189): the interaction of the complex of the examined ligand with $BF(OMe)_2$ or $SnCl_4 \cdot NEt_3$ [411].

The interactions of compounds containing the oxime group with coordinated organonitriles are very important in template syntheses [384,413–416]. These reactions take place as a nucleophilic addition and lead to the formation of complexes with unusual iminoacyl ligands. The iminoacylation reaction was studied in detail for various oximes and organonitriles, coordinated to $PtCl_4$ [384,413–416]. Thus, the template transformation (3.192) of the discussed type in case of the oximes ***744*** takes place in acetonitrile or chloroform and yields complexes of the type ***745*** [413a]:

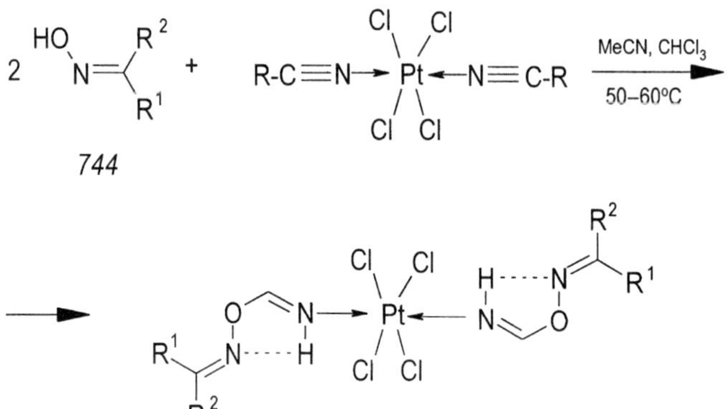

(3.192)

R = Me, Ph, CH_2Ph; $R^1\text{-}R^2 = C_4H_8\text{-}cyclo$, $C_5H_{10}\text{-}cyclo$, $-CHC_6H_4OH\text{-}o$

Similar reactions with participation of aromatic chloro- and amido-oximes take place in CH_2Cl_2 and yield complexes $[PtCl_4\{HN{=}C(Me)ON{=}CRR^1\}_2]$, where $R/R^1 = Cl$, Ph, $Cl/C_6H_4\text{-}p\text{-}Me$, $Cl/C_6H_4\text{-}p\text{-}NO_2$, H_2N/Ph [414]. The reaction (3.193) with dioximes takes place in the same synthetic conditions with yields of final products ***746*** $\sim$ 80–90% [415]:

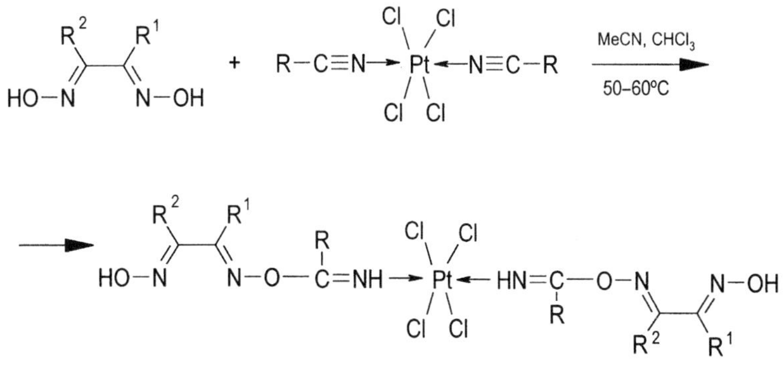

746

(3.193)

The described transformation is interesting because the vicinal dioxime groups do not participate in substitution of organonitrile groups, when traditional chelate compounds would be formed, but join the organonitriles of one of their OH groups, yielding addition products [415].

The interaction (3.194) of oximes and nitriles, coordinated to $RhCl_3$, takes place differently in comparison with the described reactions (3.192) and (3.193) [416a]:

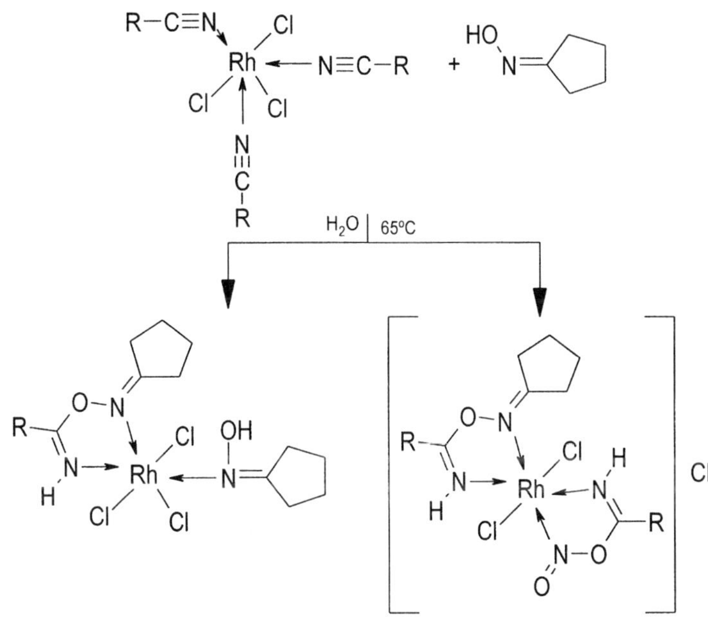

(3.194)

The chelate structures of both obtained complexes were proved by x-ray diffraction data [416a]. At the same time, it is necessary to pay attention to low yields (3–5%) of reaction (3.194) products, which considerably complicate its use for preparative goals. The interaction between the complex $[P+Cl_4(RCN)_2]$ and hydroxame acids, existing in oxime form, was reported in [416b].

Among other template transformations, related to modification of ligand systems, it is necessary to indicate reactions of deoxygenation of sulfoxide complexes, for instance (3.195) [417]:

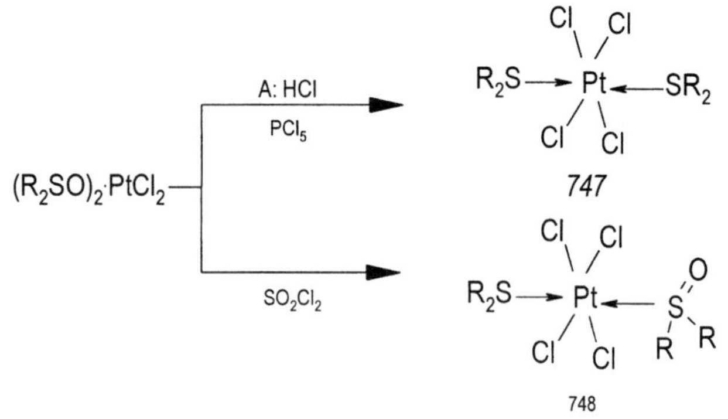

(3.195)

Additionally to the above reducing agents, ROH, Zn, NaI, AlI_3, and $M[BH_4]$ (M = Li, Na) are used. In our opinion, the possibility of obtaining mixed-ligand sulfide–sulfoxide complexes ***748*** as a result of these syntheses is of the highest interest.

Wide synthetic possibilities for modification of coordinated ligands are opened up by the classic reactions of electrophilic and nucleophilic substitution in complexes of aliphatic, aromatic, and heterocyclic compounds [314,359,418–422]. For example, the transformations (3.196) were known long ago [419]:

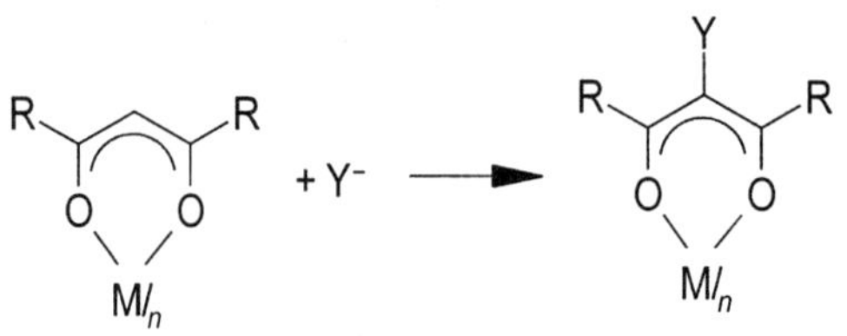

(3.196)

R = Alk, Ar
Y = Hal, SCN, SAr, NO_2, COR^1 (R^1 = H, Alk, Ar), D;
M = Be, Al, Co, Fe, Rh, Cu, Eu

The reactions of electrophilic substitution in metallocene structures of the type ***3*** (Sec. 1.1) are numerous and various [423–433]. The reaction capacity of coordinated cyclopentadienyl ligands in ferrocene ***749*** is especially studied in detail (3.197) [423–426,433]. It is significant that already in 1968 more than 300 such transformations had been reported [424], for example (3.197):

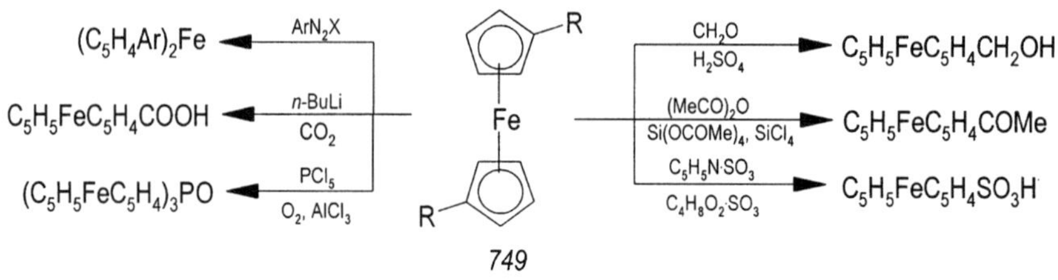

(3.197)

Among C(R)-substituted ferrocenes, various derivatives, containing cyano, amino, dialkylamino, amino(dialkylamino)methyl, azide, aldehyde, ketone, rhodanide, and other coordinatively active groups, have been obtained. They are metal-containing ligands for preparation of di- and polynuclear structures (see Sec. 3.3.2.3).

Coordinated heterocycles take part in reactions of reduction, protonation, alkylation, acylation, haloidation, nucleophilic and cycloaddition [359].

EXPERIMENTAL PROCEDURES

Example 1

Synthesis of ***739*** (3.190) [402]

Cyclohexane-1,2-dione dioxime (0.43 g, 3 mmol), 0.35 g (2 mmol) of ferrocenylboric acid, and 0.20 g (1 mmol) of $FeCl_2 \cdot 4H_2O$ were dissolved in 10 mL of methanol and stirred for 1 hr. The red–orange precipitate was filtrated and washed with methanol, diethyl ether, and pentane. The product was recrystallized from chloroform and dried in vacuo. Yield 0.65 g (71%).

Example 2

Synthesis of ***745*** (3.192) [413a]

In a typical experiment, $[PtCl_4(MeCN)_2]$ (0.15 g, 0.36 mmol) was suspended in acetonitrile or chloroform (10 mL) at room temperature, the oxime (0.90 mmol) was added, and the reaction mixture heated with stirring at 55–60°C for 10 min until homogenization of the reaction mixture. A bright yellow solution formed was evaporated to dryness and washed with diethyl ether (five 3-mL portions) to remove the excess oxime. Yields 90–95%.

Example 3

Synthesis of $[PtCl_4\{NH{=}C(Me)ON{=}C(Cl)Ph\}_2]$ [414]

A solution of an oxime (0.096 mmol) in 2 mL of CH_2Cl_2 was added to a suspension of $[PtCl_4(MeCN)_2]$ (20 mg, 0.048 mmol) in 5 mL of CH_2Cl_2 and refluxed for 38 hr. The precipitate formed was filtered off, washed with two 5-mL portions of warm (35°C) CH_2Cl_2, two 5-mL portions of Et_2O, and dried in vacuo at room temperature. Yield 79%.

3.3.2.3 Interaction of Coordinated Ligands with Lewis Acids

This type of template transformation includes mainly reactions of immediate interaction of reagents (Sec. 3.1) – coordinatively unsaturated metal-containing ligand systems (complex compounds) with Lewis acids and leads (3.198) to obtaining di- and poly-, homo- and heteronuclear structures.

$$ML_m + pM^1X_n \xrightarrow{\text{Solv}} [(ML_m)(M^1X_n)_p] \tag{3.198}$$

Alcohols, hydrocarbons, and their halide-containing derivatives, and their mixtures, are the most frequently used solvents (Solv). The majority of these reactions take place at room temperature or with slight heating, necessary for dissolution of the reagents. The transformation (3.199) was one of the first template reactions of the examined type [434]. In this reaction, the coordinatively unsaturated platinum complex is an N-base and silver nitrate is a Lewis acid:

$$[(H_3N)_2Pt(SCN)_2] + AgNO_3 \longrightarrow [(H_3N)_2Pt(SCN)(SCN) \rightarrow Ag]^+NO_3^- \tag{3.199}$$

The azomethinic (3.200) and nitrile (3.201) groups [91], which are metal-complex N-bases, possess donor properties:

(3.200)

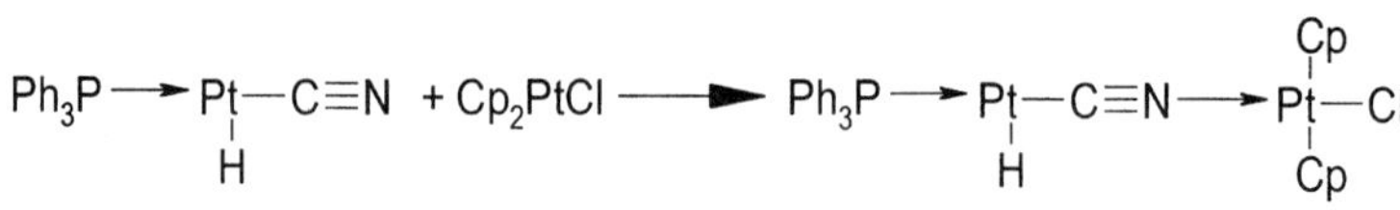

(3.201)

It was recently shown [435] that the mixed-valent complex compounds with N-coordinated bridge fragment can be obtained on the basis of the nitride rhenium complex as a result of the transformation (3.202):

$$ReNCl_2(PMe_2Ph)_3 + ReCl_4 \longrightarrow (Me_2PhP)_3Cl_2Re{\equiv}ReCl_4 \qquad (3.202)$$

Similar reactions are also described for the ligands containing the group C≡P [436].

It was established at the end of the 1960s [437,438] that the chelates of *o*-oxyazomethines and their analogues react with metal halides, forming bi- and trinuclear structures, for instance ***750*** and ***751***:

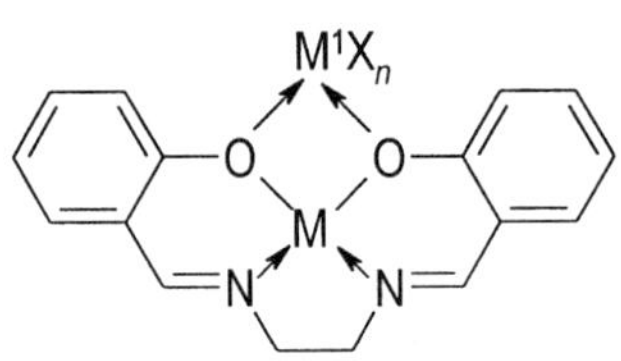

750

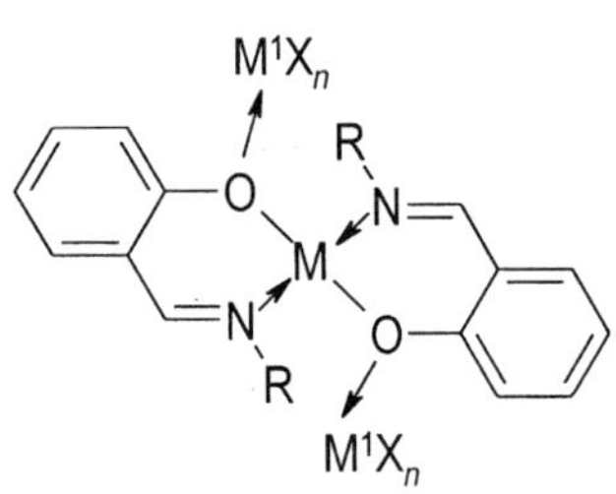

751

Further, numerous similar coordination compounds have been obtained. The results of these investigations have been generalized in reviews [53,438] and a monograph [91]. In this respect, we note the results of some studies. Thus, *bis*(salicylal)ethylenediiminates ***752*** react with metal carbonyls in THF, forming binuclear complexes ***753*** (3.203) [439]:

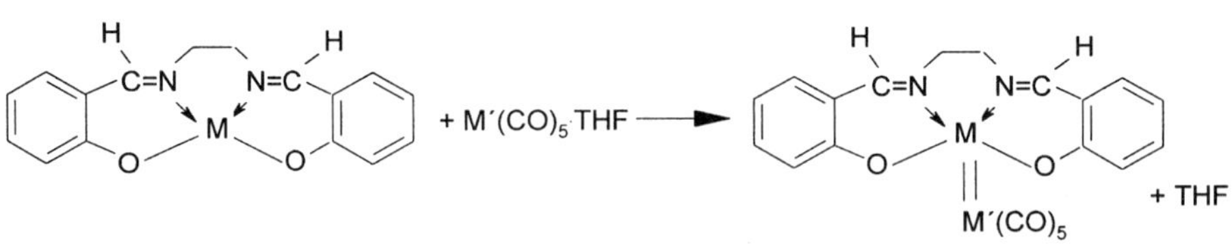

752 M = Ge, Sn, Pb; M′= Cr, W *753*

(3.203)

Additionally, to ***753***, the complexes $L_2M'(CO)_4$ (L = ***752***) can be formed. Trinuclear complexes were also obtained by coupling the salicylideneiminates or β-aminovinylketonates of bivalent metals (Co, Ni, Cu, Zn) with $M'X_3$ (M′ = B, Al; X = Cl, Br) [440]. A series of works, related to the synthesis of examined coordination compounds, was published by Sinn and coworkers [438,441–444]. These authors [442,444], as well as those of Ref. 445, showed that not only M^1X_n, but also metal β-diketonates can take part as Lewis acids, that is, such interactions take place also between chelate molecules.

The reactions with participation of *o*-oxyazomethines and organometallic Lewis acids are interesting. In particular, the following transformation was recently reported (3.204) [446]:

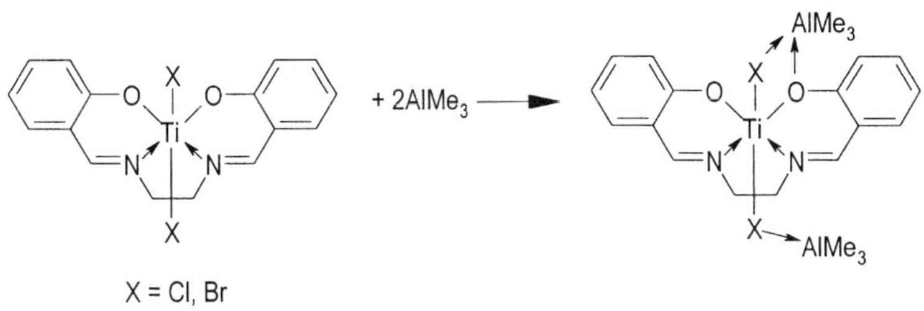

(3.204)

Template transformation, similar to (3.203), is observed in the case of *o*-hydroxy-azomethines of the heterocyclic series, for instance (3.205) [447]:

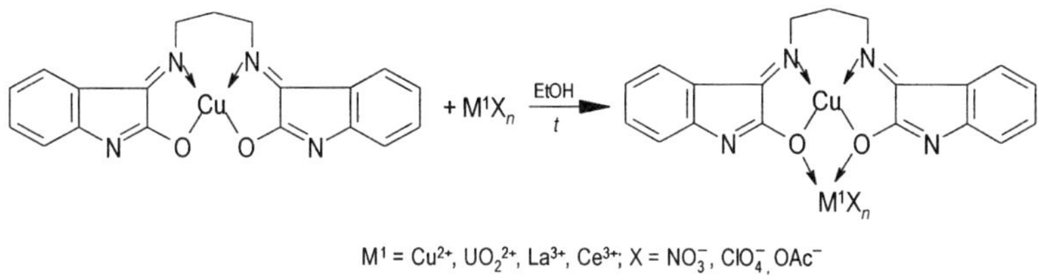

(3.205)

Additionally to the indicated chelates, other inner-complex compounds, for example those on the basis of hetarylformazanes (3.206) [448], can also take part in the examined template transformations:

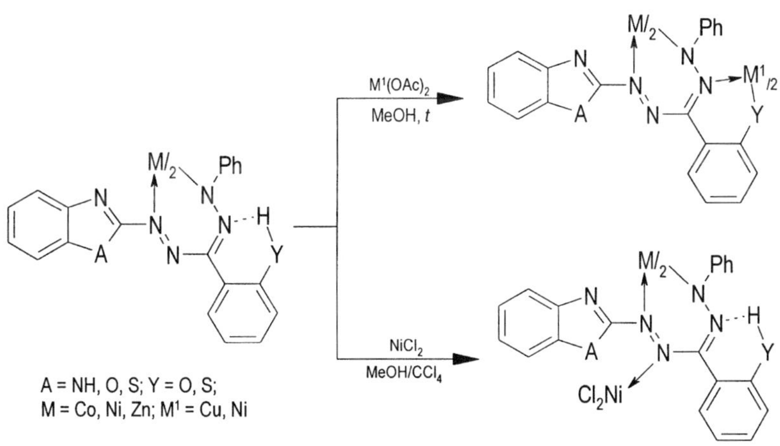

(3.206)

The anionic copper complexes of N[1],N[4]-*bis*(salicylidene)isosemicarbazidate ion (L) are used in the reactions above. As a result of their interaction with lanthanum and samarium chlorides in chloroform–methanol suspension, the heteronuclear Lu,La- or Cu,Sm-complexes are formed [449].

Organometallic compounds have long been used as ligands; their most frequent application is as metal carbonyls [74c,91,450,451]. These ligands, for instance ***754***, are interesting as ambidentate ligand systems, in which some coordination modes can be realized [91,450,451]:

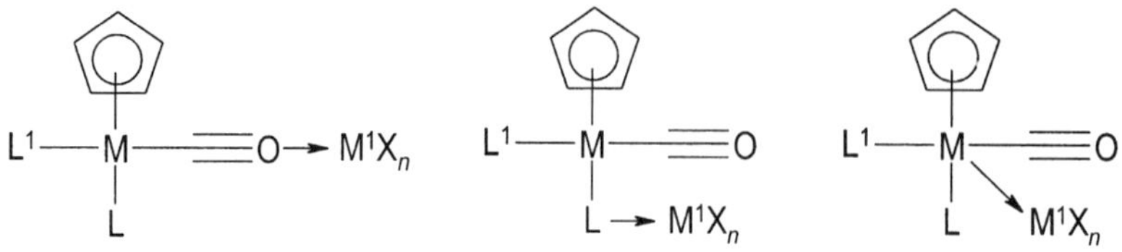

754a *754b* *754c*

L, L′= CO, NO, PPh_3; M = Mo, W; M′= Al, Sn, Ti; X = Cl; n=3-4

The coordination $M \rightarrow M^1$ (***754c***), when the M behaves as a d-base [450,451], is widely represented in complexes obtained on the basis of metal–carbonyl ligands. Thus, a binuclear compound ***755*** with Fe—Al bond is formed by interaction between $K_2[Fe(CO)_4]$ and cyclopentadienylaluminum chloride dimer [452]. The reaction takes place as follows (3.207):

$$2K_2[Fe(CO)_4] + [CpAlCl_2]_2 \longrightarrow \underset{\mathbf{755}}{2[CpAl{-}Fe(CO)_4]} + 4KCl \quad (3.207)$$

On the basis of tetrahydrofuran complexes of pentacarbonyls of Group VI metals and triphenyl derivatives of Group V elements, the binuclear complexes with M—E bond were isolated (3.208) [453]:

$$M(CO)_5(THF) + EPh_3 \xrightarrow{THF} M(CO)_5EPh_3 \quad (3.208)$$
$$M{=}Cr,\ Mo,\ W;\ E{=}Bi,\ Sb$$

The M—E bond is realized in these complexes obtained; it was proved in the case of M = Mo, E = Bi [Mo—Bi distance = 2.832(1) Å] and M = W, E = Bi [W—Bi = 2.829(1) Å] by x-ray diffraction [453].

Metal–metal bonds are formed in complexes of the type $R_m[\{M(CO)_5\}_nM'X_p]$, where R = NBu_4, PPh_4; M = Cr, Mo, W; M′ = Sn, Pb; X = Cl, Br; m = 1–2; n = 2, 3; p = 3–4 [454]. The sodium–metal–carbonyl complexes of the type $Na_2[M_2(CO)_{10}]$ participate in this case as ligands and $SnCl_2$ acts as a Lewis acid [454]. M is a bridge fragment in the case of trinuclear complexes (n = 2).

Complex compounds [455] and multinuclear iron carbonyls [456] can take part as acceptors (A) in the examined reactions with use of metal–carbonyl bases. In this respect, a series of recently reported reactions (3.209) [455] and (3.210) [456] are representative:

$$M(CO)_m(MeCN)_n + nA \xrightarrow[THF]{MeCN} M(CO)_mA_n \quad (3.209)$$

where A = $[WI_2(CO)\{PhP(CH_2)_2PPh(CH_2)_2PPh_2{-}P,\ P'\}\eta^2(RC{=}CR')]$; M = Cr, Mo, W; R, R′ = Me, Ph; m = 3–5, n = 1–3.

$$[PhC{\equiv}CRe(CO)_5]Co_2(CO)_8 + Fe_3(CO)_{12} \xrightarrow{C_6H_{12}} [PhC{\equiv}CRe(CO)_5]FeCo_2(CO)_7(\mu\text{-}CO) \quad (3.210)$$

The presence of Co—Fe [2.499(2), 2.559(2) Å] and Re → Fe [2.982(2) Å] bonds was proved by x-ray diffraction [456].

Interaction of η^6-$Cr(CO)_3$ coordinated azomethines with soft ($HgCl_2$) and hard (BF_3, $AlBr_3$, $SnCl_4$) Lewis acids leads (3.211) to adducts of the type *756* [457]:

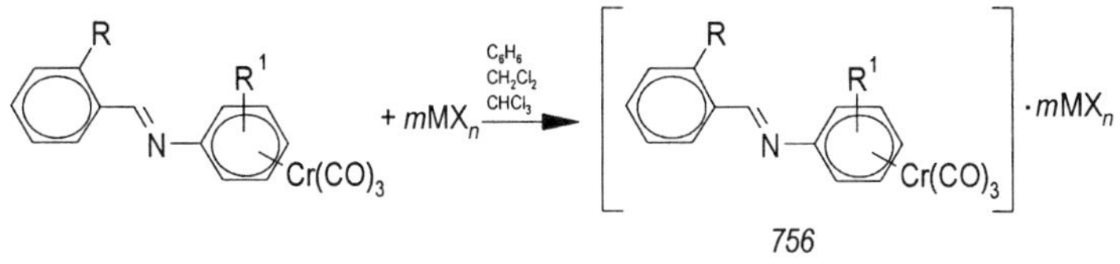

(3.211)

R = H, OH; R¹= H, Me; M = Hg, Al, Sn; X = Cl, Br; *n* = 2–4; *m* = 1, 2

It was shown on the basis of IR spectral data that the soft mercury chloride is coordinated on the chromium atom and the indicated hard acids are connected with N or O (R=OH) donor centers [457]. This conclusion required x-ray diffraction proof, since a coordination (of the type *754*) on the oxygen atom of the carbonyl group is not excluded due to the presence of the C≡O fragment [91,450,451].

The substituted ferrocenes *757* are widely used as metal-ligand systems. These compounds have a great variety of chelating groups (R): β-diketones (a) [458], semicarbazones and thiosemicarbazones (b) [458], azomethines and their analogues (c) [458–462], oximes (d), carbonic, thiocarbonic (e), and amino acids (f) [458]:

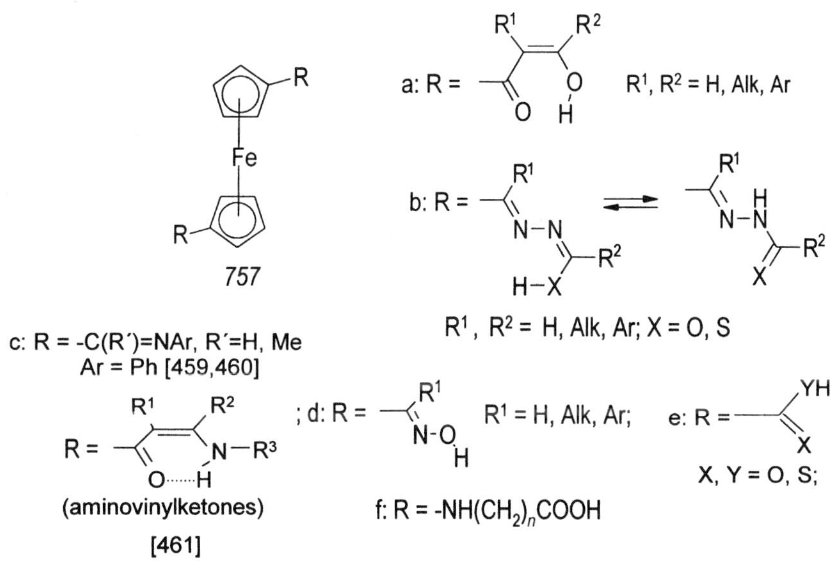

The following ferrocene derivatives, containing coordinatively unsaturated heteroaromatic N-donors as substituents in the cyclopentadienyl ring, are coordinatively active: *bis*(benzoethene)pyridine [463], aminopyridine [464], pyridine-pyrazol [465], and pyrazolylborate [466] fragments. Diphenylphosphinoferrocene [467,468], dialkylphosphinoferrocenes [469], as well as mono-, di-, and triferrocenylphosphines [470] are P-donors. Chalcogene *758* (3.212) [471] and chalcogen–nitrogen *759* (3.213) [472] ferrocene derivatives are used as metal ligands:

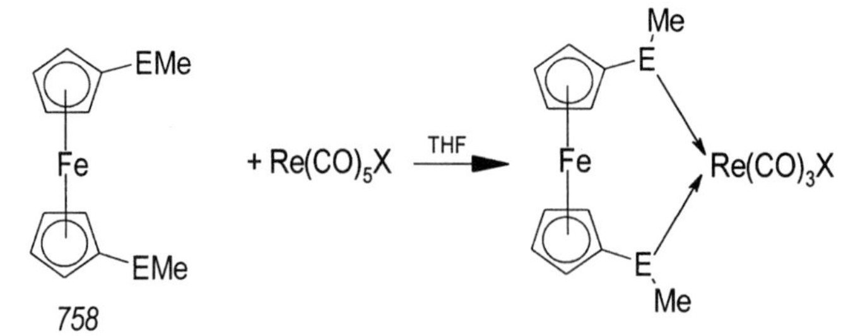

(3.212)

758

E = S, Se; X = Cl, Br, I

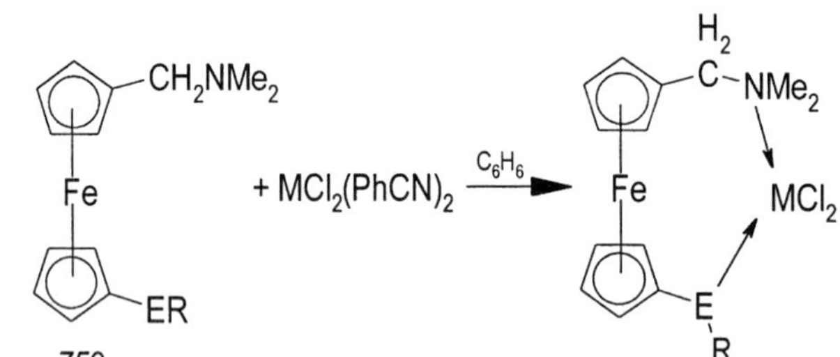

(3.213)

759

R = Alk; E = S, Se; M = Pd, Pt

Tri- and tetranuclear ferrocenyl–acetylene clusters of ruthenium and osmium are described in Ref. 473.

Together with ferrocene, the heteroferrocene ligand systems undergo modifications in the reactions with Lewis acids. Thus, interaction of azacymantrene ***760*** (E = N, R = H, L = CO) and its derivatives (for instance ***760***: L = PPh_3) with metal carbonyls and chlorides leads (3.214) to bi- ***761*** (E = N, $MX_n = M(CO)_5$, M = Cr, Mo, W; $M(CO)_2Cp$, M = Mn, Re [474–476]) and trinuclear ***762*** (E = N; M = Pd, Pt [477]) structures:

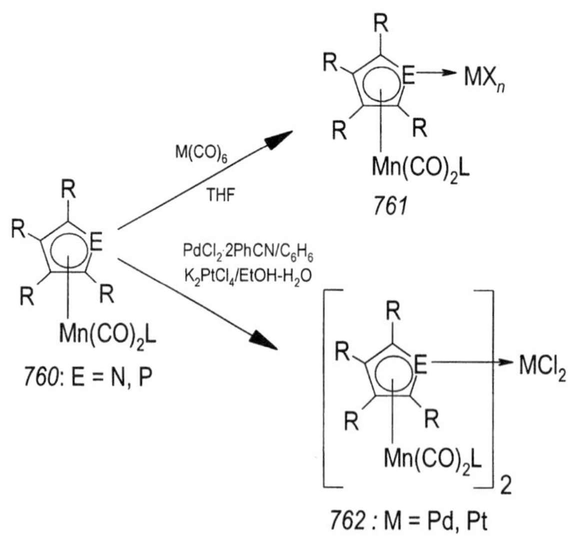

(3.214)

760: E = N, P

761

762 : M = Pd, Pt

Analogous transformations (3.215) are characteristic for azaferrocene ***763*** (R = H). They are accompanied by formation of the complex compounds ***764*** (R = H, M = Cr, X = CO, *n* = 5) [476,478] and ***765*** (M = Pd, Pt) [477]. The adducts of 2,3,4,5-tetramethylazaferrocene (***763***: R = Me) with BH_3 (***764***: R = Me, M = B, X = H, *n* = 3) were isolated (3.215) [479]:

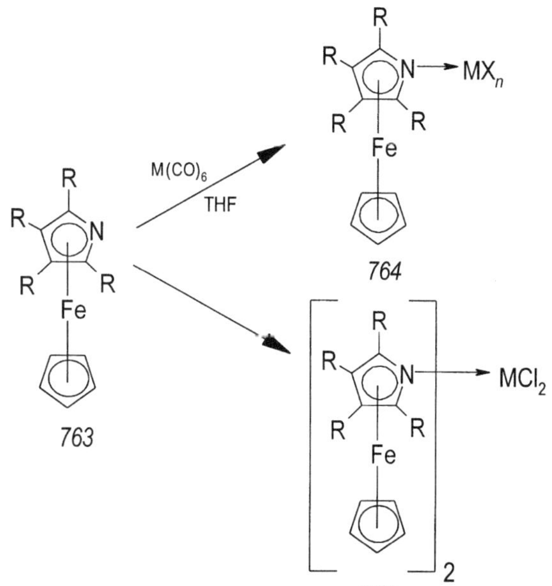

(3.215)

Octamethyl-1,1′-diazaferrocene ***766*** behaves as a bidentate N,N′-base [480,481]. It reacts (3.216) with $AgBF_4$ in methanol, forming a tetranuclear cyclic complex ***767***, whose structure was proved by x-ray diffraction [480]. ***766*** reacts analogously with BH_3, BF_3, and $HgCl_2$ [481]; however, the structures of obtained adducts have not yet been strictly proved.

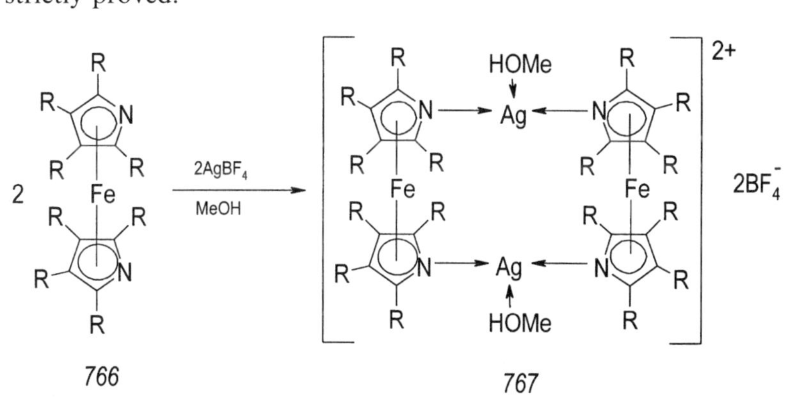

(3.216)

Not only N, but also C atoms can take part as donor centers in azacymantrene ***760*** (E = N, R = H) and azaferrocene ***763*** (R = H). In this respect, the formation of a trinuclear osmium–carbonyl adduct ***768*** is representative, which takes place in the reaction of the indicated azacenes with acetonitrile complex of triosmium decarbonyl [482]:

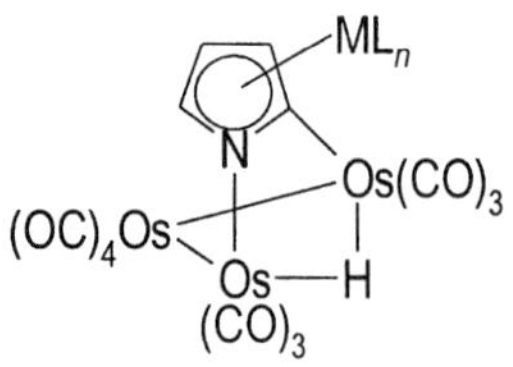

768

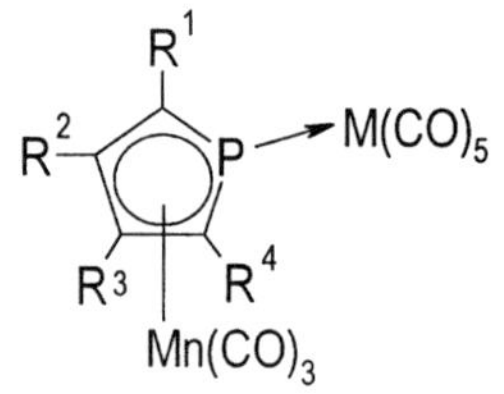

769

Phosphacymantrene *760* (E = P, R = H), as well as its R-substituted derivatives, participate as P-donors and, interacting with metal carbonyls in THF, form binuclear complexes, for instance *769* (R = H) [483]. A tetranuclear dimanganese–dipalladium chloride complex is formed as a result of the interaction of 2-phenyl-3,4-dimethylphosphacymantrene (*760*: E = P; R^1 = H, R^2 = Ph, R^3 = R^4 = Me; L = CO) with $[PdCl_2(PhCN)_2]$ in CH_2Cl_2 (25°C, 4 hr) [484]. Reaction of the same ligand with $[Pd(DBA)_2]$, where DBA is dibenzylideneacetone, under the same conditions leads to the tetramanganesetrichloride complex of phosphacymantrene. Complex structures of these compounds, as well as those of similar clusters, were proved by x-ray diffraction and do not provoke doubts [484]. Phosphanyl derivatives of phosphaferrocene *770* are P,P-chelating ligands and form (3.217) metal-cyclic structures, for example *771* [485]. Structure *771* (R = Ph, M = Mo) was proved by x-ray diffraction [485]:

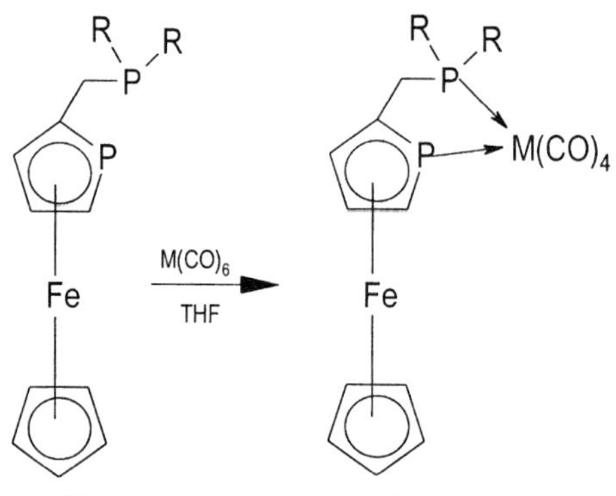

770 771

R = Cy (cyclohexyl), Ph; M = Cr, Mo

(3.217)

A series of diheteronuclear complexes was isolated (3.218) on the basis of metal *bis*-phospholyldichlorides (tetrahydroborates) [486]. They are bidentate P-donor chelating ligands and form metal-cyclic structures. The chelates of the type *772* [486] are typical examples of such compounds:

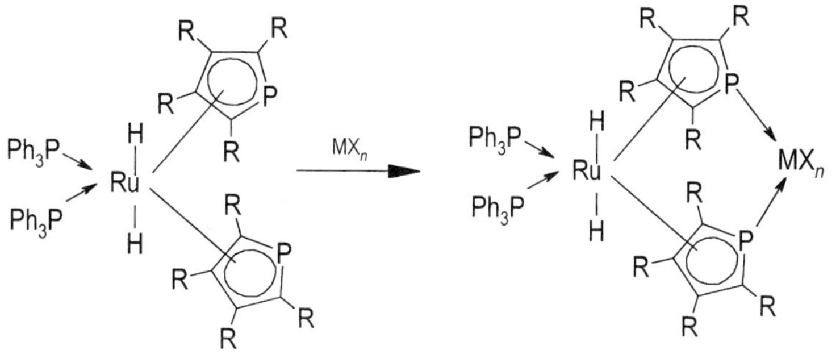

772

R = H, Me; M = Zr, Yb, U; X = Cl, BH_4

(3.218)

Interaction of phosphacymantrene (L) with $[Rh(COD)Cl]_2$ in CH_2Cl_2 at room temperature leads to tetra- and pentanuclear complexes $[L_3RhCl]$ and $[L_4Rh]^+(BF_4)^-$ [487].

As a result of the titled transformation, the binuclear complexes of thiophene (3.129) [488] and selenophene (3.220) [489] were synthesized.

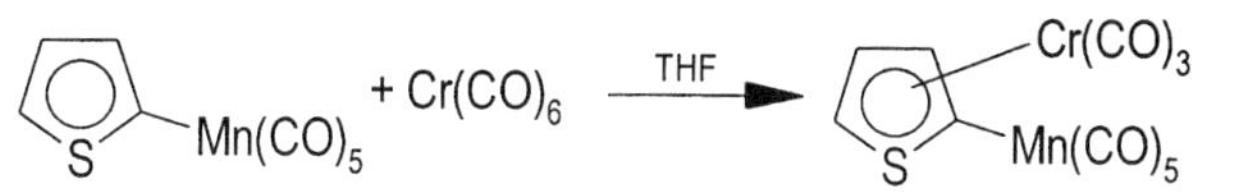

(3.219)

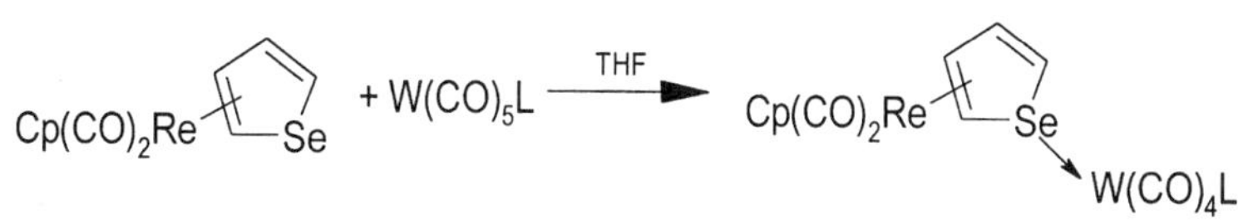

(3.220)

Use of coordinated azole and azine system containing donor-active nitrogen atoms as metal ligands open wide possibilities for obtaining homo- and heteronuclear complexes. Thus, such reactions are known where the complexes of coordinatively unsaturated pyrazole (3.221) [490] and triazole (3.222) [491] take part:

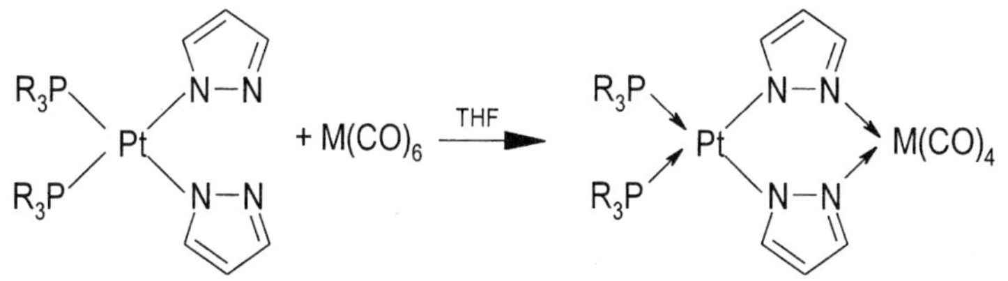

M = Cr, Mo, W

(3.221)

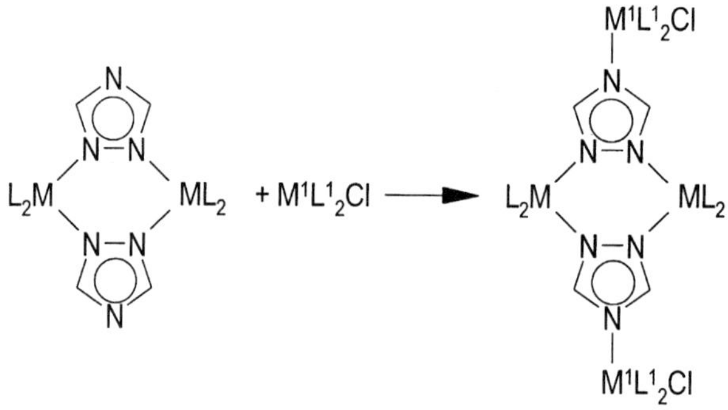

(3.222)

This interaction is carried out with participation of $[M'(\mu\text{-Cl})(cod)]_2$ (M = Rh, Ir) or $[Rh(\mu\text{-Cl})(CO)_2]_2$ and leads to a series of products, in which M = M′ = Rh, $L_2 = L_2' =$ cod, $(CO)_2$; M = Rh, M′ = Ir, $L_2 = L_2' =$ cod; M = Pd, M′ = Rh, $L_2 = \eta^3\text{-}C_3H_5$, $\eta^3\text{-}C_4H_7$, $L_2' = (CO)_2$ [491].

Modification of ligand systems in azine complexes was studied for the example of reactions (3.223) of η^6-coordinated 2,4,6-collidinechromiumtricarbonyl *773* (L = CO) [294] and its monotriphenylphosphine derivative *773* (L = PPh_3) [492] with Lewis acids ($AlBr_3$, $SnCl_4$ $TiCl_4$), leading to binuclear structures *774*:

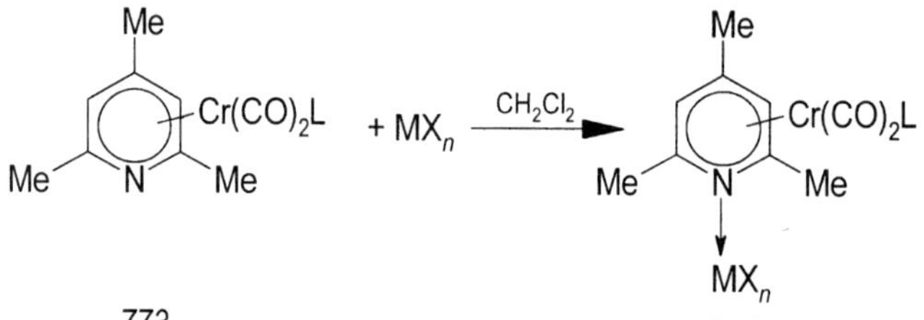

(3.223)

773 *774*

Use of noncoordinated donor centers of diazines, for instance in *775* [493], in reactions with Lewis acids (3.224) is widespread. A bridge joining the metal–carbonyl group, for example *776*, as well as formation of the metal-cycle *777* was observed in these processes:

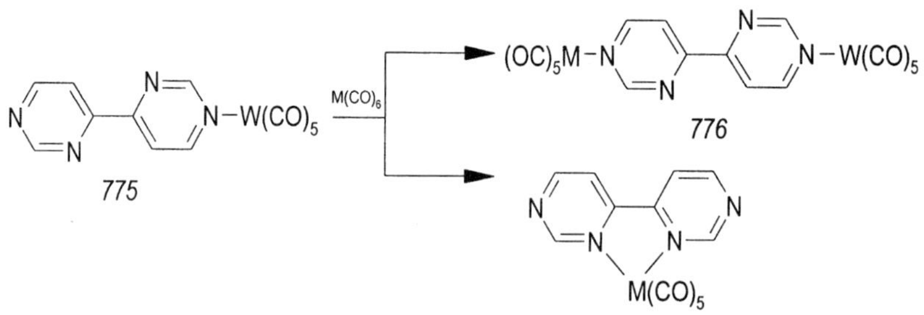

775 *776* *777*

(3.224)

The same donor center is contained in the tricarbonyliron complex of 2-diphenylphosphinopyridine *778*, on whose basis a series of binculear heterometallic complexes with or without $M—M^1$ bond, for instance *779* and *780*, respectively, was isolated (3.225) [130a,494,495]:

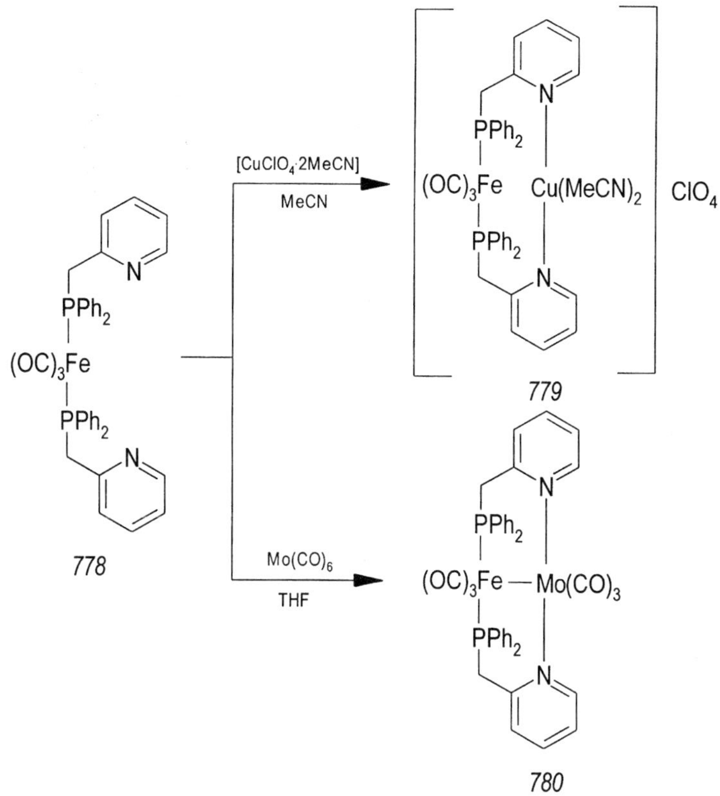

(3.225)

Practically all transformations discussed in this section represent reactions of molecular complexes (adducts) formation. At the same time, it is necessary to take into account that, in some cases, more complex processes can take place, for example cyclometallation (3.226) [460], (3.227) [496], and (3.228) [497]:

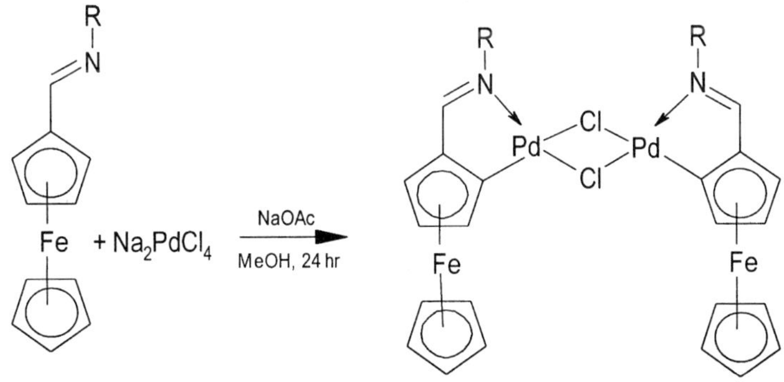

(3.226)

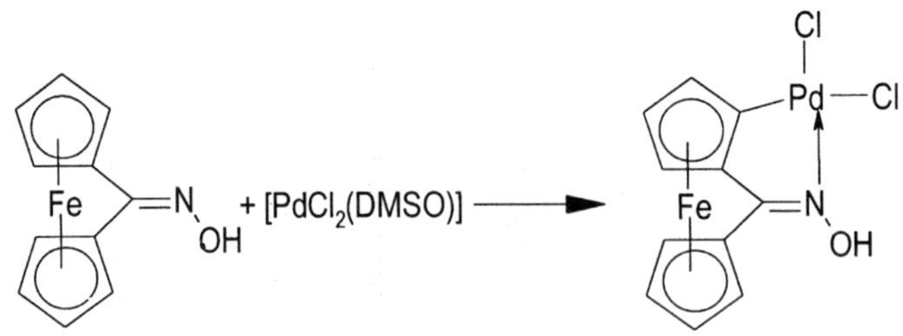

(3.227)

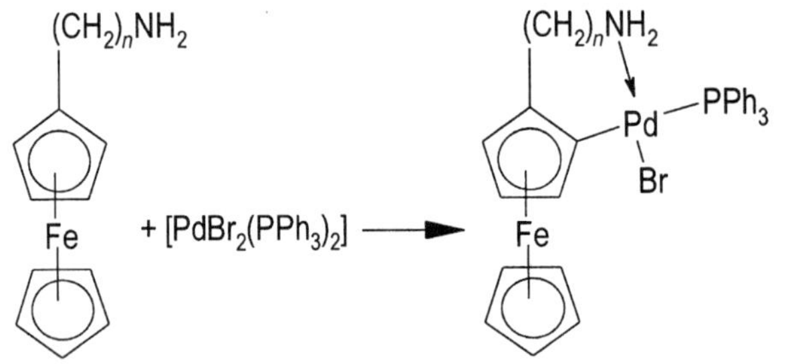

(3.228)

The cyclopalladation reactions of cobalt tetraphenylcyclobutadienepentadienyl-oxazoline are described in Ref. 498.

To conclude this section, we note that an enormous amount of material on the use of the variant of template synthesis, examined in this section, for obtaining heterometallic clusters has been accumulated [499,500]. It is emphasized that the use of clusters as metal ligands is the most suitable approach in order to synthesize such compounds with programmed structure and composition. Especially, many such reactions have been carried out at the expense of sulfur atoms of ligand cluster structures, for instance those reported in Refs. 501 and 502.

EXPERIMENTAL PROCEDURES

Example 1

Synthesis of [Cu(salabza)Gd(hfa)$_3$] and [Cu(salabza)Lu(hfa)$_3$] [503a]

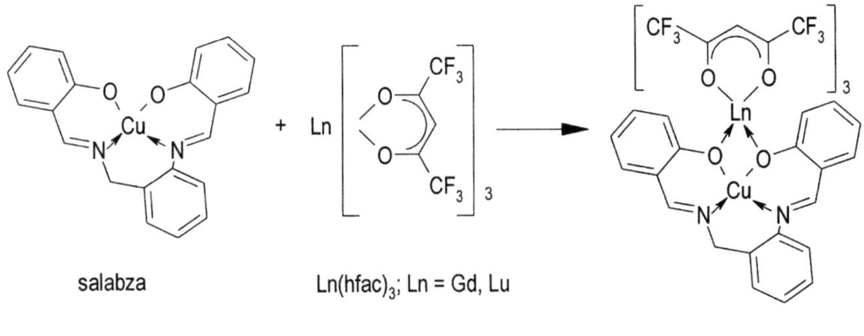

[Cu(salabza)Gd(hfa)$_3$]

The complex [Gd(hfa)$_3$]·$2H_2O$ (0.408 g, 0.5 mmol) in methanol (1 mL) was added to an equivalent amount of [Cu(salabza)] (0.196 g, 0.5 mmol) in chloroform (40 mL). After the mixture was refluxed for 2 hr and filtered, the filtrate was allowed to stand at room temperature for 5 days to give dark green crystals. Yield 0.369 g (63%).

[Cu(salabza)Lu(hfa)$_3$]

The synthetic method was similar to that for [Cu(salabza)Gd(hfa)$_3$], except for the use of [Lu(hfa)$_3 \cdot 2H_2O$] instead of [Gd(hfa)$_3 \cdot 2H_2O$]. Yield 0.327 g (55%).

Example 2

Synthesis of {[η^6-(N-Arylideneaniline)]tricarbonylchromium}tin(IV) Chloride ***756*** (3.211) [457]

A solution of 0.33 mmol of $SnCl_4$ in 2 mL of dry benezene was added to the solution of 0.33 mmol of corresponding azomethine in 5 mL of the same solvent under stirring at 0–5°C. After 5 min reaction the mixture was dissolved in light petroleum, and the precipitate was filtered off, washed with benzene, light petroleum, and dried in vacuo.

Example 3

Synthesis of η^5-2,5-Diphenyl-1-phosphacyclopentadienylmanganesetricarbonyl-P)pentacarbonyltungsten ***769*** ($R^1 = R^4 = Ph$, $R^2 = R^3 = H$, $M = W$) [483]

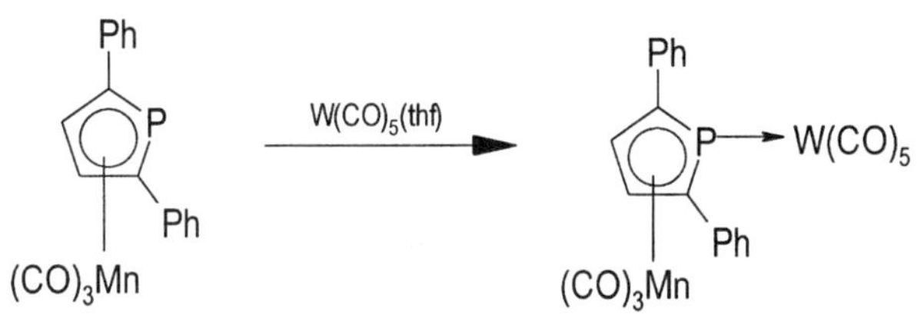

A solution of $W(CO)_6$ (0.176 g, 0.5 mmol) in 30 mL of THF was UV-irradiated for 3–4 hr at 5°C under inert atmosphere. After finishing the irradiation the manganese complex (0.187 g, 5 mmol) was added to the resultant solution, and the mixture was stirred overnight. Solvent was removed, and dark yellow residue was recrystallized from hexane resulting in light yellow crystals. Yield 0.22 g (60%).

Example 4

Synthesis of ***773*** (L = CO) (3.223) [294]

2,4,6-Collidinechromiumtricarbonyl 773 (L = CO)

The solution of 1.4 g (6.4 mmol) of $Cr(CO)_6$ and 3 g (24.8 mmol) of 2,4,6-collidine in dibutylether/dioxane (THF) mixture (20:1) in an Ofele flask was refluxed in an oil bath for 12 hr under argon atmosphere. After that the cooled solution was filtered, and the filtrate was evaporated in vacuum. An oil-like residue was washed with light petroleum until crystallization began. Crystals were filtered off and recrystallized from hexane. Yield 10–12%.

Adduct 774 (L = CO, $MX_n = SnCl_4$)

A solution of 0.33 mmol of $SnCl_4$ in 2 mL of CH_2Cl_2 was added to the solution of 2,4,6-collidinechromiumtricarbonyl (0.33 mmol) in the same solvent under continuous stirring. After some minutes an excess of light petroleum was added, and the high-unstable precipitate was filtered off, washed with light petroleum, and dried in vacuo.

Example 5

Synthesis of ***525*** [503b]

Benzoisotellurazole (0.2 g, 0.87 mmol) and tri-iron dodecacarbonyl (0.44 g, 0.87 mmol) were dissolved in 35 mL of dry toluene under an atmosphere of argon, and the reaction mixture was refluxed for 3 hr. After the reaction mixture was cooled, it was filtered to remove a black residue and the filtrate was concentrated under vacuum, yielding a dark brown solid. The solid was chromatographed on a silica gel column (2 × 12 cm) with hexane/CH_2Cl_2 (2:1) as the eluting solvent, thus revealing a purple band succeeded by a green band. Collection of the two fractions followed by evaporation of the solvent yielded dark purple crystals from the first eluate (15 mg), which were shown to be the cluster compound $Fe_3Te_2(CO)_9$, and a dark green solid from the second eluate, shown to be the trimeric compound $(C_6H_4CHNTe)_2$ $[Fe_3(CO)_7]$. Yield 55 mg (15.3% based on benzoisotellurazole).

Example 6

Synthesis of (salen)Ge=Cr(CO)$_5$ [salen = 2,2′-N,N′-*bis*(salicylidene) ethylenediamine] [439]

Method 1

A mixture of (salen)Ge (0.51 g, 1.50 mmol) and $Cr(CO)_6$ (0.33 g, 1.50 mmol) in THF (50 mL) was UV-irradiated for 2 hr, during which time a red color developed. The solvent was then removed under reduced pressure and the residual solid treated with pentane. The resulting suspension was filtered and the collected solid dried in vacuo to afford the product. Yield 0.64 g (80%).

Method 2

A solution of $Cr(CO)_6$ (0.33 g, 1.50 mmol) in THF (30 mL) was irradiated for 1.5 hr. The solution was then purged of CO by passing a stream of nitrogen through it for 15 min, and then a suspension of (salen)Ge (0.51 g, 1.50 mmol) in THF (20 mL) was added. The product was obtained by the above procedure. Yield 0.65 g (83%).

3.4 "DIRECT" SYNTHESIS OF COORDINATION COMPOUNDS

The above conventional synthetic methods are based on the use of metal salts or carbonyls as complex-formers. At the same time, as far back as at the end of the nineteenth century [504], the possibility of use of compact elemental metals for obtaining complex compounds was shown. This circumstance served as a basis for development of the electrochemical [10,24,201,202,206,505–507], gas-phase [201,202,508–512], and liquid-phase [201,202,513] syntheses of metal complexes using "zero-valent" metals. All these syntheses are united as "direct synthesis of metal complexes" [201,202,513]. Much literature is devoted to this area, generalized in a series of reviews [505–507,510–513] and monographs [10,201,202,206,508]. In this respect, only principal aspects of the direct synthesis and the most recent achievements of its application for obtaining various types of coordination and organometallic compounds will be discussed in this section.

3.4.1 Gas-Phase Synthesis

The gas-phase synthesis, described in detail in the monographs of Klabunde [514,515] and in recent books [201,202], is a variety of the method of direct (immediate) interaction of reagents (Sec. 3.1) with participation of vaporized metals and vaporized or frozen ligands. The syntheses of this type are carried out in various types of vacuum equipment (one of them is presented in Fig. 3.1), described in Refs. 201,202,514, and 516. The preparative possibilities of one-step gas-phase synthesis are limited by comparatively low yields of products in the majority of these reactions, which are frequently accompanied by decomposition of formed coordination compounds and/or transformation of ligands [508].

In this respect, the highest efficiency of gas-phase synthesis is achieved by the cocondensation of vapors of a metal and a ligand at low temperatures (cryosynthesis) [516–521]. The interval of working temperatures usually varies from 10 to 273 K, although sometimes the syntheses are carried out [509,511] at higher (295–325 K) [522] or lower (for example, in liquid helium) [523] temperatures.

Evaporation technology of metals is well developed both for low-boiling ($<1000^\circ$C) and high-boiling ($>2500^\circ$C) metals [201,202,517–519]. Such effective complex-formers as Zn, Cd, and Pb belong to the first group; the second one includes Re, Nb, Mo, W, Os, Ir, etc. The metals which are the most frequently used in cryosynthesis (Co, Fe, Cr, Ni, Pd, La, Ce, Lu) are evaporated at 1400–1700°C.

The ligands in gas-phase syntheses are represented by various inorganic and organic molecules. Thus, a large group of metal carbonyls was isolated according to Scheme (3.229) [517–519]:

$$M^0 + nCO \xrightarrow{t} M(CO)_n \quad (3.229)$$

M = Co, Mn, Cr, Fe, Ni, Pd,
Pt, Rh, Cu, Ag, Ir, Eu, Nd

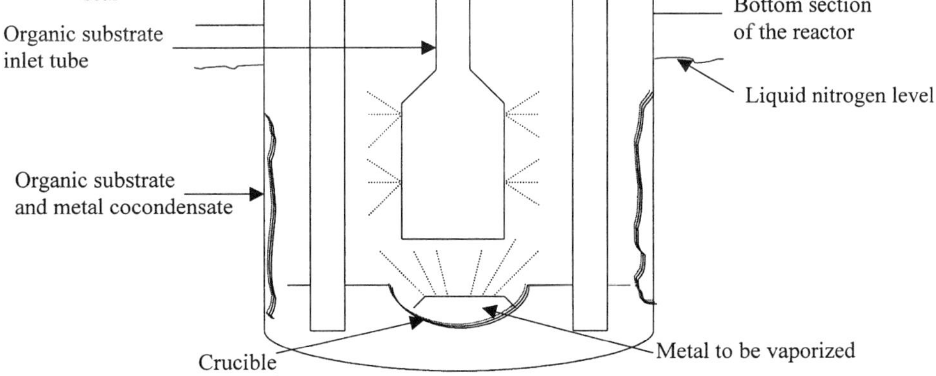

Figure 3.1 Macroscale stationary cocondensation apparatus for investigations of metal atom chemistry. (From Ref. 514, reproduced with permission.)

The yields of these complexes are low (~ 10%); in some cases, their formation has only been registered in argon matrix [517–519,524]. At the same time, a similar reaction with phosphorus trifluoride (3.230) [516,525] can be used for preparative goals due to higher yields obtained: Fe (25%), Co (50%), Cr (65%), Pd (70%), and Ni (~ 100%):

$$M^0 + nPF_3 \longrightarrow M(PF_3)_n \qquad (3.230)$$

$M = Co, Ni, Pd, n = 4$
$M = Fe, n = 5$
$M = Cr, n = 6$

Gas-phase syntheses with use of CS, CS_2, and NO were carried out; the mixed-ligand complexes, containing N_2—O_2, N_2—CO, NO—CO, and NO—PF_3 as ligands, were also reported [510–512]. The cryosyntheses of η^5-cyclopentadienyl derivatives ***3*** of various metals (M = V, Co, Mn, Fe, Cr, Mo, W) are reviewed in detail [510–512]. As is known (Sec. 3.2.2), such complexes can also be obtained by conventional chemical methods of metal exchange [3,4].

In case of η^6-arene complexes, for example dibenzenechromium ***4*** (M = Cr), these compounds are formed by conventional methods as a result of many-step syntheses [11]. In this respect, an enormous number of such compounds have been isolated by cryosynthetic route (3.231) on the basis of aromatic hydrocarbons, containing various substituents in the phenyl cycle [510–512,526,527]:

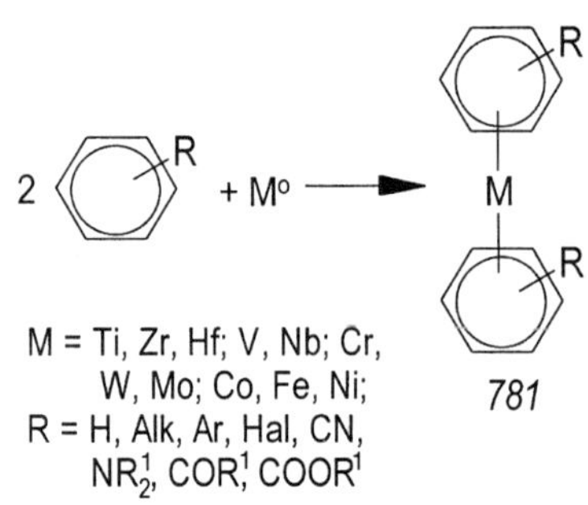

(3.231)

The yields of the obtained compounds vary in a wide range (from 2 to 60%) [201,202,512] and depend on the nature of metals and substituents. For instance, in case of ***781*** (R = Me) the yields are 20% (M = Ti), 30% (V), 40% (Mo), 60% (Cr); for ***781*** (R = 1,3,5-*t*-Bu) 20% (W), 30% (Cr, Mo) [528], 40% (Ti, Zr, Hf) [529]. The yield of the complex with M = Cr and R = *o*-$ClC_6H_4CF_3$ is 33% [530,531]. Metal *bis*-arene complexes of the type ***4*** with naphthalene (43%) [532,533] and antracene [534] were also prepared by gas-phase synthesis.

This method is the only one possible for production (3.232) of nitrogen-containing analogues of ***781***: complexes of η^6-coordinated pyridine and its C-substituted derivatives (***782***) [5,510–512,535–537a]:

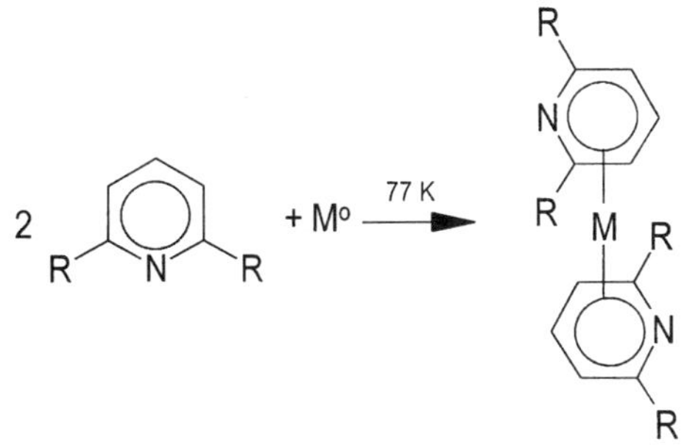

782

(3.232)

The yields of all the complexes are extremely low, for example 2% for ***782*** (R = Me) [535]. The chromium sandwich of nonsubstituted pyridine ***782*** (R = H) was prepared from ***782*** (R = $SiMe_3$) [537a]. Among phosphorus-containing complexes, the phosphaorganometallic sandwich compounds [Co(η^5-P_3C_2(t-Bu)$_2$)(η^4-P_2C_2(t-Bu)$_2$)], [Co(η^5-P_2C_3(t-Bu)$_3$)(η^4-P_2C_2(t-Bu)$_2$)], and the protonated tetraphosphabarrelene complex [Co(η^4-P_4C_4(t-Bu)$_4$H)(η^4-P_2C_2(t-Bu)$_2$)], the latter as its [W(CO)$_5$] adduct, were isolated via reaction of cobalt atoms with the phospha-alkyne t-BuCP [537b]. Cryosynthesis was also used for isolation of η^6-coordinated 2,4,6-tri(t-butyl)-phosphabenzene ***783*** [538]. The yield of ***783*** is 45%, indicating sufficiently high preparative possibilities of the transformation (3.233):

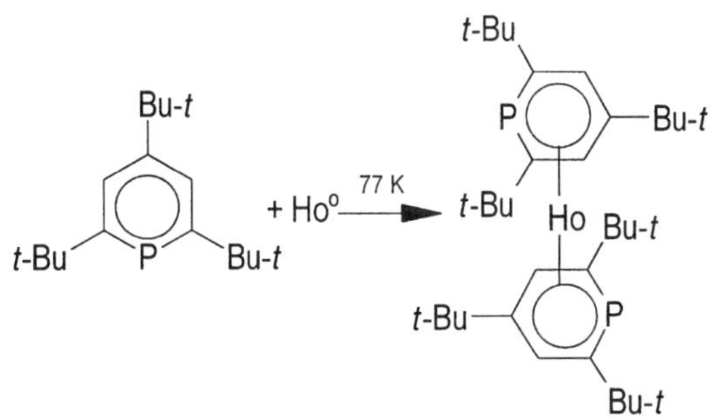

783

(3.233)

Among the chelates, the gas-phase syntheses of β-di- (tri- and tetra-) ketonates of transition and nontransition metals of the type ***2*** have been described in detail [516,517,539,540]. A lot of metals (Al, Sn, Pb, Ti, Zr, Hf, Co, Mn, Cr, Fe, Ni, Pd, Cu, Du, Ho, and Eu) were used. Special equipment for β-diketonate production from elemental metals was reported [202,540], allowing us to obtain these complexes with almost quantitative yields (90–100%) for some metals (Ti, Zr, Hf, Co, Cr, Ni, Cu) [540].

The perspective of use of cryosynthesis procedures for obtaining other metal chelates is of high interest. Thus, synthetic reactions (3.234) with use of nitrogen-containing chelating ligands were carried out [541]:

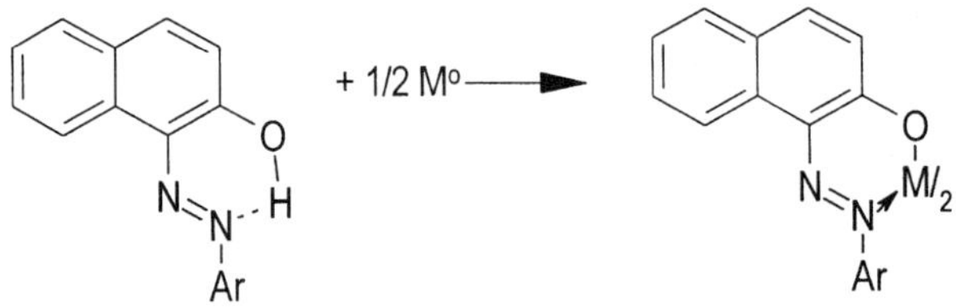

784 785 (3.234)

Nickel and copper were used as a source of complex-former in the syntheses above. Additionally to *o*-hydroxyazo compounds, other N,O-donors were applied (Sec. 2.2.5.3), in particular *o*-hydroxyazomethines, 8-hydroxyquinoline, and 2-hydroxyphenylbenzoxazol, which form ICC ***422***, ***445***, and ***446***, respectively. Yields of the products are from 10 to 60% [541].

Some examples of polynuclear polymeric metal–arene complexes, obtained by cryosynthetic reactions from tin, titanium, vanadium, chromium, and molybdenum, are represented [202,542,543]. However, the preparative possibilities of such syntheses are not yet clear [201,202].

The material presented in this section and in Refs. 201, 202, and 509 to 512 testifies that the gas-phase reactions should be effectively used for obtaining metal complexes, hardly accessible by traditional routes, such as, for instance, ***781–783***. At the same time, synthetic possibilities and advantages of this method in obtaining molecular complexes (Sec. 3.1.1.1) and chelates (Sec. 3.1.1.2) require an additional study. The first complexes $[MX_nL_m]$ can be prepared, in our opinion, in the reaction (3.235) between vapors of metals (M^0), ligands (L), and halogens:

$$mL + M^0 + n/_2X_2 \longrightarrow [MX_nL_m] \qquad (3.235)$$

However, it is necessary to take into account that this reaction can be accompanied by haloidation of the ligand system. Moreover, the adducts themselves can be decomposed to initial components easier in comparison with π-complexes or metal chelates.

Wider use of different types of chelating ligands is especially appropriate in cryosynthesis of ICC. Thus, in gas-phase conditions it is possible to create an antioxidant medium and exclude processes of oxidation of ligands, which prevent isolation of metal chelates from organic chalcogen-containing compounds with free XH groups (X = S, Se, Te), especially those containing N,S- and N,Se-donors (Secs. 2.2.5.4 and 2.2.5.5). Moreover, the examined synthetic technique allows us to obtain anion-free ICC (ICC with anions are frequently formed using metal salts, first of all acetates and halides).

The tridentate ligand systems, for example ***640***, should occupy an important place among chelating ligands. Synthetic reactions (3.74) in gas phase with use of vaporous "zero-valent" metals can open a route for obtaining a large group of binuclear structures, not containing coordinated molecules of solvents (***641***), in which similar liquid-phase transformations are usually carried out.

EXPERIMENTAL PROCEDURES (See More in Ref. 202)

Example 1

Interaction of Metal Atoms and Simple Inorganic Ligands: Reaction of Ni Atoms with PH_3 and PF_3 [544]

An equimolar mixture of PH_3 and PF_3 condensed with nickel at 77 K formed two volatile compounds $Ni(PF_3)_2(PH_3)_2$ and $Ni(PF_3)_2PH_3$, which were separated from $Ni(PF_3)_4$ by low-temperature distillation. Some hydrogen was also evolved when the condensate was warmed from 77 K. The compound $Ni(PF_3)_2(PH_3)_2$ decomposed slowly above 273 K, evolving hydrogen. When allowed to warm to room temperature in the presence of PF_3, it was converted quantitatively to $Ni(PF_3)_2PH_3$.

Example 2

Interaction of Metal Atoms and Alkenes: Reaction Between Co Atoms and C_2H_4 [545]

Cobalt vapor was generated by directly heating a 0.01-in. ribbon filament of the metal and cocondensed with C_2H_4/Ar matrices at 12–15 K. A series of mono- and binuclear cobalt–ethylene complexes $Co(C_2H_4)_l$ (where $l = 1, 2$) and $Co_2(C_2H_4)_m$ (where $m = 1, 2$) as well as a suspected tetranuclear species $Co_4(C_2H_4)_n$ have been detected by spectroscopic methods.

Example 3

Interaction of Metal Atoms and Dienes: Reaction of W Atoms with 1,3-Butadiene [546]

Tungsten atoms were generated by resistive heating of the metal wire. Cocondensation of the atoms with 1,3-butadiene (1:100 molar ratio) at liquid nitrogen temperature produces a yellow matrix. After warm-up the yellow–brown liquid was siphoned from the flask under an inert atmosphere and the excess 1,3-butadiene was pumped off, leaving a residue which could be purified by sublimation (10^{-3} Torr, 323–333 K) or by recrystallization from heptane. Yields are 50–60% based on the amount of metal deposited in the butadiene matrix. A typical 1-hr run yields about 200 mg of pure product [*tris*-(butadiene)tungsten], which is white, is in the hexagonal system, decomposes at 408 K, is soluble in organic solvents, and is stable in air (slight darkening noticeable after 2 weeks).

Example 4

Interaction of Metal Atoms and Arenes: Reaction of V Atoms and Benzotrifluoride [531]

Vanadium was vaporized from a tungsten boat. Simultaneously benzotrifluoride vapor was codeposited with the vanadium vapor on the liquid nitrogen-cooled walls of the glass reactor. Codeposition continued for 10 min (35.4 mg, 0.70 mg-atom of vanadium; 10 mL of benzotrifluoride). After completion of the reaction the cold matrix was red–black. It was allowed to warm slowly to room temperature and then pressurized with argon. With argon flushing, the solution was taken out by syringe and transferred to a sublimer. The excess ligand was slowly pumped off, and then the complex was sublimed to yield 51.2 mg (22%) of *bis*(benzotrifluoride)vanadium(0).

Example 5

Interaction of Metal Atoms and Hetarenes: Reaction of Cr Atoms and 2,6-Dimethylpyridine [536]

Bis(2,6-dimethylpyridine)chromium was prepared by cocondensing chromium atoms with the ligand at 77 K, the latter substance being present in excess. The frozen matrix of ligand-containing metal atoms undergoes the typical color change observed in other syntheses of this type (colorless→dark) as it warms to room temperature. The final reaction mixture is a red–brown solution in excess ligand containing dispersed unreacted metal. Excess ligand was removed in high vacuum, and the dark residue was sublimed to give a red–brown substance, as the sole product, in 2% yield (based upon chromium) which proved to be *bis*(2,6-dimethylpyridine)chromium (m.p. 79–80°C).

Example 6

Cocondensation of Iron with Benzene and $HSiCl_3$ [547]

Iron vapor (1.2 g) was codeposited with benzene (about 90 g) at 77 K. $HSiCl_3$ (about 10 cm^3) was distilled in and the matrix warmed to room temperature over 5–6 hr. The solution was stirred overnight. The reaction mixture was filtered through Celite under argon. The volatiles were removed under reduced pressure, the resultant yellow solid was dissolved in benzene, and the solution was layered with hexanes at room temperature. Yellow prismatic crystals of the (η^6-benzene)$Fe(H)_2(SiCl_3)_2$ were formed on the wall of the Schlenk tube within 2 weeks (1–2% yield based on Fe vaporized).

Example 7

Interaction of Metal Atoms and Alkynes: Synthesis of Cu–Acetylene Complexes [548]

Monoatomic copper vapor was generated by directly heating a tungsten-rod assembly around which copper wire was wrapped. The rate of metal atom deposition (10–12 K) was continuously monitored with use of an in situ quartz-crystal microbalance assembly. $Cu(C_2H_2)$ and $Cu(C_2H_2)_2$ were detected by spectroscopic methods.

Example 8

Interaction of Metal Atoms and Organic Halide Derivatives: Reaction of Ni Atoms and Allyl Bromide (Allyl Chloride) [549]

About 1.5 g of nickel was vaporized at 1823 K over 30 min from a resistively heated alumina-coated molybdenum wire spiral inside an evacuated 200-mm diameter glass vessel which was partly immersed in liquid nitrogen. About 20 g of the allyl halide was simultaneously vaporized into the vessel and condensed with the nickel vapor on the cold walls. During this cocondensation, the pressure in the vessel was below 2×10^{-4} Torr so that few gas-phase intermolecular collisions occurred.

When the nickel had vaporized, the vessel was warmed to room temperature and the excess allyl halide was pumped off. The vessel was then warmed to 343 K and the volatile π-allyl nickel halide pumped out into an adjoining cooled trap. 2.7 g of the π-allyl nickel bromide was isolated, a 60% yield based on nickel vapor. In case of π-allyl nickel chloride, a 75% yield was obtained, but the product was contaminated by 1–2% of a complex mixture of C_{10}–C_{15} hydrocarbons.

Example 9

Interaction of Metal Atoms and Oxygen-Containing Compounds: Reaction of Al Atoms and Ethers [550]

The interaction between aluminum atoms in their ground state and a series of ethers was carried out in an inert hydrocarbon matrix (adamantane) at 77 K in the rotating cryostate. Al atoms were deposited at a rate of ca. 0.06 g/hr by resistively heating the furnace to ca. 800°C. Dimethyl ether, diethyl ether, ethylene oxide, and 1,3,5-trioxane were used as ligands. The following products have been obtained, among others: CH_3AlOCH_3, $C_2H_5AlOC_2H_5$, CH_3OCH_2AlH, etc.

Example 10

Synthesis of Metal β-Diketonates [540]

Atoms of a series of transition and *f*-metals (Cu, Zn, Ni, Cr, Fe, Sn, Pd, Er, Dy, Ho) were cocondensed with acetylacetone previously frozen in the reactor walls at 77 K. The yields were 10–32%.

Using small metal particles of Cr, Fe, Co, Ni, Ti, Zr, Hf (instead of metal atoms) carried by inert gas in a reactor, reported in Ref. 202, it is possible to increase yields of metal β-diketonates to 90–100% [540]. If the metal particles are in the 0.1–0.3 mm size range and their specific surface is between 50 and 100 m^2/g, the time for full reaction is reduced to a fraction of a second. The product output increases sharply (up to 10–60 g of β-diketonate/min) when the β-diketone is introduced to the reactor as an aerosol.

3.4.2 Electrochemical Synthesis

Direct electrochemical synthesis is an accessible and mostly one-step method for obtaining coordination compounds [10,201,202,206,509,513,551–554]. The processes are described by a simple scheme (3.236) [555]:

$$M^+ + L + e^- \longrightarrow M \cdots L \tag{3.236}$$

In the case of use of coordinating solvents, the same reaction can be represented as follows (3.237) [556]:

$$M - me^- \xrightarrow{\text{Solv}} [M(Solv)_n]^{m+} + pL + me^- \longrightarrow [ML_p] \tag{3.237}$$

For the proton-donor ligands, this process can be written (3.238):

$$\begin{aligned} &K{:}\ LH + ne^- \longrightarrow nL^- + n/0.5nH_2 \\ &A{:}\ M - ne^- \longrightarrow M^{n+};\ M^{n+} + nL^- \longrightarrow [ML_n] \end{aligned} \tag{3.238}$$

The electrosyntheses of coordination compounds are carried out mostly using sacrificial anodes of metal complex-formers. According to Ref. 202, the following metals of the Periodic Table have been used: Group I – Cu, Ag, Au; II – Mg, Ca, Zn, Cd, Hg; III – Al, In, Ga, Tl; IV – Sn, Pb, Ti, Zr, Hf; V – Sb, V, Nb, Ta; VI – Cr, Mo, W; VII – Mn, Re; VIII – Co, Ni, Pd; actinides – Th, U. Platinum (sometimes mercury) serves as cathode [202].

The processes are carried out in various solvents with dielectric constant (ε) from 3.5 to 64 [10,206]. However, the most frequently used solvents are acetonitrile ($\varepsilon = 38$) and alcohols ($\varepsilon \sim 30$). To increase the electroconductivity of solvents, the supporting electrolites (LiCl, $LiClO_4$, $NaBF_4$, R_4NX, where R = Alk, X = Br, ClO_4, BF_4) are introduced into a reaction mixture.

Various types of electrochemical cells, for instance the cell produced by Bioanalytical Systems, Inc. (BAS) (Fig. 3.2), have been elaborated for electrosynthetic procedures [202]. At the same time, these processes can be carried out in the usual electrolytic cell, better made from polyethylene with divided anodic and cathodic space [551], in inert atmosphere at room temperature. The duration of such synthetic reactions varies from some minutes to some hours.

For a combination of electrochemical processes and simultaneous ultrasonic treatment of the reaction system, the sonoelectrochemical cell (Fig. 3.3) is used for electrosynthesis and voltammetric studies [557].

The modern state of electrochemical synthetic method [10,201,202,206] allows us to obtain all types of coordination compounds. At the same time, numbers of reported data for different types of complexes are considerably different one from another.

Molecular complexes, obtained using the electrochemical method, are represented by adducts of inorganic and organic N,P,O-donors with metal halides (pseudohalides) and organohalides [202,506,507]. The following ligands are generally used: N-donors (amines, acetonitrile, pyridine, 2,2′-bipy, 1,10-phen, azoles), P-donors (phosphines and diphosphines), O-donors (alcohols, ethers, dioxane, DMF). Among the electrochemically synthesized coordination compounds with N-donor ligands [506,507], the complexes of amines $[PdX_2(NH_2R)]$, pyridine $[PdX_2(py)_2]$, *o-bis*-azines $[PdX_2 \cdot bipy]$ and $[PdX_2 \cdot phen]$ (X = Cl, Br) [558] should be mentioned. These compounds were prepared by electrolysis in 0.2 and 0.5 M HX in the presence of corresponding ligands.

Use of the electrosynthesis allows us to isolate, together with neutral complexes, cationic and anionic compounds [506,551,559]. The most widespread principle of electrochemical synthesis of cationic complexes is based on the anodic dissolution of a metal in an organic solvent (which frequently acts as a ligand), containing a mineral acid [551]. Such syntheses, in particular (3.239) [551,560], were carried out under dissolution of metals (Cd, Zn, In, Ti, V, Mn, Co, Cr, Fe) in CH_3CN or DMSO (L), containing an equal volume of 48% solution of HBF_4:

$$M^0 + mL + nHBF_4(H_2O) \xrightarrow[10\text{–}20\ \text{hr}]{I = 0.1\text{–}0.3\ \text{A}} [ML_m][BF_4]_n \quad m = 4, 6;\ n = 2, 3 \tag{3.239}$$

The transformation above can also take place with use of other ligands (which are not the same as solvents), for example (3.240) [561]:

$$Co^0 + 1,10\text{-phen} + NEt_4ClO_4 \xrightarrow{\text{MeCN}} [CoCO_3(1,10\text{-phen})_2]ClO_4 \tag{3.240}$$

Many examples of electrochemically prepared anionic complexes are known [202,506,507,551]. Among them, the coordination compounds with ammonium cations are of interest [506,507]. In these syntheses, it is possible to vary the composition of complexes of the type $[(R_4N)_mMX_n]$. Thus, similar complexes (R = Me, X = SCN, m = 2, 3, 4; m = 4, 6) were obtained by dissolution of metals (Mn, Fe, Co, Cu, Zn, Cd) in aqueous medium, containing $(Me_4N)SCN$ [561]. Analogous syntheses with use of pyridine, 2,2′-bipy, and 1,10-phen (L) lead to neutral complexes, for instance (3.241) [561]:

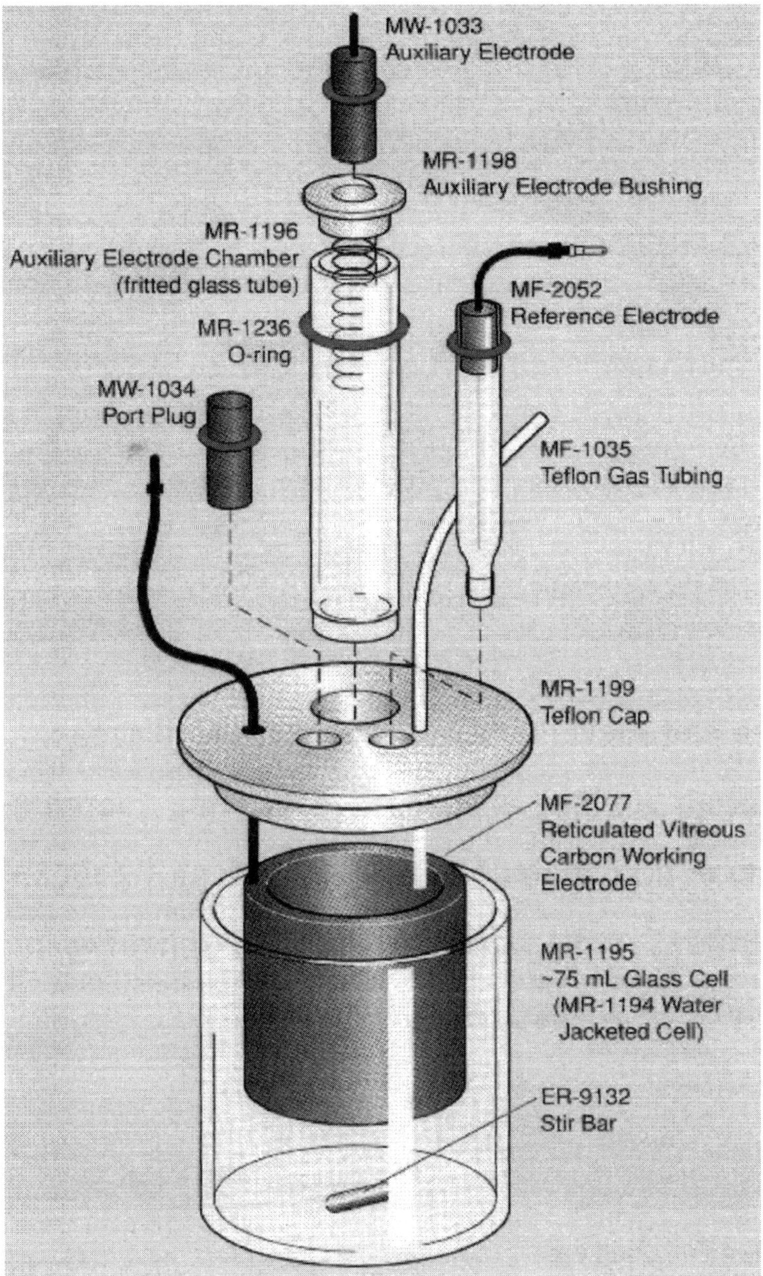

Figure 3.2 The divided electrochemical cell for electrosynthesis. (From *Bioanalytical Systems, A Handbook of Electroanalytical Products*, Bioanalytical Systems, Inc., reproduced with permission.)

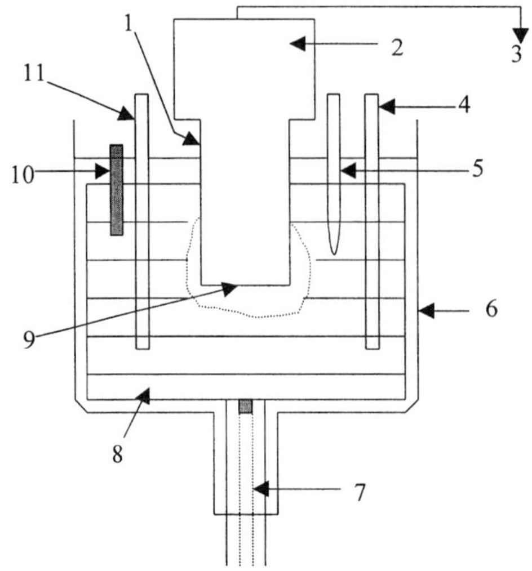

Figure 3.3 Sonoelectrochemical cell used for electrosynthesis and voltammetric studies. 1, Sonic horn; 2, transducer; 3, to control unit of sonic horn; 4, graphite counter electrode; 5, argon inlet for degassing; 6, Pyrex reservoir; 7, platinum-disk macro- or microelectrode; 8, copper cooling coil connected to the thermostatted water bath; 9, titanium tip; 10, platinum resistance thermocouple; 11, SCE reference. (From Ref. 557, reproduced with permission.)

$$M^0 + 2L + 2NH_4NCS \xrightarrow[H_2O/EtOH]{H_2O} [ML_2(NCS)_2] \qquad (3.241)$$

$$M = Co, Mn, Fe, Ni, Zn, Cd$$

Other examples of reported molecular complexes are tabulated in Refs. 202,506, and 507. However, it is necessary to note that the participation in these syntheses of acids, ammonium salts, halogens, and alkyl halides [506,507,551,562] provokes doubts in respect of determining the role of the electrochemical processes, since the reagents above can themselves dissolve elemental metals in various liquid media, forming coordination compounds (Sec. 3.4.3).

The electrochemical technique has been relatively rarely applied to obtain the metal π-complexes [202,506,507]. Thus, the electrosyntheses of complexes of copper with cyclo-octadiene and nickel with cyclo-octatetraene were reported [202,506,507]. The electrochemical reactions (3.242) are of especial interest, since they allow us to produce *bis*-cyclopentadienyl complexes [563–565]:

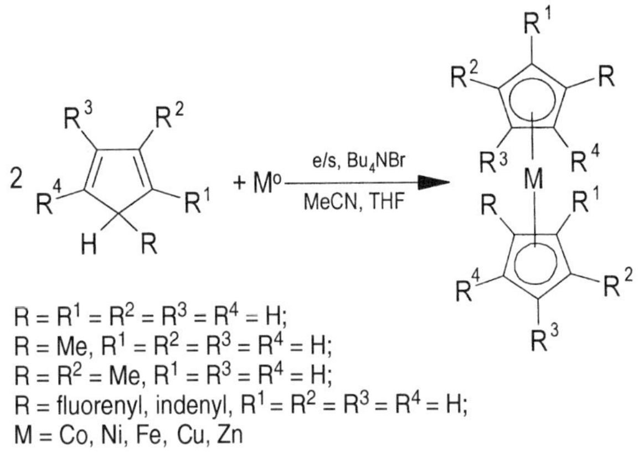

(3.242)

In the presence of KPF_6 in the reaction mixture ($M^0 = Co$) in THF, the cationic complex $[Co(Me_5C_5)_2](PF_6)$ is formed (3.243) [566]:

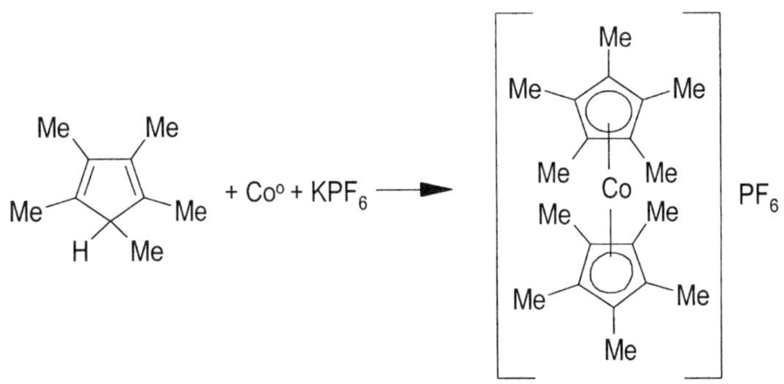

(3.243)

The electrosynthesis has been used most frequently for obtaining metal chelates [202,507,512]. In particular, at the beginning of the last century Russian chemist L.A. Chugaev obtained the first electrochemically prepared chelate – nickel *bis*-dimethylglyoximate ***736*** (R = R′ = Me) [10,206,505,513]. At present, especial interest is paid to the application of the electrosynthetic technique to create complexes with the coordination unit MN_n on the basis of ligands with low-acid NH group. Substitution of the hydrogen atom in this group is usually carried out under action of highly active (mainly alkaline) metals. So, in the traditional conventional syntheses of complex compounds of transition metals, the method of exchange of metals (Sec. 3.2.2) is used. Also, an increase in acidity of the NH group is achieved, for example, by its tosylation [Sec. 3.1.1.2, reactions (3.27), (3.28), and (3.30)]. It was shown in 1973 [567] that the mobiity of the proton of an NH group in reactions of substitution for a metal could be increased by the electrochemical method with use of complexes of azoles (LH) – imidazole and its substituted derivatives ***786***, and benzotriazole ***787***:

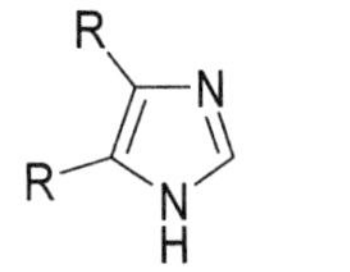

786: R = H, Ph, C_4H_4-*cyclo*

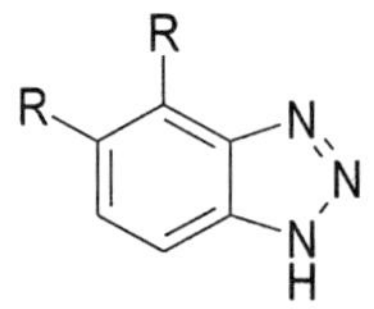

787: R = H, C_4H_4-*cyclo*

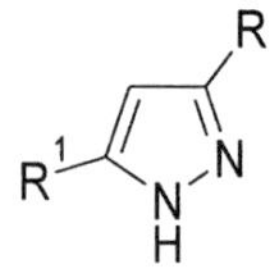

788: R, R¹ = Me

These syntheses were carried out in methanol medium at azole concentration from 0.002 to 0.4 mol/L; the yields were 12–75% (on azoles) and 15–95% (on current) [567,568]. Further, the electrosynthesis was used to produce complexes of methylpyrazoles ***788*** with yields (on ligands) 18–100%, depending on the solvent and metal nature [551,569]. The highest yield was observed in alcohols (especially in methanol: Cd, 75%; Fe, 84%; Ni, 96.7%; Co, 99%; Zn, 100%), the lowest one in acetonitrile (Zn, 18%; Co, 20%; Cd, 30%; Ni, 74%). The chelates with five-member metal-cycles ***789*** and ***790*** [567,568] were isolated in analogous conditions; such compounds cannot be obtained from ligands and metal salts:

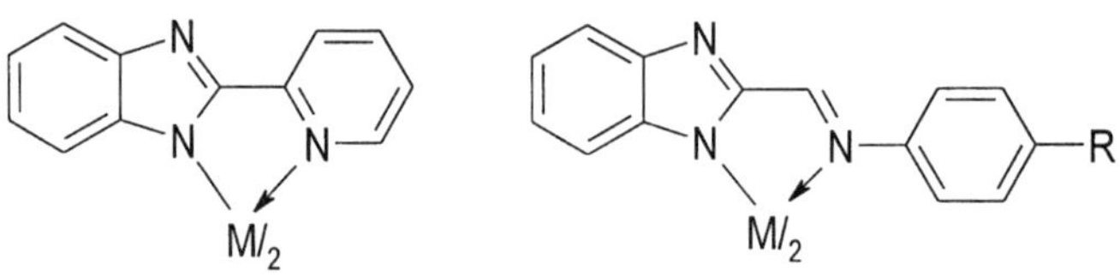

789 *790*

M = Co (82.4%), Ni, Cu (89%)

Similarly, the complexes of 2,2′-bipyridineamine (LH) with the composition ML (M = Cu, Ag) and ML_2 (M = Zn, Cd) were prepared in acetonitrile [570]. Unfortunately, the yields of formed complexes are not reported, so the preparative possibilities of this synthesis are unclear.

The complexes with M—NTs coordination can be prepared on the basis of the ligands having HNTs fragment under immediate interaction of those and metal salts (Sec. 3.1.1.2). Nevertheless, in recent years electrosynthesis has also been used in their synthesis. Thus, the complexes with coordination unit MN_4 were electrosynthesized from 2-ethylsulfamidopyridine (3.244) [571], 2-tosylaminobenzylidene-O-methyloxime (3.245) [572], and 2-tosylamino-(N-pyridyl-2)benzylideneimine (3.246) [573] as ligands:

M = Ni, Cu (3.244)

M = Co, Ni, Cu, Zn (3.245)

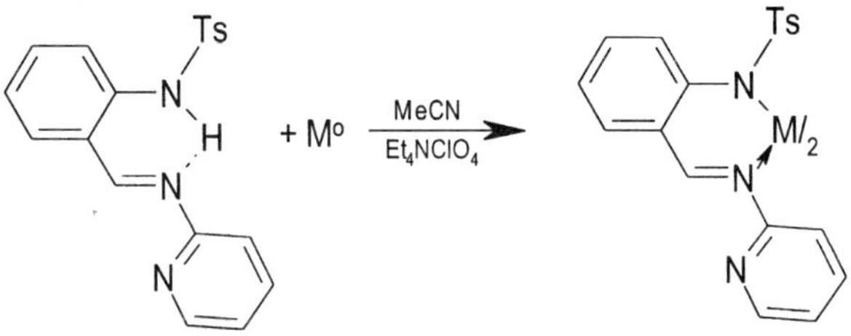

M = Co, Ni, Cu, Zn, Cd (3.246)

An analogous coordination unit exists (3.247) in complexes prepared from 2-tosylaminopyrrolylmethyleneimines [574,575]:

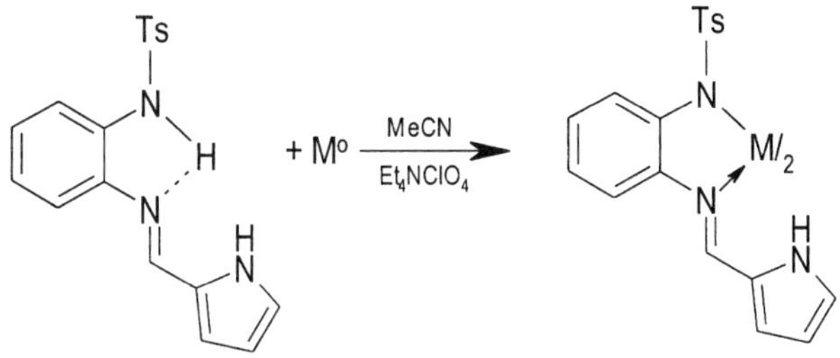

M = Cu, Ni, Zn, Cd *791* (3.247)

The peculiarity of the structure of the complex ***791***, established by x-ray single crystal diffraction [574], is that the pyrrol fragment does not take part in coordination with the metal. This situation is explained, evidently, by the higher acidity of the HNTs fragment as compared to the NH group of pyrrol. To confirm this suggestion, the electrochemical synthesis (3.248) leads to a substitution of the proton of the NH group of the pyrrol fragment to the metal [576]:

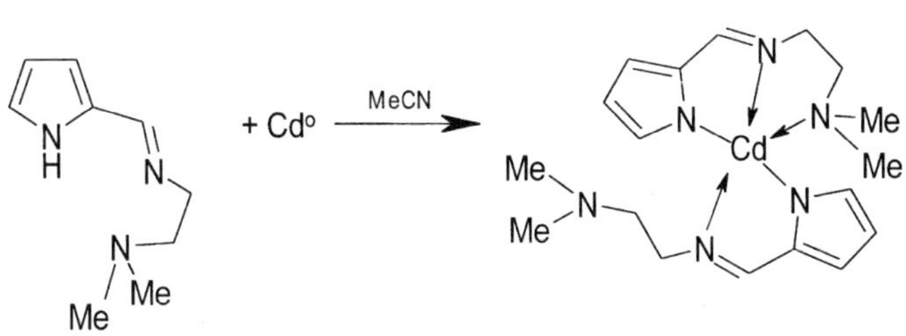

(3.248)

The coordination unit MN_6 exists in adducts of the complexes of tosylated 2-aminopyridine with pyridine, 2,2′-bipy, and 1,10-phen (L), obtained as a result of one-step electrochemical synthesis (3.249) [577]:

$m = 1$ (bipy, o-phen), 2 (py)

(3.249)

Among the O-donor ligands, the β-diketones (especially acetylacetone) are widely used (3.250) in electrosynthesis [202,509,540,551,578–584]:

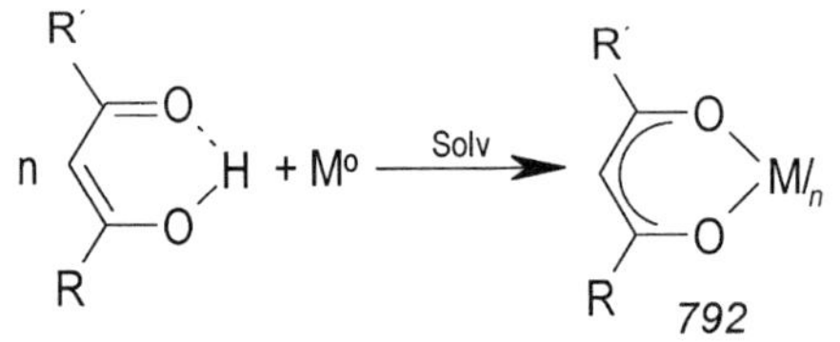

792

$n = 2–4$; M = Cu; Cd, Zn; Al, In; Ti, Zr, Hf; V, Nb, Ta; Cr, Mo; Mn; Co, Ni, Fe; Ce, Dy, U, Th. Solv = MeCN, ROH, THF, Py, diglym

(3.250)

LiCl, LiBr, $LiClO_4$, KCl, NH_4Cl, R_4NClO_4, and R_4NBF_4 were used as supporting electrolytes. The majority of complexes of the examined type were obtained with high yields, for instance acetylacetonates in dependence on the solvent and the metal nature: Ni, 68–93%; Co, 63–89%; Mn, 85% [551]. Also, benzoylacetone ***792*** (R = Me, R′ = Ph) and dibenzoylmethane ***792*** (R = R′ = Ph) [202] are used as similar ligands, as well as 1,1,1-trifluoromethylheptane-2,4-dione [585].

An oxygen-containing environment is realized in electrochemically obtained chelates of *o*-diphenol ***793***, 2-methyl-3-hydroxy-4-pyrone ***794*** [202,586], and salicylaldehyde ***795*** [579]:

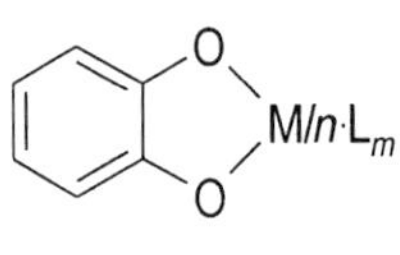

793

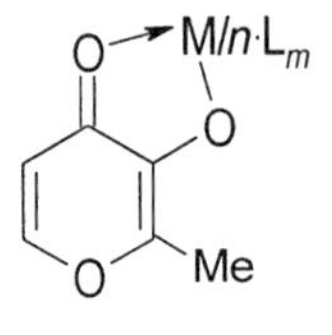

794

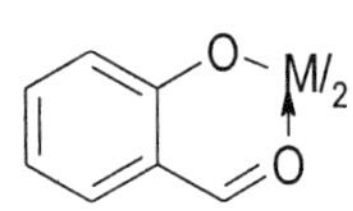

795

Among the sulfur-containing chelating ligands used in the electrosynthesis, it is necessary to mention dialkyldithiocarbamates and dialkyldithiophosphates, on whose basis complexes of the types ***796*** and ***797***, respectively, were prepared [554,587,588]:

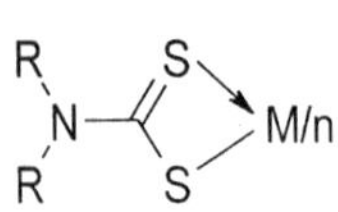

796

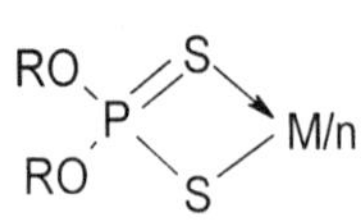

797

Considerable progress has been achieved in the electrochemical synthesis of the ICC of N,O-donor ligands. The first chelates with the coordination unit MN_2O_2 were obtained electrochemically at the beginning of the 1970s. These compounds were the following: complexes of 2-salicylideneimino-1-ethylbenzimidazole ***798*** and 2-hydroxyphenyl-4,5-diphenylimidazole ***799*** [567,568], C-hetaryl-substituted monosugars ***800*** [589,590], and 1-vinyl-2-hydroxymethylbenzimidazole ***801*** [591]:

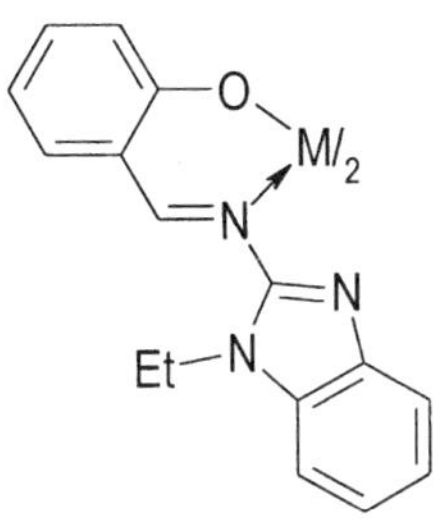

798

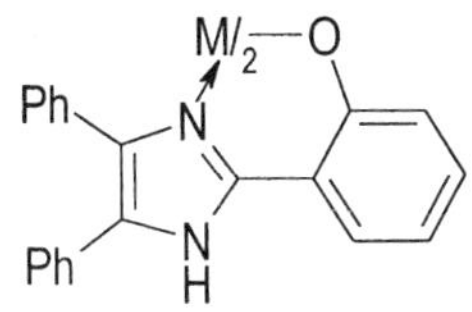

799

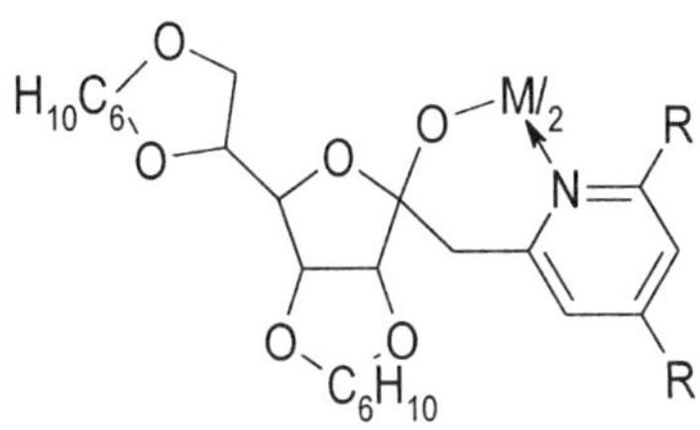

800

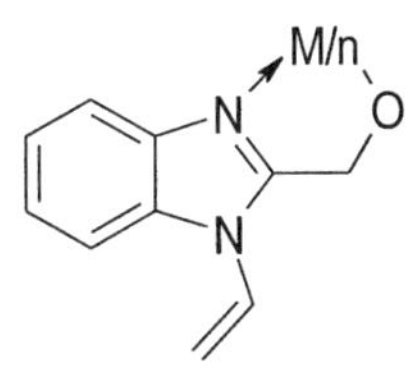

801

The syntheses of complexes of the type ***798*** and ***799*** [567,568] were carried out in methanol. The yields (on ligand) of the first-mentioned chelates are 50% (Co), 84% (Ni), and 83% (Cu), for the second ones 29% (Ni) and 83% (Cu). The yields of the chelates ***800*** are about 90% [589,590]; those of ***801*** were not reported, so there is no possibility of evaluating their preparative value.

In the 1990s, electrosynthesis was used to obtain complexes of 2-pyrrol-N-(*o*-hydroxyphenyl)aldimines ***802*** (3.251) – compounds of the type ***803*** [592,593]:

802 *803*

R = H, Me, Cl;
M = Co, Ni, Cu, Zn, Cd

(3.251)

Despite the fact that the compounds of the type ***802*** are formally tridentate ligands, a substitution of the proton of the NH group to the metal does not take place even under electrosynthesis conditions. This was confirmed by ^{1}H NMR data of zinc and cadmium complexes and x-ray diffraction data of ***803*** (M = Zn, R = 2,4-dimethyl) [593].

In a difference from ***802***, the compounds ***640*** behave under conditions of both electrosynthesis [202] and chemical synthesis (Sec. 3.1.1.2) as tridentate ligands, forming complexes of the types ***804*** and ***805*** (3.252):

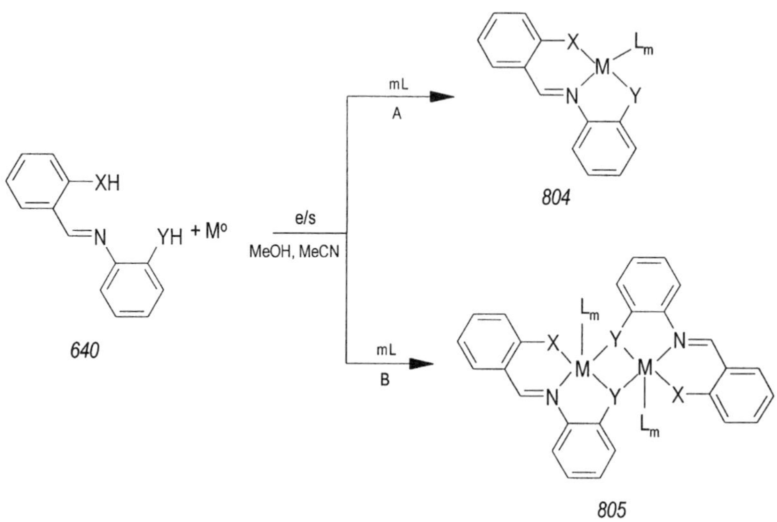

(3.252)

The majority of electrosyntheses have been carried out according to route A, for example, those reported in Refs. 594 to 597. In this respect, the electrochemical syntheses (3.252) with participation of 2-tosylamino-N-salicylideneaniline ***640*** (X = O, Y = NTS), 1-methyl-2-aminobenzimidazole (L), cobalt, copper, and zinc (M^0) were carried out in acetonitrile at room temperature (10 mA, 20 V, 4 hr) in the presence of Et_4NClO_4 [594,595]. The structures of copper [594] and cobalt [595] complexes ***804*** were proved by x-ray diffraction.

The possibility of route B was supposed on the basis of the elemental analysis data of the complexes ***805*** [596] and was strictly proved for the example of the structure ***805*** (M = Cu, X = NTs, Y = O, L = 1,10-phen, $m = 1$) [597]. The causes of successful electrosynthesis of this compound (in comparison with the chelates of the type ***804***, which are usually obtained in analogous synthetic conditions) are still unclear. However, there are reasons to suppose that one of them is the chelate coordination (proved by x-ray diffraction [597]) of two 1,10-phen molecules, stabilizing the molecule ***805***. We note that similar binuclear complexes ***805*** (M = Ni, X = NTs, Y = O, L = MeOH, $m = 2$) with structures proved by x-ray diffraction were synthesized earlier on the basis of the same ligand by conventional chemical methods [596,598].

The electrochemical synthesis was also used for production of bi(oligo)nuclear complexes on the basis of the derivatives of salicyloylhydrazones ***806*** [599] and ***807*** [600]. Other types of electrochemically synthesized di- and polynuclear complexes are described in Refs. 202 and 509.

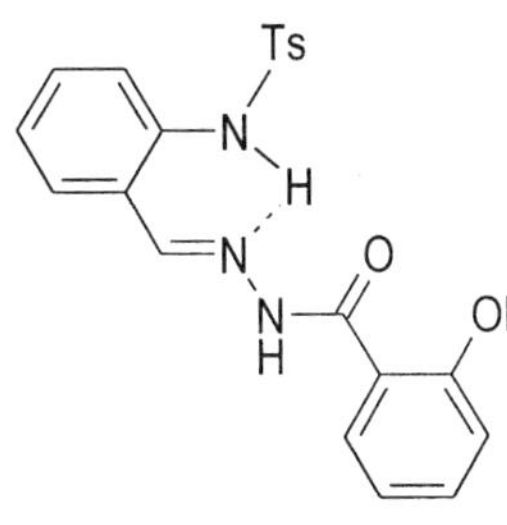

806

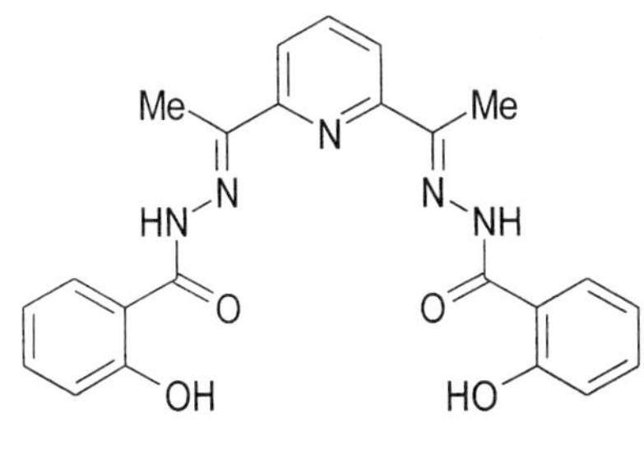

807

On the basis of ligand ***808***, the attractive complexes of the type ***809*** were recently electrosynthesized (3.253) [601]:

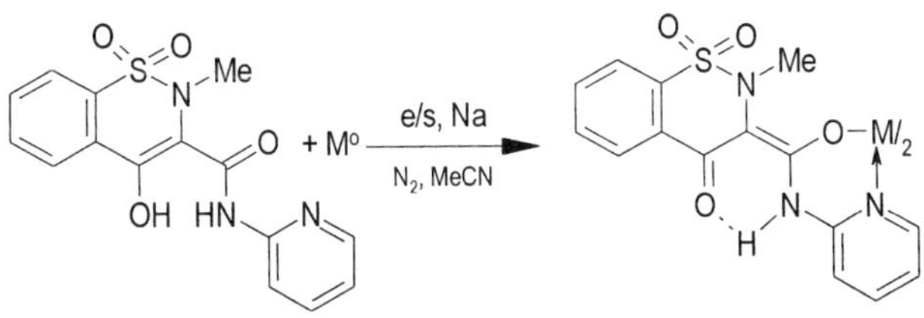

808 809: M = Cu, Zn, Cd

(3.253)

According to Ref. 601, the synthesis was carried out using sodium salts of the ligands ***808*** without supporting electrolyte (which was generated in situ). The structrures of the formed complexes ***809*** seem unproved, since the authors do not take into account the ambidentate [6] character of the ligand ***808*** and the possibility of formation of not only structures with coordination unit MN_2O_2 – ***809***, but also the β-diketone one with MO_4. Precisely such a structure, according to the preliminary data on the basis of IR and PMR study, is characteristic for similar complexes [602].

The electrosynthetic method has played a principal role in obtaining coordination compounds with N,O,S- and N,O,Se-ligand environment ***810*** [202,512,513]. The reaction of electrochemical cleavage of the S—S and Se—Se bonds, discovered by Tuck and coworkers [554a,603,604], was used for this goal. In particular, the complexes ***811*** (X = S [605,606] and Se [607]) were obtained electrochemically (3.254) on the basis of the formally tetradentate ligand ***810***:

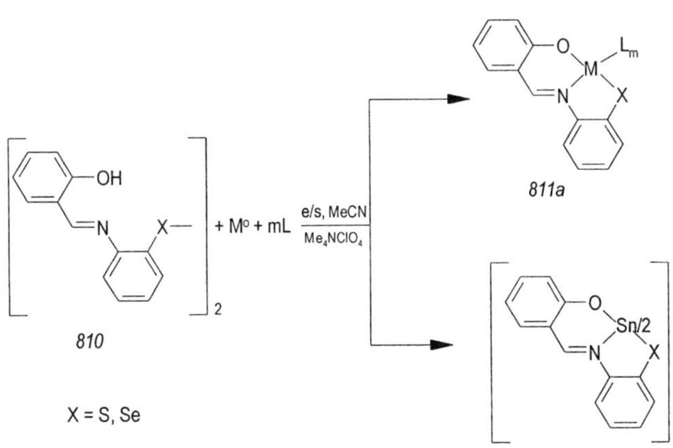

(3.254)

811b

The composition of the complexes above depends on the metal nature and its oxidation state. Thus, divalent copper, zinc, and cadmium form complexes of the type ***811a*** (L = py, bipy, 1,10-phen; $m = 1, 2$) [605], tetravalent tin forms chelates ***811b*** (X = S [606], Se [607]). The electrochemical cleavage of the S—S bond is observed not only in ligands with aromatic bridge ***810***, but also in those with aliphatic one, for example $(CH_2)_n$ in ***812*** [608]:

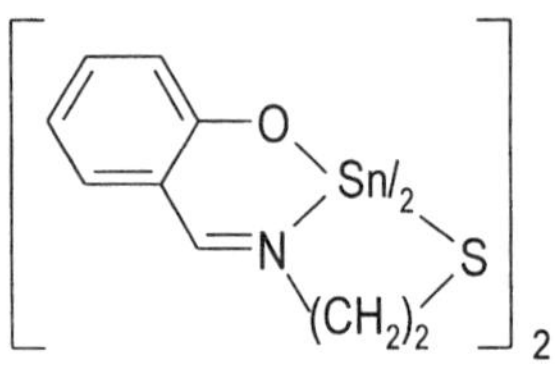

812

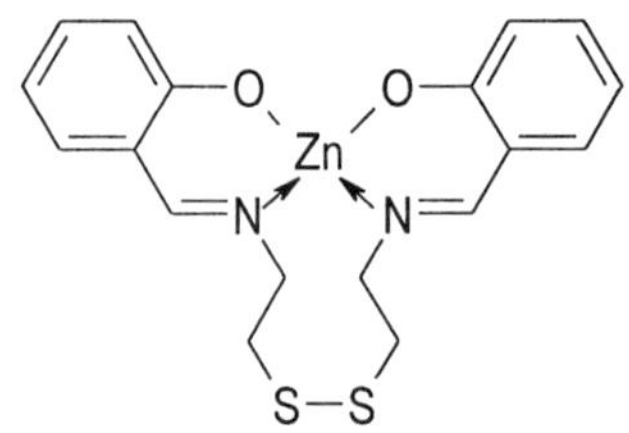

813

At the same time, divalent zinc forms under electrosynthesis conditions the tetrahedral complex ***813*** with nondestroyed S—S bond and coordination unit ZnN_2O_2 [609].

The tellurium-containing complexes, analogous to ***811*** (X = Te), have not yet been isolated, although a series of reported data are known, for example Refs. 265 and 610 to 612, where the destruction of the Te—Te bond is described. Among other electrochemically synthesized complexes, the chelates of arenephosphinothiols ***814*** [613] and thiosalicylaldehyde ***815*** [614] should be mentioned:

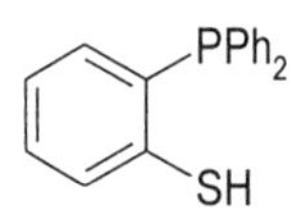

814

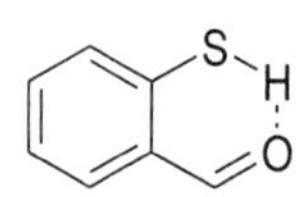

815

The coordination compounds obtained on their basis in acetonitrile using zinc and cadmium (***814***), and copper in the presence of PPh_3 (***815***), contain, respectively, P,S- and P,O,S-ligand environments. Their structure was proved by x-ray diffraction.

The direct template electrosynthesis of some metal chelates has been developed [615,616]. The electrochemical reactions were carried out in mild conditions that allowed isolation of not only the expected final products, but also complexes of initial ligand systems. In particular, the template reaction (3.255) takes place according to the following scheme [615]:

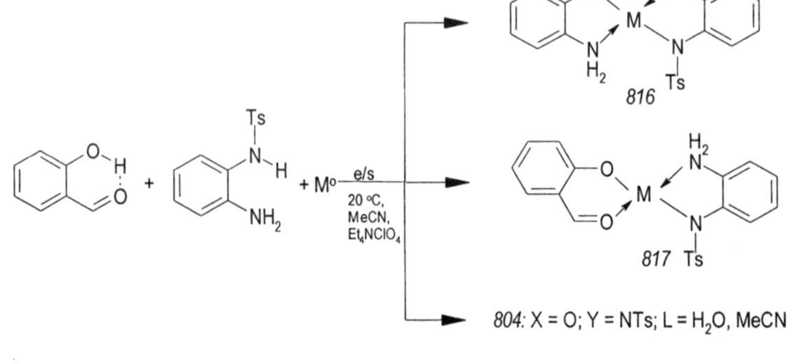

(3.255)

All the complexes were isolated and characterized by elemental analysis data, IR and ^1H NMR spectroscopy (Zn chelate), and magnetochemical data (Ni, Cu chelates). It was established that the type of formed products depends on the metal nature: all three complexes ***804***, ***816***, and ***817*** were isolated for M = Ni, two compounds ***804*** and ***816*** for Cu and Cd, and only one complex ***804*** was prepared in case of M = Zn.

The template synthesis (3.256) led to only one complex ***818*** [616], which was also obtained by a chemical method from the same components and copper acetate under boiling of the reaction mixture:

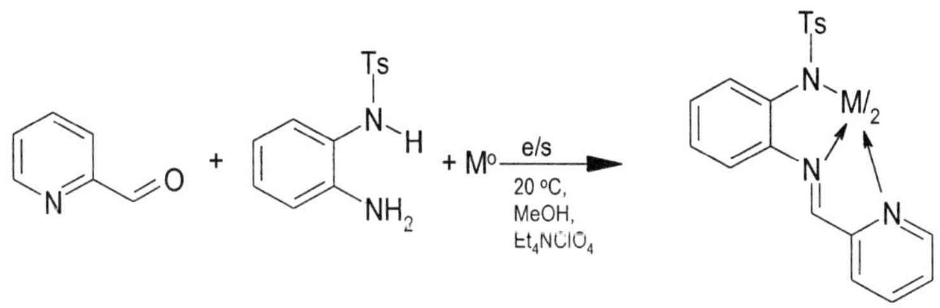

(3.256)

M = Co, Ni, Cu, Zn, Cd

818

Template electrosynthesis was applied to obtain lanthanide complexes (3.257) [617]:

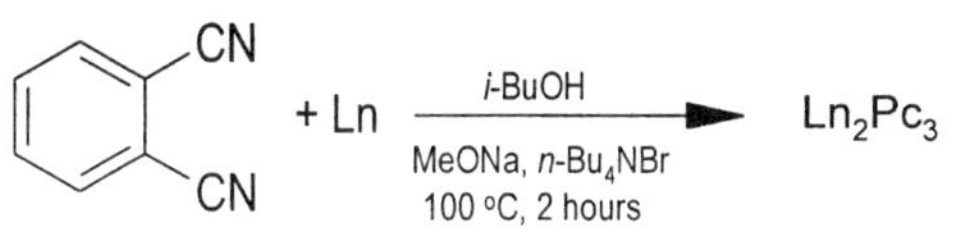

(3.257)

Ln = La, Sm, Nd, Pr

In this respect, we note that also earlier [318,618] the template electrosynthesis was used to produce phthalocyanine complexes: elemental metals or dissolved metal salts were used as a source of metal center (see details in Sec. 5.1).

To conclude this section, we note that a series of other types of electrosynthesis of coordination compounds with participation of metal salts or complexes are known [10,206,551]. They include reaction of synthesis of metal complexes, change of an oxidation number of the central ion, and nature of ligands (mostly coordinated organic compounds). In particular, the electrosynthesis of chromium carbonyls was carried out from $Cr(acac)_3$, $[Cr(py)_3Cl_2]$, and $[Cr(py)_2Cl_3]$ in dry solvents, the highest yield was registered in pyridine (up to 80%) [551,619]. The reactions of electrochemical substitution of CO groups to other ligands (py, PR_3, THF) in metal carbonyls was reported [202,620–622].

Among the electrochemical syntheses related to the change of metal oxidation number, we emphasize obtaining acetylacetonates of divalent iron, cobalt, and nickel [551,623]. The method of alternating-current electrochemical synthesis was applied to isolate π-complexes of monovalent copper with allylamines, allylimines, and allylurea from the salts of divalent copper [624–628]. We note that the same method was used for preparation of analogous π-complexes with copper(II) halides (X = Cl, Br) [629a]. Other electrochemical syntheses with participation of metal salts and complexes are described in monographs [201,202] and literature cited therein.

The data described in the present section allow us to consider electrosynthesis as a modern, rapidly developing [629b–g], and effective preparative technique for obtaining coordination compounds.

EXPERIMENTAL PROCEDURES (See More in Ref. 202)

Electrosynthesis of Adducts and Related Compounds

Example 1

Electrochemical Preparation of Group IVB Halide Complexes $MX_4 \cdot nCH_3CN$ [630]

Each metal was used as an anode of a cell containing a nonaqueous phase with a platinum wire as the auxiliary electrode. In case of Ti/Br complex the experimental conditions were the following: solution composition (volume in milliliters) CH_3CN 50; Br_2 2; $V = 10$ V; $I = 200$ mA; $t = 4.5$ hr; anode loss = 1.34 g. According to the analytical results the adduct formula was $TiBr_4 \cdot 2CH_3CN$.

The electrochemical technique gave the possibility of obtain the *bis*-acetonitrile adduct of the tetrachlorides and bromides, except in those cases where gram quantities of tetraethylammonium halide were present when the appropriate anionic halogen complexes were obtained. Thionyl chloride proved to be the most satisfactory solvent for the preparation of the anionic chlorocomplexes.

Example 2

Electrosynthesis of Manganese Ethylate [631]

Ethylcellosolve (30 mL) was added to an ethanol solution of NaBr (2.3%, 165 mL). The electrolysis (Mn anode and rotating steel cathode) was carried out at 55–60°C and anodic current density 5.1–6.0 A/dm^2 during 6.5 hr (nitrogen atmosphere). Then the solution was put (in N_2 atmosphere) into the closed flask. In 2 days the electrolyte was decanted from the dense solid, absolute ethanol (150 mL) was added. In 2 days this procedure was repeated. The last fraction of washing ethanol practically does not contain NaBr or ethylcellosolve. The ethanol rests were removed by pumping at 25–30°C. The product's weight became constant after 12–13 hr of its stay in vacuum. The formed product (56.2 g, 97.5%) was a gray crumbly powder.

Electrosynthesis of π-Complexes

Example 3

Electrochemical Synthesis of Metallocenes and Polymetallocenes [632]

A stout platinum sheet (80 mm × 4 mm) formed the cathode; metals (Fe, Co, Ni, Cu, Zn, Ag, Au) were cleaned with 5 mol/L HCl. They were finely polished with a fine emery cloth. Some reagents were used as supplied and others were adequately treated. The electrolyte (ca. 200 mg) was added to a 200-mL round-bottomed Schlenk flask into which THF and acetonitrile (typically 3:1 ratio) were distilled under Ar. Monomeric cyclopentadiene (ca. 4 mL) was then condensed into it (similarly for methylcyclopentadiene). The flask was fitted with a tight rubber septum and maintained at $-80^\circ C$. Alternatively 4 mL of Me_5C_5H was added to the solvent system at room temperature.

The electrolytic cell was a 200-mL tall-form beaker with a bottom outlet as described elsewhere. The process was carried out at voltages of 40–50 V at room temperature and a voltage of 50 V produced an initial current of 15–30 mA. Electrolysis over 1–3 hr dissolved approximately 100 mg of metal. In the case of metallocenes, the organometallic was extracted with hexane, the solvent removed, and the complex sublimed in vacuo at 80°C/13 Pa. In most cases solutions eventually turned brown. Complexes precipitated in each electrolytic process were collected, washed with ether, and then dried in vacuo or over P_4O_{10}. Some complexes obtained from different solution compositions are the following:

$Fe(Cp)_2$	(I_{in} = 150 mA, t = 50 min, anode loss = 100 mg, yield 70%)
$Fe(MeCp)_2$	(I_{in} = 15 mA, t = 55 min, anode loss = 12 mg, yield 66%)
$Ni(MeCp)_2$	(I_{in} = 24 mA, t = 50 min, anode loss = 25 mg, yield 90%)
$Ni(Cp)_2$	(I_{in} = 12 mA, t = 135 min, anode loss = 29 mg, yield 50%)

Electrosynthesis of Metal Chelates

Example 4

First Example of Template Electrosynthesis of Metal Chelates (Preparation of *804*, *816*, and *817* by Interaction of Ni, Cu, Zn, Cd, Salicylaldehyde and 2-N-Tosylaminoaniline) (3.255) [615]

An acetonitrile solution, containing 0.122 g (1 mmol) of salicylaldehyde and 0.262 g (1 mmol) of 2-tosylaminoaniline, was electrolyzed at room temperature (I = 10 mA, V = 20 V) with use of Et_4NClO_4 as supporting electrolyte. Copper, nickel, zinc, and cadmium sheets ($2 \times 2\,cm^2$) were used as anodes and platinum wire was used as cathode. In case of nickel, after 1.5 hr of electrolysis the green solid of ***817*** was formed. After retaining the mother liquor for 2 hr, a brown solid ***816*** was formed. The third product (complex ***804***) was isolated after evaporation of the mother liquor to a third of its volume.

For copper and cadmium, two products ***816*** and ***817*** only were formed; the first directly in the electrochemical cell and the second isolated after evaporation of the liquor to 50% of the initial volume. Template electrosynthesis with use of zinc led to the formation of the chelate ***804***, which was precipitated as in the electrochemical cell from partially evaporated and cooled mother liquor.

Example 5

Electrosynthesis of Uranium β-Diketonates [581,584]

The electrolysis of β-diketone solutions in ethanol (0.4–0.6 M) was carried out using a uranium anode and nickel cathode for 3–5 hr at 45–60 mA in an inert atmosphere (Ar, He) as in an oxidative one (O_2, dry air) with agitation. LiCl was used as supporting electrolyte (0.1 mol/L). β-Diketones [acetylacetone (Hacac) and benzoylacetone (Hba)] were distilled before use.

The following products were isolated: (1) $U(acac)_4$ (gray, yield 60%), $U(ba)_4$ (red–brown, 72%) (these solids were formed in inert atmosphere); (2) $UO_2(acac)_2 \cdot Hacac$ (orange, 59%) (the product was isolated after evaporation of ethanol, treatment of formed solid by ether, and its further evaporation); $UO_2(ba)_2 \cdot 2Hba$ (red, yield 30%) [the same isolation procedure as for $UO_2(acac)_2 \cdot$ Hacac]. In the presence of H_2O_2 (30% aqueous solution), added in ratio 1:25 to the initial electrolyte, a 45% yield of $UO_2(ba)_2 \cdot 2$Hba can be obtained after electrolysis and partial evaporation of the solvent [95].

Example 6

Synthesis of 2-Tosylaminobenzylidene-O-methyloxime Complexes (3.245) [572]

Direct electrochemical synthesis is carried out according to the next procedure. Sheets of copper, nickel, or zinc are used as sacrificial anodes, and platinum is used as the cathode. Methanol is used as a solvent and $LiClO_4$ as a supporting electrolyte. The ligand (0.5 g) is dissolved in methanol (30 mL) by heating and then the obtained solution is cooled to room temperature. The electrolysis is carried out for 1 hr (current 20 mA; applied voltage 20–30 V). The formed solid is filtered, washed with hot methanol (3 × 5 mL), and dried in air.

Example 7

Synthesis of ***804*** (3.252) [594,595]

Chelate cobalt complex was synthesized electrochemically by reacting 2-N-tosylamino (2′-hydroxybenzylidene)aniline with 2-amino-1-methylbenzimidazole on a cobalt anode ($2 \times 2\,cm^2$ plate) at room temperature. A working acetonitrile solution (50 mL) contained azomethine (0.250 g), N-base (0.185 g), and Et_4NClO_4 (0.01 g). Electrolysis was carried out in a special-purpose electrolytic cell at a direct current of 10 mA with a voltage of 20 V for 4 hr. The resulting green powder was filtered off and recrystallized from ethanol giving brown crystals.

Example 8

Synthesis of *tris*(Pyridine-2-selenolato)indium(III) [633]

The electrochemical procedure for the synthesis of the complex was similar to those described by Tuck [559]. The cell was a 100-cm^3 tall-form beaker fitted with a rubber bung through which the electrochemical leads enter the cell. The indium anode was suspended from a platinum wire and the cathode was a platinum wire. $[(pySe)_2]$ (0.31 g) was dissolved in acetonitrile and a small amount of tetraethylammonium perchlorate was added to the solution as a current carrier. An applied voltage of 10 V produced a current of 15 mA. During the electrolysis nitrogen gas was bubbled through the solution to provide an inert atmosphere and also to stir the solution phase. After 1 hr of reaction 70 mg of metal had been dissolved from the anode ($E_f = 1.1$).

Example 9

Synthesis of ***818*** (3.256) [616]

Zn, Cd, Co, Ni, and Cu sheets ($2 \times 2\,cm^2$) were used as sacrificial anodes and platinum wire was used as inert cathode. Methanol (50 mL) was used as solvent and Me_4NClO_4 (10 mg) as a supporting electrolyte. The electrolysis was carried out at $I = 30$ mA and initial $V = 20$ V for 1.5 hr at 25°C in an argon stream. 2-Tosylaminoaniline, pyridine-2-carbaldehyde, or 2-(N-2-tosylaminophenyl)aldiminopyridine (1 mmol) were used as ligands.

3.4.3 Other Kinds of Direct Synthesis of Coordination Compounds

Additionally to cryosynthesis and electrosynthesis, coordination compounds can be obtained by other types of direct liquid- and solid-phase syntheses, whose application for synthetic purposes is still limited.

3.4.3.1 Synthesis by Immediate Interaction of Ligands and Metals

This method is sufficiently common for preparation of coordination compounds from ligands with acidic XH groups (especially X = O) [509,513,634]. Although the examples of its successful use were known long ago (see the material generalized in Ref. 513), little attention is paid to this technique at present. In our opinion, this situation is provoked by insuffient activity in these syntheses of main complex-formers, first of all *d*- and *f*-elements. Also, the nature of ligands has an important role, since the following factors should be taken into account: high mobility of the proton of the XH group (acidity) and such a mutual position of the donor centers at which the possibility of formation of metallocyclic structures can be realized (chelate effect).

Precisely the last condition explains the fact that mainly ICC have been obtained by the immediate interaction of ligands and zero-valent metals. Thus, a large series of metal β-diketonates was synthesized in the absence of a solvent [513,634–638], for example, iron *bis*- and *tris*-acetylacetonates [635]. It was shown that other ligands can serve as activators or promoters in these processes. In particular, the introduction of α,α′ or γ,γ′-bipy into the reaction mixture gives the possibility of isolating copper acetylacetonates and adducts of similar complexes of cobalt and nickel [636], meanwhile the β-diketonates of the metals above are not formed under conditions similar to those reported in Ref. 635. Under dissolution of more active metallic barium in the mixture of another β-diketone – dipivaloylmethane (DPM) – with dyglime (DG) or tetraglime (TG) in absolute pentane, the mononuclear complex $[Ba(DPM)_2(TG)]$ and binuclear complex $[Ba_2(DPM)_4(\mu\text{-}H_2O)(DG)]$ were isolated and structurally characterized [637].

Chelate-forming ligands, containing the azomethinic and OH group in ortho position to each other, for instance *o*-hydroxyazomethines ***23***, 2-hydroxyphenylbenzazoles (the compound ***446*** is an example of their complexes), 3-hydroxyphenyl-1,2,4-oxadiazole ***819***, and hydroxyphenyltri- and oxa(thia)diazoles ***820***, react highly effectively with elemental metals in various solvents [513,639]:

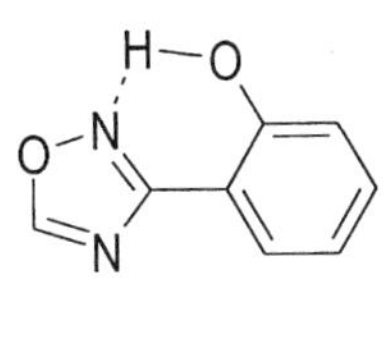

819

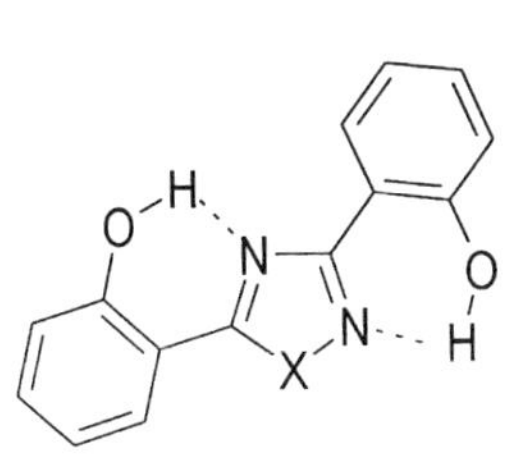

820: X = NR, O, S; R = H, Ph

The complexes of α-pyridinecarbonic acid were prepared with high yields (3.258) by its dissolution in water, containing powdered iron [640]:

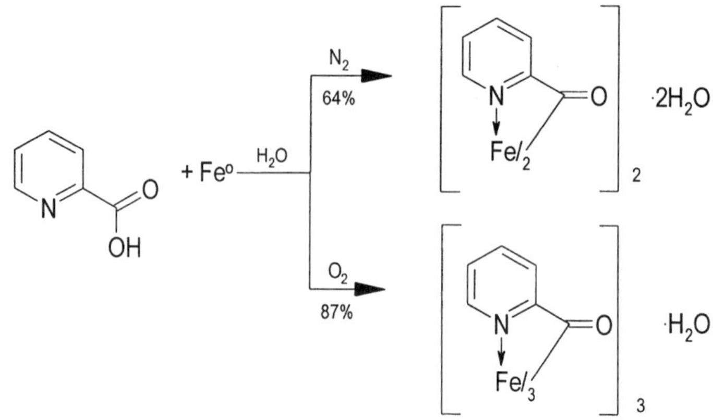

(3.258)

A "nonstandard" reaction (3.259) is of undoubted interest, and the product ***821*** was characterized by x-ray diffraction [641]:

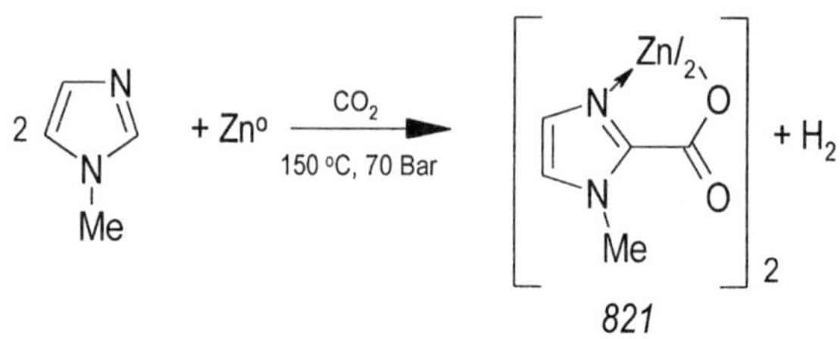

(3.259)

o-Quinones (Sec. 5.2) are highly active ligands for direct synthesis of the discussed type, especially those containing *t*-butyl substituents [509,513]. As a result of these reactions, the complexes of *o*-semiquinones and pyrocatechines having a radical nature are formed. Formation of a series of such complexes in nonaqueous solvents (hexane, toluene, chloroform, THF, and dimethoxyethane) was detected by EPR [509]. In some cases, the products were isolated in crystalline form [642–645]. In particular, a structurally characterized triradical gallium complex of 3,5-di-*t*-butylbenzosemiquinone anion ***822*** was obtained as a result of the transformation (3.260) [645]:

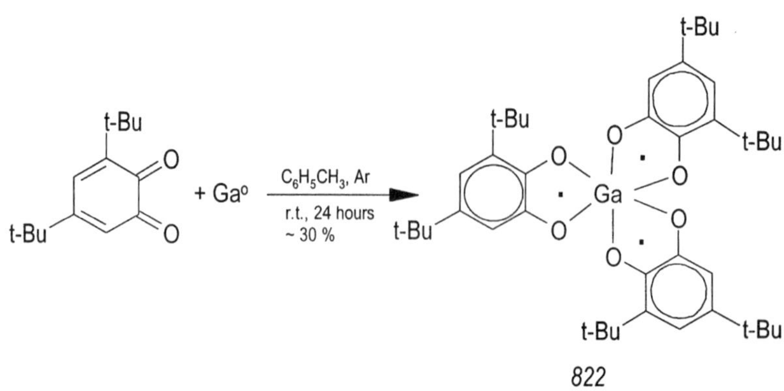

(3.260)

On the basis of the same ligand, in analogous conditions the complexes of divalent metals ML_2 (M = Hg, Zn, Cd) were isolated. The adducts ML_2L' are formed in the presence of α,α'-bipy and 1,10-phen (L′) with yields of 50 and 90%, respectively

[644]. More detailed information on metal-quinone complexes is presented in Sec. 5.2.

Such active metals as aluminum can participate in formation (3.261) of polynuclear structures, for example ***823*** [646]:

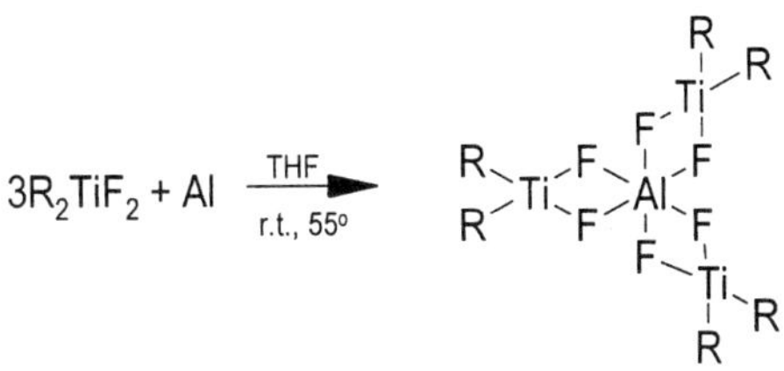

823 (3.261)

Polynuclear complexes are also obtained by boiling of chelate-bridge double bidentate ligands of the type ***820*** and metal powders (Co, Ni, Cu) in DMF [639]. The yields of these processes are almost quantitative.

Template syntheses can also be carried out by immediate interaction of the components of ligand systems and metals. A typical example of such processes is the synthesis of metal phthalocyanines, described in detail in Sec. 5.1.

3.4.3.2 Synthesis from Ligands and Chemically Activated Metals

This method is based on the introduction of oxidants into a liquid-phase reaction medium. Free halogens, HX, and halogen-substituted hydrocarbons are used as oxidants in such transformations [201,202,509,513,647].

Long ago oxidation by halogens was mainly related to reactions of ligands, metals, and molecular iodine or bromine in acetonitrile or DMF [513]. The oldest example of these reactions is the transformation (3.262) [648]:

$$3Fe^0 + 4Hal_2 + 6MeCN \xrightarrow{N_2} Fe^{II}[Fe^{III}Hal_4] \cdot 6MeCN \qquad (3.262)$$

The syntheses with use of CCI_2 were carried out long ago and led to various chelates [649,650a]. The interest in these reactions has not disappeared at present [650b]. Thus, the transformations (3.263) and (3.264) were carried out [651]:

$$Sm^0 + PhN{=}NPh + 1/2I_2 \xrightarrow[50^\circ C,\ 12\ hr]{THF} Sm(PhN{=}NPh) \cdot (THF)_n \qquad (3.263)$$

$$Ln + RS{-}SR + I_2 \xrightarrow{THF} Ln(SR)_n \cdot THF \qquad (3.264)$$

Ln=Sm, Eu, Yb

The reaction (3.265) was carried out with 73% yield of the final product [652]:

$$Me_2NCOH + 2Pd + 3I_2 \xrightarrow{DMF} [(Me_2NH_2)_2]^{2+}[Pd_2I_6(Me_2NCOH)_2]^{2-} \qquad (3.265)$$

Evidently, the coordinated iodine can play an activating role, for example in the process (3.266) [653]:

$$L_2I_4 + Co \xrightarrow{Et_2O} [CoL_2I_2] \qquad (3.266)$$

$L = (PhCH_2)_2PCH_2CH_2P(CH_2Ph)_2$

The syntheses with use of halogen-substituted hydrocarbons allows us to reach sufficient yields [202,509,513]. Among them, CCl_4 is used the most frequently. Among other metals, copper, cobalt, nickel, iron, and zinc have been used. These syntheses are especially attractive due to the possibility of dissolving gold and silver.

Acetonitrile and benzonitrile, aliphatic amines, pyridine, amides (especially DMF), alcohols, and propylene–carbonate are used as solvents in the syntheses of this type. All of them have properties of ligands. Moreover, such classic N-donors as 2,2′-bipy and 1,10-phen were used as ligands, as well as compounds with P-(triphenylphosphine) and O- donor centers (urea, pyridine-N-oxide, triphenylphosphinoxide) and chelating ligand systems (β-diketones, 8-hydroxyquinoline, dimethylglioxime) [202].

Complex compounds obtained by oxidative dissolution of elemental metals in the systems "M^0-ligand–solvent–HX (or halogen-substituted hydrocarbon)," are represented by all types of complexes [509]. However, their ratio is not equal: they are mainly molecular complexes and metal chelates; the π-complexes and di(poly)-nuclear coordination compounds are not numerous. A large number of molecular (neutral and ionic) complexes, chelates, and their adducts, mostly with N-bases, are tabulated in monographs [201,202].

Some examples [(3.267)–(3.269)] of such syntheses are represented below. They are interesting, since they open up the possibility of obtaining complexes of noble low-active metals with participation of HBr (3.267) [654] and $CHBr_3$ (3.268) [655a], (3.269) [655b]:

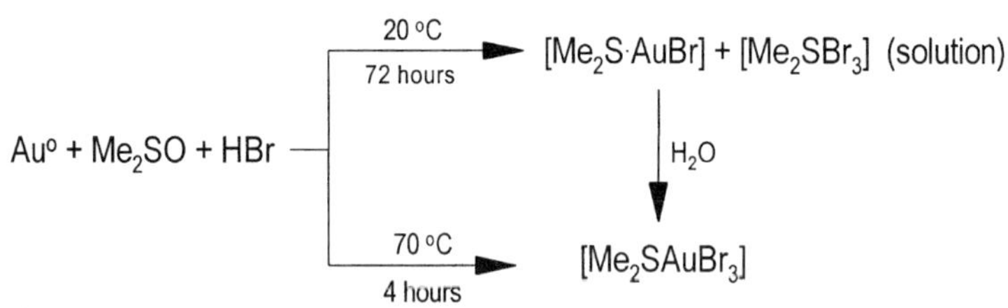

(3.267)

The structure of products was proved by x-ray diffraction and testifies that, as a result of the process, deoxygenation of DMSO takes place (compare with Ref. 417). The interaction (3.268) with use of palladium and pyridine-N-oxide produces $PdPy_2Br_2$ with yield 72.5% [655a]:

$$Pd^0 + PyO + CHBr_3 \xrightarrow[2\ hr]{135°C} PdPy_2Br_2 \qquad (3.268)$$

The synthetic reaction (3.269) takes place with sufficiently high yield (56%) [655b]:

$$Rh^0 + [Ph_3PBu]^+Br^- + CHBr_3 \longrightarrow [PPh_3Bu]_2[RhBr_5(CO)] \qquad (3.269)$$

The structure of the above-described complexes was proved by x-ray diffraction.

Copper, as a more active metal, is oxidized by CCl_4 in milder conditions (in formamide FA or methylformamide MFA at room temperature or 70°C) and forms a series of complexes $[MeNH_3][Cu(HCOO)_3]$, $[Cu(HCOO)_2(FA)_2]$, $[Cu_3Cl(HCOO)_5(FA)_6]$, and $[Cu_2Cl(HCOO)_3]\cdot 2H_2O$ with yields 72, 41, 89, and 23%, respectively [656].

In relation to the discussed syntheses, we emphasize the systematic study of the kinetics of metal dissolution [202,657–659]. Also, the oxidation of Cu—Zn alloy was studied [658], which opens a route to simple preparation of heteronuclear complexes.

The interactions with use of various ligands, metals, and ammonium salts (ammonia method) are well developed [202,509,660–663] and represent a considerable contribution in the dissolution of metals in nonaqueous media. The basis of this simple and accessible method of obtaining complexes from bulk metals is the following reaction:

$$M^0 + mNH_4X + nL + (p/2)O_2 \xrightarrow{\text{Solv}} MX_mL_n + mNH_3 + pH_2O \qquad (3.270)$$

The reactions are usually carried out [201,202] in nonaqueous solvents (Solv) (alcohols, acetonitrile, DMF, and DMSO) which take part mainly as ligands in formed molecular complexes. Additionally to them, the nitrogen-containing bases (pyridine, bipy, 1,10-phen) were used. Co, Ni, Cu, Zn, Cd, Pb have mainly been applied as a source of complex-former. The ammonium salts are represented by halides and pseudohalides, in particular rhodanides, which have been used especially widely [664–675].

The majority of synthetic reactions give high yields (from 70 to 95%) and can be used for preparative goals [202]. Thus, the interaction of copper powder, ammonium thiocyanate in monoethanolamine (LH) at 50–60°C led to the formation of [CuL(HL)SCN] with 81% yield [673]. A similar yield (77%) was reached in the synthesis of $[Sr(SCN)_2 \cdot 4HMPA]_n$ from Sr^0, NH_4SCN, and hexamethylphosphamide (HMPA) [668], as well as in the reaction to obtain the pentanuclear cluster $[Cu^{II}(DMF)_4][Cu^I{}_4(SCN)(CN)_2]$ (yield 80%), which was carried out by heating at 50–60°C a mixture of Cu^0, Me_2NH_2SCN in DMF [675].

Ammonia methods with use of metal oxides are of undoubted interest. The yields of formed complex compounds are 60–90% [202]. For example, the reaction (3.271) of diethyleneamine (L) with a powder of ZnO and ammonium thiocyanate in acetonitrile leads to complex ***824a*** (yield 90%) [676]:

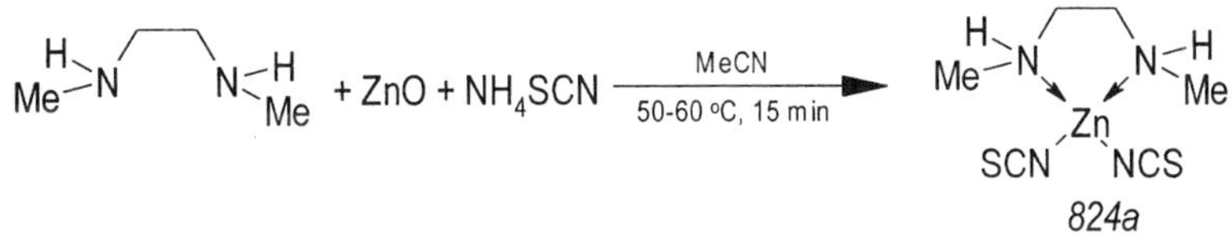

(3.271)

The structure of ***824a*** was proved by x-ray diffraction [676]. The structurally characterized complex ***824b*** was isolated (3.272) with 62% yield [677]:

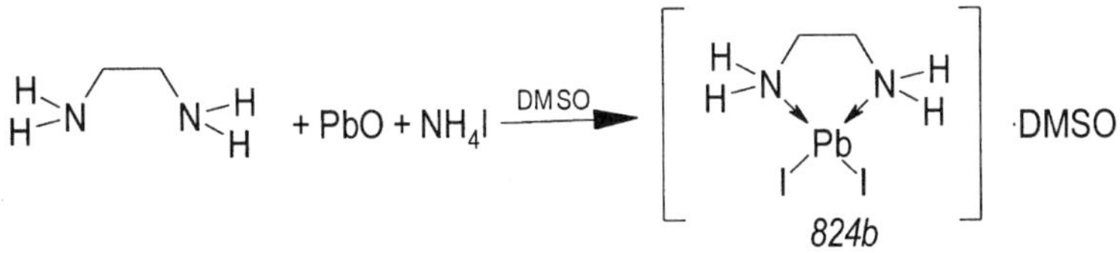

(3.272)

At the same time, a transformation, similar to (3.272), with participation of ethylenediamine (en), PbO, NH_4SCN, and 2-dimethylaminoethanol is accompanied by the formation (82%) of structurally characterized trinuclear complex $Pb_3(en)_6(SCN)_6$ [678].

The ammonia syntheses of heteronuclear complexes with use of elemental copper, lead halides, and aminoalcohols (3.273) [679–681] are significant, since they open up the possibility of preparing complexes with different oxidation states of a complex-former [681b]:

$$2Cu + Zn + 3NH_4I + 3HMe_2EA + 1.5\,O_2 \longrightarrow Cu_2Zn(NH_3)I_3(Me_2EA)_3 + 2NH_3 + 3H_2O \quad HMe_2EA - \text{2-dimethylaminoethanol} \tag{3.273}$$

These variants of ammonia syntheses represent an especial interest due to the possibility of obtaining, on their basis, in some cases, complex compounds with unusual or rare structures [202]. In this respect, it is necessary to mention the synthesis (3.274), as a result of which the anionic complex ***825*** (84% yield) with nontypical coordination (Sec. 2.2.3.5, structure ***113***) for the rhodanide group $Cu_2S{-}CN{-}Cu$ was isolated [674]:

$$Cu^0 + Me_3NHSCN \xrightarrow[70-80°C,\ 4-5\ hr]{MeOH} \underset{825}{[Me_3NH]^+[Cu_2(SCN)_3]^-} \tag{3.274}$$

The complex compound ***825***, whose structure was proved by x-ray diffraction, has a polymeric structure [674].

The reactions to obtain complexes of the type $(MX_nL_m)_p$ ($p = 1\ldots$) by ammonia method with use of NH_4X ($X = SCN$, Hal, NO_3, BF_4, PF_6), M (alkali and transition metals, lanthanides), L (diamines, hexamethylformamide, and other ligands) [682–688] should be noted. Not only M^0 [668] can serve as a source of complex-formers, but also MH or MR, for example (3.275) [688]:

$$NH_4SCN + n\text{-}BuLi + Me_2N(CH_2)_3NMe_2 \xrightarrow{C_6H_5Me} [LiNCS \cdot Me_2N(CH_2)_3NMe_2] \tag{3.275}$$

Other oxidants (HBF_4, AsF_5, $AgNO_3$, $AgClO_4$, $NOClO_4$, $NOBF_4$, Ph_3CBF_4, etc.) [10,206] have also been used in the reactions of direct synthesis of coordination compounds from elemental metals. Such transformations allow us to produce complexes of low-active metals, for example silver (3.276) [689]:

$$Ag^0 + NOX \xrightarrow{MeCN} [Ag(MeCN)_4]X \tag{3.276}$$

Various nontraditional oxidants are applied to oxidize coordination compounds [10,206]. These compounds are the following: salts of nitrosonium, nitronium, aryl-diasonium, triphenylcarbenium, tropilium, amino-oxides, some protic acids (for example, HBF_4, HPF_6, CF_3SO_3H, CF_3COOH), halides of nonmetals (PCl_5, AsF_5, $SOCl_2$, SO_2Cl_2, $PhICl_2$), a series of selective organic solvents (tetracyanoethylene, tetracyanoquinodimethane, 1,2-dioxo-3,4,5,6-chrorobenzene, or tetrachloro-1,2-benzoquinone).

Other important methods of synthesis of coordination compounds are discussed in detail [1,3,10,11,53,201,202,206,207,316,318,322,690]. In this respect, we emphasize the synthesis of metal-polymers [690,691] and preparation of complexes in the solid phase (mechano- or tribosynthesis) [10,201,202,206]. Additionally to the above-described techniques, the general methods and principles of synthesis of coordination compounds are used to obtain metal-polymers (immediate interaction of polymer ligands and metal salts, template electrosynthesis, polymer-analogous transformations). The last method consists of the polymerization of metal-monomers (metal-containing monomers) and fixation of metal complexes on the polymer matrices through the ligand [690,692]. A lot of reported data, including for understanding and specifying achievements in the area of polynuclear metal complexes, is contained in a number of monographs [690–694] and reviews [695–697], and does not require additional examination.

The solid-phase reactions starting from metal powders in conditions of friction (Examples 8 and 9) are described in two recent monographs [201,202] (see the scheme of the mechanochemical reactor in Ref. 202). In this respect, we note the possibility of a solid-state synthesis with the use of metal salts. In particular, mechanochemical syntheses are widely used for obtaining acetylacetonates [698–701d] by the method of ligand exchange, for example (3.277) [701a]:

$$PrCl_3 + 3M(acac) \xrightarrow[M = Li, Na]{} Pr(acac)_3 + 3MCl \quad (3.277)$$

The same technique was used for the preparation of europium and terbium complexes $Ln(NO_3)_3 \cdot 2L$ (L = bipy, 1,10-phen, and diphenylguanidine) [701e]. However, the preparative possibilities of (3.277) are still unclear because, in the majority of experiments, the yields of final products are limited.

EXPERIMENTAL PROCEDURES

Example 1

Synthesis of ***821*** {*bis*[1-Methyl-2-carboxyimidazole]zinc(II)} (3.259) [641]

A suspension of zinc dust (1.0 g, 15.3 mmol) in 1-methylimidazole (10 mL, 126 mmol) was stirred for 48 hr under carbon dioxide atmosphere at 150°C in a 100-mL stainless steel autoclave; a maximum value of 70 bar was noted for the pressure at 150°C. Over the course of the reaction, we obtained a clear, yellowish solution. The zinc reacted completely in this reaction. The 1-methylimidazole was distilled in vacuo (50°C at 0.1 bar) until a white precipitate formed. The white solid was vigorously stirred with diethyl ether. After filtration, the zinc carboxylate was dissolved in THF and layered with *n*-hexane. After a few days at room temperature, colorless prismatic crystals were collected.

Example 2

Synthesis of Bromo(dimethylsulfide)gold(I) [$(Me_2S)AuBr$] (3.267) [654]

Conc. HBr 10 mL, (0.07 mol) and 10 mL (0.14 mol) of DMSO were added to 0.3 g (0.0015 mol) of gold powder. The color of the mixture changed to dark red. The mixture was allowed to stand for 3 days at room temperature, and transparent, colorless, long thin needles of [$(Me_2S)AuBr$] were precipitated. The reaction mixture was heated to 70°C and kept at this temperature for 20 min, till Au was dissolved and the complex precipitated more completely. The precipitate was filtered off, washed with a mixture of ether and alcohol, and dried in an argon stream.

Example 3

Synthesis of ***825*** {[Trirhodanodicopper(I)ato][trimethylammonium]} (3.274) [674]

Cu (0.63 g, 10 mmol), $(CH_3)_3NHSCN$ (1.76 g, 15 mmol), and acetonitrile (20 mL) were placed in a flask in the above order. The mixture was heated to 70–80°C and stirred until complete dissolution of copper (4–5 hr). After cooling of the solution the colorless precipitate obtained was filtered and recrystallized from acetonitrile. Yield 1.25 g (78%).

Example 4

Synthesis of $[Cu(H_2NCH_2CH_2O)(H_2NCH_2CH_2OH)]SCN$ 3.278 [672]

$$H_2N(CH_2)_2OH + Cu + NH_4SCN \longrightarrow [Cu(H_2NCH_2CH_2O)(H_2NCH_2CH_2OH)]SCN \quad (3.278)$$

Copper powder (0.64 g, (0.01 mol), 0.76 g (0.01 mol) of ammonium thiocyanate, and 5 mL of monoethanolamine were placed into the reactor and heated at 50–60°C for 10–15 min under stirring. After cooling, *i*-propanol was added to the solution and the precipitate of the complex was filtered off, washed, and dried. Yield 1.97 g (81%).

Example 5

Synthesis of NaNCS · 2HMPTA [685]

Addition of HMPTA (3.58 g, 20 mmol) to a mixture of NH_4SCN (0.76 g, 10 mmol) and Na (0.23 g, 10 mmol) solids in toluene (12 mL) at room temperature gave an immediate blue coloration. Vigorous gas evolution occurred forthwith, the color changing to blue–green, then (after stirring at 25°C for 1.5 days) pale yellow; at this final stage, no solids remained. The solution was warmed and further toluene (10 mL) added before filtration. Cooling to 25°C and addition of hexane (18 mL) afforded crystals of NaNCS · 2HMPTA.

Example 6

Synthesis of $[CH_3NH_3][Cu(HCOO)_3]$ [656]

Copper powder (0.4 g, 0.006 mol), N-methylformamide (6 mL), and CCl_4 (12 mL) were placed in a flask and the mixture heated to 70°C, refluxed, and stirred until total dissolution of copper was observed (2 hr). The green solution was filtered and allowed to stand at room temperature for 3 days, after which turquoise long needles of $[CH_3NH_3][Cu(HCOO)_3]$ separated. These were filtered out, washed with a mixture of EtOH and Et_2O, and finally dried at room temperature. Yield 1.07 g (72%).

Example 7

Synthesis of $Cu(Ph_3PO)_2Cl_2$ [702]

Copper powder (0.127 g, 2 mmol), triphenylphosphine (1.39 g, 5 mmol), CCl_4 (2 mL), and dry methylethylketone (8 mL) were placed in a flask and the mixture was boiled, refluxed, and stirred for 3 hr. After slow cooling yellow crystals of the product precipitated from the solution. These were filtered out and washed with cold methylethylketone.

Example 8

Synthesis of the First Chiral Complexes Dichloro[(−)-sparteine-N,N′]copper(II) and Dichloro[(−)-sparteine-N,N′]zinc(II) [650b]

Metal powder (0.1 g, 1.5 mmol), (−)-sparteine (0.36 mL, 1.5 mmol), DMSO (3 mL), and CCl_4 (3 mL) were placed in a flask and the mixture heated at 65°C with magnetic stirring until total dissolution of the metal was observed (0.5–2 hr). The solution was filtered and allowed to stand at room temperature for 1–12 hr, after which the crystals were formed. Sparteine complexes were filtered off and dried at room temperature. Yield 94% (Cu complex) and 91% (Zn complex).

Example 9

Mechanosynthesis of Nickel(II) Salicylalanilinate [202]

A 1% ethanol solution (10 cm^3) of salicylalaniline was treated in the reactor for 40 min at a velocity of relative sliding of 0.8 m/sec and a loading of 50 kg/cm^2 using a nickel (sample)–steel (counterbody) friction pair. The green solution formed was filtered and evaporated. Yield 15%.

Example 10

Mechanosynthesis of Copper(II) Salicylal-*p*-chloroanilinate [202]

A 1% solution (10 cm^3) of salicylal-*p*-chloroaniline in EtOH was treated in the reactor similarly to the above method, with a copper (sample)–bronze (counterbody) friction pair. Yield 20%.

Example 11

Solid-phase Synthesis of Copper Trifluoracetylacetonate [$Cu(TFA)_2$] by Mechanical Activation [701d]

The mixture of $CuCl_2$ (0.2618 g, 1.947 mmol) and sodium trifluoroacetylacetonate (NaTFA) in the presence of steel balls (d = 12.3 mm, 150 g) was treated mechanically for 45 min in a steel reactor (volume 80 mL, vibration 12 Hz). Then 0.2520 g of the reaction mixture was used to isolate the product by sublimation (temperature of external heating 150–160°C, pressure ~ 0.1 mm Hg). The obtained product was isolated in the form of blue crystals. Yield 0.1729 g (90%).

Example 12

Solid-phase Synthesis of $Ln(NO_3)_2 \cdot 2D$ (Ln=Eu, Tb); D=bipy, phen, diphenylguanidine [701e]

The lanthanide salt hydrates and D were taken in the ratio Ln:D 1:(2–3). Their mixture and 12 ceramic balls (d = 8–10 mm) were put into a corundum thick-wall reactor (volume 150 mL) and exposed to the vibration (50 Hz, 1.5 kW) for 3–15 min. Then small amounts of solvents (ethanol, water, hexane, 3–10 mL) were added. The formed solid was recrystallized from acetone. Yields 85–92%.

3.5 SYNTHESIS IN THE CONDITIONS OF MICROWAVE AND ULTRASONIC TREATMENT

Microwave (MW) irradiation as a "nonconventional reaction condition" [703] has been applied in various areas of chemistry and technology to produce or destroy diverse materials and chemical compounds, as well as to accelerate chemical processes. The advantages of its use are the following [704]:

1. Rapid heating is frequently achieved.
2. Energy is accumulated within a material without surface limits.

3. Economy of energy applies due to the absence of the necessity to heat the environment.
4. Electromagnetic heating does not produce pollution.
5. There is no contact between the energy source and the material.
6. The suitability of heating and the possibility for automation.

The substances or materials have different capacity to be heated by microwave irradiation, which depends on the substance nature and its temperature. Generally, chemical reactions are accelerated in microwave fields [704], as well as those by ultrasonic (US) treatment, although the nature of these two techniques is completely distinct.

Microwave treatment is widely used to prepare various refractory inorganic compounds and materials (double oxides, nitrides, carbides, semiconductors, glasses, ceramics, etc.) [705], as well as in organic processes [706,707]: pyrolysis, esterification, and condensation reactions. Microwave syntheses of coordination and organometallic compounds, discussed in this chapter, are presented in a relatively small number of papers in the available literature. As is seen, the use of microwaves in coordination chemistry began not long ago and, due to the highly limited number of results, these works can be considered as a careful pioneer experimentation, in order to establish the suitability of this technique for synthetic coordination chemistry.

Contrasting with the microwaves, ultrasound is applied much more in synthetic coordination and organometallic chemistry, and has now become a classic conventional synthetic tool. This approach is used, in particular, for the activation of elemental metals in organic synthesis. A recent monograph [708] and previously published books [709–711] contain a complete description of the possibilities of ultrasonic treatment for obtaining metal complexes, so, in the present monograph, we give only selected applications.

3.5.1 Physical Principles of Microwave Irradiation and Equipment

Microwave heating is a physical process where the energy is transferred to the material through electromagnetic waves. Frequencies of microwaves are higher than 500 MHz. It is known that a nonconductive substance can be heated by an electric field, which polarizes its charges without rapid reversion of the electric field. For some given frequencies, the current component, resulting in the phase with electric field, produces a dissipation of the potency within the dielectric material. Due to this effect, a dielectric can be heated through the redistribution of charges under the influence of external electric fields. The potency dissipated within the material depends on the established electric field within the material. This potency is diminished as the electromagnetic field penetrates the dielectric.

The most common microwave application is that of multimode type, which accepts a broad range of thermal charges with problems of microwave uniformity. The application of multimode type is given in a closed metallic box with dimensions of various wavelengths and which supports a large number of resonance modes in a given range or frequencies.

A resonance cavity or heater consists of a metallic compartment that contains a microwave signal with polarization of the electromagnetic field; it has many reflections in preferential directions. The superposition of the incident and reflected waves gives rise to a combination of stationary waves. If the configuration of the electric field is known precisely, the material to be treated can be put into a position of electric field maximum for an optimal transference of electromagnetic energy.

Typical microwave equipment consists of a magnetron tube (Fig. 3.4) [704]. Just as for other vacuum tubes, the anode has a higher potential with respect to the cathode (source of electrons). So, the electrons are accelerated to the anode in the electric field. The cathode is heated till the high temperature expulses electrons. Generally, the anode is close to earth potential and the cathode has a high negative potential.

The difference between the magnetron and other vacuum tubes is that the electron flow passes along a spiral; this route is created by an external magnetic field **B** (Fig. 3.4). The electron cloud produces resonance cavities several times in its trip to the anode. These cavities work as Helmholtz resonators and produce oscillations of fixed frequency, determined by the cavity dimensions: small cavities produce higher frequencies, large cavities smaller frequencies. The antenna in the right zone collects the oscillations.

A new microwave reactor for batchwise organic synthesis is described in Ref. 712 (Fig. 3.5). Its use permits us to carry out synthetic works or kinetic studies on the 20–100 mL scale, with upper operating limits of 260°C and 10 MPa (100 atm). Microwave-assisted organic reactions can be conducted safely and conveniently, for lengthy periods when required, and in volatile organic solvents. The use of water as a solvent is also explored [712].

A typical reactor used for organic and/or organometallic syntheses is presented in Fig. 3.6 [713], which can be easily implemented using a domestic microwave oven. Due to some problems occurring during microwave treatment, for example, related to the use of volatile liquids (they need an external cooling system via copper ports), original solutions to these problems are frequently found in the reported literature. Thus, properties of solid CO_2 as a substance transparent to microwaves [714], which is not sublimed during 4 min in a microwave oven,

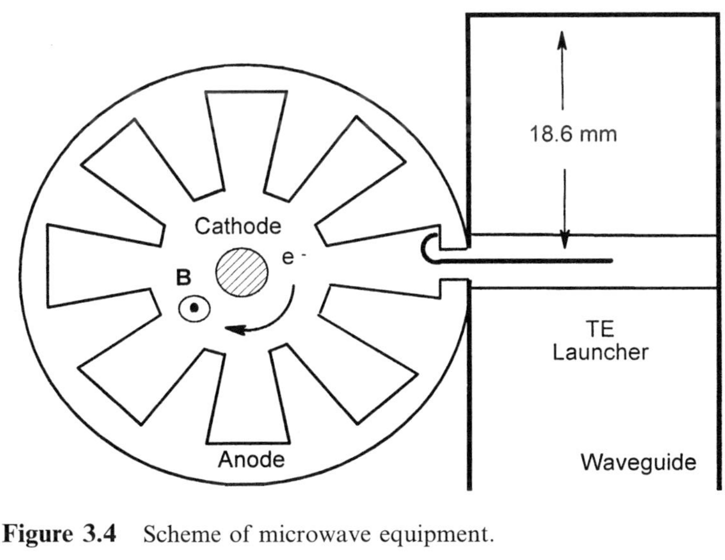

Figure 3.4 Scheme of microwave equipment.

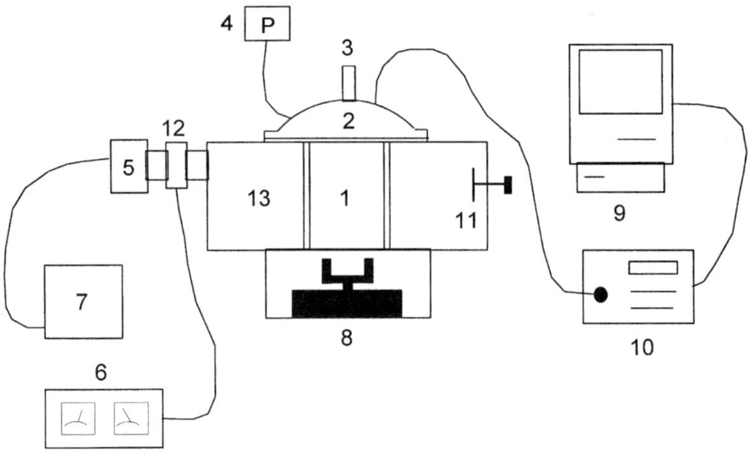

Figure 3.5 Reactor for batchwise organic synthesis. 1. Reaction vessel; 2, top flange; 3, cold finger; 4, pressure meter; 5, magnetron; 6, forward/reverse power meters; 7, magnetron power supply; 8, magnetic stirrer; 9, computer; 10, optic fiber thermometer; 11, load matching device; 12, waveguide; 13, multimodal cavity (applicator). (From Ref. 712, reproduced with permission.)

allowed its use as a coolant for volatile solvents such as benzene. The extremely simple apparatus consisting of a 250-mL beaker (as the reaction vessel) and a 150-mL beaker with a 2-cm flanged lip placed as a cover over the larger beaker, acts as a "cold finger" [714].

A combination of different techniques can frequently improve yields of final compounds or synthetic conditions, for example a reunion of direct electrochemical synthesis and simultaneous ultrasonic treatment of the reaction system [715]. Reunion of microwave and ultrasonic treatment was an aim to construct an original microwave–ultrasound reactor suitable for organic synthesis (pyrolysis and esterification) (Fig. 3.7) [716]. The US system is a cup horn type; the emission of ultrasound waves occurs at the bottom of the reactor. The US probe is not in direct contact with the reactive mixture. It is placed a distance from the electromagnetic field in order to avoid interactions and short circuits. The propagation of the US waves into the reactor occurs by means of decalin introduced into the double jacket. This liquid was chosen by the authors of Ref. 716 because of its low viscosity that induces good propagation of ultrasonic waves and inertia towards microwaves.

These "standard" and "nonstandard" reactors mentioned above have been widely used for promotion of various organic and inorganic reactions and processes [717–722]: dehydration of crystal hydrates [723–726], optimization of catalytic processes [704], activation of elemental metals [720], synthesis of inorganic compounds, materials [719,727a], nanoparticles [727b], etc. From the point of view of the author of Ref. 728, microwave radiation has become a catalyst for chemical reactions. Microwave use for the preparation of some coordination and organo-

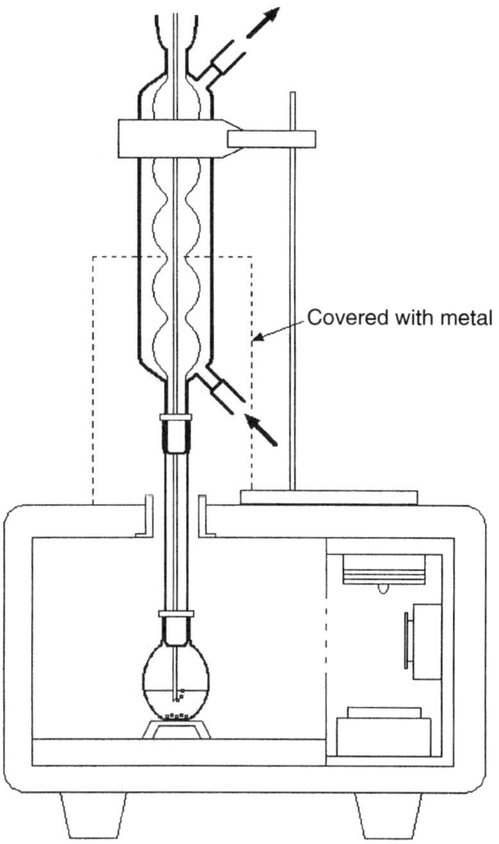

Figure 3.6 Typical MW reactor for organic and/or organometallic synthesis. (From Ref. 713, reproduced with permission.)

metallic compounds has been reported during the last 10 years and described in detail below.

3.5.2 Microwave Synthesis of Metal Complexes

Among other advantages of microwave use in organic–organometallic reactions, the possibility to carry out "*dry*" synthesis (without any solvent used) is a great contribution. Thus, according to the conventional techniques, Claisen–Schmidt template reactions of acetylferrocene and ferrocene carboxaldehyde (3.279), (3.280) are usually performed under classic homogeneous conditions in ethanol [729 and references cited therein]. Using microwave treatment of the reaction system, it became possible to prepare ferrocenylenones ***826*** and ***827*** without solvent in the presence of solid KOH with higher yields in comparison with those reported earlier (see examples below). It is noted that the reactions may be accelerated efficiently by microwave irradiation [729].

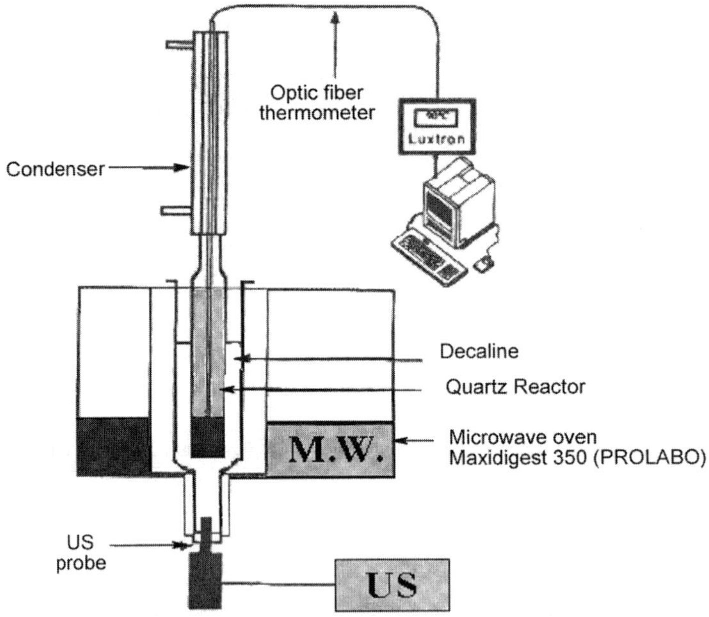

Figure 3.7 Combined MW–US reactor. (From Ref. 716, reproduced with permission.)

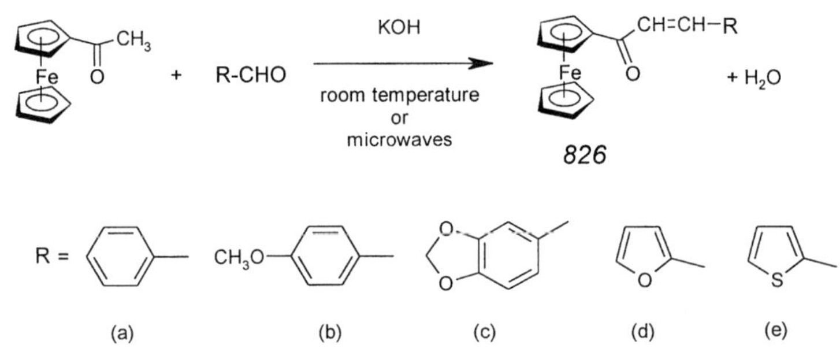

(3.279)

Condensation of acetylferrocene [729]

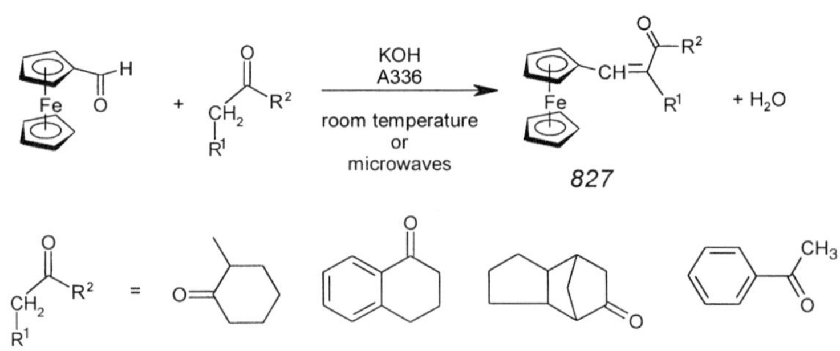

(3.280)

Condensation of ferrocene carboxaldehyde [729]

Similar iron sandwich [Fe(η-C_5H_5)(η-arene)][PF_6] [730,731] and manganese–arene [(η-arene)$(CO)_3$Mn][PF_6] [731] complexes were isolated and characterized. The compounds are synthesized after less than 4 min of microwave irradiation instead of 4–6 hr by conventional methods. It is noted that "only a 2–3 fold excess of the arene is needed, which is important if the arene used is expensive" [730].

It is known that thiophene can react with elemental iron in the form of metal atoms in cryosynthesis conditions [508] or its carbonyls [732], carrying out the desulfurization of the ligand. In reactions with iron carbonyls, the use of microwave heating evidently leads to acceleration of reported reactions of thiophene and its tellurium analogue and its derivatives with $[Fe_3(CO)_{12}]$ [732a]. The following dechalcogenation reactions take place, forming binuclear complexes ***828a,b*** (3.281). Among other organometallic compounds, prepared in this way, it is necessary to mention chromium, molybdenum, and tungsten carbonyls [732b].

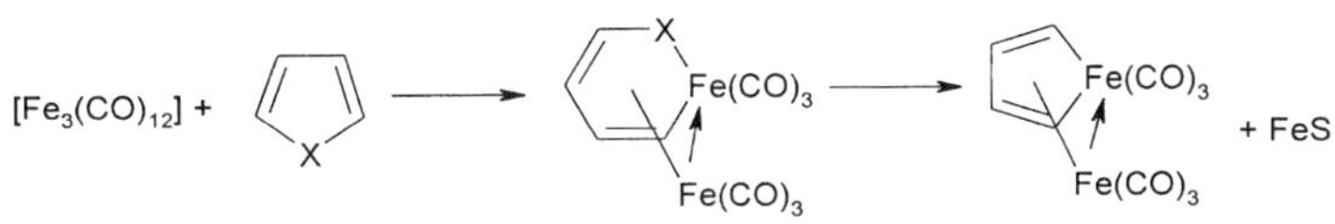

(3.281)

X = S, Te

828a *828b*

Another type of coordination compound, molecular adducts of composition $VOPO_4 \cdot C_nH_{2n+1}OH$ (1-alkanols, $n = 1$–18) and $VOPO_4 \cdot C_nH_{2n}(OH)_2$ (1,ω-alkanediols, $n = 2$–10) were prepared by the direct reaction of various liquid alcohols with solid and finely ground $VOPO_4 \cdot 2H_2O$ in a microwave field. According to x-ray diffraction data, the structures of all these polycrystalline complexes obtained retain the original layers of $(VOPO_4)_\infty$. Alcohol molecules are placed between the host layers in a bimolecular way, being anchored to them by donor–acceptor bonds between the oxygen atom of an OH group and a vanadium atom as well as by hydrogen bonds [733]. Other adducts, $[(n\text{-Bu})_4N][TlMS_4]$ (M = Mo, W), were also prepared under conditions of microwave treatment and their nonlinear optical properties studied [734].

A remarkably rapid synthesis of various ruthenium(II) polypyridine complexes from $RuCl_3 \cdot 3H_2O$, ligand L, and ethylene glycol as a solvent by microwave irradiation was carried out by the authors of Ref. 713 using the reactor presented in Fig. 3.6. After further treatment of the reaction mixture with KPF_6 or $NaClO_4$, the formed products have a general formula $[RuL_3]L'_2$, where L′ is PF_6 or ClO_4.

A large class of coordination compounds, metal chelates, is represented in relation to microwave treatment by a relatively small number of reported data, mainly β-diketonates. Thus, volatile copper(II) acetylacetonate was used for the preparation of copper thin films in $Ar—H_2$ atmosphere at ambient temperature by microwave plasma-enhanced chemical vapor deposition (CVD) [735a]. The formed pure copper films with a resistance of 2–3 μΩ cm were deposited on Si substrates. It is noted that oxygen atoms were never detected in the deposited material since Cu—O intramolecular bonds are totally broken by microwave plasma-assisted decomposition of the copper complex. Another acetylacetonate, $Zr(acac)_4$, was prepared from its hydrate $Zr(acac)_4 \cdot 10H_2O$ by microwave dehydration of the latter [726]. It is shown [704] that microwave treatment is an effective dehydration technique for various compounds and materials. Use of microwave irradiation in the synthesis of some transition metal phthalocyanines is reported in Sec. 5.1.1. Their relatives – porphyrins – were also obtained in this way [735b].

Precursors of an important class of metal chelates, N,N-dialkylthiocarbamates, were used to study rearrangements in them [736]. The reactions were carried out (3.282) in a microwave oven using various supports, in particular graphite, with yields 30–90%.

$$\text{ArO—C(S)—N(CH}_3)_2 \rightarrow \text{ArS—C(O)—N(CH}_3)_2 \qquad (3.282)$$

EXPERIMENTAL PROCEDURES

Example 1

Synthesis of the Adducts $VOPO_4 \cdot C_nH_{2n+1}OH$ (1-alkanols, $n = 1$–18) and $VOPO_4 \cdot C_nH_{2n}(OH)_2$ (1,ω-alkanediols, $n = 2$–10) [733]

Vanadyl phosphate dihydrate $VOPO_4 \cdot 2H_2O$ was obtained by long-term boiling of a V_2O_5 suspension in aqueous phosphoric acid. The layered complexes with alcohols and diols were prepared by suspending microcrystalline $VOPO_4 \cdot 2H_2O$ (ca. 1 g, grain size 0.01–0.08 mm) in dry liquid alcohol, and subsequent short exposure (0.5–5 min) to a microwave field (frequency 2450 ± 30 MHz, total generator output 800 W). The reaction mixture was placed in a 15-cm^3 glass flask equipped with a reflux and put into the waveguide of a microwave generator with stirring and heating. After cooling, the solid product formed was filtered off. When the starting material was a solid alcohol, the suspension was separated from the melt by hot filtration.

Example 2

Reactions of Tri-iron Dodecacarbonyl with Thiophene (3.281) [732a]

The reaction was carried out by sealing $[Fe_3(CO)_{12}]$ (1.0 g), thiophene (15 cm^3), and Fe_3O_4 (0.5 g) in a Teflon container and heating the contents in a Sharp Carousel microwave oven for 50 min on a high power setting. Work-up afforded the products $C_4H_4Fe_2(CO)_6$ (5%) and $C_4H_4Fe(CO)_2Fe_2(CO)_6$ (trace). A similar microwave experiment using benzothiophene gave a 49% yield of benzothiaferrole after 50 min, identical to that obtained in Ref. 737 after 18 hr; however, microwave heating was not successful in promoting a desulfurization reaction.

Example 3

Synthesis of $[(\eta\text{-Arene})(CO)_3Mn][PF_6]$ (3.283) [731]

$$\text{Arene} + \text{BrMn(CO)}_5 \rightarrow [(\eta\text{-arene})(\text{CO})_3\text{Mn}][\text{PF}_6] \qquad (3.283)$$

Bromopentacarbonyl manganese (0.50 g, 1.82 mmol), mesytilene (0.5 g, 4.16 mmol), and Al powder (0.5 g, 18.5 mmol) were mixed with 1,2,4-trichlorobenzene (TCB, 5 g) in a small reaction beaker. Finely ground $AlCl_3$ (0.5 g, 3.76 mmol) was added and the whole thoroughly mixed. The cold finger beaker with solid CO_2 was set in place and the whole microwaved for 3 min on a medium setting. A small beaker of water was placed in the oven prior to use to absorb excess microwave radiation. On cooling, the mixture was carefully treated with ice water (30 mL) and filtered to give a yellow aqueous phase. This phase was extracted with toluene, separated, and 60% HPF_6 (0.5 g, 2.00 mmol) added. The flocculent yellow precipitate was filtered off, washed with distilled water, and air-dried to give the product (0.33 g) in 45% yield.

Example 4

Synthesis of Ferrocenylenones: (a) 1-Ferrocenyl-3-piperonylprop-2-ene-1-one and (b) 2-Ferrocenylmethylene-6-methylcyclohexanone [729]

A corresponding aldehyde [1.1 mmol, piperonal (a), ferrocene carboxaldehyde (b)] and ketone [1.0 mmol, acetylferrocene (a), 2-methylcyclohexanone (b)] were stirred in an Ehrlenmeyer flask, 2 mmol of powdered KOH added, and the mixture stirred after addition of one drop of Aliquat 336 [see (3.279) and (3.280)]. The Ehrlenmeyer flask was irradiated with microwaves [(a) 3 min, 140 W; (b) 2 min, 140 W] and the final product dissolved in dichloromethane. After filtration on Celite, the solvent was removed and the residue chromatographed on a preparative chromatographic layer of silica (AcOEt/cyclohexane 20/80). The condensation product was further purified by crystallization in ether. Yield: (a) 73% (lit. 37%), red solid; (b) 63%, red solid.

Example 5

Synthesis of $[Fe(\eta\text{-}C_5H_5)(\eta\text{-arene})][PF_6]$ (3.284) [730]

$$\text{Arene} + Fe(C_5H_5)_2 \xrightarrow[\text{(i) } AlCl_3/Al;\ \text{(ii) } HPF_6]{} [Fe(\eta\text{-}C_5H_5)(\eta\text{-arene})][PF_6] \tag{3.284}$$

Ferrocene (0.01 mol), Al powder (0.01 mol), and $AlCl_3$ (0.2 mol) were ground together in a mortar. The arene (benzene, toluene, mesitylene, xylenes, etc.) was added and the mixture reground. The intimate mixture was then rapidly transferred to the microwave apparatus (Sharp Easy Chef R5A53, 850 W) and heated for 3–4 min with a beaker of water (60 mL) placed alongside the reaction vessel to absorb excess microwave radiation. The complexes were isolated by the conventional work-up of adding water and precipitating with HPF_6 from the aqueous solution.

Example 6

Synthesis of $[Ru(bipy)_3](ClO_4)_2 \cdot 3H_2O$ [713]

A domestic microwave oven (Fig. 3.6) was used. The starting material, $RuCl_3 \cdot 3H_2O$, was prepared from RuO_2. Ethylene glycol was used as solvent. The mixture of $RuCl_3 \cdot 3H_2O$ (0.2 g, 0.76 mmol) and 2,2′-bipy (0.36 g, 2.13 mmol) in ethylene glycol (20 mL) was refluxed for 20 min under microwave irradiation. The green solution turned to orange red. The solution was cooled to room temperature and filtered. The filtrate was poured into a saturated aqueous solution of $NaClO_4$ to obtain a red–orange precipitate, $[Ru(bipy)_3](ClO_4)_2 \cdot 3H_2O$. Then an acetonitrile solution of the precipitate was poured into diethyl ether for recrystallization. Yield 86%.

Example 7

Synthesis of $[(n\text{-}Bu)_4N][TlMoS_4]$ (M = Mo) [734]

A well-ground mixture of $[NH_4]_2[MoS_4]$ (0.25 g, 1.0 mmol), TlBr (0.28 g, 1.0 mmol), and $[(n\text{-}Bu)_4N]Br$ (0.96 g, 3.0 mmol) was placed in an open Pyrex glass tube which was stored in a large beaker in a domestic microwave oven equipped with a power control [National IEC-705: Hi (700 W), MedHi (500 W), Med (300 W), MedLo (100 W), Lo (50 W)]. The irradiation power and time were set in the following order: MedHi, 15 sec; MedLo, 10 sec; Lo, 10 sec; MedHi, 5 sec; Lo, 10 sec. During the microwave irradiation, an argon stream was introduced into the reaction tube for safe removal of the NH_3 and H_2S gases evolved. [When the high-power irradiation (Hi) was applied, the reaction mixture melted immediately and began to bubble violently within 20 sec.] After irradiation, the resulting dark red mixture was allowed to cool to room temperature and then extracted with DMF (3 mL) and filtered. Dark red needles of the product formed from the filtrate on standing at ambient temperature. Yield 0.17 g (25.33%).

3.5.3 Physical Principles of Ultrasonic Treatment and Equipment

Ultrasound is a part of the sound spectra having frequency of ~ 16 kHz, which is out of the normal range of human hearing. The effects produced by ultrasound are derived from the creation, expansion, and destruction of small bubbles, which appear under US-irradiation of a liquid phase. After creation, these bubbles grow to 2–10 times the initial diameter of a few microns and undergo radial vibration. The circulating liquid cools these unstable cavities, so, after a few cycles of rarefaction and compression, they collapse violently in 10^{-5}–10^{-7} sec [708]. This phenomenon, named "*cavitation*", produces high temperatures and pressures in the liquid. The temperature of cavitation varies from 1000 to 10,000 K, more frequently in the range 4500–5500 K.

It should be noted that acoustic irradiation is a mechanical energy (no quantum), which is transformed to thermal energy. Contrary to photochemical processes, this energy is not absorbed by molecules. Due to the extensive range of cavitation frequencies, many reactions are not well reproducible. Therefore, each publication related to the use of US generally contains a detailed description of equipment (dimensions, frequency used, intensity of US, etc.) [709]. Sonochemical reactions are usually marked)))), in accordance with internationally accepted usage [708].

For successful application of US, the influence of various factors can be summarized as follows [710]:

1. *Frequency*. Increase of frequency leads to the decrease of production and intensity of cavitation in liquids. This fact can be explained as follows: at high frequencies, the time necessary for a bubble to appear as a result of cavitation and to have sufficient size to affect the liquid phase is too low.
2. *Solvent*. Cavitation produces considerably minor effects in viscous liquids or those with higher surface pressures.
3. *Temperature*. Increase of temperature allows the cavitation to be performed at lower acoustic intensities. This is a consequence of increasing vapor pressure of the solvent with increasing temperature.
4. *Application of gases*. In the case of application of gases (poor or very soluble), the intensity of cavitation decreases due to the formation of a large number of additional nuclei in the system.
5. *External pressure*. Increase of external pressure leads to the increase of intensity of destruction of cavitation bubbles, that is, the effects of US in this case are more rapid and violent in comparison with normal pressure.
6. *Intensity*. In general, the increase in intensity of US intensifies the produced effects.

With respect to metal surfaces, the US action can be described as follows [710]:

1. The acoustic *flow* is a movement of the liquid, induced by the sound wave (a conversion of the sound to kinetic energy), and is not a cavitation effect.

2. The formation of *assymetric cavities* on the metal surface is a direct result of destruction of the low-life bubbles near the surface. As a result of the cavitation, the deformation of surface takes place, together with fragmentation and decrease of size of appearing particles.

A high-intensity US processor, usually applied in experiments in a laboratory scale, is presented in Fig. 3.8 (Ace Glass). Weaker US sources, used for cleaning laboratory glassware, can also be used for synthetic purposes, where a stronger US irradiation is not desirable [715]. Moreover, a tendency to unite different techniques has appeared in recent years, leading to the creation of "mixed" equipment, in particular to carry out electrochemical processes with simultaneous application of US treatment (Sec. 3.4.2, Fig. 3.2) and microwave heating of the reaction system with ultrasonic agitation (Sec. 3.5.2, Fig. 3.7).

3.5.4 Synthesis of Metal Complexes in the Conditions of Ultrasonic Treatment

There is little reported data on ultrasonic synthesis of metal chelates [738–740]. Interaction between metallic copper or nickel with azomethinic ligands in ethanol, dioxane, or various lubricant oils (10% solution) was studied in Ref. 738. After 4 hr of US irradiation of the reaction mixture, the formed complexes ***829*** (R = Ph, CH_3, H, Cl, NO_2; M = Ni, Cu) were isolated with yields of 10–25%, in dependence on metal and ligand nature:

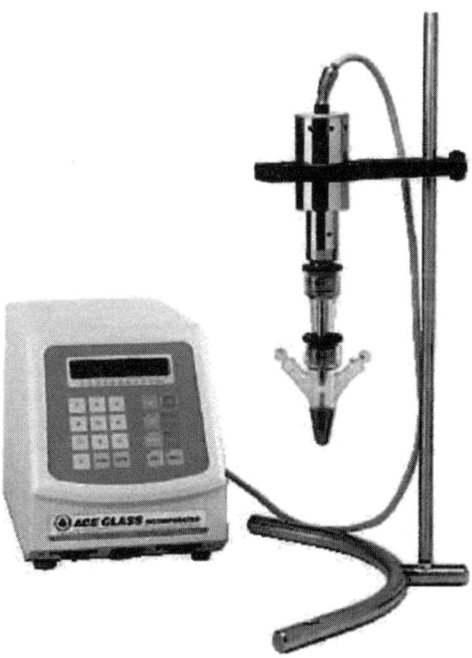

Figure 3.8 High-intensity US reactor.

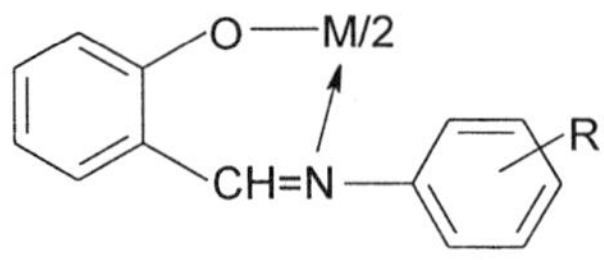

829

It was observed that, under equal conditions, the yields of copper complexes are always higher in comparison with those of nickel. An increase in donor force of the solvent applied leads to more rapid formation of complexes; an increase in viscosity leads to its delay. According to the physical–chemical study, the formed products are the same as those prepared by conventional methods from corresponding metal salts and ligands. It was established that a multimolecular layer of crystalline product is formed in the border "metal-solution." Diffusion of metal atoms takes place through this layer due to cavitation processes [738]. Another application of ultrasonic treatment for optimization of traditional synthetic methods is presented in the Experimental Procedures at the end of this section.

The US treatment is widely used for the synthesis of various organometallic compounds. In comparison with conventional techniques, the synthesis conditions are more simple, the process duration is shorter, and the yields are higher. For example, to prepare Li- and Mg-organic compounds starting from alkyl or aryl halides, it is necessary to carry out processes in dry atmosphere and to use dried ether and THF, as well as additives for initiation (I_2, CH_3I). However, on an industrial scale these reactions are dangerous, complicated, and poorly reproducible. To carry out these interactions with simultaneous US treatment, the final products are obtained rapidly, without preliminary drying of initial reagents, and in the absence of additives [741]. Using technical ether with water traces, the $Mg(OH)_2$ is precipitated together with the final product and can easily be separated. Reactions of synthesis of Mg-organic compounds are initiated immediately with the beginning of US application and the yields of final products are 60–95% [741]. The Li- and Mg-organic complexes, prepared by the "ultrasonic" method, can be used as precursors in obtaining various organic compounds (Barbié reactions) [710].

The ultrasonic field considerably accelerates processes and increases yields of Al-organic complexes with low-active alkyl halides, complexes of sodium, naphthaline, benzoquinoline, and other aromatic compounds [742]. Use of US treatment in Ulmann condensation [742] in the presence of copper powder (3.285) allows acceleration of the reaction rate 50 times and a yield of 81% in comparison with $<1.5\%$ without US:

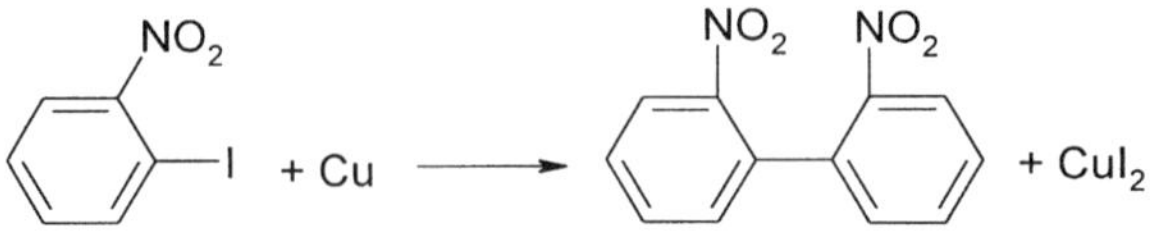

(3.285)

The H-acidic atom in organic compounds ***830*** can easily be substituted by a metal forming ***831*** (3.286) as a result of US treatment [743,744]:

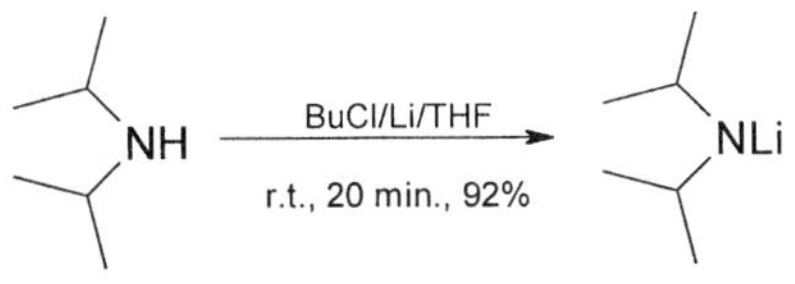

830 831

(3.286)

Organometallic compounds, obtained in a US field and having "nonstandard" structures, were reported [745,746]. Thus, organic halides produce, with aluminum and palladium, the following products ***832*** [Scheme (3.287)]:

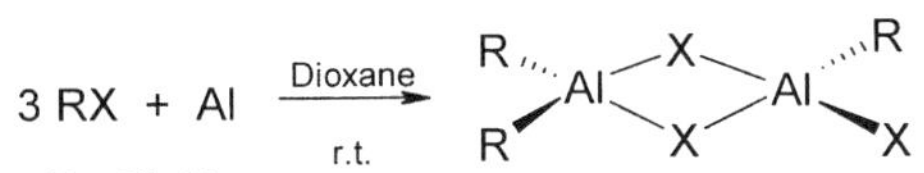

X= Cl, Br

832

(3.287)

The ruthenium π-complex ***833*** was isolated after US irradiation of the reaction system containing ruthenium trichloride, zinc powder, methanol, and 1,5-cyclo-octadiene (3.288) [747]:

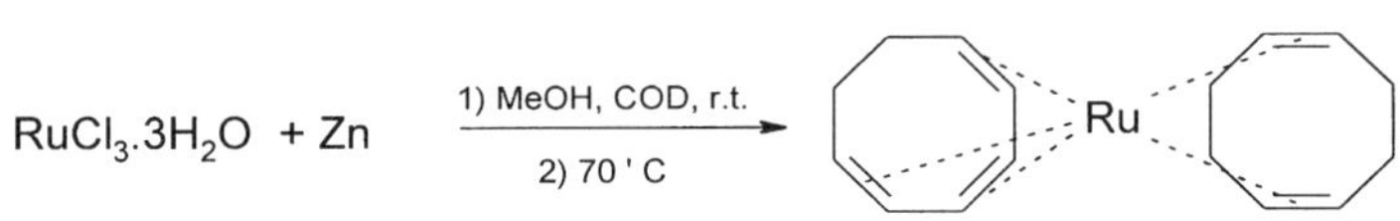

833

(3.288)

As a result, the (η^6-1,3,5-cyclo-octatriene)(η^2-1,5-cyclo-octadiene)ruthenium is formed with yields of 93 and 35% with and without US treatment, respectively. Among other π-complexes synthesized in a US field, the $Fe(CO)_2(C_5Me_5)(C_2H_5)$ ***834*** was isolated as a result of the two-step process (3.289) [748,749]:

$$Fe(CO)_2(C_5Me_5)Br \xrightarrow[C_2H_5I]{US,\ K,\ THF,\ r.t.} Fe(CO)_2(C_5Me_5)(C_2H_5) \quad (3.289)$$

834

The mercury molecular complex ***835*** was prepared from elemental metal and α,α′-dibromacetone [742] (3.290):

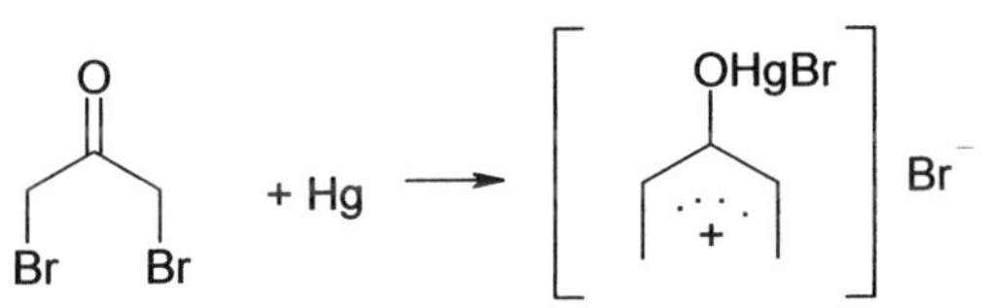

835

(3.290)

High pure metal powders can be prepared from metal carbonyls. Thus, US irradiation provokes the irreversible destruction of $Fe(CO)_5$ and for5mation of $Fe_3(CO)_{12}$ ***836*** [Scheme (3.291)] [750]:

$$Fe(CO)_5 \rightarrow Fe_3(CO)_{12} + Fe + CO \quad (3.291)$$

836

This process occurs simultaneously with the formation of amorphous iron with traces of carbon and oxygen [751], the cluster $Fe_3(CO)_{12}$ is formed, which cannot be obtained by thermal destruction of $Fe(CO)_5$ (metallic iron is produced), by its photolysis [$Fe_3(CO)_9$ is formed], or by action of other physical methods [750]. On the other hand, metal carbonyls can be prepared from transition metal halides and alkali metals in a US field. For example, $VCl_3(THF)_3$ reacts with Na in THF at CO pressure of 4.4 atm (instead of 200 atm without US application) forming $V(CO)_6^-$ with 35% yield [752]. Metal carbonyls can be prepared from their halides, for instance the interaction of WCl_6 with Na, CO in THF at 10°C produces $W_2(CO)_{10}^{2-}$ with 40% yield [708].

Finally, it is necessary to mention the classic, very active magnesium–antracene electron-transfer complex ***837*** reported in all monographic publications on ultrasound chemistry [753]. This highly reactive reducing catalyst, obtained by US treatment of magnesium, THF, and antracene, is an excellent precursor for reduction of metal salts, synthesis of π-complexes (for example, metal η^5-cyclopentadienyls), and Grignard reagents [754].

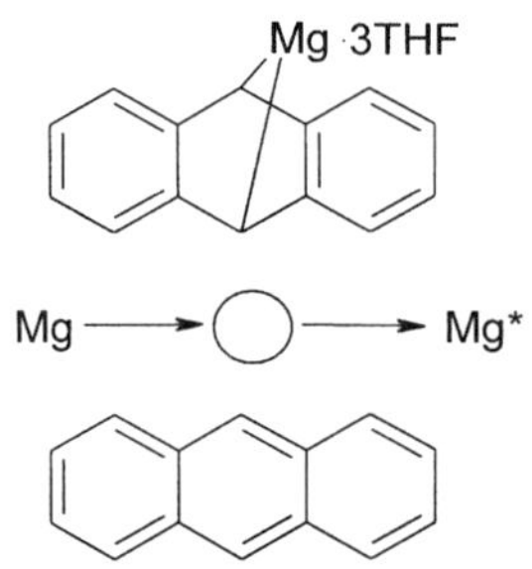

837

Precisely this approach was used to prepare a variety of metallocenes and π-allyl complexes, for example ***838*** (3.292) [722]:

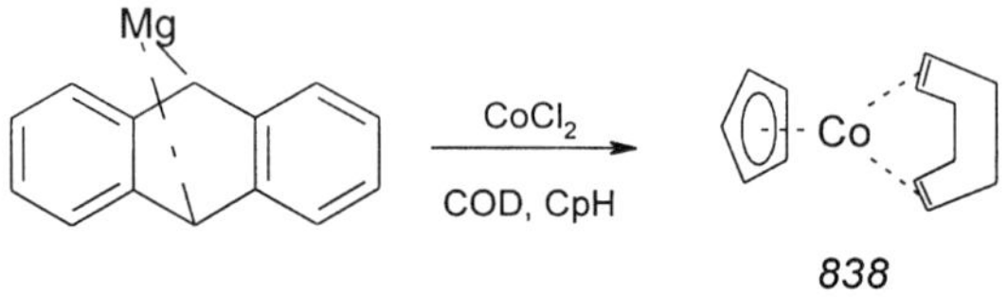

838 (3.292)

Another classic system – metallic Li and naphthalene (or biphenyl) as "electron-carrier" – is also prepared in N,N,N′,N′-tetramethylethylenediamine (tmeda) and toluene mixture by US treatment. This system serves for obtaining highly active Rieke metals [708].

Thus, use of ultrasound is very common in chemical processes, first of all in the optimization of syntheses of organic compounds, which are out of the area of the present book. Classic coordination compounds, such as, for example, metal chelates, synthesized by this technique, are not sufficiently represented in the available literature. More attention is paid to US preparation of σ- and π-organometallic compounds, mainly *s*- and *p*-metals (Li, Mg, Al), which frequently serve as precursors for organic synthesis reactions.

EXPERIMENTAL PROCEDURES

Example 1

Interaction between Metallic Copper and $[(CH_3)_2NCS_2—]_2$ (TMTD) with Simultaneous Application of Ultrasound [715]

The electrochemical synthesis (see Sec. 3.4.2) was carried out according to the technique described by Tuck and coworkers [755]. The electrochemical cell was a 100-mL tall-form beaker. The anode was a copper sheet (1 g); the cathode was a platinum foil. The electrolytic process in each case was applied for 2 hr. n-Bu_4NBr (~ 0.05 g) was used as supporting electrolyte in the case of acetonitrile and EtOH–toluene solutions (100 mL). The current was 20 mA and agitation 40 rev/min in all experiments. The isolated products were filtered, washed several times by small amounts of dry acetonitrile, and dried in air. The simultaneous ultrasonic treatment of solutions of TMTD was made in some experiments (Tables 3.1 and 3.2) during the synthesis processes using a weak source of ultrasound (ultrasonic cleaner BRANSONIC 12). More strong sources of ultrasound have not been used deliberately in order to prevent an undesirable, too rapid metal dissolution, turbulent processes, and superheating of the reaction zone.

Example 2

Interaction of Metallic Copper and Nickel with Azomethinic Ligands under Ultrasonic Treatment

Metal powder was put into a flask with ligand solution (10%) in EtOH or dioxane. The ultrasonic treatment was carried out using equipment UZDN-1 (35 kHz) during 1–3 hr. X-ray powder diffraction was used for identification of formed products ***829*** (M = Cu, Ni) [738]. Reactions of metal powders and azomethinic ligands take place also without ultrasonic treatment (with very small yields); however, it is impossible to identify the products using x-ray powder diffraction.

Table 3.1 Experimental Conditions for the Oxidative and Electrochemical Dissolution of Copper in Nonaqueous Solutions of $[Me_2N—C(S)—S—]_2$ (TMTD)

Experiment				Use of ultrasound	Voltage[c]
(ch)[a]	(el)[b]	TMTD (g)	Solvent	(ch) or (el)	(el)
A	I	1.25	AN	−	37
B	J	1.25	AN	+	33
C	K	1.25	EtOH + tol.	−	28
D	L	1.25	EtOH + tol.	+	31
E	M	1.30	DMF	−	35
F	N	1.30	DMF	+	33
G	O	1.30	DMSO	−	25
H	P	1.30	DMSO	+	29

[a]Chemical synthesis (oxidative dissolution).
[b]Electrochemical synthesis.
[c]To produce initial current of 20 mA.
In all experiments we used: 1 g [mCu(init.)] of copper; volume of solvent 100 mL; agitation 40 rev/min; $t = 2$ hr.

Table 3.2 Comparative Results Obtained in the Synthesis Processes Applied

Experiment	Solid Color	Solid m.p. (°C)	Cu (f/c)[a] (%)	*m*Cu(diss.) (g)	[*m*Cu(diss.)/*m*Cu(init.)] ×100%
A	Brown	262	20.31/20.91	0.040	4.0
B	Brown	265	20.42/20.91	0.123	12.3
C	Brown	260	19.98/20.91	0.048	4.8
D	Brown	263	20.17/20.91	0.149	14.9
E	Black	320	13.62/14.11	0.053	5.3
F	Black	323	13.55/14.11	0.138	13.8
G	Black	327	13.38/13.81	0.045	4.5
H	Black	325	13.40/13.81	0.150	13.0
I	Brown	260	20.17/20.91	0.062	6.2
J	Brown	262	20.03/20.91	0.180	18.0
K	Brown	264	20.14/20.91	0.070	7.0
L	Brown	261	19.90/20.91	0.171	17.1
M	Black	323	13.70/14.11	0.064	6.4
N	Black	325	13.59/14.11	0.175	17.5
O	Black	327	13.29/13.81	0.081	8.1
P	Black	327	13.35/13.81	0.195	19.5

[a] f = found, c = calculated.

According to studies reported in Ref. 738, a multimolecular layer of the product is formed on the metal surface. Since for its formation the presence of metal atoms or ions on the border between liquid and solid phases is needed, a diffusion of metal atoms through the compound layer is a necessary condition for such layer formation. The cavitation processes on the surface contribute to this. Since an energetic barrier should be mastered in the reaction route, a cavitation ultrasonic action has the same importance as triboplasma formed by metal friction [756].

It is necessary to mention that the synthetic approaches described in this chapter are used for the preparation of both coordination and organometallic compounds. For obtaining organometallic compounds having the σ- and π-M—C bond, we should especially mention the gas-phase synthesis [cryosynthesis, Sec. 3.4.1, Schemes (3.229), (3.231)–(3.233), Examples 2, 3, 5], synthesis of isonitrile and nitrile complexes (Secs. 2.2.4.1 and 2.2.5.1), method of ligand substitution [Sec. 3.2.1, Schemes (3.93), (3.99a), (3.119)–(3.129), (3.133)–(3.137), Examples 1, 2, 8, 9], method of metal substitution [Sec. 3.2.2, Schemes (3.147)–(3.150), Examples 2–7], cyclometallation reactions [Sec. 2.2.5.1, Scheme (2.8); Sec. 3.3.2.3, Schemes (3.226)–(3.228)]. The direct electrochemical synthesis was also applied for preparation of organometallic compounds, in particular *bis*-cyclopentadienyls [Sec. 3.4.2, Schemes (3.242) and (3.243)].

Classic oxidative addition reactions are also important in obtaining the σ-organometallic compounds. Among them should be mentioned the additions on the bonds C—H [Scheme (3.293)] [757], C—C [Scheme (3.294)] [758], and C—Hal [Scheme (3.295)] [759]:

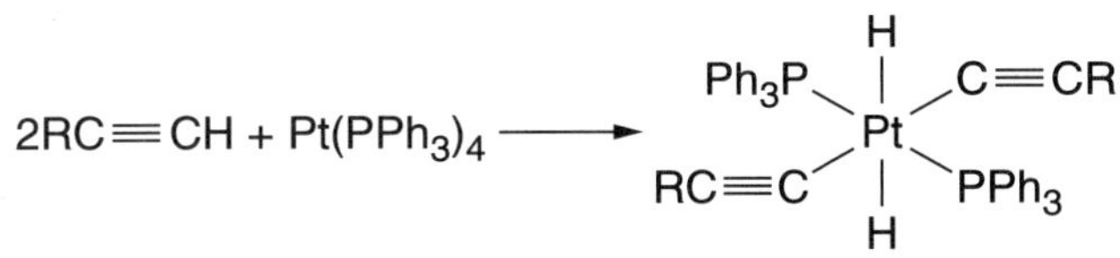

(3.293)

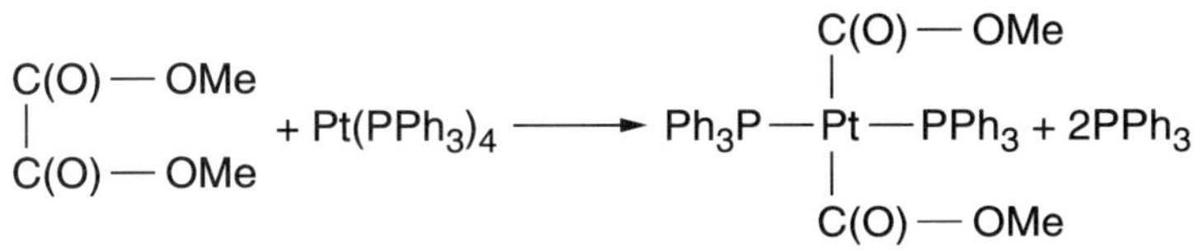

(3.294)

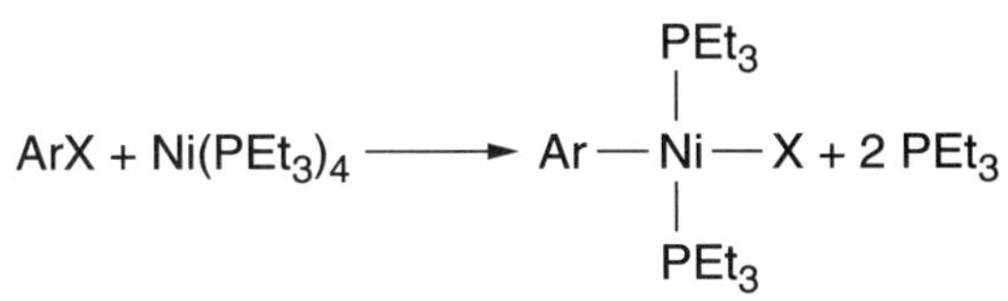

(3.295)

$$M{-}L + L' \rightarrow M{-}L'{-}L \quad (L'{=}CO,\ CO_2,\ {-}C{=}C{-},\ {-}C{\equiv}C{-},\ RNC) \tag{3.296}$$

Reactions on the metal–ligand bond are of definite interest [760,761] [Scheme (3.296)]. Other methods used in organometallic chemistry are described in Refs. 74,762,763, and references to Sec. 2.2.4.1.

ACKNOWLEDGMENTS

The authors are very grateful to Bioanalytical Systems, Inc. and Professor Richard G. Compton (Oxford University) for permission to reproduce the schemes of electrochemical cells and to Professor Ubaldo Ortiz Méndez (Universidad Autónoma de Nuevo León, Monterrey, Mexico), Professor Igor V. Melikhov, and Professor Sergey S. Berdonosov (both from Moscow State University, Russia) for useful suggestions and comments in preparation of this section, as well as to Professors Takeko Matumura (Japan), Christopher R. Strauss (Australia), and Martine Poux (France) for permission to reproduce the schemes of microwave equipment.

REFERENCES

1. *Comprehensive Coordination Chemistry*. (Edit. Wilkinson, G.). Pergamon Press: Oxford, 1987, vols. 1–7.
2. Martin, D.F. In: *Synthesis of Inorganic Compounds*. (Edit. Jolly, W.L.). *Preparative Inorganic Reactions*. (Edit. Jolly, W.L.). John Wiley: New York, London, Sydney, 1964, vol. 1.
3. *Synthetic Methods of Oranometallic and Inorganic Chemistry*. (Edit. Herrmann, W.A.). Theime: Stuttgart, New York, 1996, vols. 1–3, 6; 1997, vols. 4, 5, 7, 8.
4. *Comprehensive Organometallic Chemistry*. (Edit. Wilkinson, G.; Stone, F.G.A.; Abel, E.W.). Pergamon Press: Oxford, New York, London, 1st Edition, 1982, vols. 1–9; 2nd Edition, 1995, vols. 1–13.
5. Garnovskii, A.D.; Sadimenko, A.P. *Adv. Heterocycl. Chem.* **72**, 1 (1999).
6. Von Zelewsky, A. *Stereochemistry of Coordination Compounds*. John Wiley: Chichester, New York, 1996.

7. Black, D.St.; Hartshorn, A.J. *Coord. Chem. Rev.* **9**, 219 (1973).
8. Dyukarev, S.S.; Morozov, I.V.; Reshetova, L.N.; Guz, O.V.; Arkhangelskii, I.V.; Korenev, Yu.M.; Spiridonov, F.M. *Zhurn. Neorg. Khim.* **44**, 945 (1999).
9. Abramenko, V.L.; Garnovskii, A.D.; Surpina, L.V. *Koord. Khim.* **13**, 1232 (1987).
10. Davies, J.A.; Hockensmith, C.M.; Kukushkin, V.Yu.; Kukushin, Yu.N. *Synthetic Coordination Chemistry: Principles and Practice*. World Scientific: Singapore, London, 1992, 452 pp.
11. *Handbuch der Preparative Anorganische Chemie*. (Herausgegeben von Brauer G.). Ferdinant Enke Verlag: Stuttgart, 1981, Bd. 3.
12. Dichakjians, S.; Farago, M. *Inorg. Chim. Acta* **108**, 247 (1985).
13. Hanusa, T.P. *Chem. Rev.* **93**, 1023 (1993).
14. (a) Sergienko, V.S.; Abramenko, V.L.; Ilyujhin, A.B. *Zhurn. Neorg. Khim.* **40**, 1807 (1995); (b) Alvarez Boo, P.O.; Freijanes, E.; García Martínez, G.; Casas, J.C.; Sordo, J. *Synth. React. Inorg. Met.-Org. Chem.* **25**(1), 115 (1995); (c) Ishankhodzhaeva, M.M.; Khusenov, K.Sh.; Umarov, B.B.; Parpiev, N.A.; Alexandrov, G.G. *Zhurn. Neorg. Khim.* **43**, 1837 (1998); (d) Maekawa, M.; Munakata, M.; Kuroda-Sowa, T.; Suenaga, Y.; Sugimoto, K. *Inorg. Chim. Acta* **290**(2), 153 (1999); (e) Brisdon, B.G. *Inorg. Chem.* **6**, 1791 (1967).
15. Ribas, J.; Monfort, M.; Resino, I.; Kumar, G.; Solans, X.; Font-Bardia, M. *Polyhedron* **17**, 1735 (1998).
16. Garnovskii, A.D.; Lukova, O.A.; Alexeenko, V.A.; Komissarov, V.N.; Litvinov, V.V.; Kharabaev, N.N. *Zhurn. Neorg. Khim.* **37**, 1299 (1992).
17. Panova, G.V.; Potapov, V.M.; Ovchinnikova, O.V.; Kosina, A.N. *Zhurn. Obsh. Khim.* **57**, 2346 (1987).
18. Vasilchenko, I.S.; Antsishkina, A.S.; Burlov, A.S.; Sadikov, G.G.; Uraev, A.I.; Nivorozhkin, A.L.; Garnovskii, D.A.; Sergienko, V.S.; Korshunov, O.Yu.; Garnovskii, A.D. *Zhurn. Neorg. Khim.* **44**, 1278 (1999).
19. Sergienko, V.S.; Abramenko, V.L.; Porai-Koshitz, M.A.; Garnovskii, A.D. *Zhurn. Strukt. Khim.* **31**(1), 47 (1990).
20. Ruttimann, S.; Piguet, C.; Bernardinelli, G.; Bocquet, B.; Williams, A.F. *J. Am. Chem. Soc.* **114**(11), 4230 (1992).
21. Lu, O.; Wang, S.; Zhu, Y.; Cui, Y.; Wang, L.; Cuo, Q. *J. Chem. Soc., Dalton Trans.* No. 17, 2523 (1994).
22. Lomova, T.N.; Zaitseva, S.V.; Molodkina, O.V.; Ageeva, T.A. *Koord. Khim.* **25**, 424 (1999).
23. Sergienko, V.S.; Kovalenko, Yu. V.; Ilyukhin, A.B.; Abramenko, V.L.: *Zh. Neorgan. Khim.*, **39**, 1320 (1994).
24. Abramenko, V.L.; Sergirenko, V.S. *Koord. Khim.* **23**, 359 (1997).
25. Hiraoka, M. *Crown Compounds. Their Characteristics & Applications*. Kadansha: Tokyo, Amsterdam, Oxford, New York, 1982.
26. Ovchinnikov, Yu.A.; Ivanov, V.T.; Shkrob, A.M. *Membrane-Active Complexones*. Science: Moscow, 1974; Ovchinnikov, Yu.A.; Ivanov, V.T.; Shkrob, A.M. *Membrane-Active Complexones*. Elsevier Science: Amsterdam, Oxford, New York, 1974.
27. Bott, S.G.; Elgamal, H.; Atwood, J.L. *J. Am. Chem. Soc.* **107**(6), 1796 (1985).
28. Prinz, H.; Bott, S.G.; Atwood, J.L. *J. Am. Chem. Soc.* **1088**, 2113 (1986).
29. Belskii, V.K.; Bulichev, B.M.; Streltsova, N.R.; Storozhenko, P.A.; Ivakina, L.V. *Dokl. Akad. Nauk SSSR* **303**, 1137 (1988).
30. Frenzen, G.; Massa, W.; Ernst, T.; Dehnicke, K. *Z. Naturforsch. B* **45**(10), 1393 (1990).
31. Schafer, M.; Dehnicke, K. *Z. Anorg. Allg. Chem.* **514**, 21 (1992); **515**, 127 (1993).
32. Willey, G.R.; Palin, J.; Alcock, N.W. *J. Chem. Soc., Dalton Trans.* No. 7, 1117 (1992).
33. Belskii, V.K.; Bulichev, B.M.; Streltsova, N.R. *Zhurn. Neorg. Khim.* **37**, 1531 (1992).

34. Belskii, V.K.; Streltsova, N.R. *Zhurn. Neorg. Khim.* **38**, 72 (1993).
35. Bulichev, B.M.; Belskii, V.K. *Zhurn. Neorg. Khim.* **40**, 1834 (1995).
36. Bulichev, B.M.; Belskii, V.K. *Zhurn. Neorg. Khim.* **42**, 260 (1997).
37. Costes, R.M.; Folcher, G.; Plurien, P.; Rigny, P. *Inorg. Nucl. Chem. Lett.* **12**, 13 (1976).
38. Williams, D.L.; Deason, L.E. *J. Inorg. Nucl. Chem.* **39**, 1079 (1977).
39. Belomestnii, V.I.; Sveshnikova, L.B.; Kanisheva, A.S.; Gorbunova, Yu.G.; Mikhailov, Yu.N. *Zhurn. Neorg. Khim.* **42**, 247 (1997).
40. Hassaballa, H.S.J.; Steed, J.W.; Junk, P.C. *J. Chem. Soc., Chem. Commun.* No. 5, 577 (1998).
41. Ryabov, A.D. *Usp. Khim.* **44**, 253 (1985).
42. Omae, I. *Organometallic Intramolecular-Coordination Compounds*. Elsevier Science: Amsterdam, New York, 1986.
43. Newkome, G.R.; Puckett, W.E.; Gupta, V.K.; Kiefer, G.E. *Chem. Rev.* **86**, 451 (1986).
44. Dunina, V.V.; Zalevskaya, O.A.; Potapov, V.M. *Usp. Khim.* **57**, 434 (1988).
45. Pavlowski, V.T.; Poznyak, A.L. *Z. Chem.* **29**, 6 (1989).
46. Evans, D.W.; Baker, G.R.; Newkome, G.R. *Coord. Chem. Rev.* **93**, 155 (1986).
47. Ryabov, A.D. *Chem. Rev.* **90**(2), 403 (1990).
48. Hudson, S.A.; Maitlis, P.M. *Chem. Rev.* **93**, 861 (1993).
49. Krilova, L.F.; Lukianova, I.G.; Dikanskaya, L.D.; Podoplelov, A.V. *Metalloorg. Khim.* **1**, 1067 (1988).
50. Black, D. St. C.; Deacon, G.B.; Edwards, G.L.; Gaterhouse, B.M. *Aust. J. Chem.* **46**, 1323 (1993).
51. Atwood, J.D. *Inorganic and Organometallic Reaction Mechanisms*. VCH: Weinheim, New York, 1997.
52. Rawson, J.M.; Winpenny, R.E.P. *Coord. Chem. Rev.* **139**, 313 (1995).
53. Garnovskii, A.D.; Nivorozhkin, A.L.; Minkin, V.I. *Coord. Chem. Rev.* **126**, 1 (1993).
54. Garnovskii, A.D. *Russ. J. Coord. Chem.* **19**, 368 (1993).
55. Garnovskii, A.D.; Osipov, O.A.; Minkin, V.I.; Orlova, L.V.; Kuznetsova, L.I. *Dokl. Akad. Nauk SSSR* **178**, 598 (1968).
56. Garnovskii, A.D.; Antsishkina, A.S.; Sadikov, G.G.; Uraev, A.I.; Porai-Koshits, M.A. *Zhurn. Neorg. Khim.* **40**, 1977 (1995).
57. Watkinson, M.; Fondo, M.; Bermejo, M.R.; Sousa, A.; McAuliffe, C.A.; Pritchard, R.G.; Jaiboon, N.; Aurangzev, N.; Naeem, M. *J. Chem. Soc., Dalton Trans.* No. 1, 31 (1999).
58. Shagisultanova, G.A.; Ardasheva, L.P. *Koord. Khim.* **25**, 439 (1999).
59. Garnovskii, A.D.; Burlov, A.S.; Antsyshkina, A.S.; Divaeva, L.N. *Zhurn. Neorg. Khim.* **41**, 90 (1996).
60. (a) Burlov, A.S.; Kuznetsova, L.I.; Volbushko, N.V.; Korshunov, O.Yu.; Garnovskii, A.D. *Zhurn. Obsh. Khim.* **68**, 496 (1998); (b) Burlov, A.S.; Kharisov, B.I.; Blanco, L.M.; Kuznetsova, L.I.; Garnovskii, D.A.; Volbushko, N.B.; Garnovskii, A.D. *Rev. Soc. Quim. Mex.* **43**, 143 (1999).
61. Schmidt, C.; Dalkner, J.; Schollmeyer, D.; Singer, H. *Inorg. Chim. Acta* **257**(2), 269 (1997).
62. Kuzmina, L.G.; Grandberg, K.I.; Ilyina, L.G. *Zhurn. Neorg. Khim.* **40**, 1827 (1995).
63. García-Tojal, J.; García-Orad, A.; Serra, J.L.; Pizarro, J.L.; Lezama, L.; Arriortua, M.I.; Rojo, T. *J. Inorg. Biochem.* **75**(1), 45 (1999).
64. Kawamoto, T.; Kuma, H., Kushi, Y. *Bull. Chem. Soc. Jpn.* **70**(7), 1599 (1997).
65. Kochin, S.P.; Garnovskii, A.D.; Kogan, V.A.; Sushko, T.G. *Zhurn. Neorg. Khim.* **14**, 1428 (1969).
66. Bouman, E.; Henderson, R.K.; Reedyk, J.; Veldman, N.; Spek, A.L. *Inorg. Chim. Acta* **287**, 105 (1999).
67. Marini, P.J.; Murray, K.S.; West, B.O.; *J. Chem. Soc., Dalton Trans.* 143 (1983).

68. (a) Nivorozhkin, L.E.; Nivorozhkin, A.L.; Korobov, M.S.; Konstantinovskii, L.E.; Minkin, V.I. *Polyhedron* **4**, 1701 (1985); (b) Nivorozhkin, A.L.; Korobov, M.S.; Konstantinovskii, L.E.; Minkin, V.I. *Zh. Obshch. Khim.* **55**, 849 (1985)..
69. Chikina, N.L.; Abakarov, G.M.; Shneider, A.A.; Kuren, S.G.; Litvinov, V.V.; Sadekov, I.D.; Garnovskii, A.D. *Zhurn. Obsh. Khim.* **58**, 2496 (1988).
70. (a) Singh, A.K.; Srivastava, V.; Khandelwal, B.L. *Polyhedron* **9**(6), 851 (1990). (b) Kogan, V.A.; Kochin, S.G.; Antsyshkina, A.S.; Sadikov, G.G.; Garnovskii, A.D. *Mendeleev Commun.* 82 (1999).
71. Kuz'mina, N.P.; Kupriyanova, G.N.; Troyanov, S.I. *Koord. Khim.* **26**, 390 (2000).
72. Fuchita, Y.; Ieda, H.; Yasutake, M. *J. Chem. Soc., Dalton Trans.* No. 3, 271 (2000).
73. (a) Takhirov, T.G.; Diyachenko, O.A.; Tagiev, D.V.; Nivorozhkin, A.L.; Nivorozhkin, L.E.; Minkin, V.I. *Koord. Khim.* **17**, 505 (1991); (b) Burlov, A.S.; Antsishkina, A.S.; Sadikov, G.G.; Divaeva, L.N.; Garnovskii, A.D.; Sergienko, V.S. *Koord. Khim.* **26**, 691 (2000); (c) Chiou, S.-J.; Innocent, J.; Riordan, C.G.; Lam, K.-C.; Liable-Sands, L.; Rheingold, A.L. *Inorg. Chem.* **39**, 4347 (2000).
74. (a) Toreki, R. *The Organometallic HyperTextbook*. Intenet site http://www.ilpi.com; (b) Cotton, F.A.; Wilkinson, G.; Murillo, C.A.; Bochmann, M. *Advanced Inorganic Chemistry*. John Wiley: New York, 6th Edition, 1999; (c) Collman, J.P.; Hegedus, L.S.; Norton, J.N.; Finke, R.G. *Principles and Applications of Organotransition Metal Chemistry*. University Science Books: Sausalito, CA, 1987.
75. (a) Herberhold, M. *Metal π-Complexes. Complexes with Mono-olefinic Ligands.* Elsevier Science: Amsterdam, 1972; (b) Casey, C.P.; Wang, Y.; Tanke, R.S.; Hazin, P.N.; Rutter, Jr., E.W. *New J. Chem.* **18**(1), 43 (1994); (c) Yamamoto, A. *Organotransition Metal Chemistry*. John Wiley: New York, Chichester, Brisbane, Toronto, Singapore, 1986, 455 pp.; (d) Marr, G.; Rockett, B.W. In: *Chemistry of the Metal–Carbon Bond.* (Edit. Hartley, F.R.; Patai, S.). John Wiley: New York, 1982, vol. 1, 10; (e) Sweet, J.R.; Graham, W.A.G. *Organometallics* **2**, 135 (1983); (f) Borkowski, S.L.; Baenziger, N.C.; Jordan, R.F. *Organometallics* **12**, 486 (1993); (g) Gross, C.L.; Girolami, G.S. *Organometallics* **15**, 5359 (1996); (h) Lin, W.; Wilson, S.R.; Girolami, G.S. *Organometallics* **16**, 2356 (1997); (i) Shelby, Q.D.; Lin, W.; Girolami, G.S. *Organometallics* **18**, 1904 (1999); (j) Drew, D.; Doyle, J.R. In: *Inorganic Syntheses.* (Edit. Cotton, F.A.). McGraw-Hill: New York, 1972, vol. xiii, pp. 61 62; (k) Valderrama, M.; Arancibia, V.; Contreras, R.; Soto, C. *J. Coord. Chem.* **54**, 389 (2001) (special issue "*Contributions from Latin American authors*"); (l) Ursini, C.V.; Dias, G.H.M.; Horner, M.; Bortoluzzi, A.J.; Morigaki, M.J. *Polyhedron* **19**(22/23), 2261 (2000); (m) Prini, G.; Balboni, D.; Camurati, I.; Resconi, L.; Traverso, O. *Inorg. Chem. Commun.* **2**(2), 41 (1999); (n) Koutsantonis, G. *Laboratory Manual. Inorganic Chemistry.* http:/www.chem.uwa.edu.au/enrolled_students/chem300_Lab _Manual.html.2001; (o) Pasinskii, A.A.; Torubaev, Yu.V.; Lyakina, A.Yu.; Valiullina, R.I.; Lysenko, K.A. *Zhurn. Neorg. Khim.* **45**(6), 934 (2000); (p) Pasinski, A.A.; Torubaev, Yu.V.; Lyakina, A.Yu. *Koord. Khim.* **26**(2), 112 (2000).
76. Osechkin, S.I.; Miskiv, M.G.; Sobolev, A.I. *Metallorg Khim.* **4**, 997 (1991).
77. (a) Mikhalichko, B.M.; Miskiv, M.G.; Goreshnik, E.A. *Koord. Khim.* **25**, 65 (1999); (b) Yam, V.W-W.; Lee, W-K; Cheung, K.K.; Lee, H-K.; Leung, W-P. *J. Chem. Soc., Dalton Trans.* 2889 (1996).
78. Koide, Y.; Francis, J.A.; Bott, S.G.; Barron, A.R. *Polyhedron* **17**(5/6), 983 (1998).
79. Piguet, C.; Bernardinelli, G.; Williams, A.F. *Inorg. Chem.* **28**(15), 2920 (1989).
80. Belih, L.B.; Cherenkova, T.V.; Schmidt, F.K. *Koord. Khim.* **25**, 446 (1999).
81. Sergienko, V.S.; Kovalenko, Yu.V.; Ilyukhin, A.B.; Abramenko, V.L. *Zhurn. Neorg. Khim.* **39**, 1320 (1994).
82. Komaei, S.A.; Van Albada, G.A.; Haasnoot, J.G.; Kooijman, H.; Spek, A.L.; Reedijk, J. *Inorg. Chim. Acta* **286**, 24 (1999).

83. Komaei, S.A.; Van Albada, G.A.; Matikainen, P.; Turpeinen, U.; Reedijk, J. *Polyhedron* **18**(14), 1991 (1999).
84. Goodgame, D.M.L.; Hussain, I.; White, A.J.P.; Williams, D.J. *J. Chem. Soc., Dalton Trans.* No. 17, 2899 (1999).
85. Kalinnikov, V.T.; Rakitin, Yu.V. *Introduction in Magnetochemistry*. Science: Moscow, 1980, 301 pp.
86. Rakitin, Yu.V. *Results of Science & Technology Molecule Structure and Chemical Bond*. VINITI: Moscow, 1986, vol. 10, p. 95.
87. Velayutham, M.; Varghese, B.; Subramaniam, S. *Inorg. Chem.* **37**(6), 1336 (1998).
88. Albert, J.; González, A.; Granell, J.; Moragas, R.; Solans, X.; Font-Bardia, M. *J. Chem. Soc., Dalton Trans.* No. 11, 1781 (1998).
89. Laine, T.V.; Klinga, M.; Leskelä, M. *Eur. J. Inorg. Chem.* **6**, 959 (1999).
90. *The Chemistry of Pseudohalides*. (Edit. Golub, A.M.; Keler, H.; Skopenko, V.V.). Elseiver Science: Lausanne, Amsterdam, New York, Tokyo, 1986.
91. Garnovskii, A.D.; Sadimenko, A.P.; Osipov, O.A.; Tsdintsadze, G.V. *Hard–Soft Interactions in Coordination Chemistry*. Rostov University Press: Rostov-on-Don, 1986, 272 pp.
92. Holm, R.H.; Everett, G.W.; Chakravorty, A. *Progr. Inorg. Chem.* **7**, 83 (1966).
93. Kogan, V.A.; Zelentsov, V.V.; Osipov, O.A.; Burlov, A.S. *Usp. Khim.* **48**, 1208 (1979).
94. Burlov, A.S.; Antsishkina, A.S.; Romero, J.; Garnovskii, D.A.; García-Vázquez, A.; Sousa, A.; Garnovskii, A.D. *Russ. J. Inorg. Chem.* **40**, 1427 (1995).
95. Garnovskii, A.D.; Burlov, A.S.; Lukov, V.V.; Zaletov, V.G.; Asmaev, O.T.; Levchenkov, S.I.; Amarskii, E.G.; Garnovskii, D.A. *Koord. Khim.* **22**, 838 (1996).
96. Garnovskii, A.D.; Burlov, A.S.; Garnovskii, D.A.; Vasilchenko, I.S.; Antsishkina, A.S.; Sadikov, G.G.; Sousa, A.; García-Vázquez, A.; Romero, J. *Koord. Khim.* **23**, 399 (1997).
97. Burlov, A.S.; Garnovskii, D.A.; Kuznetsova, I.I.; Volbushko, N.V.; Korshunov, O.Yu.; Asmaev, O.T.; Kharisov, B.I.; Blanco, L.M.; Garnovskii, A.D. *Koord. Khim.* **24**, 915 (1998).
98. Garnovskii, A.D.; Burlov, A.S.; Garnovskii, D.A.; Vasilchenko, I.S.; Antsishkina, A.S.; Sadikov, G.G.; Sousa, A.; García-Vázquez, A.; Romero, J.; Durán, M.L.; Sousa-Pedrares, A.; Gómez, C. *Polyhedron* **178**, 863 (1999).
99. McNelis, B.J.; Nathan, L.C.; Clark, C.J. *J. Chem. Soc., Dalton Trans.* No. 11, 1831 (1999).
100. Yoshida, N.; Oshio, H.; Ito, T. *J. Chem. Soc., Perkin Trans.* 2, 975 (1999).
101. Kogan, V.A.; Zelentsov, V.V.; Larin, G.M.; Lukov, V.V. *Complexes of Transition Metals with Hydrazones. Physico-Chemical Properties and Structures*. Nauka: Moscow, 1990.
102. Lukov, V.V.; Levchenkov, S.I.; Kogan, V.A. *Zhurn. Neorg. Khim.* **42**, 606 (1997).
103. Bogatyreva, E.V.; Kogan, V.A.; Lokshin, V.A. *Zhurn. Neorg. Khim.* **35**, 2010 (1990).
104. Lukov, V.V.; Kogan, V.A.; Levchenkov, S.I.; Zavada, L.V.; Lisenko, K.A.; Shishkin, O.V. *Zhurn. Neorg. Khim.* **43**, 421 (1998).
105. Sakamoto, M.; Itose, S.; Ishimori, T.; Matsumoto, N.; Okawa, H.; Kida, S. *Bull. Chem. Soc. Jpn.* **63**, 1830 (1990).
106. Epifantsev, V.P.; Kogan, V.A.; Lukov, V.V.; Lokshin, V.A. *Zhurn. Neorg. Khim.* **37**, 2215 (1992).
107. Levchenkov, S.I.; Kogan, V.A.; Lukov, V.V. *Zhurn. Neorg. Khim.* **38**, 1992 (1993).
108. Lukov, V.V.; Levchenkov, S.I.; Kogan, V.A. *Koord. Khim.* **21**, 402 (1995).
109. Chan, S.C.; Koh, L.L.; Leung, P.-H.; Ranford, J.D.; Sim, K.Y. *Inorg. Chim. Acta* **236**(1/2), 101 (1995).
110. Lukov, V.V.; Levchenkov, S.I.; Kogan, V.A. *Zhurn. Neorg. Khim.* **41**, 1332 (1996).
111. Levchenkov, S.I.; Lukov, V.V.; Kogan, V.A. *Koord. Khim.* **23**, 380 (1997).

112. Levchenkov, S.I.; Lukov, V.V.; Scherbakov, I.N.; Kogan, V.A. *Koord. Khim.* **24**, 354 (1998).
113. Lukov, V.V.; Levchenkov, S.I.; Kogan, V.A. *Koord. Khim.* **24**, 946 (1998).
114. Ainscough, E.W.; Brodie, A.N.; Dobbs, A.; Randorf, J.D.; Waters, J.M. *Inorg. Chim. Acta* **267**(1), 27 (1998).
115. Koh, L.L.; Kon, K.W.; Loh, K.W.; Long, V.C.; Ranford, J.D.; Tan, A.L.C.; Tjan, V.V. *J. Inorg. Biochem. 72*, 155 (1998).
116. Randorf, J.D.; Vittal, J.J.; Wang, V.M. *Inorg. Chem.* **37**, 1226 (1998).
117. Kogan, V.A.; Lukov, V.V.; Levchenkov, S.I.; Antipin, M.Yu.; Shishkin, O.V. *Mendeleev Commun.* 145 (1998).
118. Negoyu, D.; Kalinesku, M.; Emandi, A.; Badau, E. *Koord. Khim.* **25**, 40 (1999).
119. Burlov, A.S.; Garnovskii, A.D.; Alexeenko, V.A.; Mistryukov, A.E.; Sergienko, V.S.; Zaletov, V.G.; Lukov, V.V.; Khokhlov, A.V.; Porai-Koshits, M.A. *Koord. Khim.* **18**, 859 (1992).
120. Burlov, A.S.; Kuznetsova, L.I.; Adamova, S.M.; Kurbatov, V.P.; Bondarenko, G.I.; Garnovskii, A.D. *Zhurn. Obsh. Khim.* **70**, 804 2000).
121. Yusman, T.A.; Kochin, S.G.; Lukov, V.V.; Zaletov, V.G.; Garnovskii, A.D. *Zhurn. Neorg. Khim.* **38**, 863 (1993).
122. Levchenkov, S.I.; Kogan, V.A.; Lukov, V.V. *Zhurn. Neorg. Khim.* **38**, 1687 (1993).
123. Lukov, V.V.; Levchenkov, S.I.; Kogan, V.A. *Koord. Khim.* **25**, 51 (1999).
124. Zhovmir, F.K.; Simonov, Yu.A.; Zelentsov, V.V.; Gebreleu, N.V.; Revenko, M.D.; Stroesku, A.K.; Sobolev, A.N. *Zhurn. Neorg, Khim.* **33**, 2180 (1988).
125. Lukov, V.V. Ph.D. Thesis, Rostov State University, 2000.
126. Kogan, V.A.; Lukov, V.V.; Nedzvetskii, V.S.; Abramova, N.A. *Zhurn. Neorg. Khim.* **32**, 1158 (1987).
127. Kogan, V.A.; Lukov, V.V. *Koord. Khim.* **23**, 18 (1997).
128. Kersting, B. *Eur. J. Inorg. Chem.* **8**, 1071 (1998).
129. Zelentsov, V.V.; Suvorova, K.M. *Koord. Khim.* **25**, 47 (1999).
130. (a) Garnovskii, A.D.; Sadimenko, A.P.; Uraev, A.I.; Vasilchenko, I.S.; Garnovskii, D.A. *Russ. J. Coord. Chem.* **26**, 301 (2000); (b) Fagalde, F.; Katz, N.E. *J. Coord. Chem.* **54**, 367 (2001); (c) Comba, P.; Katsichtis, C.; Nuber, B.; Protzkow, H. *Eur. J. Inorg. Chem.* No. 5, 777 (1999); (d) Brunner, H.; Kubicki, M.M.; Leblanc, J.-C.; Meiler, W.; Moise, C.; Sadorge, A.; Stubenhofer, B.; Watcher, J.; Wanninger, R. *Eur. J. Inorg. Chem.* No. 5, 843 (1999); (e) Nikonov, G.I.; Kuzmina, L.G.; Lorberth, J.; Howard, J.A.K. *Eur. J. Inorg. Chem.* No. 5, 825 (1999); (f) Fenton, D.E.; Nájera, B.A. *J. Coord. Chem.* **54**(3/4), 239 (2001); (g) Ara, I.; Forniés, J.; Gómez, J.; Ialinde, E.; Merino, R.I.; Moreno, M.T. *Inorg. Chem. Commun.* **2**(2), 62 (1999); (h) Wong, W.-Y.; Lam, H.-Y.; Lee, S.-M. *J. Organomet. Chem.* **595**, 70 (2000); (i) Cotton, F.A.; Dikarev, E.V.; Petrukhina, M.A.; Stiviba, S.E. *Organometallics* **19**(7), 1402 (2000).
131. *Encylopedia of Inorganic Chemistry*. (Edit. King, R.B.). John Wiley: Chichester, 1994.
132. Housecroft, C.E. *Coord. Chem. Rev.* **152**, 87 (1996).
133. Cortez, S.M.; Paptis, R.G. *Coord. Chem. Rev.* **169**, 363 (1998).
134. Housecroft, C.E. *Coord. Chem. Rev.* **142**, 101 (1995).
135. Smith, D.R. *Coord. Chem. Rev.* **172**, 457 (1998).
136. Wong, C.V.; Woollins, J.D. *Coord. Chem. Rev.* **130**, 243 (1994).
137. Corain, B.; Tapparo, A.; Sheikh-Osman, A.A.; Bombi, G.G.; Zatta, P.; Favarato, M. *Coord. Chem. Rev.* **112**, 19 (1992).
138. Corain, B.; Nicolini, M.; Zatta, P. *Coord. Chem. Rev.* **112**, 33 (1992).
139. Sauls, F.C.; Interrante, L.V. *Coord. Chem. Rev.* **128**(1/2), 193 (1993).
140. *Coord. Chem. Rev.* **149** (1996) (complete issue).
141. Wells, R.L. *Coord. Chem. Rev.* **112**, 273 (1992).
142. Housecroft, C.E. *Coord. Chem. Rev.* **152**, 141 (1996).

143. Hannon, M.J. *Coord. Chem. Rev.* **152**, 393 (1996).
144. Ouchi, M.; Hakushi, T. *Coord. Chem. Rev.* **148**, 171 (1996).
145. Housecroft, C.E. *Coord. Chem. Rev.* **142**, 147 (1995).
146. Meehan, P.R.; Aris, D.R.; Willey, G.R. *Coord. Chem. Rev.* **181**, 121 (1999).
147. Pikramenou, Z. *Coord. Chem. Rev.* **172**, 99 (1998).
148. Wong, C.Y.; Woollins, J.D. *Coord. Chem. Rev.* **130**, 175 (1994).
149. Hampden-Smith, M.J.; Wark, T.A.; Brinker, C.J. *Coord. Chem. Rev.* **112**, 81 (1992).
150. (a) Draper, S.M.; Twamley, B. *Coord. Chem. Rev.* **134**, 189 (1994); (b) Page, E.M.; Wass, S.A. *Coord. Chem. Rev.* **152**, 411 (1996).
151. Byrne, J.J.; Draper, S.M. *Coord. Chem. Rev.* **134**, 171 (1994).
152. Leigh, G.J.; de Souza, J.S. *Coord. Chem. Rev.* **154**, 71 (1996).
153. Padge, E.M. *Coord. Chem. Rev.* **172**, 111 (1998).
154. Kee, T.P. *Coord. Chem. Rev.* **169**, 129 (1998).
155. Zingano, E. *Coord. Chem. Rev.* **129**, 195 (1995).
156. Geiger, D.K. *Coord. Chem. Rev.* **152**, 359 (1996).
157. Geiger, D.K. *Coord. Chem. Rev.* **172**, 157 (1998).
158. Morris, M.J. *Coord. Chem. Rev.* **152**, 309 (1996).
159. Wedd, A.G. *Coord. Chem. Rev.* **154**, 5 (1996).
160. Wong, E.H. *Coord. Chem. Rev.* **152**, 251 (1996).
161. Vites, J.C.; Lynam, M.M. *Coord. Chem. Rev.* **172**, 319 (1998).
162. Housecroft, C.E. *Coord. Chem. Rev.* **142**, 21 (1995).
163. Housecroft, C.E. *Coord. Chem. Rev.* **169**, 187 (1998).
164. Vites, J.C.; Lynam, M.M. *Coord. Chem. Rev.* **142**, 1 (1995).
165. Vites, J.C.; Lynam, M.M. *Coord. Chem. Rev.* **169**, 201 (1998).
166. Vites, J.C.; Lynam, M.M. *Coord. Chem. Rev.* **172**, 357 (1998).
167. Nordlander, E.; Whalen, A.M. *Coord. Chem. Rev.* **134**, 91 (1994).
168. Powell, A.K. *Coord. Chem. Rev.* **142**, 257 (1995).
169. Sima, J.; Makanova, J. *Coord. Chem. Rev.* **160**, 161 (1997).
170. Nordlander, E.; Thapper, A.; King, J.; Lorber, C.; Carlsson, H.; Prestopino, F.; Focci, N. *Coord. Chem. Rev.* **172**, 3 (1998).
171. Weit, S.K.; Ferraudi, G.; Grutsch, P.A.; Kutal, C. *Coord. Chem. Rev.* **128**(1/2), 225 (1993).
172. Davies, M.B. *Coord. Chem. Rev.* **152**, 1 (1996).
173. Davies, M.B. *Coord. Chem. Rev.* **169**, 237 (1998).
174. Foulds, G.A. *Coord. Chem. Rev.* **142**, 153 (1995).
175. Higgins, S.J. *Coord. Chem. Rev.* **152**, 175 (1996).
176. Foulds, G. *Coord. Chem. Rev.* **169**, 3 (1998).
177. Richmond, M.G. *Coord. Chem. Rev.* **141**, 63 (1995).
178. Housecroft, C.E. *Coord. Chem. Rev.* **152**, 107 (1996).
179. Simpson, M.C.; Cole-Hamilton, D.J. *Coord. Chem. Rev.* **155**, 163 (1996).
180. Hui, J.W-S.; Wong, W.T. *Coord. Chem. Rev.* **172**, 389 (1998).
181. Pregosin, P.S.; Salzmann, R. *Coord. Chem. Rev.* **155**, 35 (1996).
182. Thompson, A.M.W.G. *Coord. Chem. Rev.* **152**, 157 (1996).
183. Charmant, J.P.H. *Coord. Chem. Rev.* **172**, 437 (1998).
184. Byers, P.K. *Coord. Chem. Rev.* **142**, 123 (1995).
185. Kilimann, U.; Edelmann, F.T. *Coord. Chem. Rev.* **141**, 1 (1995).
186. Richter, J.; Edelmann, F.T. *Coord. Chem. Rev.* **147**, 373 (1996).
187. Hegerheide, M.P.; Boersma, J.; Van Koten, G. *Coord. Chem. Rev.* **155**, 87 (1996).
188. Gunrko, Yu.K.; Edelmann, F.T. *Coord. Chem. Rev.* **156**, 1 (1996).
189. Pedersen, K.D.; Frensdorf, H.K. *Angew. Chem., Int. Ed. Engl.* **11**, 16 (1972).
190. *Synthetic Multidentate Macrocyclic Compounds.* (Edit. Izatt, R.M.; Christensen, J.J.). Academic Press: New York, 1978.

191. *Progress in Macrocyclic Chemistry*. (Edit. Izatt, R.M.; Christensen, J.J.). John Wiley: New York, vol. 1, 1979; vol. 2, 1981.
192. *Coordination Chemistry of Macrocyclic Compounds*. (Edit. Melson, G.A.). Plenum Press: New York, 1979.
193. *Host–Guest Complexes Chemistry*. Springer-Verlag Chemie: Berlin, 1981.
194. Gokel, G.W.; Korzeniovski, S.H. *Macrocyclic Polyether Synthesis*. Verlag Chemie: Berlin, 1982.
195. Izatt, R.M.; Brandshaw, J.S.; Nielsen, S.A. *Chem. Rev*. **85**, 271 (1985).
196. Bogatskii, A.V.; Lukianenko, N.G.; Kirichenko, T.I. *Zhurn. Vses. Khim. Obsh. Mendeleeva* **30**, 487 (1985).
197. Tsivadze, A.Yu.; Varnek, A.A.; Khutorskii, V.E. *Coordination Compounds of Metals with Crown Ethers*. Science: Moscow, 1991.
198. Tsivadze, A.Yu. *Mendeleev Chem. J*. **40**(4/5), Part I, 62 (1996).
199. Kratz, F.; Nuber, B.; Weiss, J.; Keppler, B.K. *Polyhedron* **11**, 487 (1992).
200. Zamias, S.; Raptopoulou, C.P.; Tetris, A.; Zafiropulos, T.F. *Inorg. Chem. Commun*. **2**, 48 (1999).
201. *Direct Synthesis of Coordination Compounds*. (Edit. Skopenko, V.V.). Ventury: Kiev, 1997, 172 pp.
202. *Direct Synthesis of Coordination and Organometallic Compounds*. (Edit. Garnovskii, A.D.; Kharisov, B.I.). Elsevier Science: Lausanne, Amsterdam, London, New York, 1999, 244 pp.
203. Skopenko, V.V.; Kokozay, V.N.; Pavlenko, V.A. *Ukr. Khim. Zhurn*. **59**, 485 (1993).
204. Skopenko, V.V.; Kokozay, V.N.; Vasilyeva, O.Yu. *Koord. Khim*. **24**, 180 (1998).
205. McCarthy, P.J.; Hovey, R.J.; Ueno, K.; Martell, A.E. *J. Am. Chem. Soc*. **77**, 5820 (1955).
206. Kukushkin, V.Yu.; Kukushkin, Yu.N. *Theory and Practice of the Synthesis of Coordination Compounds*. Nauka: Leningrad, 1990, 260 pp.
207. Goryachev, I.G.; Zaitsev, B.E.; Klyuchnikova, N.G.; Gromov, D.N. *Handbook on Inorganic Synthesis*. Chemistry: Moscow, 1997, 275 pp.
208. Gutman, V. *Coordination Chemistry in Non-Aqueous Solutions*. Springer-Verlag: Wien, New York, 1968.
209. (a) Krestov, G.A.; Afanasiev, V.N.; Agafonov, A.V. *Complex-Formation in Non-Aqueous Solutions*. Moscow: Science, 1989, 256 pp.; (b) Berezin, B.D.; Golubchikov, O.A. *Coordination Chemistry of Solvento-Complexes of Transition Metal Salts*. Moscow: Science, 1992, 236 pp.
210. Gutmann, V.; Resch, G. *Lecture Notes on Solution Chemistry*. World Scientific: Singapore, 1995, 242 pp.
211. Garnovskii, A.D.; Garnovskii, D.A.; Burlov, A.S.; Vasilchenko, I.S. *Mendeleev Chem*. J. **40**(4/5), Part I, 19 (1996).
212. Davies, J.A.; Hartley, F.R. *Chem. Rev*. **81**, 79 (1981).
213. Davies, J.A. *Adv. Inorg. Chem. Radiochem*. **24**, 115 (1981).
214. Berezin, B.D. *Zhurn. Vses. Khim. Obsh. Mendeleeva* **29**, 514 (1984).
215. Beck, W.; Sunkel, K. *Chem. Rev*. **88**(7), 1405 (1988).
216. Panina, N.S.; Yakovlev, V.N.; Kukushkin, V.Yu. *Koord. Khim*. **14**, 723 (1988).
217. Laerance, G.A. *Adv. Inorg. Chem. Radiochem*. **34**, 145 (1989).
218. Berezin, B.D. *Russ. Chem. Rev*. **60**, 996 (1990).
219. Berezin, V.D. *Koord. Khim*. **17**, 597 (1991).
220. Strauss, S.H. *Chem. Rev*. **93**, 927 (1993).
221. Garnovskii, A.D.; Burlov, A.S.; Kharisov, B.I.; Gójon-Zorrilla, G. *Rev. Soc. Quím. Méx*. **41**, 205 (1997).
222. Osipov, O.A.; Minkin, V.I.; Garnovskii, A.D. *Guide on Dipol Moments*. High School: Moscow, 1971, 414 pp.

223. Parker, A. *Usp. Khim.* **32**, 1270 (1963).
224. Parker, A.J. *Quart. Rev.* (London) **16**, 163 (1962); *Adv. Org. Chem.* **5**, 1 (1965); *Chem. Rev.* **69**, 1 (1969); *Pure & Appl. Chem.* **25**, 345 (1971).
225. Mayer, U.; Gutmann, V.; Gerger, W. *Monatsch. Chem.* **106**, 1235 (1975); **108**, 489 (1977).
226. Mayer, U. *Pure & Appl. Chem.* **51**, 1697 (1979).
227. Gordon, J.E. *The Organic Chemistry of Electrolyte Solutions.* John Wiley: New York, London, Sydney, Toronto, 1977.
228. Fialkov, Yu.Ya. *Solvent as a Medium for Controlling a Chemical Process.* Chemistry: Leningrad, 1990, 160 pp.
229. Reichardt, Ch. *Solvent Effects in Organic Chemistry.* Verlag Chemie: Weinheim, 1988.
230. Abramenko, Yu.V.; Garnovskii, A.D.; Sergienko, V.S.; Abramenko, V.L. *Koord. Khim.* **37**, 2689 (1992).
231. Lavrenova, L.G.; Larionov, S.V.; Grankina, Z.A.; Ikorskii, V.N. *Izv. Sib. Otd. Akad. Nauk SSSR, Ser. Khim.* 81 (1983).
232. Pon, G.; Willet, R.D.; Prince, V.A.; Robinson, W.T.; Turnbull, M.M. *Inorg. Chim. Acta* **255**(2), 325 (1997).
233. Yatsimirsky, K.B.; Kostromina, N.A.; Sheka, Z.A.; Davidenko, N.K.; Kriss, E.E.; Ermolenko, V.I. *Chemistry of Complex Compounds of Rare-Earth Elements.* Naukova Dumka: Kiev, 1966, 433 pp.
234. Doretti, I.; Sitran, S.; Falechini, S. *J. Inorg. Nucl. Chem.* **38**, 2123 (1976).
235. Mandral, S.; Das, G.; Singh, R.; Shukla, R.; Bharadwaj, P.K. *Coord. Chem. Rev.* **160**, 191 (1997).
236. Nivorozhkin, A.L.; Uraev, A.I.; Burlov, A.S.; Garnovskii, A.D. *Mendeleev Chem. J.* **40**(4/5), 255 (1996).
237. Toftlund, H.; Becher, J.; Olsen, P.H.; Pedersen, J. *Israel J. Chem.* **25**, 56 (1985).
238. Malachowski, M.R.; Adams, M.; Elia, N.; Rheingold, A.L.; Kelly, R.S. *J. Chem. Soc., Dalton Trans.* No. 13, 2177 (1999).
239. Dobosz, A.; Dudarenko, N.M.; Fritsky, I.O.; Glowiak, T.; Karaczyn, A.; Kozlowski, H.; Sliva, T.Yu.; Swiatek-Kozlowska, J.J. *J. Chem. Soc., Dalton Trans.* No. 13, 743 (1999).
240. Botsivali, M.; Evans, D.F.; Missen, P.H.; Upton, M.W. *J. Chem. Soc., Dalton Trans.* 1147 (1985).
241. Evans, F.E.; Jakubovic, D.A. *Polyhedron* **7**, 2723 (1988).
242. Dutta, S.; Chakravorty, A. *Polyhedron* **13**(12), 1811 (1994).
243. Kurbatov, V.P.; Antsishkina, A.S.; Kukharicheva, E.S.; Garnovskii, A.D. *Zhurn. Neorg. Khim.* **41**, 1315 (1996).
244. Kuznetsova, L.I.; Burlov, A.S.; Volbushko, N.V.; Korshunov, O.Yu.; Zaletov, V.G.; Garnovskii, A.D. *Zhurn. Obsh. Khim.* **68**, 1338 (1998).
245. Kharisov, B.I.; Blanco, L.M.; Garnovskii, A.D.; Burlov, A.S.; Kuznetsova, L.I.; Korovina, L.V.; Garnovskii, D.A.; Dieck, T. *Polyhedron* **17**, 381 (1998).
246. Litvinov, V.V.; Sergienko, V.S.; Knyazhanskii, M.I.; Porai-Koshitz, M.A.; Garnovskii, A.D.; Mistryukov, A.E. *Koord. Khim. 16*, 1360 (1990).
247. Fomicheva, E.B.; Belskii, V.G.; Panova, G.V. *Koord. Khim.* **15**, 209 (1989).
248. Jirgi, B.S.; Tembe, G.L.; Sonar, M.H.; Murti, A.S. *Trans. Met. Chem.* **18**, 171 (1993).
249. Da Silva, A.S.; De Silva, M.A.A.; Carvalho, C.E.M.; Antunes, O.A.C.; Herrera, J.O.M.; Brinn, I.M..; Mangrich, A.S. *Inorg. Chim. Acta* **292**(1) 1 (1999).
250. Shepelenko, E.N.; Minkina, L.S.; Kochin, S.G.; Khokhlov, A.V.; Bren, V.A.; Garnovskii, A.D. *Zhurn. Obsh. Khim.* **57**, 2342 (1987).
251. Shepelenko, E.N.; Rybalkin, V.P.; Minkina, L.S.; Kochin, S.G.; Khokhlov, A.V.; Bren, V.A.; Garnovskii, A.D. *Zhurn. Obsh. Khim.* **60**, 1835 (1990).
252. Elmali, A.; Atakol, O.; Svoboda, I.; Fuess, H. Acta Crystallogr. C **51**, 2520 (1995).

253. Abramenko, V.L.; Sergienko, V.S. *Koord. Khim.* **23**, 359 (1997).
254. Glez, A.; Drozdov, A.A.; Trofimov, S.I. *Koord. Khim.* **20**, 922 (1994).
255. Afzaletdinova, N.G.; Murinov, Yu.I.; Valyamova, F.G.; Vasilyeva, E.V. *Zhurn. Neorg. Khim.* **44**, 226 (1999).
256. (a) Yusman, T.A.; Kochin, S.G.; Lukov, V.V.; Zaletov, V.S.; Garnovskii, A.D. *Zhurn. Neorg. Khim.* **38**, 863 (1993); (b) Sakamoto, M.; Itose, S.; Ishimori, T.; Matsuomoto, N.; Okawa, H.; Kida, S. *J. Chem. Soc., Dalton Trans.* 2083 (1989); (c) Virovets, A.V.; Podberezskaya, N.V.; Lavrenova, L.G.; Bikzhanova, G.A. *Polyhedron* **13**(20), 2929 (1994); (d) Palkina, K.K.; Kuzmina, N.E.; Lyakhov, A.S.; Gaponik, P.N.; Bogatikov, A.N.; Govorova, A.A.; Ivashkevich, L.S. *Zhurn. Neorg. Khim.* **46**(10), 1649 (2001).
257. Langford, C.H.; Gray, H.B. *Ligand Substitution Processes*. W.A. Biniamin: New York, 1965.
258. Basolo, F.; Pearson, P. *Mechanisms of Inorganic Reactions*. John Wiley: New York, London, Sydney, 1967.
259. *Coordination Chemistry*. (Edit. Martell, A.E.). American Chemical Society: Washington, 1978.
260. *Mechanisms of Inorganic and Organometallic Reactions*. (Edit. Twigg, M.V.). Plenum Publishing Co.: New York, London, 1985, 519 pp.
261. Basolo, F. *Coord. Chem. Rev.* **100**, 47 (1990).
262. *Inorganic Reactions and Mechanisms*. (Edit. Atwood, J.D.; Zuckerman, J.J.). Wiley VCH: New York, 1998, 688 pp.
263. Weiss, J.K.; Beck, W. *J. Organomet. Chem.* **44**, 325 (1972).
264. Garnovskii, A.D.; Kolobova, N.E.; Zlotina, I.B.; Mitina, G.K.; Kolodyazhnii, Yu.V.; Osipov, O.A.; Anisimov, K.N. *Zhurn. Obsh. Khim.* **42**, 929 (1972).
265. Sadekov, I.D.; Uraev, A.I.; Garnovskii, A.D. *Russ. Chem. Rev.* **68**, 415 (1999).
266. Sadekov, I.D.; Minkin, V.I. *Sulfur Rep.* **19**, 285 (1997).
267. Sadekov, I.D.; Maximenko, A.A.; Nivorozhkin, V.L. *Russ. Chem. Rev.* **67**, 193 (1998).
268. Pasinskii, A.A.; Torubaev, Yu.K.; Drukovskii, A.G.; Eremenko, I.L.; Vegini, D.; Krasilnikova, E.V.; Privalov, V.I.; Yanovskii, A.I.; Struchkov, Yu.T. *Zhurn. Neorg. Khim.* **42**, 42 (1997).
269. Kaur, R.; Singh, H.B.; Butcher, R.J. *Organometallics* **14**(10), 4755 (1995).
270. Thaler, G.; Wurst, K.; Sladky, F. *Organometallics* **15**(21), 4639 (1996).
271. Minoura, M.; Kawashima, T.; Kokitoh, N.; Okazaki, R. *J. Chem. Soc., Chem. Commun.* No. 2, 123 (1996).
272. Pasinskii, A.A.; Torubaev, Yu.V.; Eremenko, I.L.; Krasilnikova, E.V.; Yanovskii, A.I.; Struchkov, Yu.T. *Zhurn. Neorg. Khim.* **42**, 730 (1997).
273. Kuhn, N.; Schumann, H. *J. Organomet. Chem.* **287**, 345 (1985).
274. Hou, Z.; Fujita, A.; Yamasaki, H.; Wakatsuki, Y. *J. Am. Chem. Soc.* **118**(33), 7843 (1996).
275. Hou, Z.; Wakatsuki, Y. *Chem. Eur. J.* **3**, 1005 (1997).
276. Ryabov, A.D. *Ross. Khim. Zhurn.* **40**(4/5), 33 (1996).
277. Kukushkin, V.Yu. *Coord. Chem. Rev.* **139**, 375 (1995).
278. Gavrilov, K.N.; Mikhel, I.S. *Zhurn. Neorg. Khim.* **39**, 1364 (1994).
279. Nyholm, R.S.; Sutton, G.J. *J. Chem. Soc. (London)* 560 (1958).
280. Bowden, F.L.; Fergusson, D.J. *J. Chem. Soc., Dalton Trans.* 460 (1974).
281. Doretti, R.; Madalosso, F.; Sitran, S.; Feleshini, S. *Inorg. Nucl. Chem. Lett.* **12**, 817 (1976).
282. Burlov, A.S.; Kuznetsova, L.I.; Volbushko, N.V.; Korshunov, O.Yu.; Garnovskii, A.D. *Zhurn. Obsh. Khim.* **68**, 496 (1998).
283. Liu, Z.-H.; Duan, C.-Y.; Hu, J.; You, X.-Z. *Inorg. Chem.* **38**(8), 1719 (1999).
284. Sudha, N.; Singh, H.B. *Coord. Chem. Rev.* **135/136**, 469 (1994).

285. Hennig, H. *Z. Chem.* **11**, 81 (1971).
286. Hennig, H.; Daute, I. *Z. Anorg. Allg. Chem.* **371**, 212 (1969).
287. Sergienko, V.S.; Abramenko, V.L.; Ilyukhin, A.B.; Porai-Koshitz, M.A.; Garnovskii, A.D. *Koord. Khim.* **18**, 585 (1992).
288. Sui, K.; Peng, S.-M.; Bhattacharya, S. *Polyhedron* **18**(5), 631 (1999).
289. Rao, N.S.; Mishra, D.D.; Maurya, R.C.; Rao, N.N. *Polyhedron* **13**(18), 2653 (1994).
290. Stelzig, L.; Kötte, S.; Krebs, B. *J. Chem. Soc., Dalton Trans.* No. 17, 2921 (1998).
291. Bansse, W.; Fliegner, J.; Sawusch, S.; Schilde, U.; Uhlemann, E. *Inorg. Chim. Acta* **269**(2), 350 (1998).
292. *Comprehensive Heterocyclic Chemistry*. (Edit. Katriitzky, A.R.; Rees, C.W.; Skrives, E.F.V.). Pergamon Press: Oxford, 1st Edition, 1984. vols. 2–6; 2nd Edition, 1996, vols. 2–9.
293. Biedermann, H.G.; Öfele, K.; Schuhbauer, N.; Tajtelbaum, J. *Angew. Chem.* **87**, 634 (1975).
294. Uraev, A.I.; Garnovskii, A.D.; Sadimenko, A.P. *Metalloorg. Khim.* **5**, 934 (1992).
295. Ryabov, A.D.; Yatsimirsky, A.K. *Inorg. Chem.* **23**, 789 (1984).
296. Battistuzzi, G.; Corradi, A.B.; Dallari, D.; Saladini, M.; Battistuzzi, R. *Polyhedron* **18**, 57 (1999).
297. Fornies, J.; Martin, R.; Navarro, M.R.; Sicilia, V.; Villarroya, P. *Organometallics* **15**(7), 1826 (1996).
298. Deberitz, J.; Nöth, H. *J. Organomet. Chem.* **61**, 271 (1973).
299. Garnovskii, A.D.; Uraev, A.I.; Antsyshkina, A.S.; Sadekov, I.D.; Minkin, V.I. *Koord. Khim.* **22**, 512 (1996).
300. Garnovskii, A.D.; Sadekov, I.D.; Antsyshkina, A.S.; Sadikov, G.G.; Borodkin, G.S.; Uraev, A.I.; Maksimenko, A.A.; Vasilchenko, I.S.; Borodkina, I.G.; Sergienko, V.S.; Minkin, V.I. *Russ. J. Coord. Chem.* **25**, 767 (1999).
301. (a) Oilunkaniemi, R.; Laitinen, R.S.; Ahlgren, M. *Inorg. Chem. Commun.* **3**, 8 (2000); (b) Trebbe, R.; Goddard, R.; Rufinska, A.; Seevogel, K.; Porschke, K.R. *Organometallics* **18**(13), 2466 (1999); (c) Wright, L.L.; Moore, S.J.; McIntyre, J.A.; Dowling, S.M.; Overcast, J.D.; Grisel, G.R.; Nelson, K.K.; Rupert, P.R.; Pennington, W.T. *Organometallics* **18**(22), 4558 (1999); (d) Leonova, T.G.; Glinskaya, L.A.; Klevtsova, R.F.; Batrachenko, N.I.; Larionov, S.V. *Koord. Khim.* **26**(9), 657 (2000).
302. (a) Kuzmina, L.G.; Kukina, G.A.; Leeis, Ya.E.; Grap, S.R.; Bankovskii, Yu.A.; Efimenko, I.A. *Zhurn. Neorg. Khim.* **41**, 209 (1996); (b) Kuzmina, L.G.; Kukina, G.A.; Leeis, Ya.E.; Grap, S.R.; Bankovskii, Yu.A.; Efimenko, I.A. *Zhurn. Neorg. Khim.* **41**, 215 (1996).
303. Olekhnovich, L.P.; Kurbatov, V.P.; Osipov, O.A.; Minkina, L.S.; Minkin, V.I. *Zhurn. Obsh. Khim.* **38**, 2593 (1968).
304. Nivorozhkin, L.E.; Kurbatov, V.P.; Minkina, L.S.; Korobov, M.S.; Olekhnovich, L.P.; Osipov, O.A.; Minkin, V.I. *Zhurn. Obsh. Khim.* **41**, 2726 (1971).
305. Mugesh, G.; Singh, H.B.; Butcher, R.J. *Eur. J. Inorg. Chem.* No. 8, 1229 (1999).
306. Mugesh, G.; Singh, H.B.; Patel, R.P.; Butcher, R.J. *Inorg. Chem.* **37**(11), 2663 (1998).
307. Mugesh, G.; Singh, H.B.; Butcher, R.J. *J. Organomet. Chem.* **577**(2), 243 (1999).
308. Sadekov, I.D.; Abakarov, G.M.; Kuren, S.G.; Garnovskii, A.D.; Minkin, V.I. *Zhurn. Obsh. Khim.* **56**, 2168 (1986).
309. Bonasia, Ph.J.; Arnold, J. *J. Organomet. Chem.* **449**(1/2), 147 (1993).
310. Tom, G.H.; Ludvig, R. *The Dithiocarbamates and Related Compounds*. Elsevier Science: Amsterdam, 1981.
311. Kato, S.; Niyomura, O.; Kawahara, V.; Kanda, T. *J. Chem. Soc., Dalton Trans.* No. 10, 1677 (1999).

312. Halcrow, M.A.; Chia, L.M.L.; Liu, X.; McInnes, E.J.L.; Yellowlees, L.J.; Mabbs, F.E.; Scowen, I.J.; McPartlin, M.; Davies, J.E. *J. Chem. Soc., Dalton Trans.* No. 11, 1753 (1999).
313. Perera, J.R.; Heeg, M.J.; Schlegel, H.B.; Winter, C.H. *J. Am. Chem. Soc.* **121**(18), 4536 (1999).
314. (a) Hipp, C.J.; Busch, D.H. In: *Coordination Chemistry*. (Edit. Martell, A.E.). American Chemical Society: Washington, 1978, vol. 2; (b) Eisch, J.J.; Shi, X.; Owuor, F.A. *Organometallics* **17**(24), 5219 (1998); (c) Evans, W.J.; Johnston, M.A.; Greci, M.A.; Ziller, J.W. *Organometallics* **18**(8), 1460 (1999); (d) Avis, M.W.; Vrieze, K.; Ernsting, J.M.; Elsevier, K.J.; Veldman, N.; Spek, A.L.; Katti, K.V.; Barnes, C.L. *Organometallics* **15**(9), 2376 (1996).
315. Candlin, J.P.; Taylor, K.A.; Thompson, D.T. *Reactions of Transition Metal-Complexes*. Elsevier: Amsterdam, London, New York, 1968.
316. Gebreleu, N.V. *Matrix Reactions*. Stiintsa: Kishinev, 1980, 183 pp.
317. Yatsimirski, K.B.; Kolchinskii, A.G.; Pavlishuk, V.V.; Talanova, G.G. *Synthesis of Macrocyclic Compounds*. Nauka Dumka: Kiev, 1987, 279 pp.
318. Gebreleu, N.V.; Arion, V.B. *Template Synthesis of Macrocyclic Compounds*. Stiintsa: Kishinev, 1990, 373 pp.
319. Busch, D.A. *Werner Commemoration Volume*. Verlag Helvetica Chimica Acta, 1967, p. 174.
320. Kukushkin, Yu.N. *Koord. Khim.* **7**, 323 (1981).
321. Kukushkin, Yu.N. *Reaction Capacity of Coordination Compounds*. Chemistry: Leningrad, 1987, 288 pp.
322. Gerbeleu, N.V.; Arion, V.B.; Burgess, J. *Template Synthesis*. John Wiley: New York, 1999, 565 pp.
323. Fritsky, I.O.; Kozlowski, H.; Sadler, P.J.; Yefetova, O.P.; Swatek-Kozlowska, J.; Kalibabchuk, V.A.; Glowiak, T. *J. Chem. Soc., Dalton Trans.* No. 21, 3269 (1998).
324. Doble, D.M.J.; Benison, C.H.; Blake, A.J.; Fenske, D.; Jackson, M.S.; Kay, R.D.; Li, W.-Sh.; Schröder, M. *Angew. Chem., Int. Ed. Engl.* **38**(13/14), 1915 (1999).
325. Kruger, P.E.; Edwards, P.G.; Launau, F.; McKee, V. *Chem. Commun.* No. 7, 639 (1999).
326. Johnson, D. *Thermodynamic Aspects of Inorganic Chemistry*. Cambridge University Press: Cambridge, London, New York, 1982.
327. Burger, K. *Solvatation, Ionic and Complexes-Formation Reactions in Non-Aqueous Solvents*. Academiai Kiado: Budapest, 1983.
328. *Templated Organic Synthesis*. (Edit. Diederich, F.; Stang, P.J.). Wiley-VCH: Weinheim, 1999, 411 pp.
329. Linstead, R.P. *J. Chem. Soc.* 1016 (1934).
330. Byrne, G.T.; Linstead, R.P.; Lowe, A.R. *J. Chem. Soc.* 1017 (1934).
331. Linstead, R.P., Lowe, A.R. *J. Chem. Soc.* 1022 (1934).
332. Thomas, A.L. *Phtalocyanines. Research & Applications*. CRC Press: Boca Raton, FL, 1990.
333. *Phthalocyanines. Properties & Application*. (Edit. Leznoff, C.C.; Lever, A.B.P.). Verlag Chemie: Berlin, 1993.
334. Rodríguez-Morgade, S.; Hanack, M. *Chem. Eur. J.* **3**, 1042 (1997).
335. Kulinich, V.P.; Vasiliev, S.I.; Shaposhnikov, G.P.; Maislish, V.E.; Doroshina, O.A.; Smirnov, R.P. *Koord. Khim.* **23**, 785 (1997).
336. *Vitamin B_{12} and B_{12}-Proteins*. (Edit. Kräutler, B.; Golding, B.T.; Arigoni, D.). John Wiley: New York, 1998.
337. Garnovskii, A.D.; Alexeenko, V.A.; Kogan, V.A.; Bolotin, B.M.; Osipov, O.A.; Yusman, T.A.; Chernova, N.I. *Koord. Khim.* **3**, 500 (1977).

338. Garnovskii, A.D.; Alexeenko, V.A.; Burlov, A.S.; Lukov, V.V.; Nedzvetskii, V.S. *Koord. Khim.* **16**, 879 (1990).
339. Alexeenko, V.A.; Burlov, A.S.; Bolotin, B.M.; Lukova, O.A.; Nedzvetskii, V.S.; Garnovskii, A.D. *Koord, Khim.* **17**, 76 (1991).
340. Palii, S.P.; Simonov, Yu.A.; Bourosh, P.N.; Gebreleu, N.B.; Dvorkin, A.A. *Koord. Khim.* **22**, 138 (1996).
341. Gebreleu, N.V.; Arion, V.B. *Zhurn. Neorg. Khim.* **30**, 2833 (1985).
342. Gebreleu, N.V.; Simonov, Yu.A.; Arion, V.B.; Zavodnik, V.E.; Gredinaru, D.I.; Indrichan, K.M. *Zhurn. Neorg. Khim.* **36**, 906 (1991).
343. Jeyasubramanian, K.; Thambidurai, S.; Ramalingam, S.K.; Murugesan, R. *J. Inorg. Biochem.* **72**(3/4), 101 (1998).
344. Lukianenko, N.G. *Ukr. Khim. Zhurn.* **65**, 17 (1999).
345. Glinskaja, L.A.; Zemskova, S.M.; Klevzova, R.F. *Zhurn. Strukt. Khim.* **40**, 1206 (1999).
346. Edema, J.J.H.; Stock, H.T.; Buter, J.; Kellog, R.M.; Steets, J.J.; Spek, A.L.; Van Bolhuis, F. *Angew Chem.* **105**, 448 (1993); *Int. Ed. Engl.* **32**(3), 436 (1993).
347. Gebreleu, N.V.; Revenko, M.D. *Zhurn. Neorg. Khim.* **16**, 1046 (1971); **17**, 2176 (1972).
348. Gebreleu, N.V.; Revenko, M.D.; Ablov, A.V.; Polyakov, V.A. *Dokl. Akad. Nauk SSR* **208**, 599 (1973).
349. Ablov, A.V.; Gebreleu, N.V.; Revenko, M.D. *Zhurn. Neorg. Khim.* **19**, 2761 (1974).
350. Gebreleu, N.V.; Revenko, M.D.; Indrichan, K.M.; Subbotin, B.S. *Koord. Khim.* **2**, 386 (1976).
351. Garnovskii, A.D.; Osipov, O.A.; Orlova, L.V.; Minkin, V.I. *Zh. Neorg. Khim.* **10**, 2821 (1965).
352. Garnovskii, A.D.; Orlova, L.V.; Osipov, O.A. *Zh. Obsh. Khim.* **37**, 308 (1967).
353. Korobov, M.S.; Minkin, V.I.; Nivorozhkin, L.E. *Zhurn. Org. Khim.* **11**, 836 (1975).
354. Christensen, A.; Jensen, H.S.; McKee, V.; McKenzie, C.J.; Munch, M. *Inorg. Chem.* **36**(26), 6080 (1997).
355. Goswami, N.; Eichorn, D.M. *Inorg. Chim. Acta* **303**(2), 271 (2000).
356. Gairns, C.; Nelson, S.M.; Drew, M.G.B. *J. Chem. Soc., Dalton Trans.* 1965 (1981).
357. Bailey, N.A.; Eddy, M.M.; Fenton, D.E.; Moss, S.; Mudhopadhyay, A.; Jones, G. *J. Chem. Soc., Dalton Trans.* 2281 (1984).
358. Yam, V.W.-W.; Pui, Yu.-L.; Li, W.-L.; Lo, K.K-W.; Cheung, K.-K. *J. Chem. Soc., Dalton Trans.* No. 21, 3615 (1998).
359. Blackman, A. *Adv. Heterocycl. Chem.* **58**, 123 (1993).
360. Sadimenko, A.P.; Garnovskii, A.D.; Retta, N. *Coord. Chem. Rev.* **126**, 237 (1993).
361. Vasilchenko, I.S.; Kuzmenko, V.V.; Zaletov, V.G.; Indrichan, K.M.; Garnovskii, A.D. *Koord. Khim.* **15**, 1615 (1989).
362. Sergienko, V.S.; Abramenko, V.L.; Ilyukhin, A.B.; Porai-Koshits, M.A.; Garnovskii, A.D. *Koord. Khim.* **18**, 585 (1992).
363. Garnovskii, A.D.; Sergienko, V.S.; Bren, V.A.; Vasilchenko, I.S.; Mistryukov, A.E.; Ribalkin, V.P.; Minkina, L.S.; Kochin, S.G.; Porai-Koshitz, M.A. *Koord. Khim.* **38**, 252 (1993).
364. Jäger, E.G. *Z. Chem.* **1**, 30 (1968).
365. Jäger, E.G.; Kirchhof, B.; Schmidt, E. *Z. Anorg. Allg. Chem.* **485**, 141 (1982).
366. Jäger, E.G.; Schlenfogt, G.; Kirchhof, B. *Z. Anorg. Allg. Chem.* **485**, 173 (1982).
367. Hoyer, E.; Lorenz, B. *Z. Chem.* **1**, 28 (1968).
368. Mistryukov, A.E.; Vasilchenko, I.S.; Sergienko, V.S.; Nivorozhkin, A.L.; Kochin, S.G.; Porai-Koshitz, M.A.; Nivorozhkin, L.E.; Garnovskii, A.D. *Mendeleev Commun.* 30 (1992).
369. Garnovskii, A.D.; Gerbeleu, N.V.; Kochin, S.G.; Arion, V.B.; Minkina, L.S.; Indrichan, K.M.; Vasilchenko, I.S.; Revenko, M.D.; Konstantinovskii, L.E. *Dokl. Akad. Nauk SSSR* **306**, 872 (1989).

370. Garnovskii, A.D.; Kochin, S.G.; Minkina, L.S.; Vasilchenko, I.S.; Konstantinovskii, L.E. *Koord. Khim.* **15**, 258 (1989).
371. Vasilchenko, I.S.; Mistryukov, A.E.; Kochin, S.G.; Sergienko, V.S.; Porai-Koshitz, M.A.; Garnovskii, A.D. *Zhurn. Neorg. Khim.* **37**, 1047 (1992).
372. Garnovskii, A.D.; Mistryukov, A.E.; Kochin, S.G.; Sergienko, V.S.; Vasilchenko, I.S.; Minkina, L.S.; Porai-Koshitz, M.A. *Mendeleev Commun.* 110 (1991).
373. Garnovskii, A.D.; Gerbeleu, N.B.; Kochin, S.G.; Minkina, L.S.; Vasilchenko, I.S.; Arion, V.B.; Indrichan, K.M. *Koord. Khim.* **15**, 1330 (1989).
374. Garnovskii, A.D.; Kochin, S.G. *Zhurn. Obsh. Khim.* **65**, 829 (1995).
375. Thompson, M.C.; Busch, D. *J. Am. Chem. Soc.* **86**, 3651 (1964).
376. Jones, C.J.; McCleverty, J.A. *J. Chem. Soc. A*, 2819 (1970).
377. Sellmann, D.; Frank, P. *Angew. Chem., Int. Ed. Engl.* **25**, 1107 (1986).
378. Tschugaeff, L. *Chem. Ber.* **38**, 2520 (1905).
379. Chakravorty, A.*Coord. Chem. Rev.* **13**, 1 (1974).
380. Keeney, M.E.; Osseo-Assere, K.; Woode, K.A. *Coord. Chem. Rev.* **59**, 141 (1984).
381. Singh, R.P.; Singh, N.K. *Asian J. Chem. Rev.* **2**, 62 (1991).
382. Gerasimchuk, N.N.; Bowman-James, K. In: *Encyclopedia of Inorganic Chemistry*. John Wiley: Chichester, 1994, vol. 5, p. 2254.
383. Lalia-Kantouri, M.; Hartohulles, M.; Jannakoudakis, P.D.; Voutsas, G.P. *Z. Anorg. Allg. Chem.* **621**(4), 645 (1995).
384. Kukushkin, V.Yu.; Pombeiro, A.J.L. *Coord. Chem. Rev.* **156**, 333 (1996).
385. Mehrotra, R.C. In: Ref. 1, vol. 2, p. 269.
386. Kukushkin, V.Yu.; Tudeeva, D.; Isotova, Yu.A.; Belsky, V.K.; Stash, A.I. *Inorg. Chem.* **35**, 4929 (1996).
387. Kukushkin, V.Yu.; Pombeiro, A.J.L. *Coord. Chem. Rev.* **181**, 147 (1999).
388. Burlov, A.S.; Kuznetsova, L.I.; Garnovskii, A.D.; Kharisov, B.I.; Blanco, L.M.; Lukov, V.V.; Garnovskii, A.D. *J. Coord. Chem.* **47**, 467 (1999).
389. Kostromina, N.A.; Voloshin, Ya.Z.; Nazarenko, A.Yu. *Klathrochelates: Synthesis, Structure, Properties*. Naukova Dumka: Kiev, 1992, 234 pp.
390. Voloshin, Ya.Z. *Ross. Khim. Zhurn.***42**(1/2), 5 (1998).
391. Boston, D.R.; Rose, N.J. *J. Am. Chem. Soc.* **90**, 6859 (1968); **95**, 4163 (1973).
392. Kukushkin, V.Yu.; Belsky, V.K.; Tudela, D. *Inorg. Chem.* **35**, 510 (1996).
393. Grigg, J.; Collison, D.; Garner, C.D.; Helliwell, M.; Tasker, P.A.; Thorpe, J.M. *J. Chem. Soc., Chem. Commun.* 1807 (1993).
394. Schrauzer, G.N. *Chem. Ber.* **95**, 1438 (1962).
395. Umland, F.; Thierig, D. *Angew. Chem.* **74**, 388 (1962).
396. Jackels, S.C.; Zektzer, J.; Rose, N.J. *Inorg. Synth.* **17**, 139 (1978).
397. Rai, N.S.; Lena, A.K.; Sahoo, B. *Inorg. Chim. Acta* **35**, 29 (1979).
398. Nazarenko, A.I.; Voloshin, Ya.Z. *Zhurn. Neorg. Khim.* **29**, 1776 (1984).
399. Voloshin, Ya.Z.; Kostromina, N.A.; Nazarenko, A.Yu.; Polshin, E.V. *Ukr. Khim. Zhurn.* **55**, 7 (1989).
400. Voloshin, Ya.Z.; Kostromina, N.A.; Nazarenko, A.Yu.; Shuman, V.N. *Ukr. Khim. Zhurn.* **56**, 1238 (1990).
401. Voloshin, Ya.Z.; Kostromina, N.A.; Nazarenko, A.Yu. *Inorg. Chim. Acta* **170**, 181 (1990).
402. Voloshin, Ya.Z.; Kron, T.E.; Belsky, V.K.; Zavodnik, V.E.; Maletin, Yu.A.; Kozachkov, S.G. *J. Organomet. Chem.* **536/537**(1/2), 207 (1997).
403. (a) Voloshin, Ya.Z.; Varzatskii, O.A.; Palchik, A.V.; Polshin, E.V.; Maletin, Yu.A.; Strizhakova, N.G. *Polyhedron* **17**, 4315 (1998); (b) Voloshin, Ya.Z.; Varzatskii, O.A.; Palchik, A.V.; Strach, A.I.; Belsky, V.K. *New J. Chem.* **23**(4), 355 (1999); (c) Voloshin, Ya.Z.; Varzatskii, O.A.; Kron, T.E.; Belsky, V.K.; Zavodnik, V.E.; Strizhakova, N.G.; Palchik, A.V. *Inorg. Chem.* **39**, 1907 (2000); (d) Voloshin, Ya.Z.; Zavodnik, V.E.;

Varzatskii, O.A.; Belsky, V.K.; Vorontsov, I.I.; Antipin, M.Yu. *Inorg. Chim. Acta* **321**, 116 (2001); (e) Voloshin, Ya.Z.; Varzatskii, O.A.; Stash, A.J.; Belsky, V.K.; Bubnov, Yu.N.; Vorontsov, I.I.; Potekhin, K.A.; Antipin, M.Yu.; Polshin, E.V. *Polyhedron* **20**, 2721 (2001); (f) Voloshin, Y.Z.; Zavodnik, V.E.; Varzatskii, O.A.; Belsky, V.K.; Palchik, A.V.; Strizhakova, N.G.; Voronstov, I.I.; Antipin, M. Yu. *J. Chem. Soc., Dalton Trans.*, 1193 (2002).

404. (a) Voloshin, Ya.Z.; Polshin, E.V. *Polyhedron* **11**(4), 457 (1992); (b) Voloshin, Ya.Z.; Varzatskii, O.A.; Strizhakova, N.G.; Tkachenko, E.Yu. *Inorg. Chim. Acta* **299**, 104 (2000).
405. Voloshin, Ya.Z.; Polshin, E.V. *Polyhedron* **11**(4), 457 (1992).
406. Voloshin, Ya.Z.; Kostromina, N.A.; Nazarenko, A.Yu.; Polshin, E.V. *Inorg. Chim. Acta* **185**(1), 83 (1991).
407. Voloshin, Ya.Z.; Belsky, V.K.; Trachevskii, V.V. *Polyhedron* **11**(15), 1939 (1992).
408. Voloshin, Ya.Z.; Trachevskii, V.V. *J. Coord. Chem.* **31**(2), 147 (1994).
409. Voloshin, Ya.Z.; Trachevskii, V.V.; Polshin, E.V. *Polish J. Chem.* **71**(4), 428 (1997).
410. Voloshin, Ya.Z.; Varzatskii, O.A.; Korobko, S.V.; Maletin, V.A. *Inorg. Chem. Commun.* **1**, 328 (1998).
411. Voloshin, Ya.Z.; Varzatskii, O.A.; Tkachenko, E.Yu.; Maletin, V.A.; Degdyarev, S.P.; Kochubey, D.I. *Inorg. Chim. Acta* **255**(2), 255 (1997).
412. Voloshin, Ya.Z.; Stash, A.I.; Varzatskii, O.A.; Belsky, V.K.; Maletin, V.A.; Strizhakova, N.G. *Inorg. Chim. Acta* **284**, 180 (1999).
413. (a) Kukushkin, V.Yu.; Belsky, V.K.; Alexandrova, E.A.; Konovalov, V.E.; Kozokozyan, G.A. *Inorg. Chem.* **31**, 3836 (1992); (b) Kukushkin, V.Yu.; Pakhomova, T.B.; Kukushkin, Yu.N.; Herrmann, R.; Wagner, G.; Pombeiro, A.L. *Inorg. Chem.* **37**, 6511 (1998); (c) Wagner, G.; Pombeiro, A.J.L.; Kukushkin, Yu.N.; Pakhomova, T.B.; Ryabov, A.D.; Kukushkin, V.Yu. *Inorg. Chim. Acta* **292**, 272 (1999); (d) Kuznetsov, M.L.; Bokach, N.A.; Kukushkin, V.Yu.; Pakkanen, T.; Wagner, G.; Pombeiro, A.J.L. *J. Chem. Soc., Dalton Trans.* 4683 (2000).
414. Garnovskii, D.A.; Guerdes da Silva, M.F.C.G.; Pakhomova, T.B.; Wagner, G.; Duarte, M.T.; Frausta da Silva, J.J.R.; Pombeiro, A.J.L.; Kukushkin, V.Yu. *Inorg. Chim. Acta* **300–302**, 499 (2000).
415. Kukushkin, V.Yu.; Pakhomova, T.V.; Bokach, N.A.; Wagner, G.; Kuznetsov, M.L.; Golanski, M.; Pombeiro, A.J.L. *Inorg. Chem.* **39**, 216 (2000).
416. (a) Kukushkin, V.Yu.; Ilyichev, I.V.; Wagner, G.; Frausta da Silva, J.J.R.; Pombeiro, A.J.L. *J. Chem. Soc., Dalton Trans.* No. 17, 3047 (1999); (b) Luzyanin, I.; Kukushkin, V. Yu.; Kuznetsov, M.L.; Garnovskii, D.A.; Haukka, M.; Pombeiro, A.J.L. *Inorg. Chem.* **41**, 2981 (2002).
417. Kukushkin, V.Yu. *Coord. Chem. Rev.* **139**, 375 (1995).
418. Chen, H.; Hodges, L.M.; Liu, R.; Stevens, W.C.; Sabat, M.; Harman, W.D. *J. Am. Chem. Soc.* **116**(12), 5499 (1994).
419. *Reactions of Coordinated Ligands.* (Edit. Braterman, P.S.). Harcover: New York, London, 1986, vol. 1; Plenum Press: New York, 1989, vol. 2.
420. Constable, E.C. *Metal and Ligand Reactivity: An Introduction to the Organic Chemistry of Metal Complexes.* Verlag Chemie: New York, Weinheim, 1995.
421. Black, D.St. In: Ref. 1, vol. 1, p. 415.
422. Loft, M.S.; Mowlem, T.J.; Widdowson, D.A.; Williams, D.J. *J. Chem. Soc., Perkin Trans. I*, No. 2, 105 (1995).
423. Rosenblum, M. *Chemistry of Iron Group Metallocenes: Ferrocene, Ruthenocene, Osmocene.* John Wiley: New York, London, 1965.
424. Nesmeyanov, A.N. *Chemistry of Ferrocene.* Nauka: Moscow, 1969.
425. Wilkinson, G. *J. Organomet. Chem.* **100**, 273 (1975).

426. Nesmeyanov, A.N. *Chemistry of σ- and π-Complexes of Iron Manganese, and Rhenium*. Nauka: Moscow, 1980.
427. Pearson, A.J. *Metallo-Organic Chemistry*. John Wiley: New York, 1985.
428. O'Connor, J.M.; Casey, C.P. *Chem. Rev.* **87**, 307 (1987).
429. Ernst, R.D. *Chem. Rev.* **88**, 1255 (1988).
430. Strelets, V.V. *Coord. Chem. Rev.* **114**(1), 1 (1992).
431. Coville, N.J.; Du Plooy, K.E.; Pickl, W. *Coord. Chem. Rev.* **116**, 1 (1992).
432. Kolle, U. *Chem. Rev.* **98**, 1313 (1998).
433. *Metallocenes*. (Edit. Togni, A.; Halterman, R.L.) John Wiley: New York, 1997, vol. 1; 1998, vol. 2.
434. Grinberg, A.A. *Introduction in Chemistry of Complex Compounds*. Chemistry: Moscow, Leningrd, 1971, 631 pp.
435. Haud, A.; Strähle, J. *Z. Anorg. Allg. Chem.* **624**(6), 931 (1998).
436. Konze, W.V.; Young, V.G.; Angelici, R.C. *Organometallics* **18**(2), 258 (1999).
437. Kogan, V.A.; Sokolov, V.P.; Kochin, S.G. *Zh. Obsh. Khim.* **36**, 1857 (1966).
438. Sinn, E.; Harris, S.M. *Coord. Chem. Rev.* **4**, 391 (1969).
439. Agustin, D.; Rima, G.; Gornitzka, H.; Barrau, J. *Eur. J. Inorg. Chem.* 693 (2000).
440. Tupinashvili, N.A.; Pirukhalava, N.I.; Garnovskii, A.D. *Zhurn. Obsh. Khim.* **47**, 2586 (1977).
441. Kokot, S.M.; Harris, C.M.; Sinn, E. *Aust. J. Chem.* **25**, 45 (1972).
442. O'Connor, C.J.O.; Frenberg, D.P.; Sinn, E. *Inorg. Chem.* **18**, 1077 (1979).
443. Sinn, E. In: *Biochemical and Inorganic Copper Chemistry*. (Edit. Karlin, K.D.; Zubieta, J.). Academic Press: New York, 1985, vol. 2, p. 195.
444. Thanyasiri, T.; Sinn, E. *J. Chem. Soc., Dalton Trans.* No. 6, 1187 (1989).
445. (a) Ramade, I.; Kahn, O.; Jeannin, Y.; Robert, F. *Inorg. Chem.* **36**, 930 (1997); (b) Domínguez-Vera, J.M.; Camara, F.; Moreno, J.M.; Calacio, E.; Stoecti-Evans, H. *Inorg. Chem.* **37**, 3046 (1998); (c) Alikhanyan, A.S.; Malkerova, I.P.; Kuzmina, N.P.; Glez, A.; Khulve, M.; Sanz, G.L.; Eremenko, I.L. *Zhurn. Neorg. Khim.* **44**, 969 (1999); (d) Kuzmina, N.P.; Rogachev, A.Yu.; Spiridonov, F.M.; Dedlovskaya, E.M.; Kezko, V.A.; Glez, A.; Battiston, J. *Zhurn. Neorg. Khim.* **45**, 1468 (2000); (e) Ryazanov, M.V.; Troyanov, S.I.; Malkerova, A.S.; Alikhanyan, A.S.; Kuz'mina, N.P. *Russ. J. Inorg. Chem.* **46**, 208/2001); (f) Kuz'mina, N.P.; Ryazanov, M.V.; Ketsko, V.A.; Gleizes, A.N. *Russ. J. Inorg. Chem.* **47**, 26 (2002).
446. Coles, S.J.; Hursthouse, M.B.; Kelly, D.C.; Toner, A.J.; Walker, N.S. *J. Organomet. Chem.* **580**(2), 304 (1999).
447. Hassan, A.M.A.; Abu-Al-Nasr, A.K.; Halifa, M.A. *Koord. Khim.* **23**, 387 (1997).
448. Kuznetsova, L.I.; Burlov, A.S.; Lipunova, G.N.; Garnovskii, A.D. *Koord. Khim.* **16**, 417 (1990).
449. Samus, N.M.; Zakotovenko, S.V.; Khoroshun, I.V.; Tsapkov, V.I. *Zhurn. Neorg. Khim.* **40**, 1486 (1995).
450. (a) Shriver, D.F. *Acc. Chem. Res.* **3**, 231 (1970); (b) Horwitz, C.P.; Shriver, D.F. *Adv. Organomet. Chem.* **23**, 219 (1984).
451. Lokshin, B.V.; Ginzburg, A.G.; Nazarova, E.B. *Usp. Khim.* **49**, 223 (1980).
452. Weiss, J.; Stetzkamp, D.; Nuber, B.; Fisher, R.A.; Bochme, C.; Frenking, G. *Angew. Chem.* **109**, 95 (1997); *Int. Ed. Engl.* **36**(1/2), 70 (1997).
453. Holmes, N.J.; Lewason, W.; Webster, M. *J. Organomet. Chem.* **545/546**, 111 (1997).
454. Kircher, P.; Huttner, G.; Schiemenz, B. *Chem. Ber.* **130**(6), 687 (1997).
455. Baker, P.K.; Meehan, M.M. *J. Organomet. Chem.* **535**, 129 (1997).
456. Shaposhnikova, A.D.; Kamalov, G.L.; Stadnichenko, R.A.; Pasinskii, A.A.; Yanovskii, A.I.; Dolgushin, F.B.; Struchkov, Yu.T. *Zhurn. Neorg. Khim.* **43**, 402 (1998).
457. Uraev, A.I.; Garnovskii, A.D. *Zhurn. Obsh. Khim.* **66**, 1025 (1996).

458. (a) Kogan, V.A.; Morozov, A.N. *Mendeleev Chem. J.* **40**(4/5), 141 (1996); (b) Freisleben, D.; Polborn, K.; Rubl, C.; Sunkel, K.; Beck, W. *Can. J. Chem.* **73**, 1164 (1995).
459. Wu, Ya.J.; Cui, X.L.; Liu, Y.H.J. *J. Organomet. Chem.* **543**(1/2), 63 (1997).
460. Zhao, G.; Wang, O.-G.; Mak, T.C.W. *Polyhedron* **17**, 577 (1999).
461. Briel, O.; Fehn, A.; Polborn, K.; Beck, W. *Polyhedron* **18**, 225 (1999).
462. Kadkin, O.N.; Galyametdinov, Yu.G.; Rakhmatullin, A.I.; Mavrin, V.Yu. *Izv. RAN, Ser. Khim.* 381 (1999).
463. Li, J.T.; Wu, J.J.; Ch.-Sh. *Organometallics* **15**, 5028 (1996).
464. Spannenberg, A.; Arndt, P.; Oberhuz, M.; Kempe, R. *Z. Anorg. Allg. Chem.* **623**(3), 389 (1997).
465. Less, R.J.; Wick, J.L.M.; Chatterton, N.P.; Dewey, M.J.; Cromhout, N.L.; Halerow, A.M.; Davies, J.E. *J. Chem. Soc., Dalton Trans.* No. 21, 4055 (1996).
466. Jäkle, F.; Polborn, K.; Wagner, M. *Chem. Ber.* **129**(6), 603 (1996).
467. Chalhorda, M.J.; Canales, F.; Gimeno, M.C.; Jemenez, J.; Jones, P.C.; Laguna, A.; Veiros, L.F. *Organometallics* **16**(17), 3837 (1997).
468. Ma, J.-F.; Yamamoto, Ya. *J. Organomet. Chem.* **545/546**, 577 (1167).
469. Kuwano, R.; Sawamura, M.; Okuda, S. *Bull. Chem. Soc. Jpn.* **70**(11), 2807 (1997).
470. Vilchevskaya, V.D.; Baranetskaya, N.K.; Peredutov, A.S.; Makhaev, V.D.; Krilova, A.I. *Koord. Khim.* **24**, 327 (1998).
471. Abel, E.W.; Long, N.J.; Orrell, K.G.; Osborn, A.; Sik, V.; Bates, P.A.; Hursthouse, M. *J. Organomet. Chem.* **383**(1/3), 253 (1990).
472. Nairini, A.A.; Lai, C.K.; Ward, D.L.; Brubaker, C.H. *J. Organomet. Chem.* **390**(1), 73 (1990).
473. Koridze, A.A.; Zdanovich, V.I.; Sheloumov, A.M.; Lagunova, V.I.; Peredugov, A.S.; Dolgushin, G.I.; Yanovskii, A.I. *Organometallics* **16**(11), 2285 (1997).
474. Pyshnograeva, N.I.; Setkina, V.N.; Andrianov, V.G.; Struchkov, Yu.T.; Kursanov, D.N. *J. Organomet. Chem.* **157**, 431 (1978).
475. Pyshnograeva, N.I.; Setkina, V.N.; Kursanov, D.N. *Izv. AN SSSR, Ser. Khim.* 654 (1981).
476. Zagorevskii, D.V.; Nekrasov, Yus.S.; Yakushin, S.O.; Pyshnograeva, N.I.; Mazunov, V.A.; Muftakhov, M.V. *Metalloorgan. Khim.* **4**, 1157 (1991).
477. Pyshnograeva, N.I.; Setkina, V.N.; Kursanov, D.N. *Izv. AN SSSR, Ser. Khim.* 2778 (1984).
478. (a) Pyshnograeva, N.I.; Batsanov, A.S.; Struchkov, Yu.T.; Ginsburg, A.G.; Setkina, V.M. *J. Organomet. Chem.* **297**, 69 (1985); (b) Peterleiter, M.G.; Denisovich, L.I.; Pyshnograeva, N.I.; Kravtsov, D.N. *Metalloorg. Khim.* **3**, 581 (1990).
470. Kuhn, N.; Schulter, M.; Zauder, E. *Chem. Ber.* **122**, 1891 (1989).
480. Kuhn, N.; Horn, E.M.; Boese, R.; Bläser, D. *Chem. Ber.* **122**, 2275 (1989).
481. Kuhn, N.; Kuhn, A.; Lampe, E.-M. *Chem. Ber.* **124**(5), 997 (1991).
482. Best, S.P.; Clark, R.J.H.; Deeming, A.J.; McQueln, R.C.S.; Powell, N.L.; Acuna, C.; Arece, A.J.; De Santis, Y. *J. Chem. Soc.*, Dalton Trans. 1111 (1991).
483. Ginsburg, A.G.; Batsanov, A.S.; Struchkov, Yu.T. *Metalloorgan. Khim.* **4**, 854 (1991).
484. Brunnet, L.; Mercier, F.; Ricard, L. *Angew. Chem.* **106**, 812 (1994).
485. Ganter, C.; Brassat, L.; Ganter, B. *Chem. Ber.* **130**, 1771 (1997).
486. Desmurs, P.; Visslaux, M.; Baudry, D.; Dermond, A.; Nief, F.; Ricard, L. *Organometallics* **15**, 4178 (1996).
487. Sava, X.; Mezailles, N.; Richard, L.; Mathey, F.; Le Floch, H. *Organometallics* **18**, 807 (1999).
488. Waldbach, T.A.; Van Rooyen, P.H.; Lotz, S. *Angew. Chem., Int. Ed. Engl.* **32**, 710 (1993).
489. Choi, M.G.; Angelici, R.J. *J. Am. Chem. Soc.* **112**, 7811 (1993).

490. Minghetti, G.; Banditelli, G.; Bonati, F. *J. Chem. Soc., Dalton Trans.* 1851 (1979).
491. Oro, L.A.; Pinollas, M.T.; Tejel, C. *J. Chem. Soc., Dalton Trans.* 1087 (1986); 1927 (1988).
492. Uraev, A.I.; Garnovskii, A.D.; Magomedova, N.S.; Belskii, V.K. *Zhurn. Obsh. Khim.* **65**, 1251 (1995).
493. Erust, S.; Kaim, W. *J. Am. Chem. Soc.* **108**, 3578 (1986).
494. Zhang, Z.Z.; Cheng, H. *Coord. Chem. Rev.* **147**, 1 (1996).
495. Li, S.K.; Mak, T.C.W., Zhang, Z.Z. *J. Chem. Soc., Dalton Trans.* No. 16, 3475 (1996).
496. Ryabov, A.D. *Mendeleev Chem.* J. **40**(4/5), Part II, 46 (1996).
407. Bosque, R.; López, C. *Polyhedron* **18**, 135 (1999).
498. Stevens, A.M.; Richards, C.J. *Organometallics* **18**(7), 1346 (1999).
499. Pasinskii, A.A.; Eremenko, I.L. *Zhurn. Vses. Khim. Obsh. Mendeleeva* **32**(1), 88 (1987).
500. Gubin, S.P. *Koord. Khim.* **20**, 403 (1994).
501. Richter, E.; Vahrenkampf, H. *Angew. Chem., Int. Ed. Engl.* **18**, 531 (1979).
502. Pasinskii, A.A.; Eremenko, N.L.; Gasanov, G.Sh.; Orazsakhatov, B.; Kalinnikov, V.T.; Shklover, V.E.; Struchkov, Yu.T. *Koord. Khim.* **10**, 634 (1984).
503. (a) Sasaki, M.; Manseki, K.; Horiuchi, H.; Kamagai, M.; Sakamoto, M.; Sakiyama, H.; Nishida, Y.; Sakai, M.; Sadaoka, Y.; Ohba, M.; Okawa, H. *J. Chem. Soc., Dalton Trans.* No. 3, 259 (2000); (b) Badyal, K.; McWhinnie, W.R.; Hammor, T.A.; Chen, H. *Organometallics* **16**(14), 3194 (1997).
504. Gedes, B. *J. Prakt. Chem.* **26**, 257 (1892).
505. Konev, V.A.; Kukushkin, V.Yu.; Kukushkin, Yu.N. *Zhurn. Neorg. Khim.* **31**, 1466 (1996).
506. Chakravorty, M.C.; Subrahmanyam, C.V.P. *Coord. Chem. Rev.* **135/136**, 64 (1994).
507. Garnovskii, A.D.; Blanco, L.M.; Kharisov, B.I.; Garnovskii, D.A.; Burlov, A.S. *J. Coord. Chem.* **48**, 219 (1999).
508. Blackborow, J.R.; Young, D. *Vapor Synthesis in Organometallic Chemistry*. Springer-Verlag: Berlin, Heidelberg, New York, 1979.
509. Garnovskii, A.D.; Kharisov, B.I.; Gójon-Zorrilla, G.; Garnovskii, D.A. *Russ. Chem. Rev.* **64**(3), 201 (1995).
510. Gójon-Zorrilla, G.; Kharisov, B.I.; Garnovskii, A.D. *Rev. Soc. Quím. Méx.* **40**, 131 (1996).
511. Garnovskii, A.D.; Kharisov, B.I.; Gójon-Zorrilla, G.; Garnovskii, D.A.; Burlov, A.S. *Russ. J. Coord. Chem.* **23**, 223 (1997).
512. Kharisov, B.I.; Garnovskii, A.D.; Blanco, L.M.; Burlov, A.S.; García-Luna, A. *J. Coord. Chem.* **49**, 113 (1999).
513. Garnovskii, A.D.; Ryabukhin, Yu.I., Kuzharov, A.S. *Koord. Khim.***10**, 1011 (1984).
514. Klabunde, K.J. *Chemistry of Free Atoms and Particles*, Academic Press: New York, 1980, 238 pp.
515. Klabunde, K.J. *Free Atoms, Clusters, and Nanoscale Particles*. Academic Press: San Diego, CA, 1994, 311 pp.
516. Domrachev, G.A.; Zinoviev, V.D. *Usp. Khim.* **47**, 672 (1978).
517. Timms, P.L. *Adv. Inorg. Chem. Radiochem.* **14**, 121 (1972).
518. Timms, P.L. *Acc. Chem. Res.* **6**, 118 (1973).
519. Timms, P.L. *Angew. Chem., Int. Ed. Engl.* **14**, 273 (1975).
520. *Cryochemistry* (Edit. Moskovits, M.; Ozin, G.A.). John Wiley: New York, 1976.
521. Sergeev, G.B. *Zurn. Vses. Khim. Obsh. Mendeleeva* **35**, 566 (1990).
522. Carroll, J.J.; Hang, K.L.; Weisshaar, J.C. *J. Am. Chem. Soc.* **115**, 6962 (1993).
523. Ritter, D.; Weisshaar, J.C. *J. Am. Chem. Soc.* **112**, 6425 (1990).
524. Blitz, M.A.; Mitchell, S.A.; Hackett, P.A. *J. Phys. Chem.* **95**(72), 8719 (1991).
525. Razuvaev, G.A.; Domrachev, G.A.; Zinoviev, V.D. *Dokl. AN SSSR* **223**, 617 (1975).
526. Graves, V.; Lagowskii, J.J. *Inorg. Chem.* **15**, 577 (1976).

527. Yurieva, L.P.; Zaitseva, N.N.; Uralets, I.A.; Peregudova, S.I.; Kravtsov, D.N.; Vasilkov, A.Yu.; Sergeev, V.A. *Metalloorgan. Khim.* **3**, 783 (1990).
528. Cloke, F.G.N.; Lappert, N.F.; Lewless, G.A.; Swain, A.C. *Chem. Commun.* 1667 (1987).
529. Cloke, F.G.N.; Courtney, K.A.E.; Sameh, A.A.; Swain, A.C. *Polyhedron* **8**(13/14), 1641 (1989).
530. Klabunde, K.J. *New Synth. Meth.* **3**, 135 (1975).
531. Klabunde, K.J.; Efner, H.F. *Inorg. Chem.* **14**(4), 789 (1975).
532. Kündig, E.P.; Timms, P.L. *J. Chem. Soc., Chem. Commun.* 912 (1977); *J. Chem. Soc., Dalton Trans.* 991 (1980).
533. Hawker, P.N.; Timms, P.L. *J. Chem. Soc., Chem. Commun.* 730 (1978); *J. Chem. Soc., Dalton Trans.* 1123 (1983).
534. Cintas, P. *Activated Metals in Organic Synthesis*. CRC Press: Boca Raton, FL, 1993, pp. 12–44.
535. Wücherer, E.J.; Muetterties, E.L. *Organometallics* **6**, 1691 (1987).
536. Simmons, L.H.; Riley, P.E.; Davies, R.F.; Lagowskii, J.J. *J. Am. Chem. Soc.* **98**, 1044 (1976).
537. (a) Eischenbroich, C.; Koch, J.; Kroker, J.; Wünsch, M.; Massa, W.; Baum, G.; Stork, G. *Chem. Ber.* **121**, 1983); (b) Cloke, F.G.; Hitchcock, P.B.; Nixon, J.F.; Vickers, D.M. *J. Organomet. Chem.* **635**(1/2), 212 (2001).
538. Arnold, P.L.; Cloke, F.G.N.; Hitchcock, P.B. *J. Chem. Soc., Chem. Commun.* No. 5, 481 (1997).
539. Von Gustov, E.A.K.; Jenicke, O.; Wolfbeis, O.; Eady, C.R. *Angew. Chem., Int. Ed. Engl.* **14**, 278 (1975).
540. Mazurenko, E.A. Dr. Hab. Thesis (Chemistry), Kiev, 1987.
541. Kuzharov, A.S.; Kutkov, A.A.; Suchkov. V.V. *Koord. Khim.* **6**, 1123 (1980).
542. Collin, G.F.; Timms, P.L. *J. Chem. Soc., Dalton Trans.* **14**, 1401 (1980).
543. Ozin, G.A.; Andrews, M.P.; West, R. *Inorg. Chem.* **25**, 580 (1986).
544. Timms, P.L. *J. Chem. Soc. A* **13**, 2526 (1970).
545. Hanlan, A.J.L., Ozin, G.A.; Power, W.I. *Inorg. Chem.* **17**, 3648 (1978).
546. Skell, P.S.; Vam Dam, E.M.; Silvon, M.P. *J. Am. Chem. Soc.* **96**, 626 (1974).
547. Yao, Z.; Klabunde, K.J.; Asirvatham, A.S. *Inorg. Chem.* **34**(21), 5289 (1995).
548. Ozin, G.A.; McIntosh, D.F.; Power, W.J.; Messmer, R.P. *Inorg. Chem.* **20**, 1782 (1981).
549. Piper, M.J.; Timms, P.L. *J. Chem. Soc., Chem. Commun.* 50 (1972).
550. Chenier, J.H.B.; Howard, J.A.; Joly, H.A.; LeDuc, M.; Mile, B. *J. Chem. Soc., Faraday Trans.* **86**(19), 3321 (1990).
551. Tomilov, A.P.; Chernih, I.N.; Kargin, Yu.M. *Electrochemistry of Elemento-Organic Compounds. Elements of I, II, and III Groups of the Periodic System and Transition Metals*. Science: Moscow, 1985, 254 pp.
552. Grobe, I. *Comm. Inorg. Chem.* **9**, 149 (1990).
553. *Electrochemical Synthesis*. (Edit. Little, R.D.; Weinberg, N.L.). Marcel Dekker: New York, 1991.
554. (a) Tuck, D.G. In: *Molecular Electrochemistry of Inorganic Bioinorganic, and Organometallic Compounds*. Dortrecht: Kluwer Academic, 1993, pp. 15–31; (b) García-Vázquez, J.A.; Romero, J.; Sousa, A. *Coord. Chem. Rev.* **193–195**, 691 (1999); (c) Uraev, A.I.; Vlasenko, V.G.; Kharisov, B.I.; Blanco, L.M.; Shuvaev, A.T.; Vasilchenko, I.S.; Garnovskii, A.D.; Elizondo, N.V. *Polyhedron* **19**, 2361 (2000); (d) Long, G.; Willert, R.D. *Inorg. Chim. Acta* **313**, 1 (2001); (e) Bicherov, A.V.; Kharisov, B.I.; Blanco, L.M.; Korshunov, O.Yu.; Koroleva, E.L.; Burlov, A.S.; Borodkin, G.S.; Kurbatov, V.P.; Uflyand, I.E.; Garnovskii, A.D. *J. Coord. Chem.* **54**, 337 (2001).
555. Lehmkuhl, H. *Synthesis* **7**, 377 (1973).

556. Grobe, J.; Keil, M.; Schneider, B.; Zimmermann, H. *Z. Naturforsch. B* **35**, 428 (1980).
557. Compton, R.G.; Eklund, J.C.; Page, S.D.; Rebbit, T.O. *J. Chem. Soc., Dalton Trans.* No. 3, 389 (1995).
558. Chakravorty, M.C.; Subrahmanyam, G.V.B. *Polyhedron* **11**, 3191 (1992).
559. Tuck, D.G. *Pure & Appl. Chem.* **51**, 2005 (1979).
560. Habeeb, J.J.; Said, F.F.; Tuck, D.G. *J. Chem. Soc., Dalton Trans.* 118 (1981).
561. Chakravorty, M.C.; Subrahmanyam, G.V.B. *Can. J. Chem.* **70**, 836 (1992).
562. Habeeb, J.J.; Said, F.F.; Tuck, D.G. *Can. J. Chem.* **55**, 3882 (1977).
563. Oldekop, Yu.A.; Maier, N.A.; Shirokii, V.L.; Chernishov, E.A.; Bakhtiarov, A.V.; Kabanov, B.K.; Rodnikov, I.A.; Tomilov, A.P. *Izv. AN BSSR, Ser. Khim.* **3**, 89 (1981).
564. Shirokii, V.L.; Sutormin, A.B.; Maier, N.A.; Oldekop, Yu.A. *Zhurn. Obsh. Khim.* **53**, 1892 (1983).
565. Casey, A.T.; Vecchio, A.M. *Appl. Organomet. Chem.* **4**, 513 (1990).
566. Kumar, R.; Tuck, D.G. *J. Organomet. Chem.* **281**, 47 (1985).
567. Bogdashev, N.N.; Garnovskii, A.D.; Grigoriev, V.P. *Izv. Sev.-Kavkaz. Nauchn. Tsentra Vissh. Shkoli.* **2**, 28 (1973).
568. Bogdashev, N.N.; Garnovskii, A.D.; Osipov, O.A.; Grigoriev, V.P.; Gontmakher, N.M. *Zhurn. Obsh. Khim.* **46**, 675 (1976).
569. Arakelyan, N.M.; Papyan, S.A.; Isabekyan, S.E.; Darbinyan, E.G. *Am. Khim. Zhurn.* **35**, 406 (1982); *Elektrokhim.* **19**, 940 (1983).
570. Kumar, R.; Tuck, D.G. *Inorg. Chim. Acta* **157**, 51 (1989).
571. Durán, M.L.; García-Vázquez, J.A.; Romero, J.; Castiñeiras, A.; Sousa, A.; Garnovskii, A.D.; Garnovskii, D.A. *Polyhedron* **16**, 1707 (1997).
572. Burlov, A.S.; Kuznetsova, L.I.; Garnovskii, D.A.; Kharisov, B.I.; Blanco, L.M.; Lukov, V.V.; Garnovskii, A.D. *J. Coord. Chem.* **47**, 467 (1999).
573. García-Vázquez, J.A.; Romero, J.; Durán, M.L.; Sousa, A.; Garnovskii, A.D.; Burlov, A.S.; Garnovskii, D.A. *Polyhedron* **17**, 1547 (1998).
574. Romero, J.; García-Vázquez, J.A.; Durán, M.L.; Castiñeiras, A.; Sousa, A.; Garnovskii, A.D.; Garnovskii, D.A. *Acta Chem. Scand.* **51**, 672 (1997).
575. Garnovskii, D.A., Romero, J.; García-Vázquez, J.A.; Durán, M.L.; Castiñeiras, A.; Sousa, A.; Burlov, A.S.; Garnovskii, A.D. *Koord. Khim.* **24**, 392 (1998).
576. Castro, J.A.; Romero, J.; García-Vázquez, J.A.; Sousa, A.; Castellano, F.E.; Zukerman-Spektor, J. *J. Coord. Chem.* **28**, 125 (1993).
577. Caballero, S.; Castro, J.A.; Vázquez-López, E.; García-Vázquez, J.A.; Romero, J.; Sousa, A. *Polyhedron* **18**, 1669 (1999).
578. Lehmkuhl, H.; Eisenbach, W. *Lieb. Ann.* **4**, 672 (1975).
579. Habeeb, J.J.; Tuck, D.G.; Walters, F.N. *J. Coord. Chem.* **8**, 27 (1978).
580. Matassa, L.; Kumar, N.; Tuck, D.G. *Inorg. Chim. Acta* **109**, 19 (1985).
581. Kostyuk, N.N.; Kolevich, T.A.; Shirokii, V.L.; Umbreiko, D.S. *Koord. Khim.* **15**, 1704 (1989).
582. Kostyuk, N.N.; Shirokii, V.L.; Dik, T.A.; Vinokurov, I.I.; Umreiko, D.S. *Koord. Khim.* **17**, 1573 (1991).
583. Kostyuk, N.N.; Shirokii, V.L.; Vinokurov, I.I.; Dik, T.A.; Umreiko, D.S.; Edman, A.A. *Zhurn. Neorg. Khim.* **37**, 68 (1992).
584. Kostyuk, N.N.; Shirokii, V.L.; Vinokurov, I.I.; Maier, N.A. *Zhurn. Obsh. Khim.* **64**, 1432 (1994).
585. Kumar, N.; Tuck, D.G. *Can. J. Chem.* **60**, 25 (1982).
586. Annan, T.A.; Peppe, C., Tuck, D.G. *Can. J. Chem.* **68**, 423 and 1598 (1990).
587. Geloso, C.; Kumar, R.; López-Grado, J.; Tuck, D.G. *Can. J. Chem.* **65**, 928 (1987).
588. Kharisov, B.I.; Blanco, L.M.; Salinas, M.V.; Garnovskii, A.D. *J. Coord. Chem.* **47**, 135 (1999).

589. Zhdanov, Yu.A.; Osipov, O.A.; Gontmakher, N.M.; Grigoriev, V.P.; Garnovskii, A.D.; Nechaeva, O.N.; Bogdashev, N.N.; Alexeev, Yu.E.; Alexeeva, V.G.; Yusman, T.A. *Dokl. AN SSSR* **227**, 882 (1976).
590. Zhdanov, Yu.A.; Osipov, O.A.; Grigoriev, V.P.; Garnovskii, A.D.; Alexeev, Yu.E.; Gontmakher, N.M.; Alexeeva, V.G.; Yusman, T.A.; Zaletov, V.G.; Bogdashev, N.N.; Nechaeva, O.N. *Koord. Khim.* **3**, 3 (1977).
591. Svyatkina, L.I., Baikalova, L.V.; Domnina, E.S. *Koord. Khim.* **21**, 496 (1995).
592. Castro, J.A.; Romero, J.; García-Vázquez, J.A.; Castineiras, A.; Sousa, A. *Z. Anorg. Allg. Chem.* **619**, 601 (1993).
593. Castro, J.A.; Romero, J.; García-Vázquez, J.A.; Macias, A.; Sousa, A.; Englert, U. *Polyhedron* **12**, 1391 (1993).
594. Garnovskii, D.A.; Sousa, A.; Antsishkina, A.S.; Sadikov, G.G.; Vasilchenko, I.S.; Sigeikin, S.G.; Burlov, A.S.; Castiñeiras, A.; Garnovskii, A.D. *Russ. Chem. Bull.* **45**, 1988 (1996).
595. Garnovskii, D.A.; Antsishkina, A.S.; Sadikov, G.G.; Sousa, A.; Burlov, A.S.; Vasilchenko, I.S.; Garnovskii, A.D. *Russ. J. Inorg. Chem.* **43**, 1723 (1998).
596. Kharisov, B.I.; Garnovskii, D.A.; Blanco, L.M.; Burlov, A.S.; Vasilchenko, I.S.; Garnovskii, A.D. *Polyhedron* **18**, 985 (1999).
597. Bernal, M.; García-Vázquez, J.A.; Romero, J.; Gómez, C.; Durán, M.L.; Rose, D.J.; Maresca, K.P.; Zubieta, J. *Inorg. Chim. Acta* **295**, 39 (1999).
598. Garnovskii, A.D.; Burlov, A.S.; Garnovskii, D.A.; Vasilchenko, I.S.; Antsishkina, A.S.; Sadikov, G.G.; Sousa, A.; García-Vázquez, J.A.; Romero, J. *Koord. Khim.* **23**, 399 (1997).
599. Blanco, L.M.; Kharisov, B.I.; Burlov, A.S.; Garnovskii, A.D.; Kuznetsova, L.I.; Vasilchenko, I.S. *Curr. Sep.* **18**, 41 (1999).
600. Bermejo, M.R.; Fondo, M.; Gonzáles, A.M.; Hoyos, O.L.; Sousa, A.; McAuliffe, C.A.; Hussain, W.; Pritchard, R.; Novotorsev, V.M. *J. Chem. Soc., Dalton Trans.* 2211 (1999).
601. Méndez-Rojas, M.A.; Cordova-Lozano, F.; Gójon-Zorrilla, G.; González-Vergara, E.; Quiroz, M.A. *Polyhedron* **18**, 2651 (1999).
602. Cini, A.; Giorgi, G.; Cintquanti, G.; Rossi, C.; Sabat, M. *Inorg. Chem.* **29**. 5197 (1990).
603. Said, F.F.; Tuck, D.G. *Inorg. Chim. Acta* **59**, 1 (1982).
604. Hencher, J.L.; Khan, M.; Said, F.F.; Sieler, R.; Tuck, D.G. *Inorg. Chem.* **21**, 2787 (1982).
605. Bastida, R.; Bermejo, M.R.; Louro, M.S.; Romero, J.; Sousa, A.; Fenton, D.E. *Inorg. Chim. Acta* **145**, 167 (1988).
606. Labisbal, E.; De Blas, A.; García-Vázquez, J.A.; Romero, J.; Durán, M.L.; Sousa, A.; Bailey, N.A.; Fenton, D.E.; Leeson, P.B.; Parish, R.V. *Polyhedron* **11**, 227 (1992).
607. Labisbal, E.; Romero, J.; Durán, M.L.; García-Vázquez, J.A.; Sousa, A.; Russo, U.; Pritchard, R.; Renson, M.J. *J. Chem. Soc., Dalton Trans.* 755 (1993).
608. Labisbal, E.; García-Vázquez, J.A.; Romero, J.; Sousa, A.; Castiñeiras, A.; Maichle-Mossmer, C.; Russo, U. *Inorg. Chim. Acta* **223**, 87 (1994).
609. Labisbal, E.; Romero, J.; De Blas, A.; García-Vázquez, J.A.; Durán, M.L.; Castiñeiras, A.; Sousa, A.; Fenton, D.E. *Polyhedron* **11**, 53 (1992).
610. Lukova, G.V.; Bumber, A.A.; Burlov, A.S.; Okekhnovich, E.P.; Garnovskii, A.D. *Zhurn. Obsh. Khim.* **68**, 637 (1998).
611. Lukova, G.V.; Olekhnovich, E.P.; Kletskii, M.E.; Bumber, A.A.; Garnovskii, A.D. *Izv. RAN., Ser. Khim.* 690 (1999).
612. Lukova, G.V.; Garnovskii, A.D. *Izv. RAN, Ser. Khim.* 1552 (1999); *Russ. Chem. Bull.* 1503 (1999).
613. Perez-Lourido, P.; Romero, J.; García-Vázquez, J.A.; Sousa, A.; Maresca, K.P.; Zubieta, J. *Inorg. Chem.* **38**, 3700 (1999).

614. Bott, R.C.; Haely, P.C.; Sagatys, D.S. *J. Chem. Soc., Chem. Commun.* 2403 (1998).
615. Garnovskii, D.A.; Burlov, A.S.; Garnovskii, A.D.; Vasilchenko, I.S.; Sousa, A. *Zhurn. Obsh. Khim.* **60**, 1546 (1996).
616. Garnovskii, D.A.; Vasilchenko, I.S.; Eliseeva, A.Yu.; Sousa, A.; Sadimenko, A.P.; Garnovskii, A.D. *Zhurn. Obsh. Khim.* **69**, 1533 (1999).
617. Kharisov, B.I.; Blanco-Jerez, L.M.; García-Luna, A. *Rev. Soc. Quím. Méx.* **43**, 50 (1999).
618. Yang, C.H.; Lin, S.F.; Chen, H.L.; Chang, C.T. *Inorg. Chem.* **19**, 3541 (1980).
619. Guainazzi, M.; Silvestri, G.; Gambino, S.; Filardo, G. *J. Chem. Soc., Dalton Trans.* 927 (1972).
620. Abers, M.O.; Coville, N.J. *Coord. Chem. Rev.* **53**, 227 (1984).
621. Riger, P.H. *Electrochemistry*. London: Chapman and Hall, 1993.
622. Grobe, J.; Zimmermann, H. *Z. Naturforsh. B.* **36**, 301 and 482 (1981).
623. Sock, O.; Lemoine, P.; Cross, M. *Electrochim. Acta* **26**, 99 (1981).
624. Miskiv, M.G.; Fayad, H.; Zavodnik, V.E. *Metalloorgan. Khim.* **4**, 415 (1991).
625. Miskiv, M.G.; Oleinik, V.V.; Akselrod, L.G. *Koord. Khim.* **20**, 697 (1994).
626. Filinchuk, Ya.E.; Shollmaier, D.; Oleinik, V.V.; Goreshnik, E.A.; Miskiv, M.G. *Koord. Khim.* **22**, 870 (1996).
627. Oleinik, V.V.; Miskiv, M.G. *Koord. Khim.* **22**, 876 (1996).
628. Oleinik, V.V.; Mikhalichko, B.M.; Shollmaier, D.; Miskiv, M.G. *Koord. Khim.* **23**, 383 (1997).
629. (a) Filipchuk, Ya.E.; Miskiv, M.G. *Koord. Khim.* **25**, 55 and 466 (1999); (b) Gu, J.-S.; Ma, M.-H.; Shen, L.-M.; Yu, K.-B. *Acta Chim. Sin.* **58**(10), 1270 (2000); (c) Davies, J.A.; Eagle, C.T.; Burckel, P. *Inorg. Chim. Acta* **300–302**, 645 (2000); (d) Labisbal, E.; Sousa, A.; West, D.X. *Z. Naturforsch. B* **55**(2), 162 (2000); (e) Denisovich, L.I.; Peterleitner, M.G.; Peregudova, S.M. *Russ. Chem. Bull.* **49**(2), 375 (2000); (f) Alzuet, G.; Real, J.A.; Borras, J.; Santiago-García, R.; García-Granda, S. *Inorg. Chem.* **40**(10), 2420 (2001); (g) Sanchez-Pizo, J.; García-Vazquez, J.A.; Romero, J.; Duran, M.L.; Sousa-Pedrares, A.; Labisbal, E.; Nascimento, O.R. *Inorg. Chim. Acta* 328(1), 111 (2002).
630. Habeeb, J.J.; Said, F.F.; Tuck, D.G. *Can. J. Chem.* **55**, 3383 (1977).
631. Kaabak, L.V.; Tomilov, A.P. *Zhurn. Obsh. Khim.* **67**(2), 341 (1997).
632. Casey, A.T.; Veccio, A.M. *Organomet. Chem.* **4**, 513 (1990).
633. Romero, J.; Duran, M.; García-Vázquez, J.A.; Castineiras, A.; Sousa, A.; Cristiansen, L.; Zubieta, J. *Inorg. Chim. Acta* **255**, 307 (1997).
634. Kharisov, B.I.; Garnovskii, A.D.; Gójon-Zorrilla, G.; Berdonosov, S.S. *Rev. Soc. Quím. Méx.* **40**, 173 (1996).
635. Sato, K.; Kammoni, O. *Bull. Chem. Soc. Jpn.* **42**, 2278 (1969).
636. Kurskov, S.N.; Levlev, I.N.; Khidekel, M.L. *Izv. AN SSSR* 1708 (1977).
637. Drozdov, A.A.; Troyanov, S.I. *Koord. Khim.* **20**, 171 (1994).
638. Kurskov, S.N.; Varlamov, N.V.; Labunskaya, V.I.; Koblova, O.E.; Bolshakov, A.F. *Koord. Khim.* **20**, 70 (1994).
639. Ryabukhin, Yu.I.; Shibaeva, N.V.; Kuzharov, A.S.; Korobkova, V.G.; Khokhlov, A.V.; Garnovskii, A.D. *Koord. Khim.* **13**, 869 (1987).
640. Shova, S.G.; Brashovyanu, N.V.; Turte, K.I.; Mazus, M.D. *Koord. Khim.* **22**, 466 (1996).
641. Zevago, T.A.; Trotzki, R.; Görls, H.; Dinjus, E. *Inorg. Chem. Commun.* **1**, 30 (1998).
642. Sofen, R.; Ware, D.C.; Cooper, S.R.; Raymond, K.N. *Inorg. Chem.* **18**, 234 (1979).
643. Prokofiev, A.I.; Malisheva, N.A.; Bubnov, N.N.; Solodovnikov, S.P.; Kabachnik, M.I. *Dokl. AN SSSR* **370**, 252 (1980).
644. Ozarowskii, A.; Carvey, B.R.; Peppe, C.; Tuck, D.G. *J. Am. Chem. Soc.* **113**, 3288 (1991).

645. Adams, D.M.; Reingold, A.L.; Dei, A.; Hendrickson, D.N. *Angew. Chem.* **105**, 434 (1993).
646. Liu, F.; Gornitzka, H.; Stalke, D.; Roesky, H.W. *Angew. Chem.* **105**, 447 (1993).
647. Lavrentiev, I.P.; Khidekel, M.L. *Usp. Khim.* **52**, 596 (1983).
648. Hathaway, B.J.; Holan, D.G. *J. Chem. Soc.* 2408 (1964).
649. Kurskov, S.N.; Levleva, I.N.; Lavrentiev, I.P.; Khidekel, M.L. *Izv. AN SSSR* 1708 (1977).
650. (a) Kurskov, S.N. *Koord. Khim.* **13**, 1082 (1987); (b) Gutiérrez, R.; Vázquez, J.; Vázquez, R.A.; Reyes, Y.; Toscano, R.A.; Martínez, M.; Álvarez, C. *J. Coord. Chem.* **54**, 313 (2001).
651. Nakamura, A.; Kakajama, Y.; Shibahara, T.; Mashima, K. 31st ICCC, Vancouver, Canada, 1996, p. 3L14.
652. Buchikhin, E.P.; Kuznetsov, A.Yu.; Chekmarev, A.M.; Lavrentiev, I.P. *Koord. Khim* **25**, 444 (1999).
653. Naifon, P.T.; McAuliffe, C.A.; Mackie, A.G.; Pritchard, R.G. *Inorg. Chim. Acta* **282**, 25 (1998).
654. Nifontova, G.A.; Krasochka, O.N.; Lavrentiev, I.P.; Makitov, D.D.; Atovmyan, L.O.; Khidekel, M.L. *Izv. AN SSSR, Ser. Khim.* 450 (1988).
655. (a) Makitov, D.D.; Roshchupkina, O.S.; Lavrentiev, I.P.; Atovmyan, L.O.; Khidekel, M.L. *Koord. Khim.* **14**, 394 (1988); (b) Makitov, D.D.; Krasochka, O.V.; Atovmyan, L.O.; Letuchii, Ya.A.; Roshchupkina, O.S.; Lavrentiev, I.P.; Khidekel, M.L. *Koord. Khim.* **13**, 1650 (1987).
656. Nifontova, G.A.; Filipenko, O.S.; Astakhova, A.S.; Lavrentiev, I.P.; Atovmyan, L.O. *Koord. Khim.* **16**, 218 (1990).
657. Modinova, L.I.; Sergeeva, V.P.; Meleshilova, N.P.; Maslennikov, V.P. *Zhurn. Obsh. Khim.* **68**, 20 (1998).
658. Modinova, L.I.; Sergeeva, V.P.; Maslennikov, V.P. *Zhurn. Obsh. Khim.* **68**, 1610 (1998).
659. Modinova, L.I.; Sergeeva, V.P.; Maslennikov, V.P. *Zhurn. Obsh. Khim.* **69**, 2024 (1999).
660. Pavlenko, V.A.; Kokozay, V.N.; Skopenko, V.V. *Dokl. AN UkrSSR, Ser. B* No. 7, 42 (1985).
661. Kokozay, V.N.; Dvorkin, A.A.; Vasilieva, O.Yu.; Sinkevich, A.V.; Rebrova, O.N. *Zhurn. Neorg. Khim.* **36**, 1446 (1991).
662. Skopenko, V.V.; Kokozay, V.N.; Pavlenko, V.A. *Ukr. Khim. Zhurn.* **59**, 485 (1993).
663. (a) Skopenko, V.V.; Kokozay, V.N.; Vasilieva, O.Yu. *Koord. Khim.* **24**, 180 (1998); (b) Kokozay, V.N.; Vassilyeva, O.Yu. In: *Education in Advanced Chemistry*. Vol. 7. *Perspective in Coordination Chemistry*. Poznan: Wroclaw, 2000, p. 272; (c) Vinogradova, E.A.; Vassilyeva, O. Yu.; Kokozay, V.N.; Skelton, B.M.Z. *Naturforsch* **57b**, 319 (2002).
664. Pavlenko, V.A.; Kokozay, V.N.; Skopenko, V.V. *Ukr. Khim. Zhurn.* **51**, 342 (1985).
665. Kokozay, V.N.; Nevesenko, N.D.; Skopenko, V.V. *Dokl. AN UkrSSR, Ser. B* No. 2, 40 (1987).
666. Dvorkin, A.A.; Kokozay, V.N.; Petrusenko, S.R.; Sinkevich, A.V. *Dokl. AN UkrSSR, Ser. B* No. 10, 30 (1989).
667. Sobolev, A.I.; Shvelashvili, A.E.; Kokozay, V.N.; Petrusenko, S.R.; Kapshuk, A.A. *Soob. AN GSSR* **136**, 326 (1989).
668. Barr, D.; Brooker, A.T.; Doyle, M.J.; Drake, S.R.; Raithby, P.R.; Snaith, R.; Wright, D.S. *J. Chem. Soc., Chem. Commun.* 893 (1989).
669. Pavlenko, V.N.; Kokozay, V.N. *Zhurn. Prikl. Khim.* **63**, 216 (1990).
670. Kokozay, V.N.; Simonov, Yu.A.; Petrusenko, S.R.; Dvorkin, A.A. *Zhurn. Neorg. Khim.* **35**, 2839 (1990).
671. Dvorkin, A.A.; Kokozay, V.N.; Petrusenko, S.R.; Simonov, Yu.A. *Ukr. Khim. Zhurn.* **57**, 5 (1991).

672. Vasilieva, O.Yu.; Kokozay, V.N.; Skopenko, V.V. *Ukr. Khim. Zhurn.* **60**, 227 (1994).
673. Sevostyanova, A.F.; Kokozay, V.N. *Ukr. Khim. Zhurn.* **61**, 3 (1995).
674. Babich, O.A.; Kokozay, V.N.; Pavlenko, V.N. *Polyhedron* **15**, 2727 (1996).
675. Kovbasyuk, L.A.; Babich, O.A.; Kokozay, V.N. *Polyhedron* **16**, 161 (1997).
676. Kokozay, V.N.; Petrusenko, S.R.; Sinkevich, A.V.; Somonov, Yu.A. *Ukr. Khim. Zhurn.* **59**, 236 (1993).
677. Kokozay, V.N.; Polyakov, V.R.; Simonov, Yu.A. *J. Coord. Chem.* **28**, 191 (1993).
678. Kokozay, V.N.; Sinkevich, A.V. *Polyhedron* **13**, 1427 (1994).
679. Kovbasyuk, L.A.; Vasilieva, O.Yu.; Kokozay, V.N.; Linert, W. *Z. Naturforsch. B.* **52**, 337 (1997).
680. Vasilieva, O.Yu.; Kovbasyuk, L.A.; Kokozay, V.N.; Skelton, B.W.; Linert, W. *Polyhedron* **17**, 85 (1998).
681. (a) Vasilieva, O.Yu.; Kovbasyuk, L.A.; Kokozay, V.N.; Linert, W. *Z. Kristallogr.* **213**, 437 (1998); (b) Vinogradova, E.A.; Vasillyeva, O. Yu.; Kokozay, V.N. *Inorg. Chem. Comm.* **5**, 19 (2002).
682. Barr, D.; Snaith, R.; Wright, D.C.; Wade, K. *J. Am. Chem. Soc.* **109**, 7891 (1987).
683. Barr, D.; Doule, M.J.; Drake, S.R.; Raithby, P.R.; Snaith, R.; Wright, D.C. *J. Chem. Soc., Chem. Commun* 1415 (1988).
684. Barr, D.; Doule, M.J.; Raithby, P.R.; Snaith, R.; Wright, D.C.; Mulvey, R.E.; Reed, D. *J. Chem. Soc., Chem. Commun.* 318 (1989).
685. Barr, D.; Doule, M.J.; Drake, S.R.; Raithby, P.R.; Snaith, R.; Wright, D.C. *Polyhedron* **8**, 215 (1989).
686. Barr, D.; Doule, M.J.; Drake, S.R.; Raithby, P.R.; Snaith, R.; Wright, D.C. *Inorg. Chem.* **28**, 1768 (1989).
687. Barr, D.; Brooker, A.T.; Doule, M.J.; Drake, S.R.; Raithby, P.R.; Snaith, R.; Wright, D.C. *Angew. Chem.* **102**, 300 (1990); *Int. Ed. Engl.* **29**, 285 (1990).
688. Armstrong, D.R.; Khandelwal, A.H.; Raithby, P.R.; Snaith, R.; Stalke, D.; Wright, D.C. *Inorg. Chem.* **32**, 2132 (1993).
689. Morkovnik, A.S., Morkovnik, Z.S.; Bessonov, K.N. *Zhurn. Neorg. Khim.* **32**, 3106 (1987).
690. Pomogailo, A.D.; Ufyand, I.E. *Macromolecular Chelates.* Chemistry: Moscow, 1991, 303 pp.
691. Pomogailo, A.D.; Savostianov, V.S. *Metal-Containing Monomers and Polymers.* Chemistry: Moscow, 1988, 384 pp.
692. Pomogailo, A.D. *Polymeric Immobilized Metal-Complex Catalysts.* Science: Moscow, 1988, 303 pp.
693. Bailor, G. In: *Metalloorganic Polymers.* (Edit. Karraer, Ch.; Shits, G.). Mir: Moscow, 1981, p. 315.
694. Pomogailo, A.D.; Savastyanov, V.S. *Synthesis and Polymerization of Metal-Containing Polymers.* CRC Press: Boca Raton, Ann Arbor, London, Tokyo, 1994, 164 pp.
695. Ufland, I.E.; Pomogailo, A.D. *J. Coord. Chem.* **24**, 183 (1991).
696. Ufland, I.E.; Pomogailo, A.D.; Vainshtein, E.F. *Usp. Khim.* **64**, 913 (1995).
697. Pomogailo, A.D.; Ufland, I.E. *Ross. Khim. Zhurn.* **40**(4/5), 55 (1996).
698. Trembovetskii, G.B.; Martinenko, L.I.; Muravieva, I.A. *Zhurn. Neorg. Khim.* **30**, 1163 (1985).
699. Borisov, A.P.; Petrova, L.A.; Makhaev, V.D. *Zhurn. Obsh. Khim.* **62**, 15 (1992).
700. Borisov, A.P.; Petrova, L.A.; Karpova, T.P.; Makhaev, V.D. *Zhurn. Neorg. Khim.* **41**, 411 (1996).
701. (a) Zaitseva, I.G.; Kuzmina, N.P.; Martinenko, L.I.; Makhaev, V.D.; Borisov, A.P. *Zhurn. Neorg. Khim.* **43**, 805 (1998); (b) Petrova, L.A.; Borisov, A.P.; Aleshin, V.V.; Makhaev, V.D. *Zhurn. Neorg. Khim.* **44**, 1316 (1999); (c) Petrova, L.A.; Borisov, A.P.;

Aleshin, V.V.; Makhaev, V.D. *Zhurn. Neorg. Khim.* **45**, 390 (2000); (d) Petrova, L.A.; Borisov, A.P.; Aleshin, V.V.; Makhaev, V.D. *Zhurn. Neorg. Khim.* **46**(10), 1655 (2001); (e) Kalinovskaya, I.V.; Karasev, V.E. *Zhurn. Neorg. Khim.* **45**(9), 1488 (2000).
702. Letuchii, Ya.A.; Lavrentiev, I.P.; Khidekel, M.L. *Koord. Khim.* **8**(11), 1477 (1982).
703. Giguere, R.A. In: *Organic Synthesis: Theory and Application.* (Edit. Hudlicky, T.). JAI Press: Greenwich, CT, 1989, vol. 1, pp. 103–172.
704. Roussy, G.; Pearce, J.A. *Foundations and Industrial Applications of Microwave and Radio Frequency Fields.* John Wiley: Chichester, New York, Brisbane, Toronto, Singapore, 1995.
705. Rao, K.J.; Vaidhyanathan, B.; Ganguli, M.; Ramakrishnan, P.A. *Chem. Mater.* **11**, 882 (1999).
706. Chemat, F.; Poux, M.; Berlan, J. *J. Microwave Power & Electromag. Energ.* **31**(1), 19 (1996).
707. Kidwai, M.; Goel, Y.; Kumar, R. *Indian J. Chem.* **37B**, 174 (1998).
708. *Active Metals.* (Edit. Fuerstner, A.). VCH: Weinheim, 1996, pp. 133–190.
709. Cintas, P. *Activated Metals in Organic Synthesis.* CRC Press: Boca Raton, FL, 1993, pp. 45–70; *Ultrasound: its Chemical, Physical, and Biological Effects.* (Edit. Suslick, K.S.). VCH: Weinheim, 1988.
710. Mason, T.J.; Lorimer, J.P. *Sonochemistry: Theory, Applications and Uses of Ultrasound in Chemistry.* Ellis Horwood/John Wiley: London/New York, 1988, 251 pp.
711. Mason, T.J. *Advances in Sonochemistry.* JAI Press: London, 1990, vol. 1.
712. Raner, K.D.; Strauss, C.R.; Trainor, R.W. *J. Org. Chem.* **60**, 2456 (1995).
713. Matsumura-Inoue, T.; Tanabe, M..; Minami, T.; Ohashi, T. *Chem. Lett.* 2443 (1994).
714. Dabirmanesh, Q.; Roberts, R.M.G. *J. Organomet. Chem.* **460**, C28 (1993).
715. Kharisov, B.I.; Blanco, L.M.; Salinas, M.V.; Garnovskii, A.D. *J. Coord. Chem.* **47**, 135 (1999).
716. Chemat, F.; Poux, M.; Berlan, J. *J. Microwave Power & Electromag. Energ.* **31**(1), 19 (1996).
717. (a) Majetich, G. *J. Microwave Power & Electromag. Energ.* **30**(1), 27 (1995); (b) Jiang, H.-B.; Sun, M.-Y.; Qiu, J.; Zheng, S.-L.; You, Q.-X.; Jin, H. *Chem. Res. Appl.* **12**(5), 569 (2000).
718. Bari, S.S.; Bose, A.K.; Chaudhary, A.G.; Manhas, M.S.; Raju, V.S.; Robb, E.W. *J. Chem. Educ.* **69**(11), 938 (1992).
719. Mingos, D.P. *Adv. Mater.* **5**(11), 857 (1993).
720. Whittaker, A.G.; Mingos, D.M. *J. Chem. Soc., Dalton Trans.* 2073 (1995).
721. Gedye, R.; Smith, F.; Westaway, K. *J. Microwave Power & Electromag. Energ.* **26**(1), 3 (1991).
722. Ortiz Méndez, U.; Kharisov, B.I.; Berdonosov, S.S. *Ciencia UANL* **II**(2), 162 (1999).
723. Berdonosov, S.S.; Lebedev, Y.Ya.; Melikhov, I.V.; Kozorezova, E.V. *Zhurn. Fiz. Khim.* **69**(4), 731 (1995).
724. Berdonosov, S.S.; Prokofiev, M.A.; Lebedev, V.Ya.; Berdonosova, D.G. *Neorg. Mater.* **33**(10), 1257 (1997).
725. Shantarovich, V.P.; Kevdina, V.V.; Berdonosov, S.S. *Neorg. Mater.* **33**(8), 1007 (1997).
726. Berdonosov, S.S.; Kopilova, I.A.; Lebedev, V.Ya.; Chesnokov, D.E. *Neorg. Mater.* **28**(5), 1022 (1992).
727. (a) Rao, K.J.; Vaidhyanathan, B.; Ganguli, M.; Ramakrishnan, P.A. *Chem. Mater.* **11**, 882 (1999); (b) Jiang, Z.L.; Feng, Z.W.; Shen, X.C. *Chinese Chem. Lett.* **12**(6), 551 (2001).
728. Sridar, V. *Curr. Sci.* **74**(5), 446 (1998).
729. Villemin, D.; Martin, B.; Puciova, M.; Toma, S. *J. Organomet. Chem.* **484**, 27 (1994).
730. Dabirmanesh, Q.; Roberts, R.M.G. *J. Organomet. Chem.* **460**, C28 (1993).
731. Dabirmanesh, Q.; Roberts, R.M.G. *J. Organomet. Chem.* **542**, 99 (1997).

732. (a) Singh, K.; McWhinnie, W.R.; Li Chen, H.; Sun, M.; Hamor, T.A. *J. Chem. Soc., Dalton Trans.* 1545 (1996); (b) Van Atta, S.L.; Duclos, B.A.; Green, D.B. *Organometallics* **19**(12), 2397 (2000).
733. Beneš, L.; Melánova, K.; Zima, V.; Kalousová, J.; Votinský, J. *Inorg. Chem.* **36**, 2850 (1997).
734. Lang, J.; Tatsumi, K.; Kawaguchi, H.; Lu, J.; Ge, P.; Ji, W.; Shi, S. *Inorg. Chem.* **35**, 7924 (1996).
735. (a) Pelletier, J.; Pantel, R.; Oberlin, J.C. *J. Appl. Phys.*.**70**(7), 3862 (1991); (b) Chauhan, S.M.; Sahoo, B.B.; Srinivas, K.A. *Synth. Commun.* **31**(1), 33 (2001).
736. Villemin, D.; Hachemi, M.; Lalaoui, M. *Synth. Comm.* **26**(13), 2461 (1996).
737. Ogilvy, A.G.; Draganjac, M.; Rauchfuss, T.B.; Wilson, S.R. *Organometallics* **7**, 1171 (1988).
738. Kuzharov, A.S.; Suchkov, V.V.; Vlasenko, L.A. *Zhurn. Fiz. Khim.* **53**(8), 2064 (1979).
739. Kuzharov, A.S.; Khentov, V.Ya. *Koord. Khim.* **5**, 601 (1979).
740. (a) Kuzharov, A.S.; Vlasenko, L.A.; Suchkov, V.V. *Zhurn. Fiz. Khim.* **58**, 894 (1984); (b) Ni, J.; Geng, J.-L.; Qi, W.-B.; Wang, Z.-L.; Luo, Q.-H.; Lu, G.-Y.; Feng, R. *Chinese J. Inorg. Chem.* **17**(5), 713 (2001).
741. Luche, J.-L.; Damiano, J.-C. *J. Am. Chem. Soc.* **102**, 7926 (1980).
742. Lindley, J.; Lorimer, J.P.; Mason, T.J. *Ultrasonics* **24**, 292 (1986).
743. Einhorn, J.; Luche, J.-L. *J. Org. Chem.* **52**, 4124 (1987).
744. De Nicola, A.; Einhorn, J.; Luche, J.-L. *J. Chem. Res. (S)* 278 (1991).
745. Kuchin, A.V.; Nurushev, R.A.; Tolstikov, G.A. *Zhurn. Obsh. Khim.* **53**, 2519 (1983).
746. Luche, J.-L.; Petrier, C.; Lansard, C.; Greene, A.E. *J. Org. Chem.* **48**, 3837 (1983).
747. Itoh, K.; Nagashima, H.; Ohshima, T.; Ohshima, N.; Nishiyama, H. *J. Organomet. Chem.* **272**, 179 (1984).
748. Roger, C.; Tudoret, M.J.; Guerchais, V.; Lapinte, C. *J. Organomet, Chem.* **365**, 347 (1989).
749. Roger, C.; Marseille, P.; Salus, C.; Hamon, J.R.; Lapinte, C. *J. Organomet. Chem.* **336**, C13 (1987).
750. Suslick, K.S.; Schubert, P.F.; Goodale, J.W. *J. Am. Chem. Soc.* **103**, 7342 (1981).
751. Suslick, K.S.; Choe, S.-B.; Cichowlas, A.A.; Grinstaff, M.W. *Nature* **353**, 414 (1991).
752. Suslick, K.S.; Johnson, R.E. *J. Am. Chem. Soc.* **106**, 6856 (1984).
753. Bonnermann, H.; Bogdanovic, B.; Brinkman, R.; He, D.W.; Splietlioff, B. *Angew Chem. Int. Ed. Engl.* **22**, 728 (1983).
754. Oppolzer, W.; Schneider, P. *Tetrahedron Lett.* **25**, 3305 (1984).
755. Geloso, C.; Kumar, R.; López-Grado, J.; Tuck, D.G. *Can. J. Chem.* **65**, 928 (1987).
756. Heinicke, G.; Fleischer, G. *Technik.* **31**, 458 (1976).
757. Tsuji, J. *Organic Synthesis by Means of Transition Metal Complexes*. Springer-Verlag: Berlin, 1975.
758. Deeming, A.J. *Inorg. Chem.* **9**, 271 (1974).
759. Tsou, T.T.; Kochi, J.K. *J. Am. Chem. Soc.* **101**, 6319 (1979).
760. Sokolov, V.I.; Reutov, O.A. *Coord. Chem. Rev.* **27**, 89 (1978).
761. Kukushkin, Yu.N.; Danilova, L.E. *Koord. Khim.* **7**, 163 (1981).
762. *The Chemistry of Metal–Carbon Bond.* (Edit. Hartley, F.R.). Wiley Interscience: New York, London, 1982.
763. Gubin, S.P.; Shulpin, G.B. *Chemistry of the Complexes with Metal–Carbon Bonds.* Nauka: Novosibirsk, 1984, 282 pp.

4

Synthesis of Coordination Compounds with Programmed Properties

ALEXANDER D. GARNOVSKII, IGOR S. VASILCHENKO, and DMITRY A. GARNOVSKII

Rostov State University, Rostov-on-Don, Russia

This type of synthesis of coordination compounds (which could be named "controlled," "selective," "guided," or "programmed") is related with obtaining metal complexes having beforehand programmed quantitative and qualitative composition, structure, physicochemical, and practically useful properties.

The different approaches to the synthesis of complex compounds with programmed composition are based on the problems discussed above and related to the selection of ligands and metal sources, synthetic techniques, and conditions for the complex-formation reactions. In principle the same factors should be taken into account in the preparation of complexes with the properties indicated above.

At the same time, there are two aspects of this synthesis which require special discussion. These are the creation of complex compounds with definite manner of localization of the coordination bond (regioselective synthesis) and programmed structure (polyhedron-programmed synthesis).

4.1 REGIOSELECTIVE SYNTHESIS

The theory of regioselective synthesis of metal complexes is based on the ideas of competitive coordination and the principle of "hard" and "soft" acids and bases (HSAB, Sec. 1.2.2).

4.1.1 Competitive Coordination

The idea of competitive coordination as the capacity of di- and polydentate ligands to react with metals on different donor centers was introduced in 1972 [1], further developed in a monograph [2], and a series of reviews and original publications [3–18]. Its general approach is based on the following statements [1,19]:

1. Di- and polydentate ligands (di- and polyfunctional donors) have several reaction centers, and the reaction on any of them is complicated by the competition with other atoms and/or groups, which are potentially capable of interacting with a metal.
2. If most of the charge is located on one of the nucleophil centers of such ligands, then any other one is the most polarizable.
3. The reaction capacity of di- and polydentate ligands can be successfully explained by "hard–soft" properties [2] and different spatial accessibility of their donor centers.

Those donor atoms and/or groups are examined as donor centers in the problem of competitive coordination on which, due to the molecular structure of the ligand, the most favorable conditions are created for electrophilic attack by the metal, after taking into account the acceptor properties of the metal and the conditions of complex-formation reactions [19]. Such donor centers are mostly elements of a few main subgroups belonging to Groups V and VI of the Periodic Table, and also the unsaturated, aromatic, and heteroaromatic compounds which form the fundamentals of modern ligands (Chap. 2).

The polyfunctional ligands with conjugated and isolated donor centers are distinguished in the problem of competitive coordination [1,2,11,19]. We attributed the first of them to ambidentate ligands, although in some papers, for example Ref. 20a, all ligand systems having at least two donor centers are examined as ambidentate. The notion about alterdentate ligands was introduced [20b] (some examples of complexes containing such ligands are described in Ref. 20c); the difference between them and ambidentate ligands is that the donor centers in alterdentate ligands are equivalent.

From the two types of ligands described above, the ambidentate ones are the most studied [1,7–10,12–14,17,19]. Specifically, the earlier discussed ligand systems (Chap. 2), capable of forming complex compounds with various types of localization of the coordination bond, can be attributed to them: di- and triatomic molecules (Sec. 2.2.2), inorganic anions (Sec. 2.2.3), and compounds with various homo- and heteroatomic donor centers (Secs. 2.2.4 and 2.2.5). All chelate-forming ligands, capable of coordinating metals simultaneously on two and/or more donor centers, should be formally excluded from these ligand systems [20a,21,22]. However, it will be shown below that such chelating compounds are also capable of behaving as monodentate ligands, i.e., exposing ambidentate coordinating properties.

The majority of metal ions discussed before (Sec. 3.1.2) are included as acceptors into the problem of competitive coordination [1,2,19]. To be applied to the problem of competitive coordination, the HSAB principle (Sec. 1.2.2) [1,2,11,19] emphasizes preferential binding of hard metals by hard donor centers and soft electrophilic reagents (metal centers) by soft nucleophilic atoms or groups (mainly unsaturated, aromatic, or heteroaromatic systems). This statement applies only if some other conditions are met (one of which is equal spatial accessibility of donor center in the ligand for metal center). Spatial blocking of nucleophil centers provokes a change of localization mode in the coordination bond, i.e., complex formation contrary to the expected process according to the HSAB principle.

The reaction control should be emphasized amongst the conditions of reactions of competitive complex formation [19,23]. It is necessary to take into account that it is possible to determine, and frequently predict, the direction of the electrophilic attack to the donor center of di- and polyfunctional donors (ligands) only in the case when the thermodynamically stable products are formed under conditions of kinetic control. Thus, the thermodynamic stability of complexes is discussed, when the bond between the metal and di- and polydentate ligands is localized in the place of primary attack on one of any of the donor centers by the electrophilic reagent, without further change of coordination mode in the reaction of complex formation.

This statement is very important since, in a series of cases, the end products of the examined reactions may have a structure which is different, in comparison with the expected one, on the basis of general ideas and those presented below. In this respect, we note that it is necessary to take into account the possibility of various regroupings of isomerizations which easily take place under conditions of complex formation with participation of compounds (ligands), capable of di- and polycenter interaction [19,23,24].

4.1.2 Examples of Regioselective Synthesis of Coordination Compounds

Notions about ligands containing "hard" and "soft" donor centers and various manners of their binding with metals are given above (Secs. 2.2.2 and 2.2.5). Numerous literature sources, cited therein, allow us to evaluate possibilities of regioselective synthesis of complex compounds of ligands with N,P-, N,S-, N,Se-, N,Te-, O,S-, O,Se- and O,Te-donor centers. So, in this section we pay attention to the most clear examples of obtaining metal complexes on the basis of ambidentate hard–soft ligands.

Among these ligand systems, the pseudohalide ions NCX^- ($X = S$, Se) are of permanent interest [1,5,7,11,19,21,25–41]. To explain coordination modes in the complexes of these ligands, the HSAB principle (details in Sec. 1.2.2) was first applied [42,43]: the hard Pearson acids mostly bind the hard nitrogen atom and the soft Pearson acids bind the soft sulfur or selenium atoms. This approach, related with hard–soft interactions, was also used later on [1,2,6,7,11].

The most widespread method of synthesis of complex compounds of pseudohalide ions is through the immediate interaction of ligands with their corresponding metal salts (4.1) [44]:

$$m\mathrm{L} + \mathrm{M(NCX)}_n \longrightarrow [\mathrm{L}_m\mathrm{M(NCX)}_n] \qquad (4.1)$$

This transformation takes place in various solvents (for organic L mostly in nonaqueous solvents) or with no solvent in case of using liquid ligands. In particular, synthesis without a solvent is typical for pyridine, meanwhile alcohols (mainly ethanol) [44] are chosen as solvents when *bis*-azines (2,2′-bipy, *o*-phen, and other similar chelating ligand systems) are used as ligands.

Other synthetic techniques, described below, are applied to produce pseudohalide complexes, for example reactions of exchange of perchlorate anion to pseudohalide groups [44]. This transformation has not yet lost its synthetic preparative valor at present. The synthesis (4.2) of the complexes of divalent iron with bridge NCX group (X = S [45,46] and Se [47]) with chelating ligands (L) 2,2′-*bis*-thiazoline ***839*** and 2,2′-bipyrimidine is given by:

$$L + [Fe(H_2O)_6](ClO_4)_2 + KNCX \longrightarrow [FeL(NCX)_2]_2L \qquad (4.2)$$

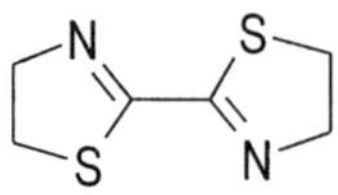

839

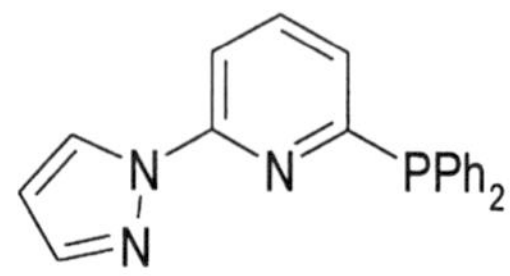

840

Developing the study of binuclear copper complexes with NCS bridge fragments [48], the transformation (4.3) was carried out [49]:

$$[Cu_2(\mu\text{-}L)_2(MeCN)](ClO_4)_2 + KNCS \xrightarrow[\text{r.t.}]{\text{MeCN}} [Cu_2(\mu\text{-}L)_2(\mu\text{-}1,1-NCS)](ClO_4) \qquad (4.3)$$

where L = 2-diphenylphosphino-6-(pyrasol-1-yl)-pyridine ***840***. The ligand-exchange method (4.4) is the basis for obtaining rhodanide complexes of divalent iron with aminoalkyldi(pyridylmethyl)amines (L) ***841*** [50], and tri(methylenepyridine)amine ***842*** [51]:

$$2L + [FePy_4(NCS)_2] \xrightarrow[\text{r.t.}]{\text{MeOH}} [FeL_2(NCS)_2] \qquad (4.4)$$

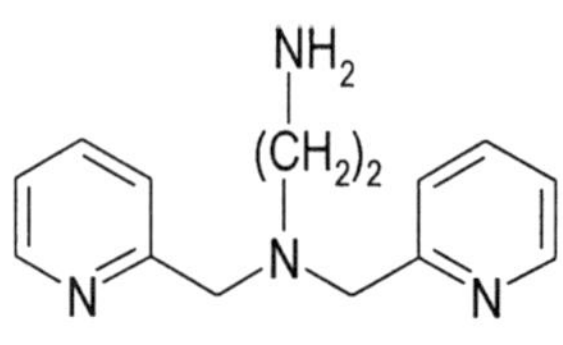

841

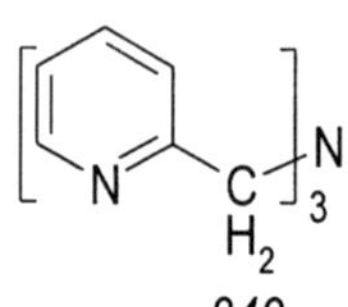

842

An interesting cycle of transformations, (4.5) and (4.6), including preparation of pseudohalide complexes and their use as metal-containing ligands (Sec. 3.3.2.3) was reported [51]:

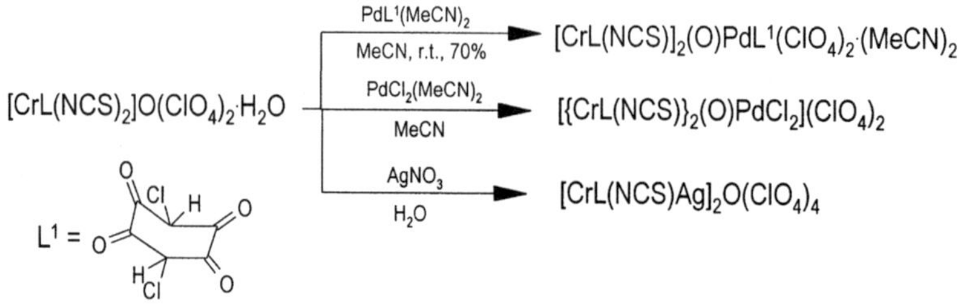

(4.5)

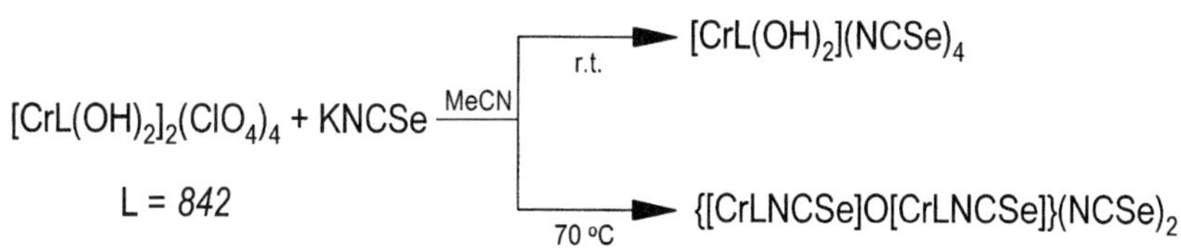

(4.6)

As a result of the syntheses (4.7) [52,53] and (4.8) [54], complexes ***843*** and ***844*** were obtained. They contain the thiocyanate ion in the inner coordination sphere.

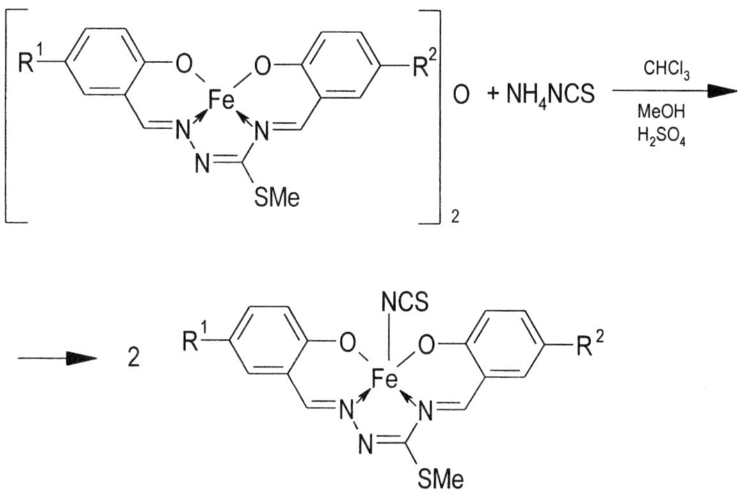

843

(4.7)

R^1 = H, OMe, Cl, Br;
R^2 = H, OMe, NO_2, Cl, Br

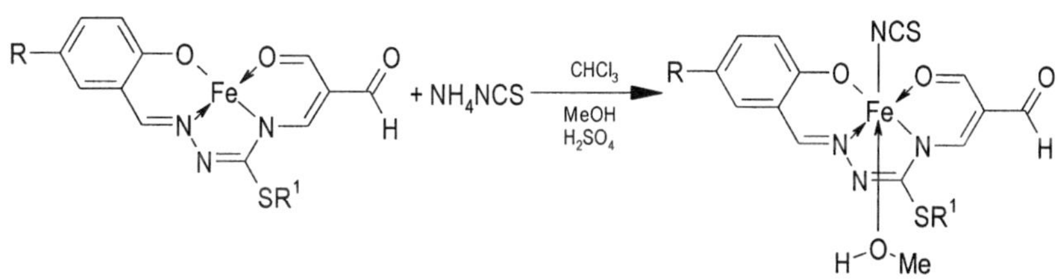

844

(4.8)

As a result of interaction of ***843*** and pyridine, the adduct ***845*** is formed [53]. The structures of coordination compounds ***844*** and ***845*** were proved by x-ray diffraction. As shown above (Sec. 3.4.3.2), the direct "ammonia" synthesis [55,56] with participation of various ligands (especially aliphatic, aromatic, and heterocyclic amines, aminoalcohols), elemental metals (or their oxides), and NH_4SCN in mostly non-aqueous media, opens definite possibilities for obtaining thiocyanate complexes. In this respect, transformation (4.9) should be mentioned [57]:

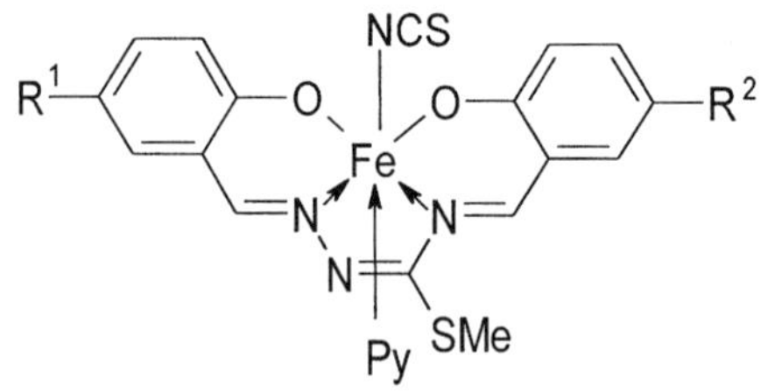

845

(4.9)

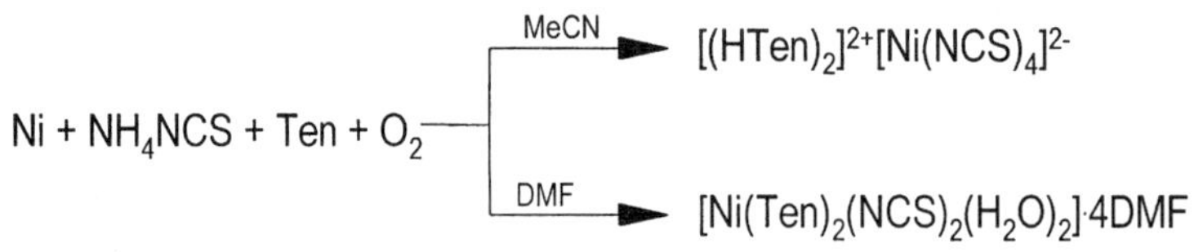

Ten – triethylenediamine

Highly specific methods for the preparation of complexes of thiocyanate ions are known, for example transformation (4.10) [58]:

$$[Cu(PPh_3)_2N_3] + CS_2 \xrightarrow[CHCl_3]{h\nu} Cu(PPh_3)_2NCS] + N_2 + S \quad (4.10)$$

Infrared and NMR-spectral analysis, and x-ray diffraction data, testify [42–54] that in case of complexes of the already discussed pseudohalide ions, the competitive coordination can be explained by the HSAB principle: "hard" Pearson acids are bound with "hard" N-center, and "soft" acids with "soft" X- donor (S, Se) centers. This situation allows us to obtain directly the coordination compounds of pseudohalides with a definite localization mode of the coordination bond, i.e., to carry out thc regioselective synthesis on the basis of the higher stability of complexes which are obtained as a result of "hard–hard" or "soft–soft" interactions [2].

At the same time, it is necessary to take into account that the approach described has a number of exceptions, related for example to the nature of other ligands forming pseudohalide complexes. A series of classic examples of inversion of the bond $M—N \rightarrow M—S \rightarrow M—N$ have been reported [6,8,11,42–44,59] and are presented in Sec. 2.2.3.5. In this respect, we especially emphasize the capacity of other ligands for "soft" or "hard" metals, related with "symbiotic" [60] and "antisymbiotic" [61] effects. Thus, Pearson [61] emphasized that "soft" ligands, which are placed in a trans position to SCN ion, contribute to N-binding of thiocyanate ions, and hard bases contribute to S-coordination of these ambidentate ligands. Metal oxidation number (Table 1.4) is important in this problem and it regulates soft–hard properties of complex-formers.

The HSAB principle very clearly explains the competitive reactions of organosulfoxides, amongst which DMSO is the most commonly used as ligand [1,2,19]. The coordination compounds of these ligands are described in monographs [62,63] and also in Sec. 3.1.1 [Eqs. (3.21)–(3.24)]. The main reaction of these syntheses is represented by a simple equation (4.11):

$$m\text{DMSO} + MX_n \longrightarrow [MX_n(\text{DMSO})_m] \quad (4.11)$$

In the vast majority of cases, such syntheses are carried out by the immediate interaction of liquid DMSO with metal salts in absence of solvents. In this respect we note the adducts of DMSO with yttrium and neodymium *tris*-acetylacetonates $[M(acac)_3(DMSO)(H_2O)]$, obtained by a lingering contact of solid acetylacetonates of the metals above with DMSO [64–68].

At the same time, various solvents are used in a series of syntheses of the examined type. Thus, the following compounds were obtained in acetone: $[Ag(DMSO)_2][ClO_4]$, $[Zn(DMSO)_n][ClO_4]_2$, $[Cd(DMSO)_6][ClO_4]_2$; $[Sc(DMSO)_6][ClO_4]_3$, $[M(DMSO)_6][ClO_4]_3$, where M = Ga, In; $[Mn(DMSO)_6]X_2$, where X = Br, I, ClO_4, BF_4. Methanol was used as a solvent in the synthesis of $[Ln(DMSO)_n][ClO_4]_3$, ethanol in preparation of $[Ru(DMSO)_6]X_n$, where X = Cl, ClO_4, $n = 2$, 3 [63].

The synthesis of DMSO complexes from elemental metals (4.12) should also be mentioned [69]:

$$Mn^0 + DMSO + HBF_4 \longrightarrow Mn[(DMSO)_6][BF_4]_2 \tag{4.12}$$

Also, coordination compounds $[V(DMSO)_6][BF_4]_2$, $[Cr(DMSO)_6][BF_4]_3$, $[Co(DMSO)_6][BF_4]_3$, and $[Ni(DMSO)_6][BF_4]_2$ were obtained [69] under electrosynthesis conditions (Sec. 3.4.2) with high yields (80–90%).

Coordination in DMSO complexes is carried out according to the HSAB principle [1,2,11,64–69]: "hard" acids are bound with the oxygen atom and "soft" acids with the sulfur atom, which have been strictly proven in many cases by x-ray diffraction. Thus, the DMSO molecule is bound with the metal through a "hard" oxygen in the adduct discussed above of "hard" yttrium acetylacetonate [65]. On the contrary, DMSO is coordinated with a "soft" platinum atom through a "soft" sulfur atom in the complex *cis*-$[Pt(DMSO)_2Br_2]$ [70]. We note that, in some cases, the syntheses with participation of DMSO are accompanied by deoxygenation of the ligand and lead to complexes of Me_2S [71].

Many other examples testifying to the possibility of regioprogrammed synthesis by selection of "hard–soft" ligands and metals have been reported [1,2,7,11,19]. Amongst them, we emphasize nitrogen-containing heteroaromatic ligands, in which the N atom can be examined as "hard," and the η^5 or η^6-π-system as "soft" donor center [4,11,15]. The general tendencies in competitive coordination of such ligand systems are based on the fact that the fundamental five-member heterocycles (π-excessive systems) are coordinated mostly on the π-center, while π-deficit heterocycles are coordinated on the N atom. In this respect, it was long accepted that only σ-N complexes can be obtained on the basis of pyridine and azoles [3,5,15]. At present, and due to the success in the synthesis of coordination compounds, the competitive reactions are controlled both on N atoms and π-systems of these nitrogen-containing heterocycles. As shown above (Sec. 3.4.1), the gas-phase synthesis had a decisive role in obtaining π-complexes of pyridine and its derivatives.

On the basis of a large amount of experimental material, it was shown that there are much more complex compounds of N-coordinated azines [synthetic method ***A***, transformation (4.13)] in comparison with complexes with analogous π-coordinated ligands (method ***B***). The mentioned reactions (4.13) represent a modern state of competitive coordination of pyridine and its C-substituted derivatives [11,15,55,56]:

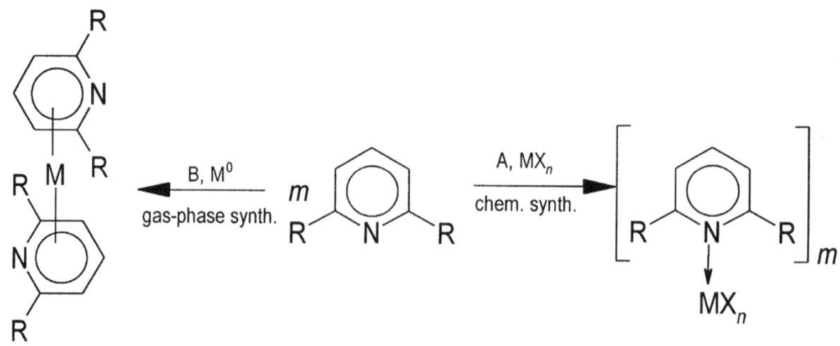

(4.13)

We note that π-complexes with blockading methylsubstituents and electron-donor R fragments can also be obtained by ligand-exchange reactions with the use of metal carbonyls, for instance (4.14) [15,72–74]:

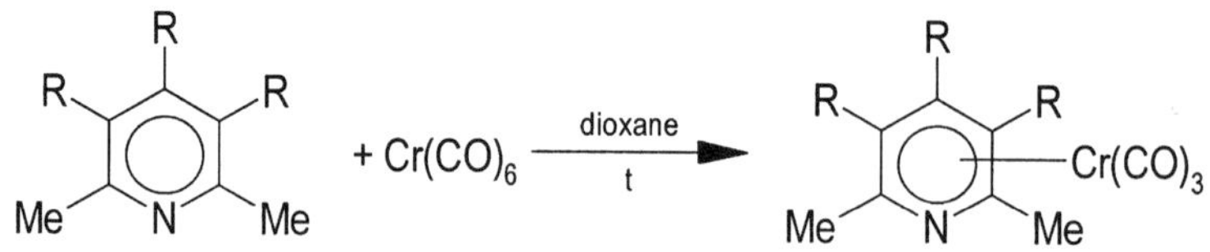

(4.14)

The first π-complexes of azole series ***711*** were isolated recently (3.147) [75]. However, the perspectives of Scheme (4.15) for obtaining other η^5-complexes of azole ligands are unclear because of the absence of reported data on the gas-phase syntheses of azole complexes.

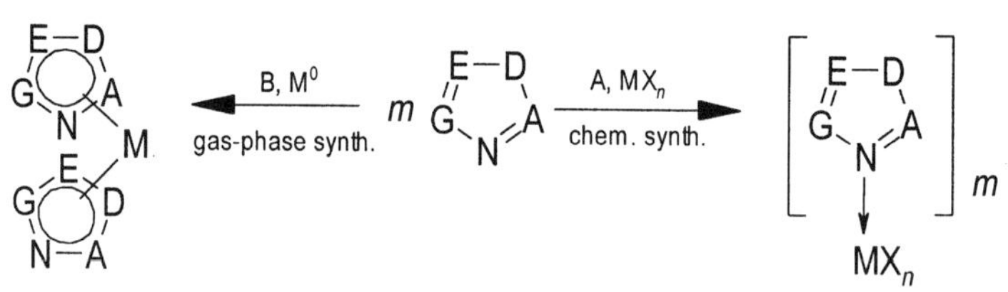

(4.15)

It is necessary to consider that the competitive reactions are characteristic also for ligands having two or more hard and soft donor centers. The first of them are NCO^-, NO_2^-, NO_3^-, and other anions [2,7,44]. The synthesis of cyanate complexes is mainly carried out by the method of immediate interaction of ligands and metal cyanates [(3.274), X = O] or by exchange of ligands, for example (4.16) [44]:

$$[MX_nL_m] + nAgNCO \longrightarrow [M(NCO)_nL_m] + nAgX \qquad (4.16)$$

Routes to control regioselectivity in such syntheses are unclear. Not in vain are the structures with N,O-coordinated bridge NCO groups, frequently observed among these complexes [44]. At the same time, due to higher softness of nitrogen in comparison to oxygen, it is possible to expect preferential N-coordination of "soft" metals, which is observed for soft Ag(I) [44] and Cu(I) [76]. It is worth noting that the Cu—O coordination is characteristic for the harder (Table 1.4) Cu(II) ion [44]. In this respect, we note that soft metals (Ag) can also form fulminate M—CNO structures [44], in agreement with the HSAB principle.

Complexes with some coordination modes are characteristic for nitrite (Sec. 2.2.3.3). In this respect, the data reported in Ref. 77 are representative, where the preparation (4.17) of nitrito-N- and nitrito-O-linkage isomers ***846*** and ***847*** of nickel complex of 1-(2-aminoethyl)piperidine in the conditions of the same experiment yields:

$$2L + K_2[Ni(NO_2)_4] \longrightarrow \begin{cases} [NiL_2(NO_2)_2] & \textit{846} \\ [NiL_2(ONO)_2] & \textit{847} \end{cases} \qquad (4.17)$$

After leaving the reaction mixture rest for a few days, a mixture of crystals were precipitated. After their mechanic separation, the existence of nitrito-N (***846***) and nitrito-O (***847***) coordinated isomers was proved by x-ray diffraction.

The nitro group has various coordination modes [7,78–81]. Synthesis of complexes of metal nitrates is carried out, in the majority of cases, by the method of immediate interaction of these salts with ligands in aqueous-organic (mainly, alcohols) media. Also, the interactions of metal oxides, ligands, and nitric acid are applied, for example (4.18) [82,83]. The structure of mixed-ligand complex ***848***, containing chelate NO_3^-- group, was proved by x-ray diffraction [82,83].

$$1,10\text{-Phen} + \text{Hacac} + Dy_2O_3 + HNO_3 \xrightarrow[H_2O]{EtOH} [Dy(Phen)_2(acac)(NO_3)_2]\cdot H_2O$$

848 (4.18)

Regioselective syntheses are of high interest in a series of complexes of chelating ligands which, as is usually accepted, form exclusively ICC (Secs. 1.2.2.2, 1.3.2.2) [22,84]. It is well known that β-diketones form β-diketonates ***849*** (4.19) in reactions with metal oxides and salts in various solvents (Sec. 3.1.1.2), as well as under gas-phase conditions (Sec. 3.4.1) and electrochemical synthesis. (Sec. 3.4.2).

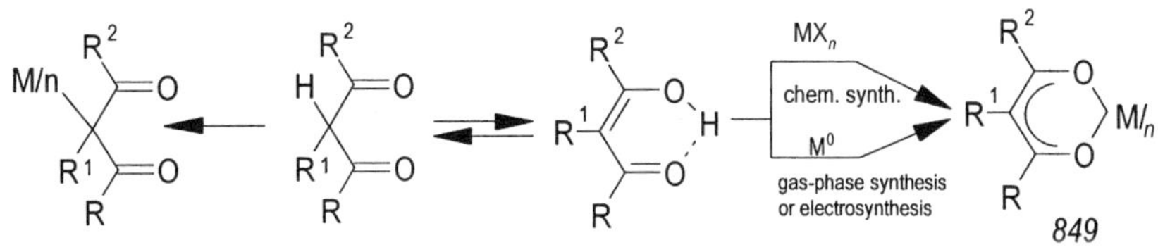

850

R, R1, R2 = Alk, Ar, Het; *n* = 2–4

(4.19)

It is less well known that the synthetic methods of β-diketonate complexes have been elaborated, where ligands of the type discussed are bound with the metal through a C-donor atom ***850*** [6,11,12,14,85–88]. Such mercury complexes, for example ***851***, were obtained as a result of transformation (4.20) [89]:

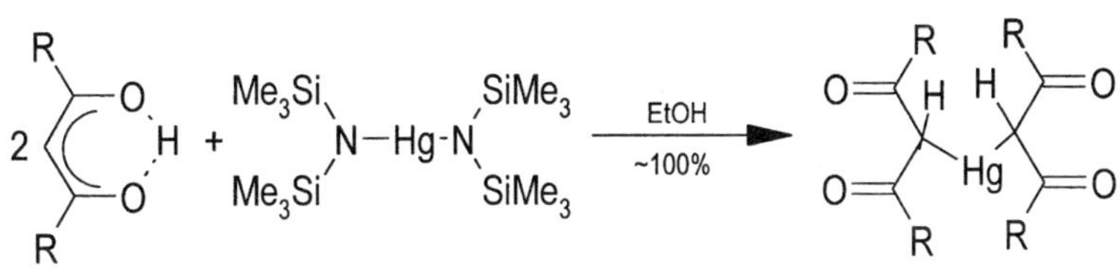

(4.20)

R = Me, CMe3

851

In addition to ***851***, coordination compound ***852*** was synthesized (4.21) and structurally characterized [90]:

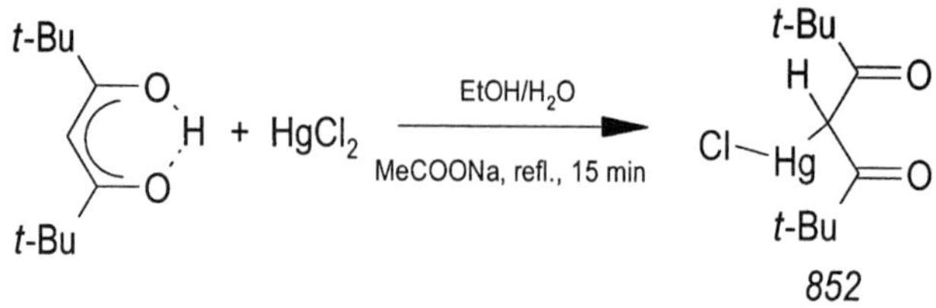

(4.21)

The mixed-ligand complex ***853***, containing C- and O-bound acetylacetonate anions, was isolated as a result of interaction (4.22) of palladium *bis*-(acetylacetonate) with triphenylphosphine [91]:

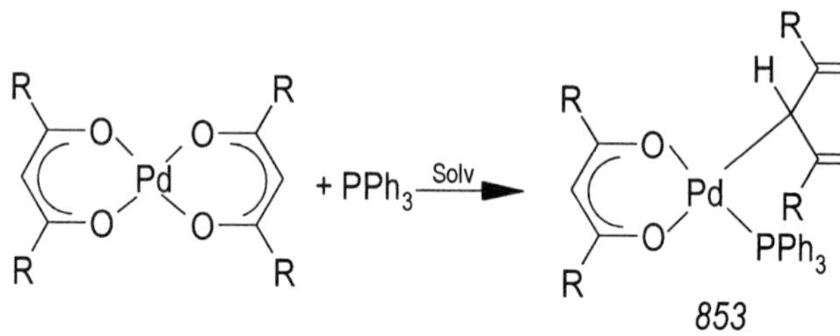

(4.22)

The structure of ***853*** was proved by x-ray diffraction [92]. The transformations given by (4.22) are widely used at present for obtaining complexes of the type ***853*** (Solv = C_6H_6; R = Bu, Ph) [93–96].

The synthetic technique of preparation of C-tri(phenylphosphine)gold-substituted β-diketones (4.23) [97] led to a series of structurally characterized coordination compounds of the type ***854*** [98,99]:

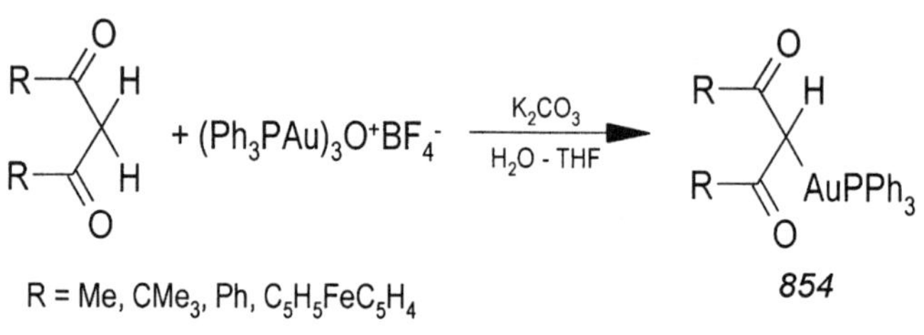

(4.23)

After analyzing the data on C-coordinated β-diketonates, it became obvious that all these complexes have been obtained from "soft" metals. It should be considered that the HSAB principle is quite suitable for regioselective synthesis of β-diketonates with soft or hard Pearson acids (Table 1.3) if it is taken into account that β-diketones are ambidentate C,O-donors, where the C atom is a "soft" and the O atom is a "hard" donor center.

o-Hydroxyazomethines ***23*** are the classic chelating ligands, on the basis of which a wide variety of ICC ***855*** with coordination units MN_nO_n (n is mostly 2) have been obtained (4.24) (Sec. 2.2.5.3):

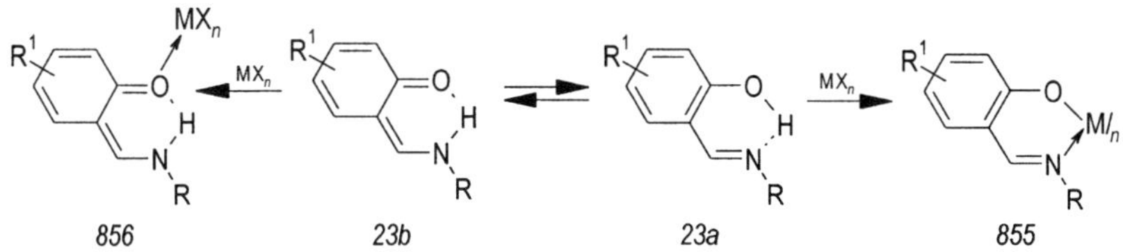

(4.24)

Together with more elaborated methods for the isolation of chelates ***855***, at present the routes of regioselective synthesis of molecular adducts ***856*** are available. These ligands participate as monodentate O-donors in ***856*** [10–12,17,100].

The most favorable conditions for obtaining molecular complexes of *o*-hydroxyazomethines of type ***856*** are created by use of strong Lewis acids, for example halides of metals of Groups III–V of the Periodic Table, and aprotic organic solvents with a low dielectric constant (hydrocarbons and their halogen-substituted derivatives). An example of such synthesis is the preparation of complexes of salicylideneimines with titanium and tin tetrachlorides (4.25) [101]:

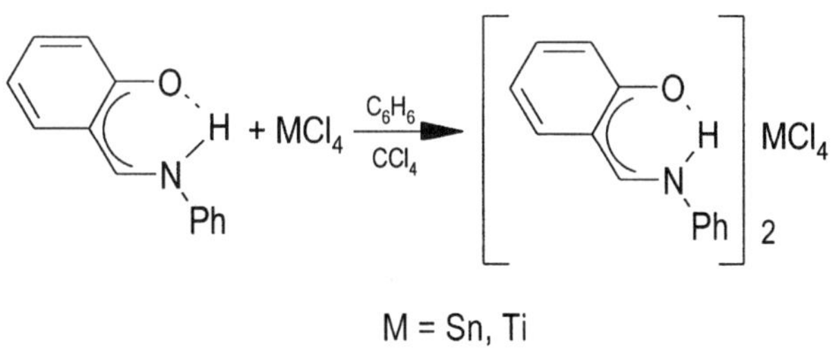

(4.25)

Absolute ether was successfully applied in the similar reaction syntheses (4.26) [102] and (4.27) [103]:

(4.26)

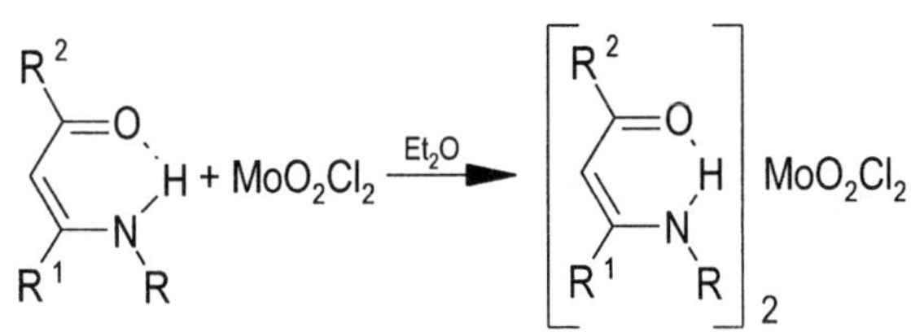

(4.27)

In some cases the syntheses of the type discussed can be carried out in ethanol (4.28) [104] or in a mixture of benzene and methanol (4.29) [105]:

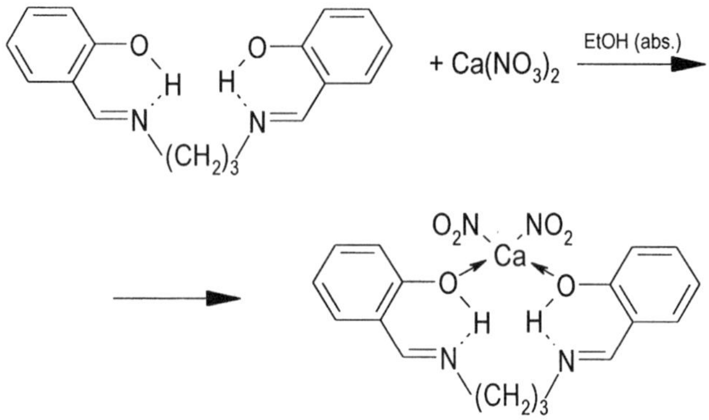

(4.28)

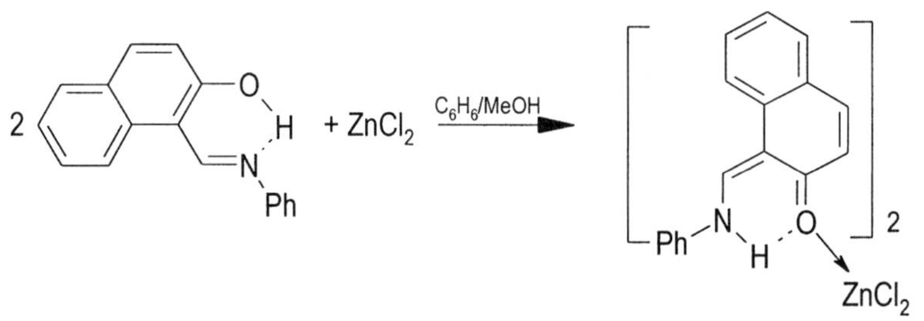

(4.29)

Recently [106], the transformation (4.30) was performed with obtaining the molecular and chelate complexes whose structures were established by x-ray structural analysis. The syntheses of molecular complexes have also been carried out using the method of ligand exchange [107,108], for instance (4.31) [108].

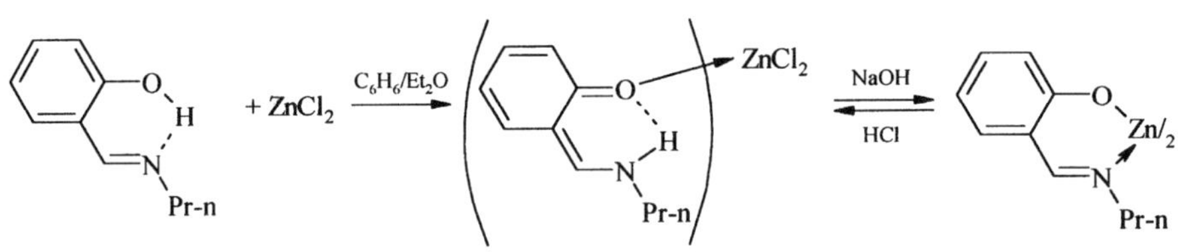

(4.30)

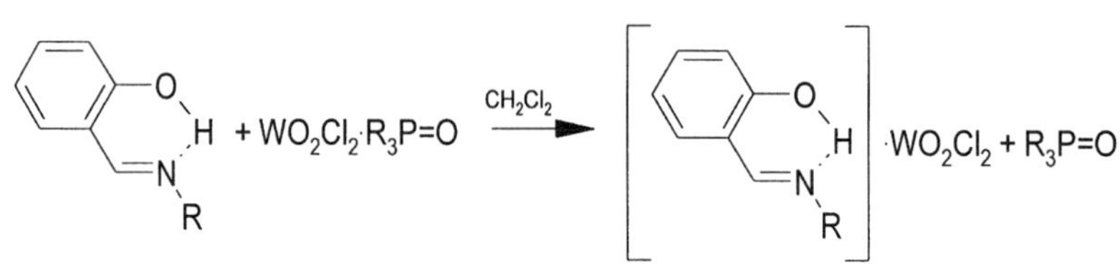

(4.31)

Template transformations, for instance (4.32) [109], are useful to carry out the kind of syntheses examined:

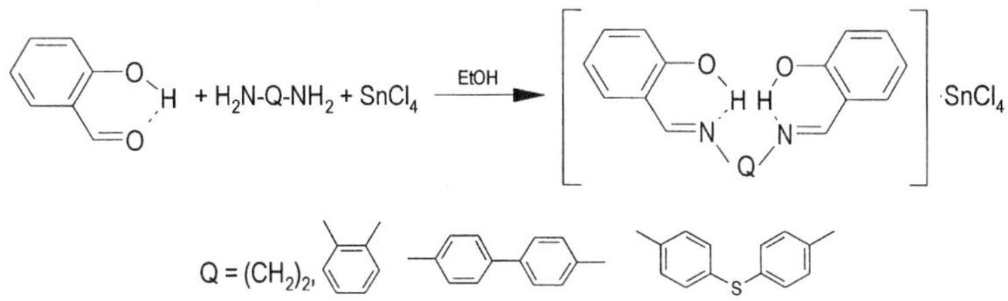

(4.32)

In this case the nature of MX_n plays a decisive role: use of metal acetates in template syntheses leads to the ICC of type ***855***, while the application of stronger Lewis acids allows the isolation of molecular complexes ***856*** [11].

The method for obtaining adducts of salicylideneimines from metal chelates (4.33) was performed [110]:

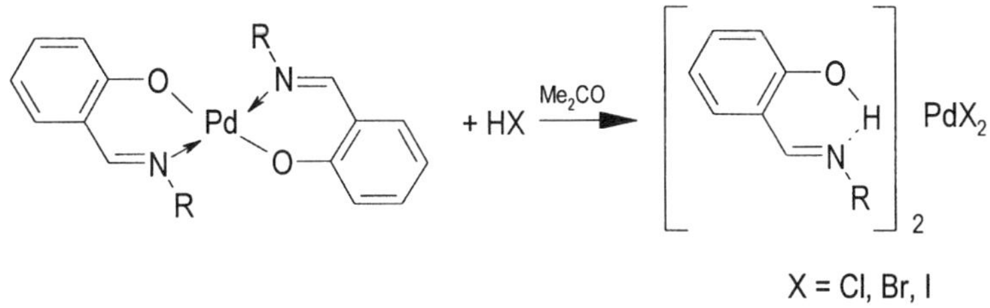

(4.33)

At the same time, preparative possibilities of the methods [Schemes (4.32) and (4.33)] are not clear due to the small number of examples about their use and the absence of data related to the yields of final products. Taking into account equations (4.24, 4.29, 4.30, and 4.33), the possibility of programmed synthesis of complexes of the types ***855*** and ***856*** and their mutual transformation should be considered as proved.

The above situation is characteristic for the complexes of β-aminovinylketones, for example (4.34) [100]:

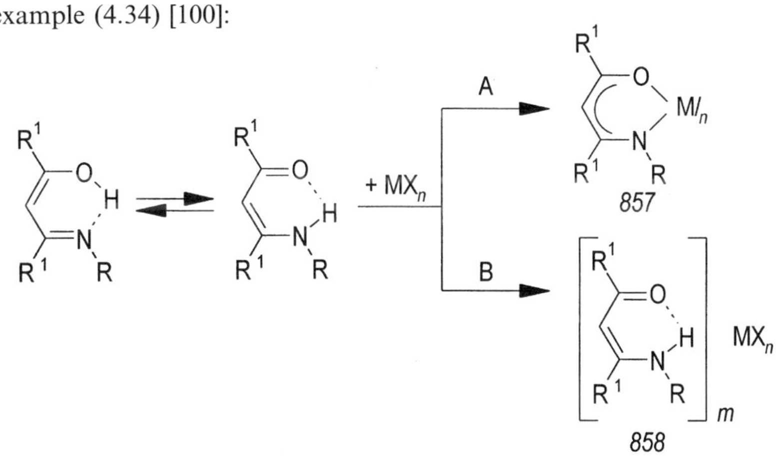

(4.34)

Complexes of these ligands are represented mostly by ICC ***857*** [(4.34), route *A*]. However, it was shown that, as a result of the interaction of N-phenylacetylacetoneimine with molybdenum dioxychloride in ethylacetate, the adduct ***858*** (R = Ph, $R^1 = R^2 = Me$, $MX_n = MoO_2Cl_2$, $m = 2$) is formed [111], whose structure was proved by x-ray diffraction.

Molecular complexes of acyl- and hetarylhydrazones of substituted salicyl aldehydes [112] and β-aminovinylthiones [113] were prepared by ligand exchange in diethyl ether from MoO_2Cl_2 and $WOCl_4$. However, the structure of the final products was examined on the basis of IR spectroscopy only and, in this respect, provokes doubts.

Amongst some specific reactions used in regioselective syntheses, we note the cyclometallation processes [Sec. 2.2.5.1, reaction (2.8); Sec. 3.1.1.2, reactions (3.40)–(3.44); Sec. 3.3.2.3, reactions (3.226)–(3.228)]. In this respect, we note that C,N-donor ligands form, depending on the nature of Lewis acids, two types of complexes. In the case of immediate interaction (4.35) of azomethines ***859*** with titanium or tin tetrachlorides (MCl_4), the molecular complexes with M—N bond ***860*** [101] were obtained (route ***A***), while the cyclometallation reaction with the use of palladium diacetate leads to binuclear chelates ***861***, in which the Pd—N, C metal-cycles are formed (4.35) [11,114–116]:

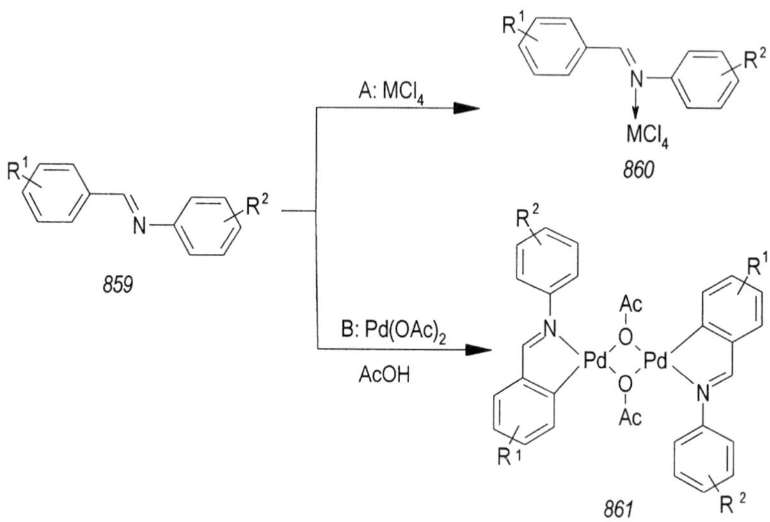

(4.35)

The solvent nature also has a regulating influence on the character of the coordination in the complexes obtained in cyclometallation reactions. Transformation (4.36) [11] is representative in this respect:

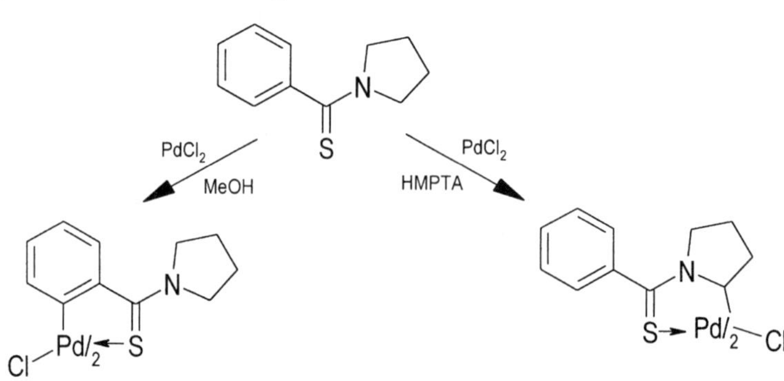

(4.36)

HMPTA –hexamethylphosphotriamide

An approach to the regioselective synthesis of metal complexes was offered based on the joint use of chemical and electrochemical methods of synthesis of coordination compounds departing from the same ligands [11], for instance transformation (4.37). As a result of these reactions, chelates with the participation of different N atoms of the imidazole ring, of the ligand in coordination, are formed:

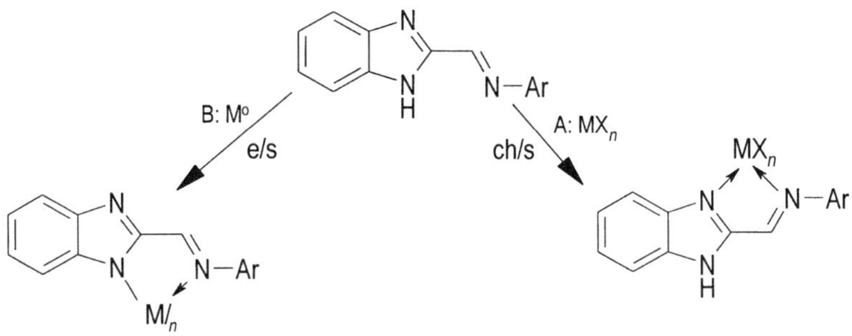

863 862 (4.37)

The chemical method (route *A*) leading to, for example, molecular metal chelates ***862*** is preferred in these [117] and other syntheses with use of azomethines of 2-formyl derivatives of N-heterocycles [118]. At the same time, the electrochemical route ***B*** allows us to obtain ICC of type ***863*** [55,56,62,63], unreachable by chemical routes.

The regioselective synthesis can be carried out in the limits of only electrochemical synthesis. In these transformations, the nature of metal complex-former (material of the anode) and of donor atom X (N, O) are the main regiocontrolling factors forming, for instance (4.38), the coordination sphere of the chelates ***864*** and ***865*** [119]:

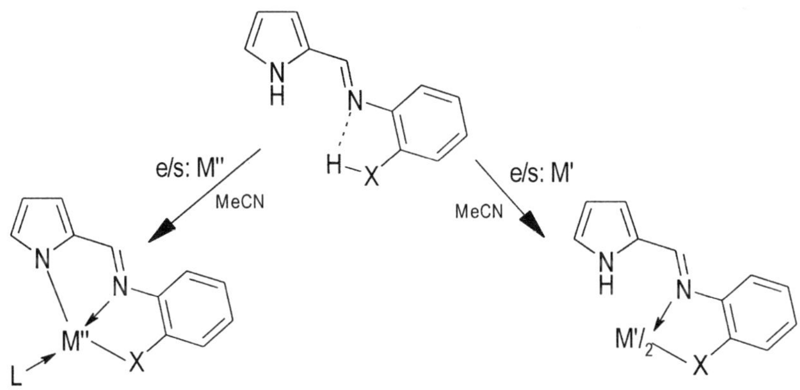

865 864 (4.38)

X = NTs, O; M' = Zn, Cd; M'' = Co, Ni, Zn

The azomethines of 2-formylpyrrole behave in this type of electrosyntheses as bi- (***864***) or tridentate (***865***) ligands. Structures of zinc complex ***864*** (X = NTs, M = Zn) [119] and its oxygen-containing analogue ***865*** (X = O, M = Zn) [120–122] were proved by x-ray diffraction.

Cyclopentadienyl derivatives with coordinative-active N-donor substituents ***866*** [123–125] are interesting ambidentate ligands for the following regioselective synthesis:

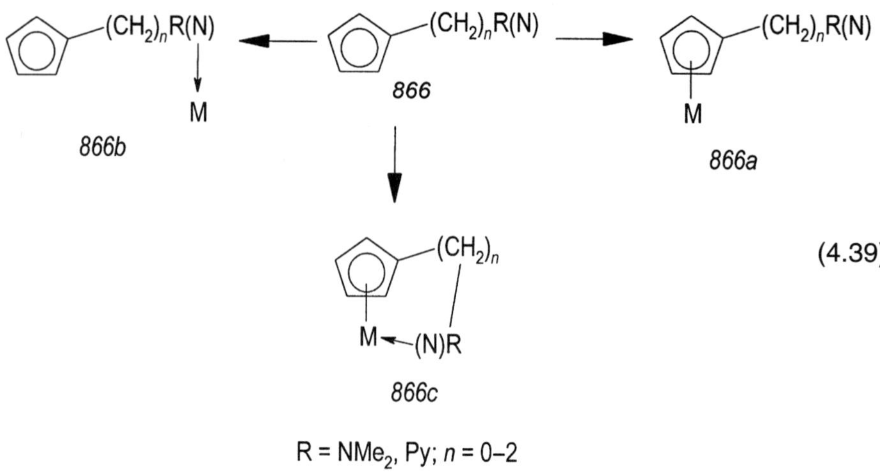

(4.39)

On their basis were isolated (4.39) the complexes with participation of a cyclopentadienyl fragment (soft donor center ***866a***), or the N atom (hard donor center ***866b***), or some other with different donor places of localization of the coordination bond.

EXPERIMENTAL PROCEDURES

Example 1

Synthesis of ***832*** (Blue), ***833*** (Brown) (4.17) [77]

The complexes $NiL_2(NO_2)_2$ (brown) and $NiL_2(ONO)_2$ (blue) were synthesized by adding a methanolic solution (5 mL) of the diamine (2 mmol) to a methanolic suspension (10 mL) of potassium hexanitronickelate(II) monohydrate (1 mmol). The resulting dark green solution was kept in a $CaCl_2$ desiccator. After a few days a mixture of brown and blue crystals was obtained, which was easily separable by tweezers (yield 81%).

Example 2

Synthesis of *bis*(acetylacetone)mercury (4.20) [89]

A solution of 12.0 g of fresh distilled acetylacetone in 50 mL of dry ether was added to the solution of 3.24 g of *bis*(hexamethylsilyl)aminomercury in 30 mL of the same solvent. Immediately a precipitate was formed which was filtered after keeping it during 1 hr, washed with ether, and dried. Yield 91%.

Example 3

Synthesis of $Pd(acac)_2PBu_3$ ***853*** (4.22) [94]

A solution of 0.726 g of tributylphosphine in 5 mL of benzene was added dropwise to the solution of 1.094 g of $Pd(acac)_2$ in 90 mL of the same solvent. The reaction mixture was stirred at room temperature for 30 min. After that the solvent was completely removed in vacuo. A yellow–orange residue was dissolved in 10 mL of pentane and cooled to −50°C. The yellow precipitate was decanted. The complex was dried in vacuum (20°C, 1 mm Hg) and packed in glass tubes. Yield 65%.

Example 4

Synthesis of *tris*(triphenylphosphinegold)oxonium Trifluoroacetylacetonate ***854*** (4.23) [97]

A mixture of 0.15 mL (1.1 mmol) of trifluoroacetylacetone and 0.5 mL of water was added to the suspension of 0.50 g (0.34 mmol) of oxonium salt and 0.15 g (1.1 mmol) of potassium carbonate in 10 mL of THF under stirring. The reaction mixture was stirred until salt was dissolved completely (15 min), after that it was dried by K_2CO_3 addition and filtered. Ether (5 mL) and heptane (10 mL) were added to the solution and the precipitate was formed. A crystal precipitate was dissolved in 2 mL of chloroform and the oxonium salt was precipitated by the ether excess. Yield 0.31 g (59%).

Example 5

Synthesis of [*bis*-1,3-salicylideneiminopropane]calcium Dinitrate (4.28) [104]

Complexes were prepared by the addition of the appropriate hydrated Ca salt [$CaCl_2 \cdot 2H_2O$ or $Ca(NO_3)_2 \cdot 4H_2O$] in absolute ethanol to a hot solution of the ligand in the same solvent. A yellow precipitate formed almost immediately. This was filtered off, washed with cold ethanol several times, and dried ($CaCl_2$).

Example 6

Synthesis of *bis*(2-hydroxy-1-naphthalideneanilinato)zinc Dichloride (4.29) [105]

A hot solution of anhydrous $ZnCl_2$ (0.68 g, 0.005 mol) in absolute methanol (5 mL) was added to the hot solution of ligand (2.47 g, 0.01 mol) in absolute benzene (30 mL). The mixture was refluxed for 5 min. A precipitate formed under cooling was filtered off and washed with hot methanol three times. Red–orange crystals were obtained after recrystallization from acetonitrile.

4.2 POLYHEDRON-PROGRAMMED SYNTHESIS

Syntheses of the complexes with a programmed geometry of coordination units are very important in modern coordination chemistry. They are based, mainly, on the factors determining stereochemistry of coordination compounds [20,126–129]. Amongst them, the nature of ligands and metals has to be emphasized. Due attention was paid to the ligand nature in two monographs [20,126]. The material in Ref. 126 was presented according to denticity of ligands, and environment of the central atom, in metal complexes with metal coordination number from 4 to 12. Reference 20 contains information, in addition to the aspects mentioned above, about slight peculiarities of structure in ligand systems: nature of donor centers, capacity of formation of metal-cycles of different size, chirality and achirality. The problems of topographical stereochemistry of mono- and polynuclear complexes also remain at the center of discussion.

The data found in a number of monographs [127–129] and reviews [86,130–135] allows the creation, for some classes of coordination, of a concrete ligand environment and facilitates the selection of metal contributing to the controlled formation of definite polyhedrons. This situation is present especially in a series of metal chelates with the most widely used coordination numbers of the central atom, i.e., 4, 5, and 6.

The preparation of azomethinic ICC, containing various types of ligand environment, will be discussed by using examples of polyhedron-programmed syntheses of complexes with tetra-, penta, and hexacoordinated structures. Choice of chelates of the examined type is related with their long-time systematic study [11,55,56,84,100,134–143], in particular by the authors of this monograph.

The data reported in a number of monographs [20,127–129] and reviews [100,130–135,144–146] allows the creation of definite classes of coordination compounds, of concrete ligand environment, and helps in the selection of metals contributing to the formation of programmed polyhedra. In order to carry out the polyhedron-programmed syntheses, it is necessary to take into account that the compounds used may have a structurally hard ligand environment, a feature which depends on the metal nature of a kind named "rigid" ligands [135]. The author of Ref. 135 relates the notions of "configurationally rigid and nonrigid ligands and metals" with stable and changing denticity and metal oxidation number, respectively. However, we will not discuss these problems, which are beyond the limits of this work.

Moreover, ligand systems which easily change the conformation of polyhedra in dependence on the nature of the metal center ("nonrigid" ligands [135]) are applied in the same type of syntheses. Such ligands can also be "rigid," i.e., mostly preserving the structure of their coordination sphere, and "nonrigid," easily changing the structure of polyhedra depending on the electronic structure of the metal ion [135].

o-Hydroxyazomethinic ligands and their analogues, having a general formula ***867***, are comparatively easily accessible compounds (especially when $X = NR^2$, O) and are widely used in coordination chemistry for obtaining ICC ***868***. The nature of donor centers (X) and R, R^1, R^2-substituents can be changed in a programmed way in these complexes.

A general scheme of chemical syntheses of complexes of the discussed ligands with six-member metal-cycles is represented by transformation (4.40):

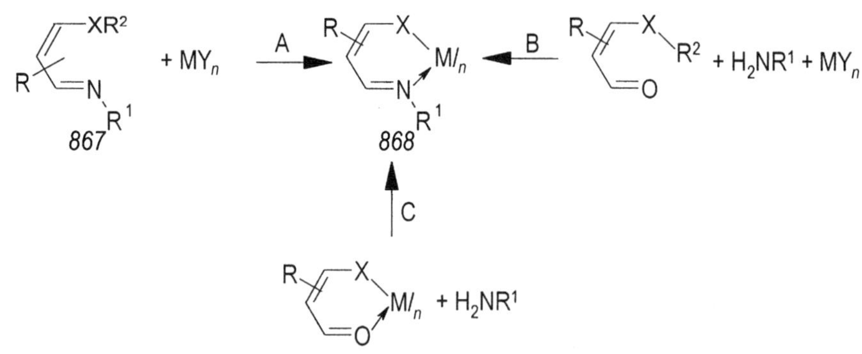

X = NR², O, S, Se; R = *cyclo*-Ar, Het; R¹ = H, Alk, Ar, Het; R² = H, M′; M′=Na, K

(4.40)

Concrete examples of this type of complex are presented in Sec. 3.2.2 [reaction (3.140)], 3.3.1 [reactions (3.169)–(3.171)], and 3.3.2 [reactions (3.176)–(3.178), (3.179), (3.181), and (3.182)]. This ICC can also be obtained by the techniques of gas-phase (Sec. 3.4.1) or electrochemical [Sec. 3.4.2, reactions (3.245)–(3.252)] synthesis.

4.2.1 Nature of Donor Atoms

Variation of the X-donor center in ligands ***867***, when R, R^1, and R^2-fragments are the same, allows the synthesis of programmed complexes with structurally different coordination units. To obtain metal chelates with $X = NR^2$, O, three synthetic routes [(4.38): ***A***, ***B***, and ***C***] are widely used, with preference for the immediate interaction (route ***A***) of azomethinic ligands with metal salts, mainly acetates (Y = OAc). An example of such synthesis is the transformation (4.41) [146]:

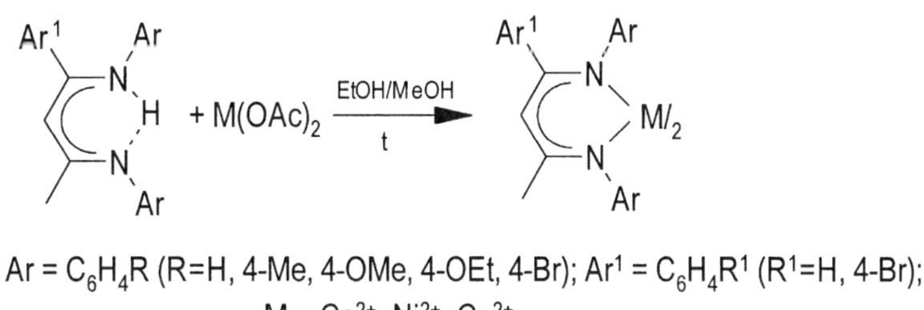

(4.41)

The immediate interaction of ligands and metal acetates is also the main method for obtaining azomethinic ICC with the coordination unit MN_2O_2, for instance (4.42) [147]:

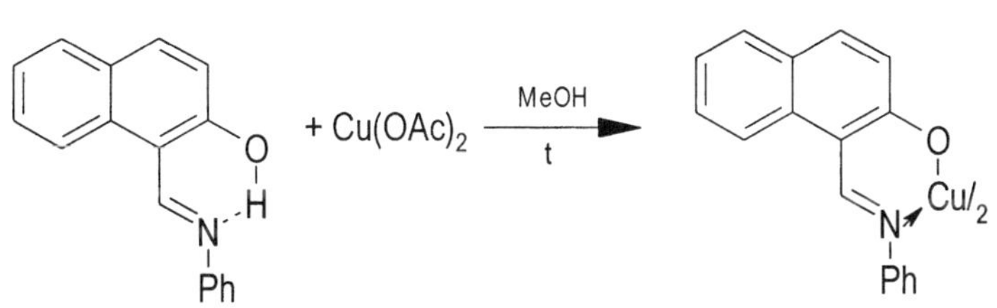

(4.42)

The azomethinic derivatives of type ***867*** with X = S, Se (R = *cyclo*-Ar) have comparatively low stability, so, in this respect, their ICC are obtained through the use of modifications on the template synthesis ***B*** or ***C*** [Scheme (4.40)]. Examples of such reactions are presented in Secs. 3.22 [reaction (3.140)], 3.3.1 [reaction (3.169)], and 3.3.2.1 [reactions (3.176), (3.178), and (3.180)–(3.182)]. In this respect, we note the recently reported transformation (4.43) [148], leading to the preparation, starting from 2,2′-dithiodibenzaldehyde ***869***, of ICC not only with the coordination unit NiN_2S_2 (***870***), but also the chelates with N_3S- (***871***) and NS_3-ligand (***872***) environment.

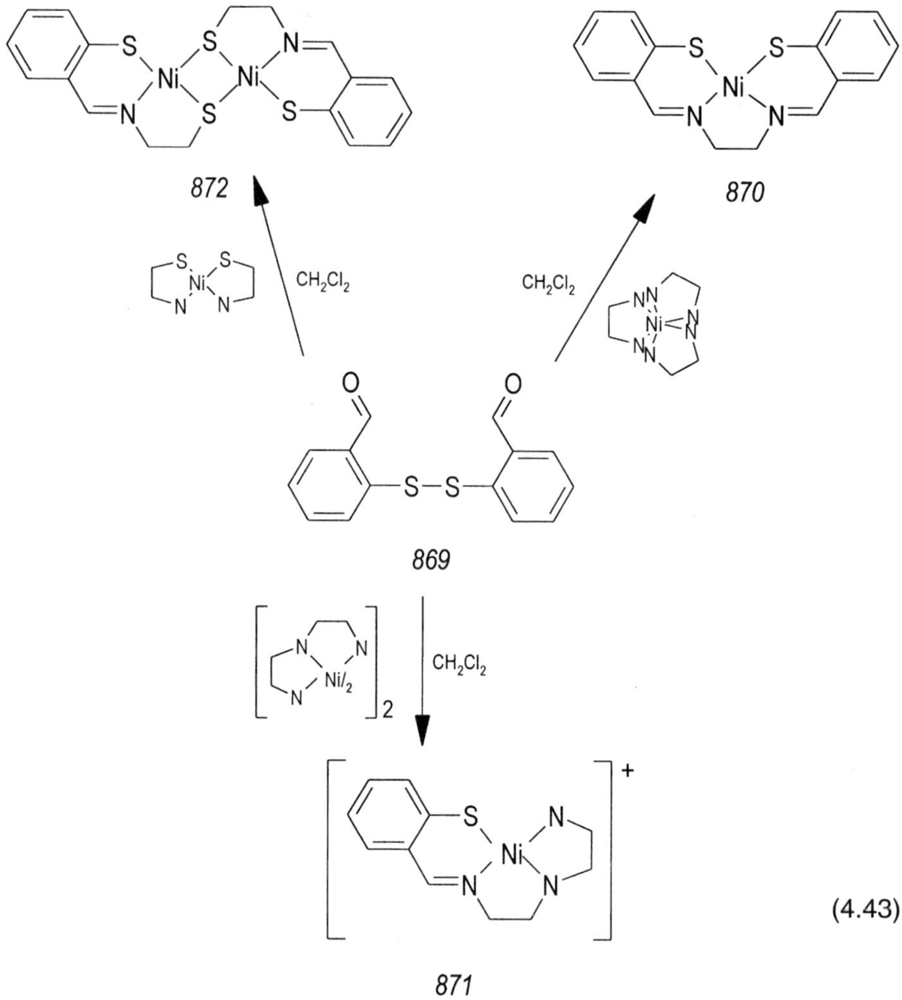

(4.43)

These template syntheses, based on the cleavage of the S—S bond of the stable compound ***869*** by nickel complexes in mixtures of methylene chloride (solvent for the disulfide) and methanol, are sufficiently promising for obtaining other ICC with N_2S_2-ligand environment. In particular, analogous reactions with coordinated aromatic and heteroaromatic amines of the types $ArNH_2$ and $HetNH_2$ are of interest. Use of these compounds in the syntheses discussed (4.43) opens a route for obtaining ICC which, for a difference of hard-plane structures ***870–872***, can yield various stereochemical forms.

The tetrahedral or tetrahedral-distorted polyhedra are preferable for azomethine chelates with coordination number 4 and coordination unit MN_4 (***868***, X = NR, M = Co, Ni, Cu) [100,130,134,135,141]. At the same time, there are some suggestions for complexes ***868*** (X = NTs), according to whether the penta- ($CN = 5$) or hexa-coordinated ($CN = 6$) structures (tetragonal pyramid, elongated octahedron, mono- or two-capped tetrahedron) exist in them. The additional ligation of tetracoordinated polyhedra takes place at the expense of binding the metal to the oxygen atoms of SO_2Ar groups [100,134,149,150].

The *trans*-planar structures, which are distorted as a result of the influence of R′-substituent, exist mostly in complexes ***868*** (X = O, M = Ni, Cu) with coordination unit MN_2O_2 [132–135,151]. Such structures are practically not observed in the case of a studied ICC of palladium, which possesses the strongest planarizing action [132–135,152,153]. Formation of tetrahedral *o*-hydroxyazomethinic ICC is characteristic exclusively for beryllium and preferably for cobalt, zinc, and mercury [132–135].

In a difference from the above, the formation of *cis*-planar structures takes place in case of chelates ***868*** (X = S, Se, M = Ni, Pd) [100,130,154–157]. The stereochemical situation, close to that above, is mainly observed in coordination compounds of azomethinic ligands containing five-member metal-cycles [100,130].

Syntheses using ***873*** (X = NR′, O, S) are carried out mostly by the method of immediate interaction of ligands and metals (4.44) resulting in chelates ***874***:

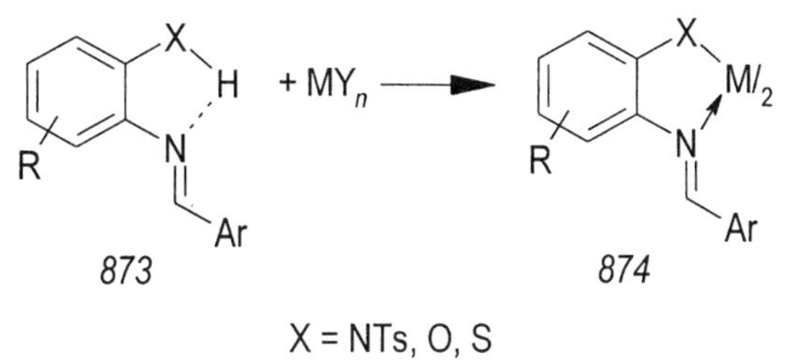

(4.44)

However, because of the low stability of ***873*** (X = Se), the template synthesis (4.45) was used for isolation of ICC of analogous selenium-containing azomethines [158]:

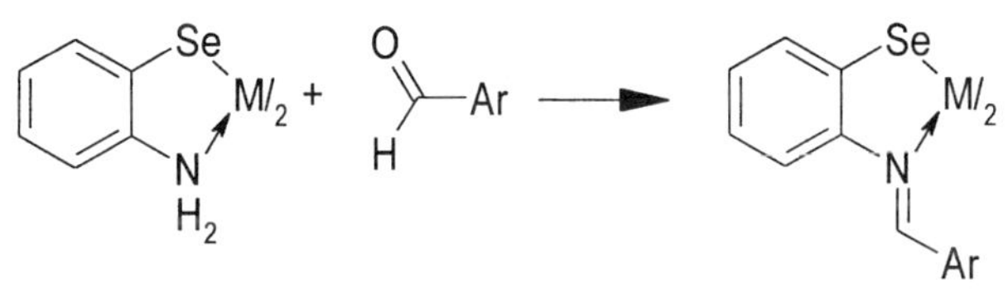

(4.45)

Attempts to isolate ICC of type ***874*** (X = Te) by using this synthetic reaction were unsuccessful [159].

The tetrahedral [160–163] and *cis*-planar [100,134,164–167] structures are characteristic for chelates of type ***874*** with coordination units NiN_4 and MN_2S_2, respectively, as well as chelates ***868*** discussed above. Original polyhedral forms were discovered by x-ray diffraction for nickel and palladium ICC of the discussed type ***874***. It is accepted that, in case of a nickel complex, the compound with a carbon–carbon bond ***875*** is formed [165,166]; formation of palladium chelates is accompanied by the cyclometallation reaction leading to tetranuclear clusters ***876***, where the tridentate ligand behaves as C,N,S-donor [168].

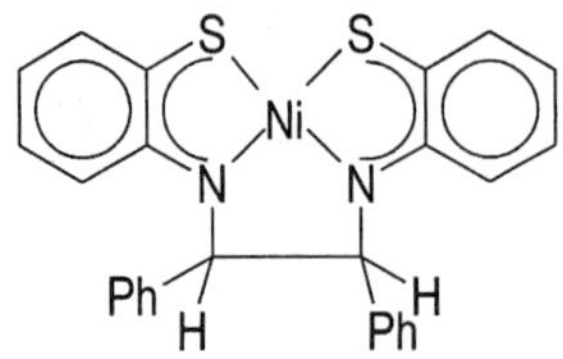

875

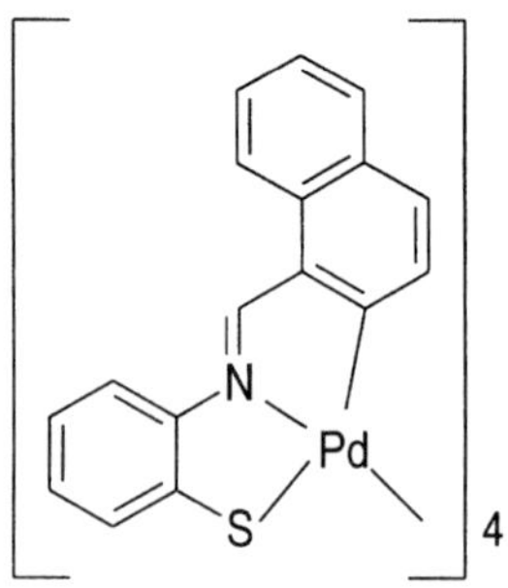

876

In a difference from above, the nickel and copper ICC ***874*** (X = O) with coordination unit MN_2O_2 evidently have tetrahedral structure, which is also characteristic for analogous cobalt and zinc complexes [169], as proved for a zinc complex by x-ray diffraction. In this respect we note a nonstandard behavior of manganese. As a result of the interaction of the ligand ***873*** (X = O, R = H, NO_2; Ar = C_6H_4R', R′ = H, OMe, NO_2) and the acetate of the metal in methanol, together with an expected ICC of type ***874*** (X = O, M = Mn)* [17], tetrameric cluster structures are formed (4.46) [171]:

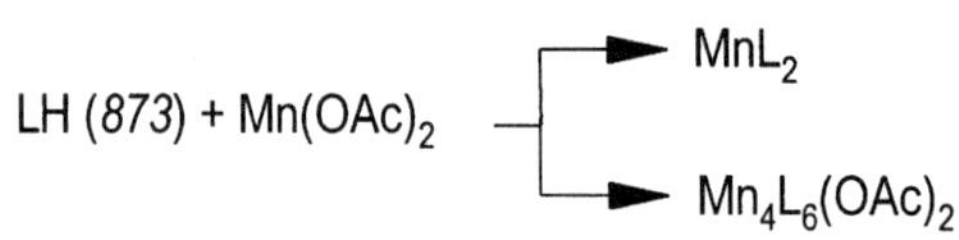

(4.46)

Amongst other ICC with five-link metal-cycles, we note *o*-indoline ***877*** (X = NTs) [160–163] and *o*-indophenol ***877*** (X = O) [163,172,173] chelates, obtained in reaction (4.47). According to the expected results, palladium complexes of the type discussed are planar, while the ICC of zinc, mercury, and lead are tetrahedral [161,172,173].

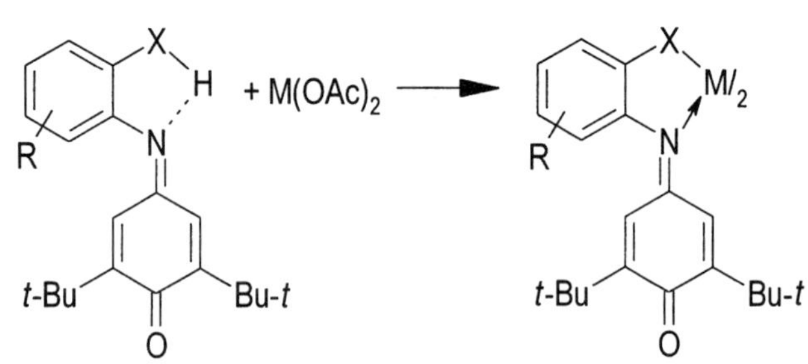

X = NTs, O

877

(4.47)

*The same complexes were obtained by template synthesis with the use of derivatives either of benzaldehyde, *o*-aminophenols, or manganese acetate, in hot methanol solution.

4.2.2 Nature of Fragments, Annelated to a Metal-Cycle

The influence of the R-substituent nature on the stereochemistry of ICC of type ***868*** is clearly displayed in complexes of hetaryl azomethinic ligands [15,100,130,134]. These chelates are mostly obtained by the method of immediate interaction of ligands and metal salts [(4.48), route ***A***], although some cases of application of template reactions for the same goals are known [(4.48), route ***B***]. In this respect, the syntheses (4.48) are representative, described in Refs. 174 and 175 for azomethines of 3-hydroxybenzo[*b*]thiophene ***878*** and their azine analogues, leading to chelates ***879***:

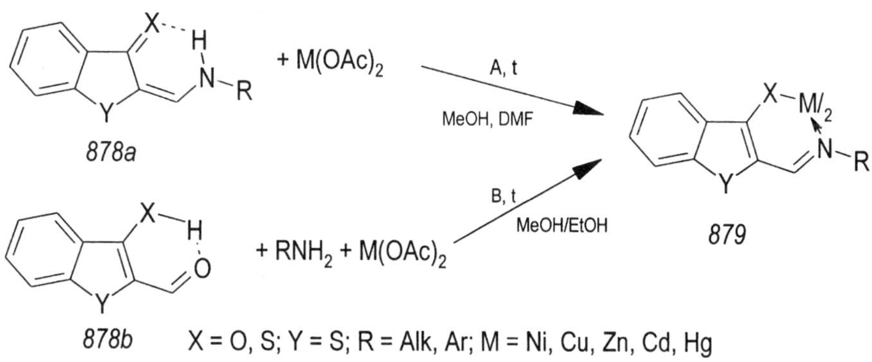

(4.48)

The ICC, synthesized on the basis of 4-aminomethylene derivatives of 5-imine-, oxo-, thio-, and selenoderivatives of pyrazole, are represented most widely. The coordination units MN_4, MN_2O_2, MN_2S_2, and MN_2Se_2, respectively, are found in them [100,134,176–186]. Thus, the methods ***A*** and ***B*** (4.49) were used for obtaining ICC of type ***868*** (X = NR, R-Het, compound ***222***) [187]:

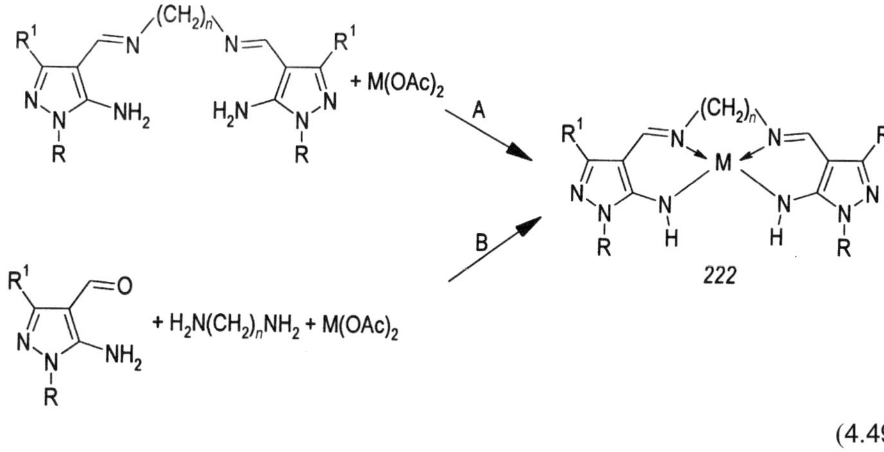

(4.49)

The immediate interaction of ligand ***471*** and metal acetates, for instance (2.13) [100,134,188–191], is practically the only method used for obtaining metal chelates of pyrazol azomethines ***468*** (Sec. 2.2.5.4). The hetarylazomethinic ligands ***880*** and metal salts form (4.50) the ICC of type ***881*** [192,193]:

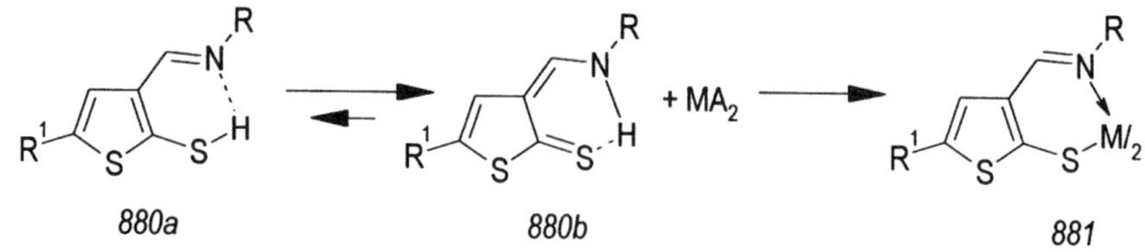

(4.50)

The structure of ***881*** was proved by x-ray diffraction [193]. The possibility of carrying out these syntheses may be explained as due to the sufficently high stability of sulfur- and selenium-containing azomethines within a heterocyclic series, in comparison with their aromatic analogues (compare the data of Sec. 3.5.2.1), which is related to the existence of ***471*** as the aminomethylene form "b" [15,194–197].

The structural aspect of the problem examined, allowing us to proceed with programmed polyhedron-controlled syntheses, is the following. The annelation of heterocyclic fragments to a metal-cycle provokes a tetrahedrization of complexes of type ***868*** (R = Het). This effect is displayed in the structures of complexes of types ***468***, ***476***, and ***881***. At the same time, there are some exceptions from this conclusion. Thus, the structure of nickel chelates ***879*** depends on the X- and Y-heteroatom nature [100,134]: the complex is tetrahedral at X = S and Y = O [198], while the planar-square configuration is found at X = Y = S [199].

The majority of metals form tetrahedral (Co, Ni, Zn, Cd, Hg) or tetrahedral-distorted (Cu) structures in the complexes of type ***468***. However, the stabilizing planar-square effect of palladium is higher than the tetrahedrizating action of annelated pyrazol fragments: complex ***468*** (M = Pd, R = *cyclo*-C_6H_{11}, R^1 = *i*-Pr, R^2 = Me) has *trans* configuration [100,200,201], while chelate ***468*** (M = Pd, R = α-pyridine, R^1 = *i*-Pr, R^2 = Me) possesses a *cis*-plane one [100,202]. The existence of the *trans* configuration, unexpected for d^8-metals in N_2S_2 environment [157], is evidently related to the action of bulk cyclohexyl substituent. This factor, controlling in some cases the structures of ICC, will be discussed later.

The described stereochemical results for nickel(II) ICC with coordination number 4 can be represented by Scheme (4.51):

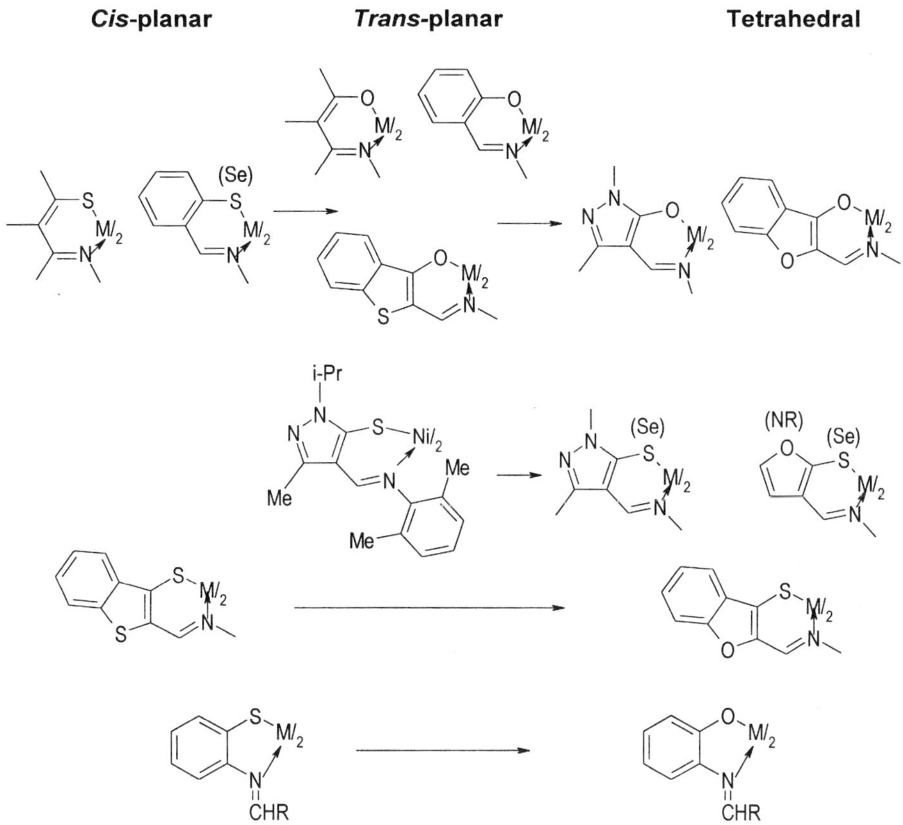

(4.51)

The tetrahedrizating influence of the heteroannelation is so high that it exceeds the *cis*-planar effect, as observed (Sec. 2.2.5.4 [133,135,157]) in nickel complexes of aromatic azomethines with N_2S_2-ligand environment.

4.2.3 Nature of Substituents in Amine Fragment

By evaluating the character of influence of an R^1-substituent on the stereochemistry of ICC of azomethinic ligands of type ***868***, the following two effects are usually emphasized: *spatial*, related to volume, and *coordinatively-active*, caused by the presence of additional donor centers. The first one causes, in the majority of cases, distortions of tetracoordinated polyhedra; the second one increases the coordination number of a metal complex-former from 4 to 5 and later to 6. Guided by these considerations, it is possible to carry out the controlled synthesis of metal complexes with a programmed geometry of the coordination unit.

The spatial effect of R-substituents has been discussed in the literature [100,134,135,139] in some detail. We examine here only a few examples of its influence on the structure of azomethinic complexes of type ***868***. Nickel chelates of *o*-hydroxyazomethines ***868*** (M = Ni, R = *cyclo*-C_4H_4, R^1 = H, Alk, Ar) in crystalline phase have, in dependence on the volume of the R^1-substituent, a planar-square or a distorted tetrahedral configuration. Meanwhile, a complicated configurational equilibrium is observed for this type of chelate in solution, including plane, tetrahedral, and octahedral (external-molecular associated) structures [100,134,135,203–206].

The tetrahedral distortions in copper ICC of the type discussed, caused by R^1-substituents with different spatial effects, are strictly proved by x-ray diffraction, as well as confirmed by EPR studies [135,207]. However, a united point of view about the structure of these complexes in solution is still absent (compare Refs. 134, 135, 205 and 206). The existence of a square-pyramidal or distorted tetrahedral configuration, as well as the equilibrium "square-tetrahedron," is accepted as possible. We emphasize that, in a series of cases, the nature of metal complex-former is the decisive factor, determining the structures of the examined ICC. Thus, practically independently of the character of the R^1-substituent, palladium chelates are planar, those of beryllium are exclusively tetrahedral, and those of cobalt, zinc, cadmium, and mercury are, in general, also tetrahedral [135].

In the past and taking into consideration the x-ray diffraction data, it was concluded (for example, Ref. 100) that the spatial effect of the R^1-substituent can influence considerably the structure of complexes of the type ***868*** (R = Het, compound ***468***). Thus, nickel chelate ***468***, containing methyl groups in *o*-positions of aryl R^1-substituent (M = Ni, X = NH, R = $C_6H_3Me_2$–2,6; R^1 = Ph, R^2 = Me) has a *trans*-planar configuration in crystalline phase (as proved by x-ray diffraction) [208]. According to ^{1}H NMR data ($CDCl_3$), the equilibrium "square-tetrahedral" configuration is observed for this complex in solution.

We note that the chain length of methylene (number *n*) is an important factor in regulating the structure of type ***882*** [100]:

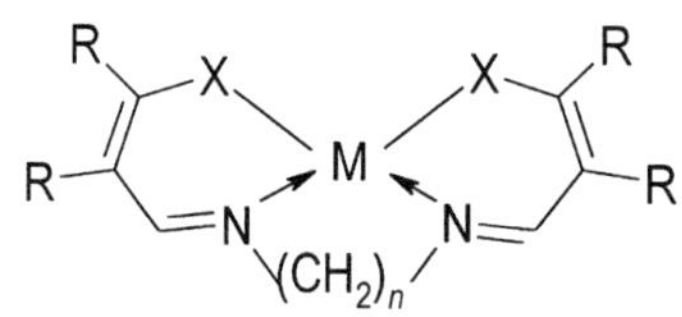

882

When *n* = 2 or 3, such structurally characterized ICC, in the majority of cases, have slightly tetrahedrizated *cis*-planar conformations [100]. The effect of the factor examined is so high that even zinc chelates ***882*** (M = Zn, R = Me, X = S, *n* = 2) show a considerably planar-distorted structure [209]. An analogous structural situation is observed for cobalt, nickel, and copper complexes containing R = *cyclo*-$(CH_2)_3$, annelated aromatic, and pyrazol fragments [100]. When *n* increases, for example *n* = 4 in the complexes capable of tetrahedrizating a structure, the influence of the examined factor decreases. This is observed in the ICC not only with X = S [100,210–212], but also for X = NH ***222*** [187].

A study of the influence of coordinatively-active R^1-substituent nature on the structure of complexes of type ***868*** was begun in the 1960s and continues to be of interest at present [100,134]. The effect of such substituents is related mainly to an increase of coordination number of the central atom of chelates ***868***, i.e., it provides a transition from square-planar or tetrahedral polyhedra to penta- or hexacoordinated ones. This circumstance allows us to carry out a programmed synthesis of ICC of the azomethinic series with pyramidal or octahedral structures.

The methods of synthesis of the examined complexes are based on a selection of azomethinic ligands, containing groups with mostly N,O,S-donor atoms at the N atom of the C=N bond. Compounds of type ***883*** are examples of such formally tridentate ligand systems:

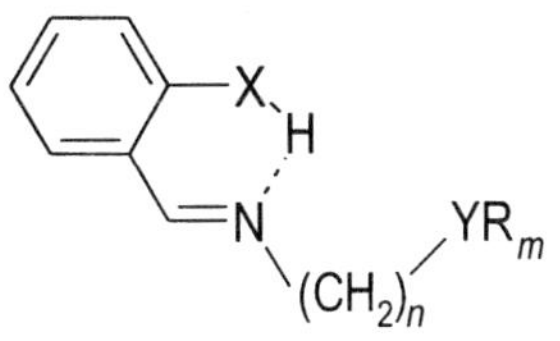

883

X = NR[1], O, S, Se; Y = N, O, S;
R = H, Alk, Ar; R[1] = H, Ts;
$m = 1, 2$; $n = 1–4$

Azomethine ligands with mono-, di-, and oligomethyleneamine (Y = N, $m = 2$), aminoalcohol (Y = O, $m = 1$), and aminomercaptoalcohol (Y = S, $m = 1$) fragments belong to these systems. Their ICC can be obtained according to Scheme (4.40) preferably by using the method of immediate interaction of ligands and metal salts. They are represented by coordination compounds of types ***884–889***, which are different in their structure ($CN = 4–6$) and number of metal nuclei [Scheme (4.52)]:

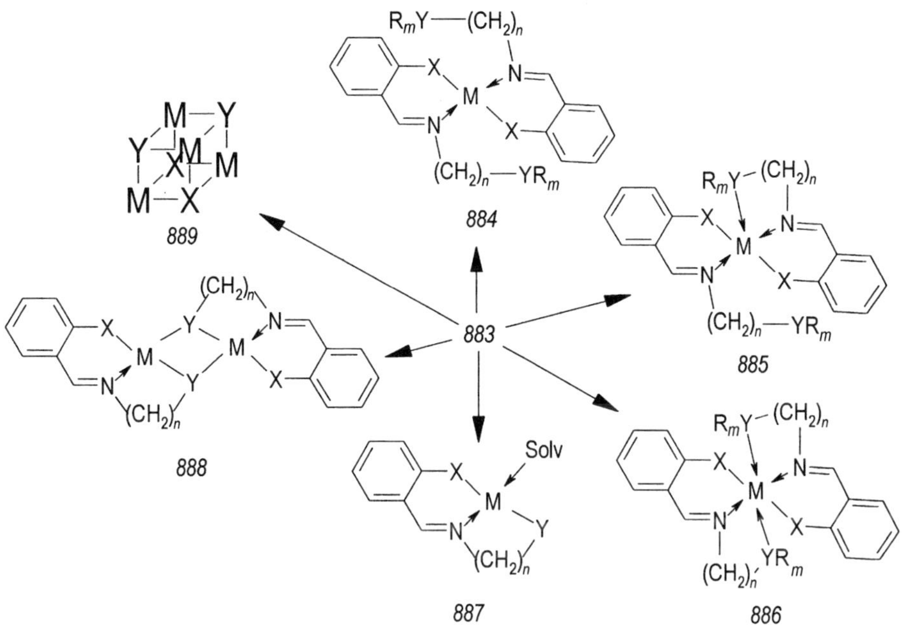

(4.52)

The ICC of the examined type with alkylamine substituents (X = O, Y = N, R = Alk, $m = 2$) have been studied for a long time [137,206]. The majority of them are represented by hexacoordinated structures ***870***. Structurally characterized copper complex ***885*** (M = Cu, X = O, Y = N, R = Me, $m = n = 2$) is an example of complexes with $CN = 5$ [213]. The indicated substituent at the imine nitrogen atom participates by coordination in structurally characterized mixed-ligand complex ***890*** obtained according to Scheme (4.53) [214]:

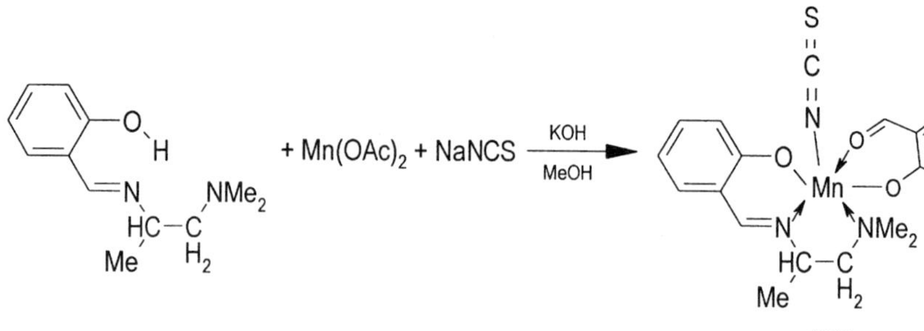

890

(4.53)

Complexes of salicylideneiminoalcohols ***884–889*** (X = Y = O, $m = 1$, $n = 1$–4) are widely known. Earlier works on the synthesis, structure, and properties of these ICC are discussed in several reviews [144,215,216] and also reported in more recent publications [217–221]. To develop this area, a series of complex compounds of aryl-substituted derivatives of salicylidene(naphthalidene)monoethanolamines ***891*** was isolated (4.54) and direct x-ray diffraction proofs of structures of type ***887*** were obtained [217]. The structure of ***891*** (R = 3,5-Cl_2) was confirmed by x-ray diffraction [217]:

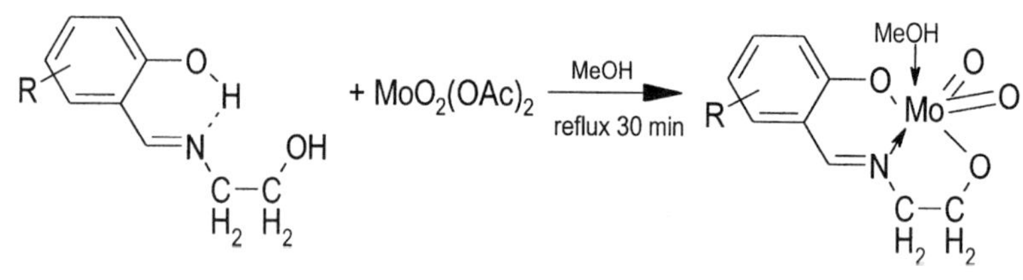

R = H, 5-Br, 5-NO_2, 3,5-Cl_2, 5,6-C_4H_4-*cyclo* 891 (4.54)

The ICC of type ***892*** containing an NTs-fragment were obtained by template synthesis (4.55) [218]:

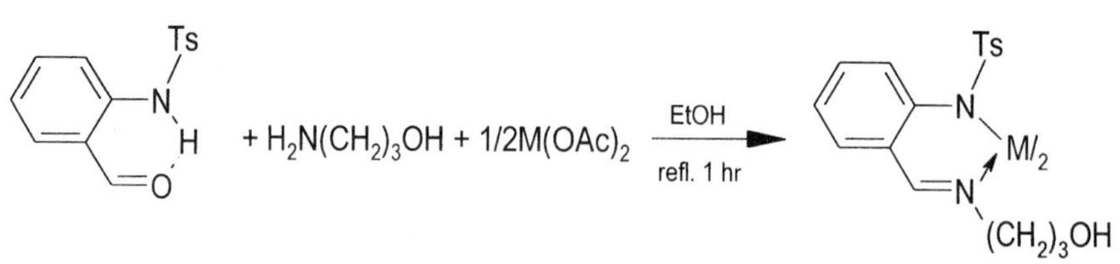

892

(4.55)

On the basis of IR and ^{1}H NMR data ($CDCl_3$), the existence of structure ***884*** was suggested [218] for the complexes ***892***, but direct structural data are still absent.

Template synthesis (4.56) was also used to obtain complexes of iminoalcohol derivatives of thiosalicylaldehyde ***893*** [221–223]:

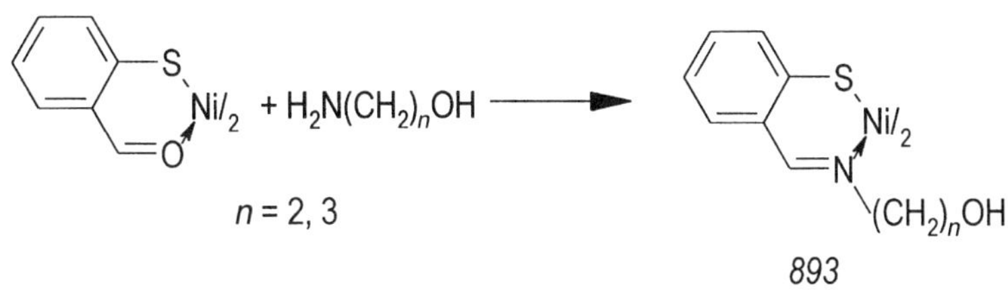

(4.56)

The structures of these ICC were not strictly established, since the available data [220–222] are synonymous: existence of both monomeric (type ***884***) and dimeric (***888***) structures is possible. The direct x-ray diffraction proofs of formation of a dimeric structure of type ***888*** are obtained for complex ***894***, isolated as a result of transformation (4.57) [224]:

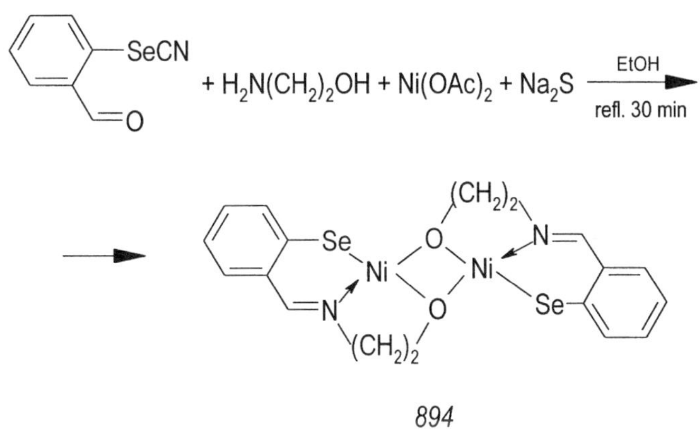

(4.57)

The examined binuclear chelate is the first example found of an ICC in which the NO_2Se-ligand environment and bridged Ni_2O_2 structure have the participation of alcohol oxygen atoms.

Various types of structures ***895*** and ***896*** were obtained with use of transformation (4.58) applied to iminoalcohol derivatives of heterocyclic carbonyl compounds used as ligands [220,225]. The structure of ***895*** [225] and ***896*** [220] was proved by x-ray analysis:

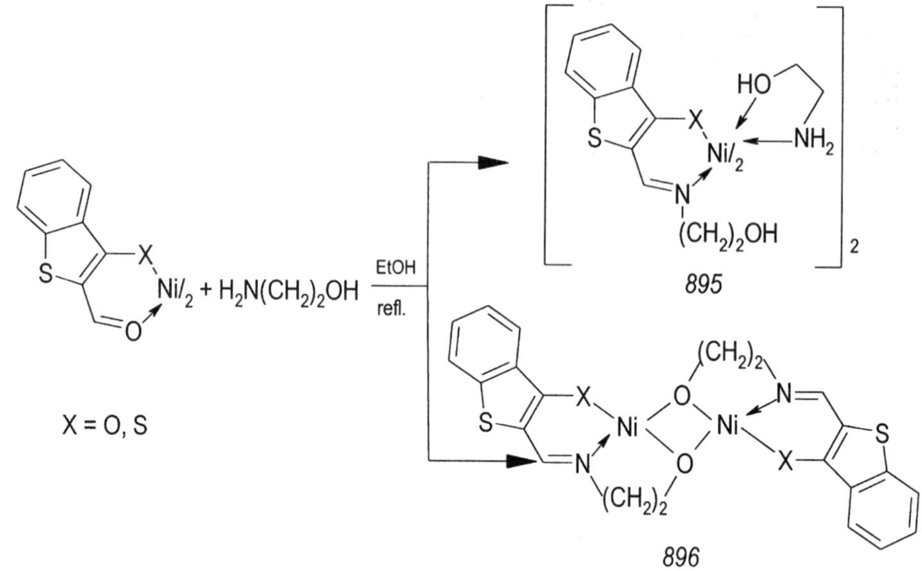

(4.58)

Since complexes of the different types **895**→**884** and **896**→**888** were synthesized under the same conditions, there is sufficient basis to consider that the factor responsible for their different structure is found in the nature of the donor atom X, which belongs to a six-member metal-cycle. Let's pay attention to structure **895**, in which the octahedral polyhedron is formed at the expense of a bidentate coordination of ethanolamine. This situation emphasizes the circumstance that, when using coordinatively-active components in template syntheses, it becomes necessary to take into consideration the possibility of independent formation of complexes which are different from the final products [compare (3.255)].

According to Ref. 220, one more type of complex **897** (compare with **896**) is formed through the reaction of immediate interaction of ligands and metal salts (4.59):

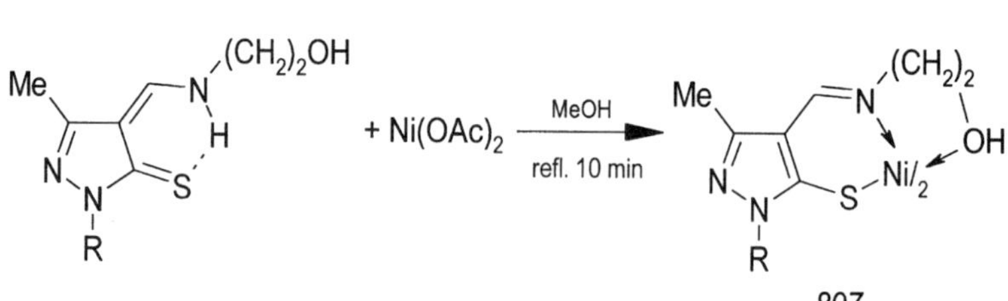

(4.59)

The octahedral structure was ascribed to these complexes on the basis of magnetochemical ($\mu_{eff} = 2.90$ and 3.01 B.M. for R = *i*-Pr and Ph, respectively) and UV-spectral studies.

A limited number of investigations are known where the ICC, obtained on the basis of mercaptoiminoalcohols ***898***, are described [226,227]. Thus, as a result of syntheses (4.60), the monomeric ***899*** [227] and dimeric ***900*** [226] complexes were synthesized, where the S atom in the mercapto group takes part in coordination with the metal:

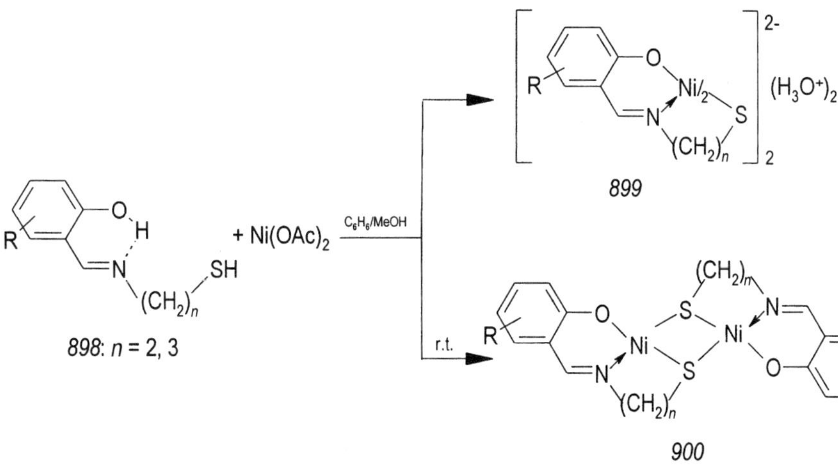

(4.60)

The formulae **899** and ***900*** are attributed to the complexes discussed only on the basis of elemental analysis and magnetochemical data and, in this respect, provoke definite doubts, especially in the case of ***899***. At the same time, it is necessary to take into account that the coordination mode of the examined ligand, as in ***899***, is quite real and was established by x-ray diffraction for the electrochemically obtained (4.61) tin(IV) complex [228]:

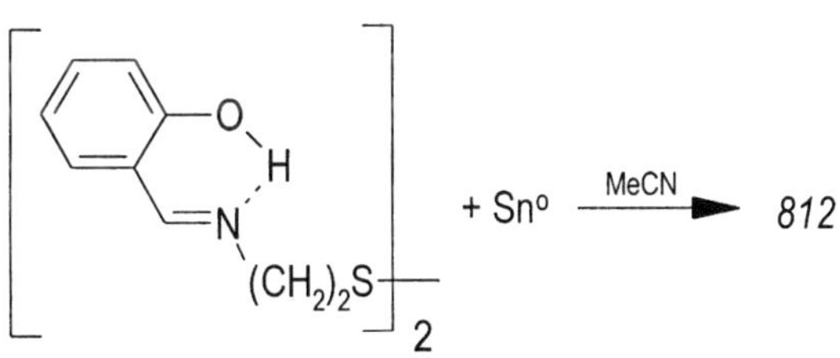

(4.61)

All the compounds ***640*** [11] should have the capacity to act as di- (***901***, ***902***) or tridentate (***903***–***905***) ligands [Scheme (4.62)]:

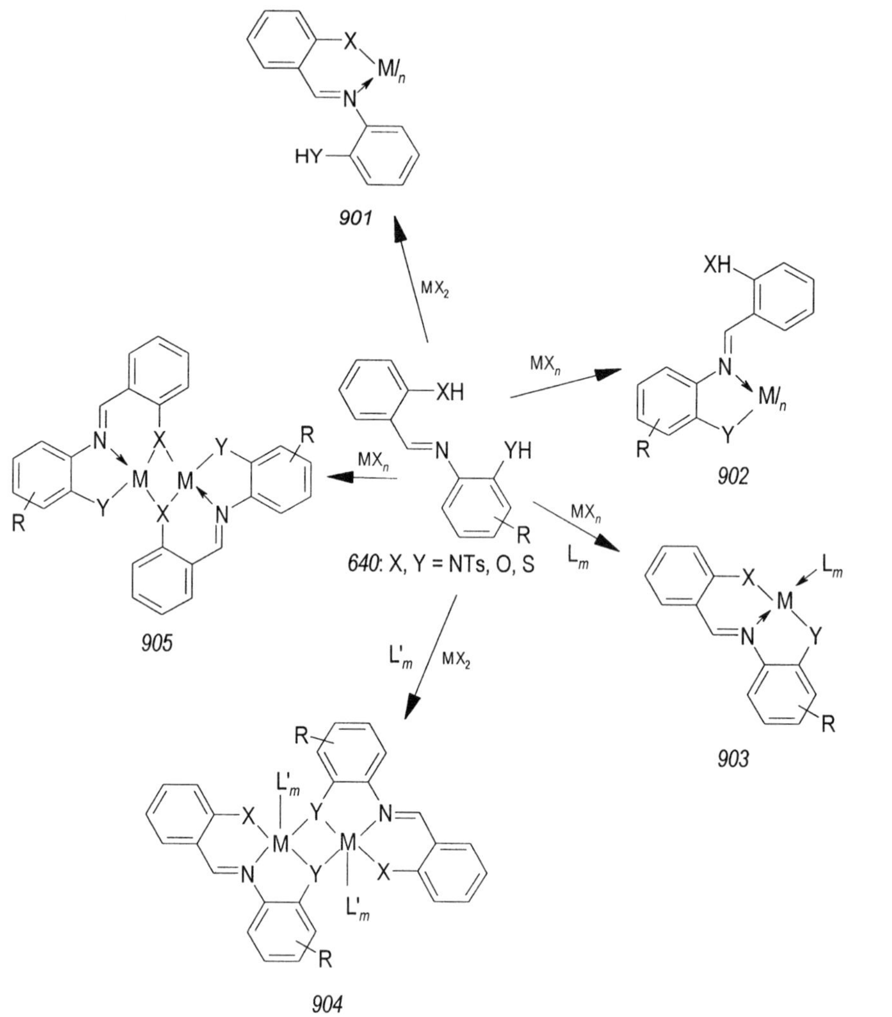

(4.62)

However, all the ICC obtained by using chemical (mainly, immediate interaction of ligands and metal salts) or electrochemical methods have the structures ***903*** or ***904***. Enormous amounts of literature related to these syntheses are mainly found in monographs [55,56] and reviews [11,229,230]. The bonding isomery, which caused the possibility of formation of structures ***904*** or ***905***, was also discussed [11,231].

The routes for controlling the synthesis of mono- (***903***) and dinuclear (***904***) structure is, evidently, related to the donor properties of additional ligands [Scheme (4.62): L and L^1]. In case of highly basic N-donors (L = py, bipy, *o*-phen), the mononuclear complexes ***903*** are generally formed [11,229,232], while the weakly basic solvents (L = MeOH [233]) lead to binuclear chelates of type ***904***. At the same time, the only example for ***805*** (L = *o*-phen) of a stable dimeric adduct with sufficiently high-basic *o*-phenanthroline is known and was shown above (Sec. 3.4.2).

It is possible to use ligands of the type ***867*** containing, as R^1-substituent, the nitrogen-containing heterocycles, annelated to the azomethinic N atom, to synthesize programmed penta- and hexacoordinated structures. However, as will be described below, in this case the linkage of the hetaryl substituent in the metal-cycle, formed as a result of an additional coordination, and the structure of the azomethinic coordination unit, both play an important role. In this respect, the situation with structures of ICC obtained on the basis of ***867***, and containing 2-pyridine [100,200–202,234–244] and 8-quinoline [182,183,245–247] substituents at the N atoms, is quite representative. Formally, tridentate ligand ***906*** reacts (4.63) with metal salts [239,240,242–244] or under electrosynthesis conditions [241], forming tetracoordinated complexes ***907*** where the N atom of the pyridine substituent does not take part in coordination with the metal:

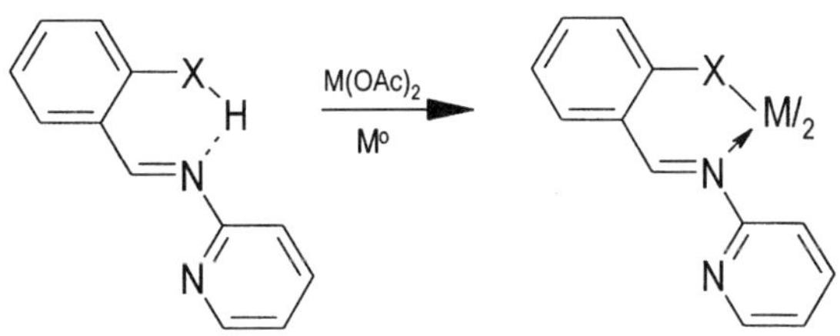

906 *907*

X = NTs, O, S

(4.63)

Complex ***907*** (X = S, M = Ni), in which according to x-ray data [242] the examined coordination is absent, was obtained by template synthesis also.

A close structural situation is observed for *cis*-planar palladium ***908*** (M = Pd) [201] and distorted tetrahedral copper ***908*** (M = Cu) [242] chelates. The N atom of pyridine substituent is turned out from the metal complex-former in these chelates.

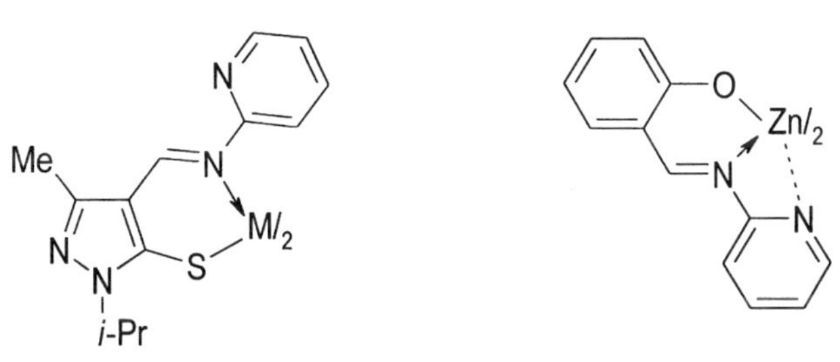

908 *909*

At the same time, according to x-ray data for zinc chelate ***909***, the nitrogen atom is turned to the side of the metal. The distance N_{py}-Zn is 2.80 Å, that allows us to consider the possible participation of the examined donor center in binding with the metal, leading to formation of a hexacoordinated structure (two-capped tetrahedron) [243]. In relation with this result, let's pay attention to the data reported in Refs. 244 and 248. The tetrahedral configuration without coordination of the nitrogen atom of pyridine is attributed to the cobalt complex ***907*** (X = NTs, M = Co), although this N atom is rotated to the side of the metal [244]. The pentacoordinated complex ***910*** is described in Ref. 248, in which only one pyridine substituent is coordinated (the distance N_{py}-Co is 2.45 Å):

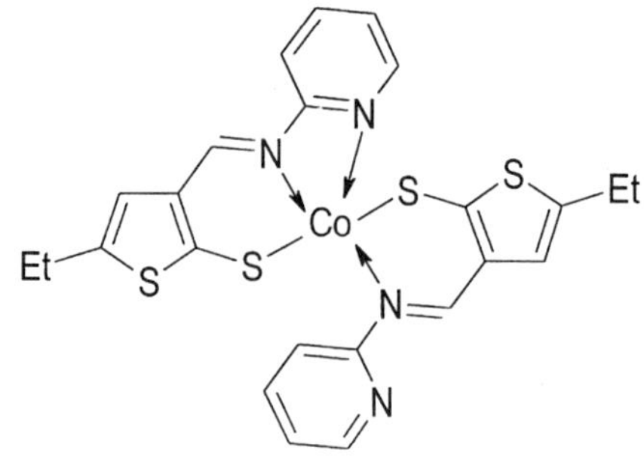

910

In a difference from above, the x-ray diffraction data clearly indicate the coordination of the N atom of the quinoline substituent in complexes ***592*** [182] and ***478*** [183]. The distances N_{quin}–Cu are 1.990 Å (***592***: X = NTs, R = H), 1.987 Å (***891***: X = O, R = 5-Cl), and 2.00 Å (***478***), in agreement with the length of a normal coordination bond. Chelates ***592*** were obtained by the method of immediate interaction of ligands and copper acetates in methanol solution [(3.314), *A*], ***478*** by interaction of pyrazol azomethine with N-quinoline substituent and $FeCl_3 \cdot 6H_2O$ in ethanol.

A comparison of data on the ICC structures of types ***592***, ***907–910*** allows us to conclude that the linkage of an additional chelate unit is the main factor determining the possibility of participation of the N atom of the heterocyclic substituent in the coordination. As expected, the four-member coordination unit (pyridine) has low stability, while the five-member (quinoline) forms stable metal-cycles.

EXPERIMENTAL PROCEDURES

Example 1

Synthesis of *879* (4.48) [174]

Method 1

An alcohol solution of 5.2 mmol of corresponding metal acetate was added to the solution of 10 mmol of ligand ***878*** in 150 mL of methanol. The mixture was refluxed in the water bath for 10–15 min. Precipitate was filtered off, washed with cold methanol, and dried over $CaCl_2$. Yield 70–90%.

Method 2

A solution of 5.2 mmol of copper or nickel acetate in 50 mL of DMF was added in small portions to the solution of 10 mmol of ligand in 50 mL of the same solvent under stirring. The mixture was heated at 100°C for 10–15 min. After cooling a precipitate of complex (or its solvate) was formed which was filtered, recrystallized twice from DMF, washed with benzene, and dried over $CaCl_2$. Yield 70–80%.

Example 2

Synthesis of Ligand *880* (R=α-Py, R^1=Et) (4.50) [192]

A solution of 5.5 g (0.058 mol) of 2-aminopyridine in 30 mL of absolute alcohol was added to the solution of 5 g (0.029 mol) of 2-mercapto-5-ethyl-3-tenylideneimine in 30 mL of the same solvent. The mixture was refluxed for 3 hr, after which the solvent was removed in vacuo;

30 mL of alcohol was added to the residue and removed under vacuum again (this procedure was repeated three times). The residue – dark oil or crystalline solid – was treated by a small amount of cold alcohol, the precipitate was filtered off, washed with water and alcohol, and dried in vacuo. Yield 4.17 g (58%).

Example 3

Synthesis of Complex ***881*** (R=α-Py, R^1=Et, M=Zn) [192]

A solution of 2.65 g (0.012 mol) of zinc acetate in 15 mL of methanol was added to the hot solution of 3 g (0.012 mol) of ligand in 50 mL of ethanol. The mixture was heated for 20 min. Formed orange precipitate was filtered off, washed with methanol, and dried. Yield 2.67 g (79%).

Example 4

Synthesis of Complex ***891*** (5.45) [217]

A solution of 0.234 g (1 mmol) of 2-(3,5-dichlorosalicylideneimino)ethanol in 20 mL of methanol was added to the solution of 0.326 g (1 mmol) of $MoO_2(acac)_2$ in 20 mL of the same solvent under stirring. The mixture was refluxed for 0.5 hr, after which it was kept refrigerated overnight. Lime-yellow crystals were filtered off, washed with cold methanol, and dried in vacuo.

Example 5

Synthesis of Complex ***895*** (4.58) [225]

Sodium sulfide (5 g) was added to the solution of 3 g of 2-formyl-3-chlorbenzo[*b*]thiophene in 80 mL of ethanol under heating, after which the mixture was refluxed for 30 min in argon atmosphere. The red–orange solution was filtered through the lay of anhydrous sodium sulfate. A solution of 1.9 g of nickel acetate in 30 mL of methanol was added to the resultant mixture. The precipitate of nickel aldehydate was filtered off, washed with methanol, and dried.

A solution of 0.15 g (2 mmol) of propanolamine in 10 mL of ethanol was added to the suspension of 0.415 g (1 mmol) of the above nickel aldehydate in 20 mL of the same solvent. The mixture was refluxed in the water bath for 2 hr. The precipitated complex was filtered off, washed with ethanol, and dried in air. Yield 70%. Complex can be recrystallized from chloroform.

4.3 SOME APPLICATIONS OF PROGRAMMED SYNTHESIS OF COORDINATION COMPOUNDS

Use of the programmed synthesis, in addition to the possibility of obtaining complexes with controlled coordination mode (regioselective synthesis, Sec. 4.1) and geometry (polyhedron-programmed synthesis, Sec. 4.2), can resolve some theoretical and practical aspects, important in coordination chemistry.

4.3.1 Theoretical Aspect

The resolution of problems of standard and nonstandard coordinations of typical chelating and heteroaromatic ligands is related in many aspects to the regioselective synthesis [6,10,11,14–16,21]. In particular, it was shown that, on the basis of β-diketones together with chelates [Sec. 3.1, Schemes (3.6), (3.36), (3.37)], the molecular complexes with nonstandard O- (***284–287***) or C- monodentate coordination Sec. 4.1.2, Schemes (4.19)–(4.21), (4.23)] of the examined ligand can be obtained. Similarly, the regioselective synthesis as it was shown above (Sec. 4.1.2) allows us to prepare [Scheme (4.24)] not only the most propagated ICC of type ***855***, but also molecular adducts ***856***, using *o*-oxyazomethines and other structurally related ligands ***857*** and ***858*** [Schemes (4.25)–(4.34)]. Regioselective syntheses in a series of heteroaromatic ligand systems are represented by complexes with differently coordinated azines [(3.232), (4.13)] and azoles [(3.147), (4.15)].

The possibility of carrying out the type-programmed synthesis (preparation of different types of coordination compounds from the same ligands) is directly related to the problem discussed [10–12,14]. On the basis of the data given above, it follows that both ICC and molecular adducts [Schemes (4.19), (4.24), and (4.34)] can be obtained from chelating ligands. Heteroaromatic ligands form [(4.13) and (4.15)] molecular and π-complexes.

Regioselective synthesis opens routes for obtaining in a controlled way complexes with fixed tautomers of ligands, and allows the correct understanding of the problem of stabilization of one of the tautomeric forms of ligand systems during coordination [18,249–255]. In this respect, most attention has long been paid to the examination of the problem of stabilization of tautomeric forms of ligands in β-diketone complexes [11,14,16,85,249,250,256]. It is accepted that the enol form of these ligands is stabilized in chelates formed by β-diketone anions [86–88,249,250]. However, such suggestions are incorrect: for β-diketonates of the types ***603*** and ***849*** with composition ML_n (LH is a ligand), according to x-ray diffraction data the bonds M—O have close lengths, so there is no reason to discuss the problem of stabilization of one of the tautomeric forms. At the same time, the stabilization of the ketone form in molecular adducts of the type $(LH)_nMA_m$, for example in complexes ***911*** [257] and ***912*** [258], is doubtless:

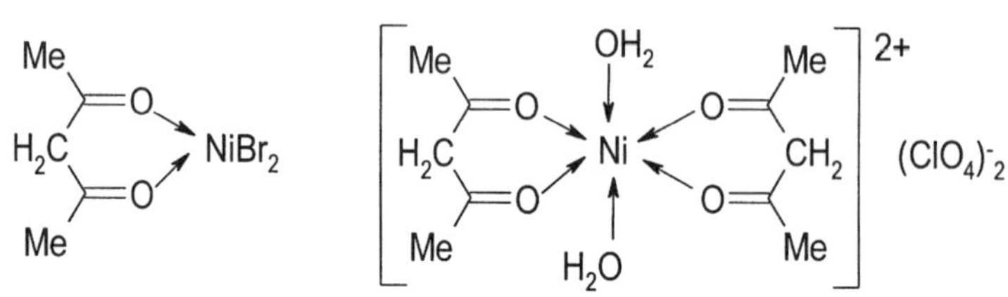

911 *912*

All the hydrogen atoms are revealed in the structure of ***912***, that clearly confirms the presence of the CH_2 group in the ketone form of the ligand [6,11,85]. The ketone form is also stabilized in C—M-derivatives of β-diketones [85–88] represented above, for example, by formulae ***850–852***.

The problem of stabilization of the enol form of β-diketone ligands, according to the available literature, is open [85]: it is accepted that such tautomers exist in coordination compounds $Cu(hfa)_2(Me_2NCH_2CH_2NMe_2)_2$ [259] and $Cu(acac)_2(H_2NCH_2CH_2NH_2) \cdot 2H_2O$ [260], where hfa is hexafluoracetylacetone. However, the x-ray diffraction study of these complexes was carried out with low precision and without localization of protons.

In the limits of the discussed problem, the topic of stabilization of tautomeric forms of ligands in complexes of *o*-hydroxyazomethines and their analogues [Sec. 2.2.5.3, formulae ***422*** and ***423***; Sec. 4.1.3, Scheme (4.24)] should be mentioned. When these ligands form ICC with the anionic form of ligand ***855***, the inner-molecular bonds become considerably smooth and, due to this, it is unlikely that we need to examine the problem of stabilization of the enol tautomer of the ligand system ***23a*** (compare, for example, Ref. 151). At the same time, the stabilization of the ketone form of these ligands was clearly proved by x-ray diffraction with localization of protons in the molecular adducts of *o*-hydroxyazomethines obtained in transformations (4.29) [105] and β-aminovinylketones (4.27) [111].

The tautomers of azole and azine derivatives, stabilized as a result of complex formation, are widely found [11,15,16,18,261–269]. Thus, the metal-cyclic structures with anionic ligands are formed in chelates of 2-hydroxy- (***436***, ***438***) and 2-mercapto-substituted (***484***, ***485***) derivatives of these nitrogen-containing heterocycles described above. At the same time, the ketone (thion) isomers are stabilized in molecular adducts. In this respect, the syntheses (4.64) [266], (4.65) [267], and (4.66) [269] are quite representative for this group:

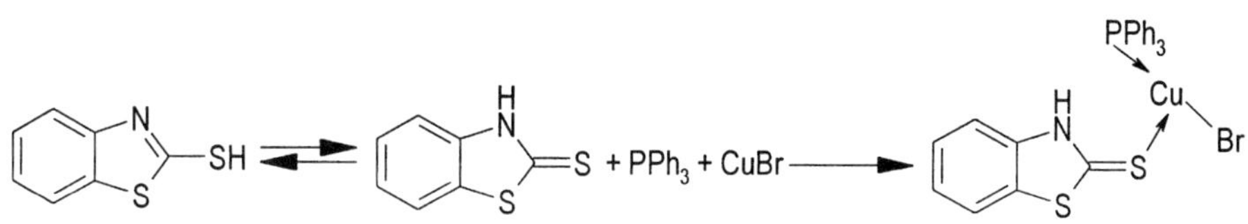

(4.64)

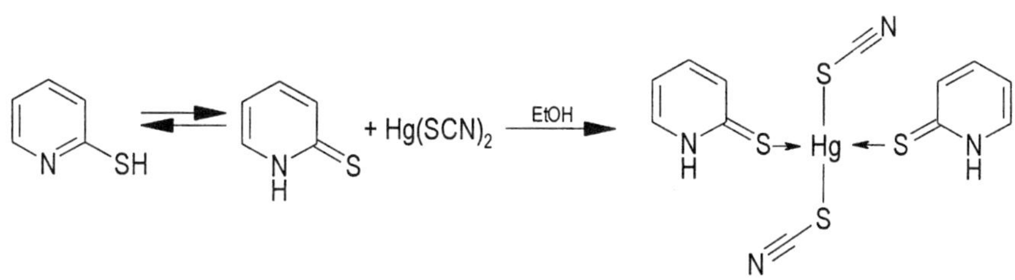

(4.65)

The thiol form of 2-mercaptopyridine is stabilized in clusters, for instance ***913***, which are obtained as a result of transformation (4.66) [269]:

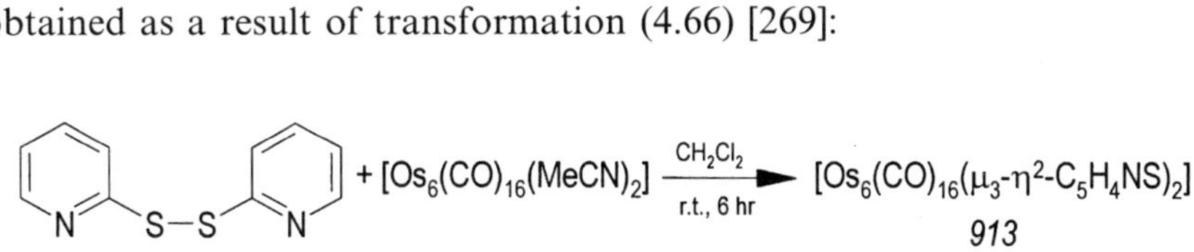

(4.66)

The formation of pyridine-2-selenolate complexes, for instance ***914***, takes place according to a closely related scheme, but under conditions of electrosynthesis (4.67) [270]:

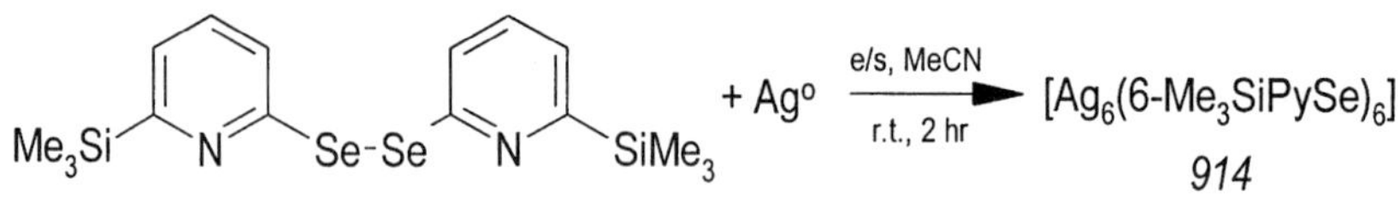

(4.67)

The structure of cluster ***914*** was proved by x-ray diffraction. The copper complexes with the same ratio M:L [Cu(PySe)] and [Cu(6-Me_3SiPySe)] were isolated according to a similar scheme, although their structure is unclear [270].

The amine tautomeric form of the ligand [15,18,271] was detected, in the majority of cases, in complexes of aminoheterocycles [Sec. 2.2.4.2 and Schemes (3.11)–(3.13)]. Only in some cases, when the electron-acceptor tosyl substituent is present at the exocyclic N atom, are the ICC formed (4.68), for example ***915***, containing anionic ligands and MN_4 metal-cycle with close lengths Ni—N (N_{py}–Ni = 2.110 Å, Ts–N–Ni = 2.095 Å) [272]:

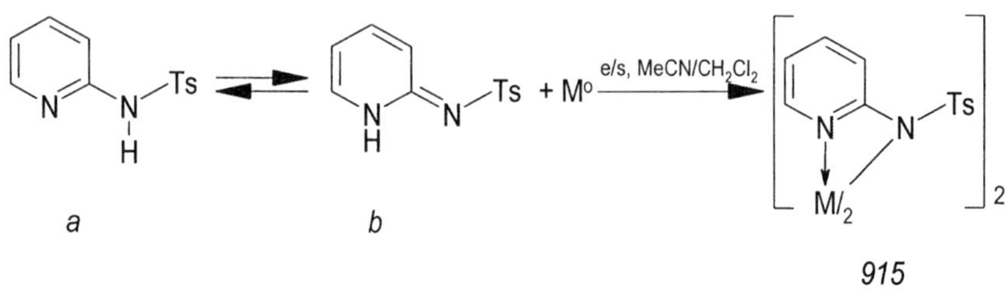

(4.68)

General concepts for obtaining complexes containing tautomeric forms of the discussed ligands, stabilized by metal, are a reaction of conditions needed for controlled synthesis of molecular adducts with completely preserved (nondeprotonated) or ICC (with deprotonated) ligand systems. These conditions, as we emphasized earlier [10,15–18], are usually created by selection of solvents and sources of metal complex-formers: for obtaining adducts – mostly aprotic solvents with low dielectric constant and metal salts (mainly halides), for obtaining chelates with the anionic form of ligand – protic solvents with $\varepsilon > 20$ and metal acetates, nitrates, or halides (in the presence of alkali agents). At the same time, some reactions are known, for example C-metallation [Schemes (4.20)–(4.23)], which lead to C-anionic complexes with fixed keto-form of monodeprotonated β-diketones.

The possibility of controlled preparation of ICC having different sizes for the metal-cycles (inner-chelate isomery), by using chelating ligands with some innercyclic donor atoms, is related to the regioselective synthesis [11,17]. This type of competitive coordination (Sec. 4.1.1) is especially characteristic for azo compounds ***4***, ***916*** [***482***, ***483***; Scheme (4.69); for details see Sec. 2.2.2.5.4].

Synthesis of complexes ***917–920*** was carried out by the method of immediate interaction of ***916*** and metal salts.

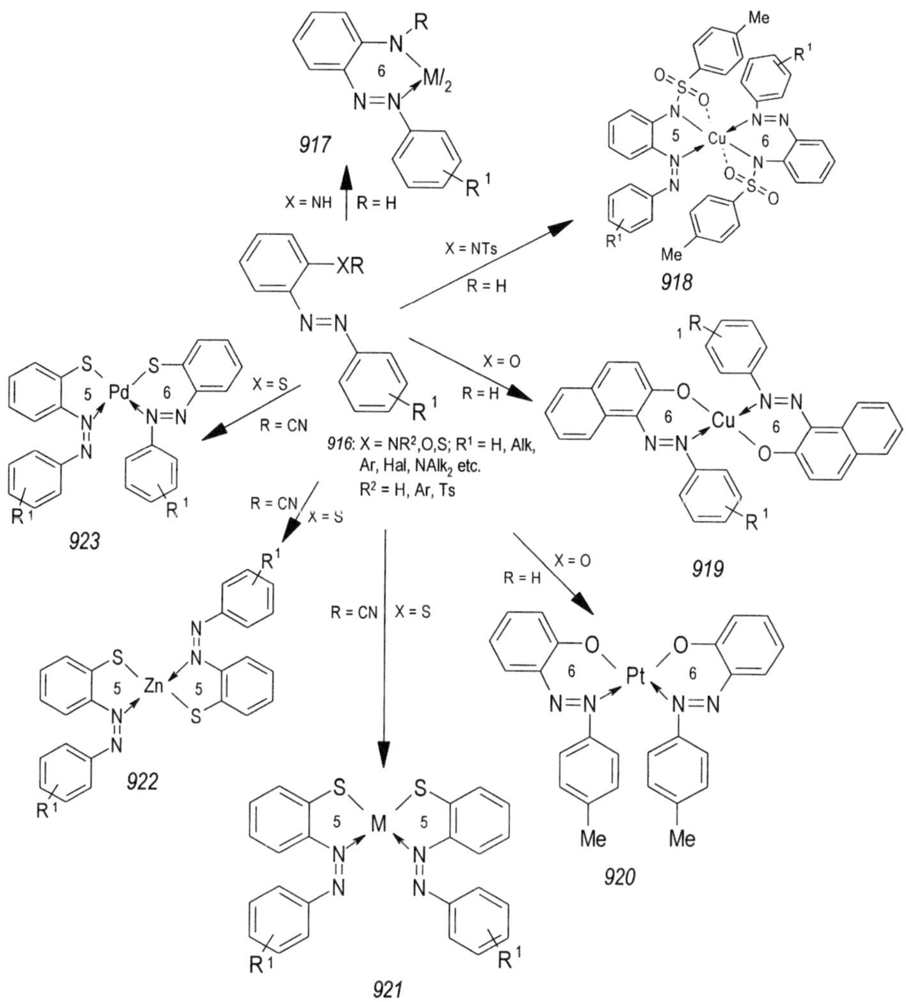

(4.69)

The ICC ***921–923*** were isolated as a result of the complicated transformation (4.70), for example ***921*** [273]:

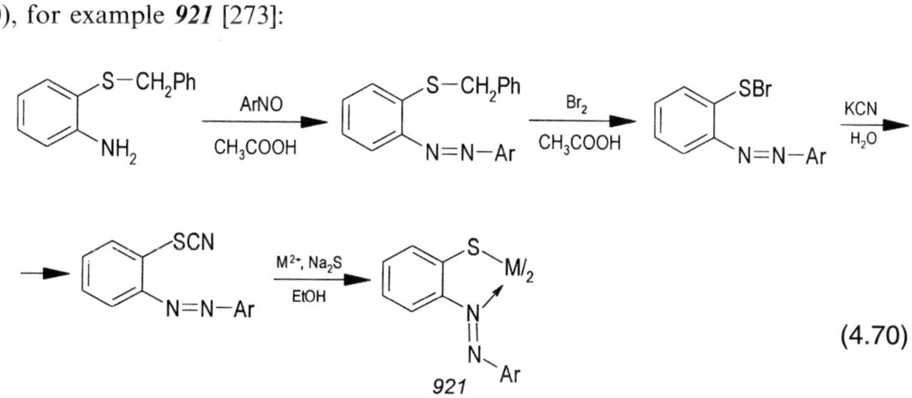

(4.70)

The nature of the donor X atom and metal (***917–923***) are factors regulating the linkage of metal-cycles in the ICC of the examined azo compounds. A comparison of structures ***228–231*** shows that the linkage of formazane chelates is determined by the character of substituents at C and N atoms ***227***. Six-member metal-cycles are formed in structures of aromatic formazanes ***228*** (except ***229***), while the ICC containing five-member metal-cycles are typical for hetarylformazanes ***230*** and ***231*** [11,274,275].

The approaches discussed above (Sec. 4.2) and some examples for obtaining complex compounds with controlled geometry of the coordination unit are only a part of the problem of regioselective synthesis. Its more general conception, especially that related to the controlled synthesis of chiral and achiral complexes, follows from the data reported in various monographs [20,127–129]. These approaches give the possibility of controlling the creation of coordination polyhedra with metal- or ligand-chiral structures (stereognostic chemistry) [20].

The examples of metal-chiral structures are mainly cationic mixed-ligand cobalt(III) complexes of ethylenediamine and its monodimethylphosphine analogue $[Co(H_2NCH_2CH_2NH_2)_3(H_2NCH_2CH_2PMe_3)_{3-n}]^{3+}$. Their synthesis, separation to enantiomers, and establishment of absolute configuration have been carried out for these compounds [276]. The binuclear cobalt(III) complexes ***924*** possess similar optical properties [277]:

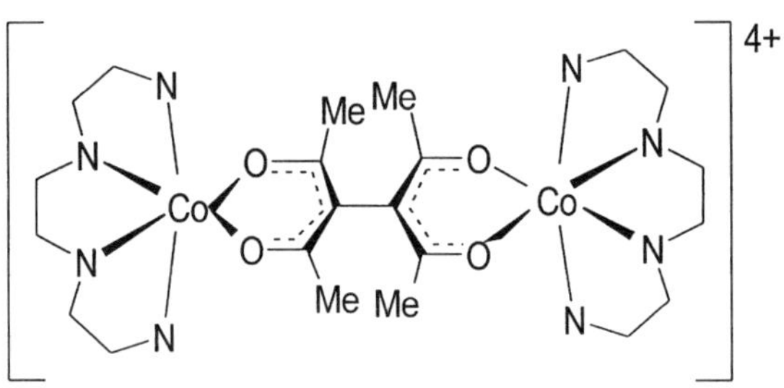

924

Chelates of optically active carbohydrates [278–282], amino and nucleine acids [283], Schiff bases [284], and *bis*-, *tris*-pyridine, and thienylpyridine [285] ligand systems have been studied in detail.

Novel aspects and perspectives of stereoselective synthesis are described in a recent publication [286].

4.3.2 Practical Aspect

Coordination compounds play an important role in modern human life [287]. In this respect, the controlled synthesis is of paramount significant in the modeling and creation of substances with physicochemical and practical properties programmed in advance, in particular those useful in medicine and biology. The examination of this problem is beyond the limits of the present monograph, but we will examine here the most important problems, in our opinion, discussed in 1990–2002.

The application of metal complexes as catalysts in various chemical and technological processes has been known for a long time. In this respect, we note a high interest displayed at present for the catalysis of hydroformylation [288–290], olefine polymerization by metallocenes [291], and hydrogenation of the carbon–carbon bond [292]. Amongst catalysts we emphasize metal clusters [293–295]. The complexes having programmed specific properties, for example magnetic anomaly [296–298], luminophor [299–302], photochromic [303,304], and liquid-crystal [305–307], have been synthesized.

Bioinorganic compounds [143,308–320] form an important area of modern coordination chemistry. The controlled synthesis in this field serves for the creation of artificial models of biologically important objects and processes, as well as for obtaining substances with valuable medical–biological properties. In this respect, the biomimetic aspect of the examined problem should be especially distinguished, related with the modeling of natural biocoordination objects [321–323]. Most attention is paid to the synthesis and study of the most important bioinorganic objects and their synthetic models: porphyrins (Sec. 2.2.4.2) [315,324–333], inophors, and crown-ethers (Sec. 2.2.4.4) [334–343]. We note that the approaches of controlled synthesis of copper(II) chelates with N,S-ligand environment (modeling active centers of "blue" copper proteins). have been elaborated in detail [344–349]. Coordination chemistry of models of various enzymes is also of great interest at present [322,323,350–353].

It is necessary to pay attention to the fact that, during recent years, the bioinorganic properties of aluminum coordination compounds have become the objects of detailed study: their general aspect [354], metabolism and toxicology [355,356], complex formation with nucleozides of di- and triphosphates and nucleozide-bound proteins [357], and x-ray analysis of biologically important complexes [358]. The data on the biocoordination chemistry of magnesium, sodium, potassium [359], and vanadium [360] ions is generalized.

Coordination compounds have become very usable in medicine [361–364]. In this respect, use of metal complexes (mostly those of lanthanides) as diagnostic [365–367] and anticancer [368–370] media should be specially emphasized. Among the last complexes, the aminoplatinum-containing compounds play an important role, so the structural study of platinum complexes as a model of nucleobases [371] is a topic of renewed interest. The new issue of *Comprehensive Coordination Chemistry II* [372] contains a wide description of nanoparticles (vols. 6 and 7), biocoordination chemistry (vol. 8), and other aspects of application of coordination compounds.

ACKNOWLEDGMENTS

The authors are very grateful for useful comments regarding Chaps. 1–4 to Professors Juan Manuel Barbarín Castillo (Universidad Autónoma de Nuevo León, Monterrey, Mexico), Kenneth J. Klabunde (Kansas State University, Kansas), Jim D. Atwood (State University of New York at Buffalo), and A. Pombeiro (Lisbon, Portugal), as well as to numerous colleagues working at Rostov State University and other Russian research centers, the collaboration with whom (see the reference lists) during many years has allowed us to write these chapters. The authors are grateful to Professor A. Sousa (Universidade de Santiago-de-Compostela, Spain) and Professor D.G. Tuck (University of Windsor, Canada) for supplying reprints of their publications.

REFERENCES

1. Garnovskii, A.D.; Osipov, O.A.; Bulgarevich, S.B. *Russ. Chem. Rev.* **41**(4), 341 (1972).
2. Garnovskii, A.D.; Sadimenko, A.P.; Osipov, O.A.; Tsintsadze, G.V. *Hard–Soft Interactions in Coordination Chemistry*. Rostov University Press: Rostov-on-Don, 1986, 272 pp.
3. Garnovskii, A.D.; Osipov, O.A.; Kuznetsova, L.N.; Bogdashev, N.N. *Russ. Chem. Rev.* **42**(2), 89 (1973).
4. Sadimenko, A.P.; Garnovskii, A.D.; Sheinker, V.N.; Osipov, O.A. *Khim. Geterotsykl. Soed.* 1299 (1983).
5. Garnovskii, A.D. Ph.D. Thesis, Rostov State University, 1972.
6. Kawaguchi, S. *Variety of Coordination Modes of Ligands in Metal Complexes. Inorganic Chemistry Concepts*. Springer-Verlag Chemie: Berlin, 1988, vol. 11.
7. Burmeiser, L. *Coord. Chem. Rev.* **105**, 65 (1990).
8. Garnovskii, A.D. *Koord. Khim.* **18**, 675 (1992).
9. Garnovskii, A.D.; Garnovskii, D.A.; Vasilchenko, I.S. *Zhurn. Neorg. Khim.* **37**, 1744 (1992).
10. Garnovskii, A.D.; Garnovskii, D.A.; Burlov, A.S.; Vasilchenko, I.S. *Mendeleev Chem. J.* **40**(4/5), Part I, 19 (1996).
11. Garnovskii, A.D.; Garnovskii, D.A.; Vasilchenko, I.S.; Burlov, A.S.; Sadimenko, A.P.; Sadekov, I.D. *Russ. Chem. Rev.* **65**(5), 389 (1997).
12. Garnovskii, A.D.; Sadimenko, A.P.; Sadimenko, M.I.; Garnovskii, D.A. *Coord. Chem. Rev.* **173**, 31 (1998).
13. Garnovskii, A.D. *Koord. Khim.* **24**, 215 (1998).
14. Garnovskii, A.D. *Russ. J. Inorg. Chem.* **43**, 1377 (1998).
15. Garnovskii, A.D.; Sadimenko, A.P. *Adv. Heterocycl. Chem.* **72**, 1 (1999).
16. Garnovskii, A.D.; Kharisov, B.I.; Blanco, L.M.; Garnovskii, D.A.; Burlov, A.S.; Vasilchenko, I.S.; Bondarenko, G.I. *J. Coord. Chem.* **46**, 365 (1999).
17. (a) Garnovskii, A.D. *Ukr. Khim. Zurn.* **65**, 27 (1999); (b) Garnovskii, A.D. *Ciencia UANL* **3**, 297 (2000).
18. Minkin, V.I.; Garnovskii, A.D.; Elquero, J.; Katritzky, A.R.; Denisko, O.V. *Adv. Het. Chem.* **76**, 157 (2000).
19. Garnovskii, A.D. *Koord. Khim.* **6**(12), 1779 (1980).
20. (a) Von Zelewsky, A. *Stereochemistry of Coordination Compounds*. John Wiley: Chichester, New York, 1996; (b) Von Zelewsky, A. *Inorg. Chem.* **20**(12), 4448 (1981); (c) Minkin, V.I.; Korobov, M.S.; Nivorozhkin, L.E.; Kompan, O.E.; Borodkin, G.S.; Olekhnovich, R.Ya. *Koord. Khim.* **24**, 164 (1998).
21. Kawaguchi, S. *Coord. Chem. Rev.* **70**, 51 (1986).
22. *Ross. Khim. Zurn.* (*Russ. J. Chem.*) **40**(4/5), 1–189 (issue dedicated to metal chelates); *Mendeleev Chem. J.* **40**(4/5), Part I, II.
23. Okhlobistin, O.Yu.; Garnovskii, A.D.; Erdineev, N.S. *Izv. Sev.-Kavkaz. Nauch. Tsentra Vissh. Shkoli. Ser. Estestv. Nauk.* No. 2, 3 (1973).
24. *Chemical Reactivity and Reaction Paths.* (Edit. Klopman, G). John Wiley: New York, London, Sydney, Toronto, 1974.
25. Kokozay, V.N.; Sienkiewicz, A.V. *Polyhedron.* **13**(9), 1427 (1994).
26. Vassilyeva, O. Yu.; Kokozay, V.N.; Skopenko, V.V. *Ukr. Khim. Zhurn.* **60**(3–4), 227 (1994).
27. Nazeeruddin, Md.K.; Zakeeruddin, S.M.; Humphry-Baker, R.; Jirousek, M.; Liska, P.; Vlachopoulos, N.; Shklover, V.; Fischer, C.-H.; Gratzel, M. *Inorg. Chem.* **38**, 6298 (1999).
28. Babich, O.A.; Kokozay, V.N.; Zub, V.Ya.; Kapshuk, A.A. *Zhurn. Neorg. Khim.* **44**, 747 (1999).

29. Vinogradova, E.A.; Vassilyeva, O. Yu.; Kokozay, V.N.; Skelton, B.M. Z. *Naturforsch.* **57b**, 319 (2002).
30. Mironov, Yu.J. *Polyhedron* **19**(4), 437 (2000).
31. Stahler, O.; Preetz, W. *Z. Anorg. Allg. Chem.* **626**(10), 2077 (2000).
32. Hehl, R.; Thiele, G. *Z. Anorg. Allg. Chem.* **626**(10), 2167 (2000).
33. Zhang, C.; Song, Y.; Xin, X. *J. Chem. Soc., Dalton Trans.* No. 8, 1317 (2000).
34. Kessenich, E.; Polborn, K.; Schulz, A. *Z. Anorg. Allg. Chem.* **627**(11), 2506 (2001).
35. Makhankova, V.G.; Vassilyeva, O.Y.; Kokozay, V.N.; Skelton, B.W.; van Albada, G.A.; Reedijk, J. *Z. Naturforsch. B* **56**(9), 931 (2001).
36. Zhang, K.L.; Chen, W.; Xu, Y.; Wang, Z.; Zhong, Z.J.; You, X. *Z. Polyhedron* **20** (15/16), 2033 (2001).
37. Besinger, J.; Fenske, D. *Z. Anorg. Allg. Chem.* **627**(7), 1487 (2001).
38. Basallote, M.G.; Duran, J.; Fernandez-Trujillo, M.J.; Manez, M.A. *Polyhedron***20**(1/2), 75 (2001).
39. Andreev, G.B.; Antipin, M.Yu.; Fedoseev, A.M.; Budanzeva, N.A. *Koord. Khim.* **27**, 231 (2001).
40. Cherkasova, T.G.; Chenskaya, V.V.; Tryasunov, B.G. *Koord. Khim.* **27**, 949 (2001).
41. Bian, Z.; Li, F.; Guo, J.; Wang, K.; Yang, M. *J. Molec. Struct.* **597**, 121 (2001).
42. Burmeister, J.L. *Inorg. Chem.* **3**, 919 (1964).
43. Porai-Koshitz, M.A.; Tsintsadze, G.V. In *Crystallochemistry* (Series "*Results of Science*"). VINITI: Moscow, 1967, p. 168.
44. (a) *Chemie der Pseudogalogeniden.* (Edit. Golub, A.M.; Koeler, Ch.; Skopenko, W.W.). VEB Deutscher Verlag der Wissenschaften: Berlin, 1979; (b) *Khimiya Psevdogalogenidov* ("*The Chemistry of Pseudohalides*") (Edit. Golub, A.M.; Koeler, Ch.; Skopenko, V.V.). Vysha Shkola: Kiev, 1981; (c) *The Chemistry of Pseudohalides.* (Edit. Golub, A.M.; Koeler, Ch.; Skopenko, V.V.). Elsevier Science: Lausanne, Amsterdam, New York, Tokyo, 1986.
45. Real, J.A.; Zarembowitch, J.; Khan, O.; Solans, X. *Inorg. Chem.* **26**, 2939 (1987).
46. Real, J.A.; Bolvin, H.; Bousseksou, A.; Dworkin, A.; Kahn, O.; Varret, F.; Zarembowitch, J. *J. Am. Chem. Soc.* **114**, 4650 (1992).
47. Real, J.A.; Castro, I.; Bousseksou, A.; Verdaguer, M.; Burriel, R.; Castro, M.; Linares, J.; Varret, F. *Inorg. Chem.* **36**, 455 (1997).
48. (a) De Munno, G.; Bruno, G.; Ficolo, F.; Jule, M.; Real, J.A. *Acta Crystallogr. C* **49**, 457 (1993); (b) Jule, M.; Verdaguer, M.; De Munno, G.; Real, J.A.; Bruno, G. *Inorg. Chem.* **32**, 5869 (1993).
49. Kuang, S.-M.; Zhang, Z.Z.; Wang, Qi-G.; Mak, C.W. *J. Chem. Soc., Dalton Trans.* 4477 (1999).
50. Matouzenko, G.S.; Roussesou, A.; Lecocg, S.; Van Konigsbruggen, P.J.; Perrin, M.; Kahu, O.; Collet, A. *Inorg. Chem.* **36**, 2975 (1997).
51. Tekut, T.F.; O'Connor, C.J.; Holwerda, R.A. *Inorg. Chem.* **32**, 324 (1993).
52. Gebreleu, N.V.; Yampolskaya, M.A.; Turte, K.I.; Kuyavskaya, B.Ya.; Sohiborov, S.S.; Yakubov, Kh.M. *Zhurn. Neorg. Khim.* **32**, 2968 (1987).
53. Yampolskaya, M.A.; Simonov, Yu.A.; Turte, K.I.; Antipin, M.Yu.; Belskii, V.K.; Struchkov, Yu.T.; Gebreleu, N.V. *Zhurn. Neorg. Khim.* **33**, 3100 (1988).
54. Gebreleu, N.V.; Yampolskaya, M.A.; Shova, S.G.; Birke, M.S.; Simonov, Yu.A.; Turte, K.I.; Spatar, F.A.; Arion, V.B.; Dvorkin, A.A. *Zurn. Neorg. Khim.* **34**, 877 (1989).
55. *Direct Synthesis of Coordination Compounds.* (Edit. Skopenko, V.V.). Ventury: Kiev, 1997, 172 pp.
56. *Direct Synthesis of Coordination and Organometallic Compounds* (Edit. Garnovskii, A.D.; Kharisov, B.I.) Elsevier Science: Lausanne, Amsterdam, London, New York, 1999, 244 pp.

57. Simonov, Yu.A.; Skopenko, V.V.; Kokozay, V.N.; Dvorkin, A.A. *Zurn. Neorg. Khim.* **36**, 626 (1991).
58. Ziolo, R.F.; Thich, J.A.; Dori, Z. *Inorg. Chem.* **4**, 640 (1972).
59. Kawaguchi, S. *Coord. Chem. Rev.* **70**, 51 (1986).
60. Jorgensen, C.K. *Inorg. Chem.* **3**, 1201 (1964).
61. Pearson, R.G. *Inorg. Chem.* **12**, 712 (1973).
62. Kukushkin, V.Yu.; Kukushkin, Yu.N. *Thoery and Practice of the Synthesis of Coordination Compounds.* Nauka: Leningrad, 1990, 260 pp.
63. Davies, J.A.; Hockensmith, C.M.; Kukushkin, V.Yu.; Kukushkin, Yu.N. *Synthetic Coordination Chemistry: Principles and Practice.* World Scientific: Singapore, London, 1992, 452 pp.
64. Dzyubenko, N.G.; Kalinichenko, Yu.B.; Martinenko, L.I. *Zhurn. Neorg. Khim.* **33**, 52 (1988).
65. Zharkova, N.Ya.; Troyanov, S.I.; Martinenko, L.I. Dzyubenko, N.G.; Chugarov, N.V. *Koord. Khim.* **24**, 60 (1998).
66. Le, Q.T.N.; Umetani, S.; Sudzuki, M.; Mazakazu, M. *J. Chem. Soc., Dalton Trans.* No. 4, 643 (1997).
67. Panina, N.S.; Kukushkin, Yu.N. *Zhurn. Neorg. Khim.* **44**, 798 (1998).
68. Nikiforov, G.B.; Alexandrov, G.G.; Il'in, E.G.; Sergienko, V.S.; Buslaev, Yu.A. *Dokl RAN* **367**, 772 (1999).
69. Leo, Q.T.H.; Umetani, S.; Suzuki, M.; Nazakazu, M. *J. Chem. Soc., Dalton Trans.* 643 (1997).
70. Lobadyuk, V.I.; Spevak, V.N.; Konovalov, L.V.; Konovalov, V.E.; Belskii, V.K. *Koord. Khim.* **23**, 389 (1997).
71. Kukushkin, V.Yu. *Coord. Chem. Rev.* **139**, 375 (1995).
72. Biedermann, H.G.; Öfele, K.; Schulbauer, N.; Teitelbaum, J. *Angew. Chem.* **87**, 634 (1975).
73. Biedermann, H.G.; Öfele, K.; Teitelbaum, J. *Z. Naturforsch. B* **31**, 321 (1976).
74. Davies, S.G.; Shipton, M.R. *J. Chem. Soc., Chem. Commun.* 995 (1989).
75. Perera, J.R.; Heeg, M.J.; Schlegel, H.D.; Winter, C.H. *J. Am. Chem. Soc.* **121**, 4536 (1999).
76. Goher, M.A.S.; Mautner, F.A. *J. Chem. Soc., Dalton Trans.* 1923 (1999).
77. Das, D.; Laskar, I.R.; Ghosh, A.; Mondal, A.; Okamoto, K.; Chaudhuri, R. *J. Chem. Soc., Dalton Trans.* 3987 (1998).
78. Kukushkin, Yu.N. *Chemistry of Coordination Compounds.* High School: Moscow, 1985, 455 pp.
79. Kostromina, N.A.; Kumok, V.N.; Skorik, N.A. *Chemistry of Coordination Compounds.* High School: Moscow, 1990, 432 pp.
80. Pandey, K.K. *Inorg. Chim. Acta* **182**, 163 (1991).
81. Virovez, A.V.; Podberezskaya, N.V.; Lavrenova, L.G.; Bikzhanova, G.A. *Polyhedron* **13**, 2929 (1994).
82. Botova, I.N.; Mirochnik, A.G.; Karasev, V.E. *Koord. Khim.* **18**, 659 (1992).
83. Karasev, V.E.; Gerasimenko, A.V.; Mirochnik, A.G.; Botova, I.N.; Polyakova, N.V. *Koord. Khim.* **25**, 945 (1999).
84. *Comprehensive Coordination Chemistry.* (Edit. Wilkinson, G.). Pergamon Press: Oxford, 1987, vols. 1–7.
85. Shkolnikova, L.M.; Porai-Koshitz, M.A. *Results of Science & Technology. Crystallochemistry.* VINITI: Moscow, 1982, vol. 15, p. 117.
86. Shugam, E.A.; Shkolnikova, L.M. In: *Results of Science & Technology. Crystallochemistry.* VINITI: Moscow, 1967, p. 5.
87. Gibson, D. *Coord. Chem. Rev.* **4**, 225 (1969).
88. Gibson, D. *Usp. Khim.* **39**, 803 (1970).

89. Allmann, R.; Flatau, K.; Musso, H. *Chem. Ber.* **105**, 3067 (1972).
90. Allmann, R.; Musso, H. *Chem. Ber.* **106**, 3001 (1973).
91. Baba, S.; Ogura, T.; Kawaguchi, S. *Inorg. Nucl. Chem. Lett.* **7**, 1195 (1971).
92. Horike, M.; Kai, V.; Yasuoka, N.; Kasai, N. *J. Organomet. Chem.* **72**, 441 (1974).
93. Ratovskii, G.V.; Burlakova, O.V.; Tkach, V.S.; Scmidt, F.K. *Koord. Khim.* **16**, 997 (1990).
94. Belih, L.B.; Dmitrieva, T.V.; Zinchenko, S.V.; Shmidt, F.K. *Koord. Khim.* **21**, 476 (1995).
95. Tyukalova, O.V.; Ratovskii, G.V.; Belih, L.B.; Shmidt, F.K. *Zhurn. Obsh. Khim.* **67**, 58 (1997).
96. Belih, L.B.; Dmitrieva, T.V.; Shmidt, F.K. *Koord. Khim.* **25**, 528 (1999).
97. Grandberg, K.I.; Kuzmina, L.G.; Smislova, E.I.; Zinin, A.I.; Krutko, D.P.; Dyadchenko, V.P. *Zhurn. Neorg. Khim.* **41**, 231 (1996).
98. Kuzmina, L.G. *Koord. Khim.* **20**, 540 (1994).
99. Kuzmina, L.G. *Koord. Khim.* **20**, 547 (1994).
100. Garnovskii, A.D.; Nivorozhkin, A.L.; Minkin, V.I. *Coord. Chem. Rev.* **126**, 1 (1993).
101. Kogan, V.A.; Osipov, O.A.; Minkin, V.I.; Sokolov, V.P. *Zhurn. Neorg. Khim.* **10**, 83 (1965).
102. Kogan, V.A.; Osipov, O.A.; Garnovskii, A.D. *Zhurn. Neorg. Khim.* **9**, 494 (1964).
103. Garnovskii, A.D.; Abramenko, V.L. *Zhurn. Obsh. Khim.* **55**, 1836 (1985).
104. Bullock, J.I.; Ladd, M.F.C.; Povey, D.C.; Taymir-Riahi, H.-A. *Acta Crystallogr. B.* **35**, 2013 (1979).
105. Sergienko, V.S.; Mistryukov, A.E.; Litvinov, V.V.; Knyazhanskii, M.I.; Garnovskii, A.D.; Porai-Koshitz, M.A. *Koord. Khim.* **16**, 168 (1990).
106. Torzilli, M.A.; Colquhoun, S.; Doucet, D.; Beer, R.H. *Polyhedron.* **21**, 697 (2002).
107. (a) Surpina, L.V.; Kolodyazhnii, Yu.V.; Garnovskii, A.D.; Osipov, O.A. *Koord. Khim.* **10**, 1459 (1978); (b) Prabhu, P.S.; Dodwad, S.S. *J. Indian Chem. Soc.* **57**, 475 (1980).
108. Yamanouchi, K.; Yamada, S. *Inorg. Chim. Acta* **11**, 223 (1974).
109. Biradar, N.S.; Kulkarni, V.N. *J. Inorg. Nucl. Chem.* **33**, 3781 (1971).
110. Andronov, E.A.; Kukushkin, Yu.N.; Murashkin, Yu.V. *Izv. Vuzov. Khim. Khim. Tekhnol.* **19**, 1479 (1976).
111. Sergienko, V.S.; Garnovskii, A.D.; Porai-Koshitz, M.A.; Abramenko, V.L. *Koord. Khim.* **10**, 1407 (1985).
112. Abramenko, V.L.; Garnovskii, A.D.; Abramenko, Yu.V. *Koord. Khim.* **20**, 39 (1994).
113. Abramenko, V.L.; Garnovskii, A.D. *Koord. Khim.* **14**, 783 (1988).
114. Ryabov, A.D. *Chem. Rev.* **90**, 403 (1990).
115. Hadson, S.A.; Maitlis, P.V. *Chem. Rev.* **93**, 861 (1993).
116. Baena, M.J.; Espinet, P.; Ros, M.B.; Serano, J.L.; Ezcurra, A. *Angew. Chem.* **105**, 1260 (1993).
117. (a) Garnovskii, A.D. *Ukr. Khim. Zhurn.* **65**, 27 (1999); (b) Teom, S.-G.; Teo, S.-B.; Lee, L.-K.; Fun, H.-K. *J. Coord. Chem.* **33**, 69 (1994).
118. Bogdashev, N.N.; Garnovskii, A.D.; Osipov, O.A.; Grigoriev, V.P.; Gontmakher, N.M. *Zhurn. Obsh. Khim.* **46**, 675 (1976).
119. Romero, J.; García-Vázquez, J.A.; Durán, M.L.; Castiñeiras, A.; Sousa, A.; Garnovskii, A.D.; Garnovskii, D.A. *Acta Chem. Scand.* **51**, 672 (1997).
120. Castro, J.A.; Romero, J.; García-Vázquez, J.A.; Sousa, A.; Castellano, F.E.; Zuckerman-Schpector, J. *Polyhedron* **12**, 31 (1993).
121. Castro, J.A.; Romero, J.; García-Vázquez, J.A.; Castineiras, A.; Sousa, A.Z. *Anorg. Allg. Chem.* **619**, 601 (1993).
122. Castro, J.A.; Romero, J.; García-Vázquez, J.A.; Macias, A.; Sousa, A.; Englert, U. *Polyhedron* **12**, 1391 (1993).
123. Jutzi, P.; Dalhuas, J. *Coord. Chem. Rev.* **137**, 179 (1994).

124. Jutzi, P.; Siemeling, U. *J. Organomet. Chem.* **500**, 175 (1995).
125. Jutzi, P.; Dalhaus, J.; Neumann, B.; Stammler, H.-G. *Organometallics* **15**, 747 (1996).
126. Kipert, D. *Inorganic Stereochemistry. Inorganic Chemistry Concepts Vol. 6*. Springer-Verlag Chemie: Berlin, Heidelberg, New York, 1982.
127. Hawkins, C.J. *Absolute Configuration of Metal Complexes*. John Wiley: New York, London, Sydney, Toronto, 1971.
128. Sokolov, V.I. *Introduction to Theoretical Stereochemistry*. Gordon & Breach: New York, 1991.
129. Müller, U. *Inorganic Structural Chemistry: A Textbook Series*. John Wiley: Chichester, 1994.
130. Calligans, M.; Nardin, G.; Randaccio, L. *Coord. Chem. Rev.* **7**, 385 (1972).
131. Maslen, H.S.; Waters, T.N. *Coord. Chem. Rev.* **17**, 137 (1975).
132. Syamal, A.; Maurya, M.R. *Coord. Chem. Rev.* **95**, 183 (1989).
133. Kharabaev, N.N. *Koord. Khim.* **17**, 579 (1991).
134. Garnovskii, A.D. *Russ. J. Coord. Chem.* **19**, 368 (1993).
135. Larin, G.M. *Russ. J. Coord. Chem.* **19**, 312 (1993).
136. Lindoy, L.F. *Chem. Soc. Rev.* **14**, 421 (1975).
137. Panova, G.V.; Vikulova, N.K.; Potpov, V.M. *Usp. Khim.* **49**, 1234 (1980).
138. Garnovskii, A.D.; Alexeenko, V.A.; Burlov, A.S.; Kurbatov, V.P.; Lipunova, G.N.; Nedzvetskii, V.S.; Sigeikin, G.I. *Izv. Vissh. Ucheb. Zaved. Ser. Khim. i Khim. Tekhn.* **34**(1), 3 (1991).
139. Garnovskii, A.D.; Alexeenko, V.A.; Burlov, A.S.; Nedzvetskii, V.S. *Zhurn. Neorg. Khim.* **36**, 886 (1996).
140. Yamada, S.; Takeuchi, A. *Coord. Chem. Rev.* **43**, 187 (1982).
141. Holm, R.H.; Ewerett, G.W.; Chakravorty, A. *Progr. Inorg. Chem.* **7**, 83 (1966).
142. Holm, R.H.; O'Connor, M.J. *Progr. Inorg. Chem.* **14**, 214 (1971).
143. Nivorozhkin, A.L.; Uraev, A.I.; Burlov, A.S.; Garnovskii, A.D. *Zhurn. Ross. Khim. Obsh. Mendeleev* **40**(4/5), 162 (1996).
144. Kogan, V.A.; Zelentsov, V.V.; Osipov, O.A.; Burlov, A.S. *Usp. Khim.* **48**, 1208 (1979).
145. Kharabaev, N.N. *Metalloorgan. Khim.* **3**, 1025 (1990).
146. Garnovskii, A.D.; Vasilchenko, I.S., Garnovskii, D.A. *Modern Aspects of Synthesis of Metallocomplexes. General Methods and Ligands*. LaPO: Rostov-on-Don, 2000.
147. Shnulin, A.N.; Mamedov, Kh.S.; Struchkov, Yu.T. *Zhurn. Strukt. Khim.* **19**, 688 (1978).
148. Coswami, N.; Eichhorn, D.M. *Inorg. Chem.* **38**, 4329 (1999).
149. Alexeenko, V.A.; Burlov, A.S.; Bolotin, B.M.; Lukova, O.A.; Nedzvetskii, V.S.; Garnovskii, A.D. *Koord. Khim.* **17**, 76 (1991).
150. Burlov, A.S.; Garnovskii, A.D.; Alexeenko, V.A.; Mistryukov, A.E.; Sergienko, V.S.; Zaletov, V.G.; Lukov, V.V.; Porai-Koshitz, M.A. *Koord. Khim.* **18**, 859 (1992).
151. Samus, I.D.; Taran, G.G.; Mazus, M.D.; Fam Ngok Fung; Tsapkov, V.I.; Popov, M.C.; Samus, N.M. *Koord. Khim.* **20**, 123 (1994).
152. Barnard, G.F.; Russel, M.J. In. Ref. 6, vol. 5, p. 1099.
153. Choloner, P.A. *Coord. Chem. Rev.* **89**, 1 (1988).
154. Olekhnovich, L.P.; Kurbatov, V.P.; Osipov, O.A.; Minkina, L.S.; Minkin, V.I. *Zhurn. Obsh. Khim.* **38**, 2593 (1968).
155. Nivorozhkin, L.E.; Kurbatov, V.P.; Minkina, L.S.; Korobov, M.S.; Olekhnovich, L.P.; Osipov, O.A.; Minkin, V.I. *Zhurn. Obsh. Khim.* **41**, 2726 (1971).
156. Korobov, M.S.; Minkin, V.I.; Nivorozhkin, L.E. *Zhurn. Org. Khim.* **11**, 836 (1975).
157. Kogan, V.A.; Kharabaev, N.N.; Osipov, O.A.; Kochin, S.G. *Zhurn. Strukt. Khim.* **22**(1), 126 (1981).
158. Garnovskii, A.D.; Minkina, L.S.; Sachashchik, L.V.; Kurbatov, V.P.; Osipov, O.A. *Zhurn. Obshch. Khim.* **41**, 1884 (1971).

159. Sadekov, I.D.; Abakarov, G.M.; Kuren, S.G.; Garnovskii, A.D.; Minkin, V.I. *Zhurn. Obsh. Khim.* **56**, 2168 (1986).
160. Garnovskii, A.D.; Olekhnovich, L.P.; Alexeenko, V.A.; Lukova, O.A.; Komissarov, V.N. *Zhurn. Obsh. Khim.* **60**, 1920 (1990).
161. Olekhnovich, R.Ya.; Korobov, M.C.; Lyubchenko, S.N.; Sukholenko, E.V.; Riskina, T.A.; Nivorozhkin, L.E.; Olekhnovich, L.P. *Zhurn. Obsh. Khim.* **62**, 901 (1992).
162. Garnovskii, A.D.; Lukova, O.A.; Alexeenko, V.A.; Komissarov, V.N.; Litvinov, V.V.; Kharabaev, N.N. *Zhurn. Neorg. Khim.* **37**, 1299 (1992).
163. Olekhnovich, I.P.; Lyubchenko, I.N.; Kurbatov, S.V.; Kogan, V.A. *Mendeleev Chem. J.* **40**(4/5), 139 (1996).
164. Kochin, S.G.; Garnovskii, A.D.; Kogan, V.A.; Sushko, T.G. *Zhurn. Neorg. Khim.* **14**, 1428 (1969).
165. D'yachenko, O.A.; Atovmyan, L.O.; Aldoshin, S.M.; Kogan, V.A.; Kochin, S.G.; Osipov, O.A. *Izv. Akad. Nauk SSSR, Ser. Khim.* 2130 (1975).
166. Kawamoto, T.; Kushi, Y. *Chem. Lett.* 893 (1992).
167. Kawamoto, T.; Kuma, H.; Kushi, Y. *J. Chem. Soc., Chem. Commun.* 2121 (1996).
168. Kawamoto, T.; Kushi, Y. *Chem. Lett. Chem. Soc. Jpn.* 1057 (1992).
169. Garnovskii, A.D.; Antsishkina, A.S.; Sadikov, G.G.; Vasilchenko, I.S.; Garnovskii, D.A.; Uraev, A.I.; Porai-Koshitz, M.A. *Zhurn. Neorg. Khim.* **40**, 1977 (1995).
170. Zelentsov, V.V.; Somova, I.K.; Kurtanidze, R.Sh.; Nikolaeva, T.B.; Rakitin, Yu.V. *Koord. Khim.* **6**, 89 (1980).
171. Belskii, V.K.; Zelentsov, V.V.; Nikolaeva, T.B. *Koord. Khim.* **15**, 1088 (1989).
172. Lyubchenko, S.N.; Litvinov, V.V.; Riskina, T.A.; Kogan, V.A.; Olekhnovich, L.P. *Zhurn. Obsh. Khim.* **60**, 1618 (1990).
173. Kompan, O.E.; Ivakhnenko, E.P.; Lyubchenko, S.N.; Olekhnovich, L.P.; Yanovskii, A.I.; Struchkov, Yu.T. *Zhurn. Obsh. Khim.* **60**, 1882 (1990).
174. Bren, V.A.; Bren, Zh.V. *Koord. Khim.* **4**, 1499 (1978).
175. Shepelenko, E.N.; Minkina, L.S.; Kochin, S.G.; Khokhlov, A.V.; Bren, V.A.; Garnovskii, A.D. *Zhurn. Obsh. Khim.* **57**, 2342 (1987).
176. Osipov, O.A.; Minkina, L.S.; Garnovskii, A.D.; Kvitko, I.Ya.; Kurbatov, V.P.; Sof'ina, E.M. *Zhurn. Obsh. Khim.* **42**, 926 (1972).
177. Toftlund, H.; Becher, J.; Olesen, P.H.; Pedersen, J. *Israel J. Chem.* **25**, 56 (1985).
178. Nivorozhkin, A.L.; Konstantinovskii, L.E.; Nivorozhkin, L.E.; Minkin, V.I.; Takhirov, T.G.; Diyachenko, O.A.; Tagiev, D.B. *Izv. Akad. Nauk SSSR, Ser. Khim.* 327 (1990).
179. Nivorozhkin, A.L.; Nivorozhkin, L.E.; Minkin, V.I.; Takhirov, T.G.; Diyachenko, O.A. *Polyhedron* **10**, 179 (1991).
180. Madal, S.; Das, G.; Singh, R.; Shukla, R.; Bharadwaj, P.K. *Coord. Chem. Rev.* **160**, 191 (1997).
181. La Cour, A.; Findeisein, M.; Hansen, K.; Hazell, R.; Hennig, L.; Olsen, C.E.; Pedersen, L. *J. Chem. Soc., Dalton Trans.* 2045 (1997).
182. Vasilchenko, I.S.; Antsyshkina, A.S.; Burlov, A.S.; Sadikov, G.G.; Uraev, A.I.; Nivorozhkin, A.L.; Garnovskii, D.A.; Sergienko, V.S.; Kurbatov, V.P.; Korshunov, O. Yu.; Garnovskii, A.D. *Russ. J. Inorg. Chem.* **44**, 1205 (1999).
183. Nivorozhkin, A.L.; Uraev, A.I.; Bondarenko, G.I.; Antsishkina, A.S.; Kurbatov, V.P.; Garnovskii, A.D.; Turte, K.I. *J. Chem. Soc., Chem. Commun.* No. 18, 1711 (1997).
184. Uraev, A.I.; Nivorozhkin, A.L.; Bondarenko, G.I.; Lisenko, K.A.; Korshunov, O.Yu.; Vlasenko, V.G.; Vasilchenko, I.S.; Shuvaev, A.T.; Kurbatov, V.P.; Antipin, M.Yu.; Garnovskii, A.D. *Dokl. Russ. Akad. Nauk.* **367**, 67 (1999).
185. Knoblach, S.; Benedix, R.; Ecke, M.; Gelbrich, T.; Sieler, J.; Somoza, F.; Hennig, H. *Eur. J. Inorg. Chem.* 1393 (1999).

186. Varnek, V.A.; Mazalov, L.N.; Uraev, A.I.; Nivorozhkin, A.L.; Garnovskii, A.D. *Zhurn. Strukt. Khim.* **41**, 427 (2000).
187. Nivorozhkin, A.L.; Uraev, A.I.; Burlov, A.S.; Garnovskii, A.D. *Mandeleev Chem. J.* **40**(4/5), 255 (1996).
188. Takhirov, T.G.; D'yachenko, O.A.; Tagiev, D.B.; Nivorozhkin, A.L.; Nivorozhkin, L.E.; Minkin, V.I. *Koord. Khim.* **14**, 237 (1988).
189. Nivorozhkin, A.L.; Sukholenko, E.V.; Nivorozhkin, L.E.; Borisenko, N.I.; Minkin, V.I.; Grishin, Yu.K.; Dyachenko, O.A.; Takhirov, T.G.; Tagiev, D.V. *Polyhedron* **8**, 569 (1989).
190. Takhirov, T.G.; Dyachenko, O.A.; Tagiev, D.B.; Nivorozhkin, A.L.; Nivorozhkin, L.E.; Minkin, V.I. *Koord. Khim.* **17**, 505 (1991).
191. Nivorozhkin, A.L.; Nivorozhkin, L.E.; Takhirov, T.G.; Dyachenko, O.A. *Polyhedron* **10**, 179 (1991).
192. (a) Goldfarb, Ya.L.; Kalik, M.A. *Khim. Geterotsikl. Soed.* 902 (1972); (b) Goldfarb, Ya.L.; Kalik, M.A. *Usp. Khim.* **41**, 679 (1972).
193. Shkol'nikova, L.M.; Zavodnik, V.E.; Il'inskii, A.L.; Kalik, M.A.; Goldfarb, Ya.L. *Khim. Geterotsiki. Saed.* **11**, 1484 (1984).
194. Bren, V.A.; Minkin, V.I. *Izv. Vissh. Ucheb. Zaved. Khim. i Khim. Tekhnol.* **25**, 663 (1982).
195. Litvinov, V.P.; Baisburg, A.F.; Mortikov, Y.Yu. *Chem. Heterocycl. Comp* **28**, 3 (1992).
196. Litvinov, V.P.; Vaisburg, A.F.; Mortikov, V.Yu. *Sulf. Rep.* **11**, 321 (1992).
197. Freidrichsen, W.; Traulsen, T.; Elguero, J.; Katritzky, A.R. *Adv. Het. Chem.* **76**, 86 (2000).
198. Bren, V.A.; Osipov, O.A.; Minkin, V.I.; Bren, Zh.V.; Minkina, L.S.; Atovmyan, L.O.; Aldoshin, S.M.; Dyachenko, O.A. *Koord. Khim.* **5**, 1058 (1979).
199. Tkachev, V.V.; Kondrashkina, E.A.; Atovmyan, L.O. *Zhurn. Strukt. Khim.* **18**, 1049 (1977).
200. Antsyshkina, A.S.; Porai-Koshitz, M.A.; Nivorozhkin, A.L.; Vasilchenko, I.S.; Nivorozhkin, L.E.; Garnovskii, A.D. *Inorg. Chim. Acta* **180**, L151 (1991).
201. Antsishkina, A.S.; Porai-Koshitz, M.A.; Nivorozhkin, A.L.; Garnovskii, A.D.; Nivorozhkin, L.E.; Ostrikova, V.N. *Zhurn. Neorg. Khim.* **35**, 154 (1991).
202. Antsishkina, A.S.; Porai-Koshitz, M.A.; Vasilchenko, I.S.; Nivorozhkin, A.L.; Garnovskii, A.D. *Dokl. AN* **330**, 54 (1993).
203. Porai-Koshitz, M.A.; Gilinskaya, E.A. *Crystallochemistry. Results of Science.* Institute of Scientific Information of Academy of Sciences of USSR: Moscow, 1966.
204. Osipov, O.A.; Garnovskii, A.D.; Minkin, V.I. *Zhurn. Strukt. Khim.* **8**, 913 (1967).
205. Osipov, O.A.; Garnovskii, A.D.; Minkin, V.I.; Kogan, V.A.; Kolodyazhnii, Yu.V.; Kurbatov, V.P.; Sheinker, V.N. *Dipol Moments in Coordination Chemistry.* Rostov University Press: Rostov-on-Don, 1976.
206. Sacconi, L.; Mani, F.; Bencini, A. In Ref. 84, vol. 5, p. 2.
207. Larin, G.M.; Kolosov, V.A.; Panova, G.V.; Vikulova, N.K. *Zhurn. Neorg. Khim.* **18**, 2868 (1973).
208. Toftlund, H.; Nivorozhkin, A.L.; La Cour, A.; Adhikary, B.; Murray, K.; Fallon, G.D.; Nivorozhkin, L.E. *Inorg. Chim. Acta* **228**, 237 (1995).
209. Cini, R.; Cinguantini, A.; Orioli, P.; Sabat, M. *Inorg. Chim. Acta* **41**, 151 (1980).
210. Hewlins, M.Y.E. *J. Chem. Soc., Dalton Trans.* 429 (1975).
211. Bereman, R.D.; Shields, G.D.; Bordner, J.; Dorfman, J.R. *Inorg. Chem.* **20**, 2165 (1981).
212. Nivorozhkin, A.L.; Konstantinovskii, L.E.; Nivorozhkin, L.E.; Minkin, V.I.; Takhirov, T.G.; Dyachenko, O.A.; Tagiev, D.B. *Izv. AN SSSR, Ser. Khim.* 327 (1990).
213. Chin, P.C.; Palenik, G.J. *Inorg. Chem.* **11**, 816 (1972).

214. Gupta, T.; Saha, M.K.; Sen, S.; Mitra, S.; Edwards, A.J.; Clegg, W. *Polyhedron* **18**, 197 (1999).
215. Casellato, U.; Vigato, P.A. *Coord. Chem. Rev.* **23**, 31 (1977).
216. Melnik, M. *Coord. Chem. Rev.* **42**, 259 (1982).
217. Sergienko, V.S.; Abramenko, V.L.; Porai-Koshitz, M.A. *Zhurn. Strukt. Khim.* **31**(5), 54 (1990).
218. Garnovskii, A.D.; Alexeenko, V.A.; Burlov, A.S.; Lukov, V.V.; Nedzvetskii, V.S. *Koord. Khim.* **16**, 879 (1990).
219. Chumakov, Yu.M.; Biyushkin, V.N.; Malinovskii, T.I.; Keleshu, S.; Tsapkov, V.I.; Popov, M.S.; Samus, N.M. *Koord. Khim.* **16**, 945 (1990).
220. Garnovskii, A.D.; Mistryukov, A.E.; Bren, V.A.; Sergienko, V.S.; Kochin, S.G.; Vasilchenko, I.S.; Minkina, L.S.; Ribalkin, V.P.; Nivorozhkin, A.L.; Porai-Koshitz, M.A.; Starikov, A.G.; Ellert, O.G. *Zhurn. Neorg. Khim.* **37**, 1513 (1992).
221. Garnovskii, A.D.; Vasilchenko, I.S.; Mistryukov, A.E.; Kochin, S.G.; Sergienko, V.S.; Kharabaev, N.N.; Porai-Koshitz, M.A. *Zhurn. Obsh. Khim.* **63**, 1144 (1993).
222. Garnovskii, A.D.; Gerbeleu, N.V.; Kochin, S.G.; Arion, V.B.; Minkina, L.S.; Indrichan, K.M.; Vasilchenko, I.S.; Revenko, M.D.; Konstantinovskii, L.E. *Dokl. Akad. Nauk SSSR* **306**, 872 (1989).
223. Garnovskii, A.D.; Kochin, S.G.; Minkina, L.S.; Vasilchenko, I.S.; Konstantinovskii, L.E. *Koord. Khim.* **15**, 258 (1989).
224. Garnovskii, A.D.; Mistryukov, A.E.; Kochin, S.G.; Sergienko, V.S.; Vasilchenko, I.S.; Minkina, L.S.; Porai-Koshits, M.A. *Zhurn. Neorg. Khim.* **37**, 1289 (1992).
225. Garnovskii, A.D.; Sergienko, V.S.; Bren, V.A.; Vasilchenko, I.S.; Mistryukov, A.E.; Ribalkin, V.P.; Minkina, L.S.; Kochin, S.G.; Porai-Koshitz, M.A. *Zhurn. Neorg. Khim.* **38**, 252 (1993).
226. Zelentsov, V.V.; Suvorova, K.M. *Zhurn. Obsh. Khim.* **38**, 480 (1968).
227. Zelentsov, V.V.; Suvorova, K.M. *Koord. Khim.* **25**, 47 (1999).
228. Labisbal, E.; García-Vázquez, J.A.; Romero, J.; Sousa, A.; Castiñeiras, A.; Maichle, C.; Russo, U. *Inorg. Chim. Acta* **223**, 87 (1994).
229. Garnovskii, A.D.; Kharisov, B.I.; Gójon-Zorrilla, G.; Garnovskii, D.A. *Russ. Chem. Rev.* **64**(3), 201 (1995).
230. Garnovskii, A.D.; Blanco, L.M.; Kharisov, B.I.; Garnovskii, D.A.; Burlov, A.S. *J. Coord. Chem.* **48**, 219 (1999).
231. Garnovskii, A.D.; Burlov, A.S.; Garnovskii, D.A.; Vasilchenko, I.S.; Antsyshkina, A.S.; Sadikov, G.G.; Sousa, A.; García-Vázquez, J.A.; Romero, J. *Koord. Khim.* **23**, 399 (1997).
232. Burlov, A.S.; Antsishkina, A.S.; Romero, J.; Garnovskii, D.A.; García-Vázquez, A.; Sousa, A.; Garnovskii, A.D. *Zhurn. Neorg. Khim.* **40**, 1427 (1995).
233. Garnovskii, A.D.; Burlov, A.S.; Garnovskii, D.A.; Vasilchenko, I.S.; Antsishkina, A.S.; Sadikov, G.G.; Sousa, A.; García-Vázquez, J.A.; Romero, J.; Durán, M.; Sousa-Pedrares, A.; Gómez, C. *Polyhedron* **18**, 863 (1999).
234. Osipov, O.A.; Minkin, V.I.; Verkhovodova, D.Sh.; Knyazhanskii, M.I. *Zhurn. Neorg. Khim.* **12**, 1549 (1967).
235. Yamada, S.; Yamanouchi, K. *Bull. Chem. Soc. Jpn.* **42**, 2562 (1969).
236. Chia, P.S.K.; Livingstone, S.E. *Aust. J. Chem.* **22**, 1613 (1969).
237. Drummond, J.; Wood, J.S. *J. Chem. Soc., Dalton Trans.* 365 (1972).
238. Kovayole, G.A. *J. Coord. Chem.* **16**, 67 (1987).
239. Parashar, K.K.; Sharma, R.C.; Kumar, A.; Mohan, G. *Inorg. Chim. Acta* **151**, 201 (1988).
240. Atherton, N.M.; Fenton, D.E.; Hewson, C.J. *J. Chem. Soc., Dalton Trans.* 1059 (1988).
241. Castiñeiras, A.; Castro, J.A.; Durán, M.; García-Vázquez, J.A.; Masas, A.; Romero, J.; Sousa, A. *Polyhedron* **8**, 2543 (1989).

242. Mistryukov, A.E.; Vasilchenko, I.S.; Sergienko, V.S.; Nivorozhkin, A.L.; Kochin, S.G.; Porai-Koshitz, M.A.; Nivorozhkin, L.E.; Garnovskii, A.D. *Mendeleev Commun.* 30 (1992).
243. Vasilchenko, I.S.; Antsishkina, A.S.; Garnovskii, D.A.; Sadikov, G.G.; Porai-Koshitz, M.A.; Sigeikin, S.G.; Garnovskii, A.D. *Koord. Khim.* **20**, 824 (1994).
244. García-Vázquez, J.A.; Romero, J.; Durán, M.L.; Sousa, A.; Garnovskii, A.D.; Burlov, A.S.; Garnovskii, D.A. *Polyhedron* **17**, 1547 (1998).
245. Vasilchenko, I.S.; Antsyshkina, A.S.; Burlov, A.S.; Sadikov, G.G.; Uraev, A.I.; Nivorozhkin, A.L.; Garnovskii, D.A.; Sergienko, V.S.; Kurbatov, V.P.; Korshunov, O.Yu.; Garnovskii, A.D. *Russ. J. Inorg. Chem.* **44**, 1205 (1999).
246. Vlasenko, V.G.; Shuvaev, A.T.; Nedoseikina, T.I.; Nivorozhkin, A.L.; Uraev, A.I.; Garnovskii, A.D.; Korshunov, O.Yu. *J. Synch. Rad.* **6**, 406 (1999).
247. Uraev, A.I.; Nivorozhkin, A.L.; Bondarenko, G.I.; Lysenko, K.A.; Korshunov, O.Yu.; Vlasenko, V.G.; Vasilchenko, I.S.; Shuvaev, A.T.; Kurbatov, V.P.; Antipin, M.Yu.; Garnovskii, A.D. *Izv. Akad. Nauk.* No. 11, 1892 (2000).
248. Shkolnikova, L.M.; Makarevich, S.S.; Zavodnik, V.E.; Kalik, M.A.; Goldfarb, Ya.L. *Koord. Khim.* **2**, 115 (1976).
249. Candlin, J.P.; Taylor, K.A.; Tompson, D.T. *Reactions of Transition Metal-Complexes.* Elsevier: Amsterdam, London, New York, 1968.
250. Kukushkin, Yu.N. *Usp. Khim.* **39**, 362 (1970).
251. Lites, D.C.; McPartlin, M.; Taskes, P.A. *J. Am. Chem. Soc.* **99**, 7704 (1977).
252. Garnovskii, A.D.; Mistryukov, A.E.; Kochin, S.G.; Sergienko, V.S.; Vasilchenko, I.S.; Porai-Koshits, M.A. *Mendeleev Commun.* 110 (1991).
253. Teo, S.-B.; Theoh, S.-G.; Okechukwu, R.C.; Fun, H.K. *J. Organomet. Chem.* **454**, 67 (1993).
254. Bravo, J.; Cordero, M.B.; Casas, J.S.; Castano, M.V.; Sanches, A.; Sordo, J. *J. Organomet. Chem.* **513**, 63 (1996).
255. Elguero, J.; Katritzky, A.R.; Denisko, O.V. *Adv. Het. Chem.* **76**, 1 (2000).
256. Porai-Koshitz, M.A.; Aslanov, L.A.; Koritnii, E.F. In *Results of Science & Technology. Crystallochemistry.* Moscow: VINITI, 1977, vol. 11, p. 5.
257. Koda, S.; Ooi, S.; Kuroya, H.; Isobe, K.; Nakamura, V.; Kawaguchi, S. *J. Chem. Soc., Chem. Commun.* 1321 (1971).
258. Cramer, R.E.; Cramer, S.W.; Cramer, K.F.; Chudyk, M.A.; Seif, K. *Inorg. Chem.* **16**, 219 (1977).
259. Busch, M.A.; Fenton, D.E.; Nyholm, R.S.; Truter, M.R. *J. Chem. Soc., Chem. Commun.* 1335 (1970).
260. Matsui, M.; Kurauchi, T.; Ooi, S.; Nakamura, V.; Kawaguchi, S.; Kuroya, H. *Inorg. Nucl. Chem. Lett.* 1041 (1973).
261. Pandey, K.K. *Coord. Chem. Rev.* **140**, 37 (1995).
262. Rawson, M.; Winpenny, R.E.P. *Coord. Chem. Rev.* **139**, 313 (1995).
263. Rapper, E. *Coord. Chem. Rev.* **61**, 115 (1985).
264. Rapper, E. *Coord. Chem. Rev.* **129**, 91 (1994).
265. Rapper, E. *Coord. Chem. Rev.* **153**, 199 (1996); **165**, 475 (1997).
266. Lang, J.; Tatsumi, K.; Yu, K. *Polyhedron* **15**, 2127 (1996).
267. Markovic-Calogovic, D.; Hasic, J.; Vicic-Topic, D. *Inorg. Chim. Acta* **285**, 208 (1999).
268. Aslanidis, P.; Karagiannidis, P.; Akrivos, P.D.; Krebs, B.; Läge, M. *Inorg. Chim. Acta* **254**, 277 (1997).
269. Leung, K.S.-Y.; Wong, W.-T. *J. Chem. Soc., Dalton Trans.* 2521 (1999).
270. Romero, J.; Duran, M.; García-Vázquez, J.A.; Castineiras, A.; Sousa, A.; Cristiansen, L.; Zubieta, J. *Inorg. Chim. Acta* **255**, 307 (1997).

271. Garnovskii, D.A.; Garnovskii, A.D.; Sadimenko, A.P.; Sigeikin, S.G. *Koord. Khim.* **20**, 83 (1994).
272. Cabeleiro, S.; Castro, J.; Vázquez-López, E.; García-Vázquez, J.A.; Romero, J.; Sousa, A. *Polyhedron* **18**, 1669 (1999).
273. Kochin, S.G.; Antsishkina, A.S.; Sadikov, G.G.; Burlov, A.S.; Kogan, V.A.; Garnovskii, A.D. *Dokl. Akad. Nauk.* **355**, 777 (1997).
274. Buzikin, B.I.; Lipunova, G.N.; Sisoeva, L.P.; Rusinova, L.I. *Chemistry of Formazanes.* Nauka: Moscow, 1992, 375 pp.
275. Lipunova, G.N.; Rusinova, L.I.; Maslakova, T.I. *Mendeleev Chem. J.* **40**(4/5), Part II, 230 (1996).
276. Kinoshita, J.; Kashiwabara, K.; Fujita, J.; Matsumo, K.; Oor, S. *Bull. Chem. Soc. Jpn.* **54**, 2863 (1981).
277. Nakano, J.; Joshikawa, J. *J. Am. Chem. Soc.* **106**, 429 (1984).
278. Zhdanov, Yu.A.; Kogan, V.A.; Alexeev, Yu.E.; Kharkovskii, V.M.; Zaletov, V.G.; Starikov, A.G. *Zurn. Neorg. Khim.* **36**, 914 (1991).
279. Whitfield, D.M.; Stojkowski, S.; Sarkaz, B. *Coord. Chem. Rev.* **122**, 171 (1993).
280. Alexeev, Yu.E.; Garnovskii, A.D.; Burlov, A.S.; Zhdanov, Yu.A. *Mendelev Chem. J.* **40**(4/5), Part II, 242 (1996).
281. Alexeev, Yu.E.; Garnovskii, A.D.; Zhdanov, Yu.A. *Russ. Chem. Rev.* **67**(8), 649 (1998).
282. (a) Gyuzcsik, B.; Nagy, L. *Coord. Chem. Rev.* **203**, 81 (2000); (b) Steinbom, D.; Junicke, H. *Chem. Rev.* **100**, 4283 (2000).
283. Aldrich-Wright, R.; Vega, R.S.; William, P.A. *Coord. Chem. Rev.* **160**, 361 (1997).
284. Yamada, S. *Coord. Chem. Rev.* **190–192**, 537 (1999).
285. Von Zelevskii, A. *Coord. Chem. Rev.* **190–192**, 811 (1999).
286. Von Zelevskii, A.; Manula, A. *J. Chem. Soc., Dalton Trans.* 219 (2000).
287. (a) Peruzzini, M.; Bertini, L. *Coord. Chem. Rev.* **185–186**, 1 (1999); (b) *XX International Chugaev's Conference on Coordination Chemistry*. Rostov University Press: Rostov-on-Don, 2001, 558 pp; (c) *The Chemistry of Metal Ions in Everyday Life: Book of Abstracts.* 33rd Conference of Coordination Chemistry, Florence, Italy, 1998; (d) Castillo-Blum, E.; Barba-Behrens, N. *Coord. Chem. Rev.* **196**, 3 (2000).
288. Ungvary, F. *Coord. Chem. Rev.* **147**, 547 (1996).
289. Ungvary, F. *Coord. Chem. Rev.* **160**, 129 (1997).
290. Ungvary, F. *Coord. Chem. Rev.* **167**, 233 (1997).
291. Heatky, G.G. *Coord. Chem. Rev.* **181**, 243 (1999).
292. Odme, G.; Classert, L.; Paetzold, E.; Meisel, R.; Drexler, R.; Fuhrmann, H. *Coord. Chem. Rev.* **185/186**, 585 (1999).
293. Pignolet, L.H.; Aubart, M.A.; Greighead, K.L.; Gould, R.A.T.; Krogstad, D.A.; Wiley, J.S. *Coord. Chem. Rev.* **143**, 219 (1995).
294. Hill, C.L.; Prosser-McCartha, C.M. *Coord. Chem. Rev.* **143**, 407 (1995).
295. Jiangliang, X.; Puddephatt, R.J. *Coord. Chem. Rev.* **143**, 457 (1995).
296. Ovcharenko, V.I.; Sagdeev, R.Z. *Usp. Khim.* **68**, 381 (1999).
297. Ouhalab, L. *Coord. Chem. Rev.* **178–180**, 1501 (2000).
298. Laget, V.; Hornic, K.C.; Rabu, P.; Drillon, M.; Ziessel, R. *Coord. Chem. Rev.* **178–180**, 1533 (2000).
299. Pettijohn, C.N.; Jochnowwitz, E.B.; Chuong, B.; Nagle, J.K.; Vogler, A. *Coord. Chem. Rev.* **171**, 85 (1998).
300. Chen, C.H.; Shi, J.; Jianmin, S. *Coord. Chem. Rev.* **171**, 161 (1998).
301. Yanagida, S.; Hasegawa, Y.; Murakoshi, K.; Wada, Y.; Nakashima, N.; Yamanaka, T. *Coord. Chem. Rev.* **177**, 461 (1998).
302. Lees, A.J. *Coord. Chem. Rev.* **171**, 3 (1998).

303. Horvath, O. *Coord. Chem. Rev.* **135/136**, 303 (1994).
304. Sima, J.; Makanova, J. *Coord. Chem. Rev.* **160**, 161 (1997).
305. (a) Galyametdinov, Yu.G.; Ivanova, G.I.; Ovchinnikov, I.V.; Binnemans, K.; Bruce, D.W. *Izw. Akad. Nauk.* 387 (1999). (b) Turanov, A.N.; Ovchinnikov, I.V.; Galyametdinov, Yu.G.; Ivanova, G.I.; Goncharov, V.A. *Izw. Akad. Nauk.* 694 (1999).
306. Hoshino, N. *Coord. Chem. Rev.* **174**, 77 (1998).
307. (a) Giroud-Godguin, A.M. *Coord. Chem. Rev.* **178–180**, 1485 (1999); (b) Molochko, V.A.; Rukk, N.S. *Koord. Khim.* **26**, 803 (2000); (c) Molochko, V.A.; Rukk, N.S. *Koord. Khim.* **26**, 883 (2000).
308. *Inorganic Biochemistry*. (Edit. Eichhon, G.). Elsevier Science: Amsterdam, Oxford, New York, 1973.
309. Williams, D. *The Metals of Life*. Van Nostrand Reinhold: London, 1971.
310. Siegel, H.; Martin, R.B. *Chem. Rev.* **82**, 385 (1982).
311. Bencini, A.; Bianchi, A.; Paoli, P. *Coord. Chem. Rev.* **120**, 51 (1992).
312. Alexander, V. *Chem. Rev.* **95**, 273 (1995).
313. Stuzhin, P.A.; Khelevina, O.G. *Coord. Chem. Rev.* **147**, 41 (1996).
314. Kudrevich, S.V.; Van Lier, J.E. *Coord. Chem. Rev.* **156**, 163 (1996).
315. Takemura, H.; Shinmyozu, T.; Inazu, T. *Coord. Chem. Rev.* **156**, 183 (1996).
316. Could, R.D. *Coord. Chem. Rev.* **156**, 237 (1996).
317. Sargeson, A.M. *Coord. Chem. Rev.* **151**, 89 (1996).
318. *Biocoordination Chemistry*. (Edit. Fenton, D.F.). Paperback Publications: Oxford, 1996.
319. Reiter, W.A.; Gerdes, A.; Lee, S.; Deffo, T.; Clifford, T.; Danby, A.; Bowman-James, K. *Coord. Chem. Rev.* **174**, 343 (1998).
320. (a) Elias, H. *Coord. Chem. Rev.* **187**, 37 (1999); (b) Chiou, S.-J.; Innocent, J.; Riordan, C.G.; Lam, K.-C.; Liable-Sands, L.; Rheingold, A.L. *Inorg. Chem.* **39**, 4347 (2000).
321. *Biomimetics*. (Edit. Starikaja, A.; Aksay, I.A.). AIP Press: New York, 1995.
322. Rose, E.; Lecas, A.; Quelguejeu, M.; Kossanyi, A.; Boitrel, B. *Coord. Chem. Rev.* **178–180**, 1407 (1999).
323. Krämer, R. *Coord. Chem. Rev.* **182**, 243 (1999).
324. *Porphyrins and Metalloporphyrins*. (Edit. Smith, K.M.). Elsevier Science: Amsterdam, 1975.
325. Chembright, P. *Coord. Chem. Rev.* **6**, 247 (1977).
326. Berezin, B.D. *Coordination Compounds of Porphyrins and Phthalocyanines*. Nauka: Moscow, 1978.
327. *The Porphyrins*. (Edit. Dolphin, D.). Academic Press: New York, 1978/79, vols. 1–7.
328. Berezin, B.D.; Koifman, O.I. *Usp. Khim.* **49**, 2389 (1980).
329. *Vitamin B_{12}*. (Edit. Dolpín, D.). Wiley Interscience: New York, 1982.
330. Smith, K.M. In *Comprehensive Organic Chemistry*. (Edit. Barton, D.; Ollis, W.D.). Oxford: Pergamon Press, 1979, vol. 4.
331. Berezin, B.D.; Enikolonov, N.S. *Metalloporphyrins*. Nauka: Moscow, 1988.
332. Gillard, D.; Kaclish, K.M. *Chem. Rev.* **88**, 1121 (1988).
333. Berezin, B.D. *Zhurn. Neorg. Khim.* **37**, 1260 (1992).
334. Pedersen, K.D.; Frensdorf, H.K. *Usp. Khim.* **42**, 492 (1973).
335. *Progress in Macrocyclic Chemistry*. (Edit. Izatt, R.M.; Christensen, J.J.). John Wiley: New York, 1979, vol. 1; 1981, vol. 2.
336. *Coordination Chemistry of Macrocyclic Compounds*. (Edit. Melson, G.A.). Plenum Press: New York, 1979.
337. *Host–Guest Complexes Chemistry*. Springer-Verlag Chemie: Berlin, 1981.
338. (a) Gokel, G.W.; Korzeniovski, S.H. *Macrocyclic Polyether Synthesis*. Verlag Chemie: Berlin, 1982, (b) *Complexation of Cationic Species by Crown Ethers* (Edit. Inouc, V.; Gokl, G.W.). New York: Marcel Dekker, 1991.

339. Hiraoka, M. *Crown Compounds. Their Characteristics & Applications*. Kadansha Ltd.: Tokyo, Amsterdam, Oxford, New York, 1982.
340. Izatt, R.M.; Brandshaw, J.S.; Nielsen, S.A. *Chem. Rev.* **85**, 271 (1985).
341. Bogatskii, A.V.; Lukianenko, N.G.; Kirichenko, T.I. *Zhurn. Vses. Khim. Obsh. Mendeleev* **30**, 487 (1985).
342. Tsivadze, A.Yu.; Varnek, A.A.; Khutorskii, V.E. *Coordination Compounds of Metals with Crown Ethers*. Science: Moscow, 1991.
343. Tsivadze, A.Yu. *Mendeleev Chem. J.* **40**(4/5), Part I, 62 (1996).
344. Kitajima, N.; Fujisawa, K., Tanaka, M.; Moto-oka, V. *J. Am. Chem. Soc.* **114**, 9232 (1992).
345. Qui, D.; Kilpantrick, L.T.; Kitajima, N.; Spiro, G. *J. Am. Chem. Soc.* **116**, 2585 (1997).
346. Czermuszewicz, R.S.; Maes, E.M. In *Education in Advanced Chemistry. Perspective in Coordination Chemistry*. Poznan: Wroclaw, 2000, p. 67.
347. Holland, P.L.; Tolman, W.B. *J. Am. Chem. Soc.* **121**, 7270 (1999).
348. Uraev, A.I.; Nivorozhkin, A.L.; Bondarenko, G.I.; Lisenko, K.A.; Korshunov, O.Yu.; Vlasenko, V.G.; Vasilchenko, I.S.; Shuvaev, A.T.; Kurbatov, V.P.; Antipin, M.Yu.; Garnovskii, A.D. *Dokl. Akad. Nauk,* **367**, 67 (1999).
349. Holland, P.L.; Tolman, W.B. *J. Am. Chem. Soc.* **122**, 6331 (2000).
350. Louloudi, M.; Hadjiliadis, N. *Coord. Chem. Rev.* **135/136**, 429 (1994).
351. Bertini, I.; Cremonini, M.A.; Ferretti, S.; Lozzi, I.; Luchinat, C.; Viezzoli, M.S. *Coord. Chem. Rev.* **151**, 145 (1996).
352. Mouucsca, J.M.; Lamotte, B. *Coord. Chem. Rev.* **178–180**, 1573 (1998).
353. Rizzarelli, E.; Vecchio, G. *Coord. Chem. Rev.* **188**(1), 343 (1999).
354. Williams, R.J.P. *Coord. Chem. Rev.* **149**(1), 1 (1996).
355. Berthon, G. *Coord. Chem. Rev.* **149**(1), 241 (1996).
356. Corain, B.; Bombi, G.G.; Tapparo, A.; Perazzolo, M.; Zatta, P. *Coord. Chem. Rev.* **149**(1), 11 (1996).
357. Nelson, D.J. *Coord. Chem. Rev.* **149**(1), 95 (1996).
358. Powell, A.K. *Coord. Chem. Rev.* **149**(1), 95 (1996).
359. Black, C.B.; Huang, H.-W.; Cowan, J.A. *Coord. Chem. Rev.* **135/136**, 165 (1994).
360. Rehder, D. *Coord. Chem. Rev.* **182**, 297 (1990).
361. Salder, P.J.; Guo, Z. *Pure & Appl. Chem.* **70**, 863 (1998).
362. Stochel, G.; Wanat, A.; Kulis, E.; Stasicka, Z. *Coord. Chem. Rev.* **171**(1), 203 (1998).
363. Schietert, C.W.; McCue, J.P. *Coord. Chem. Rev.* **184**, 67 (1999).
364. Fag, G.; Crispomi, G. *Coord. Chem. Rev.* **184**, 291 (1999).
365. Reichert, D.E.; Lewis, J.S.; Anderson, C.J. *Coord. Chem. Rev.* **184**, 3 (1999).
366. Thumus, L.; Lejeune, R. *Coord. Chem. Rev.* **184**, 125 (1999).
367. Vam, V.W.; Lo, K.K. *Coord. Chem. Rev.* **184**, 157 (1999).
368. Solovay, A.H.; Tjarks, W.; Barnum, B.A.; Rong, F.G.; Barth, R.F.; Codogni, I.M.; Wilson, J.G. *Chem. Rev.* **98**, 1515 (1998).
369. Yang, P.; Guo, M. *Coord. Chem. Rev.* **185/186**, 189 (1999).
370. Lippert, B. *Coord. Chem. Rev.* **182**, 263 (1999).
371. Zangrando, E.; Pichierri, F.; Randaccio, L.; Lippert, B. *Coord. Chem. Rev.* **156**, 275 (1996).
372. *Comprehensive Coordination Chemistry II* (Edit. Cleverty, J.A.; Meyer, T.J.). Oxford: Pergamon Press, 2003, vols. 1–10.

5

Synthesis of Selected Groups of Coordination Compounds

BORIS I. KHARISOV and LEONOR M. BLANCO JEREZ

Universidad Autónoma de Nuevo León, Monterrey, Mexico

MIGUEL A. MÉNDEZ-ROJAS

Universidad de las Américas, Puebla, Mexico

Additionally to the theoretical data and synthetic techniques for various metal complexes presented in Chaps. 2–4, we would like to pay special attention to three kinds of coordination compounds (complexes of phthalocyanines, quinones, and radioactive elements), whose syntheses, in our opinion, have been insufficiently generalized in monographs and textbooks on synthetic coordination chemistry. This choice is caused by the facts that phthalocyanines, as π-aromatic macrocyclic compounds, possess unusual thermal stability (nonstandard for organic and organometallic species); the quinone complexes have free-radical properties; and coordination and organometallic compounds of radioactive elements are interesting at least for the reasons of necessity of special precautions in their syntheses and applications in the nuclear industry and nuclear medicine. So, this chapter is dedicated to the peculiarities of structure and properties and the main synthetic procedures for the complexes above.

5.1 INFLUENCE OF SOLVENT IN THE SYNTHESIS OF PHTHALOCYANINES

5.1.1 Precursors and Techniques

The well-known metal-free phthalocyanine ***238*** (Pc or PcH_2) and its numerous metal complexes (metal phthalocyanines or, more strictly, phthalocyaninates) ***712*** (R = H) have been studied intensively since the early 1930s [1–4] and are widely used in the pigment industry. Pc can be obtained by the classic template reactions starting from diverse precursors (***925–931***), such as phthalonitrile [***925***, PN, Example 1, reaction (3.154)], *o*-cyanobenzamide (***927***, Example 2), 1,3-di-iminoisoindoline (***928***, 1,3-D), phthalimide (PM) (***930***), phthalic acid (***931***), etc., generally in high-boiling nonaqueous solvents at elevated temperatures [5–7], or electrochemically from phthalonitrile [reaction (5.1), Example 5] [8–10].

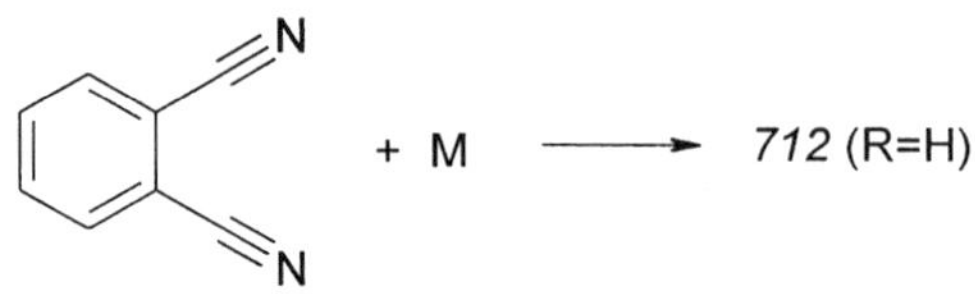

(5.1)

925
Phhtalonitrile

M = Fe, Co, Ni, Cu, Zn, Cd, Sn

Phthalocyanine precursors

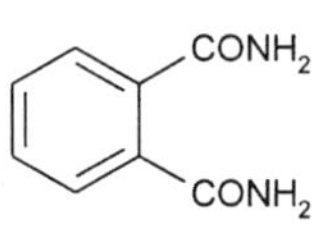

Phthalamide
926

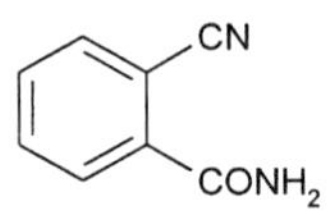

o-Cyanobenzamide
927

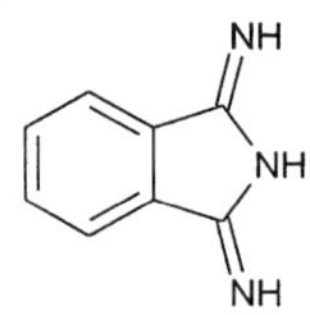

1,3-Diiminoisoindoline (1,3-D)
928

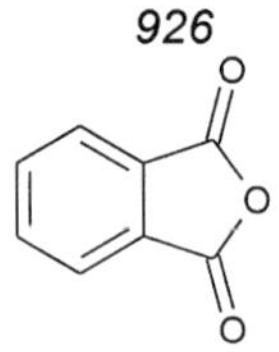

Phthalic anhydride
929

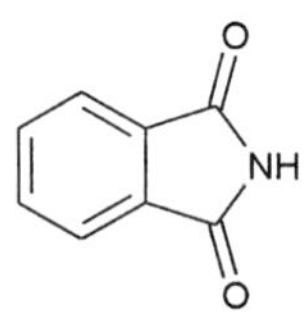

Phthalimide
930

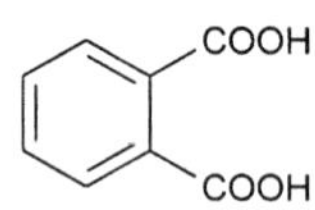

Phthalic acid
931

Metal-free phthalocyanine forms metal complexes with "strong" (for example, Fe, Cu, Ni) or "weak" (Mg, Sb) metals (according to their resistance to being eliminated from the product), which can be synthesized: (a) *chemically* from metals or their salts [2–4] or (b) *electrochemically* from the bulk metals or their salts [9,10].

The first type of reaction employs elemental metals [2–6,11–14] or their salts [5,6], the above precursors, and nonaqueous solvent. High-boiling substances such as nitrobenzene, *o*-dichloro- and trichlorobenzene, ethyleneglycol, α-methylnaphthaline, quinoline, etc., are usually used as solvents [11,15,16], although some alcohols [6,17] or benzene [18] have been applied successfully using PN as a precursor of Pc. The yields of these reactions are in the range of 90–100% [5,6,19,20a]. Evaporated solvents, for example *t*-amylbenzene [20b], were also used for some purposes (generation of an NH_3-containing gas at 200°C to carry out the transformation "phthalic acid→phthalimide" [20b]).

A series of articles is devoted to the preparation of Pc complexes from metal alloys ([21a] and references cited therein). The most important advantage of the alloys' use is an easier reaction between phthalonitrile and the alloy's component(s), due to a concentration gradient of metal particles on the alloy surface. As a consequence of such an interaction, it is possible to obtain polynuclear phthalocyanines and separate the alloy [21a]. This relatively new and intriguing field in Pc research should undoubtedly be taken into account. Another "nonstandard" technique for metal phthalocyanine production is the use of microwave irradiation under solvent-free conditions (for more details see Sec. 3.5). In particular, this method was used for obtaining phthalocyanine derivatives of Ru, Rh, Pt, and Pt [21b], Zn, Mg, Co, and Cu [21c], or Si [21d].

Metal phthalocyanines can also be obtained by an electrochemical process. The feasibility of the electrosynthesis of PcM was reported by Yang and Straughan, who obtained "PcM" of Cu, Ni, Co, Mg, and Pb using metal salts [9] or elemental Fe and Cu [9,22,23] as a source for the central atom. Furthermore, Petit's research group [10,24] studied the electrosynthesis of PcCu by electroreduction of Pc with a copper sheet or an electrodeposited layer of copper on platinum as an anode. The effect of some electrolytic parameters on the yields of the process and a mechanism involving several steps were reported in this paper. Another important result of this group is the electrosynthesis of lithium phthalocyanine radical, that is known to be a member of the class of intrinsic molecular semiconductors [25]. EtOH or its mixtures with DMA or H_2O were used as solvents; reactions were carried out in an electrochemical cell with separated anode and cathode compartments. It is shown that an increase of DMA or H_2O proportion leads to a decrease of the final product yield [10]. In all cases, the electrosyntheses of PcCu were carried out employing PN as a starting material.

The syntheses, structures, properties, and applications of metal phthalocyanines are described in numerous publications (among them, recent original papers [26a–h] and reviews [26i–m]) and well generalized in monographs [5,6,8,26n–p]. Synthetic approaches in relation to metal-free and metal-containing phthalocyanines are similar, although with some peculiarities in each area. The aim of this section is to present a comparative review and analysis of conventional and electrochemical techniques of synthesis of these compounds, taking into consideration the influence of a solvent on the reaction course.

EXPERIMENTAL PROCEDURES

Classic Chemical and Electrochemical Methods of Metal Phthalocyanine Synthesis

Example 1

Synthesis of Copper Phthalocyanine from Metallic Copper and Phthalonitrile (Classic Experiments of Linstead [3])

A mixture of phthalonitrile (4 mmol) and copper (1 mmol) was heated (in an oil bath) with stirring in a wide glass tube. A green color first formed at 190°C, the mass became pasty at 220°C, and was too stiff to be stirred at 270°C (10 min). At a bath temperature of about 220°C, the internal temperature began to mount rapidly and at times exceeded that of the bath by

45°C. The mass was left for another 5 min in the bath, cooled slightly, and ground with alcohol. The finely powdered product was repeatedly boiled with alcohol until the washings were colorless and contained no phthalonitrile; it was then dried. Yield 75–90% of the weight of phthalonitrile [4]. In more recent publications, CH_3ONa (sodium methylate, SM) and other strong bases are used in order to perform a nucleophilic attack at the cyano group of phthalonitrile (5.2) [10,19]:

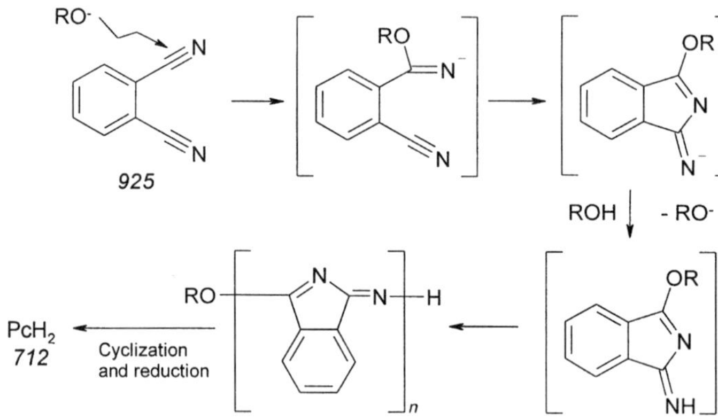

(5.2)

Example 2

Synthesis of Magnesium Phthalocyanine from Metallic Magnesium and *o*-Cyanobenzamide (Classic Experiments of Linstead [2])

Pure *o*-cyanobenzamide (10 g) was heated for 15 min at 230–240°C with 2 g of magnesium metal. The melt was powdered, extracted successively with dilute NaOH solution, 10% sulfuric acid (overnight), and hot water. After prolonged extraction with boiling alcohol, 4 g of a bright blue solid remained [2].

Example 3

Two-step Preparation of Copper Phthalocyanine [20b]

A two-step preparation of copper phthalocyanine includes: (a) treatment of phthalic acid and/or its derivatives with NH_3 and/or its gaseous mixtures in organic solvents in the presence of phthalocyanine compounds and (b) treating the reaction mixture with copper and/or its compounds and urea and/or its derivatives in the presence of catalyst. Thus, phthalic anhydride (PA, 222 parts), urea (315), CuCl (48), ammonium molybdate (2), and *t*-amylbenzene (550) were heated in an autoclave at 200°C to generate NH_3-containing gas which was added to a mixture of PA (222), copper phthalocyanine (16), ammonium molybdate (2), and *t*-amylbenzene (300) to give a suspension containing 99.5% phthalimide. The latter was treated with urea (180 parts) and *t*-amylbenzene (250) in an autoclave at 200°C to give copper phthalocyanine (223) of 95.2% purity.

Example 4

Interaction of Metal Alloys and Phthalocyanine Precursors

The alloys were mixed with phthalonitrile and pressed into pellets. The pellets were inserted into a glass ampoule, evacuated, and sealed off. The ampoule was heated at about 480 K for several days. Reaction products (SnPc from Au—Sn alloy, PbPc from Au—Pb alloy, In_2Pc_3

and SnPc from Sn—In alloy, In_2Pc_3 and Tl_2Pc from In—Tl alloy) were identified by x-ray diffraction methods on single crystals and/or powdered samples [21a].

Example 5

Electrosynthesis of Phthalocyanines [10]

The electrolysis cell (Fig. 5.1) was a divided-jacket three-electrode cell. The cathode was a gold grid (2 cm × cm, 196 mesh/cm^2), whereas the anode was twisted platinum wire.

A 100 mL solution of the chosen solvent containing LiCl (3 g, 0.07 mol) was introducted into the cell and degassed at a given temperature. The electrolytic process was started after the addition of PN in the cathodic compartment. With EtOH as the solvent, the initial colorless solution became yellow, then blue–green after the passage of 20–40 C, and finally turned into a viscous blue suspension. The electrolysis was stopped after a given amount of charge was passed. Then, the catholyte was poured into 100 mL of a 0.2 M H_2SO_4 solution. The resulting suspension was stirred for 0.5 hr and then filtered. The blue solid was adequately treated and the yields were calculated. Elemental and spectroscopic analyses indicated that the blue solid was the hydrogen phthalocyanine PcH_2 (α-form). The same electrolytic process was applied to a substituted phthalonitrile.

The electrosynthesis of PcCu was performed using a copper sheet as anode or an electrodeposited layer of copper on platinum. $CuClO_4$ was introduced into the anodic compartment.

Example 6

Electrochemical Preparation of Lithium Phthalocyanine Radical $PcLi^{\cdot}$ [25a]

The phthalocyanine radical complex of lithium ($PcLi^{\cdot}$) is a member of the class of intrinsic molecular semiconductors [27]. Its preparation is carried out by electrosynthesis at 70°C under

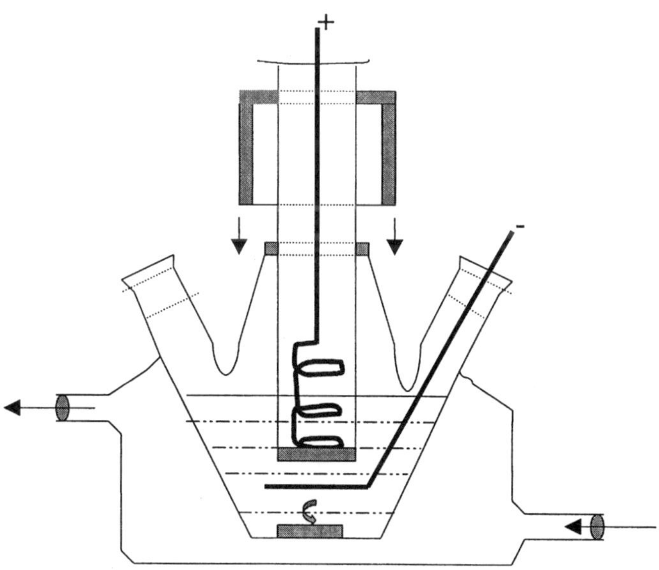

Figure 5.1 Divided electrochemical cell for phthalocyanine synthesis (a reference electrode could be added). (From Ref. 10, with permission.)

inert gas in a divided cell with two electrodes and two compartments separated by a glass-sintered disk. The cathode is a gold grid and the anode a coiled platinum wire. The electrolyzed medium is a stirred absolute ethanol solution (0.1 dm^3) containing lithium chloride (3 g, 70 mmol). Phthalonitrile (2.56 g, 20 mmol) is added to the catholyte (0.08 dm^3) and a constant current of 66 mA is applied. The catholyte rapidly darkens and the reaction is monitored using the UV–Vis spectra of solutions prepared by diluting aliquots withdrawn from the cathodic compartment into an absolute ethanol solution EtOLi (0.3 mol/dm^3). The process produces $PcLi_2$ according to (5.3):

$$4PN + 2Li^+ + 2e^- \rightarrow PcLi_2 \tag{5.3}$$

which then forms $PcLi^{\cdot}$ [(5.4)–(5.7)]:

$$EtOH + e^- \rightarrow EtO^- + 0.5H_2 \tag{5.4}$$

$$4PN \rightarrow Pc(O) \text{ (by } EtO^- \text{ action)} \tag{5.5}$$

$$Pc(O) + 2LiCl + 2e^- \rightarrow PcLi_2 + 2Cl^- \tag{5.6}$$

$$PcLi_2 + Pc(O) \rightarrow 2PcLi^{\cdot} \tag{5.7}$$

The black precipitate is isolated by filtration of the catholyte, purified with hot absolute ethanol in a Soxhlet apparatus, and then dried. $PcLi^{\cdot}$ [28a]) yield 70% (1.80 g, 3.5 mmol) is produced. More information on the electrochemical preparation and EPR studies of lithium phthalocyanine is reported in Refs. 28b and c.

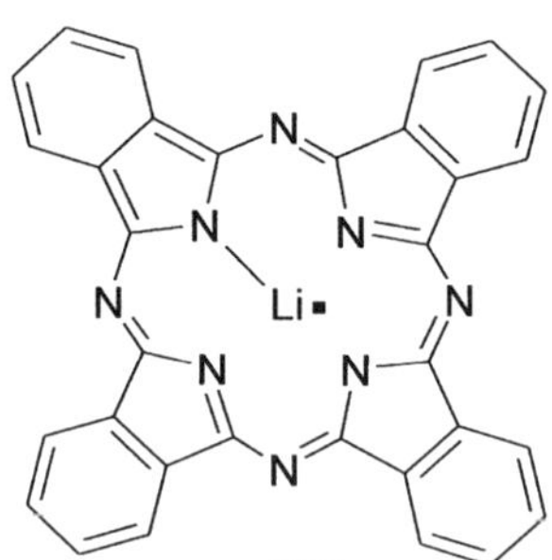

932

[28a]

Example 7

Electrosynthesis of Phthalocyaninato-Iron(II) [23]

Iron (0.1 g) is anodically dissolved (using a platinum flag) in a methanol–1,2-dichlorobenzene (80:20) mixture of 0.03 M PcH_2 and 0.05 M NEt_4Br. The latter acts as an electrolyte during the anodic dissolution. A current density of 30–50 mA/cm^2 is used for 3–9 hr and ca. 1.0 g FePc is precipitated. Analytical results confirm the +2 metal oxidation state.

Example 8

Electrosynthesis of Metal-Free and Metal-Containing Phthalocyanines Starting from Metal Salts [9]

The standard divided electrochemical cell with cation-exchange membrane between the cathode and anode compartments was used. A Hastelloy C and Pt plate of $3 \times 3\,cm^2$ served as

Table 5.1 Metal Salts and Alcohols Used as Anolytes During Electrosynthesis of Phthalocyanines

Reagent	Solvent	Product
$CuSO_4$	CH_3OH	CuPc
$NiSO_4$	CH_3OH	NiPc
$Co^{II}(OAc)_2$	CH_3OH	CoPc
$MgSO_4$	CH_3OH	MgPc
$Pb(NO_3)_2$	C_2H_5OH	PbPc
$Pb(NO_3)_2$	CH_3OH	H_2Pc
$UO_2(OAc)_2$	DMF	H_2Pc

cathode and anode, respectively. The cathode was pretreated with dilute HNO_3 before use. A total of 100 mL of catholyte containing 0.02 M tetrabutylammonium perchlorate and 0.02 M phthalonitrile was placed in the cathode compartment, whereas 100 mL of the alcoholic solution of respective metallic salts served as anolyte. The electrolyses were carried out at −1.6 V vs. SCE and 60–100 mA for 3–5 hr at room temperature. The catholyte gradually became red–brown in the course of the electrolysis. The solid formed in the cathode compartment was filtered and washed with 2 M NaOH, hot water, and acetone, and the solution was Soxhlet extracted with ethanol to remove impurities and finally sublimed in vacuo (Table 5.1).

5.1.2 Phthalonitrile as Precursor

5.1.2.1 Solvent Effect

Among all Pc precursors, PN ***925*** (example 9) [1–6,8,18,19,29] and 1,3-D ***928*** (also an intermediate product in the condensation of phthalonitrile or urea and phthalic anhyride ***929***) [30,31] are the most studied, since the preparation of metal-free Pc starting from these substances is the most simple and reproducible route. However, these precursors can mainly be used for academic purposes, since their relatively high cost makes them economically unprofitable (especially 1,3-D) for industrial production of Pc or its metal complexes (mainly PcCu).

Some organic strong bases {SM in CH_3OH (see Examples 1 and 10), 1,8-diazabicyclo[5.4.0]undece-7-ene ***933*** (DBU, Example 10), 1,5-diazabicyclo[4.3.0] non-5-ene ***934*** (DBN)} are used to perform a nucleophilic attack in the carbon atom (CN group) of PN. Various dinitriles were reacted in the presence of these and other bases {2-N,N-dimethylaminoethanol (DMAE), 1,2,3,6-tetrahydropyridine (THP), triethylamine (TEA), pyridine (Py), triphenylphosphine (TPP), 1,5,7-triazabicyclo[4.4.0]dec-5-en ***935*** (TBD), and 2-*t*-butylimino-2-diethylamino-1,3-dimethylperhydro-1,3,2-diazophosphorine ***936*** (BEMP)} in bulk or in high-boiling solvent [19,29]:

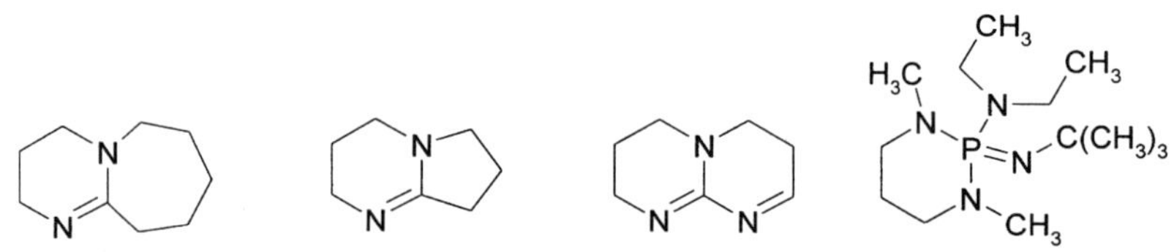

DBU 933 DBN 934 TBD 935 BEMP 936

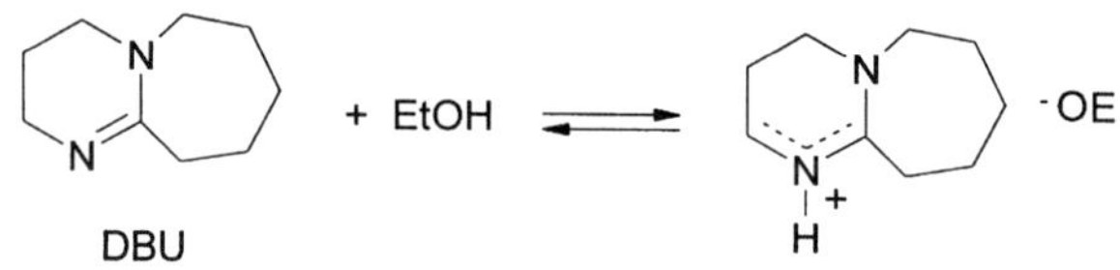

DBU 933

EXPERIMENTAL PROCEDURES

Example 9

General Procedure for the Synthesis of Phthalocyanines in Bulk [29]

Phthalonitrile (1 mmol), a little (some milligrams) basic coreactant ***933–936*** and an anhydrous metal salt (M = Cu, Co, Ni, Pb; in case of $CuCl_2$, its ratio with the phthalonitrile is 1:4) are placed in an ampoule. The ampoule is degassed by three freeze–thaw cycles and sealed in vacuo. The ampoule is heated for 4 hr at 200°C. After cooling, the ampoules are opened, and the hard, brittle product is ground in a mortar. The products are treated with ethanol in a Soxhlet extractor until the solvent is colorless. This method is valid for metal-free and metal-containing species. The products are dried over P_4O_{10} in vacuo at 150°C. Yield 64–90%.

Example 10

General Procedure for the Synthesis of Substituted Phthalocyanines Using Organic Strong Bases DBU and DBN [19,29]

The following dinitriles are used:

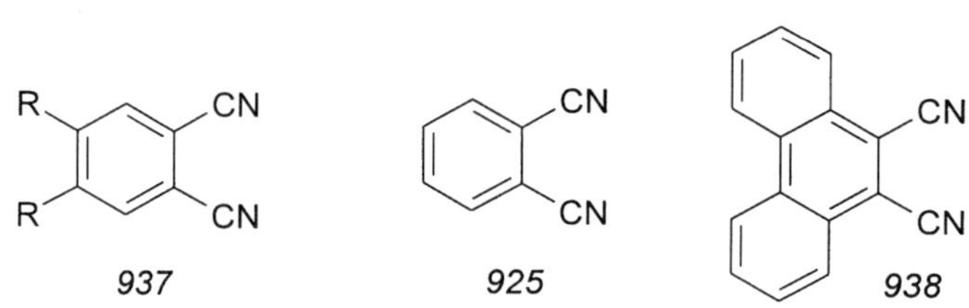

937 925 938

R = -OCH$_3$, -OC$_3$H$_6$CH$_3$
Metals used:
M = Cu, Co, Ni, Pb

Metals used:
M = Cu, Co, Ni, Pb

M = Cu

The dinitrile (5 mmol) and an equimolar amount of basic coreactant are heated in 20 mL of dry *n*-pentanol under reflux. For the reactions in the presence of metal salts, 1.25 mmol is added as an anhydrous salt together with the dinitrile. After 36 hr, methanol (50 mL) is added, and the mixture is stirred for 30 min. The isolated reaction products are washed with water and methanol and then treated with methanol in a Soxhlet extractor overnight. The products are dried over P_4O_{10} in vacuo at 150°C. Yield 82–92%. The same technique is used for the preparation of metal-free phthalocyanines.

In a recent article [32], the influence of various protic and aprotic solvents on the reaction course in the synthesis of metal-free and metal-containing phthalocyanines starting from phthalonitrile was studied. Both conventional chemical and electrochemical methods were used. In the case of aprotic solvents (DMSO, Py, dioxane, THF, DMF, acetone, nitrobenzene), no PcH_2 was obtained from phthalonitrile in the electrosynthesis conditions, as well as without electrolysis. In these experiments, different volumes of SM (0–1 mL) have been used. It was almost impossible or difficult to carry out the electrolysis in dioxane, DMF, acetone, and nitrobenzene due to unstable voltage. When SM is used, a visible interaction between the solvent and SM takes place, including formation of insoluble sediment (reactions in DMSO, DMF, THF, and acetone). The absence of SM has no influence on the reaction course. The absence of phthalocyanine formed could be explained by the impossibility of nucleophilic attack of SM to the CN group of the phthalonitrile due to the interaction between SM and the aprotic solvent. As a suggestion, instead of SM, other agents compatible with a solvent are needed to carry out a nucleophilic attack in further investigations with use of aprotic solvents.

The most reproducible results have been obtained [32] in protic solvents such as *i*-BuOH and dimethyletanolamine with almost quantitative yields (Table 5.2, Example 11). A successful synthesis in protic media is in agreement with the opinion of the authors of Ref. 10 that "a protic solvent is required for the electrosynthesis of PcH_2." Sodium methylate as a source of alkoxide anions performs a nucleophilic attack at the cyano group of PN to form 1-alkoxy-3-iminoisoindoline as an intermediate which is further reduced and cyclizated [19]. This mechanism is similar to the one described by Petit et al. [10] and Tomoda et al. [19], where metal alkoxides were used for the electrochemical preparation of Pc.

Higher yields of Pc in alcohol media with application of electrochemical procedure can be explained by the formation of RO^- ions from ROH during the process. These ions, together with the action of SM, perform a nucleophilic attack at the CN group of PN, resulting in a higher rate of Pc formation.

In contrast to N,N-dimethylethanolamine, the interaction in *i*-BuOH without electrolysis of the reaction mixture does not produce Pc. This fact could be used successfully to electrosynthesize various metal phthalocyanines, synchronizing the Pc formation on the cathode and metal anode dissolution. This could prevent the formation of mixtures of metal-free and metal phthalocyanines [33]. As will be shown below, N,N-dimethyletanolamine could also be used successfully as a "model solvent," in whose medium the formation of the metal-free Pc takes place even at room temperature under conditions of UV irradiation (see Table 5.5).

EXPERIMENTAL PROCEDURES

Example 11

Table 5.2 Protic Solvent Effect in the Electrosynthesis of PcH_2 Starting from Phthalonitrile

Experiment	Initial System	Current (mA)	Initial Voltage (V)	Time (hr)	Temp. (°C)	Yield (%)	Observations
1	*n*-BuOH TBA (0.1 g)	50	25	1.3	90	78	After addition of SM an increase of temperature is observed
2	*n*-BuOH	0	0	1.3	90	43	After addition of SM an increase of temperature is observed
3	*i*-BuOH TBA (0.1 g)	25	150	1	90	98	Simultaneous use of ultrasonic treatment stabilizes the voltage
4	*i*-BuOH	0	0	1	90	0	No product observed without electrolysis
5	*n*-$C_5H_{11}OH$ TBA (0.2 g)	25	150	1.3	95	93	Low conductivity of the solution, so it is necessary to use more TBA. No reaction observed without SM
6	*n*-$C_5H_{11}OH$	0	0	1.3	95	65	No reaction observed without SM
7	Ethylene glycol	50	10	1	100	62	High conductivity of the solution
8	Glycerine	80	37	2	100	—	Formed phthalocyanine is inseparable from the solvent, so it is impossible to evaluate the yield
9	N,N-Dimethylethanol amine TBA (0.1 g)	40	115	2	90–100	97	The conductivity increases gradually during the process
10	N,N-Dimethylethanol amine	0	0	2	90–100	66	

1: The reaction yield is calculated on the charge passed through the solution in the case of the electrosynthesis, or on the phthalocyanine formed in the conventional chemical synthesis.
2: 1 g of SM solution and 5 g of PN have been used in all experiments.
3: Tetra *n*-butyl ammonium bromide (TBA) was selected according to the dielectric constant of the corresponding solvent.
4: 100 mL of each solvent was used in all experiments.
5: Platinum sheets were used as an anode and a cathode.
Source: Ref. 32.

5.1.2.2 Temperature Effect in Pc Formation

i-BuOH and N,N-dimethyletanolamine were used as model solvents to study a temperature effect (reaction yields) of the phthalocyanine starting from phthalonitrile. Syntheses have been carried out at 55–90°C [32]. It is established that the Pc yields increase as temperature increases in both cases, as expected. In contrast to any other solvent used, an increase in conductivity of the reaction mixture is observed in N,N-dimethyletanolamine during the electrolysis (at constant temperature). This solvent can be recommended for conventional Pc synthesis (Examples 11 and 12; Tables 5.2 and 5.3). A further increase in temperature (to boiling) does not improve yields.

It is possible to carry out synthesis of Pc and its substituted derivatives even at *room temperature* using metals in primary alcohols [34]. Thus, phthalonitriles can give phthalocyanines in these conditions simply by changing the solvent to 1-octanol and using lithium 1-octanolate [34b]. The nonsubstituted phthalonitrile reacts with M (M = Ca, Mg, Zn, Fe, and Cu) in MeOH or EtOH during 20–26 days forming metal-free Pc (M = Ca) or PcM (M = Mg, Zn, Fe, and Cu) with yields from $< 1\%$ (in case of Cu) to 12% (Mg or Zn) [34a]. It is surprising that the yield is too small (traces) in case of CuPc, as it is one of the most stable metal phthalocyanines. According to the authors' opinion, "perhaps more activated forms of some metals are required" [34a]. Application of ultrasound treatment (Sec. 3.5.4) of the reaction mixture containing metallic lithium has no positive influence on yields [34a]. In our opinion, a possible use of highly active Rieke metals in appropriate solvents could help to develop this very important area of phthalocyanine synthetic chemistry. In case of using some substituted phthalonitriles [34a], M (M = Ca, Mg, Li) or LiOAlk/$Zn(OAc)_2$, and solvent (MeOH, EtOH, or 1-octanol), the yields are higher (7–60%).

EXPERIMENTAL PROCEDURE

Example 12

(a)

Table 5.3 Temperature Effect in the Electrosynthesis of Phthalocyanines Using Phthalonitrile as a Precursor

Experiment	Initial system[a]	Current (mA)	Initial voltage (V)	Time (hr)	Temp. (°C)	Yield (%)
1[b]	*i*-BuOH (100 mL) TBA (0.2 g)	25	115	2	55	3
2	*i*-BuOH (100 mL) TBA (0.2 g)	25	104	2	70	9
3	*i*-BuOH (100 mL) TBA (0.2 g)	25	97	2	75	22
4	*i*-BuOH (100 mL) TBA (0.2 g)	25	88	2	80	37
5	*i*-BuOH (100 mL) TBA (0.2 g)	25	79	2	85	66
6	*i*-BuOH (100 mL) TBA (0.2 g)	25	70	2	90	97
7[c]	Dimethylethanol amine (100 mL) TBA (0.1 g)	40	118	2	90	97
8	Dimethylethanol amine (100 mL) TBA (0.1 g)	40	127	2	80	86
9	Dimethylethanol amine (100 mL) TBA (0.1 g)	40	136	2	70	56

[a] 5 g of PN and 1 mL of SM used in all experiments.
[b] The Pc is formed starting from 52–55°C.
[c] Pc formation starts from 55°C. The conductivity is increased gradually during 2 hr of the process.
Source: Ref. 32.

(b) General Procedure for the Condensation of PN Using Different Metals at Room Temperature [34a]

PN (256 mg, 2.00 mmol) was added to the alcohol (MeOH or EtOH, 3 mL). The metal (2.20 mmol) [Ca, 88 mg as turnings; Mg, 53 mg as a powder (325 mesh); Zn, 143 mg as a powder (300 mesh); Fe, 168 mg as a powder (350 mesh)] was added and the reaction system stirred at room temperature for 20–26 days. THF (5 mL) was added and the reaction mixture subjected to flash column chromatography on silica gel using THF as eluent. The product was dispersed in ethanol–water (1:1) and collected by centrifugation. The precipitate was successively washed with acetonitrile and ethanol and collected by centrifugation. Metallated Pcs (M = Mg, Zn, Fe) were produced in 8–12% yield, while PcCu was formed but in less than 1% yield.

5.1.3 1,3-Di-iminoisoindoline as Precursor

1,3-Di-iminoisoindoline ***928*** (1,3-D) is the intermediate product in successive Pc formation and has been studied in detail [30,31], along with the mechanism of formation of the complexes derived from 1,3-D-containing ligands [35] or intermediates to metal Pc [36]. As mentioned above, its main disadvantage for industrial use, as well as that of phthalonitrile, is its relatively high cost, which makes both substances very unattractive as precursors for industrial production of Pc. The 1,3-D $\cdot$ HNO_3 whose structure can be represented by the following tautomeric formulae [30] (5.8):

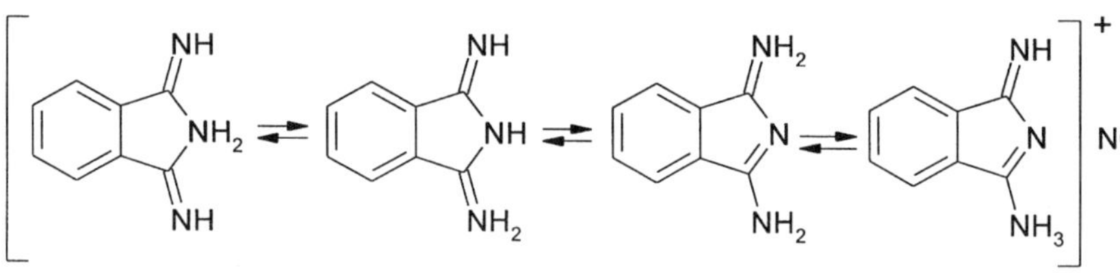

(5.8)

can be obtained starting from urea, phthalic anhydride, and NH_4NO_3 in $PhNO_2$ in the presence of catalysts such as $(NH_4)_2MoO_4$ or MoO_3 [30] with further treatment by cold NaOH to produce 1,3-D [37]. More pure 1,3-D can be obtained starting from phthalonitrile and NH_3 [38,39] (NH_3 is also used for direct preparation of PcCu from urea and phthalic acid [20b]). To synthesize Pc starting from 1,3-D, inert organic solvents such as trichlorobenzene, *o*-dichlorobenzene, etc., are generally used [1–6,31].

5.1.3.1 Solvent Effect and Ligand Concentration in Pc Formation

1,3-Di-iminoisoindoline was used as a precursor for Pc in different protic and aprotic systems, without catalysts or promoters, to study the solvent effect on the possibility of phthalocyanine formation [32]. As can be observed (Example 13), it is possible to carry out the chemical and electrochemical synthesis of metal-free Pc in aprotic solvents, such as DMF or DMSO, in contrast to the results with PN. It is surprising that the yields of Pc in ROH are comparatively small. The N,N-dimethyletanolamine is characterized by the best yields, as in the case when PN was used as precursor.

Increasing the concentration of 1,3-D in the reaction mixture leads to higher yields which are almost quantitative. With low-conducting solvents such as trichlorobenzene or *o*-dichlorobenzene, the synthesis of Pc without electrolytic conditions was carried out due to the absence of conductivity in their solutions or very low conductivity of their mixtures with DMF or DMSO.

Successful chemical and electrochemical synthesis of Pc from 1,3-D in aprotic solvents in comparison with those using PN shows that the highest influence of a solvent's nature on a reaction course takes place in the first stage of the process (1,3-D formation). For further reactions (cyclization and reduction of 1,3-D), a solvent's nature is not very important, as the results presented in Table 5.4 show. The formation of Pc from 1,3-D takes place in all the solvents used; higher yields can be achieved by optimization of the process (variation of concentration of 1,3-D, use of electrosynthesis, and/or selection of the best solvent applied) [32].

EXPERIMENTAL PROCEDURE

Example 13

Table 5.4 Electrosynthesis of Phthalocyanine Starting from 1,3-Di-iminoisoindoline

Experiment	Initial system[a]	Current (mA)	Initial voltage (V)	Time (hr)	Temp. (°C)	Yield (%)	Observations
1	EtOH TBA (0.1 g) 1,3-D (1 g)	30	150	1	75	0 (8% in case of use of 3 g of 1,3-D)	No product observed without electrolysis
2	*i*-BuOH TBA (0.15 g) 1,3-D (1 g)	15	150	1	100	0 (5% in case of use of 3 g of 1,3-D)	No product observed without electrolysis
3	N,N-Dimethyl-ethanol amine 1,3-D (1 g)	0	0	1	133	52	
4	N,N-Dimethyl-ethanol amine TBA (0.1 g) 1,3-D (1 g)	40	75	1	133	76	
5	N,N-Dimethyl-ethanol amine TBA (0.1 g) 1,3-D (3 g)	40	72	1	133	92	
6	Nitrobenzene 1,3-D (1 g)	0	0	1	190	11	The crystalline product is formed
7	Nitrobenzene TBA (0.1 g) 1,3-D (1 g)	70	150–250	1	190	44	Unstable voltage
8	DMSO 1,3-D (1 g)	0	0	1	189	5	
9	DMSO 1,3-D (1 g)	40	37	1	189	26	High conductivity of the solution
10	DMSO 1,3-D (1 g)	40	42	1	189	98	
11	*o*-Dichloro-benzene 1,3-D (1 g)	0	0	3	180	7	
12	DMF 1,3-D (1 g)	0	0	1	145	5	

Experiment	Initial system[a]	Current (mA)	Initial voltage (V)	Time (hr)	Temp. (°C)	Yield (%)	Observations
13	DMF 1,3-D (1 g)	40	35	1	145	25	Formation of Pc is observed at 120°C on the cathode surface
14	DMF 1,3-D (3 g)	40	33	1	145	94	
15	Trichlorobenzene 1,3-D (1 g)	0	0	1	180	33	
16	Trichlorobenzene 1,3-D (3 g)	0	0	1	180	95	

[a] 100 mL of solvent and different amounts of 1,3-D (1 and 3 g) and TBA have been used in the experiments.
Source: Ref. 32.

5.1.3.2 UV Irradiation Effect on Pc Formation

Tomoda et al. [40] reported that the Pc could be produced from PN even at room temperature by UV irradiation of the reaction mixture in different alcoholic solvents (Example 14). The authors concluded that the UV treatment is effective only at the initial stage of the reaction because of an absence of the product after heating the reaction mixture in the dark and further exposure to UV light [40].

EXPERIMENTAL PROCEDURE

Example 14

Synthesis of Metal-Free Phthalocyanines by UV Irradiation [40]

After sodium metal (0.115 g, 5 mmol) was dissolved in 125 mL of alcohol (CH_3OH, C_2H_5OH, n-C_3H_7OH, i-C_3H_7OH, n-C_4H_9OH, i-C_4H_9OH, s-C_4H_9OH, n-$C_5H_{11}OH$, i-$C_5H_{11}OH$), phthalonitrile (1.28 g, 10 mmol) was added to the solution. The preparation in the dark or under room light (fluorescent lamp) was carried out by heating the solution under reflux for 48 hr in nitrogen atmosphere. The photochemical preparation was carried out either at room temperature or at 70°C by irradiating internally with UV light (100 W high-pressure mercury lamp with a Pyrex filter) for 48 hr in nitrogen atmosphere. The precipitated blue product, phthalocyanine, was collected by filtration, washed with hot ethanol, and dried. Yield 0–40.3%. The formation of phthalocyanine occurred by UV irradiation even at room temperature (yield 3–16%). It was established that (a) the irradiation of the reaction mixture is effective at the initial stage of the reaction, (b) the formation of phthalocyanine is inhibited by air at room temperature, and (c) the phthalocyanine yield is better in the presence of benzene under nitrogen atmosphere.

The behavior of 1,3-D in some nonaqueous solvents under UV irradiation was studied in Ref. 32. Among other solvents used (see Table 5.5, Example 15), N,N-dimethyletanolamine has demonstrated special properties in relation to Pc formation. The Pc appears slowly even at 7°C and can be isolated (with different yields, depending on the temperature applied) at any temperature of the reaction mixture.

When PN is used as a precursor [40], free radicals of $RO^{\cdot}$ appear due to UV irradiation of the alcohol solutions of PN, contributing to the nucleophilic attack on the carbon atom of the CN group, together with the action of CH_3ONa. Probably, this is a cause of Pc formation at room temperature. In experiments using 1,3-D as a starting material, the formation of free radicals $(CH_3)_2NCH_2CH_2O^{\cdot}$ ($RO^{\cdot}$) from N,N-dimethylethanolamine probably contributes to the conglomeration of the intermediate products, which appear as a result of the transformation of 1,3-D [30,35b]. Another version which can explain the "special" properties of this solvent in all the experiments above is the possible participation of an electronic pair of nitrogen atoms of N,N-dimethylethanolamine in the nucleophilic attack.

EXPERIMENTAL PROCEDURE

Example 15

Table 5.5 Influence of UV Irradiation on the Synthesis of Phthalocyanine Starting from 1,3-D in Different Solvents

Experiment[a]	Solvent	Temperature (°C)	Observations
1	DMF	7, 20, 45, 65, 75, 85, 100	No visible changes in the system
2	DMSO	7, 20, 45, 65, 75	No visible changes in the system
		85, 100	Dark color of the solution. No phthalocyanine formed
3	Trichlorobenzene	7, 20, 45, 65, 75, 85, 100	No visible changes in the system
4	Nitrobenzene	7, 20, 45, 65, 75	No visible changes in the system
		80, 100	Dark color of the solution. Small amount of Pc (~ 0.2 g) is formed
5	N,N-Dimethyl-ethanol amine	7–100	The Pc is formed at any temperature. Intensity of the color of the solution increases with temperature. In absence of UV irradiation, the Pc is formed slowly starting from 45–50°C

[a]Solutions of 2 g of 1,3-D in 100 mL of different solvents were placed into quartz ampoules and irradiated by mercury lamp (1200 W) for 4 hr. No electrolysis was used in these experiments.
Source: Ref. 32.

5.1.3.3 Urea and Phthalic Anhydride Effects on Pc Formation

The presence of urea or PA ***929*** can also influence the reaction course of Pc starting from 1,3-D. These Pc precursors were introduced into the reaction system in some experiments in order to establish which component's abundance could affect the Pc formation. DMF was chosen as a solvent in this study [32]. It was observed (see Table 5.6, Example 16) in DMF medium that the presence of urea does not affect the Pc formation. This fact is in agreement with an earlier report [6], where urea was used as a solvent. Urea and, as will be shown below, tetramethylurea are "inert" solvents in respect of 1,3-D. On the contrary, the presence of PA decreases the reaction yield to zero. Its presence provokes competitive reactions (probably due to the attack of oxygen atoms of PA on N atoms of 1,3-D), preventing 1,3-D cyclization in the reaction system that leads to the absence of the final product.

EXPERIMENTAL PROCEDURE

Example 16

Table 5.6 Urea and Phthalic Anhydride Effects on the Formation of Pc Starting from 1,3-D

Experiment	Initial system	Current (mA)	Initial voltage (V)	Time (hr)	Temp. (°C)	Yield (%)	Observations[a]
1	*Presence of urea* DMF (100 mL) 1,3-D (1 g) Urea (1; 2; 3 g) TBA (0.1 g)	40	35–40	1	145	23; 18; 22	The product is formed without any difference in comparison with the experiments where urea is absent
2	*Presence of urea* DMF (100 mL) 1,3-D (1 g) Urea (1; 2; 3 g)	0	0	1	145	7; 9; 8.5	The product is formed without any difference in comparison with the experiments where urea is absent

[a] In the presence of phthalic anhyride (1–3 g) or its mixture with urea (3 and 4 g, respectively), no Pc was observed in electrosynthesis conditions starting from 1,3-D (1 g), as well as without electrolysis.
Source: Ref. 32.

5.1.4 Urea and Phthalic Anhyride as Precursors

5.1.4.1 Metal-Free Phthalocyanines

Industrial research in this area is devoted mainly to the synthesis of Pc or "PcM" (M = Cu, Ni, Fe, Al, etc.) starting from urea and phthalic anhydride (or its derivatives) as the most cheap precursors. A survey of the literature shows that most of the articles and patents (among them Refs. 40–49) in the "phthalocyanine" area during the last 15 years are devoted to searching for the optimal conditions for Pc or "PcM" (M = Cu, Fe, Al, etc.) preparation, as well as the study and applications of different phthalocyanine modifications [50–56], synthesis of various Pc-substituted derivatives [57–66], study of reaction mechanisms of Pc formation [9,10,18,19,29,30], and much more relevant information generalized in a recent book [67].

To carry out the interaction between urea and phthalic anhydride, molybdenum compounds such as MoO_3, Na_2MoO_4, $(NH_4)_2MoO_4$, etc., are usually employed as catalysts [41,43–48] (Example 17), as well as other metal salts such as $TiCl_4$ [42] or tungsten compounds [41]. Tetramethylurea, 1-methyl-2-pyrolidinone, and other organic compounds are added to the reaction system as promoters [41]. The reactions are carried out in high-boiling solvent such as nitrobenzene, a mixture of trihclorobenzene isomers, high alkanes, etc., at 150–250°C [6,11,41]. The central atom sources for PcM are usually the following transition metal salts: $NiCl_2$, CuCl, $CuSO_4$, $CoCl_2$, $FeSO_4 \cdot 4H_2O$, $Fe_2(SO_4)_3 \cdot 9H_2O$, $FeCl_2$, $FeCl_3$, etc. [6,43–48]. The products obtained have purity in the order of 50 to 100% and can be used as pigments after the corresponding purification. Some solvents promote the formation of by-products (for example, biphenyls [68]), or competitive reactions preventing the normal reaction course, therefore, sometimes the syntheses are carried out without any solvent [41], which could influence the morphology and crystalline structure of the formed products [69].

EXPERIMENTAL PROCEDURE

Example 17

Two-Step Synthesis of Metal-Free Phthalocyanine from Urea and Phthalic Anhydride [30,37]

Phthalic anhyride (148 g) and ammonium molybdate (1 g) are added with stirring to a hot mixture (110–120°C) of urea (300 g), ammonium nitrate (160 g), and nitrobenzene (100 mL). The mixture is heated gradually (3–4 hr) up to 165–170°C and this temperature is maintained for 15 hr. Thereafter, the mixture is heated to 180–190°C and maintained at this temperature for 2 hr with stirring. After cooling to 60°C, methanol (250 mL) is added. The solid is filtered at reduced pressure, washed with methanol, and dried in air. Yield of 1,3-D $\cdot$ HNO_3 (1,3-diiminoisoindoline nitrate) 198 g (95%) [30].

To eliminate nitrate anion, 1,3-D $\cdot$ HNO_3 (198 g) is mixed with water (910 mL) and cooled to 3–5°C. Then, a cold NaOH solution (40%, 100 g) is added. The solid formed is filtered and dried in air. Yield of 1,3-D 118 g (90% purity). Melting point 184–186°C [37].

As mentioned above, urea and PA are the most cheap phthalocyanine precursors and are produced on an industrial scale, so it is not surprising that numerous articles and patents have been dedicated to the study of their interactions [6,30,41–48]. However, only copper and some other "strong" metals (in relation to the "PcM" formation and stability) form their phthalocyanines using these precursors. There are almost no reports in the available literature about attempts to electrosynthesize PcH_2 or "PcM" starting from urea and phthalic anhydride, except for a recent work [32] where the interaction between these two precursors, as well as phthalimide, in various nonaqueous solutions by conventional chemical and electrochemical methods is studied in detail (Example 18).

EXPERIMENTAL PROCEDURE

Example 18

Attempts to Obtain Metal-Free Phthalocyanine from Urea and Phthalic Anhydride in One-Step Interaction [32]

Three series of experiments were carried out: (1) using N,N-dimethylethanolamine, nitrobenzene, trichlorobenzene, DMSO, or the mixtures "DMSO–trichlorobenzene" and "nitrobenzene–trichlorobenzene" as solvents (100 mL) and urea (4 g), PA (3 g), TBA (0.1–0.2 g), and ammonium molybdate (AM, 0.04 g); (2) the same solvents (100 mL) and urea (0; 3; 6 g), phthalimide (3 g), TBA (0.1–0.2 g), and ammonium molybdate (0.04 g); (3) mixtures of

DMSO (or nitrobenzene) (50 mL) and trichlorobenzene (50 mL), TMU, or 1-methyl-2-pyrolidinone as promoters (0.12 g), urea (4 g), PA (3 g), TBA (0.05–0.2 g), and MoO_3 (0.015 g).

As a result of all these experiments, it is impossible to obtain metal-free phthalocyanine from urea and phthalic anhyride as well as phthalimide in one-step interaction, either by conventional chemical or electrochemical methods. The presence of catalysts and promoters for CuPc (TMU or 1-methyl-2-pyrolidinone) manufacture does not influence the formation of the Pc. Similarly, the presence of small amounts of Pc introduced into the reaction mixture does not provoke a further PcH_2 formation starting from the above precursors in the solvents used [32].

According to the opinion of the authors of Ref. 32, the negative results of the experiments above do not mean that metal-free Pc could not be theoretically obtained from these precursors: additional detailed studies (combination of different techniques, such as UV irradiation, microwave treatment, use of other inert solvents, electrolysis in the systems producing free radicals, etc.) are required for successful resolution of this problem. However, under the same conditions it is possible to obtain some transition metal complexes of the Pc due to the template effects (see below).

5.1.4.2 Synthesis of Metal-Containing Phthalocyanines

Phthalocyanines of copper (Examples 19–21) and other transition (Example 21) and *p*-metals have been studied intensively [1–6,18,41–48,70–82]. Their syntheses have been carried out using all possible precursors for Pc, in different polar or nonpolar solvents or without any solvent [1–6,18,33,41–48,70–82], in the presence of NH_3 [20], N_2 or H_2 [6], using microamounts of PcH_2 [82], or applying microwave treatment [70]. Additionally to "standard" metal phthalocyanines such as PcCu, polynuclear complexes of "double sandwich" type [6,33,63] (see "*Lanthanide and actinide phthalocyanines*") or water-soluble complexes [83] have been obtained. However, the electrochemical method has been used only for the synthesis of PcCu, using a *copper anode* [9,10] or $CuSO_4$ [9], and some other metal (Ni, Co, Mg, Pb) phthalocyanines using *metal salts*, dissolved in a reaction system containing PN [9].

The authors of Ref. 32 have chosen four metals for interaction with the Pc precursors in nitrobenzene and trichlorobenzene, according to their capacity to form stable (Cu, Fe) and unstable (Mg, Sb) compounds with Pc [1–4]. As a result, the use of Cu and Fe leads to their phthalocyanines formation; the yields are considerably higher in "pure chemical" experiments (69–77%). Applying the electrosynthesis, only a small amount of CuPc (7%) is observed. Mg and Sb do not produce phthalocyanines in the above conditions.

EXPERIMENTAL PROCEDURES

Example 19

Synthesis of Copper Phthalocyanine Starting from Metallic Copper (Without Electrolysis) [41] (This Method is Widely Used in the Pigment Industry)

Urea (27 g), tetramethylurea (TMU, 0.8 g), copper powder (2.13 g), 19.95 g phthalic anhydride, 0.105 g molybdic anhydride, and trichlorobenzene (mixture of isomers, 72 g) were introduced in the above order into the reactor. The reaction mixture was heated from 20 to 170°C in 90 min with stirring and then kept at 170°C for 5 hr. The volatile portion was distilled off (at

50 mm Hg) while keeping the temperature of the external bath at 150°C. The residue was digested at 90°C for 2 hr, under stirring with 560 g of an aqueous solution of H_2SO_4 (10%). The filtrate, dried in an oven at 100°C for 24 hr, comprised 17.3 g of copper phthalocyanine in the β-crystalline form. Yield on copper 89.6%.

Example 20

Synthesis of Copper Phthalocyanine in the Presence of Tetramethylurea as a Promoter [41]

Urea (40.5 g), phthalic anhydride (29.92 g), N,N,N′,N′-tetramethylurea (TMU, 1.2 g, MoO_3 (0.155 g), CuCl (4.72 g), and trichlorobenzene (108 g) were mixed and brought up to 170°C in 90 min with stirring and kept at 170°C for 5 hr. Thereafter, the volatile portion was distilled off (at 50 mm Hg) while keeping the temperature of the external bath at 150°C. The residue was digested at 90°C for 2 hr, under stirring with 560 g of an aqueous solution of H_2SO_4 (10%). The filtrate, dried in an oven at 100°C for 24 hr, comprised 26.4 g of copper phthalocyanine in the β-crystalline form. Yield on CuCl 96.1%. Experiment without TMU gave a 8% CuPc yield.

Example 21

Interaction of Copper and Iron with Urea and Phthalic Anhydride [32]

Each experiment is carried out using metal powder (Cu, Fe, Mg, Sb, 1 g), urea (4 g), PA (3 g), TBA (0–0.2 g), TMU (0.12 mL), and MoO_3 (0.015 g). A mixture of nitrobenzene (50 mL) and trichlorobenzene (50 mL) is used as solvent. The processes are carried out during 3 hr at 170°C with or without electrolysis. Yields of CuPc or FePc 69–77%.

It is concluded [32] that it is impossible to obtain phthalocyanines electrochemically from the mentioned metals in the conditions above, except PcCu with low yields, although without electrolysis the copper and iron complexes are formed with significantly higher yields. According to Linstead [2], among other metal phthalocyanines, PcCu is the most stable due to the highest affinity of copper(II) to phthalocyanine macrocycle. Copper cannot be "extracted" from PcCu without destruction of organic matter. PcCu is formed generally in inert nonaqueous solvents or without any solvent [6], use of polar solvents (in order to create favorable conditions for electrolysis) leads to zero yields of the final product. In the experiments with use of various solvents (see Example 22), only in hysol the yield of PcCu was 85% (without electrolysis); in all other mentioned solvents no phthalocyanines were observed (except PcCu traces in mesytilene). Use of N,N-dimethyletanolamine with simultaneous UV irradiation does not produce PcCu (copper transfer from the anode to the cathode was observed in this experiment) [32].

EXPERIMENTAL PROCEDURE

Example 22

Synthesis of Copper Phthalocyanine from Urea and Phthalic Anhydride in Various Solvents

In the series of experiments, urea (4 g), PA (3 g), (0.12 g), Cu (1 g), MoO_3 (0.015 g), 1-methyl-2-pyrolidinone (0.5 mL), and 100 mL of solvent (DMEA, ethylene glycol, glycerine, DMSO, DMF, mesytilene, butylcellosolve, hysol, and xylene) were used. The reaction mixture was heated at 190°C during 3–10 hr with stirring. Yields of PcCu: in hysol, 80–85%; in mesytilene, traces; in other solvents, no phthalocyanine formed [32].

Therefore, the solvent used for successful electrosynthesis of PcCu should be "inert" in relation to PA and, on the other hand, should have electroconductivity. The compounds used as promoters [41] could theoretically serve as such solvents. Tetramethylurea (TMU) and 1-methyl-2-pyrolidinone were chosen by the authors of Ref. 32 among other promoters used in the work [41]. The first one has a nature close to that of the principal precursor (urea), and thus should not influence the reaction course negatively. The TMU has sufficient conductivity to permit electrolysis in its medium, and reasonable viscosity. The boiling point of 174–178°C is ideal for such research, since conventional syntheses of Pc from urea and PA are carried out at similar temperatures. The results of TMU use as a solvent are presented in Table 5.7. The results seem promising, and this solvent is recommended to study Pc formation in its medium in further research work. In the case of 1-methyl-2-pyrolidinone, no phthalocyanine formation was observed. No phthalocyanine was observed also in the following systems: (1) urea, PA, TBA, TMU (without copper); (2) urea, PA, TBA, TMU, Sb, or Mg (anodes; (3) TMU, urea (or without urea), phthalimide, TBA (in all cases with or without electrolysis).

EXPERIMENTAL PROCEDURE

Example 23

Table 5.7 Use of Tetramethylurea as a Solvent and Promoter Simultaneously

Experiment	Initial system[a]	Current (mA)	Initial voltage (V)	Time (hr)	Temp. (°C)	Yield (%)
(a) Interaction between urea and PA in the presence of copper						
1	PA (3 g)	20	45	3	170	35
2	Cu (1 g)	0	0	3	170	47
Interaction between phthalimide and urea in the presence of Sb or Mg						
3[b]	Phthalimide	30	50	3	170	17
4	(3 g)	0	0	3	170	33
	Sb or Mg (1 g)					

[a]TMU (100 mL), urea (4 g), TBA (0.2 g), and MoO_3 (0.015 g) have been used in all the experiments.
[b]Copper transfer from the anode to the cathode is observed. In the absence of urea, no phthalocyanine formation is observed.
Source: Ref. 32.

As mentioned earlier (Example 16, Table 5.6), an abundance of urea does not affect the reaction course of 1,3-D cyclization. Tetramethylurea can participate in similar intermediate reactions as urea due to the proximity of their nature. Theoretically, it is possible to obtain metal-free phthalocyanine by adding other catalysts and/or promoters the reaction mixture on the basis of TMU or other derivatives of urea.

5.1.5 Other Phthalocyanine Precursors

o-Cyanobenzamide ***927*** reacts readily with many metals and their compounds (Fe, its oxide and sulfide, Mg, Sb, Cr, Mn, Sn, Al, Cu, Ce, Bi_2O_3, $MnCO_3$, CaO, Sb_2O_5, etc.) at about 250°C to yield phthalocyanines. The real precursor of phthalocyanines in this case is iminophthalimidine (***939***), a derivative of *iso*-indole:

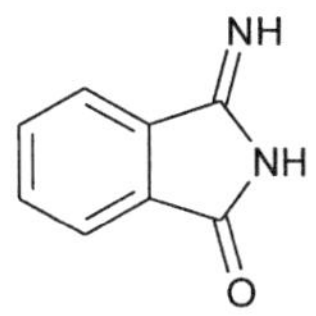

Iminophthalimidine

939

which is readily formed by isomerization of *o*-cyanobenzamide above its melting point [2]. The metals used in the preparation could often be replaced by their oxides, sulfides, or carbonates without affecting the nature of the product. The reactions with Mg, Fe, Co, Ni, and Sb are exothermic and yield phthalocyanines of the same general type, together with phthalimide and ammonia as byproducts. The reaction between *o*-cyanobenzamide and Mg (or MgO) produces magnesium phthalocyanine with ~ 40% yield (Examples 2 and 24).

EXPERIMENTAL PROCEDURE

Example 24

Synthesis of Magnesium Phthalocyanine from *o*-Cyanobenzamide [2]

o-Cyanobenzamide (crude, with phthalimide admixture, 40 g), naphthalene (15 g), and magnesium oxide (10 g) were heated for 3 hr at 230–240°C in an enameled iron pot fitted with a short air reflux and heated by an oil bath. The hard residue was powdered and extracted in a Soxhlet apparatus with hot acetone to free naphthalene, phthalimide, and a small amount of a soluble green impurity. It was then freed from excess magnesia with dilute acid, washed with water, and dried. Yield 18 g.

Phthalimide ***930*** has also been used for phthalocyanine synthesis (Example 25). In case of Mg [2], only the metal itself can be utilized. The process requires a much more careful control of temperature than that utilizing *o*-cyanobenzamide and gave only 20% of PcMg under optimum conditions.

EXPERIMENTAL PROCEDURE

Example 25

Synthesis of Magnesium Phthalocyanine from Phthalimide [2]

Phthalimide (500 g) and 20 g of magnesium were melted in a wide glass tube heated by an oil bath. The internal temperature was kept at 240–250°C and a vigorous stream of dry ammonia passed, the molten mass being stirred mechanically. A considerable quantity of phthalimide sublimed. After 8 hr heating, the viscous melt was cooled, broken up, and treated with a large bulk of 5% NaOH solution, the vessel being streamed out to remove the excess of metal, and extracted with boiling acetone in a Soxhlet apparatus. A small amount of a bright green substance was removed and 80 g of a blue powder remained. On continuous extraction of the latter with boiling quinoline the pigment was dissolved, leaving rather less than 20% of a chocolate-brown impurity. Pure PcMg was isolated in the usual way by careful crystallization of the quinoline-soluble material.

5.1.6 *f*-Metal-Phthalocyanines

5.1.6.1 Lanthanide Phthalocyanines

The lanthanide phthalocyanine complexes, obtained by conventional methods starting from metal salts at 170–290°C and phthalonitrile (Example 26), contain one or two macrocycles for each metal atom [5,6,8,63,82,84–98]. Thus, according to Refs. 6,63, and 85, the complexes having compositions $LnPc_2H$, XLnPc (X^- is halide anion), and Ln_2Pc_3 (a "super-complex") were prepared from phthalonitrile as a precursor; the ratio of the reaction products depends on the synthesis conditions and the metal nature. The ionic structure $Nd(Pc)^+Nd(Pc)_2^-$ was suggested [85] and refuted [63] for the neodymium super-complex Nd_2Pc_3; the covalent character of the donor–acceptor bonds in this compound and other lanthanide triple-decker phthalocyanines was proved by the study of dissociation conditions of these compounds [63].

The lanthanide phthalocyanines (Fig. 5.2) can also be obtained from the corresponding metal salts and metal-free phthalocyanine or 1,3-D [6]. It has been established [6,86] that "whereas a diphthalocyanine complex Pc_2LnH is mainly obtained for the heaviest lanthanides (Dy–Lu), a super-complex Pc_3Ln_2 is progressively

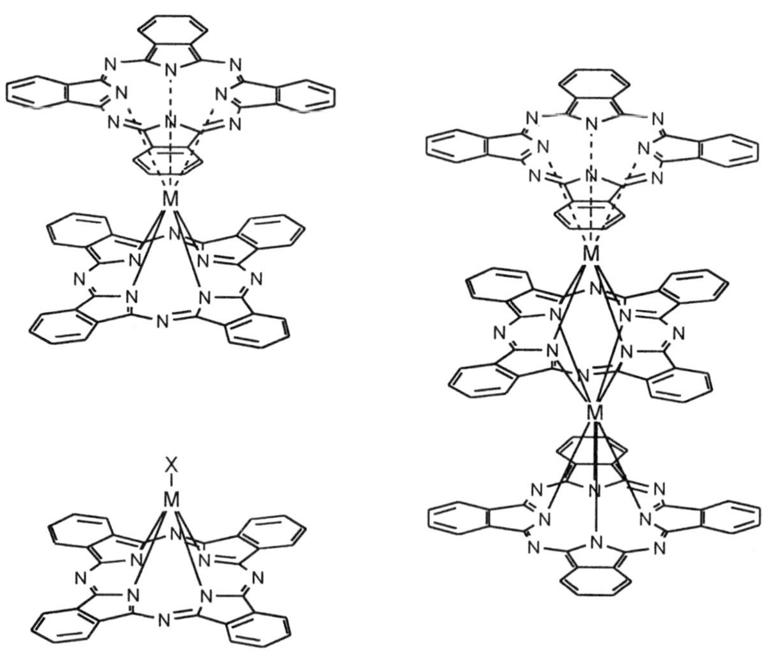

Figure 5.2 The structures of some lanthanide–phthalocyanine sandwiches.

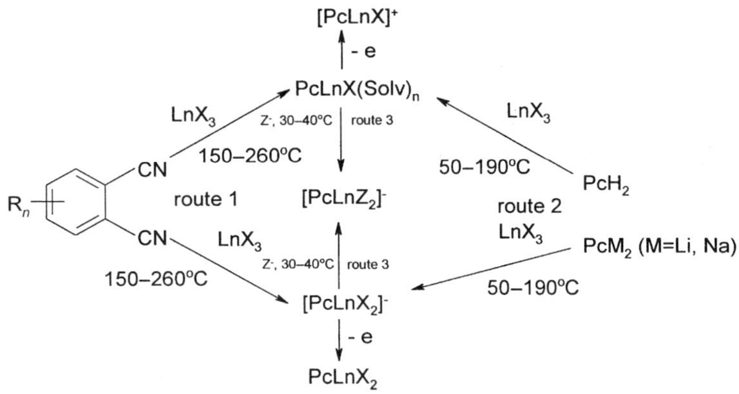

Figure 5.3 Synthetic conditions for the mixed-ligand Pc complexes, containing one Pc ligand. Route 1: interaction of *o*-dicyanobenzene(s) and their analogues with lanthanide salts. Route 2: metallation reaction of the macrocyclic ligand or its dianione by lanthanide compounds. Route 3: reactions of axial substitution in the environment of the central atom in lanthanide complexes ([96] and references cited therein). (From Ref. 96, with permission.)

formed with the lighter lanthanides, and predominates for La and Nd." However, it is impossible to avoid using metal salts and all the precursors of the phthalocyanine, with the possibility of mixed salts (XLnPc) formation [6]. We emphasize that a detailed description of conditions influencing the composition of the lanthanide–Pc complexes, as well as their spectral and electrochemical properties, are generalized in a recent review [96]. In particular, Fig. 5.3, borrowed from this work, represents the synthetic scheme for mixed-ligand complexes with one phthalocyanine ligand.

It is established that, among three possible products (LnPc, $LnPc_2H$, and Ln_2Pc_3), the Ln_2Pc_3 are produced with the best yields when the ratio of phthalonitrile:$LnCl_3$ is 6:1. Syntheses of other rare-earth element phthalocyanines are described, in particular [85,86,97–101].

EXPERIMENTAL PROCEDURE

Example 26

Synthesis of Lanthanide (Sm, Gd, Tm, Lu) Phthalocyanines [63]

Phthalonitrile (10 g) is mixed with $LnCl_3$ (Ln = Sm, Gd, Tm, Lu; 5 g) and heated at 290°C during 1 hr. Then the reaction mixture is dissolved in DMF and chromatographed in a column with Al_2O_3.

Example 27

Synthesis of Lu[15-crown-5)$_4$Pc]Pc [97]

A mixture of 0.16 g of lutetium monophthalocyaninate acetate and 0.3 g of dicyanobenzo-15-crown-5 (DCBC), dried in vacuo, was rafted (melted) in a vacuum glass ampoule, immersed in Wood alloy, at successive increases of temperature from 240 to 260°C. The melted phase was kept at 260°C for 0.5 hr. The molar ratio of the initial reagents Lu salt: DCBC was chosen as

1:5, since the DCBC was used as a reagent and a medium, allowing us to perform the reaction in melted phase. After cooling and grinding, the product was dissolved in chloroform. The reaction mixture was chromatographed on aluminum oxide with use of a chloroform–ethanol mixture (1.5 vol% EtOH) as eluent. Yield 0.063 g (14%). The synthesis of Lu(III) diphthalocyanine with crown-ether substituents is presented in Example 36.

EXPERIMENTAL PROCEDURE

Example 28

Electrosynthesis of Lanthanide (Ln, Sm, Nd, Pr) Phthalocyanines [33]

The electrochemical cell was a 100-mL glass flask with reflux. The anode was a piece of metal (Ln, Sm, Nd, Pr) and the cathode was a platinum foil. A stream of dry nitrogen served to stir the solution during all the time of electrolysis. The solution phase containing *i*-BuOH (100 mL), SM (in CH_3OH, 25% solution, 1 mL), phthalonitrile (3 g), and previously dried $(n\text{-Bu})_4NBr$ (0.1 g) as supporting electrolyte was heated to 100°C with agitation, and then the electrolysis begun (power supply PS 500-1 Sigma-Aldrich). The time of electrolysis was 2 hr (Table 5.8). After finishing the experiment, the solution was cooled and the metal phthalocyanine complex filtered and purified in a Soxhlet apparatus (absolute ethanol used as solvent). Yield 95–100%.

Simultaneous Ultrasonic Treatment

A simultaneous ultrasonic treatment of the reaction phase (with ultrasonic cleaner Bransonic 12 was carried out in all experiments using a weak source of ultrasound in order to eliminate formed products from the metal surface and thus stabilize the current in the process. Stronger sources of ultrasound have not been used to avoid turbulent processes and the uncontrolled superheating of the reaction zone.

Higher concentrations of the supporting electrolyte were not applied, in spite of the fact that the generated voltage is 55–63 V, to avoid collateral reactions, in particular, the formation of free Br_2 on the anode and its further interaction with the macrocycle molecules or the metal anode.

The electrodic reactions should be represented as follows [(5.9)–(5.10)]:

$$\text{Cathode:} \quad 4PN + 2e^- \rightarrow PcH_2 \tag{5.9a}$$

$$\text{Anode:} \quad Ln \rightarrow Ln^{3+} + 3e^- \tag{5.9b}$$

$$\text{Then:} \quad 2Ln^{3+} + 3PcH_2 \rightarrow Ln_2Pc_3 + 6H^+ \tag{5.10}$$

According to Sokolova et al. [63], the maximum yield (80–90%) of the Ln_2Pc_3 compounds (in comparison with XLnPc and $LnPc_2H$) is obtained when, in a chemical way, the molar ratio "*o*-phthalonitrile:rare-earth element salt" is 6:1 (La,

Table 5.8 Conditions of the Electrosynthesis of Lanthanide Phthalocyanines

Metal	Initial voltage (V)	Current (mA)	Yield (%)
La	60	30	97
Sm	55	30	95
Nd	63	30	92
Pr	58	30	95

Sm, Gd, Tm, Lu, initial metal salt is $LnCl_3$). Use of the electrochemical dissolution of the lanthanides allows us to obtain Ln_2Pc_3 complexes with almost 100% yields on *o*-phthalonitrile conversion, according to the elemental analysis and spectral data [33].

According to Ref. 32, there is no formation of free phthalocyanine in the system "*i*-BuOH–CH_3ONa–(*n*-Bu)$_4$NBr–*o*-phthalonitrile" without the application of electrolysis at about 100°C (Example 11), unlike some other solvents where both chemical and electrochemical formation of phthalocyanine could take place. So, this solvent was chosen by the authors of Ref. 33 in order to synchronize metal anode dissolution with the formation of free phthalocyanine on the cathode surface and to avoid obtaining a mixture of "metal-free phthalocyanine–lanthanide phthalocyanine." Unlike conventional chemical methods of preparing rare-earth metal phthalocyanines [63,85,86], where the syntheses are carried out at 170–290°C, it is possible to decrease the reaction temperature to about 100°C.

The application of simultaneous ultrasonic treatment [102–106] with electrochemical dissolution permits us to increase their efficiency, since it is not necessary to stop the electrolysis periodically to mechanically remove the formed product out of the electrode surface. As a consequence, it is possible to stabilize the voltage during the electrochemical process.

5.1.6.2 Actinide Phthalocyanines

Protactinium-233 and neptunium-239 diphthalocyanines are prepared from the corresponding thorium-232 and uranium-238 diphthalocyanines by element transformation [6]. The existence of Pa and Np di-Pcs is proven by repeated sublimation of the irradiated parent compounds using platinum gauze to retain the impurities. Neptunium di-Pc is also synthesized on the tracer scale from irradiated uranium metal, using the normal synthetic method for uranium di-Pc (Example 29) [6]. Other actinide phthalocyanines are reported [107–114]. Their structures, as well as those of 200 metal phthalocyanines and their derivatives, are classified in an excellent recent review [115]. More recent experimental data on actinide phthalocyanines are absent in the available literature.

EXPERIMENTAL PROCEDURE

Example 29

Synthesis of Th, Pa, and U Phthalocyanines [6]

Thorium phthalocyanine is prepared by heating the metal (previously etched with HCl) and *o*-phthalonitrile, 1:25, at 270 to 300°C for 5 hr. The dark blue product is cooled to room temperature, washed with benzene, and purified twice by sublimation at 520°C and 10^{-4} Torr. The protactinium-233 produced by *n*-irradiation of pure thorium phthalocyanine is separated in high purity in the residue after repeated sublimation of the thorium phthalocyanine. The thorium-231 produced by (*n*,2*n*) reaction in the thorium phthalocyanine is found to be enriched in the residue after sublimation, indicating decomposition of the phthalocyanine by irradiation. Uranyl phthalocyanine is prepared by heating a mixture of uranyl acetate and phthalonitrile at 230 to 240°C.

5.1.7 Phthalocyanine Derivatives

Most reported phthalocyanine derivatives (sulfo-, nitro-, amino-, triphenylmethyl-, polymeric, etc.) are copper complexes, although at present the synthetic chemistry of other *d*- and *f*-metal Pc derivatives is being rapidly developed (Examples 30–36) [5,6,116–118]. Some of them (in particular, copper phthalocyanine sulfonic acids) are of industrial interest because of their usefulness as dyes. Phthalocyanine sulfonic acids themselves are prepared both by urea synthesis from sulfonated phthalic anhydride and by the sulfonation of the phthalocyanine [6]. Some substituted metal phthalocyanines can be obtained by chemical or electrochemical reduction [118e]. Among a number of reported peculiarities of substituted phthalocyanines, the existence of three electronic isomers for magnesium derivative PcMn was recently confirmed [118f].

The tetrasulfo-Pc complexes of a number of metals are made by the urea melt process by heating the powdered metal, or its acetate, with triammonium-4-sulfophthalate, urea, boric acid, and ammonium molybdate. The metals or metal compounds used are those of chromium(III), manganese(II), iron(II), iron(III), cobalt(II), and zinc(II). Selected synthetic examples of sulfo- and other derivatives of metal phthalocyanines are presented below.

Recommendations on the synthesis of metal phthalocyanines. It is still difficult to evaluate real reaction mechanisms in each synthetic procedure applied. It is clear that the use of such polar protic solvents as alcohols contributes to higher yields of Pc from PN in the electrosynthesis conditions due to the ease of nucleophilic attack of the generated additional RO^-. In the further steps of Pc formation from PN or 1,3-D, a solvent's nature has no significant importance. These data about the importance of, first of all, the initial stage correspond to those reported on UV irradiation [40] of PN solutions, where such a treatment is effective only at the beginning of the process. However, in the case of the use of urea and PA, a solvent must be completely inert (or be close to urea's nature) to carry out the one-step synthesis of metal phthalocyanines, in order to exclude any negative influence on the reaction course. The fact that the yields are almost always higher in the case of direct electrosynthesis could serve as an additional confirmation about the usefulness and necessity of this technique.

It is very difficult (but, we believe, not impossible) to carry out direct *one-step* electrosynthesis of metal-free Pc starting from urea and PA. The Pc could be synthesized from PN or 1,3-D in a one-step process, according to the reported [9,10] results. Using urea and PA, the synthesis of Pc must include three steps: (a) interaction between PA, urea, and NH_4NO_3 [30], forming 1,3-D·HNO_3; (b) elimination of NO_3^- by NaOH, forming 1,3-D [37]; (c) (electro)synthesis of Pc from 1,3-D.

It is possible to improve the existing techniques of synthesis of PcCu and other metal phthalocyanines from phthalimide or urea and PA, applying electrosynthesis. These processes have many peculiarities. The solvent nature greatly affects the course of the majority of reactions of coordination compound formation [119,120]. The solvent used for PcCu preparation must be inert in order not to influence the desirable reaction course, and simultaneously must have electroconductivity to carry out electrolysis. It is recommended to use the derivatives of urea as such solvents and promoters at the same time. The use of a standard electrochemical procedure [9–14,20,121–124a] could be useful for the PcCu industry, since a typical industrial problem is the presence of initial Cu^{2+} salts in the final product. These impurities decrease the quality of the pigment and are hardly removed.

PcCu could be formed without use of any solvent [6]. It is recommended to apply electrosynthesis in melt urea or its mixtures with high-boiling conducting solvents, avoiding such typical problems as, for example, sublimation of PA. Among the other solvents used in phthalocyanine preparation, N,N-dimethyletanolamine has special properties which make it a promising solvent for further investigations in the Pc area. A recent report on the solid-phase, room-temperature synthesis of metal-free and metal complexes of substituted phthalocyanines has especial interest [124b].

A probable success could be theoretically reached using "weak" metals, such as Sb or Mg, in the chemical or electrochemical interaction with the Pc precursors. Undoubtedly, this method should be developed, since it is a possible route to prepare metal-free Pc after removing metal from the macrocycle.

Electrochemical reactions with use of a sacrificial *cathode* [121], which have not yet been reported in relation to phthalocyanine, could have theoretical success in the synthesis of "PcM." The cathode dissolution takes place due to the action of free radicals formed in solution near the cathode surface [121]; in the case of the synthesis of Pc (which is formed on the cathode surface), UV irradiation of the reaction mixture could serve as a source of free radicals.

EXPERIMENTAL PROCEDURES

Example 30

Synthesis of Copper Hexadecachlorophthalocyanine [31]

4,5,6,7-Tetrachloro-1,3-di-iminoisoindoline (11.2 g) was heated with 50 mL of trichlorobenzene and 3 g of $CuCl_2$ at 200°C with reflux and stirring during 3 hr. After finishing the reaction, trichlorobenzene was distilled with water vapor. The sediment formed was filtered, washed with small portions of chlorobenzene, benzene, acetone, and methanol, and dried. Yield of $C_{32}N_8Cl_{16}Cu$ 89%.

Example 31

Synthesis of Copper 4,4′,4″,4‴-Tetranitrophthalocyanine [31]

$CuCl_2$ (3 g) was mixed with 20 g of 5-nitro-1,3-di-iminoisoindoline and 50 mL of dry nitrobenzene and the resulting mixture heated during 1.5 hr with reflux and stirring. Then the mixture was cooled to room temperature and the product filtered, washed with hot chlorobenzene, benzene, acetone, and methanol, and dried. Yield of $C_{32}H_{12}N_{12}O_8Cu$ 89%. M.p. 452–452.5°C.

Example 32

Copper Phthalocyanine Sulfonic Acids [6]

Copper phthalocyanine (50 g) in 120 g of 6 N sulfuric acid was mixed and evaporated to dryness. The product was grinned and heated to 180–190°C for 2 hr to obtain 47 g product with a sulfur content of 8.12%. A sulfonated copper phthalocyanine containing 1.3 groups/mol is obtained, heating 32.7 g 75% 4-sulfophthalic anhydride and 25% 3-sulfophthalic anhydride mixture, 39.7 g phthalic anhydride, 250 g urea, 12 g cuprous chloride, and 12 g titanium tetrachloride in nitrobenzene at 185°C.

A 45% yield of copper 3,4,5-trichlorophthalocyanine-4′,4″,4‴-sulfonic acid is made from 3.5 g 4-sulfophthalic acid, 3.4 g tetrachlorophthalic anhydride (molar ratio 1:1), 0.5 g copper powder, 10 g urea, 1 mL titanium tetrachloride, and 0.5 g iodine, heated for 2 hr at 210°C. The yield is 3.1 g and the product is moderately water-soluble.

Example 33

Copper Phthalocyanine Nitro Compounds [6]

Copper 4,4′,4″,4‴-tetranitrophthalocyanine is prepared by treating 3-trichloromethyl-4-dichloromethyl-nitrobenzene or 3,4-*bis*(dichloromethyl)nitrobenzene with gaseous ammonia in nitrobenzene in the presence of cupric chloride for an hour at ambient temperature, and then gradually heating the mixture to 200–210°C during 6–7 hr and holding that temperature for 5 hr.

Example 34

Synthesis of Copper 1,1′,1″,1‴,4,4′,4″,4‴-Octahydroxyphthalocyanine [118a]

The above compound was prepared in 70% yield by heating 20 g 2,3-dicyanohydroquinone with 1.84 g cuprous oxide in 150 mL diethylene glycol for 3 hr at 200°C.

Example 35

Synthesis of Nonsymmetrical Copper Phthalocyanines [117]

The synthesis was performed at 280–300°C in an argon atmosphere. A mixture of equimolar amounts of tetraphenylphthalonitrile and 4-nitrophthalonitrile or 6-*t*-butylnaphthalonitrile with a twofold excess of copper(I) chloride was melted in a Wood metal–bath container for 7 hr. The cooled melt was treated with water, 5% aqueous HCl, 5% aqueous NH_4OH, water, and then dried. A mixture of nonsymmetrical reaction products was dissolved in the minimum volume of benzene and the substances chromatographed on silica gel using benzene as eluent. The yields of the formed complexes were 14% (containing 4-nitroisoindole fragments) and 36% (containing 6-*t*-butylbenzisoindole fragments), respectively.

Example 36

Synthesis of Double-decker Lutetium(III) Diphthalocyanine with Crown-Ether Substituents *bis*{tetra[(benzo-15-crown-5)-4′-yl-oxymethyl]phthalocyaninato}lutetium(III) [118d]

1-[(Benzo-15-crown-5)-4′-yl]oxymethyl-3,4-dicyanobenzene (0.212 g, 0.5 mmol), $Lu(OAc)_3 \cdot H_2O$ (0.026 g, 0.0638 mmol), DBU (37.5 mL), and *n*-hexanol (1.5 mL) were heated to 170–175°C under N_2 for 24 hr. After cooling to room temperature, the reaction mixture was diluted with EtOH (20 mL) and the dark green precipitate filtered off. It was treated with boiling EtOH and ethyl acetate several times to dissolve the unreacted starting materials and decomposition products. Then the crude product was dissolved in chloroform, filtered, and the filtrate added dropwise into ethyl acetate. The green precipitate was filtered, washed with diethyl ether, and dried in vacuo. Yield 0.030 g (13.46%).

5.2 METAL–QUINONE COMPLEXES*

At present, many coordination compounds, which contain ligands capable of participating in easy and reversible one-electron redox processes, are known. Nitroxyl radicals, spatially hindered *o*-quinones, quinoneimines, and phenoxazinone systems are examples of such ligands.

Compounds of this type obtained for all transition metals have intriguing and unique structural, magnetic, and electronic properties [125] and are of considerable interest in relation to the fact that the oxidation of the ligands in these systems, leading to paramagnetism, mimics biochemical processes such as respiration and photosynthesis. Other fields of application include their use as medicinal chemotherapeutic (antitumor) [126,127] and organ-imaging [128,129] agents, in biological intercalation studies [130], and as solid-state materials [131].

As far back as 1931, L. Michaelis suggested the formation of active forms (radical particles) of some ferments from quinones [132]. The coordination chemistry of catechols and semiquinones has developed dramatically over the past 20 years and, at present, extensive experimental data dedicated to element-organic *o*-semiquinone complexes have been accumulated and reviewed in papers by Pierpont [125,133], Tuck [134], Abakumov [135,136], and Kabachnik [137]. In this respect, a review [125] which represents an excellent generalization of the latest achievements in this area should be especially noted.

Due to some special structural and magnetic peculiarities, in particular, free-radical properties, the quinone ligands and their metal complexes are apart from the other kinds of coordination compounds [125a,138a], as will be shown below. Here we present an overview of the main methods for the synthesis of complexes containing benzoquinone, semiquinone, and catecholate ligands, and peculiarities of the products.

5.2.1 Peculiarities of Metal–Quinone and Related Complexes

5.2.1.1 Metal Oxidation Number and Redox Isomers

In general, the introduction of spatially hindered phenols into coordination compounds may produce stable free-radical forms [138b–140]. A series of metal complexes with redox ligands, containing derivatives of 2,6-di-*t*-butylphenols π- or σ-connected, or vicinal fragments in the coordination environment of the central metal atom, were synthesized in this way: π-aryl [141], π-σ-allyl [142] compounds, nitrile complexes [143], metal glioximates [144], salicylaldiminates [145,146], porphyrines [147–149], and phthalocyanines [150,151].

The ligands on the basis of spatially hindered phenols can exist in different redoxforms: diamagnetic phenol, paramagnetic phenoxyl, phenolate, and quinolate. These forms are interrelated by reversible transitions, including electron, proton and hydrogen transport (5.11):

*Written with the participation of Professor E.P. Ivakhnenko (Rostov State University, Russia).

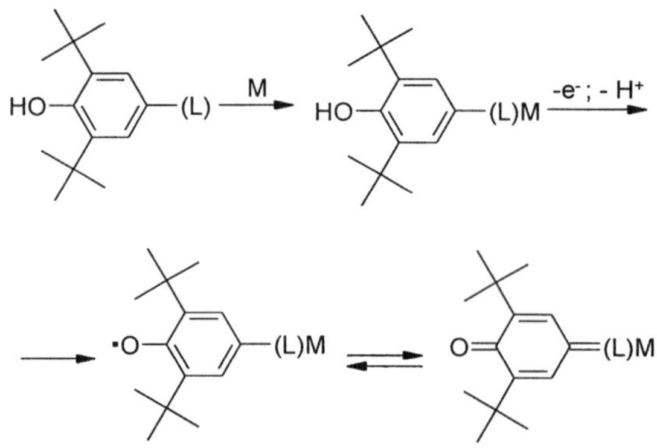

(5.11)

A detailed study of a series of coordination and organometallic compounds with ligands containing redox groups was carried out [141–151] using EPR, electronic absorption spectroscopy, electrochemical methods, and quantum-mechanical calculations. On the basis of these data, some experimental rules were established for the redox transformations, depending on the nature of the metal–ligand bond, the distance between redox fragments and metal, and the metal nature. It was shown that the introduction of a metal atom into organic radicals is an effective chemical way to stabilize them. Moreover, the transformation of ligands into their free-radical form leads to a change in reactivity of organic and organometallic compounds. According to the opinion of the authors of Ref. 152, this influence is transmitted by an innermolecular mechanism.

Studies on a series of transformations of complexes containing 2,6-di-*t*-butylphenols or corresponding phenoxyl radicals allowed us to establish that a change in the nature of the organic ligands in metal complexes, leading to their transition from diamagnetic to paramagnetic state, is a method of molecule activation. It was shown for phthalocyanines [152], containing spatially hindered phenol groups, that these compounds are polyfunctional homogeneous redox catalysts, whose activity is determined by various possible redox transformation routes.

In recent years, a considerable number of investigations have been carried out on the coordination chemistry of *o*-quinones, mainly with ligands based on 3,5-di-*t*-butyl-1,2-benzoquinone and its analogues. These compounds are capable of taking part in one-electron redox processes according to (5.12) [137]:

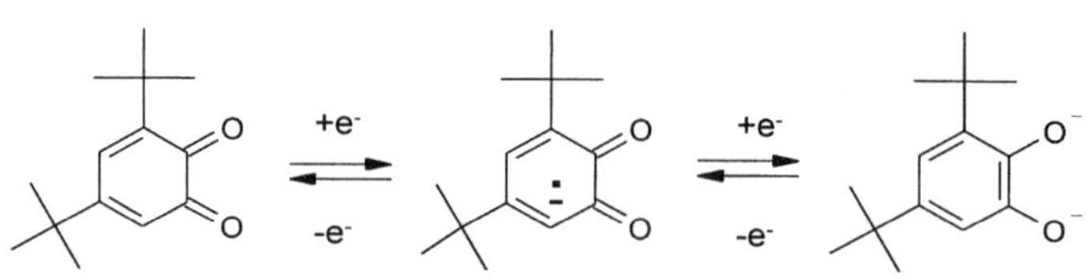

(5.12)

Their capacity to form chelating units having different oxidation states explains a variety of formed coordination compounds [133–136]. A series of novel preparative synthetic methods of *o*-semiquinolate (SQ) and catecholate complexes of transition (Mn, Re, Fe, Ru, Co, Rh, Ir, Ni, Pd, Pt, Cu, Ag, Ti, Hg), nontransition, and rare-earth metals in various oxidation states and ligand environments has been developed (see below). Application of EPR spectroscopy and x-ray singe crystal diffraction has made it possible to demonstrate the influence of electronic and spatial factors on the stability of such complexes [153–155].

In the example of thallium SQ complex [156], complex formation with *o*-quinones as neutral ligands was described, where the inner-molecular electron transition "ligand–ligand" takes place. Highly interesting redox isomery was reported [157,158]. It was discovered, for a series of mono-*o*-semiquinone complexes of rhodium, iridium, and copper, that the compounds can exist in two equilibrium isomers, differing in the place of localization of the nonpaired electron (metal or SQ ligand) (5.13). The necessary conditions for such isomers were formulated and the factors influencing the position of such an equilibrium were studied.

$$(\mathrm{COD})\mathrm{Rh}^{\mathrm{I}}\mathrm{SQ}(\mathrm{AsEt_3}) \rightleftharpoons (\mathrm{COD})\mathrm{Rh}^{\mathrm{II}}(\mathrm{Cat})(\mathrm{AsEt_3}) \tag{5.13}$$

(COD is cyclo-octadiene)

The equilibrium between metal–quinone redox isomers has been found to be extremely sensitive to the properties of nitrogen-donor coligands. The redox isomers, reported in Ref. 159, can exist (5.14) in the cobalt complexes containing semiquinolate (SQ) and catecholate (Cat) ligands derived from 3,5-di-*t*-butyl-1,2-benzoquinone (3,5-DBBQ):

$$\mathrm{Co}^{\mathrm{III}}(\mathrm{bipy})(3{,}5\text{-}\mathrm{DBSQ})(3{,}5\text{-}\mathrm{DBCat}) \rightarrow \mathrm{Co}^{\mathrm{II}}(\mathrm{bipy})(3{,}5\text{-}\mathrm{DBSQ})_2 \tag{5.14}$$

This was extended to include complexes containing various N-donor ligands and complexes prepared with 3,6-DBBQ [160–163]. According to Refs. 164 and 165, equilibrium occurs in separate electron transfer (5.15) and spin transition (5.16) steps:

$$\begin{array}{ccc} \mathrm{Co}^{\mathrm{III}}(\mathrm{N{-}N})(\mathrm{SQ})(\mathrm{Cat}) & \longleftrightarrow & \mathrm{Co}^{\mathrm{II}}(\mathrm{N{-}N})(\mathrm{SQ})_2 \\ (t_{2g})^6(e_g)^0(\pi_{Q1})^2(\pi_{Q2})^1 & & (t_{2g})^6(e_g)^1(\pi_{Q1})^1(\pi_{Q2})^1 \end{array} \tag{5.15}$$

$$\begin{array}{ccc} \mathrm{Co}^{\mathrm{II}}(\mathrm{N{-}N})(\mathrm{SQ})_2 & \longleftrightarrow & \mathrm{Co}^{\mathrm{II}}(\mathrm{N{-}N})(\mathrm{SQ})_2 \\ (t_{2g})^6(e_g)^1(\pi_{Q1})^1(\pi_{Q2})^1 & & (t_{2g})^5(e_g)^2(\pi_{Q1})^1(\pi_{Q2})^1 \end{array} \tag{5.16}$$

As a result, low-spin Co(III) is converted to high-spin Co(II). The authors of Ref. 164a noted that this process might be viewed as a charge transfer-induced spin transition. Valence tautomerism driven by pressure was found for the $[\mathrm{Co(SQ)_2(phen)}]\cdot n\mathrm{C_6H_5CH_3}$ ($n=0$, 1) complex, with pressures within 0.075–0.700 GPa for the toluene solvate and 0.10–2.5 GPa for the nonsolvated complex [164b]. The complexes $\mathrm{Co(Py_2X)(3,6\text{-}DBQ)_2}$ (X = S, Se, Te; 3,6-DBQ is 3,6-di-*t*-butyl-1,2-benzoquinone), reported here, at temperatures below 150 K are all in the $\mathrm{Co^{III}(Py_2X)(3,6\text{-}DBSQ)(3,6\text{-}DBCat)}$ isomeric form with magnetic moments $S=1/2$ due to the radical semiquinone ligand. As the sample temperature is increased, shifts to the high-spin $\mathrm{Co^{II}(Py_2X)(3,6\text{-}DBSQ)_2}$ redox isomer are observed with transition temperatures of 370, 290, and 210 K for the ligands containing S, Se, and Te bridging atoms, respectively [164a]. A similar complex with X = O, reported in the same work

[164a] and obtained from toluene, contains the $Co^{II}(Py_2O)(3,6\text{-DBSQ})_2$ redox isomer with magnetic moment 3.5–4.0μ_B and space group $P2_1/c$. In the authors' opinion, the switching properties of this compound appear to be associated with the planar/folded change in conformation of the Py_2O ligand associated with electron transfer between the metal and quinone ligand (5.17):

(5.17)

$Co^{II}(py_2O)(3,6\text{-DBSQ})_2$ $Co^{III}(py_2O)(3,6\text{-DBSQ})(3,6\text{-DBCat})$

In case of the analogous nickel complex [166], one-electron reduction (5.18) of the complex leads to one-electron oxidation of the metal:

$$Ni^{II}(3,6\text{-DBSQ})_2 + e \rightarrow [Ni^{II}(3,6\text{-DBSQ})(3,6\text{-DBCat})]^- \rightarrow [Ni^{III}(3,6\text{-DBSQ})_2]^- \quad (5.18)$$

For other metal complexes with the same ligands, such a change of metal oxidation number has also been reported: $Mn^{II}(SQ)_2 \rightarrow [Mn^{IV}(Cat)_2]^-$ [167], $V^{III}(SQ)_3 \rightarrow [V^V(Cat)_3]^-$ [168]. In all these cases reduction occurs initially at an SQ ligand [166]. Other examples are cationic porphyrin complexes of nickel which can exist as either $[Ni^{III}(porph)]^+$ or $[Ni^{II}(porph^{\cdot})]^+$ charge-localized redox isomers [169]. The oxidation state of manganese in its semiquinone and catecholate complexes is typically lower, and metal and quinone orbital energies are quite similar [125a]. The redox series of ruthenium complexes can be represented as follows (5.19) [170]:

$$[Ru^{II}(bipy)_2Q]^{3+} \xrightarrow{e^-} [Ru^{II}(bipy)_2Q]^{2+} \xrightarrow{e^-}$$
$$[Ru^{II}(bipy)_2SQ]^{+} \xrightarrow{e^-} [Ru^{II}(bipy)_2Cat] \quad (5.19)$$

It is noted [166] that the similarity in energy between metal and quinone electronic levels is responsible for shifts in charge distribution between metal and ligands. These shifts result from the change in the donor character of the ligand with reduction from SQ to Cat, and the consequential inversion in the order of localized metal and quinone electronic levels.

The change of metal oxidation number in quinone complexes can also be reached by electrochemical methods. For example, the electrochemical oxidation [(5.20), (5.21)] of $[M^{IV}(DBCat)_3]^{2-}$ (M = Mn, Tc, Re) yields products having different oxidation state of the central atom [171,172]:

$$[M^{IV}(DBCat)_3]^{2-} - e^- \rightarrow [M^{III}(DBCat)_2]^- + DBBQ \qquad M = Mn \quad (5.20)$$
$$[M^{IV}(DBCat)_3]^{2-} - e^- \Leftrightarrow [M^{V}(DBCat)_3]^- \qquad M = Tc, Re \quad (5.21)$$

Photochemistry of quinones (1,4-benzoquinone, duroquinone, 2,6-di-*t*-butylbenzoquinone, etc.) radical anions was studied in detail [173a]. According to cyclic voltammetry data in various solvents, the quinones showed typical reversible two-wave voltammograms, corresponding to two successive one-electron transfers [(5.22), (5.23)] to the radical anion then to the dianion:

$$Q + e^- \rightarrow Q^{\cdot -} \quad (5.22)$$

$$Q^{\cdot -} + e^- \rightarrow Q^{2-} \quad (5.23)$$

A molecular modeling program to calculate electron paramagnetic resonance hyperfine couplings in semiquinone anion radicals was offered [173b].

5.2.1.2 Charge Distribution

Charge distribution related to the balance in energy between frontier quinone and metal orbitals is described in Fig. 5.4. In situations where the metal orbital energy is high relative to the quinone π level, the ligand bonds as a reduced catecholate to an oxidized form of the metal. When the metal energy is low, charge is located in the metal-localized levels with ligands coordinated as partially reduced semiquinones [125a].

Periodic trends in charge distribution for transition-metal complexes containing catecholate and semiquinone ligands are discussed in detail [174,175a]. It was noted that "the most straightforward factor influencing charge distribution in the neutral $M(DBQ)_2$ and $M(DBQ)_3$ complexes is the periodic dependence of metal orbital energy" [174]. For a congeneric group of metals, charge distribution may shift from $M^{III}(SQ)_3$ for first-row metal to $M^{VI}(Cat)_3$ for the corresponding third-row metal. This shift appears clearly in the properties of the neutral complexes formed by members of the Cr, Mn, and Fe triads [125a]. Charge distributions for $M^{II}(DBSQ)_2$ (M = Mn, Co, Ni, Cu), $M^{III}(DBSQ)_3$ (M = V, Cr, Fe), $M^{VI}(DBCat)_3$ (M = Mo, W, Tc, Re) [174], and binuclear Co(III) and Cr(III) complexes with a mixed-valent semiquinone–catecholate ligand [175b] are discussed.

For some metal complexes of *o*-quinone mono-oximes, an analysis of the charge distribution within the *o*-quinone mono-oxime ligands through crystallographic data is reported in Ref. 176. It is known [177] that intermediate structures between the form ***940a*** and ***940b*** can exist within these compounds (5.24):

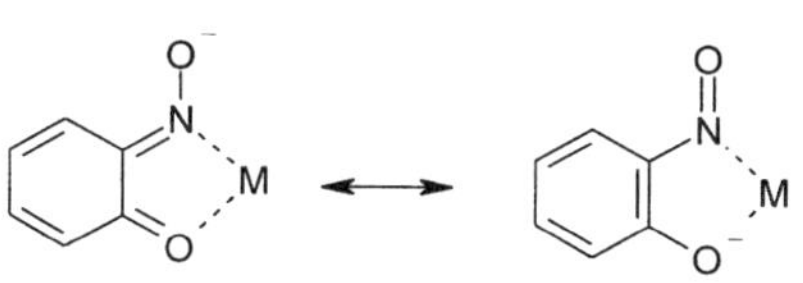

940a *940b* (5.24)

The data obtained [177] indicate that the ligands are always in an intermediate state between these limiting forms; the complexation to a *d*-metal influences the mesomeric equilibrium between these forms, causing a shift to the right.

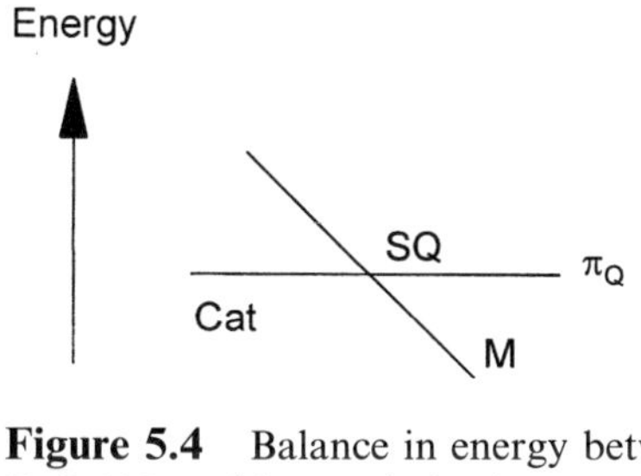

Figure 5.4 Balance in energy between frontier quinone and metal orbitals. (Adapted from Ref. 125a, with permission.)

5.2.1.3 Structural Peculiarities

As mentioned above, metal–quinone complexes frequently have some peculiarities in their structure due to the possibility of the ligands existing in different forms: quinone, semiquinone, and catecholate (5.12). The presence of substituents also affects their structure. Thus, in the reactions between $Ru_3(CO)_{12}$ with 3,6-di-*t*-butyl-1,2-benzoquinone (3,6-DBBQ) ***941*** or 2,4,6,8-tetra-*t*-butylphenoxazin-1-one (L), (L = PhenoxBQ) ***942***, leading to $Ru(CO)_2L_2$, the presence of bulky substituents at ring positions adjacent to the quinone oxygen atoms reduces the tendency for bridging interactions of both types and simplifies product isolation. *t*-Butyl substituents of the iminoquinone 2,4,6,8-tetra-*t*-butylphenoxazin-1-one may similarly block bridging interactions [178], and with the availability of these two ligands, simple quinone and iminoquinone ruthenium complexes are formed from $Ru_3(CO)_{12}$ in relatively simple reactions [179]:

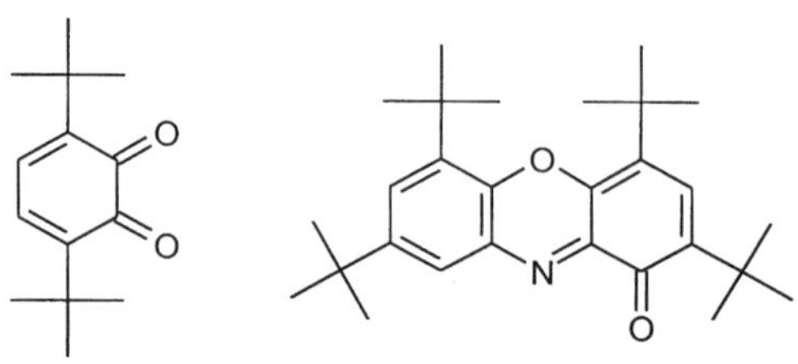

3,6-DBBQ PhenoxBQ

941 *942*

For the unique rhenium complex $Re(CO)_3(C_{14}H_{21}NO)(C_{28}H_{39}NO_2)$, prepared from $Re(CO)_5Br$ or $Re_2(CO)_{10}$, 3,5-di-*t*-butylcatechol (3,5-DBCat), 3,5-di-*t*-butyl-1,2-benzoquinone (3,5-DBBQ), and ammonium hydroxide [180], some structural features in bond lengths can be explained by delocalization of the oxygen anion and the phenoxazinyl radical throughout the chelate ring ***943***:

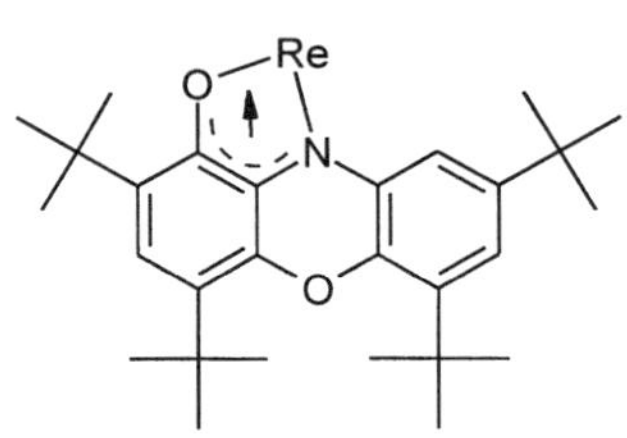

943

It is noted [181] that "mononuclear complexes of divalent rhenium are rare, and coordination of two different radical ligands to a single metal is also unique."

Template syntheses with the same ligand (3,5-di-*t*-butylcatechol) in the presence of ammonia with boron, aluminum, gallium, or strontium chloride, or with calcium or barium acetate produces, under oxidizing conditions, neutral complexes ML (M = BCl), ML_2 (M = Al, Ga, Ca, Ba), and ML_3H [181]. The ligand L [*bis*(3,5-di-*t*-butyl-1-hydroxy-2-phenyl)amine] can be in different oxidation states; the most important being the following ***944*** and ***945***:

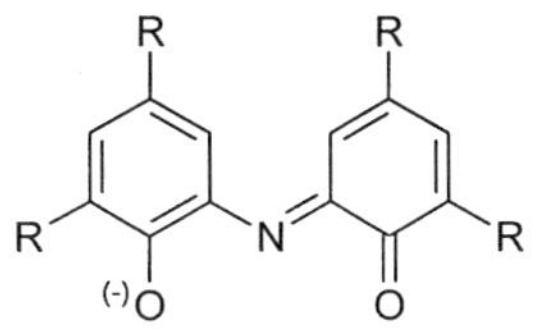

944

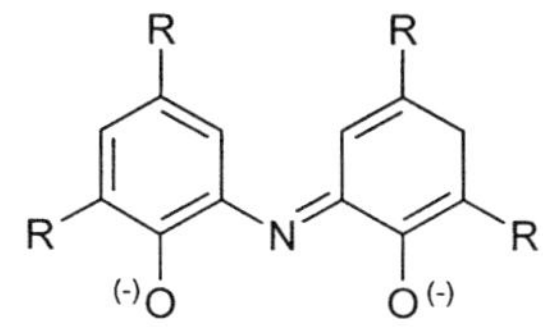

945

The proposed structure for the diamagnetic boron complex is as follows ***946*** [181c]:

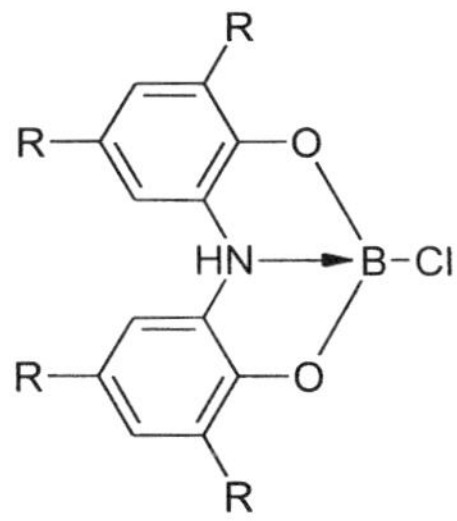

946

On the other hand, it was proposed for the complexes of Al and Ga ***947*** that they are formed by the bonding of two ligands in two different oxidation states:

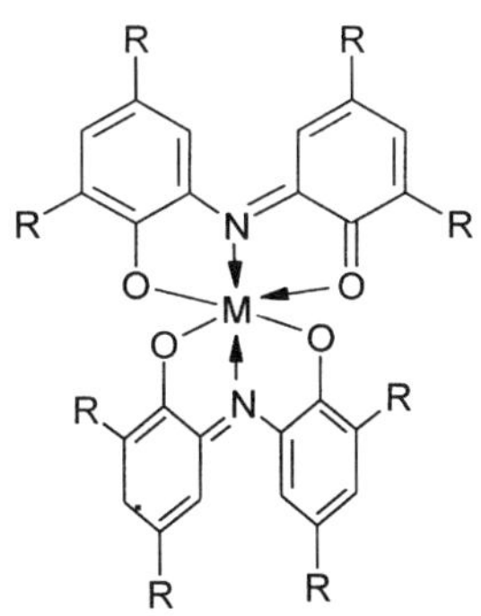

947

A series of Li and Na complexes with quinone crowns {such as [Li(NCS)(5QC-HQDME)] and [Na(NCS)(5QC-HQDME)] (5QC-HQDME = 15,17-dimethyl-16, 18-dimethoxy-3,6,9,12-tetraoxabicyclo[12.3.1]octadeca(1.14,16)-triene)} were obtained and structurally characterized [182]. The redox-active crown-ethers ***948*** used are as follows:

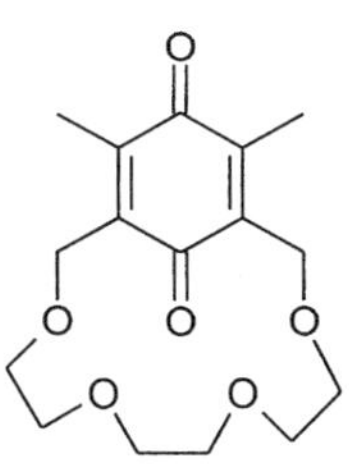

948

EPR studies demonstrated the intermolecular nature of interaction between the cations and the ligands.

Quinone derivatives such as quinone methides (the monomethylene analogues of quinones) have also been extensively studied ([183] and references cited therein) because they possess biological activity, particularly as antitumor agents. The first thermally stable quinone methide having no substituents in the methylene group ("simple quinone methide") was crystallographically characterized [183a]. It was shown that stabilization of the quinone methide can be achieved by complexation to a transition metal center (5.25):

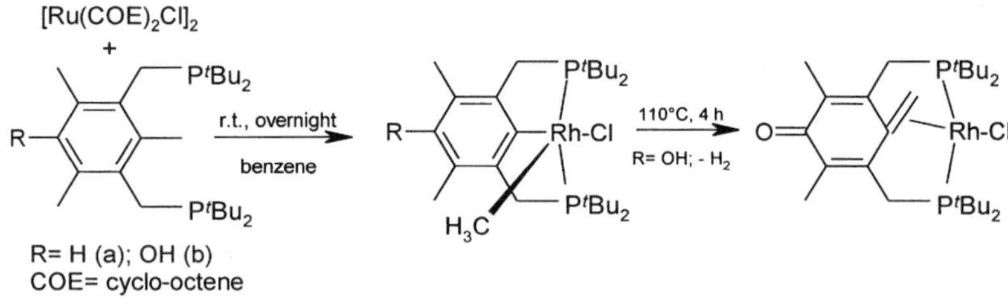

(5.25)

The synthesized complex does not react with air, carbon monoxide, or trimethylphosphine. It is noted [183a] that "the rhodium center is very strongly bound to the quinonoid ligand." Recently, Vaissermann et al. have synthesized and fully characterized (even by x-ray diffraction) the first *o*-quinone methide complex, with Ir as the metal center [183b]. This complex is highly stable at room temperature, a notable difference when it is compared with the simplest known *o*-quinone methide that is unstable above $-100°C$ [183c].

Substituted 1,4-diazabutadiene coligands complexed with the Cu(3,6-DBQ) chelate give temperature and solvent-dependent EPR and electronic spectra that indicate the following equilibrium (5.26) [125a,154,184]:

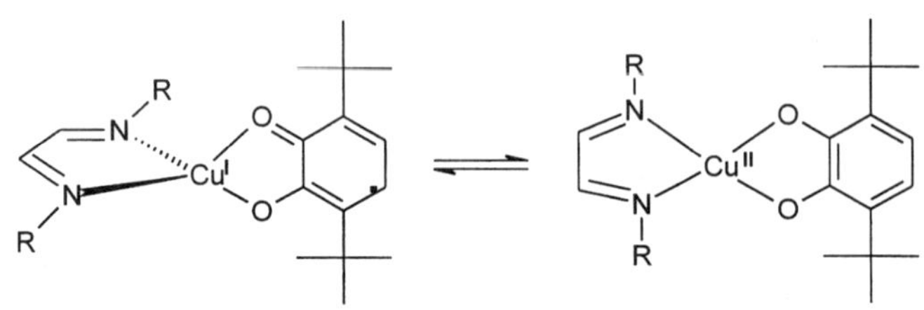

(5.26)

A series of catechol ligands ***949–951*** functionalized with N-donor chelating substituents was developed [185,186]:

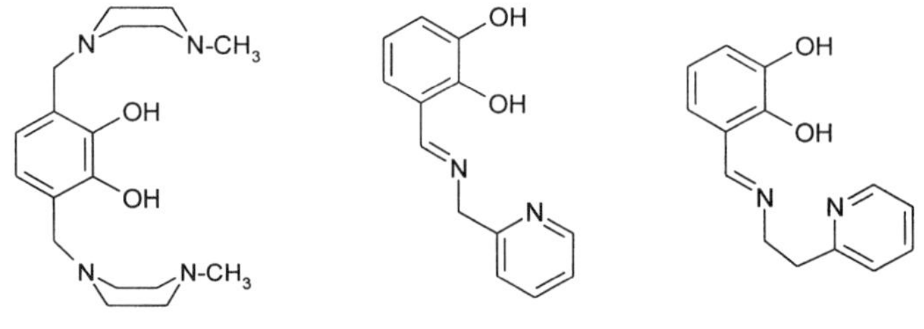

H_2L *949*　H_2L^1 *950*　H_2L^2 *951*

Multimetallic complexes (tetranuclear and pentanuclear with complicated structures) of these ligands which could have unique magnetic properties were prepared and studied. It is noted [125a] that, in the copper complex with H_2L, $Cu_5(OH)_2(L)_2(NO_3)_4$, all magnetic interactions within the complex unit H_2L appear to be antiferromagnetic, as well as in copper complexes with two other ligands, $[Cu(HL^1)(NO_3)]_4$ and $[Cu_2(L_2)(OAc)_2]_2$ [187,188].

Cationotropy (migration of alkali metal cations between equivalent positions in the ionic pairs of anion radicals) takes place because the symmetric position of a cation is not energetically beneficial. It was studied as far back as at the end of the 1960s on the *p*-benzosemiquinone anion radical and its derivatives [189,190] and later using 3,5 and 3,6-di-*t*-butyl-*o*-semiquinone complexes of metals of Groups I [191] and II [192] of the Periodic Table. The Coulombic interaction energy between the metal cation and the negative functional groups of anion radicals (for example, oxygen atoms in semiquinones) is higher in the case of the nonsymmetric position of the cation in relation to the oxygen atoms. So, a cation transferred from one position to another is related with an energetic barrier. The frequency and activation energy of the migration process depend on the cation nature, solvent used, and temperature. An EPR study of quinone complexes of alkali metals from −60°C to +60°C showed the presence of high frequencies only [191]. Studies of asymmetric complexes (5.27) of Group II metals showed [192] a superposition of EPR spectra of two forms, which are different thermodynamically and where a rapid exchange between them is absent.

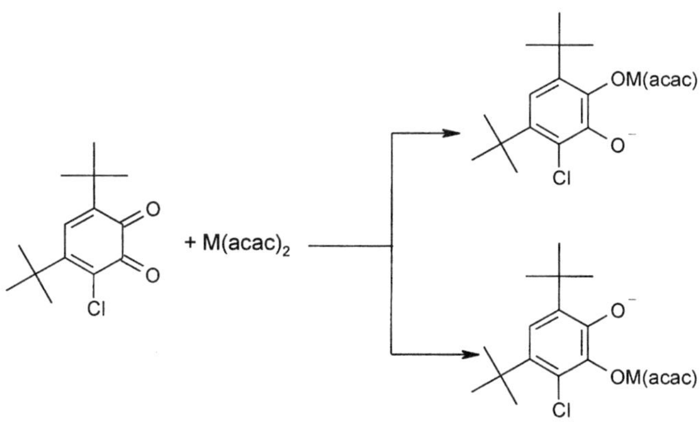

(5.27)

In case of symmetric complexes of Group II metals, their EPR spectra confirmed the following structures (5.28):

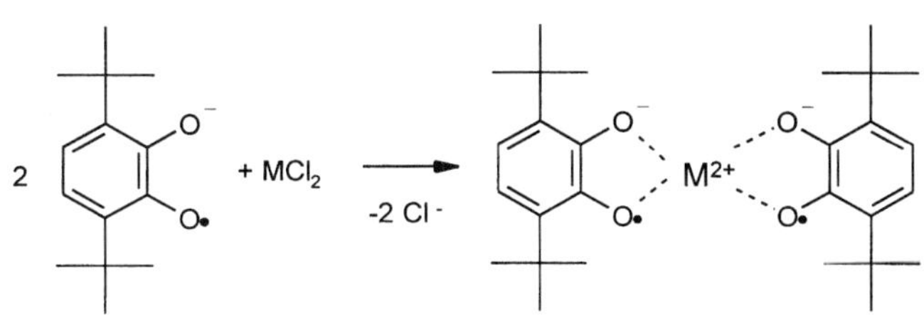

(5.28)

By using EPR data, the parameters of radical pairs (D/E) and the distance between radical centers (l) were calculated. When the metal radius is increased (from Ba to Hg), l is also increased. Radical pairs of formed complexes are tetrahedral with O atoms in the tops and metal ions in the centers. Migration of the cation in the complex containing acetylacetonate groups takes place inside the tetrahedron formed by the oxygen atoms and the acetylacetonate ligands [192].

Thus, metal–quinone complexes have a series of peculiarities, which allows us to put them into a special class of coordination compounds. As will be shown below, they can be synthesized by direct interaction between quinones and elemental metals (and also nonmetals) or from their salts or complexes.

5.2.2 Interaction of Elemental Metals and Nonmetals with *o*-Quinones

In relation to elemental metals, *o*-quinones are extremely reactive ligands [11,12,14,133,192–208], especially those containing *t*-butyl substituents. Complexes of *o*-benzoquinones, *o*-semiquinones, and catechols of this type have been obtained by direct interaction between metal powders and the corresponding ligands and reviewed [14]. It was established that thermal decomposition of such complexes of copper in solution leads to the formation of metallic copper and the initial *o*-quinone.

As shown above, these compounds are radical species and play an important role in coordination chemistry [125a,169,202,203]. In nonaqueous solvents (hexane, dimethoxyethane, toluene, THF, and chloroform), mainly the formation of complexes of types ***952–955*** (R = 3,5- or 3,6-t-Bu_2) has been observed by EPR [14]:

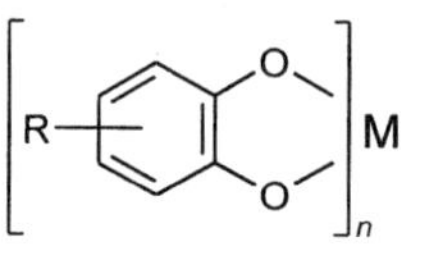

952

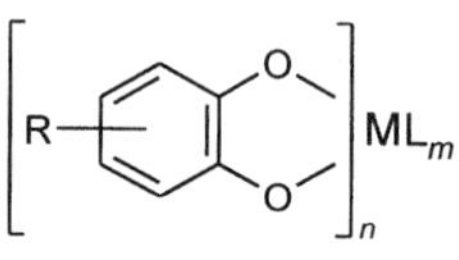

953

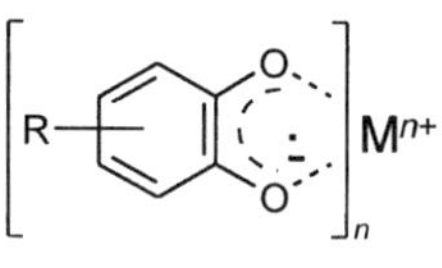

954

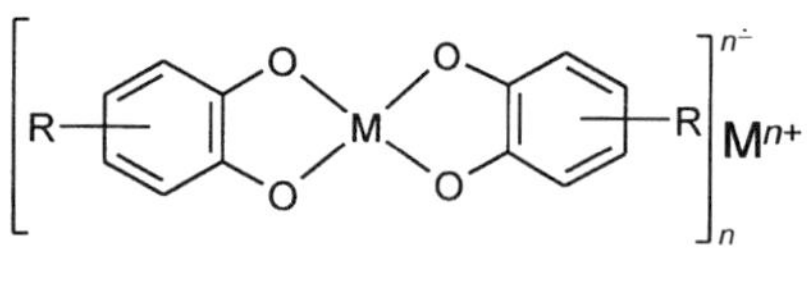

955

The formation of radical *o*-semiquinolates of type ***952*** has been detected for thallium [204], indium [198], zinc [127,196,199], cadmium [196,199,205], magnesium [199], barium [199], aluminum [196], gallium [201], and tin [195]. The adducts ***953*** are characteristic of copper and silver ($L = PPh_3$, $m = 2$, $n = 1$) [11] and also of indium (L = phen, m, $n = 1$) [198]. Data on the radical anion salts of types ***954*** and ***955*** have been presented in reviews [11,12] and other publications [195,205]. Catechol–metal complexes can also be obtained from tetrahalo-*o*-quinones, for example $Sn(X_4C_6O_2)_2$ (X = Cl, Br) [198].

In case of oxidation of mercury by *o*-quinones, no stable Hg–quinone complexes have been observed by direct oxidation of metallic Hg, with or without LiCl addition [209,210]. In the case of LiCl, *o*-quinones are reduced (5.29) by Hg to form Hg_2Cl_2 and Li–semiquinone complex ***956***:

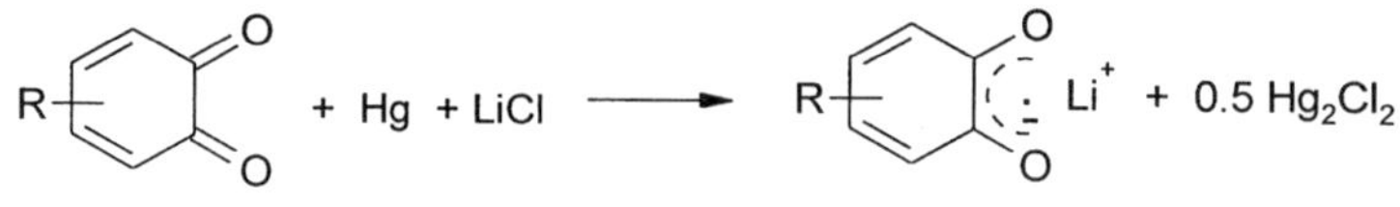

(5.29)

956

A detailed reaction mechanism of this reaction at different LiCl–quinone ratios was reported [209]. Among the small number of examples of Hg–quinone complexes obtained, those containing 3,5-DBCat with HgR and R [R = Et or Ge(*i*-Pr)$_3$] groups bound to different oxygen atoms have been prepared by treating HgR_2 with 3,5-DBBQ [211,212]. Also, the complex (3,6-DBSQ)Hg($B_{10}C_2H_{11}$) was obtained by photolysis of di-(carboran-9-yl)mercury in the presence of 3,6-DBSQ [213].

The data of Ozarowski et al. [199] and Adams et al. [201b] are of great interest for the study of the properties and structures of type ***952*** complexes. Magnesium, barium, zinc, and cadmium complexes of 3,5-di-*t*-butyl-1,2-benzoquinone (L) and their adducts with pyridine, 2,2′-bipyridine, and N,N,N′,N′-tetramethylethylenediamine (L′), having the compositions ML_2 and $ML_2 \cdot nL'$ ($n = 1, 2$), were isolated and characterized in the first of these studies. Complexes ***952*** and their adducts with L′ have been obtained by the reaction of the above elemental metals and the ligands in toluene; detailed EPR spectroscopic studies confirmed their biradical nature and made it possible to establish the conformations of the biradical ligands. Two studies on the triradical gallium complex of 3,5-di-*t*-butyl-1,2-benzoquinone ***952*** ($n = 3$) by Ozarowski et al. and Adams et al. [201] appeared almost simultaneously and presented similar conclusions. This compound was obtained by heating gallium and the ligand in boiling toluene (under an argon atmosphere) and was characterized by x-ray diffraction. The complex has the form of a three-blade propeller with the gallium atom in the center: the length of the C—C bond between the chelating oxygen atoms is 1.439(12) Å, which indicates the semiquinone nature of the ligands. It is striking that the above value differs little from the same bond (1.433 Å) in the chromium analogue, the structure of which had been characterized previously [207]. The gallium complex ***952*** ($n = 3$) exhibits ferromagnetic properties; its magnetic moment is 2.95 B.M. at 320 K and 3.58 B.M. at 9 K; it then again decreases and is 3.26 B.M. at 2 K.

Iminosemiquinone complexes of copper were prepared by treating metallic Cu with 2,4,6,8-tetra-*t*-butylphenoxazin-1-one (PhenoxBQ) ***942***. Reactions carried out with PPh_3 gave $Cu^I(PPh_3)_2$(PhenoxSQ). Reactions carried out in the absence of coligand or in the presence of a nitrogen-donor coligand gave Cu^{II}(PhenoxSQ)$_2$ [214]. Such copper complexes may serve as models for Cu–biopterin complexes found in some metalloenzymes; similarly, PhenoxSQ complexes of iron may resemble partially reduced iron–biopterin species.

The very reactive Rieke cadmium metal was allowed to react under vacuum with benzoquinone in THF. This resulted in the formation of several different paramagnetic species, $\{[Cd(C_6H_4O_2^{\cdot-})(THF)_3]^+, C_6H_4O_2^{\cdot-}\}$ or $\{[Cd(C_6H_4O_2^{\cdot-})(THF)_3]^+ + C_6H_4O_2^{\cdot-}\}$ [215]. The tetrahedral coordination sphere for each species consists of three THF molecules and an anion radical that is asymmetrically or symmetrically coordinated to the metal dication. The original apparatus for the generation of the benzoquinone anion radical via electron transfer from cadmium metal was reported [215].

Activation of elemental metals by *mechanical methods* (Sec. 3.4) in the presence of organic acceptor molecules in solid phase, described in detail [14], allows us also to obtain metal–polyradical complexes [206,216–219]. The formed radical-pair species have some unusual properties [217,218]. Compared with triplet radical pairs generated photochemically with the same donor–acceptor composition, these mechanically induced species appear to be much more stable [216]. As an example of such reactions, the synthesis of polyradical complexes in the course of a reaction of metals of Groups IIB–VB (M) with di-*t*-butyl-*o*-benzoquinones (Q) can be considered (5.30):

$$M^{n+} + NQ^{-\cdot} \rightarrow MQ_n^{\cdot} \tag{5.30}$$

where *n* indicates the valence state of the metal and $Q^{-\cdot}$ the semiquinone anion radical 3,5- or 3,6-di-*t*-butyl-*o*-semiquinone or phenoxazine [216]. Such a reaction occurs when a toluene solution of the quinone is in contact with an amalgam of the metal (Zn, Cd, Al, Ga, In, Sn) [206,219]. For the mechanochemical synthesis the mixed powder of metal and quinone is stirred in an agate or porcelain mortar at room temperature. Both synthetic techniques lead to the same di- and triradical species.

Nonmetals can also form complexes with quinones. Thus, elemental phosphorus reacts with $Cl_4C_6O_2$-*o*, providing a convenient one-pot synthesis (5.31) of the phosphorane [134,220]:

$$0.25P_4 + 2Cl_4C_6O_2\text{-}o + 0.5Br_2 \rightarrow BrP(O_2C_6Cl_4)_2 \tag{5.31}$$

Similarly, Te is oxidized (5.32) by the same quinone [221]:

$$Te + 2Cl_4C_6O_2\text{-}o \rightarrow Te(O_2C_6Cl_4)_2 \tag{5.32}$$

It will be shown below that tellurium–quinone complexes can also be obtained from Ph_2Te_2 [222]. The resulting products in the reactions above are presented in Table 5.9.

EXPERIMENTAL PROCEDURES

Example 1

Synthesis of Copper *bis*(Cyclohexyl-*i*-cyanide)-3,5-di-*t*-butylbenzosemiquinolate-1,2 [223]

Metallic copper (1.28 g, 20 mmol) was added to a solution containing 3,5-di-*t*-butylbenzoquinone-1,2 (4.4 g, 20 mmol) and cyclohexyl-*i*-cyanide (4.5 mL, 50 mmol) in toluene (50 mL). The solution was stirred for some hours at 20°C and became red–brown. A red–brown solid was formed. Solvent was reduced in vacuo at 20°C to 25 mL and the remaining solution was frozen to 0°C. The resulting solid was filtered and dried in vacuo for 2 hr at 20°C. Yield ≈ 100%, m.p. 70°C (with decomposition). Analogous derivatives of other *o*-quinones were prepared similarly.

Example 2

Synthesis of $Cu(PPh_3)_2$(PhenoxSQ) (PhenoxSQ=2,4,6,8-Tetra-*t*-butylphenoxazin-1-one anion) [214]

A mixture of PhenoxBQ (0.421 g, 1.0 mmol), copper powder (0.064 g, 1.0 mmol), and PPh_3 (0.525 g, 2.0 mmol) in 20 mL of acetonitrile was refluxed under argon for 3 hr to give a red–brown solution. The volume of the solution was reduced to give red–brown crystals of the product in 67% yield.

Table 5.9 Synthetic Methods for Obtaining Metal–Quinone Complexes

Initial system	Product	Conditions or observations	Reference
(a) Synthesis from elemental metals and nonmetals Ag, Cu, Hg (1) 3,5-di-*t*-butyl-1,2-benzoquinone (2) 3,6-di-*t*-butyl-1,2-benzoquinone + dimethoxyethane (3) phenanthrenequinone + dimethoxyethane (4) *o*-chloranil	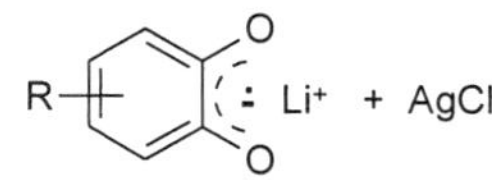		205, 210
In 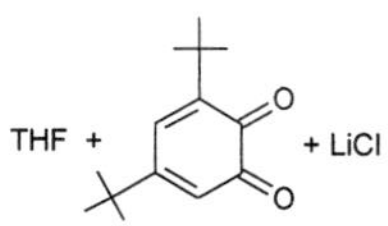	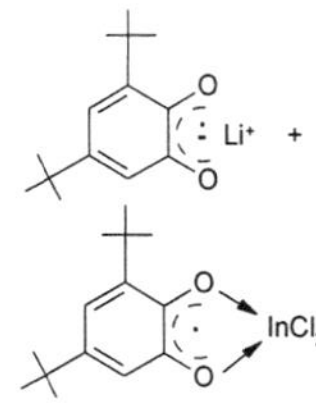 In presence of phen or Et_2O: InL·phen or $InLEt_2O$		197, 210
M = Mg, Ba, Zn, Cd 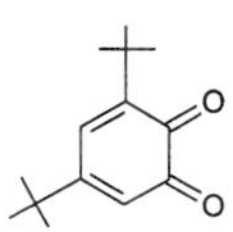 L = in toluene, in the presence of L′ = pyridine, 2,2′-bipyridine, and N,N,N′,N′-tetramethylethylenediamine	ML_2 $ML_2 \cdot nL'(n = 1, 2)$		200

Table 5.9 Continued

Initial system	Product	Conditions or observations	Reference
Cu, Hg 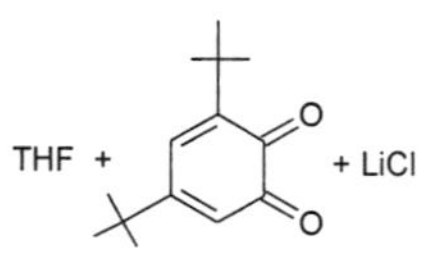	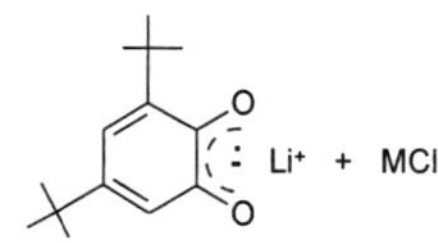		210
Cu 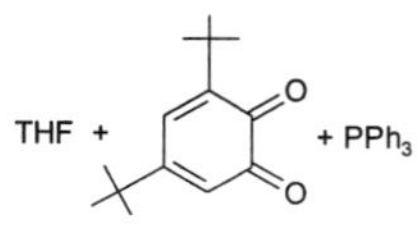	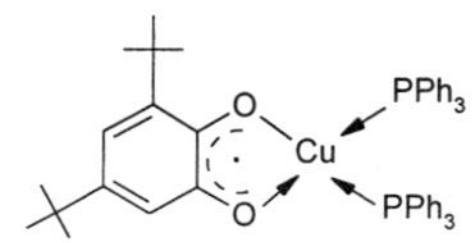		205
Cu, Ag 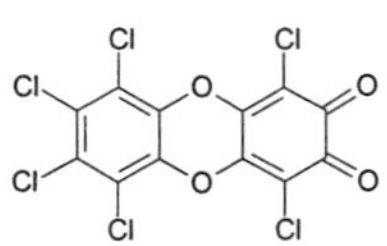	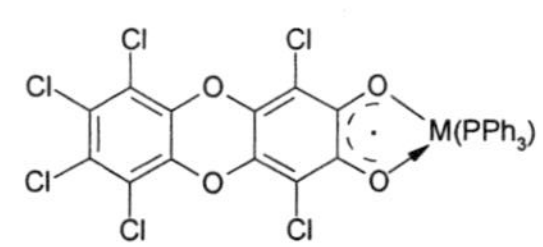		205
Tl 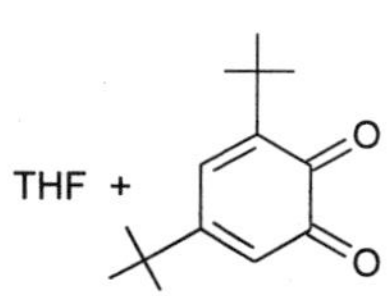	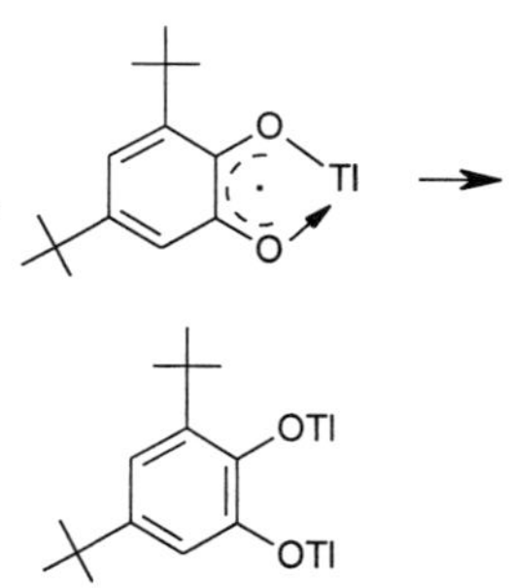		204

Tl 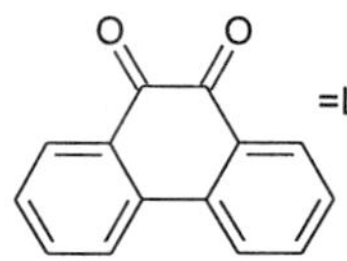 =L + THF	TlL	204
Amalgam Zn, Cd THF + 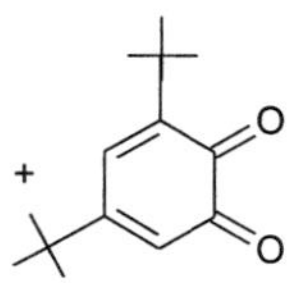	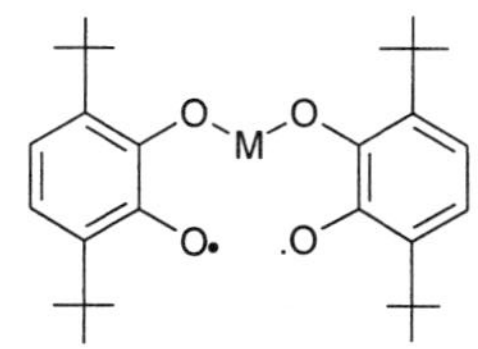	206
Zn, Cd, Al, Ga, In, Sn + 3,5- or 3,6-di-*t*-butyl-*o*-semiquinone or phenoxazine	$MQ_n^{\cdot}$	216
Sn, tetrahalogeno-*o*-benzoquinones $X_4C_6O_2$, toluene, N_2 (X = Cl, Br) in the presence of I_2 and phen	$Sn^{IV}(X_4C_6O_2)$ $SnI_2(X_4C_6O_2) \cdot$ phen	198
Sb, tetrahalogeno-*o*-benzoquinones $X_4C_6O_2$, Et_2O (X = Cl, Br)	$Sb^V(X_4C_6O_2)_{2.5} \cdot Et_2O$ X = Cl, *n* = 1.5; X = Br, *n* = 1	253
M = Ni, Pd, Pt; 1,2-benzoquinone (BQ)	$M(BQ)_2$	125a
Te, RO_2: R = Cl_4C_6, Br_4C_6, 3,5-$Bu^t_2H_2C_6$	$Te(O_2R)_2$	221

Table 5.9 Continued

Initial system	Product	Conditions or observations	Reference
(b) Synthesis from metal carbonyls			
$Co_2(CO)_8$, 3,6-di-*t*-butyl-1,2-benzoquinone (3,6-DBBQ), 2,2′-bis(pyridine) (oxy, thio, seleno, telluro) ether Py_2X, toluene	$Co(Py_2X)(3,6\text{-DBQ})_2$ (X = O, S, Se, Te)	$Co(Py_2O)$(3,6-DBSQ)(3,6-DBCat) were obtained by recrystallization from acetone. $Co(Py_2O)(3,6\text{-DBSQ})_2$ were obtained by recrystallization from toluene. 3,6-DBSQ is a semiquinolate form of 3,6-DBQ, 3,6-DBCat is a catecholate form	164
$Ni(CO)_4$, 3,6-di-*t*-butyl-1,2-benzoquinone (3,6-DBBQ), CH_2Cl_2, hexane	$Ni(3,6\text{-DBSQ})_2$	In case of 3,5-DBSQ use, the strongly paramagnetic at r.t. complex $[Ni(3,5\text{-DBSQ})_2]_4$ is formed [254]	166
$Ru_3(CO)_{12}$, toluene, 3,6-di-*t*-butyl-1,2-benzoquinone (3,6-DBBQ) or 2,4,6,8-tetra-*t*-butylphenoxazin-1-one (L)	$Ru(CO)_2L_2$		179
$Mo(CO)_6$, DBBQ	In oxygen-free atmosphere: $Mo(DBCat)_3$. In the presence of traces of O_2: $[MoO(DBCat)_2]_2$		255, 256
$M_2(CO)_{10}$ (M = Mn, Re), benzoquinone (BQ)	$M(SQ)(CO)_4$		257
$Re(CO)_5Br$ or $Re_2(CO)_{10}$, 3,5-di-*t*-butylcatechol (3,5-DBCat), 3,5-di-*t*-butyl-1,2-benzoquinone (3,5-DBBQ), NH_3, EtOH	$Re(CO)_3(C_{14}H_{21}NO)(C_{28}H_{39}NO_2)$	Reaction takes place under photolytic conditions at 0°C under N_2. The product is a complex of divalent rhenium with three carbonyl ligands, one anionic monodentate 3,5-di-*t*-butyl-2-iminophenolate ligand, and one anionic chelating 1-hydroxy-2,4,6,8-tetra-*t*-butylphenoxazinyl ligand	180

(c) Synthesis from metal salts			
In_2I_4, 3,5-di-*t*-butyl-1,2-benzoquinone (TBQ), toluene, 4-picoline (ratio In_2I_4: 2TBQ)	(TBSQ) $InI(pic)_2 \cdot C_7H_8$	When ratio In_2I_4:4TBQ is used (without 4-picoline), (TBSQ) InI_2 is isolated	225
M_2Cl_6(M = In, La, Lu, Y, Sn), 3,5-di-*t*-butyl-1,2-benzoquinone	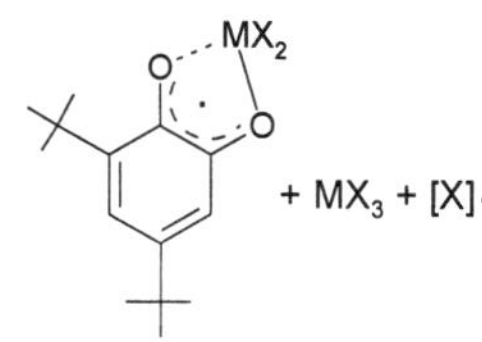		258–260
InX_3 (X = Cl, Br, I), THF, $Na^+TBSQ^{\cdot -}$ (TBSQ = 3,5-di-*t*-butyl-*o*-benzosemiquinolate anion)	Solutions of the diradicals $InX(TBSQ)_2$	Addition of pyridine or γ-picoline (L) produces $In(TBCat)XL_n$ {3,5-di-*t*-butylcatecholate–indium(III) halide adducts}	226
$TiCl_3$, VCl_3, $GeCl_4$, $SnCl_2$ in EtOH, 3,5-di-*t*-butylcatechol	$M(Cat\text{-}N\text{-}SQ)_2$	The products can be formulated as $M^{II}(Cat\text{-}N\text{-}SQ)_2$, M^{III}(Cat-N-BQ)(Cat-N-SQ) or $M^{IV}(Cat\text{-}N\text{-}SQ)_2$, where Cat-N-SQ is the dianion of the radical ligand 3,5-di-*t*-butyl-1,2-semiquinolato 1(2-hydroxy-3,5-di-*t*-butyl-phenyl)imine and Cat-N-BQ is the corresponding quinone monoanion	240
3,5-DBBQ, Hg_2Cl_2 LiCl	Li(DBSQ)		209
$RuCl_3 \cdot 3H_2O$, 3,5-di-*t*-butylcatechol, KOH, CH_3OH	$Ru(DBQ)_3$	Its osmium analogue was prepared from OsO_4 and 3,5-di-*t*-butylcatechol [126,127]	174

Table 5.9 Continued

Initial system	Product	Conditions or observations	Reference
5-Etaqo$H_2^+Cl^-$ (5-EtaqoH = 5-ethylamino-4-methyl-1,2-benzoquinone), methanol:water, K_2CO_3 or KOH	K(5-Etaquo)(5-EtaqoH) · $2H_2O$		233
(d) Synthesis from metal complexes			
Sn_2Ph_6, 3,5-di-*t*-butyl-1,2-benzoquinone (TBQ) or 9,10-phenanthrenequinone (PQ), *n*-hexane or CH_2Cl_2	$Ph_3Sn(TBSQ^{\cdot})$ or $Ph_3Sn(PSQ^{\cdot})$		224
Ph_2Te_2, $X_4C_6O_2$-*o* (X = Cl, Br), toluene	$[(X_4C_6O_2)TeC_6H_5]_2O$	$[(X_4C_6O_2)TeC_6H_5]_2O \cdot 2THF$ is formed by recrystallization of $[(X_4C_6O_2)TeC_6H_5]_2O$ from THF	222
$CoCp_2$, $Ni(3,6\text{-DBSQ})_2$, CH_2Cl_2, hexane	$(CoCp_2)[Ni(3,6\text{-DBCat})_2]$		166
$M_2(DBA)_2$ (M = Pd, Pt; DBA = dibenzylidene-acetone), 3,5-di-*t*-butyl-1,2-benzoquinone, CH_2Cl_2	$Pd(DBSQ)_2$, $Pd_2[Pd(DBSQ)_2]_2 \cdot CH_2Cl_2$, $Pt(DBSQ)_2$		229
$M(acac)_2$ (M is metal of Group II of the Periodic Table), 3,6-di-*t*-butyl-*o*-quinone, toluene or EtOH	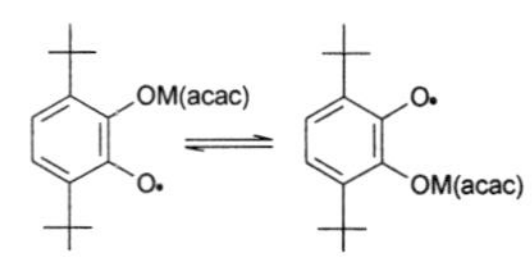		261
Cp_2ZrCl_2, 3,6-DBBQ	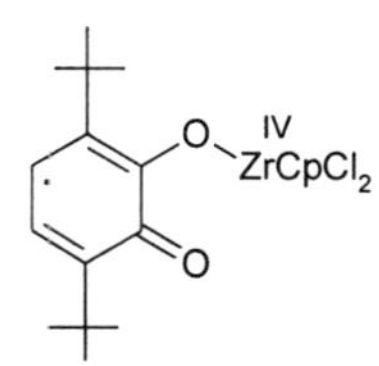	hν	262
$[Re^VO(PPh_3)_2Cl_3]$, H_2Cat/airEt_2NH	$[Re^{VII}O_2(Cat)_2]^-$	Using $H_2Cat/N_2/NEt_3$, the anion $[Re^VO(Cat)_2]^-$ is formed	125a

5.2.3 Synthesis from Metal Salts and Complexes

A series of organotin(IV) *o*-quinone complexes was prepared and characterized by Tuck et al. [224]. The primary process in the reaction of hexaphenylditin with 3,5-di-*t*-butyl-1,2-benzoquinone, phenantrene-9,10-quinone, 1,2-naphthoquinone, and tetrahalogeno-*o*-quinones was shown to involve attack by the quinone at a phenyl ligand. The intermediate thus formed decomposes to yield $Ph_3Sn(SQ^{\cdot})$, where $S(Q^{\cdot -})$ is the corresponding semiquinolate. Rearrangement of these species in solution gives rise to biradicals, while intramolecular electron transfer may lead to the formation and precipitation of $Ph_3Sn(Cat)$, where Cat^{2-} is the corresponding substituted catecholate. According to Ref. 224, the proposed reaction sequence is as follows [(5.33)–(5.36)]:

$$Q + Sn_2Ph_6 \rightarrow SQ^{\cdot}\,(Ph^{\cdot})Ph_2SnSnPh_3 \tag{5.33}$$

$$\downarrow$$

$$(SO^{\cdot})SnPh_3 + {}^{\cdot}SnPh_3 \tag{5.34}$$

followed by

$$2\,{}^{\cdot}SnPh_3 \rightarrow Sn_2Ph_6 \tag{5.35}$$

or

$$Q + {}^{\cdot}SnPh_3 \rightarrow Ph_3Sn(SQ^{\cdot}) \tag{5.36}$$

The comparable sequence for Ph_4Sn, which was also used as a precursor of tin–quinone complexes and gives related products, is (5.37):

$$Q + Ph_4Sn \rightarrow SQ^{\cdot}(Ph^{\cdot})SnPh_3 \rightarrow Ph_3Sn(SQ^{\cdot}) + Ph^{\cdot} \tag{5.37}$$

Similarly, the reaction of In_2I_4 (having the structure $In^+[InI_4]^-$ in the solid state) with substituted *o*-benzoquinones proceeds via attack of the quinone on the solvated In^+ cation which is present in solutions of In_2I_4 in toluene [225]. The formed final products have compositions $(SQ)InI_2$ or $(SQ)InI_2(pic)_2$ (pic is 4-picoline), depending on the particular *o*-quinone and on the reaction conditions. It is proposed [225] that $(SQ)InI_2$ complexes with phenantrene-9,10-quinone, tetrabromo-*o*-benzoquinone, and 1,2-naphthoquinone as ligands exist in the form of halide bridged dimers ***957***:

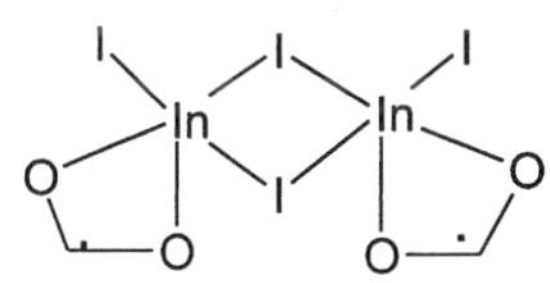

957

According to the data reported in Ref. 226, the reaction of indium(III) halides (X = Cl, Br, I) with 2 mol of $Na^+TBSQ^{\cdot -}$ (TBSQ = 3,5-di-*t*-butyl-*o*-benzosemiquinolate anion) yields (5.38) solutions of diradicals $InX(TBSQ)_2$:

$$InX_3 + 2NaTBSQ^{\cdot} \rightarrow In(TBSQ^{\cdot})_2X + 2NAx \tag{5.38}$$

Addition of pyridine or γ-picoline (L) to the complex produces $In(TBC)XL_n$ [3,5-di-*t*-butylcatecholate–indium(III) halide adducts]. The product recrystallized from DMF having the composition $[In(TBC)I(pic)_2]^{\cdot}2DMF$ belongs to the monoclinic crystal system, space group $P2_1/n$. The most interesting feature of this molecule is its dimeric nature, involving an In_2O_2 four-member ring, with In—O distances of 2.216(6) and 2.1468(5) Å. The detailed reaction mechanisms of formation of indium–quinone complexes, offered by the authors of Ref. 226, include two steps [(5.39) and (5.40)], in particular, an initial loss of a neutral ligand in toluene solution of In(TBC)XL_n and further internal electron transfer which "allows the indium(III) catecholate species to function as an indium(III) semiquinone":

$$\text{In(TBC)XL}_n \Leftrightarrow \text{In(TBC)X} + n\text{L} \tag{5.39}$$

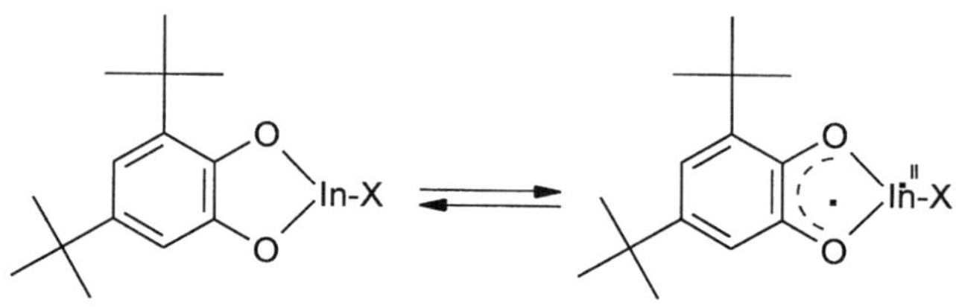

In(TBSQ)X (5.40)

A series of *o*-semiquinone (SQ, ***958–960***) complexes of Fe(III) and Cr(III) of general formula $M(SQ)_3$ is described [227]:

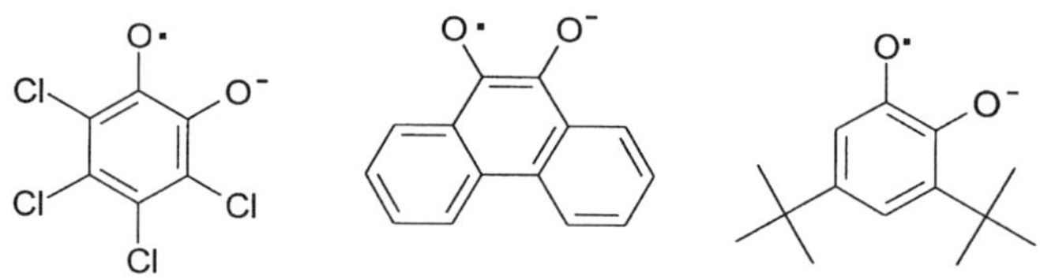

o-Cl_4SQ^- *958* phenSQ⁻ *959* 3,5-DBSQ⁻ *960*

The complexes obtained were characterized by x-ray diffraction and magnetic measurements. Iron-57 Mössbauer spectroscopy data for $Fe(SQ)_3$ complexes are consistent with the iron ion being described as high-spin iron(III) [227,228].

Reactions carried out with $M_2(DBA)_3$ (M = Pd, Pt; DBA = dibenzylideneacetone) and 3,5-di-*t*-butyl-1,2-benzquinone gave as major products the $M(DBSQ)_2$ complexes [229]. In the case of palladium, an additional product $Pd_2[Pd(DBSQ)_2]_2$ was detected, whose molecular structure consists of two planar *cis*-$Pd(DBSQ)_2$ units bridged by two Pd atoms. The Pd atoms are "sandwiched" between semiquinone rings of adjacent $Pd(DBSQ)_2$ units with three Pd—C lengths and an allyl structure ***961*** for the semiquinone rings:

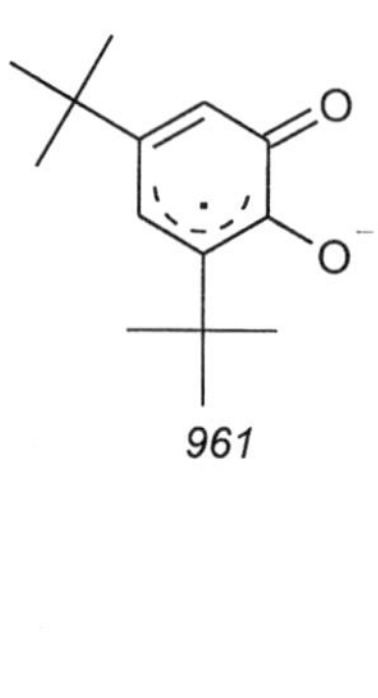

961

It was noted [229] that magnetic exchange between radical semiquinone ligands results in the near diamagnetism of $M(DBSQ)_2$ (M = Pd, Pt).

Similar chain molybdenum compounds $[Mo_2(O_2CCF_3)_4 \cdot L]_n$ (L = 9,10-antraquinone or 2,6-dimethylbenzoquinone) were prepared and characterized by x-ray structure analyses, ^{13}C-NMR spectra, and cyclic voltammetry [230,231]. In case of use of 2,6-di-*t*-butyl-*p*-benzoquinone (2,6-*t*-Bu-BQ), a *bis-p*-quinone adduct $[Mo_2(O_2CCF_3)_4(2,6\text{-}t\text{-Bu-BQ})_2]$ was isolated [231].

Stable paramagnetic binuclear complexes $\{[Ru(bipy)_2]_2(\mu\text{-L})\}^{3+}$ with L being N,O; N′,O′-coordinating 4,7-phenantroline-5,6-semidione and P,O; P′,O′-coordinating 2,5-*bis*(diphenylphosphino)-*p*-benzosemiquinone were reported [232]. *o*-Semiquinone phdo$^{\cdot -}$ complex is formed as follows (5.41):

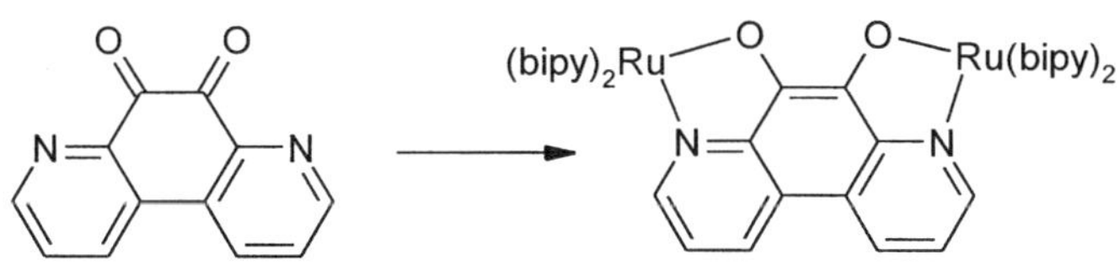

(5.41)

Spectroscopic and EPR data showed that these binuclear semiquinone complexes are situated at the borderline between anion radical complexes and metal-centered mixed-valent dimers.

Although most metal–quinone compounds are those of transition metals, alkali metal–quinone complexes have also been reported. Additionally to Li-3, 5-di-*t*-butyl-1,2-benzoquinone species, the potassium complex K(5-Etaqo)(5-EtaqoH)$\cdot 2H_2O$ (5-EtaqoH = 5-ethylamino-4-methyl-1,2-benzoquinone), obtained from potassium carbonate as starting salt, was prepared [233]. The potassium atom in this complex is coordinated to one neutral and one ionic ligand via the oxime nitrogens and the quinonoid carbonyl oxygen forming a five-membered chelate ring. The metal is therefore coordinated to seven donor atoms and is in a distorted pentagonal bipyramidal environment.

Additionally to complexes of a variety of metals and substituted *o*-benzoquinones [14], the unusual tellurium(IV) derivatives $[(X_4C_6O_2)TeC_6H_5]_2O$ (X = Cl, Br) were obtained from tetrahalogeno-*o*-benzoquinones and diphenyl ditelluride [222] [see also the reaction (5.32) with use of elemental tellurium]. The reaction pathway includes three steps: the reduction of *o*-quinone to catecholate, the oxidation of tellurium from +1 to +4 with the retention of the Te—C_6H_5 linkage, and the cleavage of the Te—Te bond of $(C_6H_5)_2Te_2$ [(5.42) and (5.43)] [222]:

$$Ph_2Te_2 \rightarrow PhTe^{\cdot} \qquad (5.42)$$

$$PhTe^{\cdot} \rightarrow (SQ^{\cdot})TePh \qquad (5.43)$$

Here the intramolecular electron transfer process (5.44) takes place; it is typical for Q/SQ/Cat ligand systems:

(5.44)

p-Quinones are also capable of producing complexes of various types with transition metals, acting as a bridge-type ligand. Thus, the reaction (5.45) of substituted *p*-

quinones with $[Rh(cod)Cl]_2$ and $AgClO_4$ in acetone at 25°C gives complex ***962*** with high yields (69–73%) [234]:

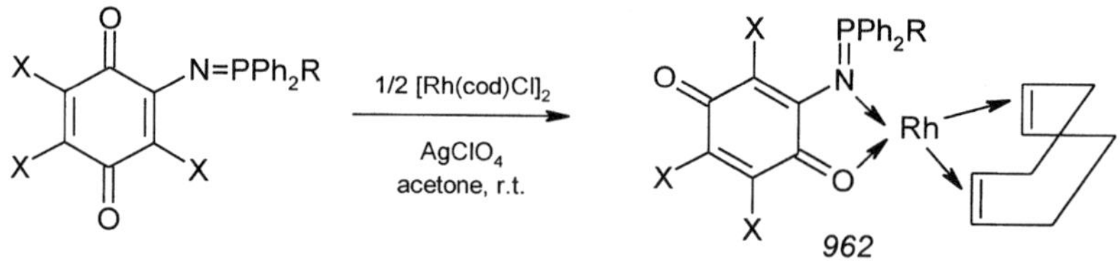

(5.45)

X= F, R= Ph; X= Cl, R= Me

The reaction of rhodium(II) pivalate dimer, $Rh_2(O_2CCMe_3)_4$, with 1,4-benzoquinone (BQ) in hexane gave a chain complex, $[Rh_2(O_2CCMe_3)_4 \cdot BQ]_n$, where the rhodium(II) pivalate dimers are connected by the bifunctional ligation of the *p*-quinone through its carbonyl oxygen or C=C double bond, i.e., the chain structure consists of two kinds of dimer units ***963*** and ***964*** [235–237]:

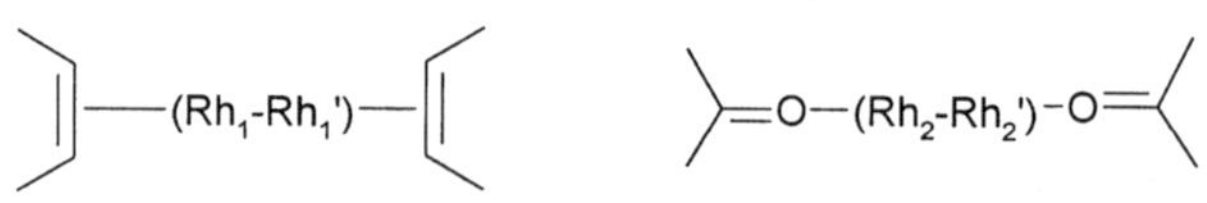

963 *964*

Rhodium complexes with the derivatives of *p*-quinone (1,4-naphthoquinone and 2,3-dimethyl-1,4-benzoquinone) were also reported [236].

The molecule Fe^{II}(salen) [salen = N,N′-ethylene-*bis*(salicylideniminate)] reacts with *p*-quinones (1,4-benzoquinone and its derivatives, 1,4-naphthoquinone, and biantrone) to give a compound $[Fe(salen)]_2Q$, where Q is the quinone moiety [238]. The compound consists of high-spin ferric ions bridged by the dianion of a hydroquinone. It was suggested that the metal ions are oxidized to M(III) and the bridge Q is the dianion of the hydroquinone. The ^{57}Fe Mössbauer data and magnetic measurements (μ_{eff} = 5.1–5.7 B.M. at 285.5 K) indicated that in all compounds obtained there is only one type of iron site and that it is most likely a high-spin Fe(III) ion. In case of biantrone complex, the following structure ***965*** was suggested for this compound:

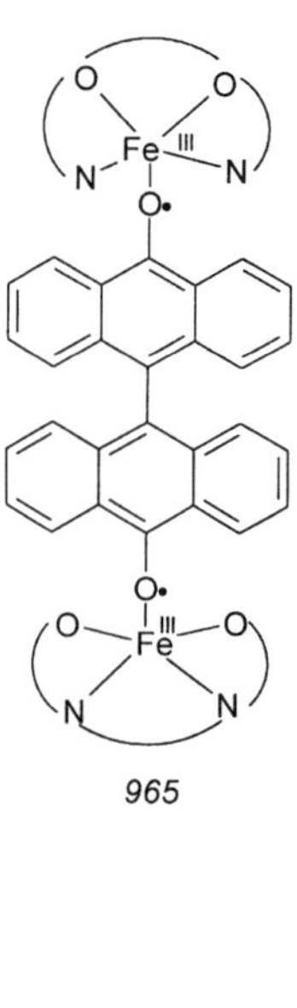

965

Among other reported *p*-quinone adducts with N,N′-ethylene-*bis*(salicylidenimi-nate), there is a cobalt one. *o*-Quinone complexes with the same ligands have the composition 1:1 [Fe(salen)Q] (Q = 9,10-phenanthrenequinone and 1,2-naphthoquinone) and $[Co(salen)(py)]_2Q$ [239].

The resulting products in the reactions above are presented in Table 5.9.

EXPERIMENTAL PROCEDURES

Example 1

Synthesis of ML_2 (M=Ti, V, Ge, Sn; L=3,5-di-*t*-butylcatechol) [240]

Concentrated aqueous ammonia (3 mL) was added to a solution of 3,5-di-*t*-butylcatechol (4 mmol) and the appropriate metal salt ($TiCl_3$, VCl_3, $GeCl_4$, $SnCl_2$, 1 mmol) in 50 mL of EtOH. The resulting mixtures were stirred in air at room temperature for 4 hr, during which time solid crystalline products separated. They were filtered, washed with EtOH, and then recrystallized from CH_2Cl_2/EtOH solutions. No yields provided.

Example 2

Synthesis of K(5-Etaqo)(5-EtaqoH) · $2H_2O$ (5-EtaqoH=5-ethylamino-4-methyl-1,2-benzoquinone) [233]

To a solution of 5-Etaqo$H_2^+Cl^-$ (5.0 g, 23 mmol) in methanol:water (4:1, 100 mL) was added potassium carbonate (9.5 g, 69 mmol). The mixture was stirred (24 hr) and then filtered to afford an orange solid. The latter was washed with water (3 × 25 mL), recrystallized from water:methanol, and dried in vacuo to afford the product (4.5 g, 90%). M.p. 180–181°C.

Example 3

Synthesis of $Co(Py_2O)(3,6\text{-}DBQ)_2$ [164]

$Co_2(CO)_8$ (86 mg, 0.25 mmol) and Py_2O (86 mg, 0.50 mmol) were combined in 30 mL of toluene under an atmosphere of Ar. The mixture was stirred for 5 min, and 3,6-DBBQ (3,6-di-*t*-butyl-1,2-benzoquinone, 220 mg, 1.00 mmol) dissolved in 30 mL of toluene was added. The solution was stirred for 2 hr at room temperature. Evaporation of the solvent gave the dark green microcrystalline product in 70% yield (235 mg).

Example 4

Synthesis of $(CoCp_2)[Ni(3,6\text{-}DBCat)_2]$ [166]

Cobaltocene (0.016 g, 0.085 mmol) was dissolved in 24 mL of hexane and transferred to a 10 mL dichloromethane solution containing $Ni(3,6\text{-}DBSQ)_2$ (0.043 g, 0.086 mmol). As the solutions mixed, 0.05 g of $(CoCp_2)[Ni(3,6\text{-}DBCat)_2]$ separated as a gray–green microcrystalline precipitate.

Example 5

Synthesis of $Ph_3Sn(BrSQ^\cdot)$ [224]

A solution of Sn_2Ph_6 (1.40 g, 2 mmol) and tetrabromo-*o*-quinone (BrQ) (1.70 g, 4 mmol) in dichloromethane (20 mL) was refluxed for 5 hr. The resultant deep brown solution was EPR active; biradical signals were also detected. Addition of *n*-hexane followed by cooling gave a red precipitate of $Ph_3Sn(BrSQ^{\cdot})$.

Example 6

Synthesis of $(TBSQ)In(pic)_2 \cdot C_7H_8$ (TBQ is 3,5-di-*t*-butyl-*o*-quinone) [225]

A solution of In_2I_4 (0.74 g, 1 mmol) in toluene (10 mL) was added to TBQ (0.44 g, 2 mmol) in the same solvent (10 mL). The resultant brown solution was stirred for 3 hr, the solvent partially removed, and excess 4-picoline added, at which point the solution became green. The air-sensitive crystals precipitated when this solution was cooled overnight were collected, washed with toluene, and dried in vacuo. Yield 0.78 g (53%).

Example 7

Synthesis of $Ru(CO)_2(3,6\text{-}DBSQ)_2$ [179]

Toluene (50 mL) was added to a mixture of $Ru_3(CO)_{12}$ (0.100 g, 0.15 mmol) and 3,6-DBBQ (0.310 g, 1.41 mmol), and the resulting mixture was refluxed under N_2. The mixture became red–violet after 30 min. After 5 hr the reaction mixture was cooled, solvent was removed under reduced pressure, and the residue was extracted with hexane. Hexane-soluble products were separated in a silica gel column using hexane as the eluent. The major product, $Ru(CO)_2(3,6\text{-}DBSQ)_2$ (60 yield), appeared first as a red–violet band; a purple band of $Ru(3,6\text{-}DBSQ)_3$ (10% yield) appeared next as a minor product. Further extraction with benzene produced a green band of unreacted 3,6-DBBQ. Crystals of $Ru(CO)_2(3,6\text{-}DBSQ)_2$ were grown from a dichloromethane/2-propanol mixture.

Example 8

Synthesis of Carbonyl(tetrachloro-*o*-catecholato)(bipyridyl)-di-μ-(di-*p*-tolyl-triazenido-N^1N^3)dirhodium $[Rh_2(CO)(o\text{-}O_2C_6Cl_4)(bipy)(\mu\text{-}RNNNR)_2]$ [241a]

To a suspension of $[Rh_2(CO)(o\text{-}O_2C_6Cl_4)(bipy)(\mu\text{-}RNNNR)_2][PF_6]$ (0.52 g, 0.42 mmol) in toluene (60 mL) was added $(Fe(\eta\text{-}C_5H_5)_2]$ (0.40 g, 2.20 mmol). After 1 hr the brown solution was filtered and concentrated to low volume in vacuo to ca. 10 mL. Slow addition of *n*-hexane (50 mL) gave a green–brown solid, which was washed with *n*-hexane. Yield 0.14 g (31%).

Example 9

Synthesis of $Li_2(SQ^{Me}—SQ^{Me})$ [241b]

4-(2-Methyl-3,4-dihydroxy-5-*t*-butylphenyl)-3-methyl-6-*t*-butyl-1,2-benzoquinone ($H_2Cat^{Me}—BQ^{Me}$) (356 mg, 1 mmol) was dissolved in 10 mL of methanol and slowly added to an aqueous solution of LiOH (48 mg, 2 mmol). The solvent was evaporated and the dark blue powder was dried in vacuo. $Li_2(SQ^{Me}—SQ^{Me})$ was obtained in 80% yield (293 mg).

5.2.4 Applications

Quinone functionalities appear as components in organic switches, and the coupled redox chemistry of quinones with transition metals may provide the basis for an organotransition metal switch [164]. A system that may exhibit light-induced switching was studied in the example of the quionone-tethered form of $Ru(bipy)_3^{2+}$ [242], but the charge-separated state that results from the $Ru(II) \rightarrow Q$ electron transfer is short-lived [164,242].

A mild triple catalytic system consisting of $Pd(OAc)_2$, hydroquinone, and a transition metal macrocycle (for example, iron phthalocyanine) was reported [243]. The catalytic effect is carried out by the interaction of Pd(II) with the substrate and the acquisition of two electrons, which are further transferred to the benzoquinone that is reduced to hydroquinone. The hydroquinone is then reorganized to benzoquinone by the O_2/metal macrocycle system. The following types of transformations were carried out in mild conditions using the developed system: 1,4-oxidation of conjugated dienes, oxidation of terminal olefins to methyl ketones, and allylic oxidation.

Vanadium (II)/catechol systems can be useful in fixation of atmospheric nitrogen, reducing it to ammonia [244–246]. A detailed EPR study concluded that the active catalyst is a trinuclear V–Cat species ***966*** [125a,246,247]:

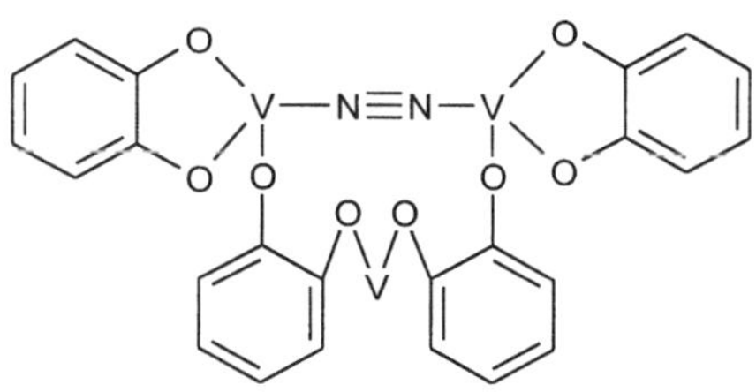

966

A simple and effective chemical method was developed for quantitatively reducing quinones, based on their reaction with metallic zinc and zinc ions [248]. Comparison of this method with conventional electrochemical reduction [249–252] revealed the chemical method to be considerably superior. A reduction reaction of vitamin K_1 and other quinones in the presence of Zn^0 and Zn^{2+} eliminates the need to apply large negative potentials and may also be performed in the absence of any applied electrochemical potential. Some quinones used, such as UQ-10, menadione, and vitamin K, of the menaquinone series (MKs 4–10) could all be reduced to their corresponding hydroquinones in these conditions.

Tc–catecholate complexes (Sec. 5.3.10) could have potential medicinal organ-imaging applications. Some of these complexes were synthesized with this purpose, in particular containing $[TcO(Cat)_2]^-$ [128] and $[TcO(Cl_4Cat)_2]^-$ [129]. If the last complex is treated with N,N-diphenylhydrazine, the dimeric anion $[Tc(NNPh_2)(Cl_4Cat)_2]_2^-$, containing a Tc—Tc bond, is formed. Other Tc–quinone complexes can be obtained by β decay of $^{99}MoO_4^{2-}$, which further reacts with catechol producing $Tc^{VI}(DBCat)_3$ [171].

Direct evidence for the formation of radical *o*-quinone (and sometimes *p*-quinone) complexes was established in the studies quoted above. Various synthetic techniques starting from elemental metals, nonmetals, metal salts, and complexes have been developed for obtaining these coordination compounds. The peculiarities of their thin structure and physical–chemical properties were investigated. The obtained products have practical applications, in particular for medical purposes. Quinone-based metal complexes have a potential applicability as cocatalysts in a wide range of reactions involving electron exchange between substrate and catalysts. Further studies in this field and on mechanisms of electron mobility between the metal center and the *o*-quinone ligands are still necessary to understand the vast and complex redox chemistry of these compounds.

5.3 COMPLEXES OF RADIOACTIVE ELEMENTS

Among the natural and artificial radioactive elements (Tc, Pm, Po, Fr, Ra, Ac, and actinides), coordination and organometallic compounds of only technetium and the actinide series (An) are well represented at the present time. The interest in their metal complexes has been motivated by the extended use of Tc, available in kilogram amounts, for medical and technical purposes, meanwhile actinides are important on their own for the nuclear industry. A lot of original papers, reviews, and chapters of some books are dedicated to Tc and An complexes [263–281]. In the present section, dedicated to the coordination and organometallic chemistry of the actinides and Tc, we intend to present the synthetic techniques for these compounds according to their ligand nature.

5.3.1 General Concepts on Actinide and Technetium Complexes

Elements with atomic numbers ranging from 90 to 103, the actinides, are members of a transition series in which the first member is actinum (atomic number 89). They are analogues to lanthanides and occupy the same part of the Periodic Table at the next period. Only four of them have been found in nature; the others are manmade elements produced by neutron irradiation or heavy-ion bombardment. All of them are radioactive [282]. Technetium (element 43), although not part of the actinides series, possesses two radioactive isotopes with long half-lives: ^{99}Tc (2.12×10^5s, has the practical use) and ^{98}Tc (1.5×10^6 years, a rhenium analogue) [283].

Electronic states of actinide atoms and ions are significantly different from those of lanthanides. In both series the successive filling of the f level proceeds up to the f^{14} configuration, but in the actinide series the filling only formally starts with Th, which has no f electrons and is the electronic analogue of Hf [282]. In contrast to lanthanides, the actinides display a wide collection of oxidation states. As An^{3+} ions they are analogues of the related Ln^{3+} ions, but as An^{4+} they resemble both Hf(IV) and Ce(IV) compounds. Actinides form various ions An^{m+} ($m = 2$–4) and AnO_2^{m+} ($m = 1$, 2) containing only f electrons. The shielding by f electrons causes the contraction of the An^{3+} ions and the magnitude of the actinide contraction along the series to be parallel to that of the lanthanide contraction. Differences in correlation between the energy of 5f and 6d, as well as 4f and 5d, levels lead to noticeable differences in the magnetic properties and electronic spectra of Ln^{m+} and An^{m+} ions [282]. Technetium ($4s^2 4p^6 4d^5 5s^2$ or $4s^2 4p^6 4d^6 5s^1$) has oxidation states from +1 to +7, however, those from +4 to +7 are the most stable [283].

Spin-orbital coupling (J) for the An^{3+} ions is very strong (2000–4000 cm^{-1}) and larger than that for the Ln^{3+} ions (ca. 1000 cm^{-1}). In contrast to lanthanides, J is comparable with the ligand field splitting and is no longer a good quantum number. The proximity of the energy of the 5f and 6d orbitals, and the population of thermally accessible excited levels, lead to the expression for effective magnetic moment $\mu_e = g[J(J+1)]^{1/2}$ being appropriate for lanthanides, but not for actinides [282,284].

Actinide organometallic complexes are compounds containing an actinide–carbon π-bond, an actinide–carbon σ-bond, or a combination of both. Actinide organometallic complexes are known for all of the early actinide elements (An) from thorium through californium [268,284b]. However, the majority of the reported data is on Th and U organometallic chemistry due to the extremely long half-lives of commercially available ^{232}Th (in the form of $ThCl_4$) and ^{238}U (as UCl_4) (1.41×10^{10} and 4.468×10^{9} years, respectively). The first well-characterized actinide organometallic compound, Cp_3UCl, was isolated by Reynolds and Wilkinson in 1956 [285].

Actinides have large metal and ionic radii and, therefore, large coordination numbers (*CN*) of up to 14 [286]. Table 5.10 contains examples of lanthanide compounds exhibiting different metal coordination number and oxidation states.

The maximum *CN* is found in the polymeric uranium borohydrides, while for bulkier ligands the *CN* decreases to 5 or even 4 as in $U(NPh_2)_4$ [282]. Like the lanthanides, the spherical An^{3+} and An^{4+} prefer high coordination numbers (usually 8 and 9), often forming isomorphous complexes with no electronically preferred coordination geometry. This type of coordination geometry is determined by the repulsion of ligands (steric effects) or packing effects [287]. For example, a square antiprism in thiocyanate complexes $Cs_4An(SCN)_8$ and $U(SCN)_4(Ph_3PO)_4$ is transformed into a dodecahedron in $Th(SCN)_4[(Me_2N)_2CO]_4$ or into a cube in $[NEt_4]_4An(SCN)_8$ [282].

The bond nature between π-donor ligands and actinide centers is discussed [263,288]. The authors emphasize the absence of a simple answer to the bond nature in organoactinide complexes. An almost exclusively ionic bonding was described for interactions between Cp^- or COT and U^{4+} [289], as well as a substantial ring metal covalence in uranocene with significant electron transfer from ligand to metal [263,290,291]. For $U(COT)_2$, it was concluded [263,292] that "the uranium 6*d* orbitals play the primary role in covalent bonding between the metal and the ligand, while the 5*f* orbitals have a secondary role."

In contrast to lanthanides, the actinides have a variety of oxidation states in aqueous solution. The stable oxidation states go from +3 for Ac to +6 for U and Np and then successively decrease to +3 for Am and succeeding elements except No (+2). The maximum and stable oxidation states coincide for Ac, Th, Pa, U, Md, and Lr. Stable states are +7 for Np and Pu, +6 for Am, +4 for Cm, Bk, Cf, Es, and Fm, and +3 for No. The unstable (except for No and Md) oxidation state +2 is known for nearly all actinides in aqueous solution [282]. The An^{2+}, An^{3+}, An^{4+}, AnO^{2+}, and AnO_2^{2+} hydrated ions are known, which act as Brönsted acids [293] (5.46):

$$An^{n+} + H_2O \rightarrow An(OH)^{(n-1)+} + H^+ \tag{5.46}$$

The An^{4+} cations are characteristic for actinides from Th through Cf (U^{4+} is readily oxidized) and in the case of Th are the only one existing in solution. Their acidity decreases in the order $Pa^{4+} \gg U^{4+} > Pu^{4+} > Np^{4+} > Th^{4+}$ [282]. The monoatomic ions exist only in very dilute solutions and tend to form polynuclear species when concentration is increased (5.47):

$$mAn^{4+} + nH_2O \rightleftarrows An_m(OH)_n^{(4m-n)+} + nH^+ \tag{5.47}$$

The acidity of the An^{n+} ions depends on the charge and radius of the central atom, so the An^{4+} and AnO_2^{2+} ions are much stronger acids than An^{3+} and AnO_2^+, respec-

Table 5.10 Coordination Numbers and Coordination Geometry of Some Actinide Complexes

Oxidation state	Coordination number	Coordination geometry	Examples of compounds
+7	6	Octahedron	$Li_5(AnO_6)$ (An = Np, Pu)
+6	6	Octahedron	$[NEt_4][PaOCl_5]$, AnF_6 (An = U, Np, Pu), *trans*-$UO_2Cl_2(OPPh_3)_2$
	7	Pentagonal bipyramid	$[UO_2Cl_2(acac)_2]\cdot Hacac$, $PuO_2(C_2O_4)\cdot 3H_2O$, $K_3UO_2F_5$
	8	Hexagonal bipyramid	$Cs_2[AnO_2(MeCO_2)_3]$ (An = Np, Pu, Am), $(UO_2)(NO_3)_2(H_2O)_2$
+5	6	Octahedron	$CsAnF_6$ (An = U, Np, Pu)
	7	Pentagonal bipyramid	$PaCl_5$
	8	Cube	$Na_3[AnF_8]$ (An = Pa, U, Np)
	9	Tricapped trigonal prism	$M_2[PaF_7]$ (M = NH_4, K, Rb, Cs)
+4	4	Tetrahedron	$U(OAr)_4\cdot[N(SiMe_3)_2]_3\cdot THF$
	5	Trigonal bipyramid	$U_2(NEt_2)_8\cdot[Li(THF)_4]\cdot[U(OAr)_5]$
	6	Octahedron	Na_2PuCl_6, *cis*-$UCl_4\cdot 2Ph_3PO$, *trans*-$UBr_4\cdot 2Ph_3PO$, $UCl_4\cdot 2HMPA$
		Trigonal prism	$U_4(dmed)_3$
	7	Pentagonal bipyramid	UBr_4, K_3UF_7
+4	7	Capped octahedron	$[UCl(tmpo)_6]Cl_3$
	8	Dodecahedron	$Np(HCOO)_4$, $Th(tta)_4$, $An(S_2CNEt_2)_4$ (An = Th, U, Np, Pu)
		Square antiprism	$K_7Th_6F_3$, $An(acac)_4$ (An = Th, U)
		Cube	$[NEt_4]_4[An(NCS)_8]$ (An = Th, Pa, U, Np)
		Bicapped trigonal prism	$Th(acac)_4$
		Bicapped octahedron	$[UCl_2(Me_2SO)_6][UCl_6]$
	9	Tricapped trigonal prism	$[NH_4]_3[ThF_7]$, $LiUF_5$, $Th(tta)_4\cdot topo$
		Capped square antiprism	Li_3ThF_7, $Th(tfa)_4\cdot 2H_2O$, $[C(NH_2)_3]_5[Th(CO_3)_3F_3]$
	10	Bicapped square antiprism	$U(C_2H_3O_2)_4\cdot 2H_2O$, $An(MeCO_2)_4$ (An = Th, U), $Na_6[Th(CO_3)_5]\cdot 2H_2O$
	11	Irregular	$An(NO_3)_4\cdot 5H_2O$ (An = Th, Pu)
	12	Icosahedron	$[PPh_4][Th(NO_3)_5(OPMe_3)_2]$, $An(BH_4)_4$ (An = Np, Pu)
	14	Bicapped hexagonal antiprism	$[An(BH_4)_4]_n$ (An = Th, U)
+3	5	Trigonal bipyramid	$[K(THF)_2]_2[U(NHAr)_5]\cdot THF$
	6	Octahedron	$[UCl_6]^{3-}$
	8	Dodecahedron	$UCl_3\cdot 3DMSO$
		Bicapped trigonal prism	$AnBr_3$ (An = Pu to Bk), AnI_3 (An = Pa to Pu)
	9	Tricapped trigonal prism	$NaPuF_4$, $AnCl_3$ (An = U to Cm), $Am(sal)_3\cdot H_2O$
+2	6	Cage structure	Th_6Br_{12}
0	8	Cube	$U(bipy)_4$

Source: Ref. 282, with permission.

tively [282]. The redox behavior of the actinides is complicated by their high radioactivity, leading, in particular, to formation of H_2O_2 in aqueous solutions.

Chloride complexation studies of the actinide ions An^{3+}, An^{4+}, AnO_2^+, and AnO_2^{2+} (An = U, Np, Pu) were reported in several comprehensive reviews [293–295]. More recent investigations on aqua and chloro complexes of UO_2^{2+}, NpO^{2+}, Np^{4+}, Pu^{3+}, etc., by x-ray absorption fine structure spectroscopy (XAFS) were reported [296,297]. In particular, it was established for U(IV) and Th(IV) aqua ions and fluoride complexes that both M(IV) aqua ions are 10-coordinate with M—O bond distances for U(IV) and Th(IV) of 2.42 ± 0.01 Å and 2.45 ± 0.01 Å, respectively [297]. Physical and chemical studies of uranium aqueous complexes are reported [298,299a]. A series of articles is dedicated to specific sequestering agents for the actinides [299b–e].

Mössbauer spectroscopy is a very useful tool to deduce the oxidation state and symmetry of the ligand environment. The gamma resonance effect is observed for ^{232}Th, ^{231}Pa, ^{238}Pu, ^{243}Am, and especially for ^{237}Np with a ^{237}U source. The isomer shifts for Np(VII) compounds are the largest (up to −70 mm/sec) and decrease to +30 mm/sec for Np(III) [300].

5.3.2 Actinide Alkoxides and Other O-Containing Complexes and Salts

Actinide alkoxides are common for uranium and trans-uranium elements; their chemistry is covered in a recent review [300]. Homoleptic alkoxides $An(OR)_n$ are known for $n = 3$ (U, Pu), 4 (Tu, U, Np, Pu), 5 (Pa, U), and 6 (U). They are widely represented by U and Th compounds and a few complexes have been obtained for Pa and Pu. For uranium, alkoxides with oxidation states +3, +4, +5, and +6 are known, as well as mixed-valence complexes such as $[U(OPh)_3(THF)]_2[UO_2(THF)_2]_2(\mu\text{-}OPh)_4(\mu_3\text{-}O)_2$ [301]. The U(IV) compounds are similar to those for Th(IV).

Actinide alkoxides are mostly oligomeric, such as $U_3O(O\text{-}t\text{-}Bu)_{10}$; derivatives of bulky alcohols and 2,6-disubstituted phenols are monomeric [263], for example $U(OCH\text{-}t\text{-}Bu_2)_4$ [302], $U(OEt)_5$ [300], $U(OMe)_6$, $U_2(OEt)_{10}$, $U(OAr)_4$, and $U(OAr)_3$, where Ar = 2,6-$Bu^tC_6H_3$ [300,303,304]. The influence of electronic factors on the structure and stability of uranium tri-*t*-butylmethoxide compounds is shown [305a]. Methoxyuranium(VI) fluorides $[UF_{6-n}(OCH_3)_n]$ ($n = 0$–5) have been studied using relativistic density functional theory. Applying the B3LYP hybrid functional and an effective core potential on uranium, equilibrium geometries have been calculated for these molecules [305b].

The monomeric alkoxides can react with each other [306] (5.48):

$$U(OEt)_4 + U(OEt)_6 \rightarrow U_2(OEt)_{10} \quad \text{(in hexane)} \tag{5.48}$$

There are also anionic alkoxide complexes such as $(Bu_4^tN)[U_2(OBu^t)_9]$ and $[Li(THF)_4][U(OAr)_5]$ [281]. Treatment of $[U(cot)(BH_4)_2]$ (cot = η-C_8H_8) with ROH (R = Et, *i*-Pr, *t*-Bu) leads to the formation of the alkoxide derivatives $[U(cot)(BH_4)(OR)]$ [307]. There is a monomer–dimer equilibrium in toluene solutions for $Th(OCH\text{-}i\text{-}Pr_2)_4$, but only the dimer, having a tri-

Table 5.11 Preparation of Actinide Alkoxides and Related O-Containing Complexes

Metal or its compound	Reaction system	Products	Conditions or observations	Reference
Synthesis from elemental metal				
U anode	Hbac, O_2 Hbac, N_2	$UO_2(bac)_2(Hbac)_{0.5}$ $UO_2(bac)_2$	Direct electrochemical synthesis	14 and references cited therein
U anode	Hacac, N_2 Hacac, O_2	$U(acac)_4$ $UO_2(acac)_2 \cdot Hacac$	Direct electrochemical synthesis	14 and references cited therein
Synthesis from metal salts				
UO_2Cl_2	$[R'_2NH_2](R'_2NCOS)$, HOR, R, R′ = Et, *n*-Pr	$[R'_2NH_2][UO_2(OSCNR'_2)_2(OR)]$		347a
$[UO_2Cl_2(THF)_2]_2$	KO(*t*-Bu), THF	$U[OCH_2C(CH_3)_3]_6$ 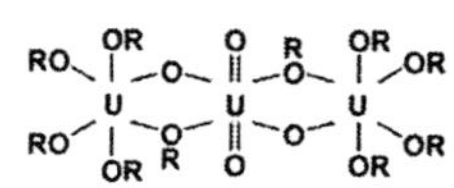 R = ${}^tBu$		347b
$ThCl_4$	NaO-*i*-Pr, HO-*i*-Pr	Th(O-*i*-Pr)-$_4$(HO-*i*-Pr)$_x$		281, 300, 348
$PaCl_5$	NaOEt, EtOH	$Pa(OEt)_5$		300, 349
$NpCl_4$	LiOR, ROH (R = Me, Et)	$Np(OR)_4$		300, 350a
$UO_2(C_2H_3O_2)_2 \cdot 2H_2O$	MX or $M(C_2H_3O_2)$ or $MClO_4$) (where A = Li, Na, K; X = Cl, Br) and crown-ethers in HCl or HBr aqueous solutions	Sandwich-type compounds $[K(18\text{-crown-}6)]_2[UO_2Cl_4]$, $[K(18\text{-crown-}6)]_2[UO_2Br_4]$, $[Na(15\text{-crown-}5)]_2[UO_2Cl_4]$, $[Na(15\text{-crown-}5)]_2[UO_2Br_4]$, $[Li(12\text{-crown-}4)]_2[UO_2Cl_4]$, and $[Li(12\text{-crown-}4)]_2[UO_2Br_4]$		350b

$UO_2(CH_3COO)_2$	L = Hacac, HDBM or HBTF	UO_2L_2		337–339, 346
Synthesis from halide adducts				
$UBr_4(CH_3CN)_4$	KOAr	$Br_2U(OAr)(THF)(4THF)$		351
$ThBr_4(THF)_4$	2,6-$Bu^tC_6H_3OK$	$ThBr_2(OAr)_2(THF)_2$	The product can be alkylated to give $Th(OAr)_2(CH_2SiMe_3)_2$	281, 352a
$UI_3(THF)_4$	$K[Zr_2\{O(i\text{-}Pr)\}_9]$	$[Zr_2\{O(i\text{-}Pr)\}_9]UI_2(THF)$	The product reacts with $K_2C_8H_8$ giving $[Zr_2\{O(i\text{-}Pr)\}_9]U(C_8H_8)$	352b
$[C_6H_5N]_2PuCl_6]$	NH_3, C_6H_6, 2-propanol	$Pu(O\text{-}i\text{-}Pr)_4$ and $Pu(O\text{-}i\text{-}Pr)_4(py)$	Recrystallization from hot 2-propanol provides $Pu(O\text{-}i\text{-}Pr)_4(HO\text{-}i\text{-}Pr)$	353
Synthesis from β-diketonates or alkoxides				
$U[OCH(CMe_3)_2]_4$	MeLi	$MeLi \cdot U[OCH(CMe_3)_2]_4$ 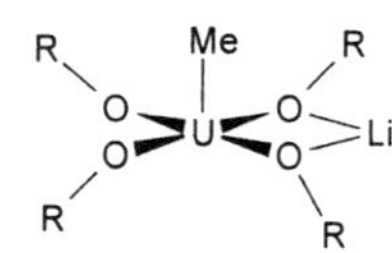		354
$Th(acac)_4$	Electrochemical reduction	$[Th^{III}(acac)_4]^-$	The product loses $acac^-$ to give $Th(acac)_3$. There are very few Th^{III} complexes	355
Synthesis from π-complexes				
$(MeCp)_3U(THF)$	Alcohols or thiols	$(MeCp)_3UR$(R = OMe, OPr-*i*, OPh, SPr-*i*)		356, 357
Cp_3UH, Cp = C_5Me_5, $C_5H_4Bu^t$, $C_5H_4SiMe_3$	EtOH	Cp_3UOEt		358

gonal bipyramidal arrangement, can be crystallized [292]. Actinide alcoxides are extremely easily hydrolyzed, for example [282] (5.49):

$$3U_2(O\text{-}t\text{-}Bu)_8(HO\text{-}t\text{-}Bu) + 2H_2O \rightarrow 2U_3O(O\text{-}t\text{-}Bu)_{10} + 7HO\text{-}t\text{-}Bu \quad \text{(in toluene)} \tag{5.49}$$

Oligomerization of homoleptic *f*-metal alkoxide species frequently takes places due to their large ionic radii, allowing high coordination numbers around the metal center [300,308,309]. As a result, dinculear ***967*** or tetranuclear ***968***, ***969*** and higher oligomers can be formed [308]. In the $[U\{O(2,6\text{-}i\text{-}Pr_2C_6H_3)\}_3]_2$ complex, dimerization occurs due to an unusual π-arene bridge [310a].

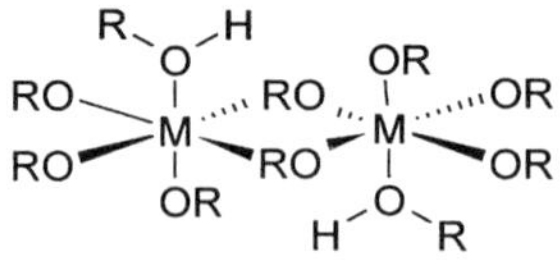

967

(M= Ce, R= *i*-Pr)

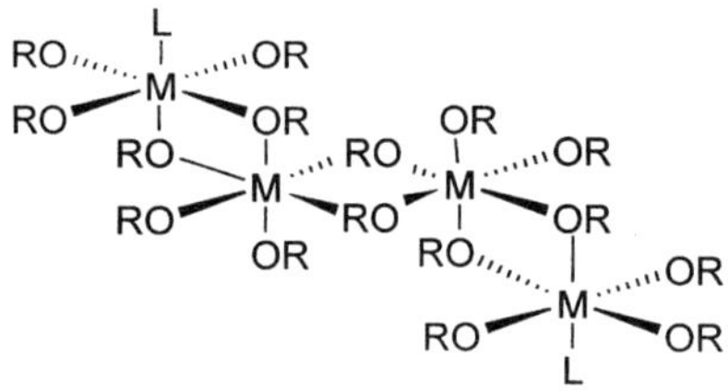

968

(M= Th, R= *i*-Pr, L= py)

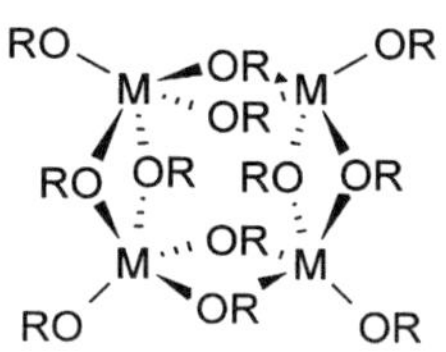

969

(M= Nd, R= CH_2-*t*-Bu)

As an example of actinide dimeric complexes, the series of compounds $[M(O\text{-}2,6\text{-}i\text{-}Pr_2C_6H_3)_3]_2$ ***970*** (M = La, Nd, Sm, Er, U), sterically demanded by *i*-propyl substituents, were isolated [308,310a]; they are held both in the solid state and in solution by π-arene interactions:

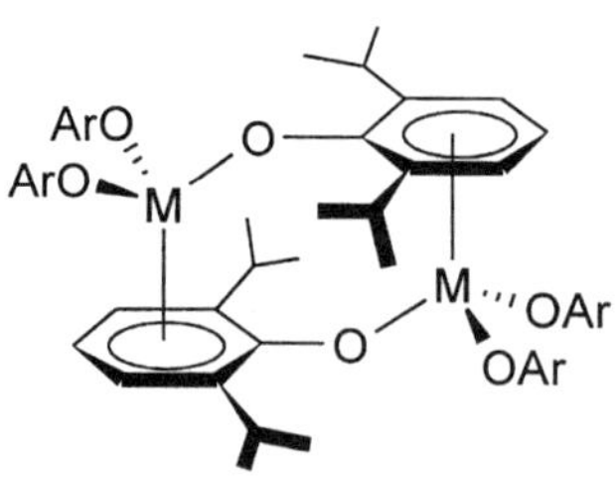

970

(M = La, Nd, Sm, Er, U)

A "nonstandard" synthesis of the first uranium cluster containing an isopolyoxometalate core was reported recently [310b]. Reduction of $[Cp'UCl_2]$ ($Cp' = 1,2,4\text{-}t\text{-}Bu_3C_5H_2$) with two equivalents of KC_8 in THF, followed by an addition of two equivalents of pyridine N-oxide, was conducted in an attempt to produce the organometallic dioxo species $[Cp'UO_2]$. However, the cluster compound $[Cp'(bipy)_2][U_6O_{13}]$ was isolated instead as the main product (54% yield) from this reaction.

Some actinide alkoxide complexes can exist in equilibrium, for example, the homoleptic complex $Th_2(OCH\text{–}t\text{-}Pr_2)_8$, obtained from a metallacyclic precursor after treatment with HOCH-*i*-Pr_2 (5.50):

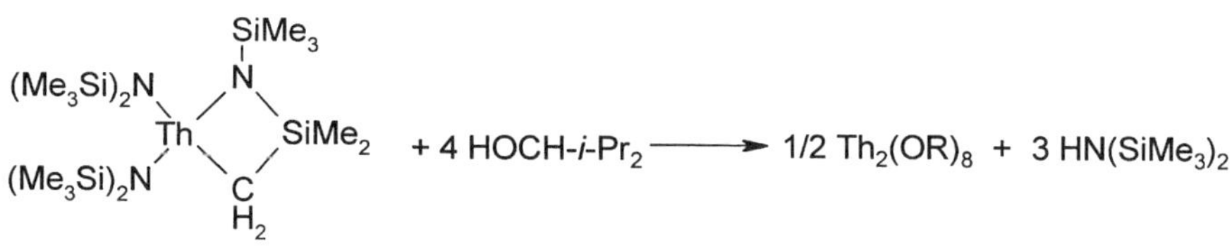

(5.50)

The homoleptic complex exists in the solid state as a dimer (*CN* of Th is 5), whose structure can be considered as two ThO_5 units (in a trigonal bipyramidal geometry) joined along a common axial–equatorial edge [281]. At room temperature, in noncoordinating solvents, there is a monomer–dimer equilibrium (5.51) [308,311]:

$$Th_2(OCH\text{-}i\text{-}Pr_2)_8 \rightleftharpoons 2Th(OCH\text{-}i\text{-}Pr_2)_4 \quad (5.51)$$

The reaction of thorium metal with two equivalents of iodine in 2-propanol produces the dimeric halide–alkoxide complex $Th_2I_4(O\text{-}i\text{-}Pr)_4(HO\text{-}i\text{-}Pr)_2$ [312] (this reaction is an example of "direct synthesis"; for details see Sec. 3.4). The complex crystallizes in a triclinic space group ($P\bar{1}$) and exhibits an edge-shared bioctahedral geometry with 2-propoxide ligands occupying bridging positions. It is isostructural to the uranium analogue $U_2I_4(O\text{-}i\text{-}Pr)_4(HO\text{-}i\text{-}Pr)_2$ [313] and structurally very similar [312] to a number of early transition metal and lanthanide alkoxide and halide–alkoxide species such as $M_2(O\text{-}i\text{-}Pr)_8(HO\text{-}i\text{-}Pr)_2$ (M = Zr, Ce) [314] and $Ti_2Cl_4(OR)_4(HOR)_2$ ($R = CH_2CH_2Cl$) [315]. It was proposed that intermediate products $[MI(O\text{-}i\text{-}Pr)_2(HO\text{-}i\text{-}Pr)_4]I$ (M = U, Th) are primarily formed and then converted to the dimeric species [312]. Other alkoxide actinide complexes have also been characterized, for example $Th_2(O\text{-}t\text{-}Bu)_8(HO\text{-}t\text{-}Bu)$ [316], $(TBA)U_2(OBu^t)_9$ (TBA = tetrabutylammonium), and $KU_2(OBu^t)_9$ [317]. For the last complex, it was shown that the K^+ cation remains associated with the dimeric anion in solution and that the structure of the molecular unit is likely to be the same in solution as in the crystalline solid. Oxidation chemistry of the uranium(III) arylalkoxide complex $U(OAr)_3$ (OAr = 2,6-di-*t*-butylphenoxide) [303] can be represented as in Fig. 5.5. As is seen, oxidation of uranium(III) arylalkoxide complexes in the presence of a suitable chalcogenide source forms the bridged binuclear uranium(IV) compounds $[(ArO)_3U]_2\text{-}\mu\text{-}X$ (X = O, S) [303].

The preparation of monocyclo-octatetraenyl uranium(IV) alkoxide complexes from reaction of uranium cyclo-octatetraene precursors with alcohols has been reported [318]. A very unusual monomeric complex, $U(OTeF_5)_6$, which can be con-

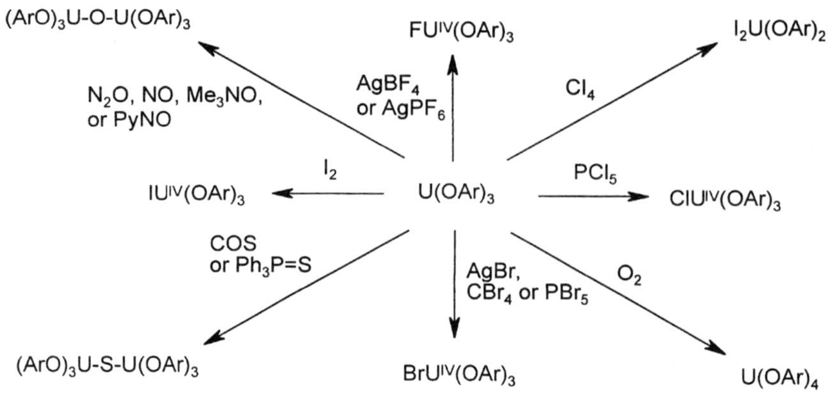

Figure 5.5 Oxidation chemistry of uranium(III) arylalkoxide complexes (OAr = 2,6-di-*t*-butylphenoxide). (From Ref. 303, with permission.)

sidered as an analogue of metal alcoholates, is sublimed at 333 K (1.3 Pa) despite its very high molecular mass [319a].

Actinide alkoxides can also be made from dialkylamides and ROH [281], as well as halide derivatives, such as, for instance, $(THF)_2ClUO_2(\mu\text{-}Cl)_2UO_2Cl(THF)_2$ [319b]. Homoleptic uranium halide complexes $UX_4(CH_3CN)_n$ can take part as precursors of mixed uranium(IV) aryloxide–halide compounds; the coordination environment of the product depends on the halide nature. Thus, $UBr_4(CH_3CN)_4$ reacts with two equivalents of KOAr to yield $Br_2U(OAr)_2(THF)(4THF)$, while UCl_4 yields the anionic complex $[K(THF)_4][UCl_3(OAr)_2]$. The mixed aryloxide–halide complexes are resistant to ligand redistribution in solution. The lighter halides of uranium(IV) are found to be superior reagents in metathesis reactions, due to the thermal instability of $UI_4(CH_3CN)_4$ [320]. A range of tetravalent (triamidoamine) uranium alkoxides and aryloxides are reported [321].

One of the routes to actinide methoxide derivatives involves the interaction of $[Cp^*_2MH_2]$ (M = U, Th) with trimethyl phosphite in pentane (5.52) [322]:

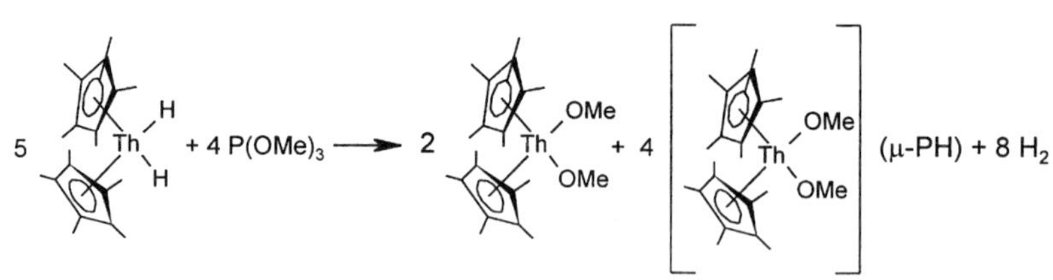

(5.52)

Other methods, for Th alkoxide complexes, involve the alcoholysis of alkyl complexes, e.g., $Cp^*_2ThMe_2$ can be obtained by reaction of $Cp^*_2ThCl_2$ with M^IOH; other reactions include insertion of ketones into Th–alkyl bonds (5.53) ([300] and references cited therein):

$$Cp^*_2ThCl(CH_3) + (CH_3)_2CO \rightarrow Cp^*_2ThCl[OC(CH_3)_3] \quad \text{(in toluene)} \qquad (5.53)$$

or via hydrogenation of η^2-acyl complexes (5.54) [300]:

$$Cp^*_2ThCl(\eta^2\text{-}OCCH_2\text{-}t\text{-}Bu) + H_2 \rightarrow Cp^*_2ThCl(\eta^2\text{-}OCH_2CH_2\text{-}t\text{-}Bu) \qquad (5.54)$$

Salts of O-containing acids and related compounds. Single crystals of $[Na_6Pu(CO_3)_5]_2 \cdot Na_2CO_3 \cdot 33H_2O$ (space group $P2_1/c$) were obtained from a 0.15 M solution of Pu(IV) in 2.6 M Na_2CO_3. In the crystal structure [323], the asymmetric unit contains a complex network consisting of $[Pu(CO_3)_5]^{6-}$ anions and Na^+ cations which are linked through interactions with CO_3^{2-} and H_2O ligands. The $[Pu(CO_3)_5]^{6-}$ ion can be viewed as a pseudohexagonal bipyramid with three CO_3^{2-} ligands in an equatorial plane and two in axial positions; the structural unit of this compound is very similar to the related $[Th(CO_3)_5]^{6-}$ anion found in the solid-state structures of $Na_6[Th(CO_3)_5] \cdot 12H_2O$ and $[C(NH_2)_3]_6[Th(CO_3)_5] \cdot 4H_2O$ ([307] and references cited therein). According to the authors' opinion [323], there is an analogy between the well-known hexagonal bipyramidal coordination polyhedron seen in $[AnO_2(CO_3)_3]^{4-}$ complexes ***971*** [324–326a] and the anion $[Pu(CO_3)_5]^{6-}$ ***972***. Among other plutonium complexes, the compound $[Pu(NO_3)_2\{2,6\text{-}[(C_6H_5)_2P(O)CH_2]_2C_5H_3NO\}_2](NO_3)_2 \cdot 1.5H_2O \cdot 0.5MeOH$ with a trifunctional ligand $2,6\text{-}[(C_6H_5)_2P(O)CH_2]_2C_5H_3NO$, coordinated to the metal through its oxygen atoms, was isolated in a mixed EtOH/MeOH solvent system [326b]. The aqua complex $[Pu^{III}(H_2O)_9][CF_3SO_3]_3$ was obtained in high yield and structurally characterized [326c].

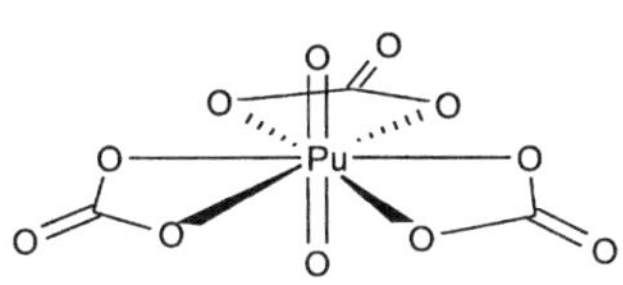

971

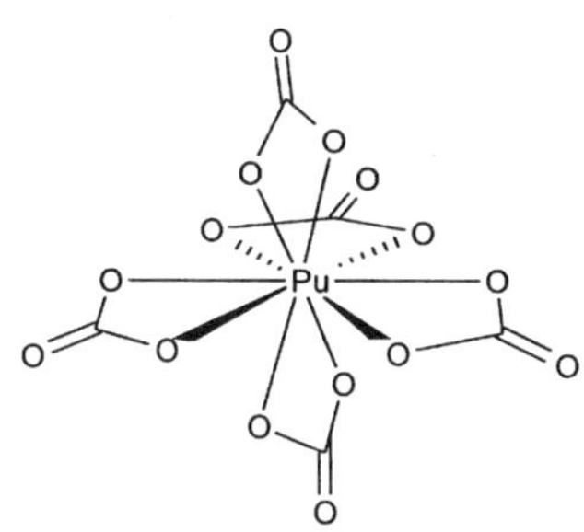

972

Studies on similar uranium or uranyl complex ions, as well as those of other actinides, formed as carbonate [327,328], sulfate [329], nitrate [330], phosphate and silicate [331], and citrate [332,333a], and other organic acid [333b,c] complexes, and other O-containing compounds were also reported. In particular, the bimetallic Pt—U amidate-bridged $[Pt\{NC(O)CH_2CH_2\}_2(PPh_3)_2]_2 \cdot (NO_3)_2$ was structurally characterized [330c]. Magnetic study of Np(V) phthalate hexahydrate $(NpO_2)_2(O_2C)_2C_6H_4 \cdot 6H_2O$ at 2–300 K showed [333c] that this compound exhibits a magnetic ordering below 4.5 K. A totality of studied magnetic properties indicates the existence of two kinds of Np lattices, one ferromagnetic and the other antiferromagnetic (metamagnetic) [333c].

The equilibrium between the U(IV) ion tetracarbonate and pentacarbonate complexes has been studied at 25°C in CO_2/HCO_3^- solutions of different ionic strength (0.5, 1.0, 2.0, 3.0 M $NaClO_4$), using UV–Vis spectrophotometry. Data have been explained by assuming the equilibrium $[U(CO_3)_4]^{4-} + CO_3^{2-} \rightarrow [U(CO_3)_5]^{6-}$ [327]. Citric acid forms a mixed-metal complex with iron and uranium, similar to the U–citrate complex; the Fe–U citrate complex is inert to biodegradation [333a]. A series of publications are dedicated to uranium peroxo complexes with various coligands: tri- and quadridentate Schiff bases [334,335], amines, or amino-carboxylic acids [336a]. Thorium(IV) complexes with neutral oxygen donor ligands are reviewed [336b].

β-Diketonates. Very strong actinide complexes with β-diketones [$An(acac)_4$ and $AnO_2(acac)_2$] are used in solvent extraction and separation of actinides. They are prepared by direct interaction of the metal or actinyl halide with the appropriate β-diketone in the presence of a base. Only fluorinated An(IV) diketonates produce adducts with Lewis bases, whereas common $AnO_2(acac)_2$ (An = Np, Pu) are stabilized by adduct formation. Fluorinated $UO_2(hfa)_2$ is a very strong Lewis acid and its adducts with H_2O and ROH can be sublimed without decomposition [282].

A detailed spectroscopic study of uranyl acetylacetonate [337–339] and other β-diketonates [339], as well as uranyl *bis*-(2-hydroxy-1-naphthaldehyde) [$UO_2(2H1N)_2$] [340] and uranyl *bis*-(2-hydroxybenzaldehyde) [341] was carried out. On the basis of the data obtained, the following structure was assigned to uranyl *bis*-(2-hydroxy-1-naphthaldehyde) [$UO_2(H1N)_2$] ***973*** [340] (with permission), whose synthesis was described earlier [342]:

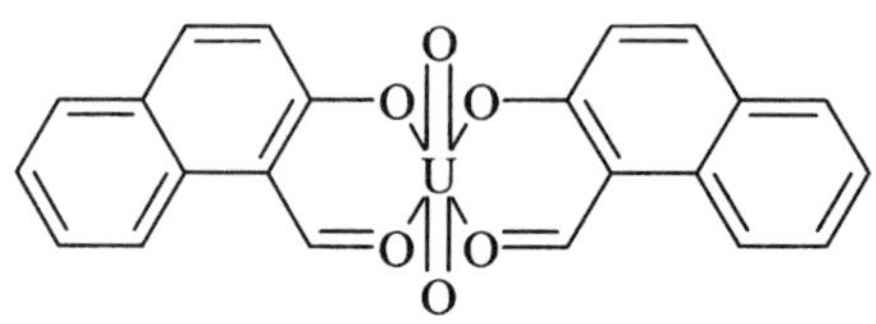

973

The Raman spectra for the studied uranyl β-diketonates (500–100 cm^{-1} and *bis*-(2-hydroxynaphthaldehyde) are presented in Figs. 5.6 and 5.7, respectively.

It is necessary to mention that uranium β-diketonates were also obtained by a direct electrochemical procedure. Thus, the electrochemical oxidation of uranium leads to chelates of the type UL_4 and UO_2L_2 (LH = diketone) [14,343–347]. In addition to these complexes, the compound having composition $UO_2L_2(HL)_{0.5}$ was also isolated [344]; the structure showed in ***974*** (compare with ***973***) was proposed on IR spectroscopy data and, in our opinion, requires a more detailed analysis.

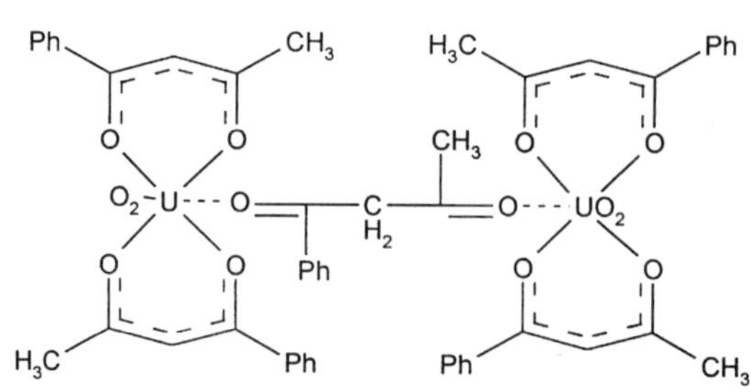

974

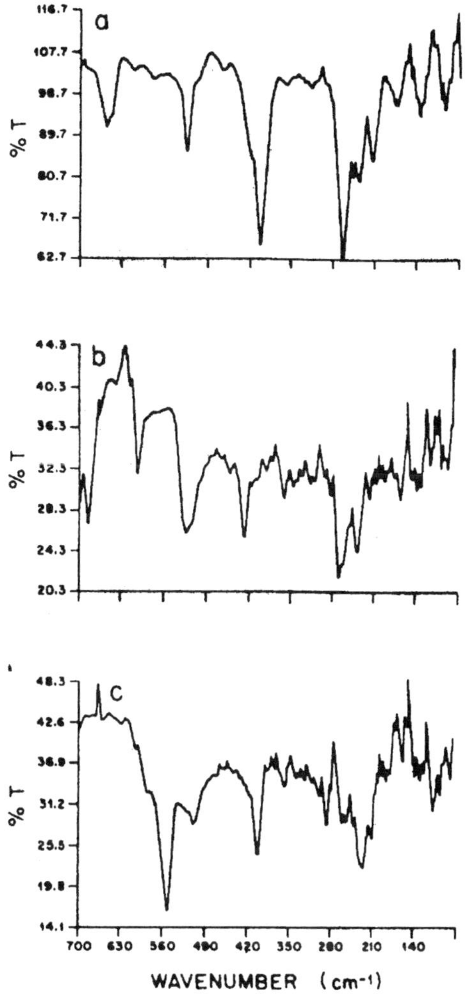

Figure 5.6 The low-region Raman spectra (500–100 cm^{-1}) for uranyl β-diketonates: (a) $UO_2(acac)_2$, (b) $UO_2(HDBM)_2$, (c) $UO_2(HBTF)_2$. The irradiance units in spectra a and c are scaled by 1.0×10^{-4} and in b by 1.0×10^{-3}. (From Ref. 339, with permission.)

EXPERIMENTAL PROCEDURES

Example 1

Synthesis of $FU(OAr)_3$ (OAr=2,6-di-*t*-butylphenoxide) [318]

To a stirred solution of 0.100 g (1.17×10^{-4} mmol) of $U(OAr)_3$ in 25 mL of hexane was added 0.023 g (1.17×10^{-4} mmol) of $AgBF_4$ dissolved in 5 mL of THF. An immediate reaction took place, resulting in a color change from brown to a black suspension, with the formation of a precipitate of silver metal. The solution was stirred for ca. 5 min and filtered through Celite, yielding a yellow solution. After the filtrate was dried under reduced pressure, the residue was redissolved in a minimum volume of hexane and the resulting solution cooled to −40°C. After 24 hr, the product was isolated by filtration in 43% yield (0.044 g).

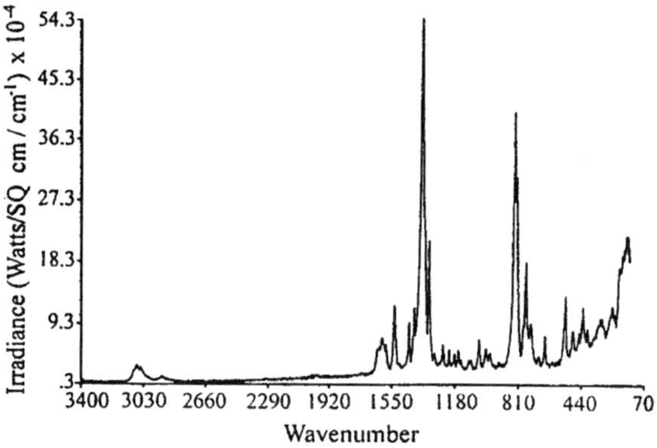

Figure 5.7 Raman spectrum of uranyl *bis*-(2-hydroxynaphthaldehyde) (cm^{-1}). (From Ref. 340, with permission.)

Example 2

Synthesis of $UO_2(HDBM)_2$ [339,346] and Uranyl *bis*-(2-Hydroxybenzaldehyde) [341]

Stoichiometrical quantities of the HDBM (1,3-diphenyl-1,3-pentanedione) and $UO_2(CH_3COO)_2$ were dissolved in tepid ethanol. The solution was concentrated and left at room temperature until crystallization. The product was washed with ether and dried at 130°C and a pressure of 10^{-7} mm Hg in an Abderhalden drying apparatus provided with P_4O_{10}. Uranyl *bis*-(2-hydroxybenzaldehyde) [79] was synthesized from stoichiometric quantities of an alcoholic solution of 2-hydroxybenzaldehyde and an aqueous solution of uranyl nitrate trihydrate, $UO_2(NO_3)_2 \cdot 3H_2O$. It is noted [79] that, for uranyl β-diketonates *bis*-(2-hydroxybenzaldehyde) and *bis*-(2-hydroxy-1-naphthaldehyde), their crystals do not exhibit the appropriate size and crystalline shape to perform polarization measurement in the solid state.

5.3.3 Actinide Halide Adducts

Light actinide metals (U, Np, Pu) react with elemental iodine or bromine in donor solvents forming trivalent AnX_3L_4 ($X = Br, I$) complexes (Table 5.12), for example (5.55) [359,360]:

$$U + 1.5I_2 + nL \rightarrow UI_3L_n,$$
$$n = 2 \text{ for } L = \text{dme and 4 for } L = \text{THF or py, at } 0°C \qquad (5.55)$$

This is a convenient, facile, and high-yield preparative route for quantitative preparation of the complexes above; no special equipment is required. $UI_3(THF)_4$ crystallizes in a $P2_1/c$ space group. It is mononuclear with pentagonal–pyramidal coordination geometry around the central uranium atom. This compound is stable until 75°C, then THF molecules are removed by steps, forming UI_3 at 162°C. Lewis base adducts of uranium tri-iodide, such as $UI_3(THF)_4$ above, are synthetically useful precursors for trivalent uranium chemistry (see Sec. 5.3.2) [360].

$UI_3(THF)_4$ can easily be prepared in large scale, which makes easy the synthesis of U(III) complexes from it. According to Ref. 360, the use of solvated complexes of other uranium halides or uranium alkoxides involves some difficulties, in

Table 5.12 Synthesis of Actinide Halide Adducts

Metal or its compound	Reaction system	Products	Conditions or observations	Reference
Synthesis from elemental metals				
U	X_2 (X = Cl, Br, I), THF	$UX_3(THF)_4$	0°C	359, 360
Th	THF, X_2 (X = Br, I)	$ThX_4(THF)_4$	0°C. The products are soluble in hydrocarbons	281
Th anode	Br_2, CH_3CN	$ThBr_4(CH_3CN)_4$	Direct electrochemical synthesis using a sacrificial Th anode. Only 0.60 g of metal is dissolved during 6 hr of electrolysis (8 V, 50 mA)	370
U	$HgCl_2$, THF	$UCl_4(THF)_3$	A lot of actinide/mercury amalgam is produced as a byproduct	371
An (U, Np, Pu)	(Br_2 or I_2) + THF	$AnX_3(THF)_4$	At 0°C. Products are air-sensitive	359
Synthesis from salts or halide adducts				
UI_4	DMSO or DIBSO	$UI_4(DMSO)_n$ (n = 6, 8)	25°C	366
UCl_4	NaH	$UCl_3 \cdot n$THF		281
$ThBr_4(THF)_4$	$KNPh_2$ or KNMePh, THF	$Th(NPh_2)_4 \cdot THF$ or $K[Th(NMePh)_5]$, respectively		372
$ThBr_4(THF)_4$	$Cp^*MgBr(THF)$	$Cp^*ThBr_3(THF)_3$	Product is a useful synthetic precursor to aryloxide and alkyl derivatives	373
$UI_3(THF)_4$	KCp^*, THF ($Cp^* = C_5Me_5$)	$Cp^*UI_2(THF)_3$	$CpAnX_2$ compounds are rare. The structure is presented in Sec. 5.3.4.	268
$AnI_3(THF)_4$ (An = U, Np, Pu)	$NaN(SiMe_3)_2$, THF	$An[N(SiMe_3)_2]_3$	25°C	359, 374

particular, formation of mixtures of products. In the case of UI_3, the trivalent oxidation state is *favored* by the iodide ligands, and its adduct $UI_3(THF)_4$ is a soluble form of UI_3 useful for further syntheses [360]. Other UX_3 adducts have been reported in the literature [361–365].

The actinide tetrahalides readily react with Lewis bases, yielding complexes containing two or four donor atoms. The adduct's composition can differ from the common $AnCl_4 \cdot 2L$ and $AnCl_4 \cdot 4L$. For example, the $AnCl_4 \cdot nL$ complexes, where $n = 2.5$ or 5, are described for L = DMA and $n = 6$ for L = DMSO or Me_3PO. The $AnCl_4 \cdot 2HMPA$ (An = Th, U) complexes are extremely volatile [282]. A trans-octahedron is the common coordination polyhedron for the $AnCl_4 \cdot 2L$ complexes [282]. Cationic uranium(IV) complexes $UX_2L_4Y_2$ [X = Cl, Br, or I; Y = ClO_4 or BPh_4; L = bulky strong neutral O-donor ligand, such as *tris*(pyrrolidine-1-yl)phosphine oxide] were reported [364].

The reactions of UI_4 with a number of sulfoxide donor ligands have been studied in nonaqueous media and compared to the behavior of UCl_4 and UBr_4 in the presence of these ligands. UL_4 is readily oxidized by DMSO and DIBSO at room temperature and the only stable complexes isolated were $UI_4(DMSO)_8$ and $UI_4(DIBSO)_6$ (Table 5.12) [366].

All actinide halides tend to accept a halogen ion, giving anionic complexes. This tendency and the stability of the complexes decreases in the order $F \gg Cl > Br \gg I$. Formal adducts of actinide trihalides are essentially ionized. Thus, uranium trichloride is crystallized from DMSO solution as the solvate $UCl_3 \cdot 3DMSO$ which is built from dodecahedral $[U(OSMe_2)_8]^+$ cations and octahedral $[UCl_6]^-$ anions. The x-ray analysis of an americium chloride hexahydrate shows an ionic structure composed of $[AmCl_2(H_2O)_6]^+$ cations and $[Cl(H_2O)_6]^-$ anions linked by hydrogen atoms [282].

Pentahalides of U and Pa form complexes of the type $AnX_5 \cdot L$ (X = Cl, Br; L = R_3PO, HMPA) and $PaX_5 \cdot 3MeCN$. Dissolution of UO_3 in thionyl chloride yields the $UCl_5 \cdot SOCl_2$ adduct, while the analogous procedure with Pa(V) hydroxide gives the ionic complex $[SO][PaCl_6]_2$. Actinide hexahalides usually do not react with Lewis bases (except the $UCl_6 \cdot$ bipy complex), whereas actinyl halides readily give complexes of the composition $AnO_2X_2 \cdot nL$ where $n = 1$, 1.5, 2, 3, and 4 [282].

Uranium fluorides UF_5 and UF_6 are shown to react with 2-fluoropyridine (F-py) or 2,2′-bipy giving adducts [367]. Thus, in CH_2Cl_2 solution UF_6 reacts with both ligands forming $UF_4 \cdot$(F-py) and $U_2F_{12} \cdot$bipy, respectively. However, for the UF_6/bipy system, the reduction of UF_6 by bipy is the dominant process, and the formation of $U_2F_{12} \cdot$bipy {which can be represented, according to the authors, as $[UF_4(bipy)_2]^{2+}[UF_7]_2^- \cdot (UF_6)$} might be a preliminary step of this reduction. In case of UF_5, the very moisture-sensitive dimorphic adduct $UF_5 \cdot$bipy and the ionic derivative $[(bipy)_2H]^+[UF_6]^-$ were obtained and structurally characterized.

The $AnX_3(THF)_4$ complexes are synthetic precursors for a range of inorganic and organometallic complexes (see Table 5.12), due in part to their favorable solubility in toluene and THF. Use of $UI_3(THF)_4$ could provide a convenient entry into the chemistry of many other trivalent uranium compounds [359]. Reactions of $AnX_3(THF)_4$ with alkali metal salts of organic ligands produce the corresponding aryloxide, amide, etc., derivatives and provide high-yield routes to both new and known trivalent actinide complexes. Thus, uranium tetrachloride UCl_4, dissolved in THF, can be reduced by NaH or Na/Hg to produce a sparingly soluble $UCl_3(THF)_n$

[368]. This compound is sometimes used as a precursor for the synthesis of uranium complexes. However, its utility is limited due to the formation of side products. For example, in the synthesis of $U[N(SiMe_3)_2]_3$ from the reaction between $UCl_3(THF)_n$ and $NaN(SiMe_3)_2$, a mixture of the desirable U(III) product with the U(IV) metallacycle [369] ***975*** is formed [359]. Adducts of uranyl salts are also frequently used as precursors of uranyl alkoxides and amides [319b]. For example, $UO_2Cl_2(THF)_3$, obtained from $UO_2Cl_2 \cdot xH_2O$ ($x = 1$–3) and $ClSiMe_3$ in THF, can be transformed to $(THF)_2ClUO_2(\mu\text{-}Cl)_2UO_2Cl(THF)_2$ (Sec. 5.3.2), which serves as an excellent precursor for the mentioned compounds [319b].

$[(Me_3Si)_2N]_2U(CH_2SiMe_2NSiMe_3)$

975

EXPERIMENTAL PROCEDURES

Example 3

Synthesis of $UF_5 \cdot$ bipy [367]

This compound was obtained from the reaction of a frozen solution of 0.50 mmol of β-UF_5 in 2 mL of CH_3CN kept at -196°C into which a solution of 1.16 mmol of bipy in 1 mL of CH_3CN was poured under vacuum. The light green, almost colorless, needle-shaped crystals that were formed upon warming to ambient temperature were decanted and pumped to dryness at this temperature.

Example 4

Synthesis of $NpI_3(THF)_4$ [359]

A 10-mL reaction vessel equipped with a Teflon-coated stir bar was charged with Np turnings (0.21 g, 0.88 mmol) and THF (ca. 5 mL). Iodine (0.33 g, 1.31 mmol, 1.5 equiv.) was slowly added to the stirred Np/THF mixture, giving a deep red–purple color. The reaction vessel was stoppered and the mixtured allowed to stir for 24 hr, during which a finely divided yellow–orange powder precipitated from solution. The solid was collected on a medium-porosity fritted filter, washed with hexane (3 × 5 mL), and then vacuum-dried to give $NpI_3(THF)_4$ (0.68 g, 0.75 mmol, 86%).

5.3.4 Actinide π-Complexes with Allyl, Cyclopolyene, Arene, and Related Ligands

Actinide complexes with the ligands above are well represented in recent literature, especially those of cyclopentadiene and its derivatives, and they have been reviewed in Ref. 268. The allyl complexes can be prepared from an allyl Grignard and $AnCl_4$ (see Table 5.13). The allyl groups are bound in an η^3-fashion, according to low-temperature NMR data [268], and may be displaced with HX or ROH to give, for example, $U(C_3H_5)_3X$ or $[U(C_3H_5)_3(OR)]_2$ [281]. The structure of *t*-butoxyallyl alkoxy-bridged dimer $[U(C_3H_5)_2(OR)_2]_2$, taken from Ref. 268, is presented in Table 5.13.

Table 5.13 Actinide π- and Related Complexes

Metal or its compound	Reaction system	Products	Conditions or observations	Reference
Allyl complexes				
$AnCl_4$	C_3H_5MgCl, Et_2O	$An(C_3H_5)_4$	The $An(\eta^3\text{-allyl})_4$ complexes are unstable above 0°C	268, 281
$U(C_3H_5)_4$	HOR, Et_2O	$[U(C_3H_5)_2(OR)_2]_2$	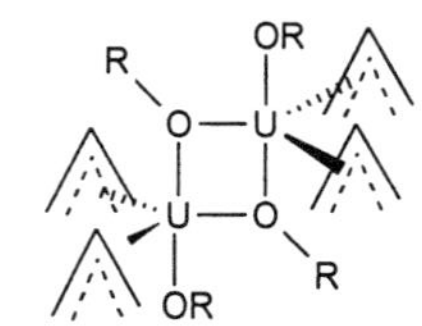	268
Cyclopentadienyl complexes				
Cp_nAn complexes				
$AnCl_n$ ($n=4$, An = Th, U, Np, Pa; $n=3$, An = Np)	Cp_2Be	Cp_nAn	Without solvent	268
U	CpH	Cp_3U		281
$AnCl_n$ (An = Th, U, Np, $n=3$ or 4)	KCp, benzene	Cp_nAn 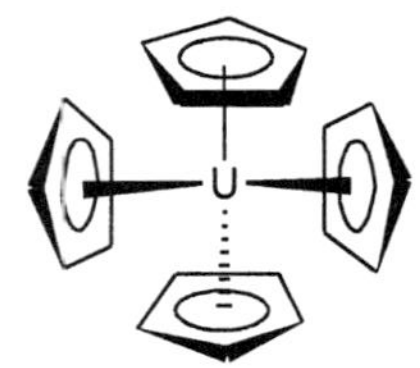	All of the Cp_4An compounds are sparingly soluble in organic solvents. U—C distance in UCp_4 is 2.81(2) Å. An-ligand bonds have covalent character	268, 281
Cp_3AnCl	$NaC_{10}H_8$ in THF or Na/Hg in THF	$Cp_3An(THF)$	Cp_3An complexes are strong Lewis acids and form adducts with a variety of Lewis bases. In case of Th, $\mu_{eff} = 0.403\mu_B$. In case of U, yield is 70%	263 and references cited therein, 402

$(Me_3Si)_2C_5H_3$ $ThCl_2$	Na, K	$(Me_3Si)_2C_5H_3Th$	The Th—C distance is 2.80(2) Å	389
Cp_3AnX complexes				
AnX_4 (An = Th, U, Np)	M′Cp DME (M′ = Na, K, Tl; X = Cl, Br, I)	Cp_3AnX 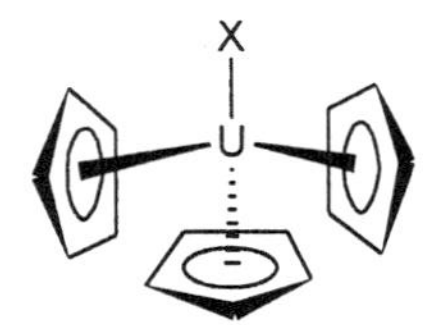	The Cp_3UCl structure is a distorted tetrahedra; the U—Cl bond is 2.559(16) Å	268, 281, 403
$(MeC_5H_4)_3U(t\text{-}Bu)$	C_2H_4 or CO (= L) in toluene	$(MeC_5H_4)_3U(L\text{-}t\text{-}Bu)$	210 psi in case of C_2H_4 use	404
Cp_3UCl	LiL (L = PPh_2, NEt_2, $SiPh_3$)	Cp_3UL		375
Cp_3UX (X = NEt_2, PPh_2, $SiPh_3$)	2,6-Dimethylphenylisocyanide	$Cp_3U[C(X){=}NC_6H_3Me_2\text{-}2,6]$ 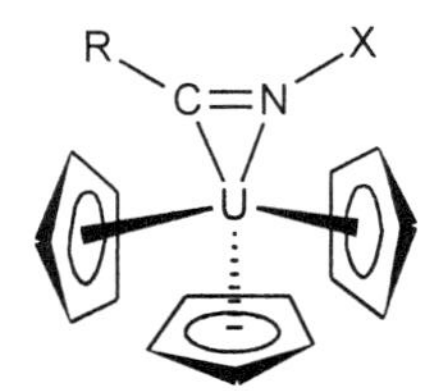	$[C(X){=}NC_6H_3Me_2\text{-}2,6]^-$ anion is η^2-coordinated	268, 405–409
Cp_3UNEt_2	$HSnPh_3$	$Cp_3U{-}SnPh_3$	Formation of U–metal bond	263 and references cited therein
$Cp_3U{=}C(H)PPh_2Me$	$HC{\equiv}CPh$	$Cp_3U{-}C{\equiv}CPh$		263

Table 5.13 Continued

Metal or its compound	Reaction system	Products	Conditions or observations	Reference
$(MeCp)_3U(THF)$	Organic azides Me_3SiN_3 or PhN_3	$(MeCp)_3U{=}NR$	Very short U—N bond distance of 2.019(6) Å; nearly linear U—N—C angle of 167.5(6)° (multiple U—N bond character)	268
Cp_3UH ($Cp = C_5Me_5$, $C_5H_4Bu^t$, $C_5H_4SiMe_3$	I_2	Cp_3UI		358
Cp_3AnCl (An = Th, U, Np)	RMgX in THF or RLi in Et_2O	Cp_3AnR	Cyclopentadienyl alkyls	268
$CpAnX_3$ complexes				
MX_4 (M = Th, U)	TiCp, DME, or THF	$CpThX_3(DME)$ and $CpUX_3(THF)_2$ 	U—Cl distance is 2.620(9) Å in $CpUX_3(THF)_2$	410
Cp_2AnX_2 complexes (X = Hal, alkyl)				
$Cp^*_2ThX_2$ (X = Cl, Br)	$CpRu(CO)_2Na$	$Cp^*_2(X)ThRu(Cp)(CO)_2$		263 and references cited therein
$ThCl_4(dmpe)$	NaCp, THF	$Cp_2ThCl_2(dmpe)$	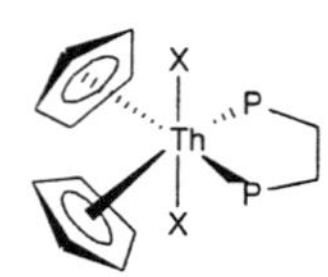	411
$Cp^*_2AnCl_2$	LiR, Et_2O	$Cp^*_2AnR_2$		268

Cp_2AnX complexes				
$(Me_3Si)_2C_5H_3UX_2$	Na—Hg, THF	$(Me_3Si)_2C_5H_3UX(THF)$	Cp_2AnX compounds are rare	268
$BkCl_3$	Cp_2Be	$[Cp_2BkCl]_2$		268
$CpAnX_2$ complexes				
$UI_3(THF)$	KCp^* in THF	$Cp^*UI_2(THF)_3$ 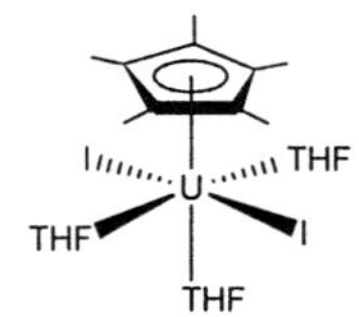	$CpAnX_2$ compounds are rare	360
Alkyl complexes				
$ThCl_4$	LiMe, tmeda, Et_2O	$[Li(tmeda)]_3ThMe_7 \cdot (tmeda)$	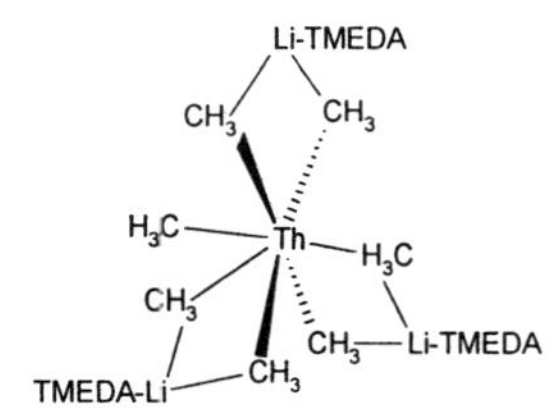	412
UCl_4	LiR (excess) + L ($L = THF, Et_2O$)	$Li_2UR_6L_8$		268

Table 5.13 Continued

Metal or its compound	Reaction system	Products	Conditions or observations	Reference
$ThCl_4$	LiMe, TMEDA, Et_2O	$[Li(TMEDA)]_3Th Me_7 \cdot TMEDA$	Six of the methyl groups [Th—C = 2.667(8)–2.765(9) Å] are coordinated in a pairwise manner to $Li(TMEDA)^+$ cations, while the seventh [Th—C = 2.571(9) Å] is not	268
Arene complexes				
UCl_4	C_6Me_6, $AlCl_3$ stoichiometric, Al	$[U_3(\mu_3\text{-}Cl)_2(\mu_2\text{-}Cl)_3(\mu_1,\eta^2\text{-}AlCl_4)_3(\eta^6\text{-}C_6Me_6)_3][AlCl_4]$		268 and references cited therein
UCl_4	Al, $AlCl_3$, C_6H_6	$(C_6H_6)U(AlCl_4)_3$	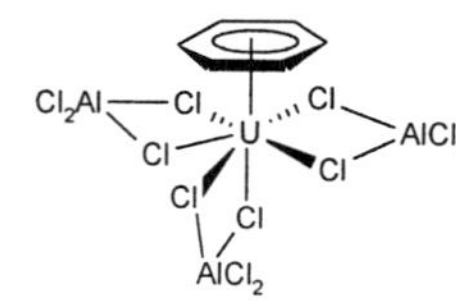	360
UCl_4	Zn, $AlCl_3$, C_6Me_6	$[(C_6Me_6)_2U_2Cl_7]^+ [AlCl_4]^-$	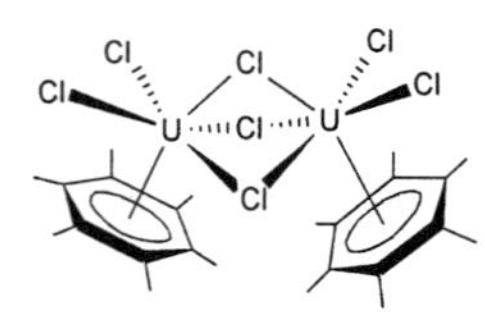	360
Cycloheptatriene complexes				
UCl_4	C_7H_8, THF	$[U(\eta\text{-}C_7H_7)_2]^-$	The C_7H_7 rings are planar	358
UX_4 (X = NEt_2 or BH_4)	$K[C_7H_9]$	$K[X_3U(\mu\text{-}\eta^7{:}\eta^7-C_7H_7)UX_5]$	Inverse cycloheptadienyl sandwich complexes	413

Cyclo-octatetraene complexes				
UCl_4	$K_2C_8H_8$	$U(C_8H_8)_2$ (uranocene)	Green, pyroforic compound having a sandwich structure with planar rings and D_{8h} symmetry	414, 415
AnX_3 (An = U, Np, Pu, Am)	$K_2(cot)$	$[K(solv)][An(cot)_2]$		416, 417
$AnCl_4$ (An = U, Th)	C_8H_8, NaH	$(cot)AnCl_2(THF)_2$	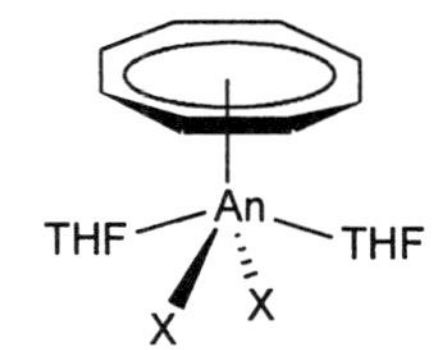	263 and references cited therein
$U(BH_4)_4$	C_8H_8	$(cot)U(BH_4)_2$		263 and references cited therein
$(PyH)_2PuCl_6$	$K_2(cot)$, THF	$Pu(cot)_2$		268
$(cot)ThCl_2(THF)_2$	$Cp^*MgCl(THF)$, PhMe	$(cot)ThCp^*Cl(THF)$	Mixed-ring cot—Cp complex	418
$UI_3(THF)_4$	(1) KCp^* in THF (2) $K_2(cot)$	$Cp^*U(cot)(THF)$	A similar uranium complex $[(Me_3Si)C_8H_7]$ $UCp^*(THF)$ has the structure: 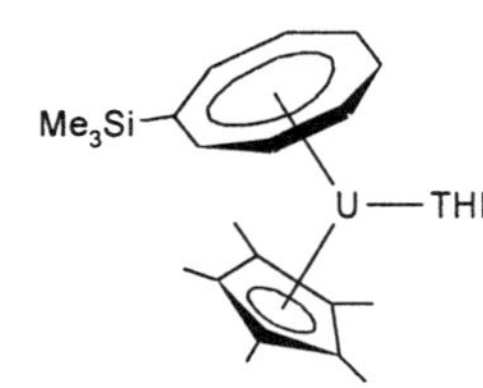	419

Table 5.13 Continued

Metal or its compound	Reaction system	Products	Conditions or observations	Reference
Carbonyls				
U (vapor)	CO, Ar	$U(CO)_x$	Cocondensation at 4 K	420
$[(Me_3Si)_2C_5H_3]U$	CO	$[(Me_3Si)_2C_5H_3]U(CO)$	Uranium-to-carbonyl π-backbonding takes place. The product dissociates reversibly	268
Cp_3MR (M = U, Th, Np)	CO	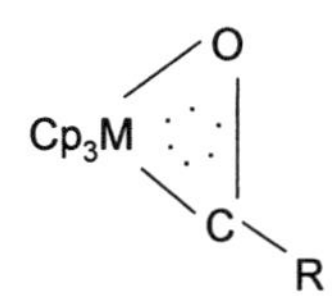	Migratory insertion reaction. The η^2-bonding of the CO insertion products is characteristic of such actinide compounds	281, 421

Among other actinide π-complexes, those of U(III,IV) and Th(III,IV) with cyclopentadiene and its derivatives are the most widespread (Table 5.13), although U(V) and U(VI) also exist [268], as well as Np(IV) and Cf(III) metallocenes [281]. All these compounds can be divided into Cp_nAn ($n=3, 4$), Cp_nAnX ($n=2, 3$), Cp_2AnX_n ($n=1, 2$), $CpAnX_n$ ($n=2, 3$) (X is generally a halide ion, although it can be R, CO, etc.) forming the corresponding cyclopentadienyl alkyls, carbonyls, alkoxides, etc. [268]. Cp_3UCl was the first synthesized organoactinide compound (made in 1956) [281]. This compound is a precursor of compounds with unbridged metal–metal bonds between, for example, uranium and iron, ruthenium $[Cp_3UMCp(CO)_2$, M = Fe, Ru], and germanium (Cp_3UGePh_3) [268,375]. The *tris*-cyclopentadienyl uranium and some of its adducts, Cp_3UL, were the first reported organouranium(III) complexes [263,376,377]. They are also used as precursors of bimetallic complexes (Fig. 5.8) [263].

Electrochemical studies of $(RCp)_3UCl$ [RCp = RC_5H_4 with R = H, CH_3, *t*-Bu, $(CH_3)_3C$, $(CH_3)_3Si$] showed evidence of disproportionation during oxidation of the complex [378]. NMR studies revealed both electron transfer and ligand exchange reactions in Cp_3UX (X = Hal, BH_4, Alk) (5.56) [263,379]:

$$\begin{aligned} &Cp_3UX \underset{-e^-}{\overset{+e^-}{\rightleftharpoons}} Cp_3UX^- \\ &Cp_3UX + Cp_3U^*(THF) \rightleftharpoons Cp_3U^*(THF) + Cp_3U^*X \\ &Cp_3UX^- + Cp_3U^*(THF) \rightleftharpoons Cp_3U^*(THF) + Cp_3U^*X^- \end{aligned} \tag{5.56}$$

For complexes with substituents in the cyclopentadiene rings, equilibrium constant determinations showed that the ligand displacement series towards $(MeCp)_3U$ is $PMe_3 > P(OMe)_3 > py >$ tetrahydrothiophene > THF > quinuclidine > CO and for $(C_5H_4SiMe_3)_3U$ is EtNC > EtCN [263,380]. This trend indicates that for π-acceptor ligands, π-backbonding from uranium is significant [263]. Electronic structure

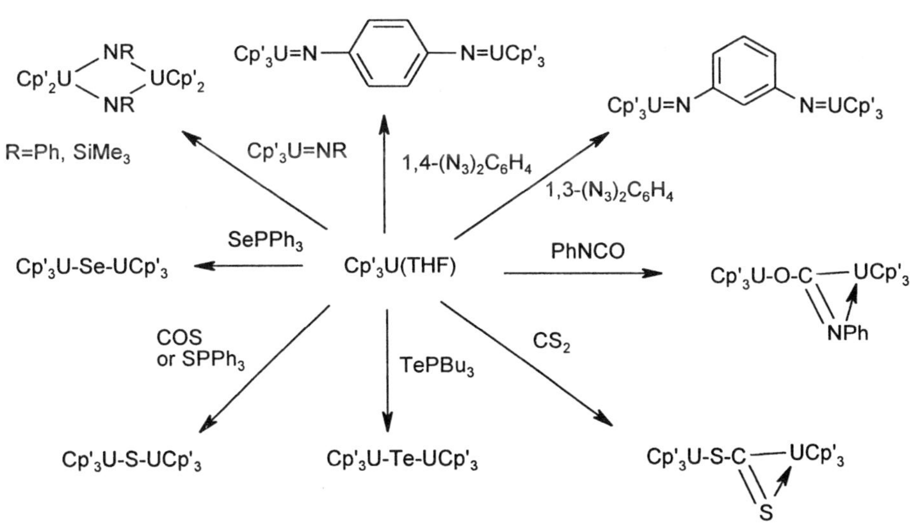

Figure 5.8 *Tris*-cyclopentadienyl uranium derivative chemistry. (From Ref. 263, with permission.)

calculations on the model complex Cp_3UCO indicated a significant U 5*f*–CO 2π backbonding which results in a stabilization of the uranium 5*f* atomic orbitals [268,381]. Uranium complexes containing CO are reported [382]. Ab initio calculations by relativistic effective core potential and gas-phase UV photoelectron spectroscopy measurements of actinide *tris*(cyclopentadienyl) complexes were carried out and reported [383,384].

Cp_3AnR complexes are exposed to insertion reactions by CO, CO_2, and CNR ligands yielding η^2-acyl ***976*** [$Cp_3An(\eta^2$-COR)], η^2-carboxylate ***977*** [$Cp_3An(\eta^2$-$O_2Cr)$], and η^2-iminoalkyl ***978*** {$Cp_3An[\eta^2$-C(R)NR′]} compounds, respectively [268]:

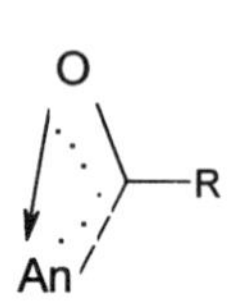

976

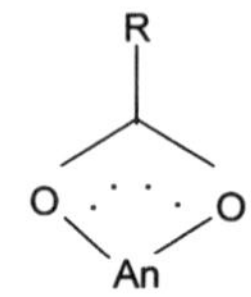

977

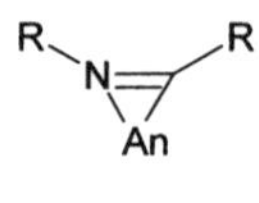

978

Cp_3UX compounds, containing U—P, U—N, U—Si, U—Sn, and U—Ge bonds, were synthesized by metathesis of the chloride ligand in Cp_3UCl [263] (Table 5.13). For complexes with U—O bonds, the x-ray crystal structures of Cp_3UOPh and $Cp_3UOSiPh_3$ revealed the short U—O distances [2.119(7) and 2.135(8) Å, respectively] and the nearly linear U—O—C and U—O—Si angles [159.4(5) and 172.6(6)°, respectively], which suggest a strong π-bonding between the U and O atoms [263]. An O-bridged bimetallic complex μ-O-*bis*[*tris*(cyclopentadienyl)uranium(IV)] was reported [385]. For complexes with U—S bond, absolute uranium–ligand bond disruption enthalpies of (UL_3—SX) complexes (L = C_5H_4Bu, $C_5H_4SiMe_3$, or $C_9H_6SiMe_3$ and X = Et or Bu) were determined [386]. Another example of a bimetallic compound is the heteroaromatic dichloro-bridged complex [$UCp_2(\mu$-$Cl)_2Li$(pmdeta)] [pmdeta = $Me_2NCH_2CH_2)_2NMe$] [387].

Ground-state configuration and electronic structure of the monomeric uranium(III) alkyl $[\eta^5$-$(CH_3)_5C_5]_2UCH[Si(CH_3)_3]_2$ were investigated and reported [388]. It was found that the $4A''[(a')^1(a'')^1(a')^1]$ state, associated with the $5f^3$ uranium configuration, represents the molecular ground state and is close in energy to the higher-lying $4A''$ ($5f^26d^1$) state. It is known ([388] and references cited therein)] that two different metal ion configurations, $5f^3$ and $5f^2d^1$, may be expected in "ligand-free" U(III) complexes. Gaseous U^{3+} ion possesses a $5f^3$ ground-state configuration. In the case of Th(III), EPR measurements on Th$[\eta^5$-$C_5H_3(SiMe_3)_2]_3$ indicated $6d^1$ ground-state configuration despite the established $5f^1$ configuration [389] of the free ion [265].

Insertion of CO_2 or CO into An—R bonds in $Cp^*_2AnMe_2$ yields insertion products ***979–981*** [390–392] or carbene-like η^2-acyl complexes ***982*** [268]:

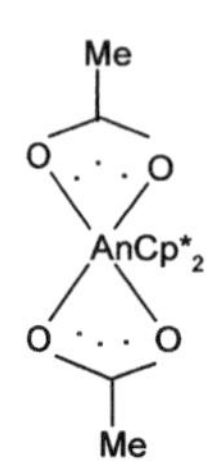

979

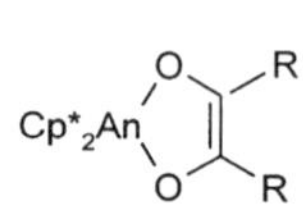

R = $CH_2C(Si)Me_3$

980

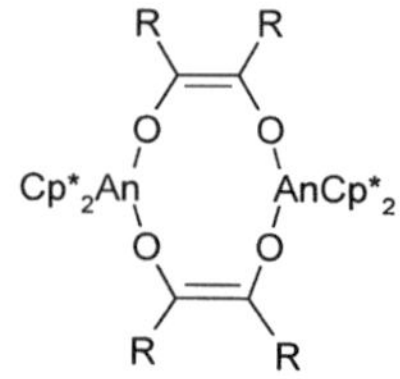

R = Me, CH_2Ph

981

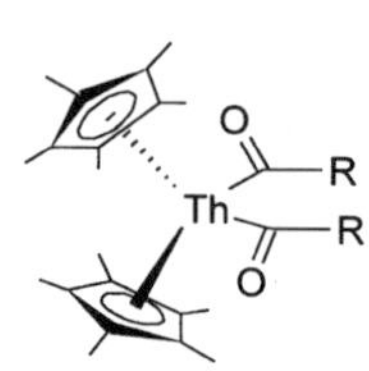

982

The dynamic behaviors of $CpUCl_3L_2$ complexes in solution were studied [263,393a]. A rapid equilibrium between two isomers was found [(5.57), from Ref. 263 with permission]:

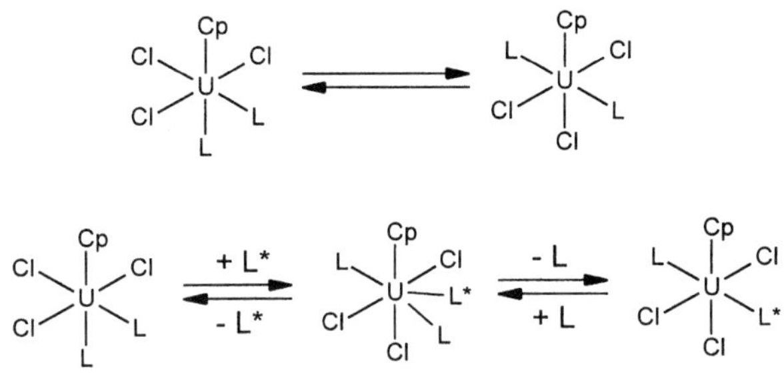

(5.57)

Structure and dynamic behavior in toluene for the complex $CpUCl_3L_2$

The uranium(III) complexes $Cp^*_2UI_2(THF)_3$ and $Cp^*UI_2(py)_3$ have been structurally characterized [393b]. The similar tribromide complex $Cp^*ThBr_3(THF)_3$ [111] provides a convenient entry into the chemistry of mono(pentamethylcyclopentadienyl) derivatives of thorium. Aryloxide and alkyl derivatives may be readily prepared by treatment of $Cp^*RhBr_3(THF)_3$ with potassium aryloxide or alkyl Grignard reagents, respectively [373], for example (5.58)–(5.60):

$$Cp^*ThBr_3(THF)_3 + KOAr \quad (\text{in THF, 18 hr}) \rightarrow Cp^*ThBr_2(OAr)(THF) + KBr \qquad (5.58)$$

$$Cp^*ThBr_3(THF)_3 + 2KOAr \quad (\text{in THF, 18 hr}) \rightarrow Cp^*ThBr_2(OAr)_2 + 2KBr \qquad (5.59)$$

$$Cp^*ThBr_2(OAr)_2MeMgBr \rightarrow Cp^*MeTh(OAr)_2 \qquad (5.60)$$

The first indenyl actinide compound, $(C_9H_7)_3UCl$, was isolated in 1971 by treatment of UCl_4 with indenyl anion [394]. In case of U(III), its indenyl complex $(C_9H_7)_3U$ was isolated from UCl_3 and C_3H_7Na [395]. The general synthetic route is the following (5.61) [268]:

$$AnX_4 + 3K(C_9H_7) \rightarrow (C_9H_7)_3AnX + 3KX \quad (\text{in THF, An} = \text{Th, U, Np}) \qquad (5.61)$$

The structure of the product $(C_9H_7)_3AnX$ is represented by ***983***:

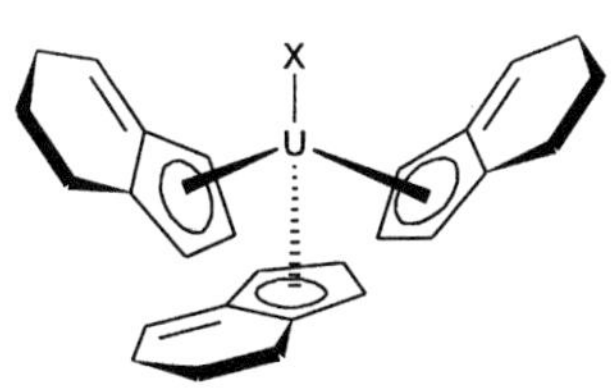

983

The uranium complexes with phospholyl ligand (formally analogous to Cp, Sec. 5.3.8) were also reported [263,396] [(5.62), from Ref. 263 with permission]:

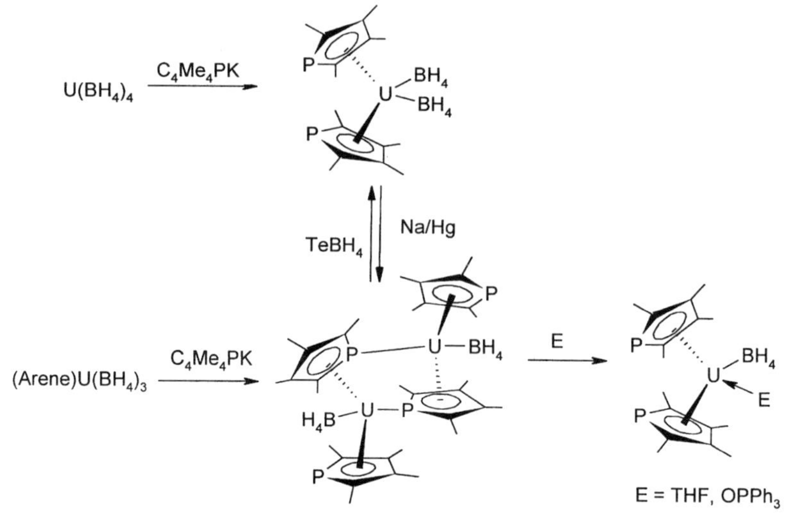

(5.62)

Phospholyl uranium complexes

The first cycloheptatrienyl sandwich compound, $[K(18\text{-crown-6})][U(\eta\text{-}C_7H_7)_2]$, was prepared and structurally characterized by Arliguie et al. [397]. In terms of the metal oxidation states, the best description of the cycloheptatrienyl ligand is formally as $C_7H_7^{3-}$, the uranium being in the +5 oxidation state [397,398]. The crystal structure of the product showed that the two C_7H_7 ligands are planar, parallel, and perpendicular to the linear axis defined by the uranium atom and the ring centroids [397].

The actinide cyclo-octatetraene complexes have general formulae $(cot)_2An$ ***984***, $(cot)AnX_2$ (X = Hal), as well as mixed-ring (cot)AnCp compounds, and may contain solvated solvent molecules, for example THF (Table 5.13) [268]:

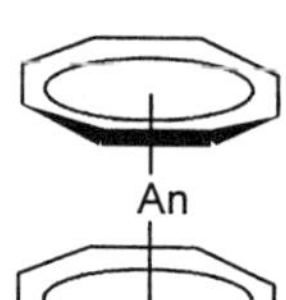

984

These complexes can be prepared by the following reactions [(5.63)–(5.66)] [268]:

$$AnCl_4 + 2K_2(cot) \rightarrow An(cot)_2 + 4KCl \quad (\text{in THF, An} = \text{Th, Pa, U, Np}) \tag{5.63}$$

$$(pyH)_2PuCl_6 + 2K_2(cot) \rightarrow Pu(cot)_2 + 4KCl + 2pyHCl \quad (\text{in THF}) \tag{5.64}$$

$$ThCl_4 + Th(cot)_2 \rightarrow 2(cot)ThCl_2(THF)_2 \quad (\text{in THF}) \tag{5.65}$$

$$UCl_4 + C_8H_8 + 2NaH \rightarrow (cot)UCl_2(THF)_2 + 2NaCl + H_2 \quad (\text{in THF}) \tag{5.66}$$

Among mixed-ligand cyclo-octatetraene complexes, the U, Th, Pa, Np, and Pu complexes $[(\eta\text{-}C_8H_8)An(\mu\text{-}SPr)_2]_2$ were reported, which can be prepared by reacting the thiol with $U(COT)(BH_4)$ [399a]. Reduction of the bulky 8-annulene thorium complex $[Th\{COT(TBS)_2\}_2]$ $[COT(TBS)_2 = —C_8H_6(t\text{-}BuMe_2Si)_2\text{-}1,4]$ by potassium yields the anionic compound $\{Th[COT(TBS)_2]_2\}K \cdot (DME)_2$, which was crystallographically characterized and is the first sandwich complex of Th(III) [399b]. The monocyclo-octatetraene uranium amide complexes $K[U(cot)(NEt_2)_2]$ ($Cot = \eta\text{-}C_8H_8$) and $[U(cot)(Net_2)_2(THF)]$ are synthesized from $[U(NEt_2)_4]$ or $[U(cot)_2]$; a series of uranium(V) derivatives are isolated after oxidation [400].

EXPERIMENTAL PROCEDURES

Example 5

Representative Procedure for One-pot Preparation of the Methyl–Aryl Complexes $Cp^*_2Th(Me)(OAr)$: $Cp^*_2Th(Me)(o\text{-}MeOC_6H_4)$ [401]

o-Arylmagnesium bromide (0.5 mL, 0.96 M in THF, 0.48 mmol) and $Cp^*_2ThCl_2$ (8225 mg, 0.39 mmol) were reacted together, in toluene (10 mL). The reaction mixture was stirred for 75 min before being taken to dryness in vacuo. The residue was extracted with toluene (15 mL) and the solution filtered to remove magnesium salts. Methylmagnesium chloride solution (126 mg, 0.37 mmol) was then added to the stirred solution of $Cp^*_2Th(o\text{-}MeOC_6H_4)X$ (X = Cl, Br). *p*-Dioxane (90 mg, 1.02 mmol) was added to the clear reaction solution, precipitating a white solid. The resulting suspension was stirred for a further 45 min, and the solvents removed in vacuo. The solid was extracted with toluene (15 mL) and the solution filtered. Toluene was then removed under reduced pressure, to yield the crude product as a white powder. Isolated solid: 240 mg (98%); purity 87%. Analytically pure product may be obtained with recrystallization from ether or toluene.

Example 6

Synthesis of $[K(18\text{-crown-}6)][U(\eta\text{-}C_7H_7)_2]$ [397]

The *bis*(cycloheptatrienyl)uranium anion $[U(\eta\text{-}C_7H_7)_2]^-$ was formed in THF (50 mL) by treating a mixture of uranium tetrachloride (212 mg) and potassium (109 mg) with an excess of cycloheptatriene (1030 mg). After stirring for 5 hr at 20°C, the solution was filtered and evaporated to dryness, leaving a green powder of $K[U(\eta\text{-}C_7H_7)_2]$ (200 mg, 52%). In the presence of 18-crown-6 (147 mg), dark green crystals of $[K(18\text{-crown-}6)][U(\eta\text{-}C_7H_7)_2]$ were isolated from THF (210 mg, 52%).

5.3.5 Actinide Hydride and Borohydride Complexes

The first organoactinide hydrides were prepared by hydrogenolysis of $Cp^*_2AnR_2$ (5.67) [268]:

$$Cp^*_2AnR_2 + 2H_2 \rightarrow Cp^*_2AnH_2 + 2RH \qquad (\text{in } C_5H_{12}) \qquad (5.67)$$

The hydride complexes of actinides usually contain a coligand, such as, for example, OR, dmpe, or Cp, and can be monomeric or oligomeric. Thus, the borohydrides $An(BH_4)_4$ produced according to (5.68) are polymeric (An = Th, U) or monomeric (An = Np, Pu), whereas the $An(BH_3Me)_4$ are all monomeric. Their volatility increases from Th to Pu, whereas their stability goes in the opposite direction [282]:

$$AnF_4 + 2Al(BH_4)_3 \rightarrow An(BH_4)_4 + 2AlF_2(BH_4) \qquad (\text{at } 0°C) \qquad (5.68)$$

Borohydrides of Pa, Th, Np, and Pu have properties similar to those of $U(BH_4)_3$ but there are some structural differences [281]. The adduct $UH(BH_4)_3(dme)$ has a U—H bond. $U(BH_4)_3[Ph_2Ppy]_2$ was also reported [281].

A trimeric thorium hydride $Th_3H_6(O-2,6\text{-}t\text{-}Bu_2C_6H_3)_6$, which represents a rare example of an early-metal hydrido complex supported exclusively by aryloxide ligation, was synthesized according to reaction (5.69), and its structure is shown in ***985*** [352]:

$$Th(OAr)_2(CH_2SiMe_3)_2 + H_2 \rightarrow 1/3[ThH_2(OAr)_2]_3 + 2SiMe_4 \qquad (5.69)$$
$$(H_2,\ \text{in benzene, 1.5 atm, 7 days})$$

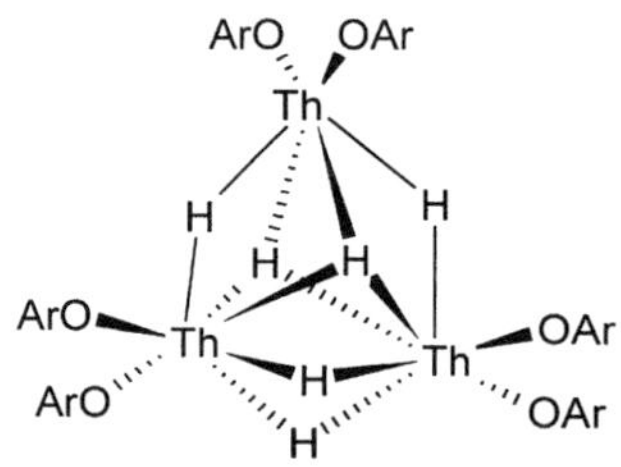

985

This complex exhibits only modest activity in the catalytic hydrogenation of 1-hexene (3 turnovers/hr at 1 atm), in a difference from other aryl oxide-based hydrido derivatives as effective arene hydrogenation catalysts [422]. Similar alkoxide–tetrahydroborate complexes of U(IV) complexes are reported [423].

The reactions of formation of $Cp^*_2AnR_2$ by hydrogenolysis (see Table 5.14) take place via σ-bond methathesis (5.70) [268]:

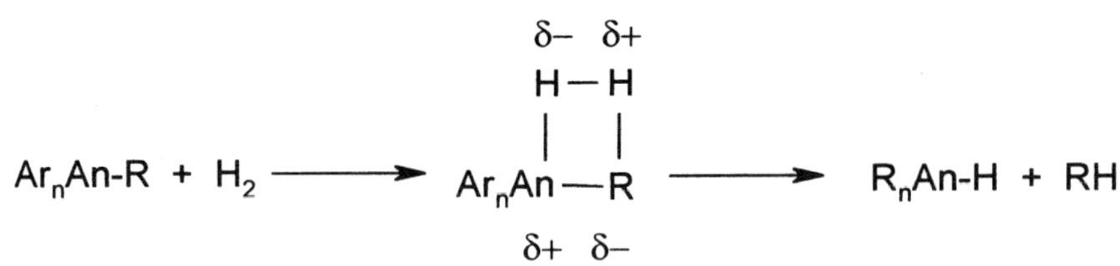

σ-Bond metathesis of actinide-alkyl bonds (5.70)

In case of thorium hydride complex $[Cp^*_2ThH_2]_2$ ***986***, the angles H—Th—Th [58(1)°] and Th—H—Th [122(4)°] and the Th···Th separation of 4.007(8) Å indicates minimal direct metal–metal interaction [268].

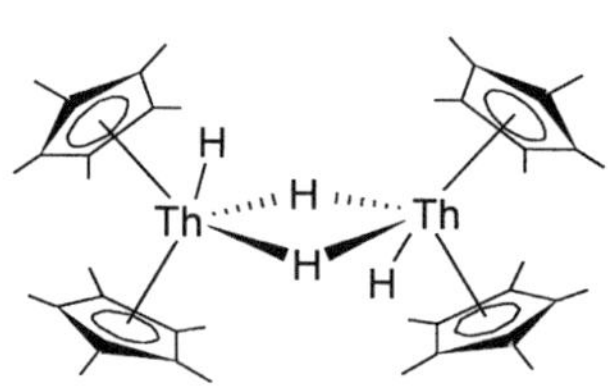

986

An unusual electron-poor cyclopentadienyl complex $(C_5H_4PPh_2,\ BH_3)_2U(BH_4)_2$ was reported [424]. It was shown that ^{31}P NMR spectra of borane adducts of diphenylphosphidocyclopentadienyl–uranium complexes and corresponding molybdenum heterobimetallics are very similar. This strongly suggests that such boranes can be used as simple models for heterobimetallics. Other related uranium borohydride complexes are reported, in particular, the structure of $[Na(THF)_6][(C_5Me_5)U(BH_4)_3]_2$ [425], the first cationic organometallic borohydride and the first organometallic dication of an *f*-element (uranium cationic cyclo-octatetraene derivative, obtained from borohydride precursors) [426b].

Some examples of hydride and borohydride complexes of actinides are presented in Table 5.14.

EXPERIMENTAL PROCEDURE

Example 7

Synthesis of $Th_3(\mu_3\text{-}H)_2(\mu_2\text{-}H)_4(O\text{-}2,6\text{-}t\text{-}Bu_2C_6H_3)_6$ [352]

Benzene (10 mL) was added to a 50-mL Kontes flask containing $Th(O\text{-}2,6\text{-}t\text{-}Bu_2C_6H_3)_2(CH_2SiMe_3)_2$ (0.325 g, 0.40 mmol), and a hydrogen atmsophere (1.5 atm) was placed over the solution. The mixture was stirred for 3 days, and then all solvent was removed in vacuo. ^1H NMR spectra of the crude reaction mixture showed the presence of unreacted reagent and two products. Benzene (8 mL) and hydrogen (1.5 atm) were again added to the flask, and the solution was stirred for an additional 4 days. All solvent was removed in vacuo, and upon addition of 1 mL of hexane, crystals were deposited. Yield 0.085 g (33%).

5.3.6 Actinide Complexes with Macrocyclic Ligands

There are some reported actinide complexes with crown-ethers and other macrocyclic ligands. Thus, diporphyrin out-of-plane sandwich complexes of Th(IV) and U(IV), $An(TPP)_2$, are produced by treatment of the amides $An(NEt_2)_4$ with free porphyrin [282]. Ultrafast electronic deactivation and vibrational dynamics of photoexcited U(IV) porphyrin sandwich complexes were reported [433]. In general, complexes of actinides with macrocyclic Schiff bases can be prepared in two ways: (1) direct complexation of a metal ion with a suitable ligand, and (2) a cyclization reaction in the presence of a metal ion. Thus, the reaction of 1,2-dicyanobenzene with anhydrous UO_2Cl_2 in DMF yields a "superphthalocyanine" uranyl complex; the condensation of ethylenediamine and 2,6-dicarbonylpyridine in the presence of UO_2^{2+} gives a hexa-aza macrocyclic complex [282]. Neutron irradiation of $TcPc_2$ gives the related Pa(IV) complex as a result of a nuclear transformation (5.71) [282]:

$$^{232}ThPc_2(n,\gamma)^{233}ThPc_2 \rightarrow {}^{233}PaPc_2 \qquad (\beta^- \text{ decay of } {}^{233}ThPc_2) \tag{5.71}$$

More information on actinide phthalocyanine complexes is given in Sec. 5.1.6.2.

Complexes with other multidentate ligands, such as complexones, serve as effective sequestrants for the actinide ions. They are soluble both in water and organic solvents [282]. The stability of these complexes grows with increasing ligand dentacity. In the An^{4+}—EDTA complexes, the ligand is hexadentate with a twist conformation [282]. Cyclic and linear catecholyl amides are used as sequestering agents for Pu and other radionuclides [434].

Only An(III) and An(IV) ions have been inserted into macrocyclic crown-ethers, e.g., in $[UCl_3(18\text{-crown-6})]_2[UO_2Cl_3(OH)H_2O]$. Hydrogen bond interactions between etheric oxygen and coordinated water molecules were only observed in the corresponding UO_2^{2+} complexes [282]. The infrared spectrum and isotope effect of uranium–crown-ether complexes were reported [435], as well as the crystal structures

Table 5.14 Actinide Hydride and Borohydride Complexes

Metal or its compound	Reaction system	Products	Conditions or observations	Reference
$U(BH_4)_4$	Thermal decomposition in solution	$U(BH_4)_3$		281
UH_3	B_2H_6, THF	$U(BH_4)_3(THF)_3$		281
$U(BH_4)_4$	THF	$U(BH_4)_4(THF)_2$		281
$U(BH_4)_4$	Et_2O	$[U(BH_4)_4(Et_2O)]_\infty$		281
ThX_4 (X = Cl, Br)	$Li_2C_2B_9H_{11}$	$[Li(THF)_4]_2[Th(\eta^5\text{-}C_2B_9H_{11})_2X_2]$	First dicarbollide complexes of Th	427
$ThBr_4(THF)_4$	CpMgBr(THF)	$CpThBr_3(THF)_3$	Product is a useful synthetic precursor to aryloxide and alkyl derivatives	428
$UBr_4(CH_3CN)_2$	$Li_2C_2B_9H_{11}$, CH_3CN	$Li_2[U^{IV}(\eta^5\text{-} C_2B_9H_{11})_2Br_2]$		429
$UI_3(THF)_4$	$Li_2C_2B_9H_{11}$, tmeda	$[Li(tmeda)][U(C_2B_9H_{11})I_2(THF)_2]$		429
$Cp_2^*UH_2$	dmpe	$Cp_2^*U(H)Cl$		268
$(Me_3XC_5H_4)_3UCl$ (X = C, Si)	$KHBEt_3$	$(Me_3XC_5H_4)_3UH$		430

Cp_3UH ($Cp = C_5Me_5$, $C_5H_4Bu^t$, $C_5H_4SiMe_3$)	Ph_3PBH_3	Cp_3UBH_4		358
$[U(\eta\text{-}C_5Me_5)_2Cl(THF)]$ or $[U(\eta\text{-}C_5H_4R)_3Cl]$ ($R = H$, Bu, or $SiMe_3$)	$[K(THF)_2][ReH_6(PPh_3)_2]$	$K[Cl(\eta\text{-}C_5Me_5)_2UH_6Re PPh_3)_2]$ or $[U(\eta\text{-}C_5H_4R)_3UH_6Re (PPh_3)_2]$	First anionic bimetallic polyhydrides	431
Cp_2MR_2 ($M = U$, Th)	H_2		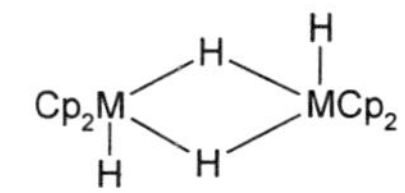	281
$Th(OAr)_2\ (CH_2SiMe_3)_2$	H_2	$Th_3(\mu_3\text{-}H)_2(\mu_2\text{-}H)_4(OAr)_6$		352
$[(\eta^5\text{-}C_2B_9H_{11})_2U^{IV} Br_2] \cdot 2[Li(THF)_4]$	Electrochemical and Na/Hg reduction	$[(\eta^5\text{-}C_2B_9H_{11})_2 U^{III}Br(THF) \cdot 2\ [Li(THF)_x]$ ($x = 2$–4)		432
$(COT)U(BH_4)_2$ (THF)	NEt_3HBPh_4, THF	$[(COT)U(BH_4)(THF)_2][BPh_4]$. It further reacts with the ammonium salt in the presence of HMPA to give $[(COT)U(HMPA)_3][BPh_4]_2$	The first cationic organometallic borohydride and the first organometallic dication of an *f* element	426b

of some of them [435,436]. Actinide crown-ether complexes can be prepared not only from their salts, but also from organometallic compounds. Thus, Cp_3UCl in THF solution of 18-crown-6 under Na/Hg treatment yields $[Cp_3UCl][Na(18\text{-crown-}6)]$ ([263] and references cited therein).

Experimental procedure (synthesis of Th, Pa, and U phthalocyanines [437]): see Sec. 5.1.6.2, Example 29.

5.3.7 Actinide Complexes with N-Containing Ligands

N-Donor bases have low affinity to actinides and usually act as proton acceptors. The An—N bond is rather weak with mono- and even bidentate N-donor ligands [282]. The only 1:1 ammonia adduct is produced with the very acidic $UO_2(hfa)_2$. Amides of actinides $An(NR_2)_4$ are generally associated like alkoxides. For example, the chelating dmed ligand forms a linear trimer $[U(dmed)_2]_3$ and a square tetramer $[U(dmed)_2]_4$, whereas the NEt_2 ligand gives a dimer [282,438]. Other dialkylamides $U[N(SiMe_3)_2]_3$, $[U(NEt_2)_3]BF_4$, $U[N(C_6H_5)_2]_4$, $UO_2[N(SiMe_3)_2]_2(OPPh_3)_2$, $[UO_2CrO_4 \cdot 2CH_3CON(C_2H_5)_2]$ are described [357,439–443].

The structures obtained by x-ray and neutron diffraction analysis of compound $[U\{N(SiMe_3)_2\}\{N(SiMe_3)(SiMe_2CH_2B(C_2B(C_6F_5)_3)\}]$ reveal the electron deficiency of the uranium atom to be effectively compensated by the formation of multicenter bonds between the U atom and Si—CH_2 units of the amido ligands; the x-ray structure of $[U\{C(Ph)(NSiMe_3)_2\}_2\{\mu_3\text{-}BH_4\}_2]$ proves unequivocally the μ_3-coordination of the BH_4 units [444]. The structure of a rare uranium(III) complex with a tripodal aromatic amine *tris*[(2,2′-bipyridin-6-yl)methyl]amine was reported [445].

Amido complexes of uranium can serve as neutral precursors to obtain cationic uranium compounds by protonolysis of the U—NEt_2 bond. Thus, complexes with the cations $[U(\eta\text{-}C_5H_5)_3(THF)]^+$, $[U(\eta\text{-}C_5Me_5)(NEt_2)_2(THF)_2]^+$, $[(U\text{-}\eta\text{-}C_5R_5)_2(NEt_2)(THF)]^+$, $[U(\eta\text{-}C_5R_5)(\eta\text{-}C_8H_8)(THF)_2]^+$ (R = H or Me) [446], $[U(NEt_2)_3]^+$, $[U(NEt_2)_2(THF)_3]^{2+}$, etc. [439] have been synthesized in this manner. Several new complexes of uranium and thorium incorporating the triply trimethylsilylated ligand $N(CH_2CH_2NSiMe_3)_3$ with further bonds from the metal to chlorine, carbon, hydrogen, and oxygen are reported. The chloro, pentamethylcyclopentadienyl, and tetrahydroborato derivatives are structurally characterized by x-ray diffraction [447]. The actinide–amide chemistry is covered in a recent review [448].

Imido complex examples are scarce: $Me_3SiN{=}U^V[N(SiMe_3)_2]_3$, $Me_3SiN{=}U^{VI}F[N(SiMe_3)_2]_3$, $Cp_2U(N\text{-}2,4,6\text{-}Bu^tC_6H_2)$ [449,450a]. For the complexes $(C_5H_4Me)_3U{=}NR$ (R = Me_3Si, Ph), both lone pairs of electrons of the N atom are involved in bonding to uranium [263,405]. In the example of imido complexes, the first example of intramolecular activation of a (C_5Me_5) methyl C—H bond in an actinide complex was discovered [450b]. Thus, thermolysis of $(C_5Me_5)_2U({=}NAd)_2$ (Ad = 1-adamantyl) in benzene or hexane results in the intramolecular C—H bond activation of a methyl group on a pentamethylcyclopentadienyl ligand across the two imido functional groups. The activation product is a reduced U(IV) metallocene *bis*(amide) complex with an N-bound methylene unit derived from the methyl group attached to one amide group [450b]. Among the simplest nitrogen ligands, the N_3^- anion gives rather stable complexes with the UO_2^{2+} cation ($K_{diss} = 5 \times 10^{-3}$), comparable with fluoride ones [282]. Heterodinculear uranium/molybdenum dinitro-

Table 5.15 Some Actinide Complexes with N-Containing Ligands

Metal or its compound	Reaction system	Products	Conditions or observations	Reference
UO_2Cl_2	$KN(SiMe_3)_2$, THF	$UO_2[N(SiMe_3)_2]_2(THF)_2$ or $[K(THF)_2][UO_2\{N(SiMe_3)_2\}_3]$	The alcoholysis of the first complex with HO-2,6-*i*-$Pr_3C_6H_3$ in the presence of py produces $UO_2(O\text{-}2,6\text{-}i\text{-}Pr_2C_6H_3)_2(py)_3$	469
$ThBr_4(THF)_4$	KNR_2	$Th(NPh_2)_4(THF)$ or $K[Th(NMePh_5)]$	The amino species can be used to make alkoxides by interaction with alcohols	281, 372
$UCl_3(THF)_3$	$NaN(SiMe_3)_2$, THF	$U[N(SiMe_3)_2]_3$	25°C	374
$U[N(C_6H_5)_2]_4$	L = py, Et_2O, THF, $(EtO)_3PO$	$U[N(C_6H_5)_2]_4 \cdot L$	The apparent trend in coordination ability is $(EtO)_3$ PO > THF ≅ py > Et_2O	441
UCl_4	$LiNEt_2$	$U(NEt_2)_4$		281
$\{[(Me_3Si)_2N]_2Th(CH_2SiMe_2NSiMe_3)\}$	HNMePh	$Th[N(SiMe_3)_2]_2(NMePh)_2$		372
$[(Me_3Si)_2N[ThN(SiMe_3)(SiMe_2CH_2)]$	Trifluoromethane sulfonic acid, then heating with CpH	$Cp[(Me_3Si)_2N]Th(\mu_2\text{-}OSO_2CF_3)_3$ $Th[N(SiMe)_3(SiMe_2CH_2)]Cp$		469
$Cp_3U{=}C(H)PPh_2Me$	$HNPh_2$	Cp_3UNPh_2		263 and references cited therein
$(C_5H_4Me)_3U(THF)$	RN_3 (R = Me_3Si, Ph)	$(C_5H_4Me)_3U{=}NR$	The first organouranium(V) complexes	405
$Cp_2U(NEt_2)_2$	RNC	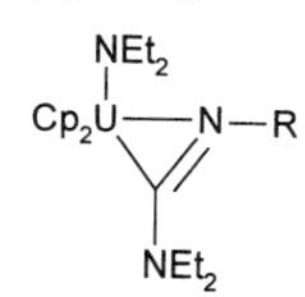		281

gen complexes are reported [451]; related infrared spectroscopic and quasirelativistic theoretical study of the coordination and activation of dinitrogen by thorium and uranium atoms was carried out almost simultaneously [452].

Mononuclear complexes $[U(C_5Me_5)_2(NHR)_2]$ (R = 2,6-dimethylphenyl, Et, or Bu) have been synthesized and structurally characterized. It was shown that in the presence of terminal alkynes and amines these complexes catalyze the intermolecular hydroamination of terminal alkynes [453]. Complex formation reactions of U(VI) with neutral N-donors in DMSO were reported [454].

An analogue of the uranyl ion was discovered in the crystal structure of the air-stable compound $PPh_4^+[UOCl_4\{NP(m\text{-toluene})_3\}]^-$ ***987*** [455]. This complex (red crystalline solid) was prepared in CH_2Cl_2 by the replacement of a chlorine atom in $PPh_4^+[UOCl_5]^-$ via the formation of Me_3SiCl (5.72):

$$\begin{aligned} &PPh_4^+[UOCl_5]^- + Me_3Si\{NP(m\text{-toluene})_3\} \\ &\rightarrow PPh_4^+[UOCl_4\{NP(m\text{-toluene})_3\}]^- + Me_3SiCl \qquad (5.72) \\ &\qquad\qquad \mathit{987} \end{aligned}$$

According to x-ray diffraction data, the ion $[UOCl_4\{NP(m\text{-toluene})_3\}]^-$ contains a linear O=U=N group with U—O and U—N distances of 176 and 190 pm, respectively, coordinated by four chlorine atoms. The U—O distance is typical of uranyl compounds {a small number of reported uranium compounds known to contain imido or phosphorane iminato groups are collected in the same publication [455]}, and the U—N distance suggests a bond order > 2. The near linearity (171.9°) of the U—N—P bond indicates full involvement of the nitrogen π-electrons in the uranium–nitrogen bond. The presence of both *f*- and *d*-valence orbitals in uranium allows each uranium oxygen bond in the uranyl ion a formal bond order of 3 [455,456a]. The electronic and geometric structures of analogues of the uranyl ion containing the —N=U=N— unit $\{[UCl_4(NPR_3)_2]$ (R = H, Me)} have been studied computationally using quasirelativistic gradient-corrected density functional theory. Other anionic actinide complexes can be obtained directly from the salts, for example, both UF_5 and UF_6 react with nitrogen bases to give molecular or ionic species, such as $[(bipy)_2H]^+[UF_6]^-$, $UF_4(2\text{-}FC_5H_4N)$, $U_2F_{12}(bipy)$ [457a]. Structural and spectral data for the Np(V)–bipy complex $[NpO_2(bipy)(H_2O)_3](NO_3)$ are reported [457b]. X-ray study of the anionic berkelium complex $(Me_4N)_2[BkCl_6]$ showed that it is isostructural with K_2PtCl_6 [458].

Thermolysis of thorium amido complex $Th(OSO_2CF_3)[N(SiMe_3)_2]_3$ with pentamethylcyclopentadiene produces the dinuclear product ***988*** (5.73) [373]:

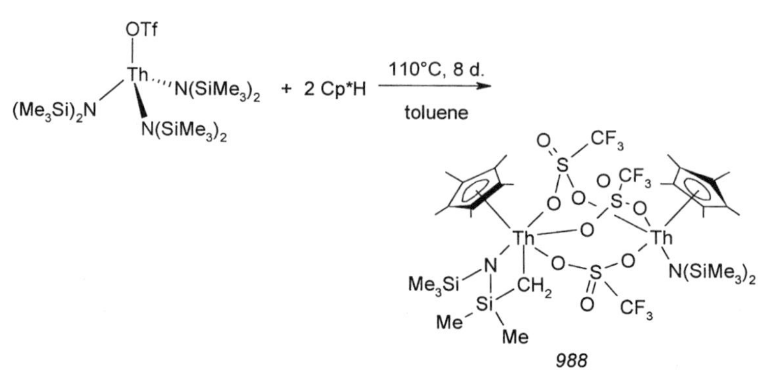

(5.73)

A red paramagnetic (1,4-diazabutadiene)thorium complex ***989*** (monoclinic, space group $P2_1/n$) was the first obtained diazabutadiene (DAB) derivative of an actinide metal [459]. It was prepared in THF at −80°C from [{(tren′)ThCl}$_2$], containing amino-*tris*{ethyl(trimethylsilyl)amido} (tren′) ligand which provides a suitable environment for the stabilization of the product (5.74):

$$[\{(\text{tren}')\text{ThCl}\}_2] + 2[\text{Li}(\text{Bu}^t_2\text{DAB})_2] \rightarrow \underset{989}{[(\text{tren}')\text{Th}(\text{Bu}^t_2\text{DAB})]} + 2[\text{Li}(\text{Bu}^t_2\text{DAB})]\text{Cl} \quad (5.74)$$

While Th(IV) compounds are diamagnetic, paramagnetic species with the sterically demanding ligand tren′ have a singly reduced 1,4-diazobutadiene ligand, i.e., $(\text{Bu}^t_2\text{DAB})^{\cdot -}$ [460]. The same (tren′) ligand was also used for the synthesis of similar uranium compounds [461,462].

Schiff base and related complexes of uranium and thorium are widely described in recent literature and covered in a review [463]. Those of U(VI) have a practical use as catalytic organic oxidants [460] or as part of a polystyrene-supported chelating resin [464,465]. Among other Schiff base precursors, salicylaldehyde [466] and triethylenetetramine [464], 3-formylsalicylic acid and *o*-hydroxybenzylamine [465], or salicylaldehyde and 1-amino-2-naphthol-4-sulfonic acid [467] were used. In the example of Schiff base complexes, kinetics of formation of U(VI) complexes and their pK_a values were studied [468].

EXPERIMENTAL PROCEDURES

Example 8

Synthesis of $PPh_4^+[UOCl_4\{NP(m\text{-toluene})_3\}]^-$, Containing a Linear O=U=N Group [455]

$Ph_4P[UOCl_5]$ (1.8 mmol) and $Me_3Si\{NP(m\text{-toluene})_3\}$ (2.0 mmol) were allowed to react in dry CH_2Cl_2 (30 mL) at room temperature for 5 hr, forming a clear red solution onto which a layer of an equal volume of toluene was added. After interdiffusion for several weeks, well-formed needles of $PPh_4^+[UOCl_4\{NP(m\text{-toluene})_3\}]^-$ were isolated in ca. 30% yield.

Example 9

Synthesis of $U[NPh_2]_4$ [441]

This compound was prepared by refluxing $HNPh_2$ (1.86 g, 11.12 mmol) with $[(Me_3Si)_2N]_2\overline{U[N(SiMe_3)(SiMe_2CH_2}\text{)]}$ (2.00 g, 2.78 mmol) in toluene (20 mL) for 6 hr. The resulting dark red solution was evaporated to a red oil in vacuo, after which hexane (5 mL) was added causing the formation of a dark red solid after 12 hr at room temperature. The solution was decanted from the solid and dried in vacuo, typically yielding 1.84 g (70%) of the product.

Example 10

Synthesis of $U[N(SiMe_3)_2]_3$ [359]

In the glovebox, 18.2 g (99.3 mmol) of $NaN(SiMe_3)_2$ was placed in a 250-mL Erlenmeyer flask and 100 mL of THF added. Then a 1-L Erlenmeyer flask was charged with 30.0 g (33.1 mmol) of $UI_3(THF)_4$, a 1-in. Teflon-coated stir bar, and 300 mL of THF. The sodium silylamide solution was added to the stirred uranium tri-iodide solution over a period of 1 hr to produce a deep red, wine-colored solution. After 12 hr of stirring, the solution was vacuum–filtered through Celite on a coasrse filter frit, and solvent was removed in vacuo to provide a purple–red microcrystalline solid. The solid was extracted into 100 mL of hexane, the extract was filtered through a fine-porosity frit, and the filtrate was cooled to −40°C to produce large purple needles, which were isolated by filtration and dried in vacuo. Reduction of the solvent volume and cooling to −40°C produced another crop of purple needles. Continued yield 19.2 g [82% based on $UI_3(THF)_4$].

5.3.8 Complexes with P-Containing Ligands

In numerous reported complexes of actinides with oxygen-containing phosphorus ligands, the metal atom is always O-bound [282]. There are a few examples of actinide phosphine complexes with An—P bonds [470]. Tetrahalides of Th(IV) and U(IV) react with pure PMe_3 giving 1:2 adducts, $MCl_4(PMe_3)_2$ [282,471a]. The eight-coordinate complex $U[Me_2PCH_2CH_2PMe_2]_4$ is isostructural with its thorium analogue [471b]. Coordination chemistry of uranium with phosphorus donors is reviewed [472].

The phospholyl ligands, mentioned in Sec. 5.3.4 as formally analogous to Cp, play a bridging function, in particular, in the dimeric U(III) complex $[\{U(\eta^5\text{-}C_4Me_4P)(\mu\text{-}\eta^5,\eta^1\text{-}C_4Me_4P)(BH_4)\}_2]$ (5.62) [263,396,473]. Reactions of the potentially tridentate diphosphinoamide ligands $N(CH_2CH_2PR_2)^{2-}$ (R = Et or *i*-Pr) as their lithium salts with uranium or thorium tetrachlorides afford the first diphosphinoamido complexes of the actinides. The ligands are shown to be versatile in that unusual complexes of uranium(IV), for example $\{UCl_2[N(CH_2CH_2PEt_2)_2]_2\}_2$, uranium(V), and thorium(IV) are isolated. These may be mono-, di-, and trisubstituted, depending upon the reaction conditions, and variations in substituents on the neutral phosphine donors radically influence the nature of the coordination complexes obtained [474].

Coordination compounds of diphosphazane dioxides with uranyl or thorium ions were synthesized [475]. The crystal structure of $[UO_2(NO_3)_2L_1]$ [$L_1 = Ph_2P(O)N(Ph)P(O)Ph_2$] reveal the bidentate chelating mode of binding of the diphosphazane dioxide to these metals. The chemistry of other uranium organophosphorus and related compounds is described [476–479]. Some of the actinide complexes are presented in Table 5.16.

Actinide complexes containing phosphines and cyclopentadienyl ligands are rare. Two unique examples are the $Cp_2ThX_2(Me_2PCH_2CH_2PMe_2)$ (X = Cl, Me, $CH_2C_6H_5$) [411] and a monomeric U(III) hydride complex $Cp^*_2U(H)[P(CH_2CH_2PMe_2)_2]$ ***990*** [480] which was synthesized by reacting the Cp^*UR_2 precursor with $[P(CH_2CH_2PMe_2)_2]$ in toluene, under a positive dihydrogen pressure, at −20°C:

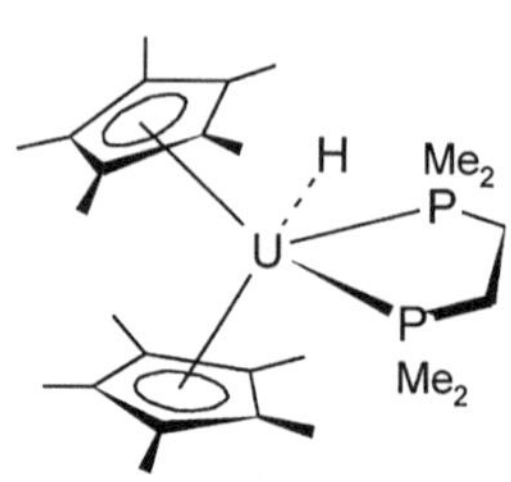

990

Table 5.16 Some Actinide P-Containing Complexes

Metal or its compound	Reaction system	Products	Conditions or observations	Reference
MCl_4 (M = Th, U)	(1) (Li or K) $P(CH_2CH_2PMe_2)_2$, toluene, THF (2) $N(CH_2CH_2PR_2)^{2-}Li_2^+$ (R = Et, *i*-Pr)	(1) $M\{P(CH_2CH_2PMe_2)_2\}_4$ (2) $\{MCl_2[N(CH_2CH_2PEt_2)_2]_2\}$		471, 474
$[UOCl_5]^-$	$Me_3Si[NP(m\text{-tol})_3]$	$[UOCl_4NP(m\text{-tol})_3]^-$	The red Ph_4P^+ salt is stable in air and soluble in CH_2Cl_2 and CH_3CN	450
$UO_2Cl_2(Ph_3PO)_2$	KO-*t*-Bu or $NaN(SiMe_3)_2$ in THF	$UO_2(O\text{-}t\text{-}Bu)_2(Ph_3PO)_2$ or $UO_2[N(SiMe_3)_2]_2(Ph_3PO)_2$		442
$Cp_2^*UCl_2$ ($Cp^* = C_5Me_5$)	$HNPPh_3$	$Cp_2^*UCl_2(HNPPh_3)$	The first *f*-element complex with a phosphine imine ligand	483a
$UO_2(NO_3)_2 \cdot 6H_2O$	HF, $(HO)(O)PHC_6H_5$	$UO_2(O_2PHC_6H_5)_2$	In a plastic beaker	483b

Actinide borohydride complexes have been synthesized by reacting $U(BH_4)_3(THF)_x$ and several diphosphine ligands, yielding $U(BH_4)_3(L)_2$ adducts (L = dimethylphosphine ethane [481] or diphenylphosphine pyridine [482a]). These compounds are, in general, fluxional on the NMR time scale, even at low temperatures. Among other actinide coordination compounds with P-containing ligands, the Np(V) complex with triphenylphosphine oxide $[NpO_2\{PO(C_6H_5)_3\}_4]ClO_4$ [482b], thorium and uranium complexes with tetraphenylimidophosphine $Th(tpip)_4$, $U(tpip)_4$, $UCl(tpip)_3$ [482c], and *bis*(diphenylphosphino)ethane dioxide [482d] were structurally characterized.

EXPERIMENTAL PROCEDURE

Example 11

Synthesis of $U\{P(CH_2CH_2PMe_2)_2\}_4$ [471]

To a stirred solution of UCl_4 in THF (0.14 M, 3 mL) at −80°C was added $KP(CH_2CH_2PMe_2)_2$ in 4:1 toluene/THF (0.1618 M, 10 mL). The mixture was stirred for 8 hr, during which time the color changed from green to black. The volatile materials were removed in vacuo. The resultant black solid was washed with cold (−80°C) light petroleum ether (2 × 50 mL). The black supernatant was evaporated to ca. 20 mL and cooled (−20°C) for 4 hr. The product crystallizes as small black prisms in good yield (0.33 g, 72%). M.p. 132–136°C.

5.3.9 Actinide S-Containing Complexes

As hard acids, actinides usually do not form stable complexes with soft-base S-donor ligands. The following few complexes, sensitive to hydrolysis, have been reported, among others: $UCl_4(dmte)_2$, $An(S_2CNEt_2)_4$ (An = Th, U, Np, Pu), $[NEt_4][Np^{III}(S_2CNEt_2)_4]$, $[NMe_4][UO_2(S_2CNEt_2)_3]$ [282], $U(SBu^n)_4$, $[U(SBu^n)_6]^{2-}$, $[U(SPh)_6]^{2-}$ [484], a unique example of a uranium(V) compound with metal–sulfur bonds $[Na(18\text{-crown-}6)(THF)][U(\eta^8\text{-}C_8H_8)(C_4H_4S_4)_2]$ [485a], structurally characterized Np(V) complex with dimethyl sulfoxide $[(NpO_2)_2\{SO(CH_3)_2\}_7(H_2O)_2](ClO_4)\cdot 2H_2O$ [486b], and similar *octakis*(dimethylsulfoxide) uranium(IV) compound [485c]. The first homoleptic dithiolate complex of an *f*-element, *tetrakis*(ethane-1,2-dithiolato)uranate(IV) $[Li(dme)]_4U(edt)_4]\cdot DME$ ***991*** (5.75) (space group $P2_1/n$) has been reported [486]:

$$UCl_4 + 4Li_2(SCH_2CH_2S) \xrightarrow{\text{in DME}} [Li(dme)]_4[U(SCH_2CH_2S)_4]\cdot DME \quad (5.75)$$

991

The nature of the U—S bonds in this complex was estimated by Hückel semiempirical calculations; it was found that U *6d*, *7s*, and *7p* orbitals are responsible for interactions with sulfur ligands; as a consequence, no significant U *5f* participation in bonding is discernible, and the U—S π-interactions are weak. The same authors [486] also presented a brief "state of the art" of the limited chemistry of actinide metal ions with S-donor ligands. Among other obtained thiolates, U(IV) tetrathiolate compounds were prepared by treatment of $U(NEt_2)_4$, $U(BH_4)_4$, or $U(SBu\text{-}n)_4$ with thiols or by oxidation of uranium metal by disulfides [487]. The U—S bond of $U(SPr\text{-}i)_4$ is found to be reactive towards H^+, I_2, and CS_2 [487].

The first organouranium complexes with bridging thiolate ligands [{U(cot)(μ-*SR*)$_2$}$_2$] (cot = η-C_8H_8) are prepared by treating [U(cot)(BH_4)$_2$] with corresponding thiol or NaSR (R = *n*-Bu, *i*-Pr, or *t*-Bu) reagent. The structure of these compounds differs from that of their alkoxide analogues that contain only two binding OR groups [488]. Among other interesting compounds, it is necessary to mention the first structurally characterized uranium tetrathiolate complex, U(SPr-*i*)$_4$[OP(NMe$_2$)$_3$]$_2$ and the first uranium–sulfur cluster, U_3(μ_3-S)(μ_3-SBu-*t*)(μ_2-SBu-*t*)$_3$(SBu-*t*)$_6$ [489].

Recently, the first U(V) heteroleptic cyclo-octatetraene–dithiolene complex ***992*** was synthesized by $AgBPh_4$ oxidation in THF of the [(COT)U^{IV}(dddt)$_2$][Na(18-crown-6)]$_2$ precursor, which was synthesized by reacting [(COT)UX_2(THF)$_n$] (X = BH_4, n = 0; X = I, n = 2) with Na$_2$dddt [490]. In general, despite the relatively small number of reported actinide S-containing complexes, they have a potentially rich coordination chemistry of such compounds of "hard" *f*-elements with "soft" S-ligands [489] and possible applications [491].

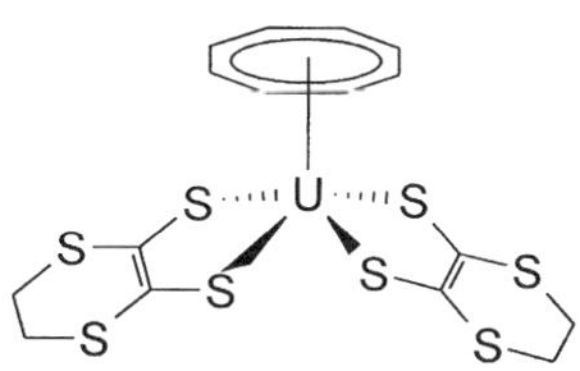

992

Some uranium and thorium complexes with the above ligands are presented in Table 5.17.

EXPERIMENTAL PROCEDURE

Example 12

Synthesis of [Li(dme)]$_4$[U(SCH$_2$CH$_2$S)$_4$]·DME [486]

A DME slurry (20 mL) of UCl_4 (0.50 g, 1.3 mmol) was carefully added to a DME suspension (15 mL) of Li$_2$edt (0.42 g, 4.0 mmol) under an argon atmosphere. The mixture was stirred for 1.5 hr at 0°C. After insoluble materials were filtered off through a medium-porosity glass frit, the bright yellow–green solution was concentrated to ca. one-fifth of its original volume. It was then cooled to −20°C for one night, during which the product crystallized as apple green prisms (0.47 g, 43% yield based on Li$_2$edt).

As mentioned in Sec. 5.3.1, at present the coordination and organometallic chemistry of actinides is represented mainly by uranium and thorium complexes with the ligands above. For other elements of this series, only some complexes have been studied in solid phase or solution [323,330b,333d,457b,458,482b,485b,493b–f], although their corresponding complex-formation processes with organic ligands have been used for extraction and separation of these metals from each other and lanthanides [493g,h]. In this respect, the recently published works of Gibson and Haire [493d–f] on Cm, Bk, and Cf organometallic chemistry should be especially noted. Here the organoberkelium and organocalifornium ions $BkCH_3^+$, BkC_2^+, AnC_2H^+, $AnCN^+$, AnC_4H^+, $An(OH)CN^+$, etc., were studied by mass spectrometry. Several of these complexes represent the first organoactinides in which Bk and Cf are

Table 5.17 Some Uranium and Thorium S-Containing Complexes

Metal or its compound	Reaction system	Products	Conditions or observations	Reference
UCl_4 or $U(BH_4)_4$	NaSR (R = *t*-Bu, Ph), THF	$[(THF)_3Na(\mu\text{-}SR)_3$ $U(\mu\text{-}SR)_3\ Na(THF)_3$	Homoleptic uranium(IV) thiolates	492
UI_4	DMSO or DIBSO	$UI_4(DMSO)_n$, n = 6,8	25°C	366
$Cp_2^*UCl_2$	Na_2dddt	Cp_2^*U(dddt)		490
$U(NEt_2)_4$	(*i*-Pr)SH	$U\{S(i\text{-}Pr)\}_4[OP(NMe_2)_3]_2$		493a

directly bonded to carbon. It is established that "both Bk and Cf exhibit a propensity to form 'monovalent' organoactinides in which the metal center forms a single σ-type bond to the organic moiety. Most earlier actinides form 'divalent' complexes having two $An^+—C$ bonds." A bonding model was applied in which the participation of one or two non-5*f* valence electrons at the An^+ center for participation in direct $An^+—C$ bonding is supposed [493e].

5.3.10 Technetium Complexes

The chemistry of technetium has become very important, especially in relation to the use of isotope ^{99m}Tc as a diagnostic agent in nuclear medicine. It has been used for many years in bone scanning and more recently in studying different diseases of the heart, brain, kidneys, liver, and other organs as well as tumor tissue. The Tc complexes are of interest for the radiopharmaceutical industry [494], and ^{99m}Tc is the radiosotope of choice for imaging in diagnostic nuclear medicine due to its ideal energy $E_\gamma = 140$ keV, lack of particulate radiation dose, half-life of 6 hr, and wide availability [495,496].

Some other applications of technetium are the following [283]: use of some of its alloys as superconductors with high critical temperature, use of ^{99}Tc in high-temperature thermocouples, construction of basic anticorrosive covers for nuclear reactors, etc.

Complexes with N-containing ligands. A series of Tc(III), Tc(II), and Tc(I) complexes with pyridine ligands was reported [497]. The interest of the authors was "to develop a coordinatively unsaturated, low-valent, *electron-rich*, Tc metal center dominated by a very weak π-acid ligand environment." The Tc^{III} complexes were prepared via the substitution chemistry of $TcCl_3(PPh_3)_2(CH_3CN)$ (see Example 12), and the Tc(II) and Tc(I) complexes were obtained by their subsequent reduction by zinc dust. The formed products {some of them are $TcCl_2(py)_4$, $TcCl_3(PPh_3)_2$(tmeda), $TcCl_3$(*t*-butyl$_3$tpy), [Tc(tpy)(py)$_3$]Cl} were characterized by electrochemical, x-ray, and spectrophotometric methods. According to the data obtained, there is a significant π-backbonding interaction in the cases of Tc(II) and Tc(I) complexes relative to Tc(III). Thus, a decrease of 0.04–0.06 Å in Tc—N bond lengths between Tc(III) and Tc(II) pyridine complexes and a decrease of 0.09 Å in Tc—N(internal) bond lengths between Tc(III) and Tc(I) terpyridine complexes take place [497]. These effects support a stabilization of the low oxidation states of the metal. The Tc(III) pyridine complexes exhibit Knight-shifted ^1H NMR spectra, transitions in the visible spectra that are tentatively assigned as charge transfer from the halide to the metal, and multiple reversible electrochemical redox couples.

The reduction of pertechnate (as well as perrhenate and molybdate) with 2-hydrazinopyridine dihydrochloride in methanol led to the preparation of a class of complexes containing the $[M(\eta^1\text{-}NNC_5H_4NH_x)(\eta^2\text{-}HNNH_yC_5H_4N)]$ (M = Tc, Re, Mo) core, represented by $[TcCl_3(NNC_5H_4NH)(HNNC_5H_4N)]$ [498]. The last compound was used to obtain $[Tc(C_5H_4NS)_2(NNC_5H_4N)(HNNC_5H_4N)]$ (Example 13, C_5H_4NS is pyridine-2-thiol), which is a precursor of the ^{99m}Tc–peptide imaging agents. Such bifunctional hydrazine ligands, used in this work, are effective and versatile linkers for labeling antibodies and protein fragments [498, ref.1].

Other organohydrazine ^{99m}Tc complexes have been synthesized. $[TcCl_2(C_8H_5N_4)(PPh_3)_2]\cdot 0.75C_7H_8$ and $[TcNCl_2(PPh_3)_2]\cdot 0.25CH_2Cl_2$ were prepared from the reaction of $[TcOCl_4]^-$ and hydrazine dihydrochloride in toluene at room temperature or by refluxing them in CH_2Cl_2, respectively [499]. A Tc(III) organohydrazine complex $[Tc(NNC_5H_4N)(PPh_3)_2Cl_2]$ ***993*** was obtained by reaction of $[Tc(MeCN)(PPh_3)_2Cl_3]$ with 2-hydrazinopyridine [500]:

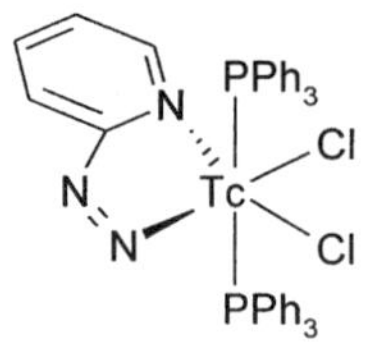

993

Complexes of Tc(III) where O-atom transfer occurs by reacting $[TcOCl_4]^-$ with several phosphine ligands in 4-picoline as solvent yield the *mer*-$[Cl_3(pic)_3Tc]$ and *mer*-$[Cl_3(pic)(PMe_2Ph)_2Tc]$ complexes. The products were characterized by x-ray diffraction and spectrophotometric methods [501].

Technetium complexes with the $[Tc{\equiv}N]^{2+}$ core are more stable in high oxidation states than the corresponding $[Tc{=}O]^{3+}$ (technetyl) complexes and their chemistry is similar to that for a technetyl core. The Tc(V) complex $[TcN(L^1)(H_2O)]\cdot 2H_2O$ ***994*** has been synthesized by a substitution reaction of $[TcNCl_2(PPh_3)_2]$ with tetraazamacrocycle H_2L^1 [502]. Another complex possessing the $[Tc{\equiv}N]^{2+}$ core was obtained using ancillary ligands such as multidentate ligands having phosphorus and nitrogen atoms $[TcNBr_4]^-$, and reacts with bipy in ethanol to yield a *cis*-octahedral $[TcNBr(bipy)_2]_2[TcBr_4]$ $[Tc{\equiv}N]^{2+}$ core complex containing a tetrahedral tetrabromo technectate(II) cation [503].

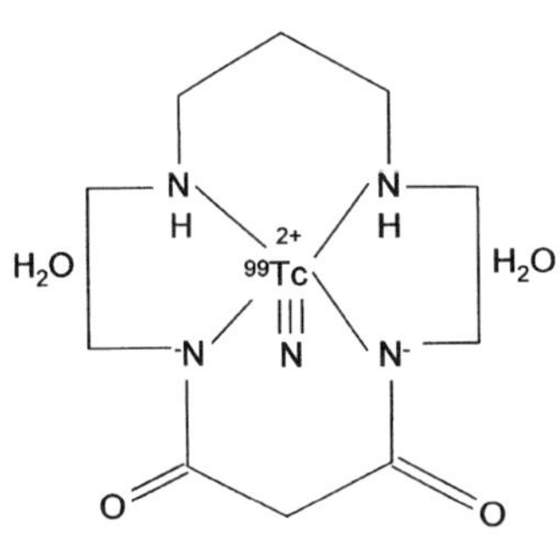

994

Other nitridotechnetium(V) chelate complexes with N_2S_2 ligands were obtained by the reaction of $[TcNCl_2(PPh_3)_2]$ with $HSCR_2CH_2NR'CH_2CH_2NR'CH_2CR_2SH$ (R = Me, Et and R′ = Me, Et) producing the corresponding chelate complex containing the $[Tc{\equiv}N]^{2+}$ core [504]. The similar chemistry of Mo^VO and $Tc^{VI}N$ cores has been explored, and mixed-ligand complexes such as $[\{TcN(S_2CNEt_2)\}_2(\mu\text{-}O)_2]$, $[\{TcN(S_2CNC_4H_8)\}_2(\mu\text{-}O)_2]$, $[AsPh_4]_2[\{TcN(CN)\}_2(\mu\text{-}O)_2]$, and $[AsPh_4]_2[\{TcN(edt)\}_2(\mu\text{-}O)_2]$ (H_2edt = ethane-1,2-dithiol) were obtained by the reaction of $[\{TcN(OH_2)_3\}_2(\mu\text{-}O)_2]^{2+}$ or $Cs_2[TcNCl_5]$ in $Na_4P_2O_7$ solution with the appropriate ligand [505].

Nitrido complexes with a ferrocene dithiocarboxylate ligand were synthesized by reacting $[TcNCl_2(PPh_3)_2]$ with the piperidinium salt of the ligand $FcCS_2$, resulting in the $Tc(N)[Fe(C_5H_4CS_2)(C_5H_5)]_2$ ***995*** complex. The two ferrocene units behave as independent redox centers bridged by the $[Tc{\equiv}N]^{2t}$ core, as seen by cyclic voltammetry [506]. Complexes with the chelating ligand $N(SPPh_2)_2$ synthesized by the reaction of $[TcN(Cl)(PPhMe_2)_3]$ or $[TcNCl_2(PPh_3)_2]$ with the sodium salt of the ligand yield $[TcNL_2]$ ***996*** and $[TcN(Cl)(PPhMe_2)L]$, respectively $[L = N(SPPh_2)_2]$ [507]:

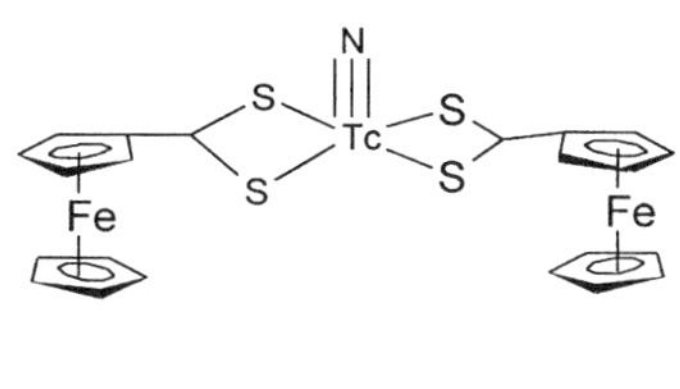

995

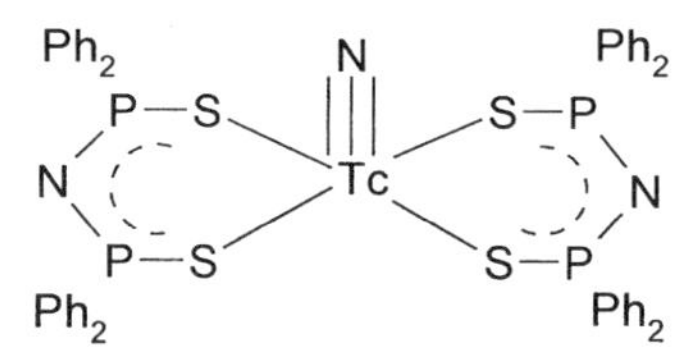

996

Tc–nitrido complexes with N-protected aminoacid derivatives with 2,5-dimethyldithiocarbazoic acid (Hdtc) were reported [508a] as compounds for radiopharmaceutical applications. $[TcN(L^n)(PPh_3)]$, where L^n = z-Gly-dtc ($n = 1$), z-Ala-dtc ($n = 2$), z-Phe-dtc ($n = 3$), z-Val-dtc ($n = 4$), and z-Leu-dtc ($n = 5$), were synthesized and characterized by spectroscopic methods and x-ray crystallography. Geometrically controlled selective formation of nitrido Tc(V) asymmetrical heterocomplexes with bidentate ligands is reported [508b].

Binuclear Tc(VII) and Tc(VI) complexes were obtained by sodium reduction of $Tc(NAr')_3I$ (Ar′ = 2,6-dimethylphenyl) or $Tc(NAr)_3I$ (Ar = 2,6-di-isopropylphenyl), resulting in the $Tc_2(NAr')_4(\mu\text{-}NAr')_2$ and $Tc_2(NAr)_6$ complexes with edge-bridged tetrahedral and "ethene-like" conformation, respectively [509]. Other binuclear complexes involving Tc(VII) and Tc(V) with the bridging ligand bptz have been prepared by reacting the pertechnetate ion and the ligand bptz·2HCl in the appropriate alcohol (X = OMe, OEt) yielding the respective product with general formula $(\mu\text{-bptz})[TcO_3X]_2$ ($X = Cl, OCH_3, OCH_2CH_3$). Mononuclear complexes of the ligand 4-phenyl-3,6-*bis*(2′pyridyl)pyridazine (pppz) were also prepared from pertechnetate and $TcOCl_4$ in ethanolic aqueous hydrochloride acid solutions [510]. A binuclear oxo-bridged Tc(III) polypyridyl complex $[(tpy)(Me_2Bpy)Tc{-}O{-}Tc(tpy)(Me_2Bpy)](OTf)_4$ (OTf^- = trifluoromethanesulfonate $CF_3SO_3^-$) was prepared from the reaction of $TcCl_3(tpy)$ and TlOTf and water present in the system [511].

Diazenido Tc-complexes were obtained by reacting $[TcCl(NNR)_2(PPh_3)_2]$ ($R = {-}C_6H_4\text{-}p\text{-}Cl$) with bidentate ligand S_2CNR_2 and maltol, yielding $[Tc(NNR)L_2(PPh_3)]$ and $[TcCl(NNR)L(PPh_3)]$ in high yield. Reaction of $[TcO_4]^-$ with arylhydrazine hydrochlorides generates a diazenide species in situ which reacts with S_2CNR_2 to give $[TcCl(NNR)_2(S_2CNR_2)_2]$ and bipy to give $[TcCl(NNR)(bipy)_2][BPh_4]$ [512].

Monocapped boronic acid adducts of Tc(III) *tris*(dioxime) (BATO) (X = Cl, Br; dioxime = dimethylglioxime, cyclohexanedione dioxime; $R = CH_3$, C_4H_9) have been prepared by template synthesis starting with pertechnetate and stannous ion $[NBu_4][TcOCl_4]$ or $M_2[TcXO]$ ($M = NH_4$, X = Cl, Br), yielding seven-coordinate monocapped boronic acid adducts of technetium *tris*(dioxime)(BATO) complexes [513].

An acetonitrile-containing Tc(III) complex $[TcCl_3(MeCN)\{PR_3\}_2]$ ($R=C_6H_5$, C_6H_4Me-3) was synthesized by the zinc reduction of $[TcCl_4(PPh_3)_2]$ in acetonitrile in the presence of PPh_3. This is a useful Tc(III) precursor to obtain other compounds. The reaction of this complex with bipy, phen, and tpy gives the Tc(III) dicationic complexes $[Tc(bipy)_3]^{2+}$, $[Tc(phen)_3]^{2+}$, and $[Tc(tpy)_3]^{2+}$ as their $[BPh_4]^-$ or $[PF_6]^-$ salts [514].

Complexes with O,S-containing ligands. Thiolato–Tc(IV) complexes can be obtained by reduction of $[Tc(OH)O(dmpe)_2]^{2+}$ in excess of H_2tdt yielding $[Tc(tdt)(dmpe)_2](PF_6)$ [515]. The product was characterized by spectroscopic methods and x-ray diffraction. The coordination geometry around the Tc atom is intermediate between octahedral and trigonal prismatic [Tc—S 2.318(6) Å, S—Tc—S angle 84.49(4)°, Tc—P, 2.902(7) Å].

The reaction of $[NBu_4][TcOCl_4]$ with naturally occurring oxazoline and thiazoline ligands [HL = 2-(2′-hydroxyphenyl)-2-oxazoline, 2-(2′-hydroxy-3-methylphenyl)-2-oxazoline, 2-(2′-hydroxyphenyl)-2-thiazoline, and 2-(2′-hydroxyphenyl)-2-benzoxazoline] yields the hexacoordinate complexes $TcOClL_2$ in refluxing alcoholic solutions (MeOH, EtOH) [516].

A mixture of $[Tc^{VI}(DBCat)_3$ and $Tc^{VI}(DBCat)_2(DBAP)]$ is produced by the reaction of 3,5-di-3,5-di-t-butylcatechol (DBCat) and ammonium pertechnate in MeOH [517a]. Schiff-base condensation of ammonia (from ammonium) and catechol is responsible for the formation of the amidophenolate (DBAP) ligand. EPR and x-ray crystallographic data are consistent with a Tc(VI) complex, which is the least common oxidation state of Tc. The catecholate ligand serves as both a reducing and a chelating agent [517a]. Trigonal bipyramidal Tc(III) mixed-ligand complexes with SES/S/P coordination [E = O, $N(CH_3)$, S] are reported [517b].

Complexes with N,S-containing ligands. Other technetium complexes, important for biology and medicine, $HTcO(cysteine)_2$ ***997*** and its barium salt $Ba[TcO(cysteine)_2]_2$ ***998***, were prepared and characterized by the authors of Ref. 518 (5.76):

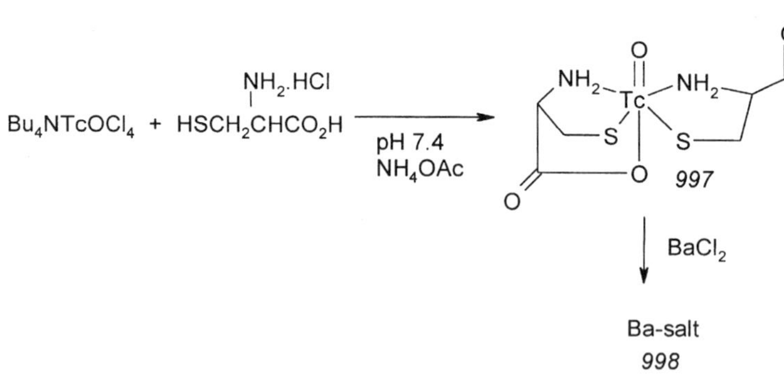

(5.76)

The same products can be obtained from NH_4TcO_4 as a precursor and cysteine. Earlier attempts to obtain technetium complexes with cysteine always resulted in the formation of a product contaminated with polymeric species [518]. Other attempts at chelating technetium with polyfunctional ligands that were accompanied by formation of undesired products are also reported [519–523]. It was postulated that the presence of excess ligand has a degrading effect on the chelate initially formed. This can be avoided by using suitable S-protecting groups (for example: benzyl, acetyl aminomethyl, and benzoyl amino methyl) that can undergo metal-induced deprotection, thus avoiding the presence of excess thiolate in the chelate mixture [518]. The obtained ^{99m}Tc analogue of oxorhenium *bis*-cysteinate has important biological properties, in particular it can be fixed into the kidney [524], which is useful for the diagnosis of the morphological status of that organ [518].

Chelate ligands containing N and S centers have been used to synthesize potential radiopharmaceuticals for diagnosing renal function. Carboxylic groups in the ligands favor the renal uptake of these compounds. Synthesis from the corresponding ammonium tetrahalo–metal oxo precursors $(NH_4)[(TcOX_4]$ and the ligand (2R,7R)-2,7-dicarboxy-3,6-diaza-1,8-octanedithiol (ECH_3) yields the $^{99}TcO(ECH_3)$ complex. Unexpected coordination of one of the carboxylate groups trans to the oxo-ligand was observed for the isostructural Re(V) analogue [525].

Other Schiff-base ligands have been chosen because of their good coordination properties toward Tc in various unusual oxidation states. The chemistry of Tc(I) complexes is relatively unexplored and only a few examples are fully characterized and identified. A Tc(I) complex was synthesized by reacting $[Tc(PPh_3)_2(CO)_3Cl]$ with the lithium salt of the Schiff base N-o-hydroxybenzylidene-2-thiazolylimine in boiling THF to yield $[Tc(PPh_3)_2(CO)_2][(C_3H_2NS)N{=}CHC_6H_4O]$. This compound has a six-coordinated distorted octahedral geometry, with *trans*-PPh_3, *cis*-CO groups and one chelate bidentate anion [526].

Complexes where the ligands are biologically important molecules (such as peptides, proteins, or antibodies) can have unique applications for target-specific diagnostic radiopharmaceuticals. The reaction of $[TcO_4]^-$ with $SnCl_2$, sodium gluconate, and RP294 (dimethyl—glycyl-L-seryl-L-cysteinyl—glycinamide) produced the ^{99}Tc(V) oxo RP294 complex which exists as the syn and anti isomers. Crystallographic resolution of the isostructural Re(V) complexes shows that the serine CH_2OH group is responsible for conferring the isometry. The isomers interconvert in solution at room temperature. The ^{99m}Tc and Re RP294 complexes have similar chemical behavior [527].

The ^{99}Tc and ^{99m}Tc complexes with new tetradentate N_2S_2 and NS_2 ligands have been synthesized by refluxing MeOH solutions of the Tc(V) precursor $[TcOCl_4][NBu_4]$ with the appropriate NS_3H_3 amido proligands to form Tc(V) species $[TcO(NS_3)][NBu_4]$ ***999*** in good yields. However, the compounds decompose over a period of hours or days. Crystal structures of the analogous Re oxo complexes indicate that the compound can be considerate as a square pyramidal complex with the oxygen at the apical position [528]:

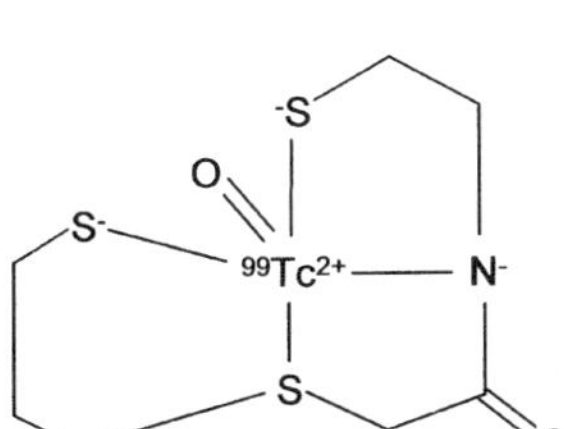

999

Complexes with P-containing ligands. The phosphine derivatives used are of interest as heart-imaging agents, and Tc complexes with them could be useful for the pharmaceutical industry. The complexes $[TcO_2(PR_3)_3](BPh_4)$ (R = Et, Pr), reported in Ref. 494, have a distorted trigonal bipyramid structure with the two oxo ligands in the trigonal plane. The salts of the cation $[TcO_2(PR_3)_3]^+$ are good starting reagents for the preparation of other dioxo mixed-ligand species (5.77):

$$[TcO_2(PR_3)_2]^+ + py \rightarrow [TcO_2(PR_3)_2(Py)_2]^+ \qquad (\text{in } CH_3OH) \qquad (5.77)$$

If R = Me, the $[TcO_2(PMe_3)_2(py)_2]^+$ complex can be prepared directly from $[TcO_4]^-$ by a one-pot method, which could probably be adapted to the commercial kits used in hospitals for the preparation of the ^{99m}Tc radiopharmaceuticals (5.78):

$$NH_4[TcO_4] + py + PMe_3 \rightarrow [TcO_2(PMe_3)_2(py)_2]^+ \qquad (\text{in } CH_3OH) \qquad (5.78)$$

The compounds containing the cation $[TcO_2(PR_3)_2(py)_2]^+$ are diamagnetic, indicating an important deformation from the ideal octahedral geometry [494].

Pertechnates react with derivatized phosphinocarboxylate ligands {L = 2-(diphenylphosphine)benzoic acid $[Ph_2P(C_6H_4COOH)]$, 3-(diphenylphosphine) propionic acid $[Ph_2P(C_2H_4COOH)]$, and (diphenylphosphine)acetic acid $[Ph_2P(CH_2COOH)]$}, producing the corresponding $[TcL_3]$ complexes. Spectroscopic and structural characterization of the compounds indicates that the complexes have a distorted-octahedral geometry with mer configuration, with two pairs of like donor atoms trans to each other and the remaining phosphorus atom trans to the oxygen atom [529]. Complexes of the short-lived isotope ^{99}Tc were prepared by similar procedures and their physical and chemical properties agreed with those of the ^{99m}Tc complexes. These compounds show significant brain uptake in biological tests.

Technetium(V) complexes with *bis*(*o*-hydroxyphenyl)phenylphosphine and (*o*-hydroxyphenyl)diphenylphosphine ligands were prepared by metathesis reactions with the appropriate Tc(V) precursor and/or by reduction/ligand-exchange reactions with ammonium pertechnetate [530]. It was expected that the combination of one soft phosphine P-donor and two hard phenolate O-donors in the chelate should stabilize Tc centers in intermediate oxidation states.

A Tc thiourea complex $[Tc(tu\text{-}S)_6]Cl_3 \cdot 4H_2O$ was used as a precursor in the preparation of $[Tc(dppe)_2(t\text{-}BuNC)_2](PF_6)$ [531a]. The preparation involves mixing both the corresponding ligand and the Tc(III) precursor in ethanol under reflux and is a more convenient synthetic route to these compounds than that using sodium amalgam, dppe, and $[TcCl_4(PPh_3)_2]$. The Tc atom has a distorted octahedral coordination geometry with the isocyanide ligands in trans-position to each other [531a]. Among other P-containing complexes, $TcCl(CO)_3[P(C_6H_5)_3]_2$ [531b] and the complexes with Tc—C multiple bonds $Tc(=C=CHPh)Cl(dppe)_2$ and $[Tc(=C—CH_2\text{-}t\text{-}Bu)Cl(dppe)_2]^+$ [531c] are reported.

Neutral Tc(III) complexes with S,P-bidentate phosphine–thiolate ligands [2-(diphenylphosphino)ethanedithiolate, 2-(diphenylphosphino)propanethiolate, and 2-(diphenylphosphino)thiophenolate] were obtained from the reaction of $[TcO_4]^-$ with an excess of the ligands [532a]. The neutral compounds have a five-coordinate trigonal bipyramidal geometry, with two phosphorus donors of two chelates coordinated mutually trans in the axial positions. Symmetric bis-substituted and asymmetric mono-substituted nitridotechnetium complexes with heterofunctionalized phosphinothiolate ligands are reported [532b].

Complexes containing M—M bond. Rhenium complexes containing multiple metal–metal bonds are "model" complexes in developing an understanding of the physical and chemical properties of such bonds between metal atoms. However, for its analogue, technetium, the development of its coordination chemistry is strongly limited by the fact that all its isotopes are radioactive. The isotope most used for academic purposes, ^{99}Tc, is a fission product of uranium-235. This isotope is a low-energy β^- emitter ($E_{max} = 0.29$ MeV) with a very long half-life (2.1×10^5 years) [497]. The reported data on Tc—Tc multiple bonds are rare ([533] and refs. 3–5 cited therein). A series of triple metal–metal bonded diamagnetic ditechnetium(II) phosphine complexes $Tc_2Cl_4(PR_3)_4$ ***1000*** ($PR_3 = PEt_3$, PPr^n_3, $PMePh_2$, PMe_2Ph) was reported in the same work [533]. These compounds are the first examples of phosphine complexes that contain a Tc—Tc multiple bond and are formed as a result of the reduction of $TcCl_4(PR_3)_2$ under heating (50–55°C) in toluene or by sonication in benzene (> 90% yield in both cases).

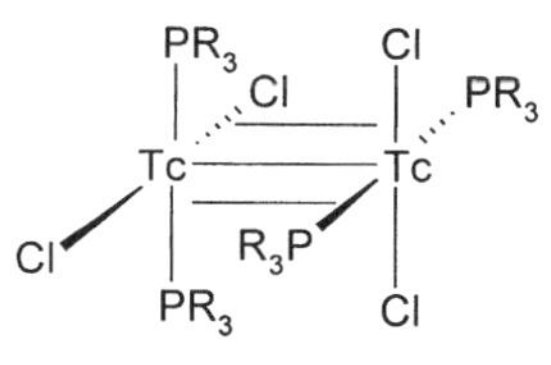

1000

The extended polymeric chain structure of $[Tc_2Cl_6]_n^{2n-}$, reported in Ref. 534, also contains such metal–metal triple bonds. Triple order Tc—Tc bonds were also found at the α and β forms of $Tc_2Cl_4(dppe)_2$ [535]. The α isomer ***1001*** has an eclipsed conformation and a Tc—Tc distance of 2.15(1) Å, while the β isomer ***1002*** has a twist angle of 35(2)° and a Tc—Tc distance of 2.117(1) Å. These last two isomers were prepared by refluxing $Tc_2Cl_4(PR_3)_4$ ($R_3 = Et_3$, Me_2Ph) in toluene with and without an excess of dppe, respectively:

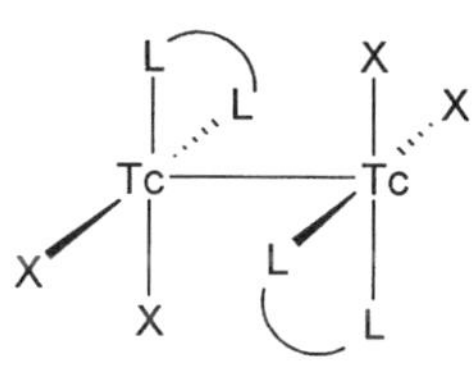

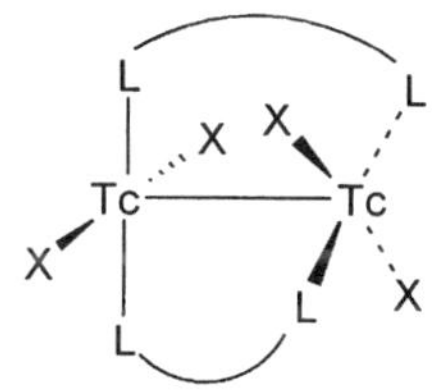

L= dppe; X= Cl

1001 *1002*

Complexes with Tc—Tc multiple bond of order 3.5 were synthesized in high yield by one-electron chemical oxidation of $Tc_2Cl_4(PMe_2Ph)_4$ with ferrocenium hexafluorophosphate in acetonitrile, producing $[Tc_2Cl_4(PMe_2PMe_2Ph)_4][PF_6]$ or neutral $[Tc_2Cl_5(PMe_2Ph)_3]$ when oxidized in the presence of *bis*(triphenylphosphine)iminium [536]. When $Tc_2Cl_4(PR_3)_4$ ($PR_3 = PEt_3$, PMe_2Ph, $PMePh_2$) react with molten formamidines (diphenylformamidine, di-*p*-tolylformamidine), they produce mixtures of tris and tetrakis-bridged formamidinate complexes of general formula $Tc_2(L)_4Cl_n$ ($n = 1$, 2) [537] in modest yield. On the other hand, triple-bonded complexes of Tc such as $[Tc_2(CH_3CN)_{10}][BF_4]_4$ [538] can be photodissociated in acetonitrile solutions

Table 5.18 Some Technetium Complexes

Metal or its compound	Reaction system	Products	Conditions or observations	Reference
Complexes with O- and S-containing ligands				
$[Tc(PPh_3)_2(CO)_3Cl]$	Lithium salt of the Schiff base N-*o*-hydroxybenzylidene-2-thiazolylimine, boiling THF. $[(LiOC_6H_4)(CH{=}N(CSNC_sH_2)]$	$[Tc(PPh_3)_2(CO)_2\{(C_3H_2NS)N{=}CHC_6H_4O\}]$		526
$KTcO_4$	3,5-di-*t*-butylcatechol, CH_3OH	*Tris*(3,5-di-*t*-butylcatecholato) technetium(VI) $[Tc(DBCat)_3]$ and *bis*(3,5-di-*t*-butylcatecholato)(di-*t*-butylamido phenolato) technetium(VI) $[Tc(DBCat)_2(DBAP)]$		540
Complexes with N-containing ligands				
$TcCl_3(PPh_3)_2(CH_3CN)$	*t*-butyl$_3$tpy, DME	$TcCl_3$(*t*-butyl$_3$tpy)		497
NH_4TcO_4	2-hydrazinopyridine · 2HCl, CH_3OH	$[TcCl_3(\eta^1\text{-}NNC_5H_4NH)(\eta^2\text{-}HNNC_5H_4N)]$		498

$[Bu_4N][TcOCl_4]$	Cysteine monohydrochloride monohydrate, aqueous solution	$[HtcO(Cys)_2]$		518
$[Tc_2Cl_4(PR_3)_4]$ ($R_3 = Et_3$, PMe_2Ph, $PMePh_2$)	CH_3CN, HBF_4	$[Tc_2^{II}(CH_3CN)_{10}]\ [BF_4]_4$	The product is an excellent precursor for the synthesis of other low-valent mono- and dinuclear technetium complexes	541
$[TcCl_6]^{2-}$	Reduction by Bu_3SnH or Zn, then treatment with $CH_3CN + HBF_4$		50% yield. Tc—Tc multiple bond is present	
Complexes with P-containing ligands				
$NH_4[TcO_4]$ or $[TcO_2(PR_3)_3]^+$ (R = Et, Pr)	PR_3, CH_3OH	$[TcO_2(PR_3)_3]^+$ or $[TcO_2(PR_3)_2(py)_2]^+$ in presence of py	Yields 60–70%. An excess of ligand is used (about 10×). No reducing agent is needed. NH_3 is produced in the reaction	494
$TcCl_3(PPh_3)_2(CH_3CN)$	tmeda, DME, toluene	$TcCl_3(PPh_3)_2(tmeda)$		497
	py	*mer*-$TcCl_3(py)_3$	Reflux	501
$TcCl_4(PPh_3)_2$	PEt_3 or $P(n\text{-}Pr)_3$, THF	$TcCl_4(PR_3)_2$		533
$TcCl_4(PEt_3)_2$	Zn, benzene	$Tc_2Cl_4(PEt_3)_4$	Sonication	533
$Tc_2Cl_4(PMe_2Ph)_4$	$[Cp_2Fe][PF_6]$	$[Tc_2Cl_4(PMe_2PH)_4][PF_6]$	Tc—Tc multiple bond	536

to give $[Tc(CH_3CN)_6]^{2+}$ ions in almost quantitative yield [539]. This deca-acetonitrile, triple-bonded Tc binuclear complex was synthesized by acidification of $Tc_2Cl_4(PR_3)_4$ with $HBF_4 \cdot Et_2O$ in a mixture of acetonitrile and methylene chloride, in very good yields [538].

Synthetic methods for the main types of Tc complexes are presented in Table 5.18.

EXPERIMENTAL PROCEDURES

Example 13

Synthesis of $[TcO_2(PR_3)_3](BPh_4)$ (R=Et, Pr) [494]

The phosphine (for R = Et, 0.1 mL and for R = Pr, 0.3 mL) was added at room temperature to a solution of 0.018 g of NH_4TcO_4 dissolved in 6 mL of methanol. After 1 hr, 0.14 g of $NaBPh_4$ dissolved in 4 mL of methanol was added to the mixture. Brown crystals started to precipitate after a few hours. Yield 60–70%. Space group $P2_1/n$ (R = Et), $P2_1/c$ (R = Pr).

Example 14

Synthesis of $TcCl_3$(*t*-butyl$_3$tpy) [497]

$TcCl_3(PPh_3)_2(CH_3CN)$ (0.15 g, 0.19 mmol) and *t*-butyl$_3$tpy (0.09 g, 0.21 mmol) were suspended in DME (10 mL). The reaction mixture was heated in an oil bath at 60°C for 2 hr and then heated to reflux for an additional 30 min. Then the reaction mixture was cooled to room temperature, and the black solid was collected, washed with DME (2 mL) and ether (2 mL), and dried in vacuo. Yield 0.077 g, 96%. Analytically pure material was obtained by suspending the solid in toluene at 50°C followed by filtration. Formed solid is insoluble in acetonitrile, acetone, or benzene, but slightly soluble in methylene chloride.

Example 15

Synthesis of $[Tc(C_5H_4NS)_2(NNC_5H_4N)(HNNC_5H_4N)]$ [498]

$[TcCl_3(\eta^1\text{-}NNC_5H_4NH)(\eta^2\text{-}HNNC_5H_4N)]$ was suspended in 3 mL of MeOH, and pyridine-2-thiol (0.009 g, 0.0814 mmol) in 1.5 mL of MeOH was added. The mixture was stirred briefly, and NEt_3 (0.017 mL, 0.122 mmol) was added. The resultant dark red solution was stirred for 1 hr, whereupon the reaction mixture was dried under vacuum to yield a mixture of $[Tc(C_5H_4NS)_2(NNC_5H_4N)(HNNC_5H_4N)]$ and $HNEt_3Cl$. Crystals were grown by diffusion of pentane into a solution of the product in CH_2Cl_2.

The complexes of radioactive elements, as seen, are part of an area of considerably developing coordination and organometallic chemistry [263,542–544a], despite the corresponding difficulties of working with radioactivity. Even such exotic compounds as actinide fullerenes are being studied [544b]. Both theoretical interest and practical applications of actinide compounds motivate its progress. For example, great attention has been paid during recent years to bimetallic complexes containing a 5*f* element, with the hope of discovering novel interesting properties and molecular structures, for example novel Si—O-based inorganic metallaheterocycles [544c,d]. In the same way, a search for uranium(V) complexes has been in progress. All this indicates continuous interest in the synthesis of novel radioactive metal complexes, in particular of nonstandard oxidation state [263]. At the same time, reports on alkyl cyclopentadienyls $Cp^*_2AnR_2$ (An = Th, U; R = H, Alk) compounds as very effective catalysts for alkene hydrogenation and polymerization (10 times more active than typical Pt/SiO_2 catalysts under the same conditions) [430,545] have led to greatly increased research in the field. Another example is the cationic actinide complex $[(Et_2N)_3U][BPh_4]$, an active catalytic precursor for the selective dimerization of terminal alkynes [Scheme (5.79)] [546]. These facts explain the interest in a wider use of actinide complexes in such classic practical areas as catalysis [544a,b] or the search for precursors of organic and organometallic synthesis.

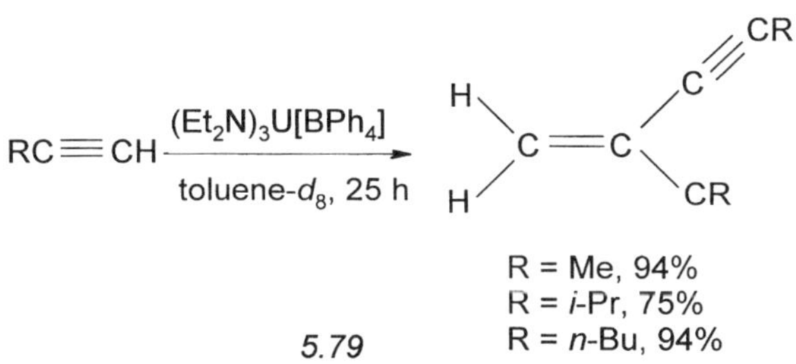

5.79 (5.79)

In the case of technetum, this is the most practically used element among non-*f* radioactive ones for medical and technical purposes [283], so the permanent interest in its coordination chemistry (in particular, the structural aspect of its compounds [547] and kinetics of substitution reactions [548]) is not surprising [549]. The theoretical interest in Tc is provoked, in particular, by the fact that this is a rhenium analogue. This element (Re) forms multiple metal–metal bond complexes and has been studied intensively in order to achieve a better understanding of the physical and chemical properties of multiple bonds between metal atoms [533].

ACKNOWLEDGMENTS

The authors are sincerely grateful to Professor Michael Klaus Engel (Dainippon Ink & Chemicals, Inc., Japan) for supplying reprints of his publications and for useful comments on phthalocyanine chemistry, and to Professor M.A. Petit (France) for permission to reproduce the scheme of an electrochemical cell for phthalocyanine synthesis. The authors are grateful to Professor E.P. Ivakhnenko (Rostov State University) for useful suggestions and critical revision of Section 5.2, to Professor Cortlandt Pierpont (University of Colorado) for permission to reproduce some experimental procedures and a scheme from his publications, to Universidad Autónoma de Nuevo León (Monterrey, Mexico) for financial support (project PAICyT-2001), and to CONACyT-Mexico (MRMA partial scholarship 134716). The authors are very grateful to Professor Michel Ephritikhine (Service de Chimie Moléculaire, France), Professor Claudio Téllez Soto (Brazil), and Professor Bob Denning (Oxford University) for permission to reproduce schemes and figures from their articles.

REFERENCES

1. Linstead, R.P. *J. Chem. Soc.* 1016 (1934).
2. Byrne, G.T.; Linstead, R.P.; Lowe, A.R. *J. Chem. Soc.* 1017 (1934).
3. Linstead, R.P.; Lowe, A.R. *J. Chem. Soc.* 1022 (1934).

4. Dent, C.E.; Linstead, R.P. *J. Chem. Soc.* 1927 (1934).
5. *Phthalocyanines. Properties and Applications.* (Edit. Leznoff, C.C.; Lever, A.B.P.). VCH: New York, vol. 1, 1990; vol. 2, 1992; vol. 3, 1993; vol. 4, 1996.
6. Thomas, A.L. *Phthalocyanines. Research and Application.* CRC Press: Boca Raton, 1990.
7. Thoma, P.F.; Habermann, W.M.; Kranz, J.L. U.S. Patent 4,145,264, 1979.
8. Moser, F.H.; Thomas, A.L. *Phthalocyanines. Properties. Vol. 1.* CRC Press: Boca Raton, 1983.
9. Yang, C.H.; Lin, S.F.; Chen, H.L.; Chang, C.T. *Inorg. Chem.* **19**, 3541 (1980).
10. Petit, M.A.; Plichon, V.; Belkacemi, H. *New J. Chem.* **13**, 459 (1989).
11. Garnovskii, A.D.; Ryabukhin, Yu.I.; Kuzharov, A.S. *Koord. Khim.* **10**(8), 1011 (1984).
12. Garnovskii, A.D.; Kharisov, B.I.; Gójon-Zorrilla, G.; Garnovskii, D.A. *Russ. Chem. Rev.* **64**(3), 201 (1995).
13. *Direct Synthesis of Coordination Compounds.* (Edit. Skopenko, V.V.). Ventury: Kiev, 1997.
14. *Direct Synthesis of Coordination and Organometallic Compounds.* (Edit. Garnovskii, A.D.; Kharisov, B.I.). Elsevier Science: Amsterdam, 1999.
15. Kirner, J.F.; Down, W.; Scheidt, W.R. *Inorg. Chem.* **15**(7), 1685 (1976).
16. Schramm, Ch.J.; Scaringe, R.P.; Stojakovic, D.R.; Hoffman, B.H.; Ibers, J.A.; Marks, T.J. *J. Am. Chem. Soc.* **102**, 6702 (1980).
17. Clarisse, C.; Riou, M.T. *Inorg. Chim. Acta* **130**, 139 (1987).
18. Borodkin, V.F. *Zhurn. Prikl. Khim.* (*Russ. J. Appl. Chem.*) **31**, 803 (1958).
19. Tomoda, H.; Saito, S.; Ogawa, S.; Shiraishi, S. *Chem. Lett.* 1277 (1980).
20. (a) Kharisov, B.I.; Méndez-Rojas, M.A.; Ganich, E.A. *Russ. J. Coord. Chem.* **26**(5), 301 (2000); (b) Tanaka, M. Japanese Patent JP 62205162 A2, 1988.
21. (a) Kubiak, R.; Janszak, J. *J. Alloys Comp.* **200**, L7–L8 (1993) (b) Shaabani, A.; Bahadoran, F.; Safari, N. *Indian J. Chem. A* **40**(2), 195 (2001); (c) Liu, L.-C.; Lee, C.-C.; Hu, A.T. *J. Porphyrins & Phthalocyanines* **5**(11), 806 (2001); (d) Davies, D.A.; Schnik, C.; Silver, J.; Sosa-Sánchez, J.L.; Riby, P.G. *J. Porphyrins & Phthalocyanines* **5**(4), 376 (2001).
22. Yang, C.H.; Chang, C.T. *J. Chem. Soc., Dalton Trans.* 2539 (1982).
23. Griffits, L.; Straughan, B.P.; Gardiner, D.J. *J. Chem. Soc., Dalton Trans.* 1193 (1983).
24. Petit, M.A.; Thami, T.; Sirlin, C.; Lelievre, D. *New J. Chem.* **15**(1), 71 (1991).
25. (a) Petit, M.A.; Thami, T.; Evan, R. *J. Chem. Soc., Chem. Commun.* 1059 (1989); (b) Petit, M.A.; Bouvet, M.; Nakache, D. *J. Chem. Soc., Chem. Commun.* 442 (1991).
26. (a) Yakubovich, T.N.; Teslenko, V.V.; Veisov, B.K.; Zub, Y.L.; Parish, R.V. *Russ. J. Coord. Chem.* **27**(8), 560 (2001); (b) Long, G.S.; Snedeker, B.; Bartosh, K.; Werner, M.L.; Sen, A. *Can. J. Chem.* **79**(5/6), 1026 (2001); (c) Zobi, F.; Stynes, N.V. *Can. J. Chem.* **79**(5/6), 795 (2001); (d) Kobayashi, N.; Muranaka, A.; Nemykin, V.N. *Tetrahedron Lett.* **42**(5), 913 (2001); (e) Kudrik, E.V.; Islyaikin, M.K.; Shaposhnikov, G.P. *Russ. J. Gen. Chem.* **70**(5), 767 (2000); (f) Sesalan, S.B.; Gul, A. *Monatsh. Chem.* **131**(11) 1191 (2000); (g) Weber, A.; Ertel, T.S.; Hanack, M. *Eur. J. Inorg. Chem.* No. 10, 2289 (2000); (h) Nemykin, V.N.; Polshina, A.E.; Kobayashi, N. *J. Chem. Soc., Dalton Trans.* No. 7, 1019 (2000); (i) Torres, T. *J. Porphyrins & Phthalocyanines* **4**(4), 325 (2000); (j) Ercolani, C. *J. Porphyrins & Phthalocyanines* **4**(4), 340 (2000); (k) Armstrong, N.R. *J. Porphyrins & Phthalocyanines* **4**(4), 414 (2000); (l) Gregory, P. *J. Porphyrins & Phthalocyanines* **4**(4), 432 (2000); (m) Becaroglu, O. *J. Porphyrins & Phthalocyanines* **4**(5), 465 (2000); (n) Milgrom, L.R. *The Colors of Life: An Introduction to the Chemistry of Porphyrins and Related Compounds.* Oxford University Press: Oxford, 1997, 249 pp; (o) McKeown, N.B. *Phthalocyanine Materials. Synthesis, Structure and Function.* Cambridge University Press: Cambridge, 1998; (p) *Handbook of Organic Conductive Molecules and Polymers.* (Edit. Nalwa, H.S.). John Wiley: Chichester, 1997, vol. 4, p. 461.

27. Sugimoto, H.; Mori, M.; Masuda, H.; Taga, T. *J. Chem. Soc., Chem. Commun.* 962 (1986).
28. (a) Turek, P.; André J.-J.; Giraudeau, A.; Simon, J. *Chem. Phys. Lett.* **134**(5), 471 (1987); (b) Ilangovan, G.; Zweier, J.L.; Kuppusamy, P. *J. Phys. Chem. B* **104**(17), 4047 (2000); (c) Ilangovan, G.; Zweier, J.L.; Kuppusamy, P. *J. Phys. Chem. B* **104**(40), 9404 (2000).
29. Wöhrle, D.; Schnurpfeil, G.; Knothe, G. *Dyes & Pigments* **18**, 91 (1992).
30. Baumann, F.; Bienert, B.; Rosch, G.; Vollmann, H.; Wolf, W. *Angew. Chem.* **68**(4), 133 (1956).
31. Borodkin, V.F.; Smirnov, R.P. *Izvest. Vissh. Ucheb. Zaved. SSSR Khim. i Khim. Tekhnol.* No. 2, 287 (1961).
32. Kharisov, B.I.; Blanco, L.M.; Torres-Martínez, L.M.; García-Luna, A. *Ind. Eng. Chem. Res.* **38**(8), 2880 (1999).
33. Kharisov, B.I.; Blanco, L.M.; García-Luna, A. *Rev. Soc. Quím. Méx.* **43**(2), 50 (1999).
34. (a) Leznoff, C.C.; D'Ascanio, A.M.; Ildis, S.Z. *J. Porphyrins & Phthalocyanines* **4**(1), 103 (2000); (b) Leznoff, C.C.; Hu, M.; Nolan, K.J.M. *J. Chem. Soc., Chem. Commun.* 1245 (1996).
35. (a) Robinson, M.A.; Trotz, S.I.; Hurley, T.J. *Inorg. Chem.* **6**(2), 392 (1967); (b) Hurley, T.J.; Robinson, M.A.; Trotz, S.I. *Inorg. Chem.* **6**(2), 389 (1967).
36. Yang, C.-H.; Chang, C.-T. *J. Chem. Soc., Dalton Trans.* 2539 (1982).
37. Krolik, L.G.; Vitkina, B.D.; Pankova, N.N. USSR Patent 170,988 (Cl. C 07d), 1965.
38. Brach, P.J.; Six, H.A. U.S. Patent 3,927,026, 1969.
39. Matsumoto, M.; Sasaki, N.; Sawano, B.; Matsumoto, M.; Sasaki, N.; Sawano, B. Japanese Patent JP 07330729 A2 951219, 1996.
40. Tomoda, H.; Hibiya, E.; Nakamura, T.; Ito, H.; Saito, S. *Chem. Lett.* 1003 (1976).
41. Bornengo, G.; Agnes, G.; Menconi, A.; Bottaccio, G.; Bozzolasco, Z.; Domenis, D. Canadian Patent CA 2031707 AA 910607, 1991.
42. Fujita, T.; Kishimoto, T.; Okayasi, H. European Patent EP 266219 A2 880504, 1988.
43. Vrana, M.; Loukota, O.; Pridal, J. Czech Patent CS 215594 B 840415, 1984.
44. Gutwirth K.; Holeci Y.; Vrana M. Czech Patent CS 202609 B 830330, 1984.
45. Watanable, A.; Takahashi, M. Japanese Patent JP 07310024 A2 951128, 1996.
46. Lin, R. Chinese Patent CN 1061609 A 920603, 1993.
47. Sawada, T. *Res. Bull. Meisei Univ., Phys. Sci. Eng.* **26**, 47 (1990).
48. Gauthier, M.V.; Ubillus, F.A.; Grados, N.Q. *Bol. Soc. Quim. Peru* **54**(1), 1 (1988).
49. Ohno, K.; Kanai, A.; Takei, T. European Patent EP 783029 A1 970709, 1997.
50. Zugenmaier, P.; Bluhn, T.L.; Hamer, G.K. *J. Mater. Sci.* **32**(20), 5561 (1997).
51. Hanack, M.; Lang, M. *Adv. Mater.* **6**(11), 819 (1994).
52. Hammond, R.B.; Roberts, K.J.; Docherty, R.; Edondson, M.; Gairns, R. *Perkin Trans.* **2**, No. 8, 1527 (1996).
53. Owens, J.W. *Inorg. Chim. Acta* **279**(2), 226 (1998).
54. Hayashi, M.; Oohashi, T. Japanese Patent JP 08176458 A2 960709, 1996.
55. Hayashi, M.; Ohashi, T.; Suzuki, S.; Fujiwara, H. Japanese Patent JP 09134024 A2 970520, 1997.
56. Qi, X.S. *Metallophthalocyanine Electrodes: Chromatographic Detection and Electrosynthesis.* Dissertation, University of Louisville, Louisville, KY, 1994.
57. Eberhardt, W.; Hanack, M. *Synthesis* **12**, 1760 (1998).
58. Pohmer, J.; Hanack, M.; Barcina, J.O. *J. Mater. Chem.* **6**(6), 957 (1996).
59. Schmid, G.; Witke, E.; Hanack, M. *J. Mater. Chem.* **5**(6), 855 (1995).
60. (a) Shirai, H.; Hanabusa, K.; Kitamura, M.; Masuda, E.; Hirabaru, O.; Hojo, N. *Makromol. Chem.* **185**(12), 2537–2532 (1984); (b) Shishkina, O.V.; Maizlish, V.E.; Shaposhnikov, G.P.; Smirnov, R.P. *Zhurn. Obshch. Khim.* **70**(1), 137–139 (2000).
61. Wu, Y.; Tian, H.; Zhu, D. *Dyes & Pigments* **3**(4), 317 (1998).

62. Jianzhuang, J.; Lau, R.L.C.; Chan, T.W.D.; Mak, T.C.W.; Ng, D.K.P. *Inorg. Chim. Acta* **255**(1), 59 (1997).
63. Sokolova, T.N.; Lomova, T.N.; Morozov, V.V.; Berezin, B.D. *Koord. Khim.* **20**(8), 637 (1994).
64. Dorr, K.; Fiedler, S.; Lin, T.; Hirsch, A.; Hannack, M. *Chem. Ber.* **130**, 1375 (1997).
65. Nabok, A.V.; Ray, A.K.; Cook, M.J. *Supramol. Sci.* **4**(3/4), 407 (1997).
66. Sommerauer, M.R.C.; Hanack, M. *J. Am. Chem. Soc.* **118**(42), 10085 (1996).
67. Polley, R.; Heckmann, H.; Hanack, M.; *Phthalocyanines and Related Compounds.* Thieme: Stuttgart, 1997, Ser. 4, vol. E-9, S717–842.
68. Delmas, C. French Patent FR 2664898 A1 920124, 1993.
69. Brinkmann, M.; Wittmann, J.C.; Andre, J.J. *Thin Solid Films* **292**(1/2), 192 (1997).
70. Filippov, E.A.; Shatalov, V.V.; Serdenko, V.A. Russian Patent RU 2045555 C1 951010, 1996.
71. Duong, V.T.; Nguen, D.Q.; Ho, C.X.; Trinh, T.D. *Tap. Chi Hoa Hoc.* **32**(3), 58 and 63 (1994).
72. Tanaka, M.; Honda, Y. Japanese Patent JP 60199066 A2, 1986.
73. Matsuura, R.; Segawa, T.; Kumada, Y.; Sawada, H. Japanese Patent JP 51034926 760325, 1976.
74. Komada, M.; Hirayama, T. Japanese Patent JP 08027388 A2, 1996.
75. Koori, Y.; Abe, Y.; Horiguchi, S. Japanese Patent JP 49097820, 1975.
76. Azuma, S.; Kano, T.; Myake, T.; Shimizu, H. Japanese Patent JP 05125289 A2 930521, 1993.
77. Butt, M.A.A.; Sheikh, Z.U.; Butt, M.A.; Jami, A. *Pak. J. Sci.* **47**(2), 52 (1995).
78. Sumimoto, C. Japanese Patent JP 57087465 A2, 1982.
79. Casas, B. U.S. Patent US 4320059 A, 1982.
80. Kimura, A.; Tanaka, M.; Kanai, A. Japanese Patent JP 08170025 A2, 1996.
81. Japanese Patent JP 57168950 A2 821018, 1983.
82. Berezin, B.D. *Coordination Compounds of Porphyrines and Phthalocyanines.* Nauka: Moscow, 1978, 280 pp.
83. Chen, F.; Xu, D. *Youji Huaxue* **10**(6), 550 (1990).
84. Moskalev, P.N. *Koord. Khim.* **16**, 147 (1990).
85. Kirin, I.S.; Moskalev, P.N.; Ivannikova, I.V. *Zhurn. Neorg. Khim.* **12**, 707 (1967).
86. M'Sadac, M.; Roncaly, J.; Garrier, F. *J. Chim. Phys. Phys.-Chim. Biol.* **83**, 211 (1986).
87. Sidorov, A.N.; Kotlyar, I.P. *Optika i Spektrosk.* **11**, 175 (1961).
88. Wang, W.; Yang, X.; Zhang, J. *J. Appl. Chem.* **36**, 58 (1986).
89. Kubota, H.; Harnoode, C.; Kitamura, F.; Ohsaka, T.; Tokuda, K. *Kidorui* **28**, 170 (1996).
90. Ostendorp, G.; Homborg, H. *Z. Anorg. Allg. Chem.* **622**, 1222 (1996).
91. Iwase, A.; Tanaka, K. *Electrochim. Acta* **35**, 1707 (1990).
92. Kolesnikova, E.E.; Shaposhnikov, G.P.; Kulinich, V.P.; Smirnov, R.P. *Izv. Vissh. Uchebn. Zaved., Khim. i Khim. Tekhnol.* **33**, 22 (1990).
93. Sugimoto, H.; Higashi, T.; Mori, M. *Chem. Lett.* **8**, 1167 (1983).
94. Moskalev, P.N. USSR Patent SU 525318 780130, 1978.
95. Moskalev, P.N.; Mishin, V.Ya.; Rubtsov, E.M.; Kirin, I.S. *Zhurn. Neorg. Khim.* **21**, 2259 (1976).
96. Nemykin, V.N.; Volkov, S.V. *Koord. Khim.* **26**(6), 465 (2000).
97. Gorbunova, Yu.G.; Tolkacheva, E.O.; Tsivadze, A.Yu. *Koord. Khim.* **22**(12), 944 (1996).
98. Tomilova, L.G.; Gorbunova, Yu.G.; Rodríguez-Méndez, M.L.; De Saja, J.A. *Mendeleev Commun.* 127 (1994).
99. Ishikawa, N.; Kaizu, Y. *Chem. Lett.* No. 2, 183 (1998).
100. Ishikawa, N.; Okubo, T.; Kaizu, Y. *Inorg. Chem.* **38**(13), 3173 (1999).

101. Nemykin, V.N.; Chernii, V.Y.; Volkov, S.V. *J. Chem. Soc., Dalton Trans.* No. 18, 2995 (1998).
102. *Ultrasound: its Chemical, Physical, and Biological Effects.* (Edit. Suslick, K.S.). VCH: Weinheim, 1988.
103. Mason, T.J. *Advances in Sonochemistry.* JAI Press: London, 1990, vol. 1.
104. Blanco, L.M.; Kharisov, B.I.; Garnovskii, A.D. *Proceedings of XXXII International Conference on Coordination Chemistry*, Santiago-de-Chile, 1997, p. 33.
105. Cintas, P. *Activated Metals in Organic Synthesis.* CRC Press: Boca Raton, 1993, pp. 61–70; Luche, J.L.; Cintas, P. *Ultrasound-Induced Activation of Metals: Principles and Applications in Organic Synthesis.* (Edit. Fuerstner, A.). VCH: Weinheim, 1996, pp. 133–190.
106. Kharisov, B.I.; Ortiz Méndez, U. *Rev. Mex. Acust.* No. 14, 6–9 (1999).
107. Nartsissov, B. U.S. Department of Commerce, Aerospace Technology Division, Library of Congress. Washington D.C. Report No. AD 629515 (microfilm), 1966, 41 pp.
108. Lux, F.; Ammentorp-Schmidt, F.; Dempf, D.; Graw, D.; Hagenberg, W. *Radiochim. Acta* **14**(2), 57 (1970).
109. Lux, F.; Dempf, D.; Graw, D. *Angew. Chem., Int. Ed. Engl.* **7**(10), 819 (1968).
110. Gieren, A.; Hoppe, W. *J. Chem. Soc., Chem. Commun.* 413 (1971).
111. Kirin, I.S.; Kolyadin, A.B.; Lychev, A.A. *Sov. J. Struct. Chem.* **15**, 415 (1974).
112. Kobayashi, T. *Bull. Inst. Chem. Res., Kyoto Univ.* **56**, 204 (1978).
113. Darovskikh, A.N.; Farnk-Kamenetskaya, O.V.; Fundamenskii, V.S.; Golynskaya, O.A. *Sov. Phys. Crystallogr.* **30**, 631 (1985).
114. Kirin, I.S.; Kolyadin, A.B.; Lychev, A.A. *Russ. J. Inorg. Chem.* **18**, 1215 (1973).
115. Engel, M.K. *Report Kawamura Inst. Chem. Res.* **8**, 11–54 (1997) (in English).
116. Subbotin, N.B.; Nemykin, V.N.; Voloshin, Y.Z. *Mendeleev Commun.* 121 (1993).
117. Nemykin, V.N.; Subbotin, N.B.; Kostromina, N.A.; Volkov, S.V. *Mendeleev Commun.* 71 (1995).
118. (a) Dabrowski, R.; Witkievicz, Z. Polish Patent 72,892, 1974; (b) Wohrle, D.; Benters, R.; Suvorova, O.; Schnurpfeil, G.; Trombach, N.; Bogdahn-Rai, T. *J. Porphyrins & Phthalocyanines* **4**(5), 491 (2000); (c) Islyaikin, M.K.; Kudrik, E.V.; Yelkin, I.A. *J. Porphyrins & Phthalocyanines* **4**(5), 573 (2000); (d) Kocak, M. *J. Porphyrins & Phthalocyanines* **4**(8), 742 (2000); (e) Chen, M.J.; Klingler, R.J.; Rathke, J.W. *J. Porphyrins & Phthalocyanines* **5**(5), 442 (2001); (f) Dolotova, O.V.; Bundina, N.I.; Kaliya, O.L.; Lukyanets, E.A. *J. Porphyrins & Phthalocyanines* **1**(4), 355 (1997); (g) Gorbunova, Y.; Kalashnikova, I.P.; Gitiérrez, N.; Rodríguez-Méndez, M.L.; Tomilova, L.G.; De Saja, J.A. In Memories of the *First International Conference on Porphyrins and Phthalocyanines*, Dijon, France, 25–30 June, 2000; Poster 544.
119. Garnovskii, A.D.; Burlov, A.S.; Kharisov, B.I.; Gójon-Zorrilla, G. *Rev. Soc. Quím. Méx.* **41**(5), 205 (1997).
120. Davies, J.A.; Hockensmith, C.M.; Kukushkin, V.Yu.; Kukushkin, Yu.N. *Synthetic Coordination Chemistry. Theory & Practice.* World Scientific Publishing: Singapore, 1996.
121. Tuck, D.G. *Pure & Appl. Chem.* **51**(9), 2005 (1979).
122. Tuck, D.G. In *Molecular Electrochemistry of Inorganic Bioinorganic and Organometallic Compounds.* Kluwer Academic: Dordrecht, 1993, pp. 15–31.
123. Chakravorty, M.C.; Subrahmaniam, G.V.P. *Coord. Chem. Rev.* **135/136**, 65 (1994).
124. (a) Garnovskii, A.D.; Blanco, L.M.; Kharisov, B.I.; Garnovskii, D.A. *J. Coord. Chem.* **48**, 219 (1999); (b) Nemykin, V.N.; Kobayashi, N.; Vokov, S.V. *Chem. Lett.* No. 5, 546 (2000).
125. (a) Pierpont, C.G.; Lange, C.W. *Progr. Inorg. Chem.* **41**, 331 (1994); (b) Pierpont, C.G. *Coord. Chem. Rev.* **219–221**, 415 (2001); (c) Pierpont, C.G.; Attia, A.S. *Collect. Czech.*

Chem. Commun. **66**(1), 33 (2001); (c) Pierpont, C.G. *Coord. Chem. Rev.* **216/217**, 99 (2001).

126. Schwartz, P.; Troy, A.L.; Gale, G.R. Presented at the *192nd National Meeting of the ACS*, Anaiheim, CA, Sept. 7–12, 1986; Abstract INOR 133.
127. Thompson, D.C.; Thompson, J.A.; Sugumaran, M.; Moldeus, P. *Chem. Biol. Interact.* **86**, 129 (1992).
128. Davidson, A.; DePamphilis, B.V.; Jones, A.G.; Franklin, K.J.; Lock, C.J.L. *Inorg. Chim. Acta* **128**, 161 (1987).
129. Abrams, M.J.; Larsen, S.K.; Zubieta, J. *Inorg. Chem.* **30**, 2031 (1991).
130. Pyle, A.M.; Barton, J.K. *Progr. Inorg. Chem.* **38**, 413 (1990).
131. Clemenson, P.I. *Coord. Chem. Rev.* **106**, 171 (1990).
132. Michaelis, L.; Smyth, C.V. *J. Biol. Chem.* **94**(2), 329 (1931).
133. Pierpont, C.; Buchanan, R. *Coord. Chem. Rev.* **38**, 45 (1981).
134. Tuck, D.G. *Coord. Chem. Rev.* **112**, 215 (1992).
135. Abakumov, T.A. *Zhurn. Vses. Khim. Obsh. Mendeleev* **24**(2), 156 (1979).
136. (a) Abakumov, G.A. *Metal Complexes with Free-Radical Ligands*. (Edit. Kabachnik, M.I.). Nauka: Moscow, 1985, p. 85; (b) Abakumov, G.A.; Cherkasov, V.K.; Nevodchikov, V.I.; Kuropatov, V.A.; Yee, G.T.; Pierpont, C.G. *Inorg. Chem.* **40**(10), 2434 (2001).
137. Kabachnik, M.N.; Bubnov, N.N.; Solodovnikov, S.P.; Prokofiev, A.I. *Usp. Khim.* **53**(3), 487 (1984).
138. (a) Kharisov, B.I.; Méndez-Rojas, M.A.; Garnovskii, A.D.; Ivakhnenko, E.P.; Ortiz-Méndez, U. *J. Coord. Chem.* (2002); **55** (7), 745 (2002); (b) Okhlobistin, O.Yu.; Ukhin, LYu.; Rubezhov, A.Z.; Ivakhnenko, E.P.; Milaeva, E.R. USSR Patent 754833 (CIC07F15/00, C07F1/08, C07F3/06), 1980.
139. Milaeva, E.R.; Rubezhov, A.Z.; Prokofiev, A.N.; Okhlobistin, O.Yu. *Usp. Khim.* **51**(10), 1698 (1982).
140. Milaeva, E.R.; Speier, G. *As. J. Chem. Rev.* **1**(2), 159 (1990).
141. Pokhodenko, V.D.; Melezhik, A.V.; Vovk, D.N. *Koord. Khim.* **8**(9), 1225 (1982).
142. Rubezhov, A.Z.; Milaeva, E.R. *Izv. AN SSSR, Ser. Khim.* No. 2, 291 (1983).
143. Rubezhov, A.Z.; Milaeva, E.R.; Prokofiev, A.I.; Kursanov, N.V.; Okhlobistin, O.Yu. *Izv. AN SSSR, Ser. Khim.* No. 5, 1143 (1984).
144. Milaev, A.G.; Panov, V.B.; Okhlobistin, O.Yu. *Zhurn. Obshch. Khim.* **52**(1), 219 (1982).
145. Medzhidov, A.A.; Kasumov, V.T.; Mamedov, Kh.S. *Koord. Khim.* **7**(1), 66 (1981).
146. Ivakhnenko, E.P.; Lyubchenko, S.N.; Kogan, V.A.; Okekhnovich, L.P.; Prokofiev, A.I. *Zhurn. Obsh. Khim.* **56**(4), 869 (1986).
147. Melezhik, A.V.; Pokhodenko, V.D. *Zhurn. Org. Khim.* **18**(5), 1054 (1982).
148. Evans, T.A.; Srivatsa, G.S.; Sawyer, D.T.; Traylor, T.G. *Inorg. Chem.* **24**(26), 4733 (1985).
149. Pokhodenko, V.D.; Platonova, E.P.; Melezhik, A.V.; Vovk, D.N. *Elektrokhim.* **20**(2), 169 (1984).
150. Milaeva, E.R.; Chechulina, I.N.; Prokofiev, A.I.; Bundel, Yu.G. *Zhurn. Obsh. Khim.* **59**(12), 2794 (1989).
151. Milaeva, E.R.; Mamaev, V.M.; Gloriozov, N.P.; Chechulina, I.N.; Prokofiev, A.I.; Bundel, Yu.G. *Dokl. AN SSSR* **306**(6), 1387 (1989).
152. Milaeva, E.R.; Szaverenyi, Z.; Simandi, L.I. *Inorg. Chim. Acta* **167**, 139 (1990).
153. Safianov, Yu.N.; Zakharov, L.N.; Struchkov, Yu.T.; Lobanov, A.V.; Cherkasov, V.K.; Abakumov, G.A. *Koord. Khim.* **15**(9), 1233 (1989).
154. Abakumov, G.A.; Cherkasov, V.K. *Metalloorg. Khim.* **3**(4), 838 (1990).
155. (a) Abakumov, G.A.; Cherkasov, V.K.; Lobanov, A.V. *Dokl. AN SSSR* **266**(2), 361 (1982); (b) Dei, A.; Gatteschi, D.; Pecaut, J.; Poussereau, S.; Sorace, L.; Vostrikova, K. *Comptes Rendus – Acad. Sci. Paris, Ser. 2, Chim. Fasc. C* **4**(2), 135 (2001); (c) Kalina,

O.G.; Bashilov, V.V.; Khodak, A.A.; Tumanskii, B.L. *Russ. Chem. Bol.* **50**(3), 566 (2001); (d) Churchill, M.R.; Keil, K.M.; Gilmartin, B.P.; Schuster, J.J.; Keister, J.B.; Jankik, T.S. *Inorg. Chem.* **40**(17), 4361 (2001); (e) Moriuchi, T.; Watanabe, T.; Ikeda, I.; Ogawa, A.; Hirao, T. *Eur. J. Org. Chem.* 277 (2001); (f) Waga, T.; Tsuge, K.; Tanaka, K. *Chem. Lett.* No. 8, 910 (2000).
156. Abakumov, G.A.; Cherkasov, V.K.; Durzhkov, N.O.; Nevodchikov, V.I.; Afanasiev, V.N.; Korovushkin, A.N. *Dokl. AN SSSR* **308**(3), 628 (1989).
157. Abakumov, G.A.; Nevodchikov, V.I.; Cherkasov, V.K. *Izv. Akad. Nauk SSSR, Ser. Khim.* No. 1, 65 (1986).
158. Abakumov, G.A.; Cherkasov, V.K.; Nevodchikov, V.I.; Bubnov, M.P. *Izv. Akad. Nauk SSSR, Ser. Khim.* No. 8, 1861 (1987).
159. Buchachan, R.M.; Pierpont, C.G. *J. Am. Chem. Soc.* **102**, 4951 (1980).
160. Jung, O.-S.; Pierpont, C.G. *J. Am. Chem. Soc.* **116**, 1127 (1994).
161. Jung, O.-S.; Pierpont, C.G. *Inorg. Chem.* **33**, 2227 (1994).
162. Abakumov, C.G.; Cherkasov, V.K.; Bubnov, M.P.; Ellert, O.G.; Dobrokvotova, Z.B.; Zakharov, L.N.; Struchkov, Yu.T. *Dokl. Akad. Nauk* **328**, 12 (1993).
163. Adams, D.M.; Dei, A.; Rheingold, A.L.; Hendrickson, D.N. *J. Am. Chem. Soc.* **115**, 8221 (1993).
164. (a) Ok-Sang, J.; Du Hwan, J.; Young-A., L.; Conklin, B.J.; Pierpont, C.G. *Inorg. Chem.* **36**, 19 (1997); (b) Roux, C.; Adams, D.M.; Itie, J.P., Polian, A.; Hendrickson, D.N.; Verdaguer, M. *Inorg. Chem.* **35**, 2846 (1996).
165. Pierpont, C.G.; Ok-Sang, J. *Inorg. Chem.* **34**, 4281 (1995).
166. Lange, C.W.; Pierpont, C.G. *Inorg. Chim. Acta* **263**, 219 (1997).
167. Larsen, S.K.; Pierpont, C.G.; DeMunno, G.; Dolcetti, G. *Inorg. Chem.* **25**, 4828 (1986).
168. Gass, M.E.; Gordon, N.R.; Pierpont, C.G. *Inorg. Chem.* **25**, 3962 (1986).
169. Nahor, G.S.; Neta, P.; Hambright, P.; Robinson, L.R. *J. Phys. Chem.* **95**, 4415 (1991).
170. Lever, A.B.P.; Masui, H.; Metcalfe, R.A.; Stufkens, D.J.; Dodsworth, E.S.; Auburn, P.R. *Coord. Chem. Rev.* **125**, 317 (1993).
171. DeLarie, L.A.; Haltiwanger, R.C.; Pierpont, C.G. *J. Am. Chem. Soc.* **111**, 4324 (1989).
172. Chin, D.-H.; Sawyer, D.T.; Schaefer, W.P.; Simmons, C.J. *Inorg. Chem.* **22**, 752 (1983).
173. (a) Eggins, B.R.; Robertson, P.K.J. *J. Chem. Soc., Faraday Trans.* **90**(15), 2249 (1994); (b) Haddy, A. *J. Chem. Educ.* **78**(9), 1206 (2001).
174. Bhattacharya, S.; Boone, S.R.; Fox, G.A.; Pierpont, C.G. *J. Am. Chem. Soc.* **112**, 1088 (1990).
175. (a) McGarvey, B.R.; Ozarowski, A.; Tuck, D.G. *Inorg. Chem.* **32**, 4474 (1993); (b) Bencini, A.; Daul, C.A.; Dei, A.; Mariotti, F.; Lee, H.; Schultz, D.A.; Sorace, L. *Inorg. Chem.* **40**(7), 1582 (2001).
176. Carugo, O.; Djinović, K.; Rizzi, M.; Castellani, C.B. *J. Chem. Soc., Dalton Trans.* 1255 (1991).
177. Carreck, P.W.; Charalambous, J.; Kensett, M.J.; McPartlin, M.; Sims, R. *Inorg. Nucl. Chem. Lett.* **10**, 749 (1974).
178. Whalen, A.M.; Bhattacharya, S.; Pierpont, C.G. *Inorg. Chem.* **33**, 347 (1994).
179. Bhattacharya, S.; Pierpont, C.G. *Inorg. Chem.* **33**, 6038 (1994).
180. DeLearie, L.A.; Haltiwanger, R.C.; Pierpont, C.G. *Inorg. Chem.* **28**, 644 (1989).
181. (a) McGarvey, B.R.; Ozarowski, A.; Tian, Z.; Tuck, D.G. *Can. J. Chem.* **73**, 1213 (1995); (b) Brown, M.A.; Castro, J.A.; McGarvey, B.R.; Tuck, D.G. *Can. J. Chem.* **77**, 502 (1999); (c) Camacho, C.C.; Merino, G.; Martínez, F.J.M.; Nöth, H.; Contreras, R. *Eur. J. Inorg. Chem.* 1021 (1999).
182. Delgado, M.; Wolf, R.E. Jr.; Hartman, J.-A.R.; McCafferty, G.; Yagbasan, R.; Rawle, S.C.; Watkin, D.J.; Cooper, S.R. *J. Am. Chem. Soc.* **114**, 8983 (1992).

183. (a) Vigalok, A.; Milstein, D. *J. Am. Chem. Soc.* **119**, 7873 (1997); (b) Amouri, H.; Besace, Y.; Le Bras, J.; Vaissermann, J. *J. Am. Chem. Soc.* **120**, 6171 (1998); (c) Amouri, H.; Vaissermann, J.; Rager, M.N.; Grotjahn, D.B. *Organometallics* **19**, 1740 (2000); (d) Di Valentin, C.; Freccero, M.; Zanaletti, R.; Sarzi-Amade, M. *J. Am. Chem. Soc.* **123**(34), 8366 (2001); (e) Amouri, H.; Vaisserman, J.; Grotjahn, D.B. *Organometallics* **19**(9), 1740 (2000).
184. Abakumov, G.A.; Garnov, V.A.; Nevodchikov, V.I.; Cherkasov, V.K. *Dokl. Akad. Nauk SSSR* **304**, 107 (1989).
185. Gojon, E.; Gaillard, J.; Latour, J.-M.; Laugier, J. *Inorg. Chem.* **26**, 2046 (1987).
186. Belorizky, E.; Fries, P.H.; Gojon, E.; Latour, J.-M. *Mol. Phys.* **61**, 661 (1987).
187. Gojon, E.; Latour, J.-M.; Greaves, S.J.; Povey, D.C.; Ramdas, V.; Smith, G.W. *J. Chem. Soc., Dalton Trans.* 2043 (1990).
188. Gojon, E.; Greaves, S.J.; Latour, J.-M.; Povey, D.C.; Smith, G.W. *Inorg. Chem.* **26**, 1457 (1987).
189. Khakhar, M.P.; Prabhananda, B.S.; Das, H.R. *J. Am. Chem. Soc.* **89**, 3100 (1967).
190. Oakes, J.; Simons, M. *Trans. Faraday Soc.* **66**, 10 (1970).
191. Prokof'ev, A.I.; Solodovnikov, S.P.; Belostotskaya, I.S.; Ershov, V.V. *Izv. Akad. Nauk SSSR, Ser. Khim.* 199 (1974).
192. Malisheva, N.A.; Prokof'ev, A.I.; Solodovnkiov, S.P.; Bubnov, N.N.; Kabachnik, M.I. *Izv. Akad. Nauk SSSR, Ser. Khim.* 997 (1979).
193. Sembiring, S.B.; Colbran, S.B.; Craig, D.C. *Inorg. Chem.* **34**, 761–762 (1995).
194. Kabanos, T.A.; Slawin, M.Z.; Williams, D.J.; Woolings, J.D. *Polyhedron* **11**, 995 (1992).
195. Prokofiev, A.I.; Prokofiev, T.I.; Bubnov, N.N.; Solodovnikov, S.P.; Belostotskaya, I.S.; Ershov, V.V.; Kabachnik, M.I. *Dokl. Akad. Nauk SSSR* **245**, 1393 (1979).
196. Volieva, V.B.; Prokofiev, A.I.; Prokofieva, T.I.; Ivanova, E.B.; Zhorin, V.A.; Ershov, V.V.; Enikolopyan, N.S. *Izv. Akad. Nauk SSSR, Ser. Khim.* 2800 (1986).
197. Annan, T.A.; McConvili, D.H.; McGarvey, B.R.; Ozarovski, A.; Tuck, D.G. *Inorg. Chem.* **28**, 1644 (1989).
198. Annan, T.A.; Tuck, D.G. *Can. J. Chem.* **67**, 1807 (1989).
199. Ozarowski, A.; McGarvey, B.R.; Peppe, C.; Tuck, D.G. *J. Am. Chem. Soc.* **113**, 3288 (1991).
200. Ceneschi, A.; Dei, A.; Gatteshi, D. *J. Chem. Soc., Chem. Commun.* 630 (1992).
201. (a) Ozarowski, A.; McGarvey, B.R.; El-Hadad, A.; Tian, Z.; Tuck, D.G.; Krovich, D.J.; DeFotis, G.C. *Inorg. Chem.* **32**, 841 (1993); (b) Adams, D.M.; Reingold, A.L.; Dei, A.; Hendrickson, D.N. *Angew. Chem.* **105**, 434 (1993).
202. Kaim, W. *Coord. Chem. Rev.* **76**, 187 (1987).
203. Pierpont, C.G.; Larsen, S.K.; Bose, S.R. *Pure Appl. Chem.* **60**, 1331 (1988).
204. Muraev, V.A.; Abakumov, G.A.; Razuvaev, G.A. *Dokl. Akad. Nauk SSSR*, **217**, 1083 (1974).
205. Muraev, V.A.; Abakumov, G.A.; Razuvaev, G.A. *Dokl. Akad. Nauk SSSR*, **236**, 620 (1977).
206. Prokofiev, A.I.; Malisheva, N.A.; Bubnov, N.N.; Solodovnikov, S.P.; Kabachnik, M.I. *Dokl. Akad. Nauk SSSR* **252**, 370 (1980).
207. Sofen, R.S.; Ware, D.C.; Copper, S.R.; Raymond, K.N. *Inorg. Chem.* **18**, 234 (1979).
208. Lange, C.W.; Pierpont, C.G. *J. Am. Chem. Soc.* **114**, 6582 (1992).
209. Muraev, V.A.; Abakumov, G.A.; Razuvaev, G.A. *Izv. Akad. Nauk SSSR, Ser. Khim.* No. 1, 159 (1975).
210. Abakumov, G.A.; Muraev, V.A.; Razuvaev, G.A. *Dokl. Akad. Nauk SSSR* **215**, 1113 (1974).
211. Abakumov, G.A.; Gladyshev, E.N.; Vyazankin, N.S.; Razuvaev, G.A.; Bayushkin, P.Y.; Muraev, V.A. *J. Organomet. Chem.* **64**, 327 (1974).

212. Razuvaev, G.A.; Abakumov, G.A.; Gladyshev, E.N.; Bayushkin, P.Y.; Tsaryapkin, V.A. *Izv. Akad. Nauk SSSR, Ser. Khim.* 2138 (1982).
213. Tumanskii, B.L.; Kampel, V.T.; Solodovnikov, S.P.; Bregadze, V.I.; Godovikov, N.N. *Izv. Akad. Nauk SSSR, Ser. Khim.* 2644 (1985).
214. Speier, G.; Whalen, A.M.; Csihony, J.; Pierpont, C.G. *Inorg. Chem.* **34**, 1355 (1995).
215. Stevenson, C.D.; Reiter, R.C.; Burton, R.D.; Halvorsen, T.D. *Inorg. Chem.* **34**, 1368 (1995).
216. Cheremisov, S.D.; Perekhodtsev, G.D.; Tipikin, D.S.; Lebedev, Ya.S.; Prokof'ev, A.I.; Aleksandrov, A.I.; Dubinskii, A.A.; Möbius, K.; Poluektov, O.G.; Schmidt, J. *J. Chem. Soc., Faraday Trans.* **92**(11), 1959 (1996).
217. Cheremisov, S.D.; Grinberg, O.Ya; Tipikin, D.S.; Lebedev, Ya.S.; Kurreck, H.; Möbius, K. *Chem. Phys. Lett.* **218**, 353 (1994).
218. Tipikin, D.S.; Lebedev, Ya.S.; Poluektov, O.G.; Schmidt, J. *Chem. Phys. Lett.* **215**, 199 (1993).
219. Prokof'ev, A.I.; Malysheva, N.A.; Bubnov, N.N.; Solodovakov, S.P.; Kabachnik, M.I. *Dokl. Akad. Nauk.* **245**, 1123 (1979).
220. Lucken, E.A.C.; Ramírez, F.; Catto, V.P.; Rhum, D.; Dershowitz, S. *Tetrahedron* **22**, 637 (1966).
221. Annan, T.A.; Tian, Z.; Tuck, D.G. *J. Chem. Soc., Dalton Trans.* 19 (1991).
222. Tian, Z.; Tuck, D.G. *J. Organomet. Chem.* **462**, 125 (1993).
223. Abakumov, G.A.; Cherkasov, V.K.; Lobanov, A.V.; Razuvaev, G.A. *Izv. Akad. Nauk SSSR, Ser. Khim.* **7**, 1610 (1984).
224. Brown, M.A.; McGarvey, B.R.; Ozarowski, A.; Tuck, D.G. *J. Organomet. Chem.* **550**, 165 (1998).
225. Brown, M.A.; McGarvey, B.R.; Ozarowski, A.; Tuck, D.G. *Inorg. Chem.* **35**, 1560 (1996).
226. Annan, T.A.; Brown, M.A.; El-Hadad, A.; McGarvey, B.R.; Ozarowski, A.; Tuck, D.G. *Inorg. Chim. Acta* **225**, 207 (1994).
227. Buchachan, R.M.; Kessel, S.L.; Downs, H.H.; Pierpont, C.G.; Hendrickson, D.N. *J. Am. Chem. Soc.* **100**(25), 7894 (1978).
228. Greenwood, N.N.; Gibb, T.C. *Mössbauer Spectroscopy*. Chapman & Hall: London, 1971, chap. 6.
229. Fox, G.A.; Pierpont, C.G. *Inorg. Chem.* **31**, 3718 (1992).
230. Handa, M.; Sono, H.; Kasamatsu, K.; Kasuga, K.; Mikuriya, M.; Ikenoue, S. *Chem. Lett.* 453 (1992).
231. Handa, M.; Matsumoto, H.; Namura, T.; Nagaoka, T.; Kasuga, K.; Mikuriya, M.; Kotera, T.; Nukada, R. *Chem. Lett.* 903 (1995).
232. Ernst, S.; Hänel, P.; Jordanov, J.; Kaim, W.; Kasack, V.; Roch, E. *J. Am. Chem. Soc.* **111**, 1733 (1989).
233. Charalambous, J.; Rees, R.G.; Thomas, T.A. *Polyhedron* **14**(17/18), 2547 (1995).
234. Li, J.; Katti, K.V.; Pinkerton, A.A.; Nar, H.; Cavell, R.G. *Can. J. Chem.* **74**, 2378 (1996).
235. Handa, M.; Takata, A.; Nakao, T.; Kasuga, K.; Mikuriya, M.; Kotera, T. *Chem. Lett.* 2085 (1992).
236. Handa, M.; Mikuriya, M.; Sato, Y.; Kotera, T.; Nukada, R.; Yoshioka, D.; Kasuga, K. *Bull. Chem. Soc. Jpn.* **69**, 3483 (1996).
237. Handa, M.; Nakao, T.; Mikuriya, M.; Kotera, T.; Nukada, R.; Kasuga, K. *Inorg. Chem.* **37**, 149 (1998).
238. Kessel, S.L.; Hendrickson, D.N. *Inorg. Chem.* **17**(9), 2630 (1978).
239. Floriani, C.; Fachinetti, G.; Calderazzo, F. *J. Chem. Soc., Dalton Trans.* 765 (1973).
240. Bruni, S.; Caneschi, A.; Cariati, F.; Delfs, C.; Dei, A.; Gatteschi, D. *J. Am. Chem. Soc.* **116**, 1388 (1994).

241. (a) Connelly, N.G.; Loyns, A.C. *J. Organomet. Chem.* **411**, 285 (1991); (b) Abakumov, G.A.; Cherkasov, V.K.; Nevodchikov, V.I.; Kuropatov, V.A.; Yee, G.T.; Pierpont, C.G. *Inorg. Chem.* **40**, 2434 (2001).
242. Goulle, V.; Harriman, A.; Lehn, J.-M. *J. Chem. Soc., Chem. Commun.* 1034 (1993).
243. Bäckvall, J.-E.; Hopkins, R.B.; Grennberg, H.; Mader, M.M.; Awasthi, A.K. *J. Am. Chem. Soc.* **112**, 5160 (1990).
244. Luneva, N.P.; Nikonova, L.A.; Shilov, A.E. *Kinet. Katal.* **21**, 1458 (1980).
245. Nikonova, L.A.; Isaeva, S.A.; Pershikova, N.; Shilov, A.E. *J. Mol. Catal.* **1**, 367 (1975).
246. Luneva, N., Moravsky, A.P.; Shilov, A.E. *Nouv. J. Chim.* **6**, 245 (1982).
247. Shilov, A.E. *J. Mol. Catal.* **41**, 221 (1987).
248. Haroon, Y.; Bacon, D.S.; Sadowski, J.A. *Biomed. Chromatogr.* **2**(1), 4 (1987).
249. Haroon, Y.; Schubert, C.A.W.; Hauchka, P.V. *J. Chromatogr. Sci.* **22**, 89 (1984).
250. Hart, J.P.; Shearer, M.J.; McCarthy, P.T.; Rahim, S. *Analyst* **109**, 477 (1984).
251. Langenberg, J.P.; Tjaden, U.R. *J. Chromatogr.* **305**, 61 (1984).
252. Ueno, T.; Suttie, J.W. *Anal. Biochem.* **133**, 62 (1983).
253. Tian, Z.; Tuck, D.G. *J. Chem. Soc., Dalton Trans.* 1381 (1993).
254. Buchachan, R.M.; Fitzgerald, B.J.; Pierpont, C.G. *Inorg. Chem.* **20**, 1038 (1981).
255. Buchachan, R.M.; Pierpont, C.G. *Inorg. Chem.* **18**, 1616 (1979).
256. Gass, M.E.; Pierpont, C.G. *Inorg. Chem.* **25**, 122 (1986).
257. DeLearie, L.A.; Pierpont, C.G. *J. Am. Chem. Soc.* **109**, 7031 (1987).
258. Abakumov, G.A.; Klimov, E.S.; Razuvaev, G.A. *Izv. Akad. Nauk SSSR, Ser. Khim.* 1827 (1971).
259. Razuvaev, G.A.; Abakumov, G.A.; Klimov, E.S. *Dokl. Akad. Nauk SSSR* **201**(1), 120 (1971).
260. Abakumov, G.A.; Klimov, E.S. *Dokl. Akad. Nauk SSSR* **202**(4), 827 (1972).
261. Kukes, S.G.; Prokof'ev, A.S.; Masalimov, A.S.; Bubnov, N.N.; Solodovnikov, S.P.; Kabachnik, M.I. *Izv. Akad. Nauk SSSR, Ser. Khim.* 1519 (1978).
262. Atkinson, J.M.; Brindley, P.B.; Davies, A.G.; Hawari, J.A.-A. *J. Organomet. Chem.* **264**, 253 (1984).
263. Ephritikhine, M. *New J. Chem.* **16**, 451 (1992).
264. (a) Marks, T.J.; Ernst, R.D. *Comprehensive Organometallic Chemistry*. (Edit. Willkinson, D.; Stone, F.G.; Abel, E.W.). Pergamon Press: Oxford, 1982, vol. 3; (b) Egelmann, F.T.; Lorenz, V. *Coord. Chem. Rev.* **200**, 99 (2000); (c) Kharisov, B.I.; Méndez-Rojas, M.A. *Russ. Chem. Rev.* **70**(10), 865 (2001).
265. Marks, T.J.; Day, V.W. In *Fundamental and Technological Aspects of Organo-f Element Chemistry*. (Edit. Marks, T.J.; Fragala, I.L.). Reidel: Dordrecht, 1985, p. 115.
266. Takats, J. In *Fundamental and Technological Aspects of Organo-f Element Chemistry*. (Edit. Marks, T.J.; Fragala, I.L.). Reidel: Dordrecht, 1985, p. 159.
267. Santos, I.; Pires de Matos, A.; Maddock, A.G. *Adv. Organomet. Chem.* **34**, 65 (1989).
268. Clark, D.L.; Sattelberger, A.P. *Actinides: Organometallic Chemistry*. In *Encyclopedia of Inorganic Chemistry*. (Edit. King, R.B.). John Wiley: Chichester, 1994, vol. 1, p. 19.
269. Bursten, B.E.; Strittmatter, R.J. *Angew. Chem., Int. Ed. Engl.* **30**(9), 1069 (1991).
270. Burns, C.J.; Bursten, B.E. *Comm. Inorg. Chem.* **9**, 61 (1989).
271. *Actinides and the Environment*. (Edit. Sterne, P.A.). NATO ASI Series. Partnership Subseries 2, Environment, 1998, vol. 41.
272. *Organometallic Compounds of the Lanthanides, Actinides, and Early Transition Metals*. (Edit. Cardin, D.J.). Chapman & Hall: London, 1985.
273. Housecroft, C.E. *Coord. Chem. Rev.* **169**, 187 (1998).
274. Schwochau, K.; Pleger, U. *Radiochim. Acta* **63**, 103 (1993).
275. Housecroft, C.E. *Coord. Chem. Rev.* **162**, 305 (1997).
276. Schoster, F.S.; Zeisler, S.K. *J. Rad. Nucl. Chem.* **220**(2), 149 (1997).

277. Vanlic-Razumenic, N.M.; Maksin, T.N.; Vuksanovic, L.P. *J. Serb. Chem. Soc.* **60**(8), 641 (1995).
278. Housecroft, C.E. *Coord. Chem. Rev.* **142**, 21 (1995).
279. Constable, E.C.; Housecroft, C.E. *Coord. Chem. Rev.* **131**, 153 (1994).
280. Mazzi, U. *Polyhedron* **8**(13/14), 1683 (1989).
281. Cotton, F.A.; Willkinson, G.; Murillo, C.A.; Bochmann, M. *Advanced Inorganic Chemistry*, 6th Ed. John Wiley: New York, 1999.
282. Soloveichik, G.L. *Actinides: Inorganic & Coordination Chemistry*. In *Encyclopedia of Inorganic Chemistry*. (Edit. King, R.B.). John Wiley: Chichester, 1994, vol. 1, p. 2.
283. Nesmeyanov, A.N. *Radiochemistry*. Chemistry: Moscow, 1978, 560 pp.
284. (a) Brooks, M.S.S.; Johansson, B.; Skriver, H.L. In *Handbook on the Physics and Chemistry of the Actinides*. (Edit. Freeman, A.J.; Lander, G.H.). North-Holland: Amsterdam, 1984, vol. 1, chap. 3, p. 153; (b) Edelmann, F.T.; Lorenz, V. *Coord. Chem. Rev.* **209**, 99 (2000).
285. Reynolds, L.T.; Wilkinson, G. *J. Inorg. Nucl. Chem.* **2**, 246 (1956).
286. Burns, J.H. *Structural Chemistry*. In *The Chemistry of Actinide Elements*. (Edit. Katz, J.J.; Seaborg, G.T.; Morris, L.R.). Chapman & Hall: London, 1986, vol. 2, chap. 20, p. 1417.
287. Kepert, D. *Inorganic Stereochemistry*. Springer-Verlag: Berlin, 1982.
288. Burns, C.J.; Bursten, B.E. *Comm. Inorg. Chem.* **9**, 61 (1989).
289. Ryykko, P.; Lohr, L.L. *Inorg. Chem.* **20**(7), 1950 (1981).
290. Rosch, N.; Strietwieser, A. *J. Am. Chem. Soc.* **105**(25), 7237 (1983).
291. Rosch, N. *Inorg. Chim. Acta* **94**(6), 297 (1984).
292. Chang, A.H.H.; Pitzer, R.M. *J. Am. Chem. Soc.* **111**(7), 2500 (1989).
293. Ahrland, S. *Solution Chemistry and Kinetics of Ionic Reactions*. In *The Chemistry of Actinide Elements*. (Edit. Katz, J.J.; Seaborg, G.T.; Morris, L.R.) Chapman & Hall: London, 1986, vol. 2, chap. 21, p. 1480.
294. Grenthe, I.; Fuger, J.; Konings, R.J.M.; Lemire, R.J.; Muller, A.B.; Nguyen-Trung, C.; Wanner, H.; Forest, H. In *Chemical Thermodynamics of Uranium*. (Edit. Wanner, H.; Forest, I.). Elsevier Science: Amsterdam, 1992, p. 193.
295. Fuger, J. In *Plutonium Chemistry*. (Edit. Carnall, W.T.; Choppin, G.R.). ACS Symposium Series 216. American Chemical Society: Washington, DC, 1983, p. 74.
296. (a) Allen, P.G.; Bucher, J.J.; Shuh, D.K.; Edelstein, N.M.; Reich, T. *Inorg. Chem.* **36**(21), 4676 (1997); (b) Allen, P.G.; Bucher, J.J.; Shuh, D.K.; Edelstein, N.M.; Craig, I. *Inorg. Chem.* **39**(3), 595 (2000).
297. Moll, H.; Denecke, M.A.; Grenthe, I. *Inorg. Chem.* **38**(8), 1795 (1999).
298. Plyasunov, A.; Fanghanel, T.; Grenthe, I. *Acta Chem. Scand.* **52**(3), 250 (1998).
299. (a) Basak, S.; Banerjea, D. *Indian J. Chem. Sect. A: Inorg.* **32**(2), 129 (1993); (b) Xu, J.; Raymond, K.N. *Inorg. Chem.* **38**(2), 308 (1999); (c) Durbin, P.W.; Kullgren, B.; Xu, J.; Raymond, K.N.; Allen, P.G.; Bucher, J.J.; Edelstein, N.M.; Shuh, D.K. *Health Phys.* **75**, 34 (1998); (d) Volf, V.; Burgada, R.; Raymond, K.N.; Durbin, P.W. *Int. J. Radiat. Biol.* **70**, 765 (1996); (e) Volf, V.; Burgada, R.; Raymond, K.N.; Durbin, P.W. *Int. J. Radiat. Biol.* **70**, 109 (1996).
300. Van der Sluys, W.G.; Sattelberger, A.P. *Chem. Rev.* **90**(6), 1027 (1990).
301. Zozulin, A.J.; Moody, D.C.; Ryan, R.R. *Inorg. Chem.* **21**(9), 3083 (1981).
302. Stewart, J.L.; Andersen, R.A. *J. Chem. Soc., Chem. Commun.* **24**, 1846 (1987).
303. Avens, L.R.; Barnhart, D.M.; Burns, C.J.; McKee, S.D.; Smith, W.H. *Inorg. Chem.* *33*(19), 4245 (1994).
304. Berg, J.M.; Sattelberger, A.P.; Morris, D.E.; Van der Sluys, W.G.; Fleig, P. *Inorg. Chem.* **32**(5), 647 (1993).
305. (a) Baudin, C.; Baudry, D.; Ephritikhine, M. *J. Organomet. Chem.* **415**(1), 59 (1991); (b) Schreckenbac, G. *Inorg. Chem.* **39**(6), 1265 (2000).

306. Jones, R.G.; Bindschadler, E.; Blume, D.; Karmas, G.; Martin, G.A. Jr.; Thirtle, J.R.; Yeoman, F.A.; Gilman, H. *J. Am. Chem. Soc.* **78**(23), 6030 (1956).
307. Arliguie, T.; Baudry, D.; Ephritikhine, M. *J. Chem. Soc., Dalton Trans.* No. 6, 1019 (1992).
308. Clark, D.I.; Huffman, J.C.; Watkin, J.G. *J. Chem. Soc., Chem. Commun.* **3**, 266 (1992).
309. Barnhart, D.M.; Clark, D.L.; Gordon, J.C.; Huffman, J.C.; Watkin, J.G.; Zwick, B.D. *Inorg. Chem.* **34**(22), 5416 (1995).
310. (a) Menrotra, R.G.; Singh, A.; Tripathi, U.M. *Chem. Rev.* **91**(6), 1287 (1991); (b) Duval, P.B.; Burns, C.J.; Clark, D.L.; Morris, D.E.; Scott, B.L.; Thompson, J.D.; Werkema, E.L.; Andersen, R.A. *Angew. Chem., Int. Ed. Engl.* **40**(18), 3357 (2001).
311. Van der Sluys, W.G.; Burns, C.J.; Huffman, J.C.; Sattelberger, A.P. *J. Am. Chem. Soc.* **110**(7), 5924 (1988).
312. Barnhart, D.M.; Frankcom, T.M.; Gordon, P.L.; Sauer, N.N.; Thompson, J.A.; Watkin, J.G. *Inorg. Chem.* **34**(19), 4862 (1995).
313. Van der Sluys, W.G.; Huffman, J.C.; Ehler, D.S.; Sauer, N.N. *Inorg. Chem.* **31**(8), 1316 (1992).
314. Vaartstra, B.A.; Huffman, J.C.; Gradeff, P.S.; Hubert-Pfalzgraf, L.G.; Daran, J.-C.; Parraud, S.; Yunlu, K.; Caulton, K.G. *Inorg. Chem.* **29**(17), 3126 (1990).
315. Winter, C.H.; Sheridan, P.H.; Heeg, M.J. *Inorg. Chem.* **30**(8), 1962 (1991).
316. Clark, D.L.; Watkin, J.G. *Inorg. Chem.* **32**(9), 1766 (1993).
317. Berg, J.M.; Sattelberger, A.P.; Morris, D.E.; Van der Sluys, W.G.; Fleig, P. *Inorg. Chem.* **32**(5), 647 (1993).
318. Arliguie, T.; Baudry, D.; Berthet, J.C. *New J. Chem.* **15**(7), 569 (1991).
319. (a) Templeton, L.K.; Templeton, D.H.; Barlett, N.; Seppelt, K. *Inorg. Chem.* **15**(11), 2720 (1976); (b) Wilkerson, M.P.; Burns, C.L.; Paine, R.T.; Scott, B.L. *Inorg. Chem.* **38**(18), 4156 (1999).
320. McKee, S.D.; Burns, C.J.; Avens, L.R. *Inorg. Chem.* **37**(10), 4040 (1998).
321. Roussel, P.; Hitchcock, P.B.; Tinker, N.D.; Scott, P. *Inorg. Chem.* **36**(25), 5716 (1997).
322. Duttera, M.R.; Day, V.W.; Marks, T.J. *J. Am. Chem. Soc.* **106**(10), 2907 (1984).
323. Clark, D.L.; Condrason, S.D.; Keogh, D.W.; Palmer, P.D.; Scott, B.L.; Tait, C.D. *Inorg. Chem.* **37**(12), 2893 (1998).
324. Anderson, A.; Chieh, C.; Irish, D.E.; Tong, J.P.K. *Can. J. Chem.* **58**(16), 1651 (1980).
325. Coda, A.; Dell Guista, A.; Tazzoli, V. *Acta Crystallogr. B***37**, 1496 (1981).
326. (a) Graini, R.; Bombieri, G.; Forsellini, E. *J. Chem. Soc., Dalton Trans.* **19**, 2059 (1972); (b) Bond, E.M.; Duesler, E.N.; Paine, R.T.; Neu, M.P.; Matonic, J.H.; Scott, B.L. *Inorg. Chem.* **39**(18), 4152 (2000); (c) Matonic, J.H.; Scott, B.L.; Neu, M.P. *Inorg. Chem.* **40**(12), 2638 (2001).
327. Bruno, J.; Grenthe, I.; Robouch, P. *Inorg. Chim. Acta* **158**(2), 221 (1989).
328. Serezhikina, L.B.; Serezhkin, V.N. *Radiochem.* **38**(2), 110 (1996).
329. Serezhkina, L.B.; Serezhkin, V.N. *Russ. J. Inorg. Chem.* **41**(3), 410 (1996).
330. (a) Serezhkina, L.B.; Serezhkin, V.N. *Russ. J. Inorg. Chem.* **41**(3), 420 (1996); (b) Henderson, W.; Oliver, A.G.; Rickard, C.E. *Inorg. Chim. Acta* **307**(1/2), 144 (2000).
331. Serezhkina, L.B.; Serezhkin, V.N. *Russ. J. Coord. Chem.* **22**(10), 738 (1996).
332. Dodge, C.J.; Francis, A.J. *Environ. Sci. Technol.* **32**(3), 379 (1998).
333. (a) Dodge, C.J.; Francis, A.J. *Environ. Sci. Technol.* **31**(11), 3062 (1997); (b) Bismondo, A.; Cassol, A.; Zanonato, P.L. *Annali di Chimica* **89**(1/2), 185 (1999); (c) Nakamoto, T.; Nakada, M.; Nakamura, A. *Solid State Commun.* **119**(8/9), 523 (2001).
334. Tarafder, M.T.H.; Khan, A.R. *J. Indian Chem. Soc.* **74**(6), 489 (1997).
335. Tarafder, M.T.H.; Khan, A.R. *Polyhedron* **10**(9), 973 (1991).
336. (a) Bhattacharjee, M.; Chaudhuri, M.K.; Dutta, R.N. *J. Chem. Soc., Dalton Trans.* **9**, 2883 (1990); (b) Agarwal, R.K.; Agarwal, H.; Arora, K. *Rev. Inorg. Chem.* **20**(1), 1 (2000).

337. Téllez, C.S.; Gómez-Lara, J. *Spectrochim. Acta* **51A**(3), 395 (1995).
338. Téllez, C.S.; Gómez-Lara, J. *Spectrosc. Lett.* **27**(2), 209 (1994).
339. Téllez, C.S.; Gómez-Lara, J.; Mondragón, M.A.; Castaño, V.M.; Mena, G.R. *Vibr. Spectrosc.* **9**, 279 (1995).
340. Téllez, C.S.; Gómez-Lara, J. *J. Braz. Chem. Soc.* **7**(6), 505 (1996).
341. Téllez, C.S.; Arisawa, M.; Gómez-Lara, J.; Mondragón, M.A. *Polyhedron* **19**(22/23), 2353 (2000).
342. Gómez-Lara, J. *Bol. Inst. Quím. UNAM* **XVI**, 27 (1964).
343. Matassa, L.; Kumar, N.; Tuck, D.G. *Inorg. Chim. Acta* **109**(1), 19 (1985).
344. Kostyuk, N.N.; Kolevich, T.A.; Shirokii, V.L.; Umreiko, D.S. *Koord. Khim.* **15**(12), 1704 (1989).
345. Kostyuk, N.N.; Shirokii, V.L.; Vinokurov, I.I.; Mayer, N.A. *Zhurn. Obsh. Khim.* **64**(9), 1432 (1994).
346. Belford, R.L.; Martell, A.E.; Calvin, M. *J. Inorg. Nucl. Chem.* **14**, 169 (1960).
347. (a) Perry, D.L. *Inorg. Chim. Acta* **48**, 117 (1981); (b) Wilkerson, M.P.; Burns, C.P.; Dewey, H.J.; Martin, J.M.; Morris, D.E.; Paine, R.T.; Scott, B.L. *Inorg. Chem.* **39**(23), 5277 (2000).
348. Bradley, D.C.; Saad, M.A.; Wardlaw, W. *J. Chem. Soc.* 1091 (1954).
349. Maddock, A.G.; Pires de Matos, A. *Radiochim. Acta* **18**, 71 (1972).
350. (a) Samulski, E.T.; Karraker, D.G. *J. Inorg. Nucl. Chem.* **29**, 993 (1967); (b) Danis, J.A.; Lin, M.R.; Scott, B.L.; Eichhorn, B.W.; Runde, W.H. *Inorg. Chem.* **40**(14), 3389 (2001).
351. McKee, S.D.; Burns, C.J.; Avens, L.R. *Inorg. Chem.* **37**(16), 4040 (1998).
352. (a) Clark, D.L.; Grumbine, S.K.; Scott, B.L.; Watkin, J.G. *Organometallics* **15**(3), 949 (1996); (b) Evans, W.J.; Nyce, G.W.; Greci, M.A.; Ziller, J.W. *Inorg. Chem.* **40**(26), 6725 (2001).
353. Bradley, D.C.; Harder, B.; Hudswell, F. *J. Chem. Soc.* **3**, 3318 (1957).
354. Van der Sluys, W.G.; Burns, C.J.; Sattelberger, A.P. *Organometallics* **8**(3), 855 (1989).
355. Vallat, A.; Laviron, E.; Dormond, A. *J. Chem. Soc., Dalton Trans.* No. 3, 921 (1990).
356. Brennan, J.G.; Andersen, R.A.; Zalkin, A. *Inorg. Chem.* **25**(10), 1756 (1986).
357. Stults, S.D.; Andersen, R.A.; Zalkin, A. *Organometallics* **9**(5), 1623 (1990).
358. Berthet, J.-C.J.; Le Marechal, J.F.; Lance, M.; Nierlich, M.; Vigner, J.; Ephritikhine, M. *J. Chem. Soc., Dalton Trans.* No. 9, 1573 (1992).
359. Avens, L.A.; Bott, S.G.; Clark, D.L.; Sattelberger, A.P.; Watkin, J.G.; Zwick, B.D. *Inorg. Chem.* **33**(10), 2248 (1994).
360. Clark, D.L.; Sattelberger, A.P.; Bott, S.G.; Vrtis, R.N. *Inorg. Chem.* **28**(10), 1771 (1989).
361. Drozdzyúski, J.; Du Preez, J.G.H. *Inorg. Chim. Acta* **218**(1/2), 203 (1994).
362. Avens, L.R.; Barnhart, D.M.; Burns, C.J., McKee, S.D. *Inorg. Chem.* **35**(2), 537 (1996).
363. Van der Sluys, W.G.; Berg, J.M.; Barnhardt, D. *Inorg. Chim. Acta* **204**(2), 251 (1993).
364. Du Preez, J.G.H.; Gouws, L.; Rohwer, H.E.; Van Brecht, J.A.M.; Zeelie, B.; Casellato, U.; Graziani, R. *J. Chem. Soc., Dalton Trans.* **10**, 2585 (1991).
365. Du Preez, J.G.H.; Rohwer, H.E.; Van Brecht, J.A.M.; Zeelve, H.; Casellato, U.; Graziani, R. *Inorg. Chim. Acta* **189**(1), 67 (1991).
366. Du Preez, J.G.H.; Zeelie, B. *Inorg. Chim. Acta* **161**(2), 187 (1989).
367. Arnaudet, L.; Bougon, R.; Buu, B.; Lance, M.; Nierlich, M.; Vinger, J. *Inorg. Chem.* **33**(20), 4510 (1994).
368. Moody, D.C.; Odom, J.D. *J. Inorg. Nucl. Chem.* **41**, 533 (1979).
369. Simpson, S.J.; Andersen, R.A. *J. Am. Chem. Soc.* **103**(14), 4063 (1981).
370. Kumar, N.; Tuck, D.G. *Inorg. Chem.* **22**(13), 1951 (1983).
371. Deacon, G.B.; Tuong, T.D. *Polyhedron* **7**(3), 249 (1988).
372. Barnhart, D.M.; Clark, D.L.; Crumbine, S.K.; Watkin, J.G. *Inorg. Chem.* **34**(7), 1695 (1995).

373. Butcher, R.J.; Clark, D.L.; Grumbine, S.K.; Scott, B.L.; Watkin, J.G. *Organometallics* **15**(5), 1488 (1996).
374. Andersen, R.A. *Inorg. Chem.* **18**(6), 1507 (1979).
375. Sternal, R.S.; Marks, T.J. *Organometallics* **6**(12), 2621 (1987).
376. Kanellokopulos, B.; Fischer, E.O.; Dornberger, E.; Baumgartner, F. *J. Organomet. Chem.* **24**(2), 507 (1970).
377. Baudry, D.; Bulot, E.; Ephritikhine, M. *J. Chem. Soc., Chem. Commun.* No. 18, 1316 (1989).
378. Clappe, C.; Hauchard, D.; Durand, G. *J. Electroanal. Chem.* **448**(1), 95 (1998).
379. Le Maréchal, J.-F.; Villers, C.; Charpin, P.; Lance, M.; Nierlich, M.; Vigner, J.; Ephritikhine, M. *J. Chem. Soc., Chem. Common.* No. 5, 308 (1989).
380. Brennan, J.G.; Stults, S.D.; Andersen, R.A.; Zalkin, A. *Inorg. Chim. Acta* **139**(1/2), 201 (1987).
381. Bursten, B.E.; Strittmatter, R.J. *J. Am. Chem. Soc.* **109**(22), 6606 (1987).
382. Ellis, J.E.; Beck, W. *Angew. Chem.* **34**(22), 2489 (1995).
383. Di Bella, S.G.; Giuseppe, A.L. *J. Phys. Chem.* **97**(45), 11673 (1993).
384. Gulina, A.C.; Di Bella, S. *Organometallics* **11**(10), 3248 (1992).
385. Kanellakopulos, B.; Spirlet, M.-R.; Powietzka, B. *Polyhedron* **15**(9), 1503 (1996).
386. Jemine, X.; Goffart, J.; Leverd, P.C. *J. Organomet. Chem.* **469**(1), 55 (1994).
387. Blake, P.C.; Lappert, M.F.; Atwood, J.L. *J. Chem. Soc., Chem. Commun.* No. 1 (21), 1436 (1988).
388. Di Bella, S.; Lanza, G.; Fragala, I.L.; Marks, T.J. *Organometallics* **15**(1), 205 (1996).
389. Kot, W.K.; Shalimov, G.V.; Edelshtein, N.M.; Edelman, M.A.; Lappert, M.F. *J. Am. Chem. Soc.* **110**(3), 986 (1988).
390. Moloy, K.G.; Marks, T.J. *J. Am. Chem. Soc.* **106**(23), 7051 (1984).
391. Moloy, K.G.; Marks, T.J. *Inorg. Chim. Acta* **110**(2), 127 (1985).
392. Tatsumi, K.; Nakamura, A.; Hoffman, P.; Stauffert, P.; Hoffmann, R. *J. Am. Chem. Soc.* **107**(15), 4440 (1985).
393. (a) Le Maréchal, J.-F.; Ephritikhine, M.; Folcher, G. *J. Organomet. Chem.* **299**(1), 85 (1986); (b) Avens, L.R.; Burns, C.J.; Zwick, B.D. *Organometallics* **19**(4), 451 (2000).
394. Burns, J.M.; Laubereau, P.G. *Inorg. Chem.* **10**(12), 2789 (1971).
395. Goffart, J.; Bettonville, S. *J. Organomet. Chem.* **361**(1), 17 (1989).
396. Baudry, D.; Ephritikhine, M.; Nief, F.; Rcard, L.; Mathey, F. *Angew. Chem., Int. Ed. Engl.* **29**(12), 1485 (1990).
397. Arliguie, T.; Lance, M.; Nierlich, M.; Vinger, J.; Ephritikhine, M. *J. Chem. Soc., Chem. Commun.* No. 2, 183 (1995).
398. Green, J.C.; Kaltsoyannis, N.; Sze, K.H.; McDonald, M. *J. Am. Chem. Soc.* **116**(5), 1994 (1994).
399. (a) Schake, A.R.; Avens, L.R.; Burns, C.J.; Clark, D.L.; Sattelberger, A.P.; Smith, W.H. *Organometallics* **12**(5), 1497 (1993); (b) Parry, J.S.; Cloke, F.G.N.; Coles, S.J.; Hursthouse, M.B. *J. Am. Chem. Soc.* **121**(29), 6867 (1999).
400. Boisson, C.; Berthet, J.C.; Ephritikhine, M. *J. Chem. Soc., Dalton Trans.* No. 6, 947 (1996).
401. England, A.F.; Burns, C.J.; Buchwald, S.L. *Organometallics* **13**(9), 3491 (1994).
402. Kanellakopulos, B.; Dornberger, E.; Baumgarther, F. *Inorg. Nucl. Chem. Lett.* **10**, 155 (1974).
403. Wong, C.; Yen, T.; Lee, T. *Acta Crystallogr.* **18**, 340 (1965).
404. Weydert, M.; Brennan, J.G.; Andersen, R.A.; Bergman, R.G. *Organometallics* **14**(8), 3942 (1995).
405. Brennan, J.G.; Andersen, R.A. *J. Am. Chem. Soc.* **107**(2), 514 (1985).
406. Paolucci, G.; Rosserto, G.; Zanella, P.; Fischer, R.G. *J. Organomet. Chem.* **284**(2), 213 (1985).

407. Ossola, F.; Rosserto, G.; Zanella, P.; Paolucci, G. *J. Organomet. Chem.* **309**(1), 55 (1986).
408. Porchia, M.; Brianese, N.; Casellato, U.; Ossola, F.; Rosserto, G.; Zanella, P.; Graziani, R. *J. Chem. Soc., Dalton Trans.* No. 4, 677 (1989).
409. Zanella, P.; Brianese, N.; Casellato, U.; Ossola, F.; Porchia, M.; Rosserto, G.; Graziani, R. *J. Chem. Soc., Dalton Trans.* No. 8, 2039 (1987).
410. Ernst, R.D.; Kennelly, W.J.; Day, C.S.; Marks, T.J. *J. Am. Chem. Soc.* **101**, 2656 (1979).
411. Zalkin, A.; Brennan, J.G.; Andersen, R.A. *Acta Crystallogr. C* **43**, 418 and 421 (1987).
412. Lauke, H.; Swepston, P.N.; Marks, T.J. *J. Am. Chem. Soc.* **106**(22), 6841 (1984).
413. Arliguie, T.; Lance, M.; Ephritikhine, M. *J. Chem. Soc., Dalton Trans.* No. 14, 2501 (1997).
414. Boussie, T.R.; Eisenberg, D.C.; Rigsbee, J.; Streitwieser, A.; Zalkin, A. *Organometallics* **10**(6), 1922 (1991).
415. Streitwieser, A.; Barros, M.T.; Wang, H.K.; Boussie, T.R. *Organometallics* **12**(12), 5023 (1993).
416. Karraker, D.G.; Stone, A. *J. Am. Chem. Soc.* **96**(22), 6885 (1974).
417. Karraker, D.G. *J. Inorg. Nucl. Chem.* **39**, 87 (1977).
418. Gilbert, T.M.; Ryan, R.R.; Sattelberger, A.P. *Organometallics* **7**(12), 2514 (1988); **8**(3), 857 (1989).
419. Achake, A.R.; Avens, L.R.; Burns, C.J.; Clark, D.L.; Sattelberger, A.P.; Smith, W.H. *Organometallics* **12**(5), 1497 (1993).
420. Ogden, J. In *Cryochemistry*. (Edit. Moskovitz, M.; Ozin, G.A.). John Wiley: New York, 1976, chap. 7.
421. Villiers, C.; Ephritikhine, M. *J. Chem. Soc., Dalton Trans.* **23**, 3397 (1994).
422. Ankianies, B.C.; Fanwick, P.E.; Rothwell, I.P. *J. Am. Chem. Soc.* **113**(12), 4710 (1991).
423. Adam, R.; Villiers, C.; Ephritikhine, M. *New J. Chem.* **17**(7), 455 (1993).
424. Hafid, A.; Dormond, A.; Baudry, D. *New J. Chem.* **18**(5), 557 (1994).
425. Ryan, R.R. *Inorg. Chim. Acta* **162**(2), 221 (1989).
426. (a) Baudry, D.; Bulot, E.; Ephritikhine, M. *J. Chem. Soc., Chem. Commun.* No. 20, 1369 (1988); (b) Cendrowski-Guillaume, S.M.; Lance, M.; Ephritikhine, M. *Organometallics* **19**(16), 3257 (2000).
427. Rabinovich, D.; Chamberlin, R.M.; Abney, K.D. *Inorg. Chem.* **36**(19), 4216 (1997).
428. Clark, D.L.; Watkin, J.G.; Butcher, R.J. *Organometallics* **15**(5), 1488 (1996).
429. Rabinovitch, D.; Haswell, C.M.; Scott, B.L.; Miller, R.L.; Nielsen, J.B.; Abney, K.D. *Inorg. Chem.* **35**(6), 1425 (1996).
430. Berthet, J.-C.; Le Maréchal, J.-F.; Ephritikhine, M. *J. Chem. Soc., Chem. Commun.* No. 6, 360 (1991).
431. Cendrowki-Guillaume, S.M., Ephritikhine, M. *J. Chem. Soc., Dalton Trans.* No. 8, 1487 (1996).
432. De Rege, F.M.; Smith, W.H.; Abney, K.D. *Inorg. Chem.* **37**(15), 3664 (1998).
433. Bilsel, O.; Milam, S.N.; Girolami, G.S. *J. Phys. Chem.* **97**(28), 7216 (1993).
434. Raymond, K.L.; Freeman, G.E.; Kappel, M.J. *Inorg. Chim. Acta* **94**(4), 193 (1984).
435. Han, Y.-D.Z.; Ying-Jie Niu, S.-Y. *Science in China, Ser. B, Chem. Life Sci.* **36**(9), 1061 (1993).
436. Rogers, R.D.; Benning, M.M.; Etzenhouser, R.D. *J. Coord. Chem.* **26**(4), 299 (1992).
437. Thomas, A.L. *Phthalocyanines. Research and Application.* CRC Press: Boca Raton, 1990.
438. Lappert, M.G.; Power, P.P.; Sanger, A.R.; Srivastava, R.C. *Metal and Metalloid Amides*. John Wiley: New York, 1980.
439. Berthet, J.C.; Boisson, C.; Lance, M.; Viegner, J.; Nierlich, M.; Ephritikhine, M. *J. Chem. Soc., Dalton Trans.* No. 18, 3019 (1995).

440. Boisson, C.; Berthet, J.C.; Lance, M.; Viegner, J.; Nierlich, M.; Ephritikhine, M. *J. Chem. Soc., Dalton Trans.* No. 6, 947 (1996).
441. Turman, S.E.; Van der Sluys, W.G. *Polyhedron* **11**(24), 3139 (1992).
442. Burns, C.J.; Smith, D.C.; Sattelberger, A.P.; Gray, H.B. *Inorg. Chem.* **31**(18), 3724 (1992).
443. Mikhailov, Yu.N.; Gorbunova, Yu.E.; Serezhkin, V.N.; *Dokl. Chem.* **358**(1/3), 21 (1998).
444. Muller, M.; Williams, V.C.; Prout, K. *Inorg. Chem.* **37**(6), 1315 (1998).
445. Wietzke, R.; Mazzanti, M.; Pecaut, J. *J. Chem. Soc., Dalton Trans.* **24**, 4087 (1998).
446. Berthet, J.-C.; Boisson, Ch.; Ephritikhine, M. *J. Chem. Soc., Dalton Trans.* No. 18, 3027 (1995).
447. Scott, P.; Hitchcock, P.B. *J. Chem. Soc., Dalton Trans.* No. 4, 603 (1995).
448. Berthet, J.C.; Ephritikhine, M. *Coord. Chem. Rev.* **178**, 83 (1998).
449. Arney, D.S.J.; Burns, C.J. *J. Am. Chem. Soc.* **117**(37), 9448 (1995).
450. (a) Brown, D.R.; Denning, R.G.; Jones, R.H. *J. Chem. Soc., Chem. Commun.* No. 22, 2601 (1994); (b) Peters, R.G.; Warner, B.P.; Scott, B.L.; Burns, C.J. *Organometallics* **18**(14), 2587 (1999).
451. Odom, A.L.; Arnold, P.L.; Cummins, C.C. *J. Am. Chem. Soc.* **120**(23), 5836 (1998).
452. Kushto, G.P.; Souter, P.F.; Andrews, L. *J. Chem. Phys.* **108**(17), 7121 (1998).
453. Straub, T.; Frank, W.; Eisen, M.S. *J. Chem. Soc., Dalton Trans.* No. 12, 2541 (1996).
454. Cassol, A.; Bernando, P.; Di Portanova, R. *Inorg. Chem.* **29**(5), 1079 (1990).
455. Brown, D.R.; Denning, R.G.; Jones, R.H. *J. Chem. Soc., Chem. Commun.* No. 22, 2601 (1994).
456. (a) Denning, R.G. *Struct. Bond.* **79**, 215 (1992); (b) Kaltsoyannis, N. *Inorg. Chem.* **39**(26), 6009 (2000).
457. (a) Arnaudet, L.; Bougon, R.; Buu, B.; Lance, M.; Nierlich, M.; Vigner, J. *Inorg. Chem.* **33**(20), 4510 (1994); (b) Andreev, G.B.; Fedosseev, A.M.; Budantseva, N.A.; Antipin, M.Y. *Mendeleev Commun.* No. 2, 58 (2001).
458. Morss, L.R.; Carnall, W.T.; Williams, C.W. *J. Less-Common Met.* **169**(1), 1 (1991).
459. Scott, P.; Hitchcock, R.B. *J. Chem. Soc., Chem. Commun.* No. 5, 579 (1995).
460. El-Hendawy, A.M.; El-Kourashy, A.; El-Ghany Shanab, M.M. *Polyhedron* **11**(5), 523 (1992).
461. Scott, P.; Hitchcock, P.B. *Polyhedron* **13**(10), 1651 (1994).
462. Scott, P.; Hitchcock, P.B. *J. Chem. Soc., Dalton Trans.* 603 (1995).
463. (a) Maurya, M.R.; Maurya, R.C. *Rev. Inorg. Chem.* **15**(1/2), 1 (1995); (b) Goyal, R.C.; Agrawal, D.D.; Arora, K. *Asian J. Chem.* **12**(4), 1181 (2000); (c) Goyal, R.C.; Arora, K.; Agrawal, D.D.; Sharma, K.P. *Asian J. Chem.* **12**(3), 919 (2000); (d) Arora, K.; Agrawal, D.D.; Goyal, R.C. *Asian J. Chem.* **12**(3), 893 (2000); (e) Moustafa, M.E.; Shama, S.A. *Anal. Lett.* **33**(8), 1635 (2000).
464. Syamal, A.; Singh, M.M. *Indian J. Chem. Sect. A: Inorg.* **37**(4), 350 (1998).
465. Syamal, A.; Singh, M.M. *Indian J. Chem. Sect. A: Inorg.* **31**(2), 110 (1992).
466. Amrallah, A.H. *Pakistan J. Sci. Ind. Res.* **37**(6/7), 231 (1994).
467. Syamal, A.; Singh, M.M. *React. Polym.* **21**(3), 149 (1993).
468. Erk, B.; Baran, Y. *Synth. React. Inorg. Met. Syst.* **21**(9), 1321 (1991).
469. (a) Clark, D.L.; Watkin, J.G.; Butcher, R.J. *Organometallics* **14**(6), 2799 (1995); (b) Burns, C.J.; Clark, D.L.; Donohoe, R.J.; Duval, P.B.; Scott, B.L.; Tait, C.D. *Inorg. Chem.* **39**(24), 5464 (2000).
470. Fryzuk, M.D.; Haddad, T.S.; Berg, D.J. *Coord. Chem. Rev.* **99**, 137 (1990).
471. (a) Edwards, P.G.; Andersen, P.A.; Zalkin, A. *J. Am. Chem. Soc.* **103**(26), 7792 (1981); (b) Edwards, P.G.; Parry, J.S.; Reed, P.W. *Organometallics* **14**(8), 3649 (1995).
472. Deshpande, S.G.; Jain, S.C. *Indian J. Chem. Sect. A: Inorg.* **30**(6), 549 (1991).
473. Nierlich, M.; Gradoz, P.; Ephritikhine, M. *J. Organomet. Chem.* **481**(1), 69 (1994).

474. Coles, S.J.; Danopoulos, A.A.; Read, P.W. *J. Chem. Soc., Dalton Trans.* No. 20, 3401 (1995).
475. Aparna, K.; Krishnamurthy, S.S.; Nethaji, M. *J. Chem. Soc., Dalton Trans.* No. 18, 2991 (1995).
476. Vasilescu-Saidel, A.; Turcanu, C.N.; Grosescu, R. *J. Radioanal. Nucl. Chem.* **203**(1), 177 (1996).
477. Vasilescu-Saidel, A.; Turcanu, C.N.; Grosescu, R. *J. Radioanal. Nucl. Chem.* **203**(1), 189 (1996).
478. Arad-Yellin, R.; Zangen, M.; Gottlieb, H. *J. Chem. Soc., Dalton Trans.* No. 7, 2081 (1990).
479. Novoderezhkina, T.L.; Borina, A.F.; Kulikovskii, B.N. *Russ. J. Coord. Chem.* **22**(10), 749 (1996).
480. Duttera, M.R.; Fagan, P.J.; Marks, T.J.; Day, V.W. *J. Am. Chem. Soc.* **104**(3), 865 (1982).
481. Wasserman, H.J.; Moody, D.C.; Ryan, R.R. *J. Chem. Soc., Chem. Commun.* No. 8, 532 (1984).
482. (a) Wasserman, H.J.; Moody, D.C.; Paine, R.T.; Ryan, R.R.; Salazar, K.V. *J. Chem. Soc., Chem. Commun.* No. 8, 533 (1984); (b) Charushnikova, I.A.; Krot, N.N.; Starikova, Z.A. *Radiochemistry* **43**(1), 24 (2001); (c) Carvalho, A.; García–Montalvo, V.; de Matos, A.P. *Polyhedron* **19**(14), 1699 (2000); (d) Bonfada, E.; Schulz-Lang, E.; Abram, U. *Z. Naturforsch.* **55**(3/4), 285 (2000).
483. (a) Cramer, R.E.; Roth, S.; Gilje, J.W. *Organometallics* **8**(10), 2327 (1989); (b) Grohol, D.; Gingl, F.; Clearfield, A. *Inorg. Chem.* **38**(4), 751 (1999).
484. Villiers, C.; Ephritikhine, M. *J. Chem. Soc., Dalton Trans.* No. 23, 3563 (1994).
485. (a) Loured, P.C.; Lance, M.; Viegner, J.; Nierlich, M.; Ephritikhine, M. *J. Chem. Soc., Dalton Trans.* No. 2, 237 (1995); (b) Charushnikova, I.A.; Krot, N.N.; Starikova, Z.A. *Radiochemistry* **42**(5), 439 (2000); (c) Kim, S.-Y.; Harada, M.; Tomiyasu, H.; Shiokawa, Y.; Ikeda, Y. *J. Nucl. Sci. Technol.* **37**(11), 999 (2000).
486. Tatsumi, K.; Matsubara, I.; Inoue, Y.; Nakamura, A.; Cramer, R.E.; Tagoshi, G.J.; Golen, J.A.; Gilje, J.W. *Inorg. Chem.* **29**(24), 4928 (1990).
487. Arliguie, T.; Fourmigue, M.; Ephritikhine, M. *Organometallics* **19**(2), 109 (2000).
488. Leverd, P.C.; Arliguie, T.; Lance, M. *J. Chem. Soc., Dalton Trans.* No. 4, 501 (1994).
489. Leverd, P.C.; Arliguie, T.; Ephritikhine, M. *New J. Chem.* **17**(12), 769 (1993).
490. Arliguie, T.; Fourmigue, M.; Ephritikhine, M. *Organometallics* **19**(2), 109 (2000).
491. Bahadur, L.; Pandey, J.P. *J. Electrochem. Soc.* **137**(12), 3755 (1990).
492. Leverd, P.C.; Lance, M.; Nierlich, M. *J. Chem. Soc., Dalton Trans.* No. 14, 2251 (1993).
493. (a) Leverd, P.C.; Arliguie, T.; Ephritikhine, M. *New J. Chem.* **17**(12), 769 (1993); (b) Yaita, T.; Tachimori, S.; Edelstein, N.M.; Bucher, J.J.; Rao, L.; Shuh, D.K.; Allen, P.G. *J. Synchrotron Radiat.* **8**(2), 663 (2001); (c) Auwer, C.D.; Charbonnel, M.C.; Thuery, P. *Inorg. Chem.* **39**(7), 1487 (2000); (d) Gibson, J.K.; Haire, R.G. *Organometallics* **18**(22), 4471 (1999); (e) Gibson, J.K.; Haire, R.G. *Radiochim. Acta* **89**(6), 363 (2001); (f) Gibson, J.K.; Haire, R.G. *Int. J. Mass Spectrom.* **203**(1–3), 127 (2000); (g) Takeishi, H.; Kitatsuji, Y.; Kimura, T.; Meguro, Y.; Yoshida, Z.; Kihara, S. *Anal. Chim. Acta* **431**(1), 69 (2001); (h) Drew, M.G.B.; Iveson, P.B.; Madic, C. *J. Chem. Soc., Dalton Trans.* No. 5, 821 (2000).
494. (a) Rochon, F.D.; Melanson, R.; Kong, P.-C. *Inorg. Chem.* **37**(1), 87 (1998); (b) Murugesan, S.; Shetty, S.J.; Srivastava, T.S.; Noronha, O.P.D.; Samuel, A.M. *Appl. Radiat. Isot.* **55**(5), 641 (2001).
495. Baum, S.; Bramlet, R. *Basic Nuclear Medinice*. Appleton-Century-Crofts: New York, 1975.
496. *Radiopharmaceuticals*. (Edit. Subramanian, G.; Rhodes, B.A.; Cooper, J.F.; Sodd, V.J. The Society of Nuclear Medicine: New York, 1975.

497. Barrera, J.; Burrel, A.K.; Bryan, J.C. *Inorg. Chem.* **35**(2), 335 (1996).
498. Rose, D.J.; Maresca, K.P.; Nicholson, T.; Davidson, A.; Jones, A.G.; Babich, J.; Fischman, A.; Graham, W.; DeBord, J.R.D.; Zubieta, J. *Inorg. Chem.* **37**(11), 2701 (1998).
499. Abrams, M.J.; Larsen, S.K.; Shalkh, S.N.; Zubieta, J. *Inorg. Chim. Acta* **185**(1), 7 (1991).
500. Hirsch-Kuchima, M.; Nicholson, T.; Davidson, A.; Davis, W.M.; Jones, A.G. *Inorg. Chem.* **36**(15), 3237 (1997).
501. Lu, J.; Yamano, A.; Clarke, M.J. *Inorg. Chem.* **29**(18), 3483 (1990).
502. Marchi, A.; Rossi, R.; Magon, L.; Duatti, A.; Casellato, V.; Grazianni, R.; Vidal, M.; Riche, F. *J. Chem. Soc., Dalton Trans.* No. 6, 1935 (1990).
503. Archer, C.M.; Diworth, J.R.; Griffiths, D.V.; McPartlin, M.; Kelly, J.D. *J. Chem. Soc., Dalton Trans.* No. 1, 183 (1992).
504. Marchi, A.; Marcelli, L.; Rossi, R.; Magon, L.; Bertolasi, V.; Ferretti, V.; Gilli, P. *J. Chem. Soc., Dalton Trans.* No. 9, 1485 (1992).
505. Baldas, J.; Boas, J.F.; Colmanet, S.F.; Williams, G.A. *J. Chem. Soc., Dalton Trans.* No. 19, 2845 (1992).
506. Bolzati, C.; Uccelli, L.; Duatti, A.; Venturini, M.; Morin, C.; Cheradame, S.; Refosco, F.; Ossolo, F.; Tisato, F. *Inorg. Chem.* **36**(17), 3582 (1997).
507. Abram, U.; Lang, E.S.; Aram, S.; Wegmann, J.; Dilworth, J.R.; Kirmse, R.; Woollins, J.D. *J. Chem. Soc., Dalton Trans.* No. 4, 623 (1997).
508. (a) Cattabriga, M.; Marchi, A.; Marvelli, L.; Rossi, R.; Vertuani, G.; Pecoraro, R.; Scattarin, A.; Bertolasi, V.; Ferretti, V. *J. Chem. Soc., Dalton Trans.* No. 9, 1453 (1998); (b) Bolzati, C.; Boschi, A.; Bandoli, G. *J. Am. Chem. Soc.* **122**(18), 4510 (2000).
509. Burrell, A.K.; Clark, D.L.; Gordon, P.L.; Sattelberger, A.P.; Bryan, J.C. *J. Am. Chem. Soc.* **116**(9), 3813 (1994).
510. Du Preez, J.G.; Gerber, T.I.A.; Gibson, M.L. *J. Coord. Chem.* **22**(1), 33 (1990).
511. Barrera, J.; Bryan, J.C. *Inorg. Chem.* **35**(7), 1825 (1996).
512. Archer, C.M.; Dilworth J.R.; Jobanputra, P.; Thompson, R.M.; McPartlin, M.; Hiller, W. *J. Chem. Soc., Dalton Trans.* **6**, 897 (1999).
513. Treher, E.N.; Francesconi, L.C.; Gougouta, J.Z.; Malley, M.F.; Nunn, A.D. *Inorg. Chem.* **28**(18), 3411 (1989).
514. Archer, C.M.; Dilworth, J.R.; Thompson, R.M.; McPartlin, M.; Povery, D.C.; Kelly, J.D. *J. Chem. Soc., Dalton Trans.* No. 3, 461 (1993).
515. Konno, T.; Kirchloff, J.R.; Heeg, M.J.; Heinemann, W.R.; Deutsch, E. *J. Chem. Soc., Dalton Trans.* No. 21, 3069 (1992).
516. Shuter, E.; Hoveyda, H.R.; Karunaratne, V.; Rettig, S.J.; Orvig, C. *Inorg. Chem.* **33**(2), 368 (1996).
517. (a) De Learie, L.A.; Haltiwangen, R.C.; Pierpont, C.G. *J. Am. Chem. Soc.* **111**(12), 4324 (1989); (b) Pietzsch, H.-J.; Tisato, F.; Refosco, F.; Leibnitz, P.; Drews, A.; Seifert, S.; Spies, H. *Inorg. Chem.* **40**(1), 59 (2001).
518. Chatterjee, M.; Achari, B.; Das, S.; Banerjee, R.; Chakrabarti, C.; Dattagupta, J.K.; Banerjee, S. *Inorg. Chem.* **37**(21), 5424 (1998).
519. Bryson, N.; Dewan, J.C.; Lister-James, J.; Jones, A.G.; Davison, A. *Inorg. Chem.* **27**(15), 2154 (1988).
520. Bryson, N.; Lister-James, J.; Jones, A.G.; Davis, W.M.; Davison, A. *Inorg. Chem.* **29**(16), 2948 (1990).
521. Yokoyama, A.; Saji, H.; Tanaka, H.; Odori, T.; Morita, R.; Mori, T.; Torizuka, K. *J. Nucl. Med.* **17**, 810 (1976).
522. Ikeda, I.; Inoue, O.; Kurata, K. *J. Nucl. Med.* **18**, 1223 (1977).
523. Brandau, W.; Bubeck, B.; Eisenhut, M.; Taylor, D.M. *Appl. Radiat. Isot.* **39**, 121 (1988).

524. Chattopadhyay, M.; Jalan, K.N.; Pal, A.K.; Benerjee, S.N. *Nucl. Med. Biol.* **15**, 535 (1988).
525. Marzilli, L.G.; Banaszcyk, M.G.; Hansen, L.; Kuklenyik, Z.; Cini, R.; Taylor, A. *Inorg. Chem.* **33**(22), 4850 (1994).
526. Rossi, R.; Marchi, A.; Magon, L.; Duatti, A.; Casellato, V.; Graziani, R. *Inorg. Chim. Acta* **160**(1), 23 (1989).
527. Wong, E.; Fauconnier, T.; Bennett, S.; Valliant, J.; Nguyen, T.; Lau, F.; Lu, L.F.L. Pollak, A.; Bell, R.A.; Thornback, J.R. *Inorg. Chem.* **36**(25), 5799 (1997).
528. Archer, C.M.; Dilworth, J.R.; Griffithi, D.V.; Al-Jebbori, M.J.; Kelly, J.D.; Lu, C.; Rosser, M.J.; Zheng, Y. *J. Chem. Soc., Dalton Trans.* No. 8, 1403 (1997).
529. Refosco, F.; Tisato, F.; Bandoli, G.; Deutsch, E. *J. Chem. Soc., Dalton Trans.* No. 19, 2901 (1993).
530. Luo, H.; Setyawati, I.; Rettig, S.J.; Orvig, C. *Inorg. Chem.* **34**, 2287 (1995).
531. (a) Kaden, L.; Pombeiro, A.J.L.; Wang, Y.; Abram, U. *Inorg. Chim. Acta* **230**(1), 189 (1995); (b) Alberto, R.; Herrmann, W.A.; Kiprof, P. *Inorg. Chem.* **31**(5), 895 (1992); (c) Burrel, A.K.; Bryan, J.C.; Kubas, G.J. *Organometallics* **13**(4), 1067 (1994).
532. (a) Tisato, F.; Refosco, F.; Bandoli, G.; Bolzati, C.; Moresco, A. *J. Chem. Soc., Dalton Trans.* **9**, 1453 (1994); (b) Bolzati, C.; Malago, E.; Bandoli, G. *New. J. Chem.* **23**(8), 807 (1999).
533. Burns, C.J.; Burrell, A.K.; Cotton, F.A.; Haefner, S.C.; Sattelberger, A.P. *Inorg. Chem.* **33**(10), 2257 (1994).
534. Cotton, F.A.; Daniels, L.M.; Falvello, L.R.; Grigoriev, M.S.; Kryuchkov, S.V. *Inorg. Chim. Acta* **189**(1), 53 (1991).
535. Cotton, F.A.; Daniels, L.M.; Haefner, S.C.; Sattelberger, A.P. *Inorg. Chim. Acta* **288**(1), 69 (1999).
536. Cotton, F.A.; Haefner, S.C.; Sattelberger, A.P. *Inorg. Chem.* **35**(7), 1831 (1996).
537. Cotton, F.A.; Haefner, S.C.; Sattelberger, A.P. *Inorg. Chem.* **35**(25), 7350 (1996).
538. Bryan, J.C.; Cotton, F.A.; Daniels, L.M.; Haefner, S.C.; Sattelberger, A.P. *Inorg. Chem.* **34**(7), 1875 (1995).
539. Cotton, F.A.; Haefner, S.C.; Sattelberger, A.P. *J. Am. Chem. Soc.* **118**(23), 5486 (1996).
540. De Learie, L.A.; Haltiwanger, R.C.; Pierpont, C.G. *J. Am. Chem. Soc.* **111**(12), 4324 (1989).
541. Cotton, F.A.; Sattelberger, A.P.; Bryan, J.C. *Inorg. Chem.* **34**(7), 1875 (1995).
542. Kilimann, U.; Edelmann, F.T. *Coord. Chem. Rev.* **141**, 1 (1995).
543. Richter, J.; Edelmann, F.T. *Coord. Chem. Rev.* **147**, 373 (1996).
544. (a) Gun'ko, Yu.K.; Edelmann, F.T. *Coord. Chem. Rev.* **156**, 1 (1996); (b) Akiyama, K.; Zhao, Y.; Sueki, K.; Tsukada, K.; Haba, H.; Nagame, Y.; Kodama, T.; Suzuki, S.; Ohtsuki, T.; Sakaguchi, M.; Kikuchi, K.; Katada, M.; Nakahara, H. *J. Am. Chem. Soc.* **123**(1), 181 (2001); (c) Lorenz, V.; Fischer, A.; Jacob, K.; Bruser, W.; Edelmann, F.T. *Chemistry* **7**(4), 848 (2001); (d) Lorenz, V.; Fischer, A.; Giemann, S.; Gilje, J.W.; Gun'ko, Y.; Jacob, K.; Edelmann, F.T. *Coord. Chem. Rev.* **206/207**, 321 (2000).
545. He, M.-Y.; Xiong, G.; Toscano, P.J.; Burwell, R.L.; Marks, T.J. *J. Am. Chem. Soc.* **107**(3), 641 (1985).
546. Wang, J.Q.; Dash, A.K.; Berthet, J.C.; Ephritikhine, M.; Eisen, M.S. *Organometallics* **18**(13), 2407 (1999).
547. Bandoli, G.; Dolmella, A.; Porchia, M.; Refosco, F.; Tisato, F. *Coord. Chem. Rev.* **214**, 43 (2001).
548. Omori, T. *Radiochemistry* **39**(3), 198 (1997).
549. Omori, T. *Reports: Fac. Sci. Shizuoka Univ.* **35**, 39 (2001).

Conclusions

ALEXANDER D. GARNOVSKII

Rostov State University, Rostov-on-Don, Russia

BORIS I. KHARISOV

Universidad Autónoma de Nuevo León, Monterrey, Mexico

The material generalized in recent monographs [1–4], special issues [5,6], and in the present book presents ideas about the most common synthetic methods for obtaining coordination compounds. Analysis of these and other numerous data, reported in modern chemical literature, shows that the methods of immediate interaction of ligands and metal salts are the most frequently used. Template syntheses are also widely applied, which allow us to form complex compounds on the metal matrix or modify coordinated ligand systems. The methods of gas-phase (cryosynthesis) and electrochemical synthesis are being developed rapidly and lead to obtaining hard-accessible and, in a series of cases unique, coordination and organometallic compounds.

The nature of the ligand is chosen as a principal factor, responsible for the type and structure (mode of localization of the coordination bond and stereochemistry) of synthesized metal complexes. Systematization of ligands, constructed taking into consideration the character and mutual position (possibility of chelation) of donor centers, and their detailed examination (Chap. 2) are undertaken in the present book with the objective of highlighting progress in this area, achieved after the publication of a number of important many-volume issues [7,8].

The material of the present book is intensified in comparison with earlier, similar subject, monographs [1–4], in the detailed description of the modern state of selection of ligands and the main synthetic methods (Chaps. 2 and 3) and the presence of data on the controlled synthesis of metal complexes (Chap. 4) and synthesis of selected groups of coordination compounds (Chap. 5). However, the chapters mentioned above contain a series of shortcomings, related to the incomplete elucidation of the enormous amount of reported data and the limited examination of general concepts of controlled creation of coordination compounds with a programmed set of properties. In this respect, it would be worthwhile to generalize in future these data and give an account of the material on such synthesis, as done in the present monograph for type-guided (Sec. 4.1) and regioselective (Sec. 4.2) reactions of complex formation.

The expansion in use of solid-state synthesis is of undoubted interest. These routes can be applied not only for obtaining complexes by the method of immediate interaction of components in the reaction of complex formation (Sec. 3.1), but also via modification of coordinated ligand systems (Sec. 3.3.2). To carry out this variant of template synthesis, it is worthwhile using the techniques of solid-state organic syntheses [9–11].

Evidently, this book requires more experimental procedures in each section, similar to those described in Refs. 4 to 6. This disadvantage is compensated, in our opinion, by the numerous schemes of complex-formation reactions, containing not only reagents, but also, in the majority of cases, process conditions. These schemes can be used as a guide to carrying out synthetic experiments.

Synthesis was and will be the basis of modern chemistry; the development of chemical science and the creation of newest science-containing technologies are impossible without an actualization of its methods and techniques.

REFERENCES

1. Kukushkin, V.Yu.; Kukushkin, Yu.N. *Theory and Practice of the Synthesis of Coordination Compounds*. Nauka: Leningrad, 1990, 260 pp.
2. Davies, J.A.; Hockensmith, C.M.; Kukushkin, V.Yu.; Kukushkin, Yu.N. *Synthetic Coordination Chemistry: Principles and Practice*. World Scientific: Singapore, London, 1992, 452 pp.
3. *Direct Synthesis of Coordination Compounds*. (Edit. Skopenko, V.V.). Ventury: Kiev, 1997, 172 pp.
4. *Direct Synthesis of Coordination and Organometallic Compounds*. (Edit. Garnovskii, A.D.; Kharisov, B.I.). Elsevier Science: Lausanne, Amsterdam, London, New York, 1999, 244 pp.
5. *Handbuch der Preparative Anorganische Chemie*. (Herausgegeben von Brauer G.). Ferdinand Enke Verlag: Stuttgart, 1981, Bd. 3.
6. *Synthetic Methods of Organometallic and Inorganic Chemistry*. (Edit. Herrmann, W.A.). Thieme: Stuttgart, New York, 1996, vols. 1–3, 6; 1997, vols. 4, 5, 7, 8.
7. *Comprehensive Coordination Chemistry*. (Edit. Wilkinson, G.). Pergamon Press: Oxford, 1987, vols. 1–7.
8. *Comprehensive Organometallic Chemistry*. (Edit. Wilkinson, G.; Stone, F.G.A.; Abel, E.W.). Pergamon Press: Oxford, New York, London, 1st Ed., 1982, vols. 1–9; 2nd Ed., 1995, vols. 1–13.
9. Hermkens, P.H.H.; Ottenheijm, H.C.J.; Rees, D.C. *Tetrahedron* **52**, 4527 (1996).
10. Hermkens, P.H.H.; Ottenheijm, H.C.J.; Rees, D.C. *Tetrahedron* **53**, 5647 (1997).
11. Booth, S.; Hermkens, P.H.H.; Ottenheijm, H.C.J.; Rees, D.C. *Tetrahedron* **54**, 15385 (1998).

Author Index

Subject Index